Linux

Linux

The Textbook, Second Edition

Syed Mansoor Sarwar and Robert M. Koretsky

CRC Press
Taylor & Francis Group
Boca Raton London New York

CRC Press is an imprint of the
Taylor & Francis Group, an **informa** business

A CHAPMAN & HALL BOOK

CRC Press
Taylor & Francis Group
6000 Broken Sound Parkway NW, Suite 300
Boca Raton, FL 33487-2742

© 2019 by Taylor & Francis Group, LLC
CRC Press is an imprint of Taylor & Francis Group, an Informa business

No claim to original U.S. Government works

Printed on acid-free paper
International Standard Book Number-13: 978-1-138-71008-5 (Hardback)

Library of Congress Cataloging-in-Publication Data

Names: Sarwar, Syed Mansoor, author. | Koretsky, Robert, author.
Title: Linux : the textbook / Syed Mansoor Sarwar, Robert M Koretsky.
Description: Second edition. | Boca Raton : Taylor & Francis, CRC Press, 2018. | Includes index.
Identifiers: LCCN 2018019196 | ISBN 9781138710085 (hardback : alk. paper)
Subjects: LCSH: Linux. | Operating systems (Computers)
Classification: LCC QA76.774.L46 S37 2018 | DDC 005.4/46--dc23
LC record available at https://lccn.loc.gov/2018019196

Visit the Taylor & Francis Web site at
http://www.taylorandfrancis.com

and the CRC Press Web site at
http://www.crcpress.com

To Abbujee with love

SMS

To my family

RMK

Contents

Preface

Authors

Together, the authors have almost 75 years of practical teaching experience at the college level. Our continuing concept for this book grew out of our unwillingness to use either the large, intractable Linux reference sources (both online and printed) or the short "nutshell" guides, to teach meaningful, complete, and relevant introductory classes on the subject. We still feel very strongly that a textbook approach, with pedagogy incorporating in-chapter tutorials and exercises, as well as useful basic and advanced problem sets at the end of each chapter, allows us to present all the important Linux topics for a classroom lecture–laboratory–homework presentation.

In this second edition, we have continued to fine-tune in a manner consistent with what we feel are optimal learning outcomes, i.e., well-thought-out sequencing of old and new topics, well-developed and timely lessons, and homework exercises/problems synchronized with the sequencing of chapters in the book. To assist our pedagogy, we have also incorporated in the earlier chapters, many forward references to later chapters, and in the later chapters, backward-references to the earlier material.

Audience

This second edition is guided by the same intentions and core principles as the first edition, particularly in terms of both its scope and its content. It is intended as a textbook on the modern, 21st-century Linux operating system. It uses an introductory pedagogic style, very similar to the explication and layout of the previous edition.

With the exception of two newly introduced chapters on system programming and two chapters on system administration, the first 14 chapters of the book can be used successfully by a complete novice, as well as by an experienced Linux system user. The text can be deployed in an informal, continuing education environment, at home or work, and in a formal, higher education classroom learning environment. Professionals interested in furthering their knowledge will also benefit from the depth and breadth this book has to offer. It is aimed at anyone with a desire to learn.

The Purposes of This Book in the Second Edition

Our primary purpose in the core chapters remains a didactic description of the Linux application user's interface (AUI). We try to do this in a way that gives the reader insight into the inner workings of the system, along with explanations of some important Linux concepts, data structures, and algorithms. Notable examples include the in-depth descriptions of the Linux file, process, and input/output (I/O) redirection concepts.

Our secondary purpose is to extensively describe the Linux application programmer's interface (API) in terms of C/C++ libraries and Linux system calls. In writing this second edition, particularly for the system programming chapters, we do assume prior basic to intermediate knowledge of C/C++ programming on the part of the reader.

Our tertiary purpose is to describe some important Linux software engineering tools for developers of C/C++ software, Python programs, and shell scripts.

Last, our ancillary purpose is to present and situate various system administration topics in a modern, 21st-century Linux environment controlled by the systemd "superkernel." These topics can be applied in a single, stand-alone Linux system environment, or in a distributed computing environment. Their subject matter can be extended from laptop or desktop systems to small-to-medium sized, server-class systems, and beyond.

Distribution Information and Additions

The Linux system distributions, or "flavors," that we use to illustrate everything in this edition are from the Debian and Red Hat (RHEL) families: Debian, Ubuntu, Linux Mint, and CentOS. We have chosen these flavors primarily because they are very easy to install and use, and, therefore, they are universally used by a large international base of ordinary Linux users. They are freely available to an ordinary user who wants to install the system on their own PC. The exemplary case of this is the free CentOS, versus the only commercially available RHEL.

Most importantly, we have chosen to use recent versions that incorporate systemd—Debian 9.1, Ubuntu 16.04, Linux Mint 18.2, and CentOS 7.4.

Whenever we illustrate a Linux command, by default we preface the command with only the Bash shell prompt "$," and we use the output that a Debian-family distribution would give. That output is generally uniform across Debian, Ubuntu, and Linux Mint, and also CentOS. Whenever the syntax, execution, and output of a command significantly differ from the default on the Debian-family distributions, particularly as seen in Appendix A on Debian and CentOS, we preface the command with a shell prompt as follows to help you differentiate the distribution the command has been executed on:

[root@debian]# for Debian

[root@centos]# for CentOS

By default, and in a majority of cases, we use the traditional text-based command line interface, in a text-based terminal window interface, as it is presented on the distributions and versions we have chosen. The use of a Graphical User Interface (GUI) desktop is also an integral part of our presentation, and we use this wherever it is warranted, and promotes expediency for the beginner.

Because of their use of systemd in the kernel, these versions are distinctly different from earlier, pre-systemd major release versions. There are many things that also make these distributions and versions superior to, as well as very different from, any other contemporary, nominally Linux distribution. They are also distinctly different from other NIX-like operating systems, such as TrueOS, UNIX, and OS X. There are many topics covered in this edition of the book that older, more traditional textbook approaches to teaching Linux, and other NIX-like systems, could not include—systemd and ZFS being prime examples of that.

The biggest difference between this edition and the previous one is the partitioning of the text into three major modules:

Chapters 1–14 comprise the basic Linux core module.

Chapters 15 and 16 comprise the system programming module.

Chapters 17 and 18 comprise the system administration module.

We also provide two appendices (A and B). Appendix A is a set of instructions for critical software installation on Debian-family and Red Hat family systems. Appendix B is a listing of further reference books. A glossary of key terms is available for download at https://www.crcpress.com/9781138710085.

How will the reader benefit from this change? The biggest benefit of partitioning the book into these modules, and all the supplements that we provide, is that the book can be used in a wider variety of learning situations. We describe these in detail in the "Pathways through the Text" section later.

The next major difference is that the basic material concerning these three major modules is found in the printed book, and advanced, supplementary material for these modules is placed at the book website.

Because Debian and Red Hat families of Linux have had many important functional additions made to the AUI since the previous edition of the book came out, and because Linux now holds a vast and widely dispersed share of the operating systems marketplace, we felt that we needed to add instructional material to the book covering these additions, including the following:

- A complete tutorial chapter on systemd and framing the presentation of everything in Linux within a systemd environment
- Two new, complete chapters on Linux system programming and the Linux API
- A new, extensive chapter on Linux system administration using systemd that details installation, maintenance, and updating/upgrading your Linux systems on your own PC
- A revised and updated chapter on networking and internetworking to bring it in line with current standards
- Complete coverage of the system call interface, files, file-related data structures in the Linux kernel, file I/O paradigms, and file manipulation API
- Extensive coverage of the concepts of Linux processes and threads, process-related kernel data structures in the Linux kernel, process management API, and signal handling
- Comprehensive coverage of interprocess communication in Linux using pipes and named pipes (FIFOs)
- Use of GUI desktop environments, such as Cinnamon on Linux Mint
- New diagrams, tables, interactive shell sessions, in-chapter tutorials, in-chapter exercises, and basic and advanced end-of-chapter questions, problems, and projects
- Coverage of many new commands and enhancing coverage of existing commands
- Up-to-date URLs for important Web resources on nearly everything in the book
- New chapter on the nano text editor
- Redesigned text layout for a more usable active learner printed document

In the advanced, supplementary chapters at the book website, all prefaced with the letter "W" to help you differentiate them from the printed book chapters:

- A complete tutorial chapter on the Python 3 programming language and its use in Linux (Chapter W19)
- Two advanced chapters on system programming to supplement the two chapters on that topic in the printed book (Chapters W20 and W21)
- A complete reference chapter on the Zettabyte File System (ZFS), the kernel-loadable user file system, with many practical Linux system administration examples (Chapter W22)
- A complete chapter on virtualization methodologies, illustrating Linux containers with LXC/LXD, cloud computing with Amazon Web Services EC2, and installation of various guest operating systems in popular host systems using VirtualBox (Chapter W23)
- An extensive and complete tutorial on the git command, and using GitHub (Chapter W24, Section W5.7)
- A complete tutorial on the more traditional Linux text editors vi, vim, gvim, and emacs, and adding many example methods used to customize those editors (Chapter W25)
- Additional advanced system administration topics, for example, extensive coverage of CUPS printing in Linux (Chapter W26)
- Additional advanced and critical systemd topics, such as cgroups, namespaces, and the clone system call (Chapter W27)
- Wayland (W28)
- TC Shell Basic and Advanced Programming (W29, W30)

The second edition also offers enhanced usability of all shell scripts, Python and C/C++ programs, and other programming code shown in the printed book, by maintaining them as correctly formatted and usable text files at a GitHub repository. This feature allows you to easily utilize a single git command, or Web browser download, to obtain these files at a local repository or directory.

Features

This edition has had many new diagrams and tables added, and there are many new in-chapter tutorials, interactive shell sessions, in-chapter exercises, and end-of-chapter questions and problems. We have partitioned the end-of-chapter questions and problems into basic and advanced, with project-level problems presented as well. These project-level problems can be done in the short term in a single week or can stretch over the course of an entire term if used in a formal learning environment.

We have added more general command descriptions whenever we introduce a new command or utility. These general command descriptions describe the exact syntax of the command (and any other pertinent variants of the basic syntax), its purpose, the output produced by the command, and its useful options and features. In addition, every chapter contains a summary of the material covered in the chapter. There is also an appendix that shows how to install key software used throughout the book for both Debian and Red Hat family systems.

Supplements

At the book website for everyone (available from the CRC website at https://www.crcpress.com/9781138710085:

Whole Chapters or Sections, prefaced with the letter "W"

Python Programming (W19)

System Programming III: Interprocess Communication (W20)

System Programming IV: Practical Considerations (W21)

Zettabyte File System (W22)

Virtualization Methodologies (W23)

git/GitHub (W24, Section W5.7)

vi/vim/emacs Tutorial and Extensions (W25)

System Administration (W26)

systemd (W27)

Wayland (W28)

TC Shell Basic and Advanced Programming (W29, W30)

Solutions to In-Chapter Exercises for all Printed Book and Online Supplementary Chapters.

A Glossary of key terms used throughout the printed book and the supplementary chapters.

At the book website, for instructors only*:

Solutions to Questions and Problems/Advanced Questions/Problems/Projects sections for core chapters (1–14)

Workbook and Solutions Manual for core chapters (1–14)

Test Bank and Solutions for core chapters (1–14)

PowerPoint Slides for core chapters (1–14)

Contact your CRC Press representative to gain access to this material.

At the author-maintained GitHub site:

1. Source code for C/C++ programs, Python code, and long shell scripts, arranged by chapter.

---

* _Note:_ Resources available to qualified instructors only.

2. Author-maintained Web resource hyperlinks to many other Linux resources on the Web.
3. Updates of version-specific content of Debian and Red Hat family Linux systems that severely impact our printed book presentations.
4. Errata.

The following are the instructions for using the author-maintained GitHub site:

a. You can use your Web browser and retrieve any part of the materials from the following GitHub repository:
   https://github.com/bobk48/linuxthetextbook
b. Or you can do the following to prepare and download these materials:
   Prepare a git repository directory on your local system, using the instructions found in Chapter W24, Section W5.7.
   To access these materials, pull from the repository using this git command:
   git pull https://github.com/bobk48/linuxthetextbook master
   You will find items 1–4 previously in your local git repository.

A Note to Instructors

The didactic structure of each chapter in this new edition follows one of two similar formats: either the shell session format or the tutorial format.

In the shell session format (used in all chapters in the printed book except 3, 17, and 18), the following outline is used:

- Learning objectives
- Introduction
- Topic discussion and background organized in sections and subsections
- Illustrative commands or topic illustrations presented as shell sessions, where the user types in commands shown and results are displayed
- In-chapter exercises that reinforce what was discussed on a topic or done interactively in a shell session
- Summary
- End-of-chapter basic and advanced questions and problems, keyed to topics presented

In the tutorial format (used in Chapters 3, 17, 18, and in all the book website chapters), the following outline is used:

- Learning objectives
- Introduction
- Topic discussions and background organized in sections and subsections
- One or several example sessions or practice session tutorials that illustrate the commands and topics of interest in any particular section or subsection
- Illustrative commands or topic illustrations presented as shell sessions, where the user types in commands shown and results are displayed
- In-chapter exercises that reinforce what was discussed in an example or practice session, on a topic, or done interactively in a shell session
- Summary
- End-of-chapter basic and advanced questions and problems keyed to topics presented

Pathways through the Text

If this book is to be used as the main text for an introductory course in Linux, Chapters 1–13 should be covered.

If the book is to be used as a companion to the main text in an operating system concepts and principles course, the coverage of chapters would be dictated by the order in which the main topics of the course are covered, but should include Chapters 4, 9, 10, 15, and 16.

For use in a C/C++ or shell programming course, Chapters 1 and 4–14, and relevant sections of Chapter 3 and W25 would be a great help to students. The extent of coverage of Chapter 14 would depend on the nature of the course—partial coverage in an introductory course and full coverage in an advanced course. Also, depending on the text editor used, relevant sections of Chapter W25 should be covered.

For use in a Linux system administration course, Chapters 4–10, 17, 18, W26, and W27, and relevant sections of Chapters 11 and 12 should be used.

In a course on introduction to system programming, Chapters 11, 15, and 16; relevant portions of Chapters 4–10; and Chapters 17, 18, W20, and W21 should be covered.

Finally, people using this book for professional development, and in informal learning environments, can select relevant chapters that meet their individual learning objectives. For example, Chapter W24, Section W5.7, can be used by anyone who wants to learn the basics of git and GitHub.

The Design of Fonts

The following typefaces have been used in the book for various types of text items:

Font
> Minion Pro, italic
> Minion Pro, bold
> Courier Std
> Courier Std, bold

Text Item
> Key terms:
> - Whatever directory you are currently in is known as the present working directory.

Files/directories/symbolic constants/menu paths:

- The directory first and the file myfile2 are now removed.
- Make the Options menu choice Save Options
- Make the pull-down menu choice File>Quit
- A socket with AF_INET address family is known as the Internet domain socket.

Commands, program code, output of commands and programs, and options:

- Use the man and whatis commands to find information about the passwd command.
- The output of the date command is Thu Apr 7 13:53:30 PKT 2016.
- You can use the −l option to display the long listing.
- The following session shows the Bash shell script in the for_demo1 file:

```
$ cat for_demo1
#!/bin/sh
for people in Debbie Jamie John Kitty Kuhn Shah
```

```
do
echo "people"
done
exit 0
$
```

Keystrokes:

- <Enter>, <Alt+V>, <F1>, a

Prompts, messages, dialogs, and windows:

- A user who runs a write or talk command sees the message Permission denied.
- The system then displays the `login:` prompt.
- In the Find `file:` dialog box that opens.
- Click the OK button in the Save window.

User input:

- $ ssh 192.168.0.8
- Password for bob@pcbsd-2467: XXX

We take full responsibility for any errors in the book. You can send your error reports and comments to us via the email address provided in the `README.md` file at the GitHub site mentioned previously. We will incorporate your feedback and fix any errors in subsequent printings.

A Note to Students and Readers

The way to use this book most effectively is to learn by doing. That is, after all, why we have provided so many command line sessions. Type in what you see in those sessions and compare it to the output you get on your Linux system.

And install the Linux system of your choice, hopefully chosen from the ones we have illustrated everything in the book with, on your own PC. Then experiment and play with Linux. Play is the foundation upon which learning is built.

Acknowledgments

First and foremost, we would like to acknowledge the outstanding stewardship and guidance provided by our editor, Randi Cohen, at CRC Press/Taylor & Francis Group. Education, in the modern world, is most importantly a project of advocacy, and she stands as our greatest advocate.

Second, we were guided along in the excellent development process leading up to the final form of the book by Eve Strillacci and Holly Omand at CRC Press/Taylor & Francis Group. They provided many insights into how best to reach our target audience, through complete and articulate reports based upon the input of numerous reviewers.

Third, we thank Todd Perry at CRC Press/Taylor & Francis Group and Paige Force, Rebecca Dunn, and Alexandra Andrejevich at codeMantra.

Personal Acknowledgments

I thank my parents, wife, children, and siblings for their love, support, and trust. They have all been a positive influence in my life, have helped me in many ways, and remain my biggest supporters. I can never pay back the grace that they have extended to me over the years. My mother, a class act in motherhood, has been fighting serious illnesses for the past several years. I pray for her good health and peace of mind.

I lost my father, Abbujee, on April 3, 2018, as I was finalizing the last of my chapters for the first draft. He was my primary inspiration for book writing. Even during the final days of his life, as he was fighting bilateral pneumonia and was placed at the intensive care unit (ICU) of a local hospital, he would ask me about the status of this book. He was elated to hear when I told him that I was about to submit my final chapter. A man out of this world, he was an avid reader even at the age of 97 and an icon of wisdom, courage, care, intellect, logical reasoning, and mental toughness, and remains my inspiration for the quest of knowledge discovery, rational thinking, and service to fellow humans. But, above all, he was the most dedicated and model father that one can envision.

After completing his masters in philosophy in 1945 from the prestigious Government College, Lahore, Pakistan, he pursued a 45-year career as a journalist and served in famous national English dailies, including over 35 years at the famous London-based *Civil and Military Gazette* and *The Pakistan Times*. He was well read in philosophy, world history (American, European, and Muslim), various religions (Judaism, Christianity, Islam, Hinduism, Buddhism, and Zoroastrianism), and world politics, and had written many thought-provoking articles on philosophy, constitution, comparative religions, and Islam. His seminal work was published as a 162-page book, *A Fresh Look at the Quran: Unmasking the Truth*, published by Amazon in 2013.

About a month before his death, my father finished rereading the following books that he had won in a declamation contest as a college freshman in 1942: *The Prince* by Niccolò Machiavelli, and *Plays* and *What Then Must We Do?* by Leo Tolstoy. He was in the middle of rereading the classic *Creative Evolution* by Henri Bergson when he was admitted to the ICU on March 18, 2018. His lead pencil remains placed between pages 276 and 277 of this book. The next book on his to-read list was *The Autobiography on an Unknown Indian* by Nirad C. Chaudhri. Without his and my wife's continuous encouragement and support, writing this book would never have been possible.

My wife, Robina, and children, Maham, Ibraheem, and Hassan, have been extremely patient and supportive during the course of this project. Thank you for your understanding and support, guys! I could not have done it without Robina, an exemplary mother and wife. At the time of writing the first edition of the book, Hassan and Maham were in elementary school, and Ibraheem was not even born. Now, Hassan is the CEO of infinione (infinione.com), a Los Angeles-based technology company that he founded 5 years ago when he was a junior at the University of Southern California, and a passionate world traveller and avid hiker, who has already been to more than a dozen countries. Winter glacier hiking in Iceland has thus far been the hallmark of his expeditions. After finishing her masters in textile design innovation from Nottingham Trent University, Nottingham, Maham is a textile designer at Republic, a top brand in fashion, and an adjunct faculty member at the Beaconhouse National University, Lahore. With the Best Designer award under her belt in a national competition, several of her designs have already been marketed through well-known fashion women's wear brands such as Beechtree. Ibraheem is a young 6'2" teenager, with a love for mathematics and science, and has the goal to become a top-notch scientist as well as a professional basketball player. He has already played at the national level at the age of 16.

My very special thanks to my sisters, Rizwana and Farhana, and brothers, Masood, Nadeem, Aqeel, and Nabeel, for their friendship and care. They are the best siblings one could possibly ask for. They are always there for me whenever I need them. I convey my sincere gratitude to them, my sisters-in-law, Maimoona, Sadaf, and Farzana, and brothers-in-law, Hamid and Abdul Ghafoor, for taking care of our father and ailing mother during the hours of their need. Folks, I will forever remain indebted to you for performing my share of service toward our parents. I hope to be able to offer a payback in some form someday.

I also thank my colleagues and friends at Punjab University College of Information Technology, Institute of Business and Information Technology, Department of Space Science, Institute of Chemistry, Department of Persian, Department of Kashmir Studies, and Institute of Geology for their help and encouragement through thick and thin as I was completing this project and handling the difficult university affairs, particularly during the past 4 years. They are too many to name here, but they know who they are. Thank you fellows!

I owe very special thanks to my coauthor, Robert Koretsky, for his friendship and out-of-the-way support throughout this project. He was there whenever I needed his help. In particular, after the death of my father, he took over the responsibility to complete my remaining work without even my request. Bob, all this meant a lot to me. I am truly indebted to you for your encouragement and support throughout. This book would never have been published in time without you.

Syed Mansoor Sarwar

Without the love, encouragement, and guidance of my wife, Kathe, this work would never have been realized. I would like to also thank my two children, Tara and Cody, for always being there for me. A special thanks to my two grandsons, Victor and Garvey.

Dr. Syed Mansoor Sarwar provides the foundations for the vision of this book as a conceptual whole. My greatest debt of gratitude goes to him for this, and many other things, in friendship and admiration.

Robert M. Koretsky

Authors

Syed Mansoor Sarwar is a professor and principal at Punjab University College of Information Technology, Lahore, Pakistan, and a former tenured associate professor at the Multnomah School of Engineering at the University of Portland (UP), Portland, Oregon. He received his MS and PhD in computer engineering from the Iowa State University (ISU), Ames, Iowa, and has over 30 years of post-PhD experience in teaching and research. He has authored over 40 research publications in international journals and conferences, and authored, as lead author, five books published by top world publishers. He was nominated for the Best Graduate Researcher Award at ISU for his PhD research, and Best Researcher and Best Teacher awards at UP. He has been learning, using, and teaching UNIX since 1986. His family is spread over two continents. He currently resides in Lahore with his wife, daughter, and younger son. His older son runs a software company in Los Angeles.

Robert M. Koretsky is a retired lecturer in mechanical engineering at the Multnomah School of Engineering at the University of Portland, Oregon. Principally educated at the Pratt Institute in Brooklyn, New York, he has also worked as an automotive engineering designer with Freightliner Corp. in Portland. He currently resides in Portland, with his wife, two children, and two grandchildren.

Together, they are the authors of four textbooks on UNIX and Linux.

1

Overview of Operating Systems

OBJECTIVES

- To explain what an operating system is
- To briefly describe operating system services
- To describe character and graphical user interfaces
- To discuss different types of operating systems
- To briefly describe the Linux operating system
- To give an overview of the structure of a contemporary system
- To briefly describe the structure of the Linux operating system
- To detail some important system setups
- To briefly describe the history of the Linux operating system
- To provide an overview of the different types of Linux systems

1.1 Introduction

Many operating systems are available today, some general enough to run on any type of computer (from handheld, Internet of things (IoT) devices, to cloud-based, warehouse-scale server clusters) and some specifically designed to run on a particular type of computer system, including real-time embedded computer systems used to control the movement of mechanical devices such as robots, IPads, and a plethora of cell phone models. In this chapter, we describe the purpose of an operating system and the different classes of operating systems. Before describing the different types of operating systems and where Linux fits in this categorization, we present a layered diagram of a contemporary computer system and discuss the basic purpose of an operating system. We then describe the different types of operating systems and the parameters used to classify them. Finally, we identify the class that Linux belongs to and briefly discuss the different members of the Linux family.

The people who use Linux comprise application developers, systems analysts, programmers, administrators, business managers, academicians, and people who just wish to read their e-mails. From its inception in the early 1990s as a hobbyist project, it was explosively developed in conjunction with the development of the Internet, via a vast community of developers, and then completely adopted for commercial uses. In the fully developed and mature versions of today, Linux has an underlying functionality that is complex but easy to learn, and extensible yet easily customized to suit a user's style of computing. One key to understanding its longevity and its heterogeneous appeal is to study the history of its evolution.

1.2 What Is an Operating System?

A computer system consists of various hardware and software resources, as shown in a layered fashion in Figure 1.1. The primary purpose of an operating system is to facilitate easy, efficient, fair, orderly, and secure use of these resources. This purpose can be conveniently described as a controlling function that ensures concurrency, virtualization, and persistence. It allows the users to employ

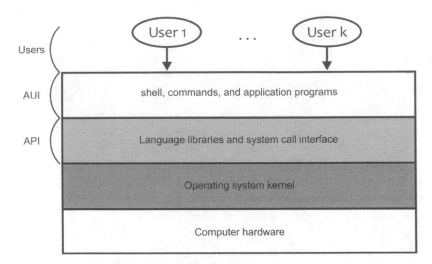

FIGURE 1.1 A layered view of a contemporary computer system.

application software—spreadsheets, word processors, Web browsers, e-mail software, and other programs. Programmers use language libraries, system calls, and program generation tools (e.g., text editors, compilers, and version control systems) to develop software. Fairness is obviously not an issue if only one user at a time is allowed to use the computer system, including single-user desktop systems, laptops, tablet computers, and cell phones. However, if multiple users are allowed to use the computer system, fairness and security are two main issues to be addressed by the operating system designers.

Hardware resources include keyboards, touch pads, display screens (may also be touch screens), main memory (commonly known as *random access memory* or RAM), disk drives, network interface cards, and central processing units (CPUs). Software resources include applications such as word processors, spreadsheets, games, graphing tools, picture- and video-processing tools, and Internet-related tools such as Web browsers. These applications, which reside at the topmost layer in the diagram, form the *application user interface* (AUI). The AUI is glued to the operating system kernel via the language libraries and the system call interface. The system call interface comprises a set of functions that can be used by the applications and library routines to execute the kernel code for a particular service, such as reading a file. The language libraries and the system call interface comprise what is commonly known as the *application programming interface* (API). The kernel is the core of an operating system, where issues such as CPU scheduling, memory management, disk scheduling, and interprocess communication (IPC) are handled. The layers in the diagram are shown in an expanded form for the Linux operating system in Figure 1.2, which are described briefly.

There are two ways to view an operating system: top down and bottom up. In the bottom-up view, an operating system can be viewed as a software system that allocates and deallocates system resources (hardware and software) in an efficient, fair, orderly, and secure manner. For example, the operating system decides how much RAM space is to be allocated to a program before it is loaded and executed. The operating system ensures that only one file is printed on a particular printer at a time, prevents an existing file on the disk from being accidentally overwritten by another file, and further guarantees that, when the execution of a program given to the CPU for processing has been completed, the program relinquishes the CPU so that other programs can be executed. Thus, in the bottom-up view, the operating system is a resource manager.

In the top-down view, which we espouse in this textbook, an operating system can be viewed as a piece of software that isolates you from the complications of hardware resources. You therefore do not have to deal with the extremely difficult (and sometimes impossible for most users) task of interacting with these resources. For example, as a user of a computer system, you don't have to write the code that allows you to save your work as a file on a hard disk, use a mouse as a point-and-click device, use a touch screen or touch pad, or print on a particular printer. Also, you do not have to write new device driver software for

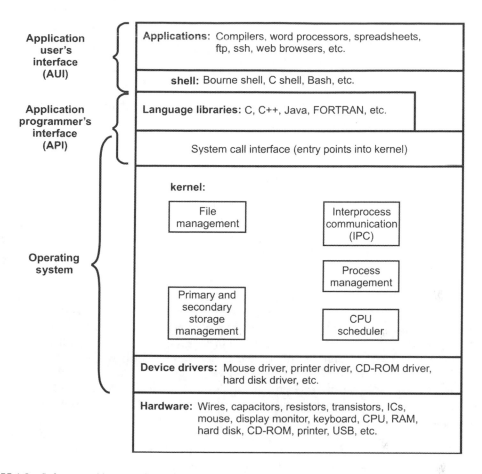

FIGURE 1.2   Software architecture of the Linux operating system.

a new device (e.g., mouse, disk drive, or DVD) that you buy and install in your system. The operating system performs the task of dealing with complicated hardware resources and gives you a comprehensive machine with a simple, ready-to-use interface. This machine allows you to use simple commands to retrieve and save files on a disk, print files on a printer, and play movies from a DVD. In a sense, the operating system provides a virtual machine that is much easier to deal with than the physical machine. You can, for example, use a command such as `cp memo letter` to copy the memo file to the letter file on the hard disk in your computer without having to worry about the location of the memo and letter files on the disk, the structure and size of the disk, the brand of the disk drive, and the number or name of the various drives (hard drive, Solid State Drive (SSD), DVD, etc.) on your system.

1.3 Operating System Services

An operating system provides many ready-made services for users. Most of these services are designed to allow you to execute your software, both application programs and program development tools, efficiently and securely. Some services are designed for housekeeping tasks, such as keeping track of the amount of time that you have used the system. The major operating system services therefore provide mechanisms for the following secure and efficient operations and processes:

* Execution of a program
* Input and output operations performed by programs

- Communication between processes
- Error detection and reporting
- Manipulation of all types of files
- Management of users and security

A detailed discussion of these services is outside the scope of this textbook, but we discuss them briefly when they are relevant to the topic being presented.

1.4 Character (Command Line) versus Graphical User Interfaces

To use a computer system, you have to give commands to its operating system. An input device, such as a keyboard, is used to issue a command. If you use the keyboard to issue commands to the operating system, the operating system has a character user interface (CUI), commonly known as the *command line interface*. If the primary input device for issuing commands to the operating system is a point-and-click device, such as a mouse, a touch screen, or a touch pad, the operating system has a graphical user interface (GUI). Most, if not all, operating systems have both CUI and GUI, and you can use either. Some have a command line as their primary interface but allow you to also run software that provides a GUI. Operating systems such as UNIX and Linux have CUIs, whereas Mac OSX and Microsoft Windows primarily offer GUIs but have the capability to allow a user to enter text in a terminal screen. Although Linux comes with a CUI as its fundamental interface, it can just as easily run the GUI-based software that uses Wayland to provide the GUI interface. We primarily discuss the Linux GUI in Chapters 17 and W28.

Although a GUI makes a computer easier to use, it gives you an automated setup with reduced flexibility. A GUI also presents an extra layer of software between you and the task that you want to perform on the computer, thereby making the task slower. That is because this GUI layer of software consumes a significantly larger part of system resources to maintain its operability. In contrast, a CUI gives you ultimate, fine-grained control of your computer system and allows you to run application programs any way you want. A CUI is also more efficient because a minimal layer of software is needed between you and your task on the computer, thereby enabling you to complete the task faster. The CUI software consumes a significantly smaller part of system resources to maintain its operability. It is also malleable and gives the user more control. Because many people are accustomed to the graphical interfaces of popular gizmos and applications, such as IoT devices, high-powered video games, and Web browsers, the character interface presents an unfamiliar, and sometimes less intuitive, and difficult style of communicating commands to the computer system. However, computer science students are usually able to meet this challenge after a few hands-on sessions.

1.5 Types of Operating Systems

Operating systems can be categorized according to the number of users who can use the system at the same time and the number of processes (executing programs) that the system can run simultaneously. These criteria lead to three types of operating systems:

- *Single-user, single-process system*: These operating systems allow only one user at a time to use the computer system, and the user can run only one process at a time. Such operating systems are commonly used for PCs. Examples of these operating systems are earlier versions of Mac OS, DOS, and many of Microsoft's Windows operating systems.
- *Single-user, multiprocess system*: As the name indicates, these operating systems allow only a single user to use the computer system, but the user can run multiple processes simultaneously. These operating systems are also used on PCs. Examples of such operating systems are OS/2, Windows XP Workstation, and batch operating systems. Batch processing is still commonly

used in mainframe computers, and most modern operating systems including UNIX, Microsoft Windows 10, Linux, and Mac OSX perform some tasks in batch mode. Even smartphone/IoT operating systems, including Android and iOS, perform tasks in batch mode.

* *Multiuser, multiprocess system*: These operating systems allow multiple users to use a computer system simultaneously, and every user can run multiple processes at the same time. These operating systems are commonly used on computers that support multiple users in organizations such as universities and large businesses. Examples of these operating systems are UNIX, Linux, Windows 10 Server, z/OS, and z/VM.

Multiuser, multiprocess systems are used to increase resource utilization in the computer system by multiplexing expensive resources such as the CPU. This capability leads to increased system throughput (the number of processes finished in a unit of time). Resource utilization increases because, in a system with several processes, when one process is performing input or output (e.g., reading input from the keyboard, capturing the event generation of a mouse click, or writing to a file on the hard disk), the CPU can be taken away from this process and given to another process—effectively running both processes simultaneously by allowing them both to make progress [one is performing input/output (I/O) and the other is using the CPU]. The mechanism of assigning the CPU to another process when the current process is performing I/O is known as *multiprogramming*. Multiprogramming is the key to all contemporary multiuser, multiprocess operating systems. In a single-process system, when the process using the CPU performs I/O, the CPU sits idle because there is no other process that can use the CPU at the same time.

Operating systems that allow users to interact with their executing programs are known as *interactive operating systems*, and the ones that do not are called batch operating systems. Batch systems are useful when programs are run without the need for human intervention, such as systems that run payroll database programs. Almost all well-known contemporary operating systems (UNIX, Linux, Windows, etc.) are interactive. UNIX and Linux also allow programs to be executed in batch mode, with programs running in the background (see Chapter 10 for details on "background process execution" in Linux). Multiuser, multiprocess, and interactive operating systems are known as *time-sharing systems*. In time-sharing systems, the CPU is switched from one process to another in quick succession. This method of operation allows all the processes in the system to make progress, giving each user the impression of sole use of the system. Examples of time-sharing operating systems are UNIX, Linux, and Windows 10 Server.

1.6 The Linux Distribution Families

In the early 1990s, the name Linux referred to a single operating system, but it is now used to refer to three basic families of distributions, or "flavors," of operating systems that are offshoots of the original in terms of their user interface. In Section 1.8, we give a brief history of some of the most popular and developmentally influential Linux systems. In Section 1.9, we sketch the very large variety and distribution-specific implementation of some of the common features in all distributions.

1.7 Linux Software Architecture

Figure 1.2 shows a layered diagram for a Linux-based computer system, identifying the system's software components and their logical proximity to the user and hardware. We briefly describe each software layer from the bottom up.

1.7.1 Device Driver Layer

The purpose of the device driver layer is to interact with various hardware devices. It contains a separate program for interacting with each device, including the hard disk driver, DVD or CD-ROM driver, keyboard driver, mouse driver, touch pad driver, and display driver. These programs execute on behalf of the Linux kernel when a user command or application needs to perform a hardware-related operation such as a file read that translates to one or more disk reads. The user doesn't have direct access to these programs and therefore can't execute them as commands.

1.7.2 Linux Kernel

The Linux kernel layer contains the actual operating system. Some of the main functions of the Linux kernel, listed in Figure 1.2, are described in this section. In addition, the kernel performs several other tasks for fair, orderly, and safe use of the computer system. These tasks include managing the CPU, printers, and other I/O devices. The kernel ensures that no user process takes over the CPU forever, that multiple files are not printed on a printer simultaneously, and that a user cannot terminate another user's process.

1.7.2.1 Process Management

This part of the kernel manages processes in terms of creating, suspending, resuming, and terminating them, and maintaining their states. It also provides various mechanisms for processes to communicate with each other and schedules the CPU to execute multiple processes simultaneously in a time-sharing system. IPC is the key to the client–server-based software that is the foundation for Internet applications, including Web browsing (HTTP), File Transfer Protocol (FTP), and remote login [secure shell (SSH)]. The Linux system provides three primary IPC mechanisms/channels:

- *Pipe*: Two or more related processes running on the same computer can use a pipe as an IPC channel. Typically, these processes have a parent–child or sibling relationship. A pipe is a temporary channel that resides in the main memory and is created by the kernel, usually on behalf of the parent process.
- *Named pipe*: A named pipe, also known as First-In First-Out (FIFO), is a permanent communication channel that resides on the disk and can be used for IPC by two or more related or unrelated processes running on the same computer.
- *Berkeley Software Distribution (BSD) socket*: A BSD socket is also a temporary channel that allows two or more processes in a network (or on the Internet) to communicate, although processes on the same computer can also use them. Sockets were originally a part of the BSD UNIX only, but they are now available on every Linux system. Internet software such as Web browsers, FTP, SSH, and electronic mailers are implemented by using sockets.

We discuss these mechanisms of IPC in detail in the major component of the book, *Linux System Programming*, in particular, Chapter W20 at the book website.

1.7.2.2 File Management

This part of the kernel manages files and directories, also known as folders. It performs all file-related tasks, including file creation and removal, directory creation and removal, setting access privileges on files and directories, and maintaining their attributes, such as file size. A file operation usually requires manipulation of a disk. In a multiuser system, a user must never be allowed to manipulate a disk directly because it contains files belonging to other users, and user access to a disk poses a security threat. Only the kernel must perform all file-related operations, such as file removal. Also, only the kernel must decide where and how much space to allocate to a file.

1.7.2.3 Main Memory Management

This part of the kernel allocates and deallocates RAM in an orderly manner so that each process has enough space to execute properly. It also ensures that part or all of the space allocated to a process does not belong to some other process. The space allocated to a process in the memory for its execution is known as its *address space*. The kernel ensures that no process accesses an area of memory that does not belong to its address space. The kernel maintains areas in the main memory that are free to be allocated to processes. The kernel code that performs this task is called the *free space manager*. When a program is to be loaded in the main memory, the free space manager allocates adequate space for it, and the *loader* loads the program into this space. The kernel also records where all the processes reside in the memory so that, when a process tries to access main memory space that does not belong to it, the kernel can terminate the process and give a meaningful message to the user. When a process terminates, the kernel deallocates the space allocated to the process and puts it back in the free space pool so that it can be reused.

1.7.2.4 Disk Management

The kernel is also responsible for maintaining free and used disk space and for the orderly and fair allocation and deallocation of disk space. It decides where and how much space to allocate to a newly created file. The kernel code that performs this task is known as the *disk storage manager*. Also, the kernel performs *disk scheduling*, deciding which request to serve next when multiple requests for file read, write, and so on arrive for the same disk.

1.7.3 System Call Interface

The system call interface layer contains entry points into the kernel code. Because the kernel manages all system resources, any user or application request that involves access to any system resource must be handled by the kernel code. However, user processes must not be given open access to the kernel code for security reasons so that user processes can invoke the execution of kernel code. Linux provides several openings, or function calls, into the kernel, known as *system calls*. There are numerous system calls that deal with the manipulation of processes, files, and other system resources. These calls are well tested, and most of them have been used for several years, so their use poses much less of a security risk than if any user code was allowed to perform the task.

1.7.4 Language Libraries

A *language library* is a set of prewritten and pretested functions in a programming language available to programmers for use with the software that they develop. The availability and use of libraries save time because programmers do not have to write these functions from scratch. This layer contains libraries for several languages, including C, C++, C#, Java, Perl, and Python. For the C language, for example, there are several libraries, including a string library (which contains functions for processing strings, such as a function for comparing two strings) and a math library (which contains functions for mathematical operations, such as finding the cosine of an angle).

As we stated earlier in this chapter, the libraries and system call interface form what is commonly known as the API. In other words, programmers who write software in a language such as C can use in their code the prewritten functions available in the various C libraries and system calls.

1.7.5 Linux Shell

The *Linux shell* is a program that starts running when you log on and interprets the commands that you enter. The most popular shells are the Bourne Again shell (bash, the default shell in all major distributions of Linux), Bourne shell (sh), C shell (csh), TC shell (tcsh), and Korn shell (ksh). We show the usage of shell commands in all of the core chapters of this book, and shell scripting in Chapters 12 and 13 for bash, and Chapters W29 and W30 for the TC shell.

1.7.6 Applications

The application layer contains all the applications (tools, commands, and utilities) that are available for your use. A typical Linux system contains hundreds of applications; we discuss the most useful and commonly used applications throughout this textbook. When an application that you're using needs to manipulate a system resource (e.g., reading a file), it needs to invoke some kernel code that performs the task. An application can find the appropriate kernel code to execute in one of two ways: (1) by using a proper library function and (2) by using a system call. Library calls constitute a higher level interface to the kernel than system calls, which makes library calls a bit easier to use. However, all library calls eventually use system calls to begin execution of the appropriate kernel code. Therefore, the use of library calls in software results in slightly slower execution. A detailed discussion of language libraries and system calls is, generally, beyond the scope of this textbook. However, we discuss and show the use of several library calls and system calls in Chapters 15 and 16, and W20 and W21, on Linux system programming.

The user can use any command or application that is available on the system. As we mentioned earlier in this chapter, this layer is commonly known as the AUI.

1.8 Historical Development of the Linux Operating System

Why has Linux achieved the status and position in the marketplace it has now?

In the 21st century, information, more than ever, is the common currency between everyone. Because Linux, as a multiuser, multiprocessing stream computer operating system that can run on various processor architectures, processors that are found on very small- to very large-scale computer hardware used by *everyone*, it is the premier expediter of information. It is used by casual, individual users on their cell phones and home computers, to participate in widely dispersed social media, all the way up to multiprocessor servers in a cloud configuration at a commercial facility that services international populations. As such, it can link your personal computing device to the Internet with a standard browser such as Opera, or Firefox, and

> Linux basically runs the Internet because of the servers that it runs on.

And what really is the difference between the major families of Linux we detail in the chapters in this book? When you say "Linux," what you can be referring to is either the kernel or a particular distribution. The kernel and its API, between distributions, or "flavors," in the Debian, Slackware, and Red Hat families are very uniform. This also depends on what release of the kernel the particular distribution, and the version of the distribution, is using. This uniformity is overseen and controlled by Linus Torvalds and a legion of developers. From our subjective perspective, given what distributions we use in this book, we describe some of the essential important differences between those distributions in Section 1.8.3.

As stated in Section 1.1, the people who use the Linux operating system are application developers, gamers, systems analysts, Web programmers, system administrators, business managers, scientists, academicians, and people who just want to read their e-mails or make phone calls on their way home from work.

From its inception in the early 1990s, as a hobbyist project, it owes its historic development to the Internet, and the developer base that has, and continues, to forge the kernel components, and the entire suite of the eight features (and more) listed earlier. And then, of course, it is highly endorsed for commercial use. Today's Linux has an underlying functionality that is complex but easy to learn because of its GUI and extensible yet easily customized to suit a user's style of computing. In our opinion, the desktop GUI is better than the equivalent on Windows or OS X machines. One of the primary keys to understanding its longevity, and its heterogeneous appeal, is to study the history of its evolution, which briefly follows next.

1.8.1 Beginnings

Before we describe the evolution of Linux, first we have to ask, "Why is this operating system so user friendly and accommodating?" Part of the answer is: This ever-evolving operating system that is accepted and used throughout the world was developed in response to the needs and activities of a heterogeneous community of developers and computer users. It grew, changed, and improved because of the work and cooperation of many diverse, and sometimes opposing, individuals and groups.

Linux has continuously grown, changed, and improved alongside the development of computer hardware, software applications, networking, and other components of the "computer revolution." The Linux project started as a personal and subjective endeavor of Linus Torvalds but, because of Internet-distributed development by thousands of programmers, has exploded into a universal and generic technical tool. Thus, its various audiences must have found some basic advantages in this tool—particularly the largest audience of ordinary, single-computer users. Separating the influences of various user groups in the development of Linux is difficult, but one thing cannot be disputed: Linus Torvalds oversees it all. Moreover, because the system is fundamentally an open software system—that is, the source code is freely distributed among the community of users—its evolution has been shaped to some extent by a populist mind-set. For example, this development model is expedited in the 21st century by development resource and source code repositories such as GitHub. See the Web Resources available at the GitHub site for the printed book for a link to GitHub. The instructions for accessing the book GitHub site are given in the Preface. And it will continue to be in the future, thanks to the pervasive use of the Internet in social media life, academic settings, and business and professional settings.

It is the underlying core functionality of Linux that brings together its diverse audiences into a community not so much in the sociological sense, but more in an independent "Do-It-Yourself," intellectual sense. As you delve into the subject matter of this textbook, you might wonder where you fit into the Linux community and how its functionality might be adapted for your uses. Essentially, it is the style of your interaction with the computer that will be the most important, invigorating, and critical aspect of your work with the Linux operating system.

The development of other contemporary operating systems is motivated and informed by completely different forces and bases (primarily commercialization) than those that motivated the inception and development of Linux (primarily a user-friendly, text-based operating system). The history of Linux is a record of how a system should be developed, regardless of how you believe that system should be structured, how you think it should function (whatever your user perspective), and whom you believe should control that development.

Table 1.1 describes the three main branches of Linux systems: Debian, Slackware, and Red Hat. We show them as they were developed from the early 1990s until the time of the writing of this book. The current representative derived systems we use in this book (Debian 9.1, Ubuntu 16.04, Linux Mint 18.2, and CentOS 7.4) are shown in the rightmost columns. The approximate dates of the initial and current development of each of the branches, and distributions, are shown at the top. Also see the Web Resources listed at the GitHub site for this book for more detailed diagram of the history of many more Linux distributions, what each was derived from, and their initial release dates.

A very important historical footnote to the development of Linux in its early years was the coding of the operating system in the high-level programming language C. Written in C, Linux is very portable. Also, C is much easier to program with than a lower level and processor architecture-specific language, which is characteristically difficult to program with.

TABLE 1.1

Schematic Linux Timeline in Years

1993	1994	1995	2000	2005	2010	2016	2017
Debian							9.1
				Ubuntu		16.04	17.10
				Mint			18.2
Slackware							14.2
	SUSE			openSUSE			43.2
	Red Hat						7.4
				CentOS			7.4

1.8.2 The History of Shells

When you install, or in some capacity, use Linux, you must realize that the Gnu's Not UNIX (Gnu) utilities the Free Software Foundation contributed to the kernel, such as the various Linux shells, are the core of the user interface to the kernel.

Development of the shell as a Linux utility parallels the development of Linux and Linux systems themselves. The first commercially available shell, the Bourne shell, was written by Stephen R. Bourne. The C shell, written in the late 1970s primarily by Bill Joy, was made available soon after in 2.0 BSD UNIX. When introduced, it provided a C-program-like programming interface for writing shell scripts. Following the development of the C shell, the Korn shell was introduced officially in Unix System V, Release 4, in 1989. Written by David Korn of Bell Laboratories, it included a superset of Bourne shell commands that had more functionality. It also included some useful features of the C shell. An additional shell derived from the C shell is the Tenex C (TC) shell. It was written and implemented in the late 1970s.

In all of the current representative Linux distributions we show in this book, the default shell is the Gnu bash shell. Bash was written by Brian Fox for the Gnu Project and was first released in 1989. It is a superset of the Bourne, Korn, and C shells, with several programming and interactive features that make it more useful for all users. These features include command line editing, unlimited command history, job control, a larger set of shell functions, and other programming and interactive advantages.

We discuss the development history of the Linux shell in more detail in Chapter 2.

The major Linux shells have slightly different features and command sets. In this textbook, we discuss common features and command sets for all Linux shells and versions. Whenever we discuss a feature or command that is particular to a shell or version, we state that specifically.

If you want to know what shells are available on your Linux system, on the default bash command line, type **more /etc/shells**

The output of this command on our Linux Mint system is as follows:

```
$ more /etc/shells
# /etc/shells: valid login shells
/bin/sh
/bin/dash
/bin/bash
/bin/rbash
/bin/csh
$
```

1.8.3 Future Developments

As we see them currently, the following three developments pose the biggest challenge to the Linux community of developers:

1. Probably the most exciting and challenging current Linux development for all distributions of Linux focuses on the incorporation of the Zettabyte File System (ZFS) into the kernel, and having the "boot," or system disk, use ZFS. ZFS was developed by Sun Microsystems in the years before its incorporation into the Solaris family in 2006. It is now a kernel-loadable module on our representative Debian-family systems, Ubuntu and Linux Mint. As a "user file system," ZFS does not preclude the use of EXT4. We give a listing of Web Resources at the GitHub site for this book for open-zfs.org.

2. Another major challenge on the horizon for Linux systems is the further development and dispersion of systemd in the kernel, most importantly, in the area of container virtualization. Examples of container virtualization are Linux Containers version C/Linux Containers version D (LXC/LXD) and Docker. Currently, systemd is a "superkernel" suite of system management daemons, API libraries, and utilities, designed as a central management and configuration

platform for Linux. systemd is used on a majority of the current implementations and official releases of the Linux kernel. It is a Linux **init** system (the process called by the Linux kernel to initialize the user space during the Linux start-up process and, most importantly, manage all processes and services afterward), thus replacing the Linux System V and BSD-style **init** daemon. The name systemd adheres to the convention of making daemons easier to distinguish by having the letter "d" as the last letter of the filename.

3. Finally, the replacement of the X Window System Protocol by various other software systems, such as Wayland, promises to yield a smaller, more effective GUI system. Wayland is a protocol that specifies the communication between a display server (called Wayland compositor) and its clients, as well as a reference implementation of the protocol in C. We give a listing of Web Resources at the GitHub site for this book for more information on Wayland.

1.9 A Basic Comparison of Linux System Distributions

As shown in Table 1.1, the general development of Linux system distributions proceeded along three main branches: Debian, Slackware, and Red Hat. And as seen in the Web Resources diagram, available at the book GitHub site, at http://futurist.se, there are a multitude of other distributions, whose history and specific implementation details run parallel to (or in some cases, diverge from) the three major branches.

There is a very large variety and distribution-specific implementation of features in all distributions shown in the Web Resources diagram. Some of these specific distribution implementation differences are listed, in somewhat of a subjective order on our part, and are briefly described as follows:

1. Whether ZFS can be easily installed and maintained on the system.
2. The version of the kernel used and the specific details of the implementation of the AUI.
3. The package management system used in any particular distribution (such as the Advanced Package Tool (APT) or the Yellow Dog Updater, Modified (YUM)).
4. The variety of application software packages available via the package management system.
5. The approximate number of available precompiled and source packages.
6. Whether packages can be installed with a graphical package management front end (such as the Software Manager in Linux Mint), and how reliable, extensive, and inclusive its GUI is.
7. The file system organization from the file system root down to the user level.
8. The reliability of the method of maintaining each release, either as a "rolling" or as a stable upgrade.
9. The device driver base, or what and how many peripheral devices actually work with the system.
10. The GUI desktop management system(s) available, if the system can be installed by default with one, and their look and feel (such as Gnome and the K Desktop Environment (KDE)).
11. Instruction set architecture support (such as x86, x86-64, and Advanced RISC Machine (ARM)).
12. The default file system type and the types of file system that can be readily accommodated (such as Btrfs, XFS, ZFS, ext2, ext3, ext4, and NFSv4).
13. Availability and version(s) of Gnu compiler collection components (such as Gnu C Compiler [gcc] and Python).
14. Availability of CD/DVD/Universal Serial Bus (USB) live media images, and if it is possible to "test drive" the system (perhaps even with "persistence") using one of these.
15. The availability of standard images for the system in popular Virtual Machine Managers (such as VirtualBox or LXC/LXD).
16. The security system used (SELinux, AppArmor).

All of these implementation differences, and many more, present personal choices to the user base. They allow for the customization of the user experience, given a wide range of use cases available to users.

1.9.1   Linux System Standardization

Distribution variety and divergence have the drawback that programs, and even commands, working on one distribution either don't exist or fail to work on another distribution. This defeats the inherent strength of user-friendliness of the system itself. Attempts have been made to standardize Linux and Linux—for example, via the IEEE Portable Operating System Interface (POSIX). This software standard not only covers Linux but also, in particular, specifies program operation and user interfaces, leaving their implementations to the developer. Several standards have been adopted, and more have been proposed. For example, adopted POSIX standards specify shell and utility standardization.

By far, the most inclusive and wide-ranging standardization mechanism is the Single UNIX Specification (SUS). It is a family of standards for computer operating systems, compliance with which is required to qualify for the name "Unix." The core specifications of the SUS are developed and maintained by the Austin Group, which is a joint working group of IEEE, ISO JTC 1 SC22, and The Open Group. See the listing of Web Resources at the GitHub site for this book in The Open Group for more information.

In addition, the Linux Standard Base (LSB) is a standardization specification protocol that promotes the compatibility between Linux distributions, by ensuring that compiled applications run on Linux systems that claim to be in conformance with the LSB. At the time of the writing of this book, LSB version 5 was the current specification. See the listing of Web Resources at the GitHub site for this book for LSB.

Summary

An operating system is software that runs on the hardware of a computer system to manage its hardware and software resources. It also gives the user of the computer system a simple, virtual machine that is easy to use. The basic services provided by an operating system offer efficient and secure program execution, I/O operations, communication between processes, error detection and reporting, and file manipulation.

Operating systems are categorized by the number of users that can use a system at the same time and the number of processes that can execute on a system simultaneously: single-user, single-process; single-user, multiprocess; and multiuser, multiprocess operating systems. Furthermore, operating systems that allow users to interact with their executing programs (processes) are known as *interactive systems*, and those that do not are called batch systems. Multiuser, multiprocess interactive systems are known as *time-sharing systems*, of which Linux is a prime example. The purpose of multiuser, multiprocess systems is to increase the utilization of system resources by switching them among concurrently executing processes. This capability leads to higher system throughput, or the number of processes finishing in unit time.

To use a computer system, the user issues commands to the operating system. If an operating system accepts commands via the keyboard, it has a CUI. If an operating system allows users to issue commands via a point-and-click device such as a mouse, it has a GUI. Although Linux comes with a CUI as its basic interface, it can run software based on the X Window System (Project Athena, MIT) that provides a GUI. Most Linux systems now have both interfaces. Mac OS X (Darwin), running on Apple products, is the most well-known GUI-based Linux system.

A computer system consists of several hardware and software components. The software components of a typical Linux system consist of several layers: applications, shell, language libraries, system call interface, Linux kernel, and device drivers. The kernel is the main part of the Linux operating system and performs all the tasks that deal with allocation and deallocation of system resources. The shell and applications layers contain what is commonly known as the AUI. The language libraries and the system call interface contain the API.

The historical development of Linux is characterized by an open systems approach, whereby the source code was freely distributed among users. Development of many versions of Linux progressed along three main branches. Compatibility releases of various versions have been aimed at standardizing the system. The POSIX and the Single UNIX Specification are related standardization effort. Three exciting and challenging new developments in the future of Linux systems, as we see them, are as follows:

1. Incorporation of the Zettabyte File System (ZFS) into the kernel as the root file system
2. Further development and dispersion of **systemd** and its "superkernel" functionality
3. Replacement of the X Window System Protocol with the Wayland Protocol.

Questions and Problems

1. What is an operating system?
2. What are the three types of operating systems? How do they differ from each other?
3. What is a time-sharing system? Be precise.
4. What are the main services provided by a typical contemporary operating system? What is the basic purpose of these services?
5. List one advantage and one disadvantage each for the CUI and the GUI.
6. What is the difference between a CUI and a GUI? What is the most popular GUI for Linux systems? Where was it developed?
7. What comprises the API and the AUI?
8. What is an operating system kernel? What are the primary tasks performed by the Linux kernel?
9. What is a system call? What is the purpose of the system call interface?
10. If you access a Linux system with the ssh command, write down the exact step-by-step procedure you go through to log on and log off. Include as many descriptive details as possible in this procedure so that if you forget how to log on, you can always refer back to this written procedure.
11. What is a shell? Name the most popular Linux shells. Log on to your Linux computer system and note the shell prompt being used.
12. How can you tell which variant from the main branches of Linux (see Table 1.1) is being used on the computer system that you log on to?
13. If you were designing a Single UNIX Specification standard, what would you include in it? You might want to research the already adopted Single Linux Specification, Release 4 standards, presented briefly earlier in the chapter and online, before answering this question.
14. What were the names of some of the systems that were the immediate predecessors of Linux? Where were these predecessors, and Linux itself, initially developed, and by whom?
15. Name the major versions and the three main branches of Linux development. Which is the commercial branch?
16. What three important characteristics of Linux during its early development helped popularize it? Explain how these characteristics apply to you as a Linux user, whatever your perspective.
17. Name the two most popular Linux systems that are the basis of most Linux systems. Where were they developed?
18. Trace the history of Linux by browsing the Web. How many Linux systems have been developed so far? How many non-Linux systems have been developed? What is the most popular Linux system for PCs? Why do you think it is so popular?
19. Name five popular members of each of the three major Linux families. What is the name of your Linux system, and which major family does it come from?

Advanced Questions and Problems

20. According to what your personal use case for both the hardware and the software of a Linux system would be, reprioritize the listing of implementation discriminators provided in Section 1.9, in a new list according to what you think would be most critical when choosing a particular distribution and version of Linux. In preparation for answering this question, define, in your own terms, what each of the 15 presented implementation discriminators in Section 1.9 is, and how you would situate each of them in a new, reprioritized list according to your own needs. For example, if you want to use your hardware and Linux system exclusively to stream media on a home network, which of the presented list items would be most important to you, and in terms of a newly prioritized listing?

Looking for more? Visit our sites for additional readings, recommended resources, and exercises.
  CRC Press e-Resource: https://www.crcpress.com/9781138710085.
  Authors' GitHub: https://github.com/bobk48/linuxthetextbook.

2

"Quick Start" into the Linux Operating System

OBJECTIVES

- To introduce the Linux character user interface (CUI) and show the generic structure of Linux commands
- To describe how to connect and log on to a computer running the Linux operating system
- To explain how to manage and maintain files and directories
- To show where to get online help for Linux commands
- To demonstrate the use of a beginner's set of utility commands
- To describe what a Linux shell is
- To briefly describe some commonly used shells
- To cover the following basic commands and operators:

  alias, biff, cal, cat, cd, chsh, cp, csh, echo, exit, hostname, login, logout, lp, lpr, ls, man, mesg, mkdir, more, mv, passwd, PATH, pg, pwd, rm, rmdir, set, ssh, su, sudo, talk, telnet, unalias, uname, whatis, whereis, who, whoami, write

2.1 Introduction

To start working productively in Linux, a beginner needs to know eight sequential topics in the order presented as follows:

1. How to type a syntactically correct command on the Linux command line. One of the most useful modes of interaction with the Linux system uses text-based, typed commands.

2. How to log in and log out of a computer running Linux, using one of the standard methods we show. Linux allows users to enter the operating system autonomously, do a combination of text- and graphics-oriented operations, and exit gracefully.

3. How to maintain and organize files in the file structure. Creating a tree-like structure of folders (also called directories) and storing files in a logical fashion in these folders is critical to working efficiently in Linux.

4. How to get help on commands and their usage. In the keyboard entry, command-based character user interface (CUI) environment, being able to find out, in a quick and easy way, how to use a command by typing it on the keyboard correctly, is imperative to working efficiently.

5. How to execute a small set of essential utility commands to set up or customize your working environment. Once a beginner is familiar with the right way to construct file maintenance commands, adding a set of utility commands makes each session more productive.

6. The essential ways to work with Linux shells, what they are, and how to find out what shell is running when you log in.

7. Ways to change your shell and what the shell environmental variables are.

8. What shell metacharacters are.

To use this chapter successfully as a springboard into the remainder of the book, you should carefully read, follow, and execute the instructions and command line sessions we provide, in the order presented. Each section in this chapter, and every subsequent chapter as well, builds on the information that precedes it. They will give you the concepts, command tools, and methods that will enable you to work and program using the Linux operating system.

Throughout this edition of the book, we illustrate everything using four representative Linux distributions: Debian-family Debian, Ubuntu, Linux Mint, and RedHat Enterprise Linux (RHEL)-family CentOS. By default, we present command line sessions and their output as seen on Debian-family Linux Mint. When a command or its output is significantly different on one of the representative systems from the default, we note this. A vast majority of commands and their output are uniform across the representative systems.

In all chapters, the major commands we want to illustrate are first defined in a "syntax box," with an abbreviated syntax description, which will clarify general components of those commands. The syntax box format and their descriptions are as follows:

Syntax: The exact syntax of how a command, its options, and its arguments are correctly typed on the command line

Purpose: The specific purpose of the command

Output: A short description of the results of executing the command

Commonly used options/features: A listing of the most popular and useful options and option arguments

---

2.2 The Structure of a Linux Command

Linux can very efficiently be used with a graphical user interface (GUI), especially by novices. But it is also essential, at critical times, to use a text-based CUI. Therefore, correctly typed syntax of Linux commands ensures subsequent correct execution of commands on all representative systems.

After a user successfully logs on to a Linux computer, a shell prompt, such as the $ character, appears on the screen. The shell prompt is simply a message from the computer system to say that it is ready to accept keystrokes on the command line that directly follows the prompt. The general syntax, or structure of a *single* Linux command (sometimes called a *simple* command), as it is typed on the command line is as follows:

```
$ command [[-]option(s)] [option argument(s)] [command argument(s)]
```

where

> $ is the command line or shell prompt from the computer;
> anything enclosed in **[ ]** is not always needed;
> **command** is the name of the valid Linux command for that shell in lowercase letters;
> **[-option(s)]** is one or more modifiers that change the behavior of command;
> **[option argument(s)]** is one or more modifiers that change the behavior of **[-option(s)]**;
> **[command argument(s)]** is one or more objects that are affected by **command**.

Note the following seven essentials:

1. A space separates command, options, option arguments, and command arguments, but no space is necessary between multiple option(s) or multiple option arguments.
2. The order of multiple options or option arguments is irrelevant.
3. A space character is optional between the option and the option argument.
4. Always press the <Enter> key to submit the command for interpretation and execution.

5. Options may be preceded by a single hyphen - or two hyphens, --, depending on the form of the option. The short form of the option is preceded by a single hyphen, the long form of the option is preceded by two hyphens. No space character should be placed between hyphen(s) and option(s).

6. A small percentage of commands (like `whoami`) take no options, option arguments, or command arguments.

7. Everything on the command line is case sensitive!

Also, it is possible and *very* common to type *multiple* Linux commands (sometimes called *compound* commands, to differentiate them from simple commands) on the same command line, before pressing the `<Enter>` key. The components of a multiple Linux command are separated with input and output redirection characters, to channel the output of one into the input of another. We show this in detail in Chapter 9.

The following are examples of single Linux commands typed on the Linux command line after the $ prompt and illustrate some of the variations of the correct syntax for a single command that may have options and arguments:

```
$ ls
$ ls -la
$ ls -la m*
$ lpr -Pspr -n 3 proposal.txt
```

The first example contains only the command. The second contains the command `ls` and two options, `l` and `a`. The third contains the command `ls`, two options, `l` and `a`, and a command argument, `m*`. The fourth contains the command `lpr`, two options, `P` and `n`, two option arguments, `spr` and `3`, and a command argument, `proposal.txt`.

You should also use the following rule of thumb: If the command executes properly, then you are returned to the shell prompt; if it does not execute properly, then you get an error message displayed on the command line, and then you are returned to the shell prompt. For example, if you type `xy` on the command line and then press `<Enter>`, usually you will get an error message saying that no such command can be found, and you are returned to the shell prompt so that you can keystroke a valid command.

This rule of thumb does not ensure that what you wanted to achieve by typing the syntactically correct command on the command line will be achieved. That is, you could execute a command and get no error messages. But the command may not have done the things you wanted it to do, simply because you used it with the wrong options or command arguments!

In addition, the following Web link is to a site that allows you to type in a single or multiple Linux command, and get a verbose explanation of the components of that command:

https://explainshell.com/

In-Chapter Exercises

1. Type the following commands on your Linux system's command line and note the results. Which ones are syntactically incorrect? Why? (The Bash prompt is shown as the $ character in each, and we assume that **file1** and **file2** exist)

```
$ la ls
$ cat
$ more -q file1
$ more file2
$ time
$ lsblk-a
```

2. How can you differentiate a Linux command from its options, option arguments, and command arguments?

3. What is the difference between a single Linux command and a multiple Linux command, as typed on the command line before pressing <Enter>?

4. If you get no error message after you enter a Linux command, how do you know that it actually accomplished what you wanted it to?

2.3 Logging On and Logging Off

How can you log on to a Linux computer and then gracefully leave?

Using one of these general ways, or a hybrid, combined version of them:

- *Stand-alone*: Use a stand-alone system, where Linux is the only operating system on the hardware.
- *Remote*: Connect to a remote computer running Linux from a computer running Linux or another operating system.
- *Virtual*: Start Linux as a guest operating system in a virtual environment, such as Linux Containers version C/Linux Containers version D (LXC/LXD) containers or VirtualBox, while another operating system is the host system on the hardware of the computer.

These general ways are the first step a user takes in a typical Linux session, gaining access to a Linux system properly, and in a secure and autonomous way. Autonomous with respect to an individual user, the operating system is seen as a virtual, concurrent, and persistent environment. It provides services "as if" she were the only user, is able to do many operations at the same time, and ensures that those operations are securely preserved over time.

A more detailed description of these ways is as follows:

1. *Stand-alone*: This way, the most common case and the methodology we deploy in the rest of this book involves sitting at a computer that can function completely on its own. This does not mean that the stand-alone computer is not hooked up to a local area network (LAN), intranet, or the Internet.

   Rather, the users' connection to Linux is *dedicated* to a single user at a time (or possibly many autonomous users that log on to the same system individually at different times) sitting at the computer and logging on to use Linux on that hardware platform exclusively.

2. *Remote*: There are several variations of using this way. Here are just two possible scenarios:

   - You sit at a computer that acts like the traditional *terminal* connected to a mainframe computer. This could also be a *thin client* (a minimally configured and capable device) connected to a server. It is connected by a high-speed communication link to another single computer or multiple computers that are interconnected with a LAN or the Internet. At the terminal, and the console or command window that appears on its screen, your interface with the operating system runs on a single, or even multiple, other computer(s). This is a shared resource method, where several users on many different terminals can share a single Linux system.

   - You sit at a stand-alone computer or device, and via software such as PuTTY, Secure Shell (SSH), or SSH X Windows forwarding (a variant of Transmission Control Protocol (TCP) port forwarding), you connect to another system over a high-speed telecommunications link. The PuTTY or SSH software then becomes your *graphical connection*, allowing you to log on and use a remote computer or system that is running Linux. This is usually a shared resource method, where several users on many different remote computers can share a single Linux system. A variation of this is using a virtual private network and its interconnection software, from a mobile device, such as a cell phone or tablet, to log onto a system remotely via the Internet.

3. *Virtual*: You have a Linux or UNIX-like operating system, such as OS X, or another operating system installed and running the computer you are sitting in front of, and you have installed a virtual environment such as VirtualBox on that computer. Then, when you want to use a Linux system, you simply switch environments so that the Linux system in the virtual environment is what you are using to interface with the computer hardware. Another variation of this is using a cloud-based operating system that exists, and is running as an "instance," on a server that you legitimately log into via a Web browser.

We do not cover virtual connections in this chapter. But in Chapter W23, titled "Virtualization Methodologies," at the book website, we cover virtual environments such as LXC/LXD and VirtualBox. We illustrate the ways of connecting to virtual machines in that chapter.

In-Chapter Exercises

5. Define autonomy in the context of a single user's login to a Linux system.
6. Which of the abovementioned three ways do you use to gain access to your Linux system? Is it a way not shown previously? Is it a hybrid way of the ones shown previously and/or something not shown?

    Write a brief description of exactly how you login to your Linux system.
7. Explain in your own words what the difference is between a stand-alone, remote, and virtual login to a Linux system?
8. Describe the exact nature and general hardware configuration of a "thin client."

In the following subsections, we present three practical, useful, easy, and popular techniques of connecting and logging on and off a computer running the Linux system, as outlined in this section.

The three techniques of connecting and logging into a Linux system we show are as follows:

1. Stand-alone login and logout.
2. Remote login via the PuTTY program from a computer running Microsoft Windows to a Linux computer.
3. Remote login via an SSH client from a Linux client computer to another remote Linux host computer.

What is common to all three of these techniques is that your first task is to identify yourself correctly as a valid and autonomous user to the Linux system. This is the primary method of keeping computer systems secure. Doing so involves typing in a valid username, or login name, consisting of a string of valid characters. You then have to type in a valid password for that username.

Before proceeding with the remainder of this chapter, you should determine which one of the preceding three ways you will use to log in to a Linux system, and then select from the three following sections that give details on how to use that way correctly. If you cannot determine this on your own, get help from your instructor or the system administrator at your site. Be aware that you may have to use some form of hybrid way of the three ways we show, to log in and out of your Linux system.

2.3.1 Stand-Alone Login Connection to Linux

The login and logout procedures shown in this section are standard and vary only slightly between all Linux and NIX-like systems. This way assumes that someone has either logged out gracefully or rebooted the computer *before* you got to it, but has not shutdown the hardware or operating system of the computer.

In this section, we also make these basic assumptions:

1. That you are logging on to an already-running computer with Linux as the operating system.

2. That you are physically sitting in front of that Linux computer.

3. As previously stated, when you log in, identifying yourself to the Linux system is your first task. Doing so involves typing in a valid username, or login name, consisting of a string of valid characters. Then you type in a valid password for that username.

Be aware that when typing on the command line, Linux is case sensitive!

2.3.1.1 Graphical Login and Logout Procedures

A login window should appear on the screen. In the login window, your username should appear in the username field. For example, in one of our representative Linux systems, Linux Mint, from the username choices presented to us, we chose a username graphically by clicking on it with the mouse. Also found in the login window, as seen in Figure 2.1, are graphical icon choices that allow you to change keyboard layout, country, and desktop system to be used during this login session. The lambda character in the upper right of Figure 2.1 allows desktop system changes and presents choices based upon what desktop system environments are available or installed.

A password entry field appears on the screen as well, as seen in Figure 2.1.

In the password field of the window, type in your password. Finally, click on the OK button as seen in Figure 2.1, and you will be logged in.

To gracefully terminate your login session with a computer running Linux Mint, for example, make the Mint Menu choice **Log out**. On many systems, this does *not* restart the computer or shut it down, it only logs you out of the system.

In-Chapter Exercises

9. Describe all of the components of the GUI login window on your Linux system, including things such as the ones shown in Figure 2.1.

10. When you logout of your Linux system, does the hardware reboot? What would be the security advantage of having the hardware reboot every time a single user logs off the system? The assumption in this question is that this applies to a stand-alone system, where only one user can be logged into the operating system and hardware at a given time.

2.3.2 Connecting via PuTTY from a Microsoft Windows Computer

In this section, we make these basic assumptions:

1. That you are sitting at a computer running Microsoft Windows and trying to connect and log on to a computer running the Linux operating system.

2. On your Microsoft Windows computer, you are connected to the Internet or an intranet LAN where you know the Internet Protocol (IP) address of the Linux computer you want to log on to. On an intranet, this is something like 192.168.0.13.

FIGURE 2.1   Linux Mint password entry window.

3. You have downloaded and installed the PuTTY program on your Microsoft Windows computer, or the system administrator has done so for you. The details of downloading this software and installing it are not given here. At the time of writing, the most current download site for the PuTTY program was

   http://www.chiark.greenend.org.uk/~sgtatham/putty/download.html

4. You are using PuTTY to make an SSH connection to a Linux computer.

5. You know a valid username and password pair that will allow you to log in to the Linux computer.

2.3.2.1 Login and Logout Procedures

Once you execute the PuTTY program, you use the valid username/password pair, and then you can type commands into a console window or terminal screen. What you type in is shown as follows in **bold** text and is always followed by pressing the <Enter> key on the keyboard.

To begin, on the Microsoft Windows computer, double click on the PuTTY program icon, or from the **Start Menu>Programs** submenu, choose PuTTY. When the PuTTY program first launches, the PuTTY configuration dialog window opens on screen, similar to Figure 2.2.

From the PuTTY configuration window, you can modify several of the parameters that control your interactive session with a remote system. Almost all of these parameters can be left at their defaults. The only two things that most users will need to do in this configuration window is type the host name (or IP address) of the Linux computer they are trying to connect and log in to (in Figure 2.2, this is 192.168.0.13), and click the protocol button for SSH, as seen in Figure 2.2. The port number is automatically set at 22 if you click on the SSH button for "Connection type." You need to know what the host name or IP address

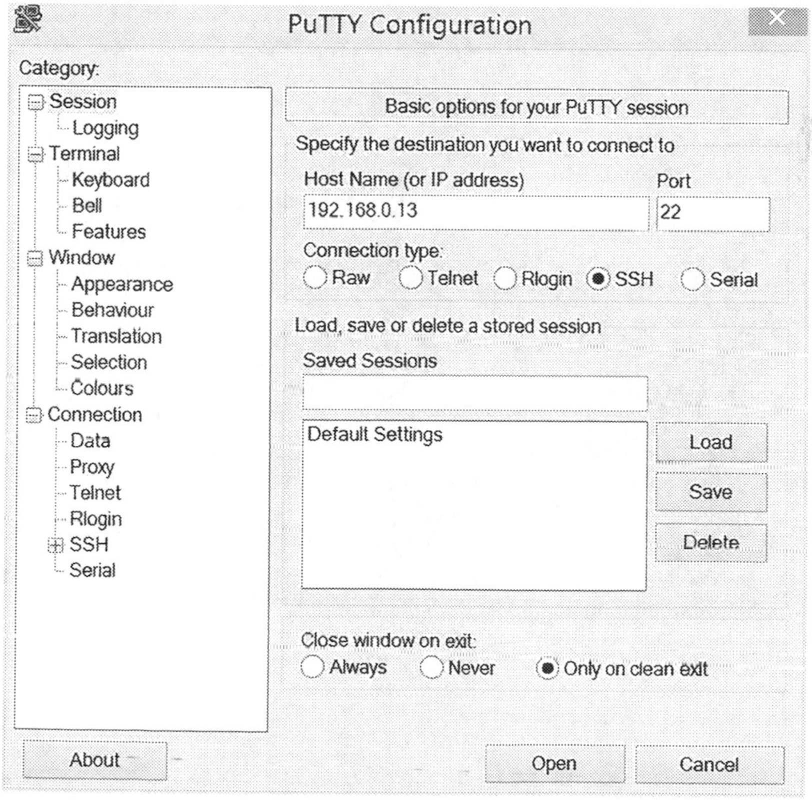

FIGURE 2.2 PuTTY configuration window.

of the Linux computer you want to log on to is. Then, click on the Open button, and a console window will open on screen, allowing you to log in to the Linux computer.

As previously stated, in the process of logging in, identifying yourself to the Linux system is your first task. Doing so involves typing in a valid *username* or *login name*, consisting of a string of valid characters. You then type a valid *password* for that username. There are both valid and invalid characters that you can use in both your username and password. See your system administrator or instructor to find out what these characters are on the Linux system you want to log in to, if they have not already told you what they are.

In response to the login: prompt, you type in your username on the Linux system, and then press <Enter> on the keyboard. In our case, the username is **bob**. Remember that Linux is case sensitive. When the Password: prompt appears, type your password on the Linux system and then press <Enter> on the keyboard.

In-Chapter Exercise

11. What are the constraints on what your password can be on your Linux system, and what is the expiration period set at (if any), as imposed by the system administrator on your Linux system? That is, number of characters allowed, 6 months, every term, etc.

To terminate your connection type logout at the command line prompt and then press <Enter> on the keyboard or on a blank line press <Ctrl+D>—that is, hold down the <Ctrl> and D keys on the keyboard at the same time. Logging out is somewhat system dependent as well as being an operation that can be tailored to a specific installation of Linux by the local system administrator. In the Bourne again shell (Bash), which is the default shell used on most Linux systems, typing the logout command is the default way of leaving the system gracefully.

If you use the Bash, Bourne, or Korn shell, holding down <Ctrl+D> or typing exit will accomplish the same thing. You will then be logged off the system, the current PuTTY session will end, and all PuTTY windows will close.

If you started a new shell during your session and didn't exit that shell before logging off, Linux will prompt Not login shell, and you will not be able to log off immediately. In this case, press <Ctrl+D> and the new shell will terminate. Also, if you started more than one shell and haven't exited from those shells before you log off, you will have to use <Ctrl+D> to terminate each shell individually. On some systems, you type exit on the command line to terminate a shell process. In either case, you will then be able to use the logout procedure previously described to leave the system.

2.3.3 Connecting via an SSH Client between Linux Machines

This way allows a user on one Linux computer to remote log in and log out of another Linux computer using the SSH protocol. As further detailed in Chapter 11, Section 8.5, SSH is an encrypted channel of communication between a client computer and a server on a LAN or the Internet. The computer you are using to log into another computer with is known as the SSH *client*. The computer you want to log into with SSH is known as the *server* or the *host* system.

Before this way can be used, both client and server systems must be able to talk to each other over the SSH channel. In other words, the server system must have the SSH server-side software package installed on it. This is usually *not* the default on most Linux systems, so you have to follow the installation instructions given for your particular flavor of Linux to accomplish this. On our Debian-family Linux Mint system, we were able to install the server-side software with the following package management command:

```
$ sudo apt-get install openssh-server
```

The client-side SSH software is installed by default on most Linux system implementations, including all of the representative Linux systems we show in this book. See Appendix A for information on how to use the package management system on any of our representative Linux systems.

Also, as previously stated, the user must know a valid username/password pair on the remote server system to be able to log in to the remote system!

We show three possible methods that can be used in this dialog. First, if the user has previously already logged into the host successfully from the client before, and the authentication keys have not changed. Second, if the user has never logged into the host successfully before from the client. Third, if the user has logged into the host before, but the authentication key on the host has changed since the last successful login. These are practical situations one might encounter any time you use this remote login method.

2.3.3.1 Login and Logout Procedures

What the user types in is shown in **bold** text:

Method 1. Having logged in before successfully:

```
$ ssh 192.168.0.8
bob@192.168.0.8's password: www
Welcome to Linux Mint 18.1 Serena (GNU/Linux 4.4.0-45-generic x86_64)

 * Documentation:  https://www.linuxmint.com
Last login: Mon Jan  9 07:25:18 2017 from 192.168.0.25
bob@bob-PowerEdge-T110 ~ $ Execute Command Line Linux Commands
bob@bob-PowerEdge-T110 ~ $ logout
Connection to 192.168.0.8 closed.
$
```

Method 2. Having never logged in before:

```
$ ssh 192.168.0.8
The authenticity of host '192.168.0.8 (192.168.0.8)' can't be established. ECDSA
key fingerprint is SHA256:uZpqi4U6uBN5SOBVFRbqbl5HspmV3eZAw/nUvPBTS5I.
Are you sure you want to continue connecting (yes/no)? yes
Warning: Permanently added '192.168.0.8' (ECDSA) to the list of known hosts.
bob@192.168.0.8's password: www
Welcome to Linux Mint 18.1 Serena (GNU/Linux 4.4.0-45-generic x86_64)

 * Documentation:  https://www.linuxmint.com
Last login: Sat Dec 24 16:54:10 2016 from 192.168.0.6
bob@bob-PowerEdge-T110 ~ $ Execute Command Line Linux Commands
bob@bob-PowerEdge-T110 ~ $ logout
Connection to 192.168.0.8 closed.
$
```

Method 3. Logged in before but host key has changed:

```
$ ssh 192.168.0.8
@@@@@@@@@@@@@@@@@@@@@@@@@@@@@@@@@@@@@@@@@@@@@@@@@@@@@@@@@@@@@@@@@
@    WARNING: REMOTE HOST IDENTIFICATION HAS CHANGED!     @
@@@@@@@@@@@@@@@@@@@@@@@@@@@@@@@@@@@@@@@@@@@@@@@@@@@@@@@@@@@@@@@@@
IT IS POSSIBLE THAT SOMEONE IS DOING SOMETHING NASTY!
Someone could be eavesdropping on you right now (man-in-the-middle attack)!
It is also possible that a host key has just been changed.
The fingerprint for the ECDSA key sent by the remote host is
43:e8:cf:33:d5:ed:dd:05:d9:e9:a5:9d:d3:18:1d:2b.
Please contact your system administrator.
Add correct host key in /home/bob/.ssh/known_hosts to get rid of this message.
Offending ECDSA key in /usr/home/bob/.ssh/known_hosts:2
ECDSA host key for 192.168.0.8 has changed and you have requested strict checking.
Host key verification failed.
$ cd /home/bob/.ssh
$ rm known_hosts
$ cd
```

```
$ ssh 192.168.0.8
The authenticity of host '192.168.0.8 (192.168.0.8)' can't be established.
ECDSA key fingerprint is 43:e8:cf:33:d5:ed:dd:05:d9:e9:a5:9d:d3:18:1d:2b.
No matching host key fingerprint found in DNS.
Are you sure you want to continue connecting (yes/no)? yes
Warning: Permanently added '192.168.0.8' (ECDSA) to the list of known hosts.
bob@192.168.0.8's password: www
Last login: Sat Dec 24 16:54:10 2016 from 192.168.0.6
Output truncated...
bob@bob-PowerEdge-T110 $ Execute Command Line Linux Commands
bob@bob-PowerEdge-T110 $ logout
Connection to 192.168.0.8 closed.
$
```

In all of the three methods, the user is assumed to have an account with the same username, and possibly password, on both client and server systems.

In method 2, the keys are generated on host and client after the user types in **yes** and presses <Enter>.

In method 3, after the first failed attempt to establish an SSH connection, the error message indicates that the authentication key has changed on the host. A very helpful component of the error message is the instruction:

Add correct host key in /home/bob/.ssh/known_hosts to get rid of this message.

Offending ECDSA key in /usr/home/bob/.ssh/known_hosts:2

So a removal of the offending key in the file **/home/bob/.ssh/known_hosts** on the client machine is done by deleting that file. Then a new key is generated, an exchange can take place, and the login can proceed.

The line in all three methods that reads "Execute Command Line Linux Commands" is where the user types in any of the valid Linux commands we show in this chapter and throughout the rest of this book. Finally, after typing **logout**, the user cuts the SSH channel connection and is returned to the command line prompt of the local client system.

In-Chapter Exercises

12. Use ssh to log into another machine on your LAN or intranet that can accept ssh logins as a server, and that you have access to with a username and password. Which of the earlier methods in this section did you have to deploy, on first ssh login, and on every subsequent ssh login, to that host?

13. Can you log into a system using ssh that is a stand-alone (or dedicated) system, and that you are already logged into? Remember that a stand-alone system can be connected to an intranet, LAN, or the Internet, but as we define it, only one user (out of a population of possibly many users) can be logged into it at any given time. Also, the assumption here is that the ssh client and server software is installed on that system. Why would you want to do this?

It is also interesting that some virtual connections, such as to Docker containers, are not configured to easily do this via a Dynamic Host Configuration Protocol (DHCP) server-assigned IP address on a LAN.

2.4 File Maintenance Commands and Help on Linux Command Usage

After your first-time login to a new Linux system using one of the three ways we described in the previous section, one of your first actions will be to construct and organize your workspace environment and the files that will be contained in it. The operation of organizing your files according to some logical scheme is known as *file maintenance*. A logical scheme used to organize your files might consist of creating *bins* for storing files according to the subject matter of the contents of the files or according to the dates of their creation. In the following sections, you will type file creation and maintenance commands that produce a structure as shown in Figure 2.3. Complete the operations shown in the following sections in the order they are presented, to get a better overview of what file maintenance really is. Also, it is critical that you review what was presented in Section 2.2 regarding the structure of a Linux command so that when you begin to type commands for file maintenance, you understand how the syntax of what you are typing conforms to the general syntax of any Linux command.

2.4.1 File and Directory Structure

When you first log in, you are working in the *home directory*, or folder, of the autonomous user associated with the username and password you used to log in. Whatever directory you are presently in is known as the *current working directory*, and there is only one current working directory active at any given time. It is helpful to visualize the structure of your files and directories using a diagram. Figure 2.3 is an example of a home directory and file structure for a user named **bob**. In this figure, directories are represented as parallelograms and plain files (e.g., files that contain text or binary instructions) are represented as rectangles. A *pathname*, or path, is simply a textual way of designating the location of a directory or file in the complete file structure of the Linux system you are working on. For example, the path to the file **myfile2** in Figure 2.3 is **/home/bob/myfile2**. The designation of the path begins at the root (/) of the entire file system, descends to the folder named **home**, and then descends again to the home directory of the user named **bob**.

As shown in Figure 2.3, the files named **myfile**, **myfile2**, and **renamed_file** are stored under or in the directory **bob**. Beneath **bob** is a *subdirectory* named **first**. In the following sections, you will create these files, and the subdirectory structure, in the home directory of the username that you have logged into your Linux system with.

In-Chapter Exercise

14. Type the following two commands on your Linux system:

```
$ cd /
$ ls
```

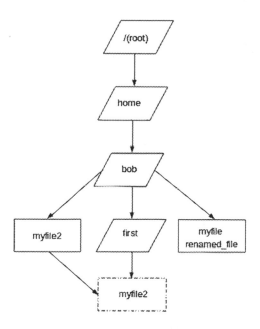

FIGURE 2.3 Example directory structure.

Similar to Figure 2.3, sketch a diagram of the directories and files whose names you see listed as the output of the second command. Save this diagram for use later.

2.4.2 Viewing the Contents of Files

To begin working with files, you can easily create a new text file by using the cat command. The syntax of the cat command is as follows:

Syntax:
   `cat [options] [file-list]`

Purpose: Join one or more files sequentially or display them in the console window

Output: Contents of the files in **file-list** displayed on the screen, one file at a time

Commonly used options/features:

+E	Displays $ at the end of each line
-n	Puts line numbers on the displayed lines
-- help	Displays the purpose of the command and a brief explanation of each option

---

The cat command, short for concatenate, allows you to join files. In the example, you will join what you type on the keyboard to a new file being created in the current working directory. This is achieved by the redirect character >, which takes what you type at the *standard input* (in this case the keyboard) and directs it into the file named **myfile**. You can consider the keyboard and the stream of information it provides as a file. As stated in Section 2.2, this usage involves the command cat but no options, option arguments, or command arguments. It simply uses the command, a redirect character, and a target, or destination, named **myfile**, where the redirection will go.

This is the very simplest example of a *multiple command* typed on the command line, as opposed to a single command, as shown and briefly described in Section 2.2. In a multiple command, you can string together single Linux commands in a chain with connecting operators, such as the redirect character shown here.

```
$ cat > myfile
This is an example of how to use the cat command to add plain text to a file
<Ctrl+D>
$
```

You can type as many lines of text, pressing <Enter> on the keyboard to distinguish between lines in the file, as you want. Then, on a new line, when you hold down <Ctrl+D>, the file is created in the current working directory, using the command you typed. You can view the contents of this file, since it is a plain text file that was created using the keyboard, by doing the following:

```
$ more myfile
This is an example of how to use the cat command to add plain text to a file
$
```

This is a simple example of the syntax of a single Linux command.

The general syntax of the more command is as follows:

Syntax:
   `more [options] [file-list]`

Purpose: Concatenate/display the files in **file-list** on the screen, one screen at a time

Output: Contents of the files in **file-list** displayed on the screen, one page at a time

Commonly used options/features:

+E/str	Starts two lines before the first line containing str
-nN	Displays N lines per screen/page
+N	Starts displaying the contents of the file at line number N

The more command shows one screenful of a file at a time by default. If the file is several pages long, you can proceed to view subsequent pages by pressing the <Space> key on the keyboard or by pressing the Q key to quit. Linux has a command named pg that accomplishes the same thing as the more command.

In-Chapter Exercise

15. Use the cat command to produce another text file named testfile. Then join the contents of myfile and testfile into one text file, named myfile3, with the cat command.

2.4.3 Creating, Deleting, and Managing Files

To copy the contents of one file into another file, use the cp command. The general syntax of the cp command is as follows:

Syntax:
  cp [options] file1 file2
Purpose: Copy **file1** to **file2**; if **file2** is a directory, make a copy of **file1** in this directory
Output: Copied files
Commonly used options/features:
  -i   If destination exists, prompt before overwriting
  -p   Preserves file access modes and modification times on copied files
  -r   Recursively copies files and subdirectories

For example, to make an exact duplicate of the file named **myfile**, with the new name **myfile2**, type the following:

```
$ cp myfile myfile2
$
```

This usage of the cp command has two required command arguments. The first argument is the source file that already exists and which you want to copy. The second argument is the destination file or the name of the file that will be the copy. Be aware that many Linux commands can take plain, ordinary, or regular files as arguments, or can take directory files as arguments. This can change the basic task accomplished by the command. It is also worth noting that not only can file names be arguments but pathnames as well. This changes the site or location, in the path structure of the file system, of operation of the command.

To change the name of a file or directory, you can use the mv command. The general syntax of the mv command is as follows:

Syntax:
  mv [options] file1 file2
  mv [options] file-list directory
Purpose: First syntax: Rename file1 to file2
         Second syntax: Move all the files in file-list to directory
Output: Renamed or relocated files
Commonly used options/features:
  -f   Forces the move regardless of the file access modes of the destination file
  -i   Prompts the user before overwriting the destination

In the following usage, the first argument to the mv command is the source file name, and the second argument is the destination name.

```
$ mv myfile2 renamed_file
$
```

It is important at this point to notice the use of spaces in Linux commands. What if you obtain a file from a Windows 10 system that has one or more spaces in one of the file names? How can you work with this file in Linux? The answer is simple. Whenever you need to use that file name in a command as an argument, enclose the file name in double quotes ("). For example, you might obtain a file that you have "detached" from an e-mail message from someone on a Windows 10 system, such as **latest revisions october.txt**.

In order to work with this file on a Linux system—that is, to use the file name as an argument in a Linux command—enclose the whole name in double quotes. The correct command to rename that file to something shorter would be

```
$ mv "latest revisions october.txt" laterevs.txt
$
```

To delete a file, you can use the rm command. The general syntax of the rm command is as follows:

Syntax:
　　`rm [options] file-list`
Purpose: Remove files in **file-list** from the file structure (and disk)
Output: Deleted files
Commonly used options/features:
　　-f　Removes regardless of the file access modes of **file-list**
　　-i　Prompts the user before removing files in **file-list**
　　-r　Recursively removes the files in **file-list** if **file-list** is a directory; use with caution!

To delete the file **renamed_file** from the current working directory, type

```
$ rm renamed_file
$
```

In-Chapter Exercise

16. Use the rm command to delete the files testfile and myfile3 you created in In-Chapter Exercise 15.

The most important command you will execute to do file maintenance is the ls command. The general syntax for the ls command is as follows:

Syntax:
　　`ls [options] [pathname-list]`
Purpose: Send the names of the files and directories in the directory specified by **pathname-list** to the display screen
Output: Names of the files and directories in the directory specified by **pathname-list** or the names only if **pathname-list** contains file names only
Commonly used options/features:
　　-F　Displays a slash character (/) after directory names, an asterisk (*) after binary executables, and an "at" character (@) after symbolic links
　　-a　Displays names of all the files, including hidden files
　　-i　Displays inode numbers
　　-l　Displays a long list that includes file access modes, link count, owner, group, file size (in bytes), and modification time

The ls command will list the names of files or folders in your current working directory or folder. In addition, as with the other commands we have used so far, if you include a complete pathname specification for the pathname-list argument to the command, then you can list the names of files and folders along that pathname list. To see the names of the files in your current working directory, type the following:

```
$ ls
Desktop      Documents    Downloads    Dropbox    Music      Pictures
Public       Templates    Linuxthetextbook2    Videos
$
```

Please note that you will probably not get a listing of the same file names as we did here, because your system will have placed some files automatically in your home directory, as in the example we used, aside from the ones we created together named **myfile** and **myfile2**. Also note that this file name listing does not include the name **renamed_file**, because we deleted that file.

The next command you will execute is actually just an alternate or modified way of executing the ls command, one that includes the command name and options. As shown in Section 2.2, a Linux command has options that can be typed on the command line along with the command to change the behavior of the basic command. In the case of the ls command, the options l and a produce a longer listing of all ordinary and system (dot) files, as well as providing other attendant information about the files. Don't forget to put the space character between the s and the dash. Remember again from Section 2.2 that spaces delimit, or partition, the components of a Linux command as it is typed on the command line.

Now, type the following command:

```
$ ls -la
total 164
drwxr-xr-x 25 bob   bob    4096 Jan   9 15:06 .
drwxr-xr-x  3 root  root   4096 Nov  21 09:48 ..
-rw-------  1 bob   bob     503 Jan  10 16:30 .bash_history
-rw-r--r--  1 bob   bob     220 Nov  21 09:48 .bash_logout
-rw-r--r--  1 bob   bob    4000 Nov  21 09:48 .bashrc
drwx------ 11 bob   bob    4096 Jan   8 21:13 .cache
drwxr-xr-x  4 bob   bob    4096 Jan   9 15:06 .cinnamon
drwxr-xr-x 16 bob   bob    4096 Jan   8 21:15 .config
drwx------  3 bob   bob    4096 Nov  21 10:00 .dbus
drwxr-xr-x  5 bob   bob    4096 Jan   8 21:14 Desktop
-rw-------  1 bob   bob      29 Jan   9 15:06 .dmrc
drwxr-xr-x  2 bob   bob    4096 Nov  21 10:00 Documents
drwxr-xr-x  2 bob   bob    4096 Jan   8 08:36 Downloads
drwx------  5 bob   bob    4096 Jan   9 15:06 .dropbox
drwx------ 30 bob   bob    4096 Jan   9 15:06 Dropbox
drwxr-xr-x  3 bob   bob    4096 Jan   4 08:40 .dropbox-dist
drwx------  2 bob   bob    4096 Jan  10 07:57 .gconf
-rw-r-----  1 bob   bob       0 Jan   8 08:32 .gksu.lock
drwx------  2 bob   bob    4096 Jan   8 20:56 .gphoto
-rw-------  1 bob   bob    4446 Jan   9 15:06 .ICEauthority
drwxr-xr-x  4 bob   bob    4096 Nov  21 10:01 .linuxmint
drwxr-xr-x  3 bob   bob    4096 Nov  21 10:00 .local
drwxr-xr-x  4 bob   bob    4096 Nov  21 10:02 .mozilla
drwxr-xr-x  2 bob   bob    4096 Nov  21 10:00 Music
drwxr-xr-x  2 bob   bob    4096 Nov  21 10:00 Pictures
-rw-r--r--  1 bob   bob     675 Nov  21 09:48 .profile
drwxr-xr-x  2 bob   bob    4096 Nov  21 10:00 Public
drwx------  2 bob   bob    4096 Jan   5 11:14 .putty
drwx------  2 bob   bob    4096 Jan   5 11:12 .ssh
-rw-r--r--  1 bob   bob       0 Nov  21 10:03 .sudo_as_admin
drwxr-xr-x  2 bob   bob    4096 Nov  21 10:00 Templates
drwxr-xr-x  2 bob   bob    4096 Nov  21 10:00 Videos
-rw-------  1 bob   bob     123 Jan   9 15:06 .Xauthority
-rw-r--r--  1 bob   bob   21022 Jan  11 07:10 .xsession-errors
```

```
-rw-r--r--  2 bob   bob     797 Jan 16 10:00 myfile
-rw-r--r--  2 bob   bob     797 Jan 16 10:00 myfile2
$
```

As you see in this screen display (which shows the listing of files in our home directory and will not be the same as the listing of files in your home directory), the information about each file in the current working directory is displayed in eight columns. The first column shows the type of file, where d stands for directory, l stands for symbolic link, and – stands for ordinary or regular file. Also in the first column, the access modes to that file for user, group, and others is shown as r, w, or x. In the second column, the number of links to that file is displayed. In the third column, the username of the owner of that file is displayed. In the fourth column, the name of the group for that file is displayed. In the fifth column, the number of bytes that the file occupies on disk is displayed. In the sixth column, the date that the file was last modified is displayed. In the seventh column, the time that the file was last modified is displayed. In the eighth and final columns, the name of the file is displayed. This way of executing the command is a good way to list more complete information about the file. Examples of using the more complete information are (1) so that you can know the byte size and be able to fit the file on some portable storage medium or (2) to display the access modes so that you can alter the access modes to a particular file or directory.

In-Chapter Exercise

17. Use the **ls -la** command to list all of the filenames in your home directory on your Linux system. How does the listing you obtain compare with the listing shown earlier? Remember that our listing was done on a Linux Mint 18.2 system.

You can also get a file listing for a single file in the current working directory, by using another variation of the ls command, as follows:

```
$ ls -la myfile
-rw-r--r--  1 bob   bob   797 Jan 16 10:00 myfile
$
```

This variation shows you a long listing with attendant information for the specific file named **myfile**. A breakdown of what you typed on the command line is (1) **ls**, the command name; (2) **-la**, the options; and (3) **myfile**, the command argument.

What if you make a mistake in your typing and misspell a command name or one of the other parts of a command? Type the following on the command line:

```
$ lx -la myfile
lx: not found
$
```

The **lx: not found** reply from Linux is an error message. There is no **lx** command in the Linux operating system, so an error message is displayed. If you had typed an option that did not exist, you would also get an error message. If you supplied a file name that was not in the current working directory, you would get an error message too. This makes an important point about the execution of Linux commands. If no error message is displayed, then the command executed correctly and the results might or might not appear on screen, depending on what the command actually does. If you get an error message displayed, you must correct the error before Linux will execute the command as you type it.

Typographic mistakes account for a large percentage of the errors that beginners make.

2.4.4 Creating, Deleting, and Managing Directories

Another critical aspect of file maintenance is the set of procedures and the related Linux commands you use to create, delete, and organize directories in your Linux account on a computer. When moving through the file system, you are either ascending or descending to reach the directory you want to use. The directory directly above the current working directory is referred to as the *parent* of the current working directory. The directory or directories immediately under the current working directory are referred to as the *children* of the current working directory. For more information on file system structure, see Chapter 4. The most common mistake for beginners is misplacing files. They cannot find the file names listed with the ls command because they have placed or created the files in a directory either above or below the current working directory in the file structure. When you create a file, if you have also created a logically organized set of directories beneath your own home directory, you will know where to store the file. In the following set of commands, we create a directory beneath the home directory and use that new directory to store a file.

To create a new directory beneath the current working directory, you use the mkdir command. The general syntax for the mkdir command is as follows:

Syntax:
    `mkdir [options] dirnames`

Purpose: Create directory or directories specified in **dirnames**

Output: New directory or directories

Commonly used options/features:

    `-m MODE`   Creates a directory with given access modes

    `-p`   Creates parent directories that don't exist in the pathnames specified in **dirnames**

To create a child, or subdirectory, named **first** under the current working directory, type the following:

```
$ mkdir first
$
```

This command has now created a new subdirectory named **first** under, or as a child of, the current working directory. Refer back to Figure 2.3 for a graphical description of the directory location of this new subdirectory.

To change the current working directory to this new subdirectory, you use the cd command. The general syntax for the cd command is as follows:

Syntax:
    `cd [directory]`

Purpose: Change the current working directory to **directory** or return to the home directory when **directory** is omitted

Output: New current working directory

To change the current working directory to **first** by descending down the path structure to the specified directory named **first**, type the following:

```
$ cd first
$
```

You can always verify what the current working directory is by using the pwd command. The general syntax of the pwd command is as follows:

Syntax:
    `pwd`

Purpose: Display the current working directory on screen

Output: Pathname of current working directory

You can verify that **first** is now the current working directory by typing the following:

```
$ pwd
/home/bob/first
$
```

The output from Linux on the command line shows the pathname to the current working directory or folder. As previously stated, this path is a textual route through the complete file structure of the computer that Linux is running on, ending in the current working directory. In this example of the output, the path starts at **/**, the root of the file system. Then it descends to the directory **home**, a major branch of the file system on the computer running Linux. Then it descends to the directory **bob**, another branch, which is the home directory name for the user. Finally, it descends the branch named **first**, the current working directory.

On some systems, depending on the default settings, another way of determining what the current working directory is can be done by simply looking at the command line prompt. This prompt may be prefaced with the complete path to the current working directory, ending in the current working directory.

You can ascend back up to the home directory, or the parent of the subdirectory **first**, by typing the following:

```
$ cd
$
```

An alternate way of doing this is to type the following, where the tilde character (~) resolves to, or is a substitute for, the specification of the complete path to the home directory:

```
$ cd ~
$
```

To verify that you have now ascended up to the home directory, type the following:

```
$ pwd
/home/bob
$
```

You can also ascend to a directory above your home directory, sometimes called the parent of your current working directory, by typing the following:

```
$ cd ..
$
```

In this command, the two periods (..), represent the parent or branch above the current working directory. Don't forget to type a space character between the d and the first period. To verify that you have ascended to the parent of your home directory, type the following:

```
$ pwd
/home
$
```

To descend to your home directory, type the following:

```
$ cd
$
```

To verify that there are two files in the home directory that begins with the letters my, type the following command:

```
$ ls my*
myfile myfile2
$
```

The asterisk following the y on the command line is known as a *metacharacter* or a character that represents a pattern; in this case, the pattern is any set of characters. When Linux interprets the command after you press the <Enter> key on the keyboard, it searches for all files in the current working directory that begin with the letters my and end in anything else.

In-Chapter Exercise

18. Use the **cd** command to ascend to the root (/) of your Linux file system, and then use it to descend down each subdirectory from the root recursively to a depth of two subdirectories, sketching a diagram of the component files found on your system. Make the named entries in the diagram as complete as possible, listing as many files as you think necessary. Retain this diagram as a useful map of your particular Linux distribution's file system.

Another aspect of organizing your directories is movement of files between directories or changing the location of files in your directories. For example, you now have the file **myfile2** in your home directory, but you would like to move it into the subdirectory named **first**. See Figure 2.3 for a graphic description to change the organization of your files at this point. To accomplish this, you can use the second syntax method illustrated for the mv file-list directory command to move the file **myfile2** down into the subdirectory named **first**. To achieve this, type the following:

```
$ mv myfile2 first
$
```

To verify that **myfile2** is indeed in the subdirectory named first, type the following:

```
$ cd first
$ ls
myfile2
$
```

You will now ascend to the home directory and attempt to remove or delete a file with the rm command. *Caution*: you should be very careful when using this command, because once a file has been deleted, the only way to recover it is from archival backups that you or the system administrator has made of the file system.

```
$ cd
$ rm myfile2
rm: myfile2: No such file or directory
$
```

You get the error message because in the home directory, the file named **myfile2** does not exist. It was moved down into the subdirectory named first.

Directory organization also includes the ability to delete empty or nonempty directories. The command that accomplishes the removal of empty directories is rmdir. The general syntax of the rmdir command is as follows:

Syntax:
  rmdir [options] dirnames
Purpose: Remove the empty directories specified in **dirnames**
Output: Remove directories
Commonly used options/features:
    -p Removes empty parent directories as well
    -r Recursively deletes files and subdirectories beneath the current directory

---

To delete an entire directory below the current working directory, type the following:

```
$ rmdir first
rmdir: first: Directory not empty
$
```

Since the file **myfile2** is still in the subdirectory named **first**, **first** is not an empty directory, and you get the error message that the rmdir command will not delete the directory. If the directory was empty, rmdir would have accomplished the deletion. One way to delete a nonempty directory is by using the rm command with the -r option. The -r option recursively descends down into the subdirectory and deletes any files in it before actually deleting the directory itself. Be cautious with this command, since you may inadvertently delete directories and files with it. To see how this command deletes a nonempty directory, type the following:

```
$ rm -r first
$
```

The directory **first** and the file **myfile2** are now removed from the file structure.

2.4.5 Obtaining Help with the man Command

A very convenient utility available on Linux systems is the online help feature, achieved via the use of the man command. The general syntax of the man command is as follows:

Syntax:
```
man [options][-s section] command-list
man -k keyword-list
```

Purpose: First syntax: Display Linux Reference Manual pages for commands in command-list one screen at a time

Second syntax: Display summaries of commands related to keywords

in **keyword-list**

Output: Manual pages one screen at a time

Commonly used options/features:

-k keyword-list	Searches for summaries of keywords in **keyword-list** in a database and display them
-s sec-num	Searches section number **sec-num** for manual pages and display them

---

To get help by using the man command, on usage and options of the ls command, for example, type the following:

```
$ man ls

LS(1)                          User Commands                          LS(1)

NAME
       ls - list directory contents

SYNOPSIS
       ls [OPTION]... [FILE]...

DESCRIPTION
       List  information  about  the FILEs (the current directory
       by default).
       Sort entries alphabetically if none of -cftuvSUX nor -sort
       is  specified.

       Mandatory  arguments  to  long  options are mandatory for
       short options too.
```

```
    -a, --all
            do not ignore entries starting with .

    -A, --almost-all
            do not list implied . and ..

    --author
Manual page ls(1) line 1 (press h for help or q to quit)
```

This output from Linux is a Linux *manual page*, or *man page*, which gives a synopsis of the command usage showing the options, and a brief description that helps you understand how the command should be used. Typing q after one page has been displayed, as seen in the example, returns you to the command line prompt. Pressing the space key on the keyboard would have shown you more of the content of the manual pages, one screen at a time, related to the ls command.

To get help in using all the Linux commands and their options, use the man man command to go to the Linux reference manual pages.

The pages themselves are organized into eight sections, depending on the topic described and the topics that are applicable to the particular system. Table 2.1 lists the sections of the manual and what they contain. Most users find the pages they need in Section 1. Software developers mostly use library and system calls and thus find the pages they need in Sections 2 and 3. Users who work on document preparation get the most help from Section 7. Administrators mostly need to refer to pages in Sections 1, 4, 5, and 8.

The manual pages comprise multipage, specially formatted, descriptive documentation for every command, system call, and library call in Linux. This format consists of seven general parts: name, synopsis, description, list of files, related information, errors, warnings, and known bugs. You can use the man command to view the manual page for a command. Because of the name of this command, the manual pages are normally referred to as Linux man pages. When you display a manual page on the screen, the top-left corner of the page has the command name with the section it belongs to in parentheses, as with LS(1), seen at the top of the output manual page.

The command used to display the manual page for the passwd command is

```
$ man passwd
```

The manual page for the passwd command now appears on the screen, but we do not show its output. Because they are multipage text documents, the manual pages for each topic take up more than one screen of text to display their entire contents. To see one screen of the manual page at a time, press the space bar on the keyboard. To quit viewing the manual page, press the Q key on the keyboard.

Now type this command:

```
$ man pwd
```

TABLE 2.1

Sections of the Linux Manual

Section	What It Describes
1	User commands
2	System calls
3	Language library calls (C, FORTRAN, etc.)
4	Devices and network interfaces
5	File formats
6	Games and demonstrations
7	Environments, tables, and macros for troff
8	System maintenance-related commands

If more than one section of the man pages has information on the same word and you are interested in the man page for a particular section, you can use the –S option. The following command line therefore displays the man page for the read system call and not the man page for the shell command read.

```
$ man -S2 read
```

The command man –S3 fopen fread strcmp sequentially displays man pages for three C library calls: **fopen**, **fread**, and **strcmp**.

Using the man command and typing the command with the –k option allows specifying a keyword that limits the search. It is equivalent to using the **apropos** command. The search then yields useful man page headers from all the man pages on the system that contain just the keyword reference. For example, the following session yields the on-screen output on our Linux system:

```
$ man -k passwd
chgpasswd (8)            - update group passwords in batch mode
chpasswd (8)             - update passwords in batch mode
Crypt::PasswdMD5 (3pm) - Provides interoperable MD5-
                          based crypt() functions
fgetpwent_r (3)         - get passwd file entry reentrantly
getpwent_r (3)          - get passwd file entry reentrantly
gpasswd (1)             - administer /etc/group
                          and /etc/gshadow
grub-mkpasswd-pbkdf2 (1) - generate hashed password for GRUB
pam_localuser (8)       - require users to be listed
                          in /etc/passwd
passwd (1)              - change user password
passwd (1ssl)           - compute password hashes
passwd (5)              - the password file
```

Output truncated…

2.4.6 Other Methods of Obtaining Help

To get a short description of what any particular Linux command does, you can use the whatis command. This is similar to the command man –f. The general syntax of the whatis command is as follows:

Syntax:
   whatis keywords
Purpose: Search the whatis database for abbreviated descriptions of each keyword
Output: Prints a one-line description of each keyword to the screen

---

The following is an illustration of how to use whatis.
   The output of the two commands are truncated.

```
$ whatis man
man (7)                 - macros to format man pages
man (1)                 - an interface to the on-line
                          reference manuals
$
```

You can also obtain short descriptions of more than one command by entering multiple arguments to the whatis command on the same command line, with spaces between each argument. The following is an illustration of this method:

```
$ whatis login set setenv
login (1)               - begin session on the system
login (3)               - write utmp and wtmp entries
setenv (3)              - change or add an environment
```

```
                                    variable
set: nothing appropriate.
$
```

The following in-chapter exercises ask you to use the man and whatis commands to find information about the passwd command. After completing the exercises, you can use what you have learned to change your login password on the Linux system that you use.

In-Chapter Exercises

19. Use the man command with the -k option to display abbreviated help on the passwd command. Doing so will give you a screen display *similar* to that obtained with the whatis command, but it will show all apropos command names that contain the characters passwd.

20. Use the whatis command to get a brief description of the passwd command shown in Exercise 19, and then note the difference between the commands whatis passwd and man -k passwd.

2.5 Utility Commands

There are several major commands that allow the beginner to be more productive when using the Linux system. A sampling of these kinds of utility commands is given in the following sections, and is organized as system setups, general utilities, and communication commands.

2.5.1 Examining System Setups

The whereis command allows you to search along certain prescribed paths to locate utility programs and commands, such as shell programs. The general syntax of the whereis command is as follows:

Syntax:
  whereis [options] filename

Purpose: Locate the binary, source, and man page files for a command

Output: The supplied names are first stripped of leading pathname components and extensions, then pathnames are displayed on screen

Commonly used options/features:

  -b Searches only for binaries

  -s Searches only for source code

For example, if you type the command whereis bash on the command line, you will see a list of the paths to the Bash shell program files themselves, as follows:

$ whereis bash

bash: /bin/bash /etc/bash.bashrc /usr/share/man/man1/bash.1.gz

Note that the paths to a "built-in," or internal, command cannot be found with the whereis command. We provide more information about internal and external shell commands in Chapter 10.

When you first log on, it is useful to be able to view a display of information about your userid, the computer or system you have logged on to, and the operating system on that computer. These tasks can be accomplished with the whoami command, which displays your userid on the screen. The general syntax of the whoami command is as follows:

Syntax:
  whoami

Purpose: Display the effective user id

Output: Displays your effective user id as a name on standard output

The following shows how our system responded to this command when we typed it on the command line.

```
$ whoami
bob
$
```

The following in-chapter exercises give you the chance to use whereis, whoami, and two other important utility commands, who and hostname, to obtain the important information about your system.

In-Chapter Exercises

21. Use the whereis command to locate binary files for the Korn shell, the Bourne shell, the Bourne again shell, the C shell, and the Z shell. Are any of these shell programs not available on your system?

22. Use the whoami command to find your username on the system that you're using. Then use the who command to see how your username is listed, along with other users of the same system. What is the on-screen format of each user's listing that you obtained with the who command? Try to identify the information in each field on the same line as your username.

23. Use the hostname command to find out what host computer you are logged on to. Can you determine from this list whether you are using a stand-alone computer or a networked computer system? Explain how you can know the difference from the list that the hostname command gives you.

2.5.2 Printing and General Utility Commands

A very useful and common task performed by every user of a computer system is the printing of text files at a printer.

This is accomplished using the configured printer(s) on the local, or a remote, system. Printers are controlled and managed with the Common UNIX Printing System. We show this utility in detail in Chapter W26, Section 5, at the book website.

The common commands that perform printing on a Linux system are lpr and lp. The general syntax of the lpr command is as follows:

Syntax:
   lpr [options] filename

Purpose: Send files to the printer

Output: Files sent to the printer queue as print jobs

Commonly used options/features:

   -P printer   Sends output to the named printer

   -# copies   Produces the number of copies indicated for each named file

---

The following lpr command accomplishes the printing of the file named **order.eps** at the printer designated on our system as **spr**. Remember from Section 2.2 that no space is necessary between the option (in this case -P) and the option argument (in this case spr).

```
$ lpr -Pspr order.eps
$
```

The following lpr command accomplishes the printing of the file named **memo1** at the default printer.

```
$ lpr memo1
$
```

The following multiple command combines the man command and the lpr command, and ties them together with the Linux *pipe* (|) redirection character, to print the man pages describing the ls command at the printer named hp1.

```
$ man ls | lpr -Php1
$
```

The following shows how to perform printing tasks using the lp command.

The general syntax of the lp command is as follows:

> **Syntax**:
>
>   lp [options][option arguments] file(s)
>
> **Purpose**: Submit files for printing on a designated system printer or alter pending print jobs
>
> **Output**: Printed files or altered print queue
>
> **Commonly used options/features**:
>
>   -d destination   Prints to the specified destination
>
>   -n copies   Sets the number of copies to print

In the first command, the file to be printed is named **file1**. In the second command, the files to be printed are named **sample** and **phones**. Note that the –d option is used to specify which printer to use. The option to specify the number of copies is –n for the lp command.

```
$ lp -d spr file1
request id is spr-983 (1 file(s))
$ lp -d spr -n 3 sample phones
request id is spr-984 (2 file(s))
$
```

Among the most useful of the general purpose, personal productivity utility commands, the cal command displays a calendar for a year or a month. The general syntax of the cal command is as follows:

> **Syntax**:
>
>   cal [[month]year]
>
> **Purpose**: Display calendar on screen as text
>
> **Output**: Displays a calendar of the month or year

The optional parameter month can be between 1 and 12, and year can be 0–9999. If no argument is specified, the command displays the calendar for the current month of the current year. If only one parameter is specified, it is taken as the year. For example, the following command displays the calendar for April 2018:

```
$ cal 4 2018
April 2018
Su  Mo  Tu  We  Th  Fr  Sa
 1   2   3   4   5   6   7
 8   9  10  11  12  13  14
15  16  17  18  19  20  21
22  23  24  25  26  27  28
29  30
```

2.5.3 Communication Commands

The write command is used to send a message to another user who is currently logged on to the system. The syntax and a brief description of the command is as follows:

Syntax:
  `write username [terminal]`
Purpose: Write on the terminal screen or console window of the user with login name username;
   the user must be logged on to the system, and the user's terminal must have write access privilege given by the `mesg` command.
Output: Message on another user's console window.

The example shown in the following command line dialog session illustrates the use of this command.
The prerequisite for executing the `write` command is execution of the `mesg y` command by both sender
(in anticipation of a reply) and receiver to allow writing to their respective terminal screens or console
windows. The `who` command is used to determine whether the person to whom you want to write is logged
on. In this case, both sender (**sarwar**) and receiver (**bobk**) are logged on to the computer **upibm7**, **sarwar**
at terminal **ttyp0** and **bobk** at terminal **ttyC2**. The receiver's screen is garbled with the message, but no
harm is caused to any work that the user is doing. Under the shell, pressing <Enter> performs the trick of
resetting the screen, and inside the `vi` editor (discussed in Chapter W25 at the book website), the screen can
be reset by pressing the <Ctrl> and R keys on the keyboard at the same time. Notice also that the sending
of the message is accomplished by holding down the <Ctrl> and D keys on the keyboard at the same time.

Sender's (sarwar) screen

```
$mesg y
$who
bob                     upibm7:ttyC2               Oct      12       13:47 :34
sarwar                  upibm7:ttyp0               Oct      12       14:20 :15
$write bob ttyC2
Bob,
        How are the new chapter revisions coming along?
Take care,
Mansoor
<Ctrl+D>
```

Receiver's (bob) screen

```
$mesg y
$
Message from sarwar@upibm7.egr.up.edu on ttyp0 at 14:26
Bob,
        How are the new chapter revisions coming along?
Take care,
Mansoor
EOF
```

The `mesg` command enables or disables real-time one-way messages and chat requests from other users
with the `write` and `talk` commands, respectively. The `mesg y` command permits others to initiate communication with you by using the `write` or `talk` command. If you think that you are bothered too often
with `write` or `talk`, you can turn off the permission by executing the `mesg n` command. When you do
so, a user who runs a `write` or `talk` command sees the message `Permission denied`. When the `mesg`
command is used without an argument, it returns the current value of permission, n or y.

    The `biff` command lets the system know whether you want to be notified immediately of an incoming
e-mail message. The system notifies you by sounding a beep on your terminal. You can use the command `biff y` to enable notification and `biff n` to disable notification. When the `biff` command is used
without an argument, it displays the current setting, n or y.

In-Chapter Exercise

  24. Display a calendar for the month of February 2020 on the command line of your Linux system.

2.6 Command Aliases

The alias command can be used to create pseudonyms, or nicknames, for commands. The alias command has one syntax in the Bourne, Korn, and Bourne again shells (Bash); the default shell in Linux shells, and another in the C shell; both forms are illustrated in the following example. The general syntax for the alias command is as follows:

> **Syntax**:
>   alias [name [=string] ...]in Bourne, Korn, Bash shells
>   alias [name [string]]in C shell
>
> **Purpose**: Create pseudonym string for the command name
>
> **Output**: Pseudonyms that can be used for commands

Nicknames are usually created for commands, but they can also be used for other items, such as naming e-mail groups. Both Bash and C shells allow you to create aliases from the command line one at a time, or put them multiply in the resource file for the particular shell.

Command aliases can be placed in the **.profile** file or the **.login** file, but they are typically placed in the **.bashrc** file (for the Bash shell) and the **.cshrc** file (for the C shell). The **.profile** or **.login** file executes when you log on, and the **.cshrc** or **.bashrc** file executes every time you start a C or Bash shell.

Table 2.2 lists some useful aliases to put in one of these files. If set in your environment by any of these means, the aliases in the session later allow you to use the names dir, rename, spr, ls, ll, and page as commands, substituting them for the actual commands given in quotes. Thus, when you type dir Linuxbook, the shell executes the ls -la Linuxbook command.

When you use the alias command without any argument, it lists all the aliases currently set by default.

The following session illustrates the use of this command with a Bourne, Korn, or Bash shell.

The aliases shown in the following command dialog are those found on our Linux Mint system, which uses the Bash shell as the default. They may not be the same as the ones defined by default on your system.

```
$ alias
alias egrep='egrep --color=auto'
alias fgrep='fgrep --color=auto'
alias grep='grep --color=auto'
alias l='ls -CF'
alias la='ls -A'
alias ll='ls -alF'
alias ls='ls --color=auto'
$
```

Running the same command under the C shell, where previously the system administrator or the individual user has already defined a set of aliases, produces the following output. Again, they may not be the same as the ones defined on your system:

TABLE 2.2

Some Useful Aliases for Various Shells

Bourne, Korn, and Bash Shells	C Shell
alias dir='ls -la \!^'	alias dir 'ls -la \!^'
alias rename='mv \!*'	alias rename 'mv \!*'
alias spr='lpr -Pspr \!*'	alias spr 'lpr -Pspr \!*'
alias ls='ls -C'	alias ls 'ls -C'
alias ll='ls -ltr'	alias ll 'ls -ltr'
alias page='more'	alias page 'more'

```
% alias
dir          ls -la
rename            mv
spr          lpr -Pspr
ls           ls -C
ll           ls -ltr
page         more
%
```

You can use the `unalias` command to remove one or more aliases from the alias list.

In the following Bash session, the first of the two `unalias` commands removes the `alias` for `ls`, and the second removes all of the aliases from the alias list. Note that the output of the first `alias` command does not contain an alias for the `ls` command after the `unalias ls` command has been executed. Use of the second `alias` command produces no output because the `unalias -a` command removes all the aliases from the alias list.

```
$ unalias ls
$ alias
dir='ls -la'
rename='mv'
spr='lpr -Pspr'
ll='ls -ltr'
page='more'
$ unalias -a
$ alias
$
```

In the following in-chapter exercises, you will use the `write`, `alias`, and `unalias` commands to practice their syntax and gain more insight into their utility. You will also examine a system file that keeps track of users that can log in.

In-Chapter Exercises

25. Use the `write` command to communicate with a friend who is logged on to the system.
26. Use the `alias` command to display the nicknames (aliases) of commands on your Linux system, if there are any. If there aren't any, create a few useful ones for yourself according to what you might use frequently and beneficially as a nicknamed command. Then, use `unalias` to remove one or more of them. Use `unalias -a` to remove all of the aliases. After you have unaliased all the defaults or defined aliases, how do you reinstate them?
27. Display the contents of the **/etc/passwd** file on your system to determine how many users can log on to the system. What do the contents of this file show you for your username?

Table 2.3 shows some useful commands for beginners.

2.7 Introduction to Linux Shells

When you log on and enter a CUI using a console window or terminal, the Linux system starts running a program that acts as an interface between you and the Linux kernel. This program, called a *Linux shell*, executes the commands that you have typed on the keyboard. When a shell starts running, it gives you a prompt and waits for your commands. When you type a command and press <Enter>, the shell interprets your command and executes it. If you type a nonexistent command, the shell tells you this, then redisplays the prompt and waits for you to type the next command. Because the primary purpose of the shell is to interpret your commands, it is also known as the *Linux command line interpreter.*

TABLE 2.3

Useful Commands for the Beginner

Command	What It Does
<Ctrl+D>	Terminates a process or command
alias	Allows you to create pseudonyms for commands
biff	Notifies you of new e-mail
cal	Displays a calendar on screen
cat	Allows joining of files
cd	Allows you to change the current working directory
cp	Allows you to copy files
exit	Ends a shell that you have started
hostname	Displays the name of the host computer that you are logged on to
login	Allows you to log on to the computer with a valid username/password pair
lpr or lp	Allows printing of text files
ls	Allows you to display the names of files and directories in the current working directory
man	Allows you to view a manual page for a command or topic
mesg	Allows or disallows writing messages to the screen
mkdir	Allows you to create a new directory
more	Allows you to view the contents of a file one screen at a time
mv	Allows you to move the path location of, or rename, files
passwd	Allows you to change your password on the computer
pg	Displays one screen of a file at a time
pwd	Allows you to see the name of the current working directory
rm	Allows you to delete a file from the file structure
rmdir	Allows you to delete directories
talk	Allows you to send real-time messages to other users
telnet	Allows you to log on to a computer on a network or the Internet
unalias	Allows you to undefine pseudonyms for commands
uname	Displays information about the operating system running the computer
whatis	Allows you to view a brief description of a command
whereis	Displays the path(s) to commands and utilities in certain key directories
who	Allows you to find out login names of users currently on the system
whoami	Displays your username
write	Allows real-time messaging between users on the system

A shell command can be internal/built-in or external. The code to execute an internal command is part of the shell process, but the code to execute an external command resides in a file in the form of a binary executable program file or a shell script. (We describe in detail how a shell executes commands in Chapter 10.)

Because the shell executes commands entered from the keyboard, it terminates when it finds out that it cannot read anything else from the keyboard. You can inform your shell of this by pressing <Ctrl+D> at the beginning of a new line. As soon as the shell receives <Ctrl+D>, if you are logged in at a text-only console, it terminates and logs you off the system. The system then displays the login: prompt again, informing you that you need to log on again in order to use it. If you are using a desktop environment, such as the Linux Mint Cinnamon desktop, and you have opened a terminal, or console, window on the desktop, the effect of pressing <Ctrl+D> is to close the terminal or console window.

The shell interprets single Linux commands that are structured according to Section 2.2—that is, by assuming that the first word in a command line is the name of the command that you want to execute. It assumes that any of the remaining characters, starting with a hyphen (–) are options (possibly followed by option arguments) and that the rest are the command arguments.

After reading your command line, the shell determines whether the command is an internal or external command. It processes all internal commands by using the corresponding code segments that are within its own code. To execute an external command, it searches several directories in the file system structure (see Chapter 4), looking for a file that has the name of the command. It then assumes that the file contains the code to be executed and runs the code.

The names of the directories that a shell searches to find the file corresponding to an external command are stored in the shell variable named PATH (or path in the C shell). Directory names are separated by colons in the Bourne, Korn, and Bash shells and by spaces in the C shell. The directory names stored in the PATH variable form what is known as the *search path* for the shell. You can view the search path for your variable by using the echo $PATH command in the Bourne, Korn, Bash, and C shells.

The following are two sample sessions run with this command in a terminal window on our Linux Mint system. The first is done in the Bash shell and the second in the C shell. Note that in the default Bash shell on our Linux systems, the search path contains the directory names separated by colons and that in the C shell the directory names are separated by spaces.

```
$ echo $PATH
/usr/local/sbin:/usr/local/bin:/usr/sbin:/usr/bin:
/sbin:/bin:/usr/games:/usr/local/games
$

% echo $path
/usr/local/sbin /usr/local/bin /usr/sbin /usr/bin
/sbin /bin /usr/games /usr/local/games
%
```

The PATH (or path) variable is defined in a hidden file (also known as a dot file) called **.profile** or **.login**. If you can't find this variable in one of those files, it is in the start-up file (also a dot file) specific to the shell that you're using. You can change the search path for your shell by changing the value of this variable. To change the search path temporarily for your current session only, you can change the value of PATH at the command line. For a permanent change, you need to change the value of this variable in the corresponding dot file.

In the following Bash shell example, the search path was augmented by two directories, **~/bin** and **.** (current directory). Moreover, the search starts with **~/bin** and ends with the current directory.

Be careful when editing or changing the PATH variable, so that you don't lose any component of the default search path set by the system administrator for all users of the system.

```
$ PATH=~/bin:$PATH:.
$
```

You can determine your login shell by using the echo $SHELL command, as described in Section 2.8.3. Each shell has several other environment variables set up in a hidden file associated with it. We describe these files in Section 2.8.4 and present a detailed discussion of Linux files in Chapter 4.

In-Chapter Exercise

28. What command prints the names of only the dot files in your home directory on your Linux system at standard output? What are the names of those dot files? What does each of the dot files accomplish? Make a list of them, and next to each entry in the list, write a short description in your own words of what that dot file accomplishes.

2.8 Various Linux Shells

All of our representative Linux systems come with the Bash shell preinstalled, and Bash is the default shell. The C, Bourne, Korn, Tennex C (TC), and Z shells are less popular. But they can be installed on

Debian-family systems using the apt package management facility, and with YUM on RHEL-family CentOS. These other shells offer advantages for certain applications and ways of working with the Linux system. When you log onto your Linux system using a non-GUI, text-based CUI, one particular type of shell starts execution. This shell is known as your *login shell*, and it is usually determined by the system administrator of your Linux system. If you want to use a different shell, you can do so by first installing that shell program with the appropriate package management system, and then running a corresponding command available on your system to execute the shell program.

For example, your login shell is Bash by default, but you want to use the C shell, you can do so by first installing the C shell on a Debian-family Linux system with the apt-get command, as shown in the following example, and then using the csh command.

```
$ sudo apt-get install csh
Output truncated...
$ csh
#
```

2.8.1 Shell Programs

Essentially the shell program itself, which is implemented in the C programming language, allows you to do interpreted programming (as opposed to compiled programming). It does this in two senses. First, so you can employ simple, single or complex, multiple Linux commands connected by redirection operators and/or utilities such as sed, awk or grep, to do common tasks. Second, via user-written script files, coded in the shell interpreted language, that automate and simplify those common, perhaps highly repetitive, tasks. This interpreted language has all the features of any other structured, high-level programming/scripting language, such as Perl, Tcl, or Python. The shell language is just not as complex as these other common scripting languages. This fact should tell you why there are so many different shells, just as there are many different high-level programming and scripting languages.

Programming languages have a tendency to evolve and grow with time, depending on the needs of users, and shell programs are typical of this evolution. Table 2.4 contains a list of the most common shells, their location, and the program names of those shells.

The locations shown in Table 2.4 are typical for most Linux systems. Consult your instructor or system administrator if you can't find the location shown for a shell on your system, or if you can't use the whereis command, as shown in Section 2.5.1.

Figure 2.4 traces the development of various shell families, and indicates the increasing functionality of each family as it appears higher in the hierarchy. The Bourne shell (sh) is the *grandmother* of the main shell families and has nearly the least level of functionality. Near the top of the hierarchy is the Korn shell (ksh), which includes all the functionality of the Bourne shell and much more. The rc and zsh shells are outliers that cannot be readily associated with any of the primary shell families.

TABLE 2.4

Shell Locations and Program Names

Shell	Default Location on Most Linux Systems (if the shell has been installed!)	Shell Program (Command) Name
rc	/bin/rc	rc
Bourne shell	/bin/sh	sh
C shell	/bin/csh	csh
Bourne again shell (Bash)	/bin/bash	Bash
Z shell	/bin/zsh	Zsh
Korn shell	/bin/ksh	Ksh
TC shell	/bin/tcsh	Tcsh

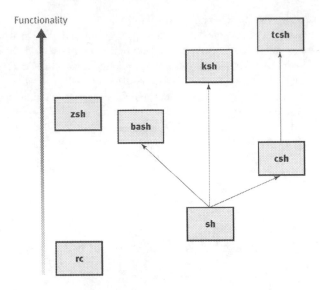

FIGURE 2.4   Development hierarchy of various Linux shells.

2.8.2 Which Shell Suits Your Needs?

Most shells perform similar functions, and knowing the details of how they do so is important in deciding which shell to use for a particular task. Also, using more than one shell during a session is a common practice, especially among shell script file programmers. For example, you might use the Bourne or Korn shell for their programming capabilities and use the C shell to execute individual commands. We discuss this example further in Section 2.8.3. The similarities of major shell functions are summarized in Table 2.5.

2.8.3 Ways to Change Your Shell

You can easily determine what your default shell is by typing echo $SHELL on the command line when you first log on to your computer system.

The question is: Why would you want to change your default shell, or for that matter, even use an additional shell? The answer is that you want the greater, or in some sense qualitatively different, functionality of another shell.

For example, your default shell might be the Bash shell (bash). A friend of yours offers you a neat and useful Z shell script that allows you to take advantage of the Z shell script programming capabilities, a script that wouldn't work if it ran under the Bash shell. You can use this script by running the Z shell at the same time you are running the default Bash shell. Because Linux is a multiprocess operating system, more than one command line interpreter at a time can be active. That doesn't mean that a single command will be interpreted multiply; it simply means that input, output, and errors are "hooked" into whatever shell process has control over them currently. [See Chapter 10 for more information about process and shell command input/output (I/O).]

TABLE 2.5

Shell Similarities

Function	Description
Execution	The ability to execute programs and commands
I/O handling	The control of program and command input and output
Programming	The ability to execute sequences of programs and commands

You can change your shell in one of two ways:

Method 1. Changing to a new default for every subsequent login session on your system, and

Method 2. Creating additional shell sessions running on top of, or concurrently with, the default shell only during the current login session.

The premise of both methods is that the shell you want to change to is available on your system. And if it isn't, you can use the apt command shown in Section 2.8 to install it.

Method 1. To change your default shell, after you have logged on as an ordinary user, type the following command:

```
$ chsh
password for bob: www
Changing the login shell for bob
Enter the new value, or press ENTER for the default
        Login Shell [/bin/bash]: /bin/csh
$
```

As shown on our Linux Mint system, you will be prompted for the name of the shell you want to change to. You type the complete path to the location of the shell you want to change to—for example, **/bin/sh** to change to the Bourne shell. Consult Table 2.4 for the complete paths to common Linux shells on your system.

If this method doesn't work on your system, consult your instructor or system administrator for more help.

Method 2. To create or run additional shells on top of your default shell, simply type the name of the shell program (see Table 2.4) on the command line whenever you want to run that shell. The following session illustrates the use of this method to change a default Bash shell, which uses the $ as the shell prompt, to a C shell, which shows the **%** as the shell prompt.

```
$ echo $SHELL
/bin/bash
$ csh
%
```

The first command line allows you to determine your default shell. In this case, the system shows you that the default setting is the Bash shell. The second command line allows you to run the C shell. The fourth line shows that you have been successful, because the default C shell prompt appears on your display. If the C shell was not available on your system or was inaccessible to you, you would get an error message after the third line. If your search path does not include /bin, you either have to type /bin/csh in place of csh, or include /bin in your shell's search path and then use the csh command.

To terminate or leave this new, temporary shell and return to your default login shell, hold down <Ctrl+D> on a blank line. If this way of terminating the new shell doesn't work, type exit on the command line and then press <Enter>. By doing so, you halt the running of the new shell, and the default shell prompt appears on your display. If you have opened a console or terminal window on your desktop, typing exit also closes this console or terminal window.

The following in-chapter exercises ask you to determine whether various shells are available on your system by using the whereis command and, for those that are available, to read the manual pages for them by using the man command.

In-Chapter Exercises

29. Using the whereis command illustrated in Section 2.5.1, verify the locations of the various shells listed in Table 2.4. Are all these shells available on your system? Where are they located if you do not find them at the locations shown in Table 2.4?

30. Using the man command illustrated in Section 2.4.5, read the manual pages for each shell listed in Table 2.4 that is on your system.

31. If the C, Bourne, Korn, TC, and Z shells are *not* installed on your Linux system, then install them using the appropriate package management commands. Look in Appendix A for descriptions of how to use the package management commands that are available on your Linux system to install these shells. Then, execute each of them, and experiment with any of the special features that they have, by doing online research, to find a comparative analysis of those shells. You can also gain some insight into their operation by doing Exercise 30.

2.8.4 Shell Start-Up Files and Environment Variables

The actions of each shell, the mechanics of how it executes commands and programs, how it handles the command and program I/O, and how it is programmed, are affected by the setting of certain *environment variables*.

Each Linux system has an initial system start-up file, usually named **.profile** or **.login**. This file contains the initial settings of important environment variables for the shell and some other utilities. In addition, hidden files for specific shells are executed when you start a particular shell. Known as the shell start-up files, they are **.bashrc** for Bash and **.cshrc** for the C shell. These hidden files are initially configured by the system administrator for secure use by all users. Table 2.6 lists some important environment variables common to Bash, Bourne, Korn, and C shells; the C shell variable name, where applicable, is in lowercase following the uppercase Bash, Bourne, and Korn shell variable names. Note that your system administrator may not have set some of these variables, such as ENV.

The following In-Chapter exercises let you view the settings of your environment variables. They assume that you are initially running the Bash shell.

In-Chapter Exercises

32. At the default login shell prompt for your system, type `set | more` and then press `<Enter>`. On our representative Linux systems, the default shell is Bash. What is displayed on your screen? Identify and list the settings for all the environment variables shown in Table 2.6.

33. If the C shell is not installed on your system, install it using the command shown in Section 2.8. Then make the C shell the current active shell for this session. Finally, type `setenv | more` and then press `<Enter>`. Identify and list the settings for all the environment variables shown in Table 2.6.

In addition to the shells, several other programs have their own hidden files. These files are used to set up and configure the operating environment within which these programs execute. We discuss some of these

TABLE 2.6

Shell Environment Variables

Environment Variable	What It Affects
CDPATH, cdpath	The alias names for directories accessed with the **cd** command
EDITOR	The default editor used in programs such as the e-mail program Elm
ENV	The path along which Linux looks to find configuration files
HOME, home	The name of the user's home directory when the user first logs on
MAIL, mail	The name of the system mailbox file
PATH, path	The directories that a shell searches to find a command or program
PS1, prompt	The shell prompt that appears on the command line
PWD, cwd	The name of the current working directory
TERM	The type of console terminal being used

hidden files in Chapters 4 and 5. They are called hidden files because when you list the names of files contained in your home directory—for example, with the `ls -l` command and option (see Chapter 4)—these files do not appear on the list. The hidden file names always start with a period (.), such as **.login**.

2.9 Shell Metacharacters

Most of the characters other than letters and digits have special meaning to the shell. These characters are called *shell metacharacters* and, therefore, cannot be used in shell commands as literal characters, without specifying them syntactically in a particular way. Thus, try not to use them in naming your files. Also, when these characters are used in commands, no space is required before or after a character. However, you can use spaces before and after a shell metacharacter for clarity. Table 2.7 contains a list of the shell metacharacters and their purposes.

The shell metacharacters allow you to specify multiple files in multiple directories in one command line. We describe the use of these characters in subsequent chapters, but we give some simple examples here to explain the meanings of some commonly used metacharacters:

- *
- ?

TABLE 2.7

Shell Metacharacters

Metacharacter	Purpose	Example
`<New Line>`	To end a command line	
`<Space>`	To separate elements on a command line	`ls /etc`
`<Tab>`	To separate elements on a command line	`ls /etc`
`#`	To start a comment	`# This is a comment line`
`"`	To quote multiple characters but allow substitution	`"$file" bak`
`$`	To end line and dereference a shell variable	`$PATH`
`&`	To provide background execution of a command	`command &`
`'`	To quote multiple characters	`'$100,000'`
`( )`	To execute a command list in a subshell	`(command1; command2)`
`*`	To match zero or more characters	`chap*.ps`
`[ ]`	To insert wild cards	`[a-s] or [1,5-9]`
`^`	To begin a line and negation symbol	`[^3-8]`
`` ` ``	To substitute a command	`PS1=`command``
`{ }`	To execute a command list in the current shell	`{command1; command2}`
`|`	To create a pipe between commands	`command1	command2`
`;`	To separate commands in sequential execution	`command1; command2`
`<`	To redirect input for a command	`command < file`
`>`	To redirect output for a command	`command > file`
`?`	To substitute a wild card for **lab.?** exactly one character	
`/`	To be used as the root directory and **/usr/bin** as a component separator in a pathname	
`\`	To escape/quote a single character; n command `arg1 \` used to quote `<New Line>` character `arg2 arg3` to allow continuation of a shell `\?` command on the following line	
C and Korn Shells Only		
`!`	To start an event specification in the history list and the current event	`!!, !$`
`%`	The C shell prompt or the starting character for specifying a job number	`% or %3`
`~`	To name home directory	`~/.profile`

- ~
- [ ]

The ?.txt string can be used for all the files that have a single character before **.txt**, such as **a.txt**, **G.txt**, **@.txt**, and **7.txt**. The [0-9].c string can be used for all the files in a directory that have a single digit before **.c**, such as **3.c** and **8.c**. The lab1\/c string stands for **lab1/c**. Note the use of backslash (\) to quote (escape) the slash character (/).

The following command prints the names of all the files in your current directory that have two-character file names and an .html extension, with the first character being a digit and the second being an uppercase or lowercase letter. The printer on which these files are printed is **spr**.

```
$ lpr -Pspr [0-9][a-zA-Z].html
$
```

Note that **[0-9]** means any digits from 0 to 9 and **[a-zA-Z]** means any lowercase or uppercase letter. The following command displays the names of all six-character-long files with **.c** extension in your current directory, with the first three characters being lab, the fourth being a digit, and the remaining being any two characters.

```
$ ls lab[0-9]??.c
lab11a.clab1a1.c lab123.clab4ab.c
$
```

2.10 The sudo and su Commands

The sudo command allows a permitted user to execute a command as the superuser, or to assume the role of another user, as specified by security policy. The su command allows an ordinary user to switch user roles or to also simulate being the superuser on the system. The superuser has file permission and access privileges to everything on the system.

In many of the operations shown in the following chapters, particularly in Chapter 17 on system administration, it will be necessary to execute the sudo or su command to accomplish the tasks shown. In order to use this command, it is necessary to know the root or superuser password.

We give a more complete explanation of the sudo command in Chapter W26, Section 9.2.1, at the book website.

An example of using the su command is as follows:

```
$ su
Password: www
/home/bob#
```

Summary

The Linux operating system is most famous for its text-based command execution, but in the 21st century it has a competitively developed GUI environment as well. This chapter serves to familiarize you with the basic structure of a CUI Linux command. It also shows you how to log in via three popular and typical login methods and how to gracefully log off.

A beginner must be able to do basic file maintenance, and a core set of CUI file maintenance commands and their options are introduced in this chapter. These commands will be useful throughout the rest of this book. Finally, we illustrate and give examples of some basic utility commands—most importantly, the commands and their options that allow you to print files and the alias command.

When you log on to a Linux computer, the system runs a program called a shell that gives you a prompt and waits for you to type commands, either as single commands or as multiple commands connected by redirection or piping operators. The shell program, coded in C, is an interpreter, and as such has the same structured programming capabilities of high-level languages. When you type a command and press <Enter>, the shell interprets and tries to execute the command, assuming that the first word in the command line is the name of the command. A shell command can be built-in or external. The shell has the code for executing a built-in command, but the code for an external command is in a file. To execute an external command, the shell searches several directories, one by one, to locate the file that contains the code for the command. If the file is found, it is executed if it contains the code (binary or shell script). The names of the directories that the shell searches to locate the file for an external command form are known as the search path. The search path is stored in a shell variable called PATH (for the Bourne, Korn, and Bash shells) or path (for the C shell). You can change the search path for your shell by adding new directory names in PATH or by deleting some existing directory names from it.

Several shells are available for you to use. These shells differ in terms of convenience of use at the command line level and features available in their programming languages. The most commonly used shells in a Linux-based system are the Bash and C shells. The Bourne shell is the oldest and has a good programming language. The C shell has a more convenient and rich command-level interface. The Korn shell has some good features of both and is a superset of the Bourne shell.

Certain characters, called shell metacharacters, have special meaning to the shell. Because the shell treats them in special ways, they should not be used in file names. If you must use them in commands, you need to quote them for the shell to treat them literally.

Questions and Problems

1. Create a directory called Linux in your home directory. What command line did you use to do this?

2. Give a command line for displaying the files **lab1**, **lab2**, **lab3**, and **lab4**. Can you give two more command lines that do the same thing? What is the command line for displaying the files **lab1.c**, **lab2.c**, **lab3.c**, and **lab4.c**? (*Hint*: use shell metacharacters.)

3. Give a command line for printing all the files in your home directory that start with the string memo and end with **.ps** on a printer called **upmpr**. What command line did you use to do this?

4. Give the command line for nicknaming the command who -H as W. Give both Bash and C shell versions. Where would you put it if you want it to execute every time you start a new shell?

5. Type the command man ls > ~/Linux/ls.man on your system. This command will put the man page for the ls command in the **ls.man** file in your **~/Linux** directory (the one you created in Problem 1). Give the command for printing two copies of this file on a printer in your lab. What command line would you use to achieve this printing?

6. What is the mesg value set to for your environment? If it is on, how would you turn it off for your current session? How would you set it off for every login?

7. What does the command lpr -Pqpr [0-9]*.jpg do? Explain your answer.

8. Use the passwd command to change your password. If you are on a network, be aware that you might have to use the yppasswd command to modify your network login password. Also, make sure you abide by the rules set up by your system administrator for coming up with good passwords!

9. Using the correct terminology (e.g., command, option, option argument, and command argument), identify the constituent parts of the following Linux single commands.

   ls -la *.exe

   lpr -Pwpr file27

   chmod g+rwx *.*

10. View the man pages for each of the useful commands listed in Table 2.3. Which part of the man pages is most descriptive for you? Which of the options shown on each of the man pages is the most useful for beginners? Explain.

11. How many users are logged on to your system at this time? What command did you use to discover this?

12. Determine the name of the operating system that your computer runs. What command did you use to discover this?

13. Give the command line for displaying manual pages for the socket, read, and connect system calls on your system.

14. What is a shell? What is its purpose?

15. What are the two types of shell commands? What are the differences between them?

16. Give the names of five Linux shells. Which are the most popular? What is a login shell?
    What do you type in to terminate the execution of a shell? How do you terminate the execution of your login shell?

17. What shells do you think are *supersets* of other shells? In other words, which shells have other shells' complete command sets plus their own? Can you find any commands in a subset shell that are not in a superset shell? Refer to Figure 2.4.

18. What is the search path for a shell? What is the name of a shell variable that is used to maintain it for the Bash, C, and Korn shells? Where (i.e., in which file) is this variable typically located?

19. What is the search path set to in your environment? How did you find out? Set your search path so that your shell searches your current and your **~/bin** directories while looking for a command that you type. In what order does your shell search the directories in your search path? Why?

20. What are hidden files? What are the names of the hidden files that are executed when you log on to your Linux systems, and how did you find this out?

21. What is a shell start-up file? What is the name of this file for the Bash shell? Where (i.e., in which directory) is this file stored?

22. What important features of each shell, as discussed on the manual pages for that shell, seem to be most important for you as either a new, intermediate, or advanced user of Linux? Explain the importance of these features to you in comparison with the other shells available and their features.

23. Suppose that your login shell is a C shell. You receive a shell script that runs with the Bourne shell. How would you execute it? Clearly write down all the steps that you would use.

Advanced Questions and Problems

24. Following is a typical /etc/profile configuration file, this particular one is from a default installation on our Linux Mint 18.2 system:

```
# /etc/profile: system-wide .profile file for the Bourne shell (sh(1))
# and Bourne compatible shells (bash(1), ksh(1), ash(1), ...).

if [ "$PS1" ]; then
  if [ "$BASH" ] && [ "$BASH" != "/bin/sh" ]; then
    # The file bash.bashrc already sets the default PS1.
    # PS1='\h:\w\$ '
    if [ -f /etc/bash.bashrc ]; then
      . /etc/bash.bashrc
    fi
```

```
    else
      if [ "`id -u`" -eq 0 ]; then
        PS1='# '
      else
        PS1='$ '
      fi
    fi
  fi

  if [ -d /etc/profile.d ]; then
    for i in /etc/profile.d/*.sh; do
      if [ -r $i ]; then
        . $i
      fi
    done
    unset i
  fi
```

Write an explanatory sentence in your own words describing exactly what you consider important lines in the file accomplish, including the comments (the lines that begin with the pound sign #). Examine this file on your Linux system. How does it compare, line-for-line, with the one above? We assume here that, by default, Bash is both the interactive and login shells on your system.

25. What is the default umask setting in an ordinary, nonprivileged account on your Linux system, from both a login and a nonlogin shell? Describe in your own words what the umask setting is, and how it is applied to newly created directories and files. Is the umask set in /etc/profile on your Linux system? If not, where can the umask be set most effectively on a persistent basis, for a particular single user, in both a login and a nonlogin shell?

26. Assume that all users, when they log into your Linux system, have Bash as their default shell. What file sets the shell prompt for them on your Linux system? Is it the file illustrated in Problem 24? Describe the lines in the file that actually specify the shell prompt, and give a short description of the components of those lines. Experiment to find out which file accomplishes the actual shell prompt setting for ordinary users (for both nonlogin and login shells), and write an explicit description of what you have discovered.

   Additionally, set the shell prompt for yourself in the current nonlogin shell, so that it contains the following:

   A display of just the date/time.

   A display of the date and time, hostname and current directory.

   A display where the entire prompt is in red text, along with hostname and current directory.

   Then make those changes persistent for yourself in both login and nonlogin shells. Finally, undo the persistent changes.

   As a follow-up, design your own shell prompt so that it contains the information you want in a useful display, given your use case(s), and make that designed prompt persistent for yourself on your Linux system.

27. Give a sequential list of the exact commands you would use to make the TC shell the default login shell for your user account on your Linux system. Is the TC shell installed by default on your Linux system? If not, how would you install it on a Debian-family or CentOS system? Give the exact commands for installation of not only the TC shell, but any of the other four major Linux shells available.

28. Execute all of the compound command examples provided at the Web link https://explainshell. com/ and then use the output shown to explain all of them in your own words. Try executing the examples with meaningful arguments on your Linux system, if possible.

Projects

Project 1

After completing Problems 24–26, gather your findings together in a summary report that details the default settings (within the scope of the files you have examined, and in the context of those problems) of the Bash environment on your Linux system. For example, which actual file takes precedence by default, and what components of the Bash environment are set in that file? What are the critical default settings in the Bash environment, and what actual files on your Linux system affect them?

Project 2

Design a scaled-down CUI command line interpreter, or shell, that is similar in function to the Bash shell. It should be an interactive, nonlogin shell that uses the same exact structure for command, options, option arguments (if you include those), and command arguments that all *single* Linux commands in the traditional shells have. It should implement a minimum of five Linux commands, such as `ls`, `cd`, `more`, `pwd`, and `rm`. Specify as carefully, and in as much detail as necessary, all of the operations of your shell program: how it is launched, its treatment of arguments, how it will handle errors in command input and execution, and how it is exited. Also specify, as completely and articulately as possible, how it can be tested on a set of valid (and invalid) arguments passed to it, and how options and option arguments will function on valid arguments.

This project can be done by an individual or in a group.

First pseudocode the software to plan how it will execute the limited set of commands on selected arguments, using a small set of options as "switches" to those commands. As an essential part of the planning process, the completeness of the specification in pseudocode is critical. For example, modularizing the program at this stage is a robust way of designing the logic of its execution.

Then, when the logic of execution is complete, convert your pseudocode into Python, C, C++, or any other high-level language you are familiar with. You must *not* simply "wrap" Bash, or any other Linux shells commands, inside of your coded program. Depending upon your knowledge of the coding language chosen, and your familiarity with Linux API system calls (many of which we detail in Chapters 15 and 16, and at the book website in Chapters W20 and W21), you can make your program as "low level" as you can.

Execute the program, and run it through the suite of test specifications you have designed, and let the test results feed back into the coded version of the program if necessary.

Finally, produce a written report of your results that details all of your design criteria, the pseudocode, the final code, and the testing results as well.

Looking for more? Visit our sites for additional readings, recommended resources, and exercises.

CRC Press e-Resource: https://www.crcpress.com/9781138710085

Authors' GitHub: https://github.com/bobk48/linuxthetextbook

3

Editing Text Files

OBJECTIVES

- To explain the general utility of editing text files on a Linux system
- To give a basic introduction to the nano text editor
- To cover the following commands and primitives:
  nano

3.1 Introduction

In this chapter, we use the following simple text editor that is commonly available on our base modern Linux systems: nano. We also comment on the general uses of text editing, and in the light of these uses, discuss other common Linux text editors. At the website for this book, we present more in-depth coverage of the vi/vim/gvim and emacs editors.

3.1.1 nano Typographic Conventions

To stress how the keyboard keys are used in the nano editor, we provide the following typographic convention reference for the keys used to execute commands or change modes:

1. Pressing the Escape key is signified as <Esc>.
2. Pressing the Enter key is signified as <Enter>.
3. Pressing the <Ctrl> key (represented in shortcut form on the nano screens as ^) in combination with another single key is signified as <Ctrl+X>, where you hold down the <Ctrl> key and press the X key (or any valid key for that combination) *at the same time*.
4. Pressing the Alt key in combination with another single key is signified as <Alt+X>, where you hold down the <Alt> key and press the X key (or any valid key for that combination) *at the same time*.
5. A variant of 3. and 4. is shown as <Ctrl+X> a [b], where you first press *and release* <Ctrl> and X simultaneously, then press the a key, and optionally press the b key (or any valid combination of single keys or strings of characters).

What you type or hold down on the keyboard is shown in **bold** text.

3.1.2 Comments on Linux Text Editors and Their General Use

Modern Linux uses both a graphical user interface (GUI), with powerful window management systems, and a character user interface (CUI). Therefore, to do useful things such as execute multiple commands from within a script file, write e-mail messages, or create C language programs, you must be familiar with one or perhaps multiple ways of entering text into a file. In addition, you must also be familiar with how to edit existing files efficiently—that is, to change their contents or otherwise modify them in some way. Text editors allow you to view a file's contents, similar to the more command, so that you can identify the key features of the file, and then read and utilize the information contained in it. For example, a file without any extension, such as **foo** (rather than **foo.txt**) might be a text file that you can view with a text editor.

The editors that we consider here are all considered full-screen display editors. That is, on the display screen or monitor that you are using to view or edit a file, you are able to see a portion of the file, which fills most or all of the window allocated to the text editor screen display. You are also able to move the cursor, or point, to any of the text you see in this full-screen display, with either the arrow keys on the keyboard or with a mouse. That text material is usually held in a temporary storage area in computer memory called the editor *buffer*. If your file is larger than one screen, the buffer contents change as you move the cursor through the file. The difference between a file, which you edit, and a buffer is crucial. For text-editing purposes, a file is stored on disk as a sequence of data. When you edit that file, you edit a copy that the editor creates, which is in the editor buffer. You make changes to the contents of the buffer—and can even manipulate several buffers at once—but when you save the buffer, you write a new sequence of data to the disk, thereby saving the file.

Another important operational feature of all Linux editors is that, traditionally, their actions are based on keystroke commands, whether they are a single keystroke or combinations of keys pressed simultaneously or sequentially. Because one of the primary input devices in Linux is the keyboard, using the correct syntax of keystroke commands is *mandatory*. But the keyboard method of input, once you have become accustomed to it, is as efficient or, for some users, even more efficient than mouse/ GUI input. Keystrokes are also more flexible, giving you more complete and customizable control over editing actions. Generally, you should choose the editor you are most comfortable with, in terms of the way you prefer to work with the computer. Of course, for most beginners, that will be nano. However, your choice of editor also depends on the complexity and quantity of text creation and manipulation that you want to do. Practically speaking, editors such as vi, vim, gvim, and Gnu emacs are capable of handling complex editing tasks in multiple windows on multiple files, and provide you with a visual software development environment, as well as document production and management capability. But to take advantage of that power, you have to learn the mechanics of the commands that are needed to perform those tasks and how they are implemented either graphically or by typing them—and retain that knowledge. The basic functions common to the text editors are listed in Table 3.1, along with a short description of each function.

For the text editors vi, vim, and gvim, you can't immediately begin to enter text into the file you are editing. You have to be in *Insert mode* to do that. vi, vim, and gvim have modes.

In nano (and Gnu emacs), you can start typing text into the file immediately. nano is a *modeless* editor, and that's why it is easy to use. That basically means that you can immediately begin entering and editing text using the keyboard and some particular pointing device, such as a mouse or keyboard arrow keys.

We present the tutorial information on nano in this chapter using typed commands.

It is very important to realize that vi, vim, and gvim generally use the same commands and have basically the same functionality. vim and gvim are not only more graphical—allowing you to work more

TABLE 3.1

Basic Text-Editing Functions

Function	Description
Cursor movement	Moving the location of the insertion point or current position in the buffer
Cut or copy, paste	"Ripping out" text blocks or duplicating text blocks, reinserting ripped or duplicated blocks
Deleting text	Deleting text at a specified location or in a specified range
Inserting text	Placing text at a specified location
Opening, starting	Opening an existing file for modification, beginning a new file
Quitting	Leaving the text editor, with or without saving the work done
Saving	Retaining the buffer as a disk file
Search, replace	Finding instances of text strings, replacing them with new strings

efficiently in GUI environments such as those on our base modern Linux systems—but they also have an improved and expanded command structure.

At the time of the writing of this book, our base Linux systems had the following text editors preinstalled:

Debian 9.1: nano, vi

Ubuntu 16.04: nano, vi

Linux Mint 18.2: nano, vi

CentOS 7.4: nano, vi, vim

The earlier listing is true if you have done a basic installation of the system, as detailed at the beginning of Chapter 17. gvim, and Gnu emacs are not preinstalled on any of our base Linux systems.

The easiest and best way to install these editors on your system is by using the package management system available on your system. We show the basics of package management systems for our base Linux systems in Chapter 17.

We also show how to install vi, vim, and emacs on our base Linux systems in Appendix A.

The most expedient way of doing an installation of these editors, if they are not already installed on your Linux system, is to use a graphical form of package management. This assumes you are interacting with your Linux system using a desktop management system GUI, such as Cinnamon in Linux Mint. For example, when we show installation of an editor on our Linux Mint system, we searched in the graphical Software Manager for vim, gvim, and emacs, and then used the Linux Mint Software Manager's facilities to install the following packages:

For vim: Vim Version 2:7.4.1689-3ubuntu1.2

For gvim: Vim-gtk 2:7.4.1689-3ubuntu1.2

For emacs: Gnu emacs 24.3.1

In addition, be aware that if you are logging into a Linux system via a terminal window, such as with PuTTY from a Windows machine, many of the graphical modes and techniques of using Linux text editors will not be available to you. But that does not prevent you from using the traditional typed commands and keyboard edits that we show for nano!

3.1.3 Other Ways of Treating Text and Files in Linux

Text editors, such as nano, vi/vim/gvim, and emacs, are not the only way of entering text into a file. As we show in later chapters, some of the other methods are by using output redirection in a multiple Linux command, and "stream editing," using traditional text processing utilities such as sed and gawk.

Output redirection, with the > and >> shell command operators in a multiple command (as opposed to a single Linux command whose structure we showed in Chapter 2) allows a quick and easy method of entering text into a file. Coupled with the echo and cat commands, which we show examples of in Chapter 9, output (and input) redirection is a viable and expedient way to create and modify text in text files.

Stream editing, simply stated, is the manipulation of text from an input stream, such as a file or standard input, using codes that produce an output stream that has the desired text content. It is traditionally and most readily accomplished using the sed (Stream Editor) or gawk commands. This allows you to append or modify text into existing files, such as configuration files, and for example, use either basic or extended Regular Expressions to construct the additions or modifications in the output stream.

For more complete documentation on sed, see the man page for sed on your Linux system. We also provide a Web reference at the GitHub site for this book, which gives a more complete description of sed, with numerous examples. gawk (the Gnu version of awk) is another stream editing method, which uses a complete and complex programming language environment. gawk allows you to process text files. For more information about gawk, see the man page for gawk on your Linux system.

3.2 A Quick Introduction to the nano Editor

The simplest and most readily available text editor in all of the Linux systems we show in this book is nano. For many users, nano would be the most efficient and adequate editor to deploy whenever a text editor is needed. Many of the tasks shown in the other standard Linux editors can be done quickly and easily by a novice user of Linux using nano.

To get started in nano, either when you only have a CUI available, or when you are interacting with Linux via some desktop GUI (such as Cinnamon or Gnome), on the Linux command line in a terminal window, type the following:

$ **nano**

You are immediately able to enter text into the editor and move the cursor around with the arrow keys available on either a physical or a "virtual" keyboard (such as is available on an iPad running the Termius terminal application). The first screen display presented by nano is shown in Figure 3.1. You can also use the Delete key to delete text on the current line, at the first character before the current position of the cursor in the on-screen display of the text.

If you typed **nano file2**, where file2 is the name of an already-existing text file, that text file is displayed on-screen, allowing you to make changes or additions to it.

There are four main sections of the editor. The top line shows the program version, the current filename being edited, and whether or not the file has been modified. Next is the main editor window showing the file being edited. The status line is the third line from the bottom and shows important messages. The bottom two lines show the most commonly used commands executed via keyboard "shortcuts" in the editor.

Referring to Figure 3.1, at the bottom of the nano screen display, there are several keyboard shortcut choices that you can make by holding down the Control (**<Ctrl>**, displayed as ∧ on the nano menu) key on your keyboard, in combination with a letter key. The two most important shortcut choices presented are as follows:

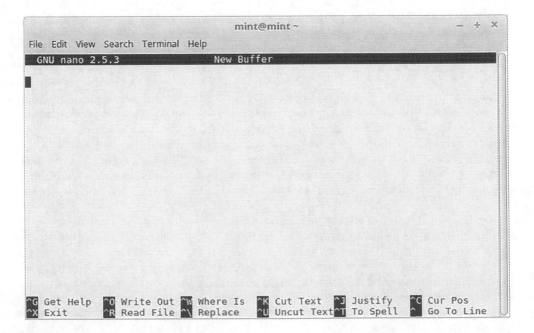

FIGURE 3.1   General appearance of the nano screen.

The command **<Ctrl+O>** allows you to write (in other words, save) the contents of what is shown on-screen to a file in the current working directory, without leaving the nano editor.

The command **<Ctrl+X>** allows you to exit from the editor and return to the Linux command prompt.

To get a more in-depth explanation of the other menu choices at the bottom of the nano screen display, and of the capabilities of nano itself, see Section 3.2.1. We also refer you to the various online help pages. A good place to start is "The Beginner's Guide to Nano, the Linux Command-Line Text Editor."

3.2.1 A Brief nano Tutorial

The following subsections illustrate and explain some of the basic operations as well as some of the features of nano.

3.2.1.1 Creating and Opening a New File

If you want to create a new file and open it using nano, use the following command(s):

```
$ nano
```

or

```
$ nano [filename]
```

The second command is also used to open an existing file, where **filename** is the name of the existing file. If you want to open a file that is not located in your current directory, then you have to use the absolute or relative path to that file. For example, /home/mint/filename.

Figure 3.2 shows a file that has been opened with the first command shown earlier.

At the bottom of the editor window, keyboard shortcuts are displayed (they are not GUI menu choices!) that let you perform some basic operations. Examples of these are Cut (and paste) Text, save (Write Out) the buffer content to a file, and exit the editor.

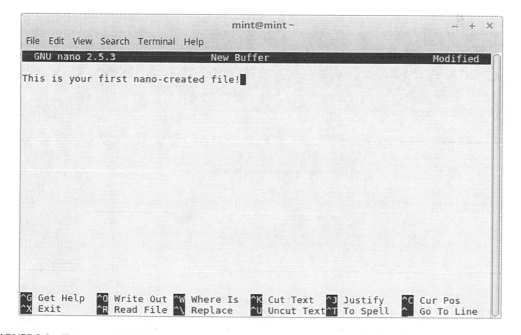

FIGURE 3.2 First opened file in nano.

3.2.1.2 How to Save a File

To save a file, use the keyboard shortcut **<Ctrl+O>**. When you use this key combination, the editor will prompt you for a filename (or confirm the name if it was already provided when the editor was started). Enter your filename and press **<Enter>** to save the file with that name.

This is illustrated in Figure 3.3.

3.2.1.3 How to Cut and Paste Text

To cut and paste a particular line of existing text that has already been entered into the buffer, first bring the cursor to any character on that line by using the arrow keys on the keyboard. Press **<Ctrl+K>** to cut that whole line out of the buffer, then position the cursor with the arrow keys to the place where you want to paste the "cut" line back into the buffer. Finally, use **<Ctrl+U>** to paste the cut line back in.

For example, in Figure 3.4, if you want to cut the first line and paste it multiply to two lines below the first. Go to the beginning of line 1 using the arrow keys on the keyboard, and then press **<Ctrl+K>**. Then, the cursor below the first line and press **<Ctrl+U>**. Repeat to do this multiply.

3.2.1.4 How to Search and Replace a Word

This feature allows you to search for a particular word in the buffer, as well as replace it with another word.

To search for a word in nano, press **<Ctrl+W>**. Then, you will be asked to enter the word which you want to search for. After typing in the word, press **<Enter>** and the tool will take you to the matched entry (See Figure 3.5).

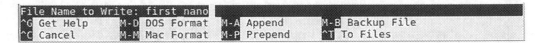

FIGURE 3.3 Writing out a file with a specific file name.

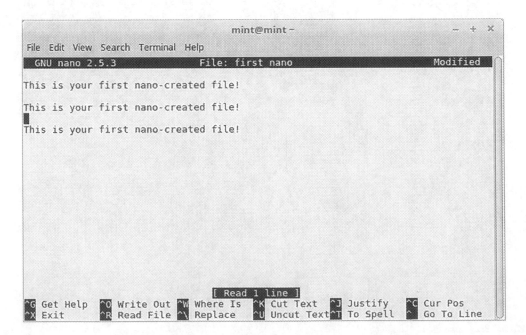

FIGURE 3.4 Cut and paste text from one line to two more lines.

FIGURE 3.5   Searching for text.

You can also replace a word (or phrase) with another by pressing **<Ctrl+\>**. When you press this key combination, nano prompts you for the word (or phrase) that you want to replace. After typing in the word (or phrase), press **<Enter>**. You will be prompted for the replacement word (or phrase). After this, you will be prompted for various ways of doing the replacement, and then are asked to confirm the changes. Once confirmed, the replacement is done.

Figures 3.6–3.9 illustrate this procedure.

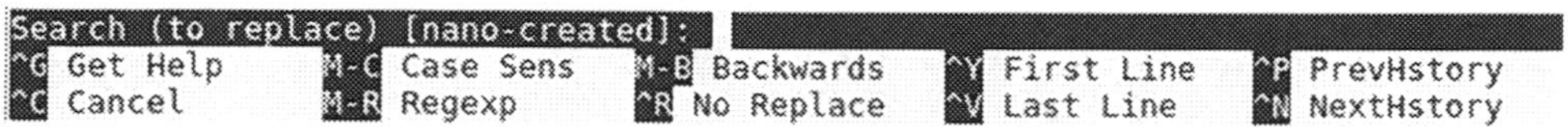

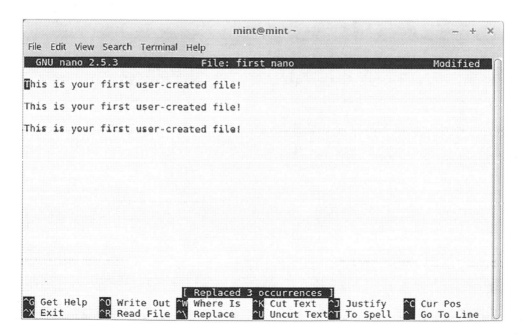

FIGURE 3.6   First replace text screen.

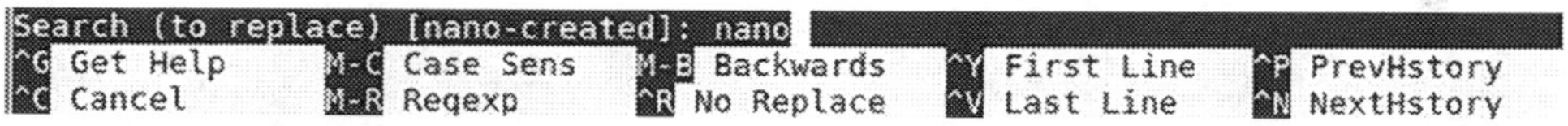

FIGURE 3.7   Second replace text screen.

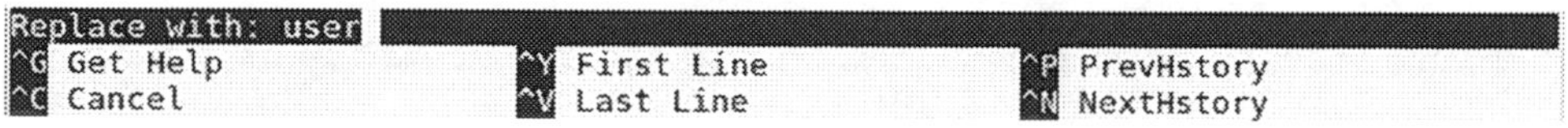

FIGURE 3.8   Third replace text screen.

FIGURE 3.9   Final replace text screen.

3.2.1.5 How to Insert Another File into the Current One

To insert text from another, already-saved file, into the buffer you are currently editing, do the following. Press **<Ctrl+R>**, and then give the complete pathname to the file which you want to insert into the current buffer, at the current position of the cursor in that buffer.

Figures 3.10 and 3.11 illustrate this procedure.

As you can see in Figure 3.11, the text of /home/mint/second_nano was inserted at the cursor position when it was at the last line in the file first_nano.

3.2.1.6 How to Show the Cursor Position

It is possible to get information about the position of your cursor in the current buffer. This can be done by pressing the **<Ctrl+C>** keyboard shortcut.

Figure 3.12 shows that after **<Ctrl+C>** was pressed, the cursor position got highlighted in the editor area, and detailed information about it showed up in the status line (the one that's highlighted or darkened in the figure—third line from the bottom of the window). It reports the cursor is at line 4 of 8, which is 50% through the file, at column 1, and character 77 of 155 characters total in the buffer.

3.2.1.7 How to Back Up a Previous Version of a File

This facility allows you to back up the previous version of the file being edited. This is done after you make changes and save the file. This feature can be accessed using the **-B** command line option.

```
$ nano -B [filename]
```

```
File to insert [from ./] : /home/mint/second nano
^G Get Help              ^X Execute Command        ^T To Files
^C Cancel               M-F New Buffer
```

FIGURE 3.10   Inserting text from a file.

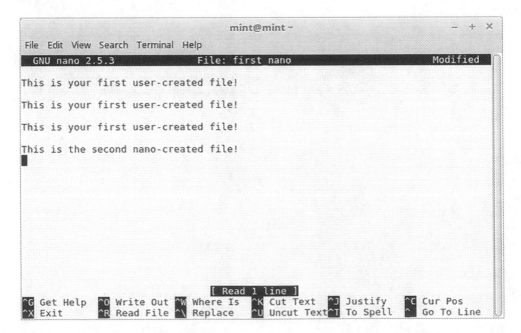

FIGURE 3.11   File after text insertion.

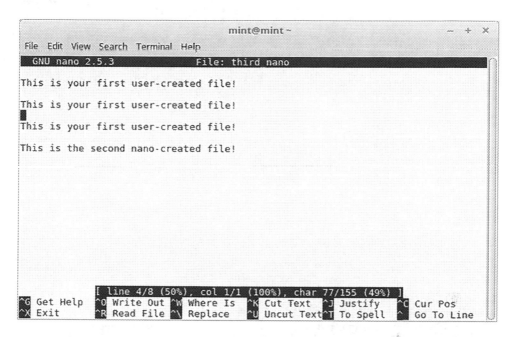

FIGURE 3.12   Cursor position report.

For example:

```
$ nano -B third_nano
```

The backup will be saved in the current directory with the same filename but suffixed with a tilde (-B). Files created for the first time cannot be backed up.

In-Chapter Exercises

1. Launch nano without specifying a file name on the command line and then enter some text into the New Buffer. Save that text into a file in your current working directory.
2. Cut some text out of a displayed file in nano, and then paste it in at another point in the display. What commands did you use to do this?
3. Replace a string of text in a displayed file in nano with another string of text. What command(s) did you use to do this? Can this be done at multiple sites repetitively in the file, and simultaneously all at once?
4. Describe, in your own words, the concept of a buffer in nano.
5. How did you select multiple characters in nano for the commands executed in the exercises earlier? If you use a single-console, text-only method of logging into and interacting with Linux, how can you select multiple characters for the operations you are asked to do in the exercises earlier?
6. (a) In only a single terminal display, how can you open multiple buffers in previously created text files in nano, and switch between editing in them? (b) What would be the major advantage of doing this?

Summary

Linux text editors are critical to the operation of CUI-based interfaces to your Linux system, and also very important even if you are using a GUI. They allow you to create text files important and critical for system operations, such as Bash shell scripts or systemd unit files, or edit existing text entities such as configuration files for applications.

This chapter covered the basics of the CUI-based nano editor and gave some important considerations you might make when choosing a more sophisticated (although more complex) editor, such as vi, vim, or emacs.

Questions and Problems

1. List ten commonly used text-editing operations you can do in nano.

2. Run nano on your Linux system. Create and edit a block of text that you want to be the body of an e-mail message explaining the basic capabilities of the nano editor. This file should be at least one page (45–50 lines of text) long. Then, save the file as **nano_doc.txt**. Insert the body of text you created in an e-mail message and send it to yourself.

3. Log on to your Linux system, and execute the **nano** program on a new, blank file.

   On the first line of the file, type your first and last name.

   On the second line of the file, type "The nano Linux text editor allows you to do simple editing on small text files efficiently."

   Use a nano command to write the file to the default directory with the name **lab51**.

   Print the file **lab51** at your Linux system line printer.

4. Do the following steps to create a file in nano-

   Step 1: At the shell prompt, type nano and then press <Enter>.

   Step 2: In the text area of the nano screen, place the cursor on the first line and type-

   This is text that I have entered on a line in the nano editor.

   Use the <Delete> and <arrow> keys to correct any typing errors you make.

   Step 3: Press <Enter > three times.

   Step 4: Type-

   This is a line of text three lines down from the first line.

   Step 5: Hold down the <Ctrl> and <O> keys at the same time (<Ctrl-O> or <^O>).

   Step 6: At the prompt File Name to Write: type **linespaced** and then press <Enter>.

   Step 7: Hold down the <Ctrl> and <X> keys at the same time (<Ctrl-X> or <X>) to return to the shell prompt.

   Step 8: At the shell prompt, type more **linespaced** and then press <Enter>.

5. Do the following steps in nano:

   Step 1: At the shell prompt, type **nano linespaced** and then press <Enter>. The linespaced file you created in Problem 4 appears in the nano screen.

   Step 2: Position the cursor at the beginning of the fourth line, at the character T in the word This, using the <arrow> keys on the keyboard.

   Step 3: Hold down the <Ctrl> and both the <Shift> and <6> keys at the same time.

   Step 4: Move the cursor with the <right arrow> key on the keyboard until you have highlighted the entire fourth line, including the period. The cursor should be one character to the right of the period at the end of the line.

Step 5: Hold down the <Ctrl> and <K> keys at the same time. This action cuts the line of text out of the current "buffer," or file that you are working on.

Step 6: Position the cursor with the <arrow> keys at the beginning of the second line of the file, directly under the line that reads

This is text that I have entered on a line in the nano editor.

Step 7: Hold down the <Ctrl> and <U> keys at the same time. This action pastes the former fourth line into the second line of the file.

Step 8: Use the <arrow> keys on the keyboard to position the cursor at the third line of the file.

Step 9: Hold down the <Ctrl> and <U> keys on the keyboard at the same time. This action pastes the former fourth line into the third line of the file.

Step 10: Now change the wording of lines 2 and 3 so that they read-

This is a line of text 1 line down from the first line.
This is a line of text 2 lines down from the first line.

How many lines are there in this file now, as far as nano is concerned?

Step 11: Hold down the <Ctrl> and <O> keys at the same time.

Step 12: At the prompt File Name to Write: type **linespaced2** and then press <Enter>.

Step 13: Hold down the <Ctrl> and <X> keys at the same time to return to the shell prompt.

Step 14: At the shell prompt, type more linespaced2 and then press <Enter>.
What do you see on screen? How many line does the more command show in this file?

6. Complete Problem 5, use nano to add two (2) more lines of text to the file named linespaced2 below lines 2 and 3, with similar content to lines 2 and 3. Then, add a line at the top of the file with your first and last name on it. Save this new file with the name linespaced3 and print it at your Linux system line printer.

7. What version of **nano** did you use in the earlier work, and how did you find this out?

8. Use the **cat** command to create a short text file named **shorty** on your UNIX system, and then read that file into **nano**, and add text to it. What command did you use to read the **cat**-created file into **nano**?

9. Execute **nano** on your Linux system using the **-m** command option. What functionality did the **-m** option give you in **nano**?

10. Repeat Problems 4–6 by launching nano using the -m option on the command line? When would it not be possible to use nano with the -m option?

Advanced Questions and Problems

vi (vim) and emacs

It is assumed here that you refer to Chapter W25 on Linux text editors at the book website, to gain familiarity with the two major, traditional Linux text editors: vi and emacs. Also, you can use vim on your Linux system in place of vi if you want to.

vi (vim)

11. Run vi on your Linux system. Create and edit a block of text that you want to be the body of an e-mail message, explaining the basic capabilities of the vi editor. For example, part of your message might describe the difference between the Insert and Command modes. This file should be at least one page (45–50 lines of text) long. Then, save the file as **vi_doc.txt**. Insert the body of text in an e-mail message and send it to yourself.

12. Run vi on your system and create a file of definitions in your own words, without looking at the textbook, for

     a.   full-screen display editor

     b.   modeless editor

     c.   file versus buffer

     d.   keystroke commands

     e.   substitute versus search

     f.   text file versus binary file

Then refer back to the relevant sections of Chapter W25 at the book website to check your definitions. Make any necessary corrections or additions. Reedit the file in vim to incorporate any corrections or additions that you made and then print out the file using the print commands available on your system.

13. Edit the file you created in Problem 12 and change the order of the text of your definitions to (d), (a), (c), and (b), using the yank, put, and D or dd commands. Print out the file using the print commands available on your Linux system.

14. Log on to your Linux system and execute the vi program on a new, blank file.

    On the first line of the file, type your first and last name.

    On the second line of the file, type "The Linux vi text editor has almost all the features of a word processor and tremendous flexibility in creating text files."

    Print the file to your Linux system line printer, from within vi, using a single vi command. How do you accomplish this, in a non-GUI, text-only environment?

15. What vi command allows you to move to the first line in the current buffer? What command allows you to move to the last line in the buffer?

16. Use the **set** command to force vi into a 30 column by 15-line display of characters so that one screen of the display shows only 15 lines, and text is automatically wrapped onto the next line after the 30th character. How did you do this? (HINT: The **set all** command shows the current status of all vi environment variables.)

17. What file in your home directory allows you to customize your vi environment variables permanently?

18. What do the following eight **vi** commands do?

    12dw, 5dd, 12o, 5O, c5b, d5,12, 12G, 5yy

19. While editing a file, how do you "escape" to the default Linux shell (on our Linux systems, Bash) while in vi, and then how do you return to the editor?

Gnu emacs

20. Using emacs, type in a paragraph of text from one of your favorite books, but without altering the size or shape of the emacs frame or using the Enter key, use the word wrap feature of emacs to format it exactly the way that it is printed in the book. Print the file at your Linux system line printer.

21. Which emacs commands move you forward and backward one character, one word, one sentence, and one paragraph?

22. Define an emacs keyboard macro that accomplishes a common editing task for you.

23. Create, edit, compile, link, and execute a short C program of your choice in emacs.

24. Try working with emacs in a text-only window and use only keystroke commands.

    To do this, you will have to launch emacs from a console or terminal window by typing **emacs -nw newfile**. The **-nw** option specifies that emacs will run in nongraphical mode. Then, in the console or terminal window, a nongraphical emacs will open on the buffer **newfile**. As stated in Section W25.3 at the book website, you can still gain access to the menu bar menus at the top of the emacs screen by pressing the escape **<Esc>** key on the keyboard and then pressing the single back quote (`) key. You can then descend through the menu bar choices by pressing the letter key of the menu choice you want to make. For example, pressing the **f** key on the keyboard gives you access to the File pull-down menu choices, and then pressing the **s** key on the keyboard allows you to save the current buffer.

Log on to your Linux system and execute the emacs program on a new, blank file, using the **-nw** command option.

On the first line of the file, type your first and last name.

On the second line of the file, type "The emacs editor is the most complex and customizable of the Linux text editors."

Print the file to your Linux system line printer from within emacs, with a single emacs command. How do you accomplish this, in a non-GUI, text-only environment?

25. Use emacs's capability of sending e-mail while you're in emacs. Send an e-mail message to one of your friends, composing the message body and sending from within emacs.

   Answer: No answer required.

26. While editing a file, how do you "escape" to the default Linux shell (on our Linux systems, Bash) while in emacs, and then how do you return to the editor?

Projects

Project 1

From within non-GUI, text-only vi, vim, and emacs sessions, create a text file that you want to print at one of the available printers on your Linux system. Then, *while still in the editor*, give a series of Common UNIX Printing Service (CUPS) commands to manage the CUPS service, either locally or on your network, that will enable you to print the text file you created in the editor. This would involve things like starting the CUPS service if it is not running by default, checking the status of the service and attached printers with system commands, or changing the name of a particular attached printer, etc. What specific commands do you use to accomplish these things, in all three editors? Create a short report organizing both a general, and specific methodologies, someone could use to manage a CUPS service, and print documents from within these editors, via the CUPS service.

Project 2

zenity is a graphical, GTK+ dialog box program that allows you to create interactive dialog boxes using Bash script files. It is installed by default on our GUI-based Debian-family and CentOS Linux systems. In this project, the zenity dialog box you will deploy will allow you to easily create new users on your Linux system. Of course, it is assumed you have the privilege to do this account creation on your system! Use emacs to create and execute a zenity-based Bash script file, following these steps:

   a. In emacs, create and save the following Bash script file, named zen1, in your home directory:

```
#!/bin/bash
    zenity --forms --title="newusers Command" --text="Add batch new user" \
        --add-entry="Username" \
        --add-password="Password" \
        --add-entry="User Number UID" \
        --add-entry="Group Number GID" \
        --add-entry="GECOS Entry" \
        --add-entry="Default Home Directory" \
        --add-entry="Default Shell" \
    >> zen_out
    sed -i -e 's/|/:/g' /home/bob/zen_out
```

   b. Make zen1 executable, then on the command line, type **./zen1**

   A zenity dialog box will open on-screen. In the GUI dialog box you will create the seven fields needed to be supplied to the **newusers** command, which we show in Chapter 17, Section 3.1,

Example 17.3, to create new users from a "batch file" on your Linux system. The seven inputs you supply to the dialog box will be written to a file named zen_out.

The seven fields, separated by the colon character (:), are the new user accounts name, password, User ID (UID), Group ID (GID), General Electric Comprehensive Operating Supervisor (GECOS) commentary, default home directory, and default shell.

For example,

hassan:QQQ:2001:2001:CFO of Accounting:/home/hassan:/bin/bash

c. Use zen1 to create a file of several new users you want to put on your Linux system. Then, put those users on your system!

Looking for more? Visit our sites for additional readings, recommended resources, and exercises.
  CRC Press e-Resource: https://www.crcpress.com/9781138710085
  Authors' GitHub: https://github.com/bobk48/linuxthetextbook

4

Files and File System Structure

4.1 Introduction

Most computer users work with the file system structure of the computer system that they use. While using a computer system, a user is constantly performing file-related operations: creating, reading, writing/modifying, or executing files. Therefore, the user needs to understand what a file is in Linux, how files can be organized and managed, how they are represented inside the operating system, and how they are stored on the disk. In this chapter, the description of file representation and storage is simplified, due to the scope of this textbook. More details on these topics are available in books on operating system concepts and principles and in books on Linux internals.

4.2 The Linux File Concept

One of the many remarkable features of the Linux operating system is its concept of files. This concept is simple, yet powerful, and results in a uniform view of all system resources. In Linux, a file is a sequence of bytes. Thus, everything, including a network interface card, a disk drive, a Universal Serial Bus (USB) flash drive, a keyboard, a printer, a simple/ordinary (text, executable, etc.) file, or a directory, is treated as a file. As a result, all input and output devices are treated as files in Linux, as described under file types and file system structure.

4.3 Types of Files

Linux supports seven types of files:

- Simple/ordinary file
- Directory
- Symbolic (soft) link
- Character special file
- Block special file
- Named pipe (also called FIFO)
- Socket

You can use the `ls -l` command to display the type of a file, as shown in Table 4.1.

4.3.1 Simple/Ordinary File

Simple/ordinary files are used to store information and data on a secondary storage device, typically a disk. An ordinary file may contain any of the following:

- Unstructured text
- Source code (or script) in a programming language such as C, C++, Java, Ruby, Python, LISP, and Bash
- An executable program that you have created by compiling (and linking) a source program
- Applications such as compilers, database tools, desktop publishing tools, and graphing software
- PostScript code
- Pictures
- Video
- Audio
- Graphics
- Etc.

TABLE 4.1

Summary of the Output of the `ls -l` Command (Fields Listed Left to Right)

Field	Meaning
First letter of first field	File type:
	- ordinary file
	b block special file
	c character special file
	d directory
	l link
	p named pipe (FIFO)
	s socket
Remaining letters of first field	Access permissions for owner, group, and others
Second field	Number of hard links
Third field	Owner's login name
Fourth field	Owner's group name (can also be a number)
Fifth field	File size in bytes
Sixth, seventh, and eighth fields	Date and time of last modification
Ninth field	File name

TABLE 4.2

Commonly Used Extensions for Some Applications

Extension	Contents of File
`.bmp, .jpg, jpeg, .gif`	Graphics
`.c`	C source code
`.C, .cpp, .cc`	C++ source code
`.class`	Java class file
`.gz`	Compressed
`.html, .htm`	File for a Web page
`.java`	Java source code
`.o`	Object code
`.pdf`	Contains data in Portable Document Format
`.txt`	Contains unformatted text that is recognized by a text editor or word processing software

Linux does not treat any one of these files differently from another. It does not give a structure or attach a meaning to a file's contents, because every file is simply a sequence of bytes. Meanings are attached to a file's contents by the application that uses/processes the file. For example, a C program file is no different to Linux from a Hypertext Markup Language (HTML) file for a Web page or a file for a video clip. However, a C compiler (e.g., cc or gcc), a Web browser (e.g., Firefox), and a video player (e.g., VLC Media Players) treat these files differently.

You can name files by following any convention that you choose to use; Linux does not impose any naming conventions on files of any type. On most Linux systems, the default maximum length of a file-name is 255 characters. You can use the getconf command on your Linux system to display or configure the maximum size of a file name in number of characters.

Although you can use any characters for file names, we strongly recommend that nonprintable characters, white spaces (spaces and tabs), and shell metacharacters (as described in Chapter 2) not be used because they are difficult to deal with as part of a file name. You can give file names any of your own or application-defined extensions, but the extensions mean nothing to the Linux system—they are treated as part of file name. For example, you can give an **.exe** extension to a document and a **.doc** extension to an executable program. Some applications require extensions, but others do not. For example, all C compilers require that C source program files have a **.c** extension, but not all Web browsers require an **.html** extension for files for Web pages. Even so, extensions should be used—it helps keep track of which files are for what purposes. Some commonly used extensions are given in Table 4.2.

4.3.2 Directory

A directory contains the names of other files and/or directories (the terms directory and subdirectory are used interchangeably). In some systems, terms such as *folder*, *drawer*, or *cabinet* are used for a directory. A directory file in any operating system consists of an array of directory entries, although the contents of a directory entry vary from one system to another. In Linux, a directory entry has the structure shown in Figure 4.1.

The *inode number* is four bytes long and is an index value for an array on the disk. An element of this array, known as an index node, more commonly called an *inode*, contains the attributes of a file such

Inode number	File name

FIGURE 4.1 Linux directory entry structure.

as file size (in bytes). When you create a new file, the Linux kernel allocates an inode to it. Thus, every unique file in Linux has a unique inode (and inode number). As discussed in Section 4.5.7 and Chapter 8, hard links to a file have the same inode, and inodes are unique per partition. This means that two different files on different partitions can have the same inode number. The details of an inode and how the kernel uses it to access a file's contents on disk are discussed in Sections 4.5.7 and 4.6.

4.3.3 Link File

A file of type link "points to" an existing file. The content of a link file is the pathname of the existing file. Thus, a link file allows you to access an existing file through another path in the file system structure and share it without duplicating its contents. The concept of a link in Linux is fully discussed in Chapter 8. But, for now, a file of type link is created by the system when a *symbolic link* is created to an existing file. Symbolic links are currently available on all versions of Linux.

4.3.4 Special (Device) File

A *special file*, also known as a *device node*, is a means of accessing hardware devices, including the keyboard, disks, tape drive, graphic cards, network cards, and printers. Each hardware device is associated with at least one special file—and a command or an application accesses a special file to access the corresponding device. Special files are divided into two types: *character special files* and *block special files*. Character special files correspond to character-oriented devices, such as a keyboard, and block special files correspond to block-oriented devices, such as a disk. Special files are typically placed in the **/dev** directory (see Section 4.4 for more details).

Applications and commands read and write peripheral device files in the same way that they read or write an ordinary file. That capability is the main reason that input and output in Linux is said to be *device independent*. Various special devices simulate physical devices and are, therefore, known as *pseudo-devices*. These devices allow you to interact with a Linux system without using the devices that are physically connected to it. These devices are becoming more and more important, because they allow use of a Linux system via a network connection or with ssh-tunneled virtual terminals in Chapter W28.

4.3.5 Named Pipe (FIFO)

Linux has several tools that enable processes to communicate with each other. These tools, which are the key to the ubiquitous client–server software paradigm, are called *interprocess communication (IPC) mechanisms* (commonly known as *IPC primitives* or *IPC channels*). These primitives are called *pipes*, *named pipes* (also called *FIFOs*), and *sockets* (systems that are strictly UNIX System V-compliant have a mechanism called *transport layer interface*). These primitives are discussed in detail under "System Programming" in Chapters 15, 16, W20, and W21. Here, we briefly mention the purpose of each so that you can appreciate the need for each mechanism and understand the need for FIFOs.

A pipe is an area in the kernel memory (a kernel buffer) that allows two processes to communicate with each other, provided the processes are running on the same computer system and are related to each other; typically, the relationship is parent–child or sibling. A FIFO is a file (of named pipe type) that allows two processes to communicate with each other if the processes are on the same computer; however, the processes do not have to be related to each other through a parent–child or sibling relationship. We illustrate the use of pipes and FIFOs at the command level in Chapter 9.

4.3.6 Socket

A socket can be used by processes on the same computer or on different computers to communicate with each other; the computers can be on a network (intranet) or on the Internet. Sockets can belong to different address families, each specifying the protocol suite to be used by processes to communicate. For example, the application layer protocols such as the Hypertext Transfer Protocol (HTTP) use sockets of address family **AF_INET** in the Transmission Control Protocol/Internet Protocol (TCP/IP) protocol suite for communication (see Chapter 11 for a detailed discussion on TCP/IP). A socket with address family **AF_INET** is also known as the *Internet domain socket*, which means that processes running on computers on the Internet can use sockets of this domain to communicate with each other. A socket with address family **AF_UNIX** or **AF_LOCAL** can be used for communication between processes that run on the same machine under a Linux operating system. This kind of socket is also known as a *UNIX domain socket*. See Chapter W20 for more details on UNIX and Internet domain sockets.

4.4 File System Structure

Three issues are related to the file system structure of an operating system. The first is how files in the system are organized from the user's point of view. The second is how files are stored on the secondary storage (usually, a hard disk). The third is how files are manipulated (read, written, etc.). In this chapter, we will address the first issue, that is, the user view of files and directories in a Linux system. As has been the case so far, our focus will be on Mint Linux.

4.4.1 File System Organization

The Linux file system is structured hierarchically and is treelike, but upside down, with the root at the top. Thus, the file system structure starts with one main directory, called the root directory, and can have any number of files and subdirectories under it, organized in any desired way. This structure leads to a parent–child relationship between a directory and its subdirectories/files. A typical Linux system contains hundreds of files and directories. For our Linux Mint system, the files and directories under the root directory, denoted as **/** in Linux terminology, are shown in the following session:

```
$ ls -l /
total 96
drwxr-xr-x     2 root root  4096 Dec 13  2016 bin
drwxr-xr-x     4 root root  4096 Jul 28  2016 boot
drwxr-xr-x     2 root root  4096 Jul 28  2016 cdrom
drwxr-xr-x    19 root root  4480 Apr  4 20:02 dev
drwxr-xr-x   151 root root 12288 Dec 14  2016 etc
drwxr-xr-x    12 root root  4096 Dec 14  2016 home
lrwxrwxrwx     1 root root    32 Jul 28  2016 initrd.img -> boot/initrd.img-4.4.0-21-generic
drwxr-xr-x    25 root root  4096 Jul 28  2016 lib
drwxr-xr-x     2 root root  4096 Jun 28  2016 lib64
drwx------     2 root root 16384 Jul 28  2016 lost+found
drwxr xr x     2 root root  4096 Jun 28  2016 media
drwxr-xr-x     2 root root  4096 Jun 28  2016 mnt
drwxr-xr-x     2 root root  4096 Jun 28  2016 opt
dr-xr-xr-x   224 root root     0 Apr  4 20:02 proc
drwx------    11 root root  4096 Jul 28  2016 root
drwxr-xr-x    32 root root  1060 Jul  4 21:45 run
drwxr-xr-x     2 root root 12288 Jul 28  2016 sbin
drwxr-xr-x     2 root root  4096 Jun 28  2016 srv
dr-xr-xr-x    13 root root     0 Jul  1 15:26 sys
drwxrwxrwt    14 root root  4096 Jul  4 22:17 tmp
drwxr-xr-x    10 root root  4096 Jun 28  2016 usr
drwxr-xr-x    11 root root  4096 Jun 28  2016 var
lrwxrwxrwx     1 root root    29 Jul 28  2016 vmlinuz -> boot/vmlinuz-4.4.0-21-generic
$
```

Note that the first line of the output (i.e., total 96) shows the size of the root directory in kilobytes. You can view the total size of the root directory in kilo (K), mega (M), or giga (G) bytes by running the ls -lh / command. The rest of the output of the command shows that the root directory has 21 directories and two symbolic links: `initrd.img` and `vmlinuz` (see Chapter 8 for symbolic links). In addition, there are the current directory (`.`) and parent of the current directory (`..`). You can display the long listing of all the files, including directories as well as the other dot files using the `-al` options. The big-picture user view of our system's files and directories is shown in Figure 4.2.

4.4.2 Home and Present Working Directories

When you log on, the Linux system puts you in a specific directory, called your home/login directory. For example, the directory called **sarwar** in Figure 4.2 is the home directory for the user with the login **sarwar**. While using the C, tcsh, Bash, or Korn shell, you can specify your home directory by using the tilde (~) character. The directory that you are in at any particular time is called your *present working directory* (also known as your *current directory*). The present working directory is also denoted as **.** (pronounced "dot"). The parent of the present working directory is denoted as **..** (pronounced "dot dot").

Later in this chapter, we describe the commands you can use to determine your home and present working directories. We also identify commands you can use to interact with the Linux file system in general.

4.4.3 Pathnames: Absolute and Relative

A file or directory in a hierarchical file system is specified by a *pathname*. Simply put, a pathname is the full name of a file. Pathnames can be specified in three ways: (1) starting with the root directory, (2) starting with the present working directory, and (3) starting with the user's home directory. When a pathname is specified starting with the root directory, it is called an *absolute pathname,* because it can be used by any user from anywhere in the file system structure. For example, **/home/faculty/sarwar/courses/ ee446** is the absolute pathname for the **ee446** directory under the user **sarwar**'s home directory. The absolute pathname for the file called **mid1** under **sarwar**'s home directory is **/home/faculty/sarwar/ courses/ee446/exams/mid1**. Note that whereas, as shown in Figure 4.2, our system partitions **/home** into subdirectories admin, students, and faculty, most Linux systems place all home directories under **/home**.

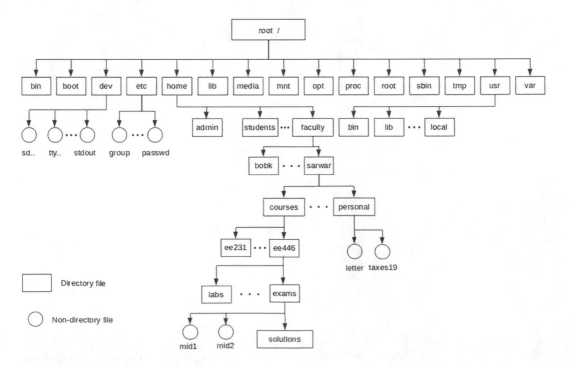

FIGURE 4.2   Typical Linux file structure.

Pathnames starting with the present working directory or a user's home directory are called **relative pathnames**. When the user **sarwar** logs on, the system puts him into his home directory, **/home/faculty/ sarwar**. While in his home directory, **sarwar** can specify the file **midl** (see Figure 4.2) by using a relative pathname, **./courses/ee446/exams/midl** or **courses/ee446/exams/midl**. The user **sarwar** (or anyone else) in the directory **ee446** can specify the same file with the relative pathname **exams/midl**. The owner (or anyone logged on as the owner) of the **midl** file can also specify it from anywhere in the file structure by using the pathname **~/courses/ee446/exams/midl** or **$HOME/courses/ee446/exams/midl**. Or, you could specify **ee446** from your personal directory as **../courses/ee446**.

A typical Linux system has several disk drives that contain user and system files, but, as a user, you do not have to worry about which disk drive contains the file that you need to access. In Linux, multiple disk drives and/or disk partitions can be *mounted* on the same file system structure, allowing their access as directories and not as named drives A:, B:, C:, and so on, as in MS-DOS and Microsoft Windows. You can access files and directories on these disks and/or partitions by specifying their pathnames as if they are part of the file structure on one disk/partition. Doing so gives a unified view of all the files and directories in the system, and you do not have to worry about remembering the names of drives and the files and directories they contain.

4.4.4 Some Standard Directories and Files

Every Linux system contains a set of standard files and directories. The standard directories contain some specific files. In this section, we discuss some of the important directories. You may like to browse through the first website listed in Web Resources (Table 4.5 at the book GitHub site) to know more about the Mint directory hierarchy.

Root directory (**/**): The root directory is at the top of the file system hierarchy and is denoted as a slash (**/**). It contains some standard files and directories and, in a sense, is the master cabinet that contains all drawers, folders, and files.

/bin: Also known as the binary directory, the **/bin** directory contains binary (i.e., executable) images of most Linux programs/commands that are fundamental to starting and repairing single-user and multiuser environments. The **/bin** directory contains only 179 files on our system, including bash, bzexe, bzdiff, bzegrep, bzgrep, bzless, bzmore, cat, chmod, chown, cp, date, echo, egrep, grep, gunzip, gzip, kill, less, ln, ls, mkdir, mknod, more, mount, mv, nano, netstat, ping, ps, pwd, rm, rmdir, sed, sh, sleep, stty, su, systemd, tar, tcsh, umount, zcat, zdiff, and zgrep. The superuser and ordinary users may use these programs.

Unlike some UNIX systems, the **/usr/bin** directory is different from **/bin** under Linux. The **/usr/bin** directory contains hundreds of executable programs that are not needed for starting or repairing the system. A few of the most commonly used are awk, diff, du, env, file, find, finger, gcc, gdb, groups, head, java, javac, less, lex, locate, lp, lpq, lpr, man, nl, pstree, sort, ssh, sudo, tail, telnet, time, top, touch, tr, uniq, vi, wc, which, who, whoami, zip, zipgrep, and zipinfo. On our Linux Mint system, **/usr/bin** contains 2,078 files.

/boot: This directory contains everything required to boot the system, including the boot loaders, except the configurations files for the boot loaders. The configuration files are placed in the **/etc** directory. For example, the GRand Unified Bootloader (GRUB) bootloader must be in **/boot** and its configuration file in **/etc**. On our system, the **/boot/grub/grub.cfg** configuration file is generated by the utility grub-mkconfig using the templates in the **/etc/grub.d** directory. Similarly, the systemd-boot loads and runs the Unified Extensible Firmware Interface (EFI or UEFI for short) boot images. The **/boot** directory also contains the Linux kernel boot image in vmlinuz-4.4.0-21-generic.

/dev: The **/dev** directory, which is also known as the device directory, contains files corresponding to the devices connected to the computer, including terminals, USB and hard disk drives,

CD-ROM drive, tape drives, modems, graphics cards, network cards, printers, and so on. These files, called special files, were described in Section 4.3.4.

This directory contains at least one file for every device connected to the computer. Each device has a name and a number, and the special file representing the device reflects both. Some example files in the **/dev** directory are as follows: **sda** is the first Serial Advanced Technology Attachment (SATA) hard drive, **sg** for CD-ROM, **dvd** is a symbolic link to **sr0**, and **tty**'s for (teletype) terminals. We discuss links in Linux, including symbolic links, in Chapter 8. The **/dev/pts** directory is used to manage pseudo-terminals.

A system may have several devices of each type—for example, 10 hard disks or partitions, 20 terminals, 100 pseudo-terminals, two solid-state disks, and so on. Our Linux Mint-based system contains a total of 223 files in the **/dev** directory. This directory may contain several hundred—even more than 1,000—files in a network-based Linux environment in a medium-to-large-sized organization.

/etc: The **/etc** directory contains commands, files, and scripts needed for system configuration and administration. A typical user cannot modify files in this directory. Some of the files and directories in this directory include **crontab**, **csh.cshrc**, **csh.login**, **csh.logout**, **group**, **inetd.conf**, **login.access**, **login.conf**, **passwd**, **printcap**, **profile**, **rc.d**, **rcp**, **shells**, **services**, **ssh**, **ssl**, and **termcap**. Discussion of most of the files in this directory is beyond the scope of this textbook. However, we briefly discuss the **/etc/passwd** file toward the end of this section.

/lib: The library directory contains a collection of related files for a given language in a single file called an archive. A typical Linux system today contains libraries for C and C++. The archive file for one of these languages can be used by applications developed in that language. The **/lib** directories contains libraries that are critical for the executable programs in the **/bin** and **/sbin** directories. The **/usr/lib** directory contains the shared and archive-type libraries (created by the ar command).

/tmp: Used by several commands and applications, the **/tmp** directory contains temporary files. You can use this directory for your own temporary files as well. All the files in this directory are deleted periodically so that the disk (or a partition of the disk) does not get filled with temporary files. The life of a file in the **/tmp** directory is set by the system administrator and varies from system to system, but it is usually only a few minutes. Files in **/tmp** may or may not exist when a system is rebooted.

/usr: The **/usr** directory contains subdirectories that hold, among other things, most of the utilities, system daemons (see Chapter 10), applications, programming tools, standard C include files, shared and archive-type language libraries, manual pages and other important documents, and source code (Berkeley Software Distribution (BSD) and third party). Two of the most important subdirectories in this directory are **bin** and **lib**, which contain binary images of most Linux commands (utilities, tools, etc.) and language libraries, respectively.

/home: Organized in some fashion, the **/home** directory is normally used to hold the home directories of all the users of the system. For example, the system administrator can create subdirectories under this directory that contain home directories for certain types of users. For instance, the diagram in Figure 4.2, which shows a university-like setup, has one subdirectory each for the home directories of members of the administration, faculty, staff, students, and so on. These subdirectories are labeled **admin**, **faculty**, and **students**. As stated earlier, most Linux systems place all home directories under **/home**.

/var: The **/var** directory contains multipurpose log, temporary, and spool files. Among several other directories, the **/var/mail** directory contains files for receiving and holding incoming e-mail messages of users. When you read your new e-mail, it comes from a file in this directory. The **/var/spool/mqueue** directory contains the undelivered mail queue and the **/var/spool/output** directory contains the line printer spooling directories. The **/var/tmp** directory contains temporary files that are kept between system reboots.

/etc/passwd: The **/etc/passwd** file contains one line for every user on the system and describes that user. Each line has seven fields, separated by colons. The following is the format of the line.

```
login_name:password:user_ID:group_ID:user_info:home_directory:login_shell
```

The login_name is the login name by which the user is known to the system and is what the user types to log in. The password field contains the dummy password x (or *). The encrypted passwords are stored in the **/etc/shadow** file. Only the superuser (i.e., **root**) has read and write access permissions for this file; nobody else can even read it. Portable Operating System Interface (POSIX) requires user_ID (UID) to be an integer type. Usually, the superuser is assigned a UID of **0**. Several other login names are also assigned UIDs that are known (or from a known range). Typically, UIDs 1–499 or 1–999 are reserved. Depending on the Linux system that you use, UIDs 1,000–32,767 or 1,000–65,536 are assigned to "normal" users like you and I. In systems that use 32-bit UIDs, this range is 1,000–4,294,967,296. The group_ID identifies the group that the user belongs to, and it is also an integer between 0 and 65,535 with, usually, 0–99 reserved integers. The user_info field contains information about the user, typically the user's full name. The home_directory field contains the absolute pathname for the user's home directory. The last field, login_shell, contains the absolute pathname for the user's login shell. The command corresponding to the pathname specified in this field is executed by the system when the user logs on. Back-to-back colons mean that the field value is missing, which is sometimes done with the user_info field. The following session shows the line from the **/etc/passwd** file on our system for the user **sarwar**:

```
$ cat /etc/passwd | grep "sarwar"
sarwar:x:1004:1008:Mansoor Sarwar,,,:/home/sarwar:/bin/bash
$
```

In this line, the login name is **sarwar**, the password field contains x, the user ID is 1004, the group ID is 1008, the personal information is the user's full name (Mansoor Sarwar), the home directory is **/home/sarwar**, and the login shell is **/bin/bash**, or the Bash shell.

The following in-chapter exercises give you practice in browsing the file system on your Linux machine and help you understand the format of the **/etc/passwd** file.

Exercise 4.1

Go to the **/dev** directory on your system and identify one character special file and one block special file.

Exercise 4.2

Determine your user ID on your Linux system by viewing the **/etc/passwd** file on it. What is your user ID?

Exercise 4.3

How many files do the **/bin** and **/usr/bin** directories contain on your Linux system. Show the commands that you used to obtain your answer.

4.5 Navigating the File Structure

Now, we describe some useful commands for browsing the Linux file system, creating files and directories, and determining file attributes, the absolute pathname for your home directory, the pathname for the present working directory, and the type of a file. The discussion is based on the file structure shown in Figure 4.2 and the user name **sarwar**.

4.5.1 Determining the Absolute Pathname for Your Home Directory

When you log on, the system puts you in your home directory. You can find the full pathname for your home directory by using the echo and pwd commands.

With no argument, the echo command displays a blank line on the screen. You can determine the absolute pathname of your home directory by using the echo command, as follows:

```
$ echo $HOME
/home/sarwar
$
```

where HOME is a shell variable (a placeholder) in the Bourne shell. The shell uses this variable to keep track of the absolute pathname of your home directory. In the C shell, the variable is home. We discuss shell variables and the echo command in detail in Chapters 12, 13, W29, and W30.

Another way to display the absolute pathname of your home directory is to use the pwd command. You use this command to determine the absolute pathname of the directory you are currently in, that is, your present working directory, also known as the current directory. This command does not require any arguments. When you log on, the Linux system puts you in your home directory. You can use the pwd command right after logging on to display the absolute pathname of your home directory as follows:

```
$ pwd
/home/sarwar
$
```

4.5.2 Browsing the File System

You can browse the file system by going from your home directory to other directories in the file system structure and displaying a directory's contents (files and subdirectories in the directory), provided that you have the *permissions* to do so. We cover file security and access permissions in detail in Chapter 5. For now, we show how you can browse your own files and directories by using the cd (change directory) and ls (list directory) commands. The following is a brief description of the cd command.

Syntax:
    `cd [directory]`

Purpose: Change the present working directory to directory, or to the home directory if no argument is specified

---

The shell variable PWD is set after each execution of the cd command. The pwd command uses the value of this variable to display the present working directory. After getting into a directory, you can view its contents (the names of files or subdirectories in it) by using the ls command. The following is a brief description of this command. The cd and ls commands are two of the most heavily used Linux commands.

Syntax:
    `ls [option] [pathname-list]`

Purpose: Send the names of the files in the directories and files specified in pathname-list to the display screen

Output: Names of the files and directories in the directory specified by pathname-list, or the names only if pathname-list contains file names only

Commonly used options/features:

-F	Display/after directories, * after binary executables, and @ after symbolic links
-a	Display names of all files, including hidden files ., .., and so on.
-h	Display output in human readable sizes (e.g., 1.1 K, 31 M, 2.6 G) when used with –l and/or –s option
-i	Display inode number and file name
-l	Display long list that includes access permissions, hard link count, owner, group, file size (in bytes), and modification time

---

If the command is used without any argument, it displays the names of files and directories in the present working directory. The following session illustrates how the ls and cd commands work with and without parameters. The pwd command displays the absolute pathname of the current directory. With the exception of hidden files, the ls command displays the name of all the files and directories in the current directory. The cd courses command is used to make the courses directory the current directory. The cd ee446/exams command makes ee446/exams the current directory. The ls ~ and ls $HOME commands display the names of the files and directories in your home directory. The cd command without any argument puts you in your home directory. In other words, it makes your home directory your current directory.

```
$ pwd
/home/sarwar
$ ls
courses   Documents   linux2e   Pictures   Templates
Desktop   Downloads   Music     Public     Videos
$ cd courses
$ ls
ee231    ee446
$ cd ee446/exams
$ pwd
/home/sarwar/courses/ee446/exams
$ ls
mid1   mid2
$ ls ~
courses   Documents   linux2e   Pictures   Templates
Desktop   Downloads   Music     Public     Videos
$ ls $HOME
courses   Documents   linux2e   Pictures   Templates
Desktop   Downloads   Music     Public     Videos
$ cd
$ ls
courses   Documents   linux2e   Pictures   Templates
Desktop   Downloads   Music     Public     Videos
$
```

We demonstrate the use of the ls command with various options in the remainder of this chapter and other chapters of the book. We use the terms *flag* and *option* interchangeably.

In a typical Linux system, you are not allowed to access all the files and directories in the system. In particular, you are typically not allowed to access many important files and directories related to system administration and files and directories belonging to other users. However, you have permissions to read a number of directories and files. The following session illustrates that we have permissions to go to and list the contents of, among many other directories, the **/** and **/usr** directories.

```
$ cd /usr
$ ls
bin   games   include   lib   local   sbin   share   src
$ cd /
$ ls
bin     dev    initrd.img   lost+found   opt    run    sys   var
boot    etc    lib          media        proc   sbin   tmp   vmlinuz
cdrom   home   lib64        mnt          root   srv    usr
$ cd
$ ls /usr
bin   games   include   lib   local   sbin   share   src
$
```

Without any option, the ls command does not show all the files and directories; in particular, it does not display the names of hidden files. Examples of these files include **.**, **..**, **.bash_history**, **.bashrc**, **.config**, **.cshrc**, **.history**, **.login**, **.mailrc**, **.profile**, **.rhosts**, **.shrc**, **.ssh**, and **.xsession**. We have

TABLE 4.3

Some Important Hidden Files and Their Purposes

File Name	Purpose
.	Present working directory
..	Parent of the present working directory
.bash_history	Contains the history of commands executed under bash
.bashrc	Setup for the Bash shell
.cshrc	Setup for the C shell
.exrc	Setup for vi
.login	Setup for shell if C or tcsh shells are the login shells; executed at login time
.mailrc	Setup and address book for mail and mailx
.profile	Setup for shell if Bourne or Korn shell is the login shell; executed at login time
.rhosts	Domain names of the trusted hosts (see Chapter 11 for details)
.shrc	Setup for shell if Bourne shell is the login shell; executed at login time
.ssh	Keys of the servers on which you would be allowed to login using the ssh command. Keys are stored when you try to establish the session on a server for the first time
.vimrc	Setup for the vim text editor
.xsession	Customized X session script

already discussed the . and .. directories. The purposes of some of the more important hidden files are summarized in Table 4.3.

You can also display the names of all the files and directories in a directory, including the hidden files, by using the ls command with the –a option. In the following session, the cd command places you in your home directory and the ls –a command displays all the files and directories, including the hidden files, in your home directory.

```
$ cd
$ ls -a
.              .cache    .dmrc          linux2e     Pictures    .xsession-errors
..             .config   Documents      .linuxmint  .profile
.bash_history  courses   Downloads      .local      Public
.bash_logout   .dbus     .ICEauthority  .mozilla    Templates
.bashrc        Desktop   .lesshst       Music       Videos
$
```

You can use shell *metacharacters* in specifying multiple files or directory parameters to the ls command. For example, the command ls ~/courses/cs446/programs/*.c displays the names of all C program files in the **~/courses/cs446/programs** directory. We discuss the use of metacharacters and *regular expressions* in detail later in this chapter and in Chapters 6 and 7.

You can display the sizes of files and directory sizes in human readable (K for kilo, M for mega, G for giga) form by using the –h option, as shown in the following session with the ls –sh and ls –lh commands.

```
$ ls -sh ch14
total 27G
12.6K all.tar           4.0K compute.h         4.0K main.h
8.0K analysis.txt       4.0K compute.o         4.0K main.o
9.6G bigdata            4.0K flat_profile.out  4.0K makefile
...
4.0K input.h            8.0K working           4.0K working.c
$ ls -lh all.tar analysis.txt bigdata bugged.c buggy
-rw-r--r-- 1 sarwar faculty  10K Jul 23  2017 all.tar
-rw-r--r-- 1 sarwar faculty 5.7K Oct  2 21:16 analysis.txt
```

```
-rw-r--r-- 1 sarwar faculty 9.6G Oct  3 10:52 bigdata
-rw-r--r-- 1 sarwar faculty  694 Oct  3 08:56 bugged.c
-rwxr-xr-x 1 sarwar faculty 8.5K Oct  6 08:27 buggy
$
```

4.5.3 Creating Files

While working on a computer system, you need to create files and directories: files to store your work and directories to organize your files more efficiently. You can create files by using various tools and applications, such as editors, and create directories by using the mkdir command. You can use any of the Linux text editors, including nano, vim, and emacs, that you can use to create files containing plain text. You can create nontext files by using various applications and tools, such as a compiler, that translates source code in a high-level language (e.g., C) and generates a file that contains the corresponding executable code.

At times, you need to create empty files or files of an arbitrary size with random data. You can do so by using the touch and fallocate commands. The touch command creates an empty file and fallocate command can be used to create a file of an arbitrary size. In the following session, we use the touch command to create two empty files, file1 and file2. We then use the fallocate command to create files file3 and file4 of sizes 10 and 20 MB each. Files created by fallocate contain random data that already resides on the disk blocks allocated to the file, without writing anything explicitly on them. This means no disk input/output (I/O) takes place when fallocate is used to create a file of any size.

```
$ touch file1
$ touch file2
$ fallocate -l 10MB file3
$ fallocate -l 20MB file4
$ ls -l
total 19536
-rw-r--r-- 1 sarwar faculty        0 Mar  1 22:40 file1
-rw-r--r-- 1 sarwar faculty        0 Mar  1 22:40 file2
-rw-r--r-- 1 sarwar faculty 10000000 Mar  1 22:41 file3
-rw-r--r-- 1 sarwar faculty 20000000 Mar  1 22:42 file4
$
```

4.5.4 Creating and Removing Directories

We briefly discussed the mkdir and rmdir commands in Chapter 2. Here, we cover these commands fully. You can create a directory by using the mkdir command. The following is a brief description of this command.

Syntax:
    mkdir [option] directory-names

Purpose: Create directories specified in directory-names

Commonly used options/features:

   -m MODE   Create directories with the access permissions specified in MODE in octal (see Chapter 5)

   -p   Create parent directories that do not exist in the pathnames specified in directory-names

---

Here, directory-names are the pathnames of the directories to be created. When you log on, you can use the following command to create a subdirectory, called memos, in your home directory. Access permissions for the newly created directories are determined by the current value of umask (see Chapter 5). You can confirm the creation of this directory by using the ls -ld memo command, as in

```
$ mkdir memos
$ ls -ld memos
drwxr-xr-x 2 sarwar faculty 4096 Jul  6 15:19 memos
$
```

Similarly, you can create a directory called **test_example** in the **/tmp** directory by using

```
$ mkdir /tmp/test_example
$
```

While in your home directory, you can create the directory **professional** and a subdirectory **letters** under it by using the mkdir command with the -p option, as in

```
$ mkdir -p professional/letters
$
```

You can use the rmdir command to remove an empty directory. If a directory is not empty, you must remove the files and subdirectories in it before removing it. To remove nonempty directories, you need to use the rm command with the -r option (see Chapter 6). The following is a brief description of the rmdir command.

Syntax:
    rmdir [option] directory-names

Purpose: Remove the empty directories specified in directory-names

Commonly used options/features:

    -p   Remove empty parent directories also

The following command removes the **letters** directory from the present working directory. If **letters** directory is not empty, the rmdir command displays an error message rmdir: letter: Directory not empty on the screen. If **letters** directory is a file, the command displays an error message rmdir: letters: Not a directory.

```
$ rmdir letters
$
```

The following command removes the directory **letters** from your present working directory and **memos** from your home directory.

```
$ rmdir ~/memos/letters
$
```

If the **~/personal** directory contains only one subdirectory, called **diary**, and it is empty, you can use the following command to remove both directories ~/personal and ~/personal/diary. We assume that you are in your home directory when you execute this command.

```
$ rmdir -p personal/diary
$
```

4.5.5 Determining File Attributes

You can determine the attributes of files by using the ls command with various options. The options can be used together, and their order does not matter. For example, you can use the –l option to get a long list of a directory that gives the attributes of files, such as the owner of the file. Similarly, you can use the –ld options to display the attributes of directories. The following examples illustrate the use of the ls command with –ld and –l options.

```
$ ls -ld courses memos personal
drwxr-xr-x 2 sarwar faculty 4096 Jul  6 10:31 courses
drwxr-xr-x 2 sarwar faculty 4096 Jul  8 15:19 memos
drwxr-xr-x 2 sarwar faculty 4096 Jul  8 15:19 personal
$ cd ~/courses/ee446/exams
$ ls -l
total 44
-rw-r--r-- 1 sarwar faculty  7878 Jul  8 15:23 mid1
-rw-r--r-- 1 sarwar faculty 32253 Jul  8 15:23 mid2
drwxr-xr-x 2 sarwar faculty  4096 Jul  8 15:22 solutions
$
```

The information displayed by the ls -l command is summarized in Table 4.1.

In the preceding two uses of the ls -l command, **courses**, **memos**, **personal**, and **solutions** are directories, and **mid1** and **mid2** are ordinary files. As stated earlier, we discuss access permissions and user types in Chapter 5. The owner of the files is **sarwar**, who belongs to the group **faculty**. The values of the remaining fields are self-explanatory.

You can use the ls command with the –i option to display the inode numbers of files and directories. To display the inode number of a directory, you need to use the ls –id command. The following examples of its use show that the inode number for the **greeting** file is 6278611, and for the directories **courses**, **memos**, and **personal**, they are 6555603, 6555456, and 6555324, respectively.

```
$ ls -i greeting
16517942 greeting
$ ls -id courses memos personal
16517083 courses   16517069 memos   16517079 personal
$
```

The ls -al command displays a long list of all the files in a directory as follows:

```
$ ls -al ~/courses/ee446/exams
total 52
drwxr-xr-x 3 sarwar faculty  4096 Jul  8 15:22 .
drwxr-xr-x 4 sarwar faculty  4096 Jul  8 15:22 ..
-rw-r--r-- 1 sarwar faculty  7878 Jul  8 15:23 mid1
-rw-r--r-- 1 sarwar faculty 32253 Jul  8 15:23 mid2
drwxr-xr-x 2 sarwar faculty  4096 Jul  8 15:22 solutions
$
```

You can use the –F option to identify directories, executable files, and symbolic links. The ls –F command displays an asterisk (*) after an executable file, a slash (/) after a directory, and an "at" symbol (@) after a symbolic link (discussed in Chapter 8), as follows:

```
$ ls -F /
bin/    dev/    initrd.img@  lost+found/  opt/    run/    sys/    var/
boot/   etc/    lib/         media/       proc/   sbin/   tmp/    vmlinuz@
cdrom/  home/   lib64/       mnt/         root/   srv/    usr/
$
```

Note that there is no executable file in the root directory. The output of the ls –F /bin command would show that all the files in the **/bin** directory are executable. You are encouraged to read the online manual pages for the ls command on your system or see the Command Appendix at the CRC Press website for this book for a detailed description of the command.

By using the shell metacharacters and regular expressions, you can specify a particular set of files and directories in the file system structure or a particular set of strings in files or directories. For example, the following command can be used in the C shell to display a long list for all the files in the **~/courses/ ee446** directory that have the **.c** extension and start with the string lab followed by zero or more characters, with the condition that the first of these characters cannot be 5.

```
$ ls -l ~/courses/ee446/lab[^5]*.c
...
$
```

Similarly, the following command can be used to display the inode numbers and names of all the files in your current directory that have four-character names and an **.html** extension. The file names must start with a letter, followed by any two characters, and end with a digit from 1 to 5.

```
$ ls -i [a-zA-Z]??[1-5].html
...
$
```

The following command under the C shell displays the names of all the files in your home directory that do not start with a digit and that end with **.c** or **.C**. In other words, the command displays the names of all the C and C++ source program files that do not start with a digit. Under the Bourne shell, you may replace the ∧ character with the ! character. Thus, the ls ~/[!0-9]*.[c,C] command would produce the same results.

```
$ ls ~/[^0-9]*.[c,C]
...
$
```

4.5.6 Determining the Type of a File's Contents

Because Linux does not support file extensions, you can use any extension name for any file. This means that you can use the **.jpg** extension for an executable program file. Thus, you cannot determine the type of content of a file by simply looking at its name. Since many software tools require the use of extensions and the user may rely on extensions, extension names are, therefore, still significant. In Linux, you can find the type of a file's contents by using the file command. Mostly, this command is used to determine whether a file contains text or binary data. Doing so is important because text files can be displayed on a terminal screen, whereas displaying the contents of a binary file shows "garbage" on your terminal screen and can also freeze your terminal, as it may interpret some of the binary values as control codes. The command has the following syntax.

Syntax:
    file [option] file-list

Purpose: Attempt to classify files in file-list

Commonly used options/features:

  -f FILE   Use FILE as a file of file-list

The following session shows a sample run of the command. In this case, the types of the contents of all the files in the root directory are displayed.

```
$ file /*
/bin:        directory
/boot:       directory
/cdrom:      directory
/dev:        directory
/etc:        directory
/home:       directory
/initrd.img: symbolic link to boot/initrd.img-4.4.0-21-generic
/lib:        directory
/lib64:      directory
/lost+found: directory
/media:      directory
```

```
/mnt:           directory
/opt:           directory
/proc:          directory
/root:          directory
/run:           directory
/sbin:          directory
/srv:           directory
/sys:           directory
/tmp:           sticky, directory
/usr:           directory
/var:           directory
/vmlinuz:       symbolic link to boot/vmlinuz-4.4.0-21-generic
$
```

The following session shows a few more types of files.

```
$ cd /bin
$ file cat tar gzip which
cat:   ELF 64-bit LSB executable, x86-64, version 1 (SYSV), dynamically linked,
interpreter /lib64/ld-linux-x86-64.so.2, for GNU/Linux 2.6.32, BuildID[sha1]=
2267d831560007f67fa4388d830192fd89861061, stripped
tar:   ELF 64-bit LSB executable, x86-64, version 1 (SYSV), dynamically linked,
interpreter /lib64/ld-linux-x86-64.so.2, for GNU/Linux 2.6.32, BuildID[sha1]=
ac316e790c0a96f05810a5d5e1c3ab8f2a2411cd, stripped
gzip:  ELF 64-bit LSB executable, x86-64, version 1 (SYSV), dynamically linked,
interpreter /lib64/ld-linux-x86-64.so.2, for GNU/Linux 2.6.32, BuildID[sha1]=
43b51e737be8684de2d196ee2f036c2cddd99c7d, stripped
which: POSIX shell script, ASCII text executable
$
```

The *executable and linkable format* (ELF) is a common standard file format for executable code, object code, shared libraries, and core dumps. Linux creates a *core dump* in a file, called **core**, when a program crashes. Programmers can use this file to identify what caused the program to crash. It contains the program state at the time of crash: data that the crashed program was accessing at the time it crashed, the state of the program stack, and the location of the program statement that caused the crash. You can use the file core command to determine the name of the program that produced the core dump.

The **cat**, tar, and **gzip** files contain ELF 64-bit executable codes, and **which** contains a POSIX shell script. Some more classifications that the file command displays are English text, C program text, Bourne shell script text, empty, nroff/troff, Perl command text, Python text, PostScript, sccs, and setgid executable. You should read the manual page for the command to learn more about the file command. The terms *SUID* and *SGID* are explained in Chapter 5.

The following in-chapter exercises familiarize you with the echo, cd, ls, and file commands and the formats of their output.

Exercise 4.4
Right after you log on, run echo ~ to determine the full pathname of your home directory.

Exercise 4.5
Use the cd command to go to the **/usr/bin** directory on your system and run the ls -F command. Identify two symbolic links and five binary files.

Exercise 4.6
Run the ls command in the same directory and write down sizes of the find and sort commands (i.e., files) in bytes and in K bytes. What commands did you use to accomplish the task? What commands would you use to obtain the sizes of these commands/files, assuming your were in your home directory?

Exercise 4.7
Run the file /etc/* command to identify types of all the files in this directory.

4.5.7 File Representation and Storage in Linux

As stated earlier, the attributes of a file are stored in a data structure on the disk, called an inode. At the time of its creation, every file is allocated a unique inode from a list (array) of inodes on the disk, called the *i-list*. The index value of the inode in the i-list is called the inode number for the inode allocated to the file, and is known as the file's inode number. The Linux kernel also maintains a table of inodes, called the *inode table*, in the main memory for all open files. When an application opens a file, an inode is allocated from the inode table and the contents of the file's inode on the disk are copied into it.

The inode number is used to index the inode table, allowing quick access to the attributes of an open file. When a file's attributes (e.g., file size) change, the inode in main memory is updated; disk copies of inodes are updated at fixed intervals. For files that are not open, their inodes reside on the disk. Some of the contents of an inode are shown in Figure 4.3.

The "link count" field specifies the number of different names the file has within the system. This count is also known as the *hard link count* (see Chapter 8 for details on links). The "file mode" field specifies what the file was opened for (read, write, etc.). The "user ID" is the ID of the owner of the file. The "access permissions" field specifies who can access the file for what type of operation (discussed in more detail in Chapter 5). The file's location on disk is specified by a number of *direct* and *indirect* pointers to disk blocks containing file data.

A typical computer system has several disk drives. Each drive consists of a number of platters with two *surfaces* (top and bottom). Each surface is logically divided into concentric circles called *tracks*, and each track is subdivided into fixed size portions called *sectors*. Tracks at the same position on both surfaces of all platters comprise a *cylinder*. Disk I/O takes place in terms of one sector, also called a disk block. For this reason, disks are known as *block devices*. Traditionally, the sector size for hard disks has been 512 bytes. Newer hard disks use 4 K-byte (i.e., 4,096-byte) sector sizes. CD-ROMs and DVD-ROMs use 2 K-byte (i.e., 2,048-byte) sector sizes.

A sector may be addressed using a four-dimensional address comprising <disk #, cylinder #, surface #, and sector #>. This four-dimensional address is translated to a *linear* (one-dimensional) block number, and most of the software in Linux deals with block addresses, because they are relatively easy to deal with. These blocks start with the sector numbered as 0 on the outermost cylinder on the topmost surface (i.e., the topmost track of the outermost cylinder), which is assigned block number 0. The block numbers increase through the rest of the tracks in that cylinder, through the rest of the cylinders on the disk, and then through the rest of the disks. The diagram shown in Figure 4.4 is a logical view of a disk system consisting of an array of disk blocks. File space is allocated in *clusters* of two, four, or eight disk blocks.

Figure 4.5 shows how an inode number for an open file can be used to access a file's attributes, including the file's contents, from the disk. It also shows contents of the directory **~/courses/ee446/labs** and how the Linux kernel maps the inode of the file **lab1.c** to its contents on disk. As previously discussed,

```
Link Count
File Mode
User ID
Time Created
Time Last Updated
Access Permissions

        •
        •
        •

File's Location on Disk
```

FIGURE 4.3 Contents of an inode.

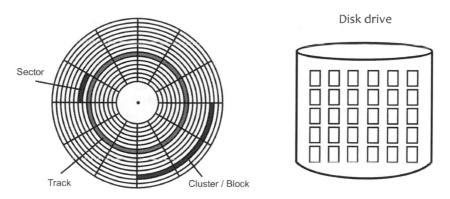

FIGURE 4.4 Physical and logical views of a disk drive in terms of tracks, sectors, clusters, and disk blocks.

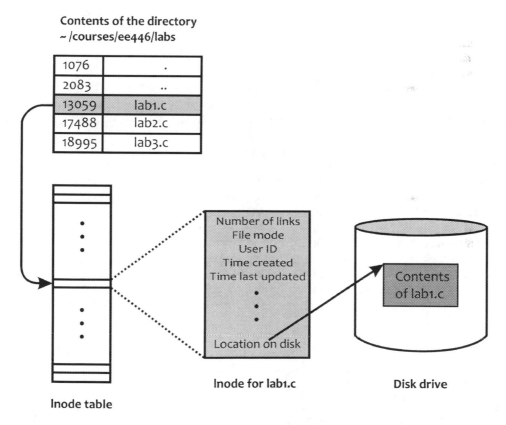

FIGURE 4.5 Relationship between the file **lab1.c** in a directory and its contents on a disk.

and as shown in the diagram, a directory consists of an array of entries <inode #, filename>. Accessing (reading or writing) the contents of **lab1.c** requires the use of its inode number to index the in-memory inode table to get to the file's inode. The inode, as previously stated, contains, among other things, the location of **lab1.c** on the disk.

The inode contains the location of **lab1.c** on the disk in terms of the numbers of the disk blocks that contain the contents of the file. The details of how exactly a Linux file's location is specified in its inode and how it is stored on the disk are beyond the scope of this textbook. These details are available in any book on Linux internals.

4.6 Standard Files and File Descriptors

When an application needs to perform an I/O operation on a file, it must first open the file and then issue the file operation (read, write, seek, etc.). Linux automatically opens three files for every command it executes. The command reads input from one of these files and sends its output and error messages to the other two files. These files are called *standard files: standard input* (**stdin**) files, *standard output* (**stdout**) files, and *standard error* (**stderr**) files. By default, these files are attached to the terminal on which the command is executed. That is, the shell makes the command input come from the terminal keyboard, and its output and error messages go to the terminal (or the console window in case of an ssh session or an xterm in a Linux system running Wayland, as discussed in detail in Chapter W28). These default files can be changed to other files by using the redirection operators: < for input redirection and > for output and error redirection.

A small integer, called a *file descriptor*, is associated with every open file in Linux. The integer values 0, 1, and 2 are the file descriptors for **stdin**, **stdout**, and **stderr**, respectively, and are also known as *standard file descriptors*. The kernel uses file descriptors to perform file operations (e.g., file read), as illustrated in Figure 4.6. The kernel uses a file descriptor to index the per-process file descriptor table to obtain a pointer to the systemwide file table. The file table, among other things, contains a pointer to the file's inode in the inode table. Once the inode for the file has been accessed, the required file operation is performed by accessing appropriate disk block(s) for the file by using the direct and indirect pointers, as described in Section 4.6.

Recall that every device, including a terminal, is represented by a file in Linux. The diagram shown in Figure 4.7 depicts the relationship between a file and its file descriptor. Here, we assume that files **lab1.c** and **lab2.c** are open for some file operations (say, file read) and have descriptors 3 and 4, respectively, that the kernel is assigned to the files when they were opened. We have described the details of this relationship in the preceding paragraph and in Section 4.6, in terms of a file table, inode table, and the storage of the file on disk; also see Figures 4.5 and 4.6.

The Linux system allows standard files to be changed to alternate files for a single execution of a command, including a shell script. This concept of changing standard files to alternate files is called *input, output,* and *error redirection*. We address input, output, and error redirection in detail in Chapter 9. We have briefly mentioned the standard files and file descriptors here because most Linux commands that explicitly require input from an outside source get it from standard input, unless it comes from a file (or list of files) that is passed to the command as a command line argument. Similarly, most Linux commands that produce output send it to a standard output. This information is important for proper understanding and use of commands in the remaining chapters.

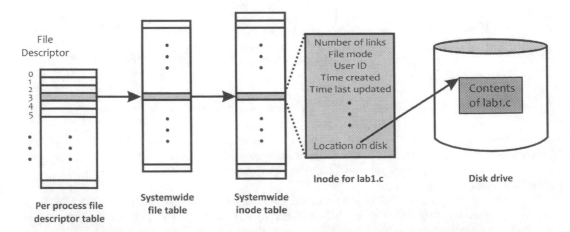

FIGURE 4.6 Relationship between file descriptors and the contents of files on disk.

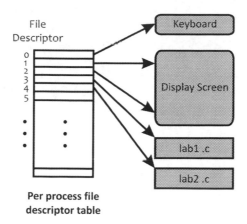

Per process file
descriptor table

FIGURE 4.7 Logical view of the relationship between file descriptors and the corresponding files.

4.7 File System

As mentioned in Section 4.2 and as detailed throughout this chapter, a file system is ordinarily thought of as a highly organized and quickly accessible collection of data, in terms of persistent storage of that data on any modern medium. That organization, the speed of accessibility, efficiency, and usefulness, are all hallmarks of any particular file system implementation. Thought of in this way, a file system is a directory hierarchy with its own root stored on a hard disk/solid-state disk (SSD) or disk partition, mounted under (glued to) a directory. The files and directories in the file system are accessed through the directory under which they are mounted. A file system may span disks, even computers connected through a network. Their home is available on any workstation, mounted as needed. All references to a file system in this textbook are to the local file system. The description of the Linux network file system (NFSv4) is beyond the scope of this textbook, but it is briefly mentioned in Chapter 11.

Over the history of Linux, various file systems have been used to provide speed, efficiency, security, and utility to an ordinary user. Three contemporary and extremely important file systems are the Extended File System (ext), XFS File System, and the Zettabyte File System (ZFS).

The most universal of these, across all major branches of Linux, is ext. Most flavors of Linux use it as the default file system. The fourth version of ext, ext4, is the current version with features such as large scalability, the ability to map to very large disk array sizes and journaling. A file system is said to use journaling if it keeps track of changes not yet committed to the file system's main part by recording the intentions of such changes in a data structure called a journal. Ext4 supports backward compatible with ext3 and ext2, block allocation until data is flushed to disk, unlimited number of subdirectories, and timestamps measured in nanoseconds.

The ZFS is a POSIX-compliant file system that was first released by Oracle in June 2006. It is a transactional file system, which means that the file system state is always consistent on disk. It is reliable, flexible, and scalable, with built-in compression and advanced features for file system backup and restoration. However, the hallmark of ZFS is its focus on maximum data integrity that protects user data against all sorts of errors, including those caused by decaying storage media, electric current spikes, accidental disk writes, and so on. It achieves such data integrity by using several techniques, including data replication. Benchmarking has shown that given multiple dedicated disks, ZFS is considerably faster than EXT4.

XFS is a high performance 64-bit journaling file system created by Silicon Graphics and ported to Linux kernel in 2001. The key feature of XFS is execution of parallel I/O operations. It is a highly scalable file system that allows tremendous scalability of I/O threads, file system bandwidth, size of files, and the size of the file system itself when spanning multiple physical storage devices.

You can use the findmnt command to display information about all mounted file systems, including their type such as ext4. This command with -t option may be used to display the file systems of a particular type. For example, you can use the findmnt -t ext4 command to display the list of all ext4 file systems mounted on your system.

Table 4.4 shows a summary comparison of Linux file systems XFS, Ext4, and ZFS. Browse through the relevant websites listed in "Web Resources" (Table 4.5 at the book GitHub site) to find out more about Ext4, XFS, and ZFS.

Exercise 4.8
Use the findmnt command to display information about all the file systems mounted on your system. How many file systems are mounted on your system?

Exercise 4.9
Use the findmnt command to display information about all ext4 file systems mounted on your system. How many of these file systems are mounted on your system?

4.7.1 Displaying Disk Usage of Files and Directories

You can display the disk usage by files and directories by using the du command. It is a useful command for new Linux users with runaway log files, hitting their quotas. Here is a brief description of this command.

Syntax:
    du [option]... file-list...

Purpose: Display a summary of the disk usage of files in file-list, recursively for directories

Commonly used options/features:

- -a   Display disk usage for all files, not only directories
- -h   Display in humanly readable format, e.g., 1.7 K, 31.9 M, and 3 G
- -s   Display only total disk usage each file in file-list

In the following session, we use the du command without any option, and with -s and -h options, to show disk usage for various files and directories in 1 K block sizes and in K, M, and G bytes.

```
$ du ~
4        /home/sarwar/.linuxmint/mintMenu/applications
34724    /home/sarwar/.linuxmint/mintMenu
...
27450108        /home/sarwar
$ du -h ~
4.0K    /home/sarwar/.linuxmint/mintMenu/applications
34M     /home/sarwar/.linuxmint/mintMenu
```

TABLE 4.4

A Summary Comparison of Linux File Systems

Feature	Ext4	ZFS	XFS
Introduced to Linux or Open Source	2006	2005	2001
Maximum file size	16 tebibytes (TiB[a])	16 exbibytes (EiB[b])	8 EiB
Maximum volume size	1 EiB	2^{128} bytes = 256 trillion yobibytes (YiB[c])	8 EiB
Maximum number of files	4 billion	Unlimited (2^{48} per directory)	16 EiB
Maximum filename length	255 bytes	255 bytes	255 bytes

[a] TiB = 2^{40} = 1024^4.

[b] EiB = 2^{60} = 1024^6.

[c] YiB = 2^{80} = 1024^8.

```
...
27G     /home/sarwar
$ du -s linux2e/ch14
27401844        linux2e/newch14
$ du -s -h linux2e/ch14
27G     linux2e/newch14
$ du -s -h /usr/bin
163M    /usr/bin
$
```

Exercise 4.10

Use the du command to display disk usage for your home directory, /lib, and /dev directories on your system in terms of 1 K blocks and in K, M, and G bytes?

4.8 End-of-File Marker

Every Linux file has an end-of-file (eof) marker. The commands that read their input from files read the eof marker when they reach the end of a file. The <Ctrl-D> on a new line is the Linux eof marker when the input file is attached to a keyboard. That is why commands such as cat while reading input from the keyboard (see Chapter 6) terminate when you press <Ctrl-D> on a new line.

Summary

In Linux, a file is a sequence of bytes. This simple, yet powerful, concept and its implementation lead to nearly everything in the system being a file; users and processes are not. Linux supports seven types of file: ordinary file, directory, symbolic link, character special file, block special file, named pipe (also known as FIFO), and socket. No file extensions are supported for files of any type, but applications running on a Linux system can require their own extensions.

Every file in Linux has several attributes associated with it, including file name, owner's name, date last modified, link count, and the file's location on disk. These attributes are stored in an area on the disk called an inode. When files are opened, their inodes are copied to a kernel area called the inode table for faster access of their attributes. Every file in a directory has an entry associated with it that comprises the file's name and its inode number. The kernel accesses an open file's attributes, including its contents, by reading the file's inode number from its directory entry and indexing the inode table with the inode number.

The Linux file structure is hierarchical, with a root directory and all the files and directories in the system under it. Every user has a directory called the user's home directory, which he or she is placed when logging on to the system. Multiple disk drives and/or disk partitions can be mounted on the same file system structure, allowing their access as directories, and not as named drives A:, B:, C:, and so on, as in MS-DOS and Microsoft Windows. This approach gives a unified view of all the files and directories in the system, and users do not have to worry about remembering the names of drives and the files (and directories) that they contain.

Directories (primarily) can be created and removed under the user's home directory. The file structure can be navigated using various commands (mkdir, rmdir, cd, ls, etc.). You can specify a file in the system by using the file's absolute or relative pathname. An absolute pathname starts with the root directory, and a relative pathname starts with a user's home directory or present working directory.

The following commands can be used to display file contents, know the type of file contents, create directories, remove directories, display the contents of a directory, create empty files, create arbitrary size files with random data, etc.: cat, cd, du, echo, fallocate, file, findmnt, getconf, ls, mkdir, more, pwd, rmdir, and touch.

You can create files on your Linux system by using several methods, including a text editor such as nano or vim, a compiler to generate the executable image of a program, the touch command to create an empty file, and the fallocate command to create an uninitialized file of any size.

Linux automatically opens three files for every command for it to read input from and send its output and error messages to. These files are called standard input (**stdin**), standard output (**stdout**), and standard error (**stderr**). By default, these files are attached to the terminal on which the command is executed; that is, the command input comes from the terminal keyboard, and the command output and error messages go to the terminal's screen display. The default files can be changed to other files by using redirection primitives: < for input redirection and > for output and error redirection.

The kernel associates a small integer with every open file. This integer is called the file descriptor. The kernel uses file descriptors to perform operations (e.g., read) on the file. The file descriptors for **stdin**, **stdout**, and **stderr** are 0, 1, and 2, respectively.

A file system is ordinarily thought of as a highly organized and quickly accessible collection of data, in terms of persistent storage of that data on any modern medium. The Extended File System, XFS, and the ZFS are the most contemporary and most important of all Linux file systems. Supporting very large size files, allowing extremely large number of directories, and maintaining data consistency are some of the hallmark features of these systems.

The du command can be used to display the disk usage for files and directories in 1 K blocks and in terms of K, M, and G bytes.

Every Linux file has an eof marker. The eof marker is <Ctrl-D> if a command reads input from the keyboard.

Questions and Problems

1. What is a file in Linux? What do character special and block special represent in Linux? How many character special and block special files reside on your system? How did you obtain your answer? Show your work.

2. Does Linux support any file types? If so, name them. Does Linux support file extensions?

3. What is a directory entry? What does it consist of?

4. What are special files in Linux? What are character special and block special files? Run the `ls /dev | wc -w` command to find the number of special files your system has.

5. What is meant by interprocess communication? Name three tools that Linux provides for interprocess communication.

6. Draw the hierarchical file structure, similar to the one shown in Figure 4.2, for your Linux machine. Show files and directories at the first two levels. Also show where your home directory is, along with files and directories under your home directory.

7. Give three commands that you can use to list the absolute pathname of your home directory.

8. Write down the line in the **/etc/passwd** file on your system that contains information about your login. What are your login shell, user ID, home directory, and group ID? Does your system contain the encrypted password in the **/etc/passwd** or **/etc/shadow** file?

9. What would happen if the last field of the line in the **/etc/passwd** file were replaced with **/usr/bin/date**? Why?

10. What are the inode numbers of the root and your home directories on your machine? Give the commands that you use to find these inode numbers.

11. Create a directory in your home directory, called **memos**. Go into this directory and create a file called **memo.james** by using one of the editors such as vi. Give three pathnames for this file.

12. Give a command for creating a subdirectory called **personal** under the **memos** directory that you created in Problem 11.

13. Make a copy of the file **memo.james** and put it in your home directory. Name the copied file **temp.memo**. Give two commands for accomplishing this task.

14. Draw a diagram like that shown in Figure 4.5 for your **memos** directory. Clearly show all directory entries, including inode numbers for all the files and directories in it.

15. Give the command for deleting the **memos** directory. How do you know that the directory has been deleted?

16. What is the purpose of the following command? What happens when you run this command on your Linux system? Does the command output make sense? Explain your answer.

    ```
    rmdir -p ~/personal/diary
    ```

17. Why does a shell process terminate when you press <Ctrl+D> at the beginning of a new line?

18. Give a command to display the types of all the files in your **~/linux2e** directory that start with the word chapter, are followed by one of the digits 1, 2, 6, 8, or 9, and end with **.eps** or **.prn**.

19. Give a command line to display the types of all the files in the personal directory in your home directory that do not start with letters a, k, G, or Q and the third letter in the name is not a digit and not a letter (uppercase or lowercase).

20. Use the **ls -i** command to display inode numbers for the **/**, **/usr**, and **~** directories on your system. Show outputs of your commands and identify the inode numbers for these directories.

21. What are ELF and core dump?

22. Where are the permissions and most other attributes for a file stored in Linux?

23. Display the absolute pathnames of your home directory by using two different methods in the Bash and TC shells. Does your system use any symbolic links in the **/** directory? If so, display those symbolic links by using a shell command.

24. Discuss possible ways to create files on Linux. Use the `fallocate` command to create files of sizes 20 and 200 GB.

25. What is a file descriptor in Linux? What are standard files? What are their descriptors?

26. What is the concept of redirection of standard files?

27. Why do you think the system-wide file table is required between the per process file descriptor table and the inode table?

28. What are sector, track, platter, and cylinder? What is the size of a sector in hard disks these days?

29. What are block and cluster?

30. What are the basic characteristics of the ZFS file system? After referring to Chapter W22 and the online resource for ZFS, is ZFS integrated into the Linux kernel, or is it a kernel-loadable module?

31. What is the maximum size (in bytes/characters) of the file name in your Linux system? What command did you use to obtain your answer?

32. Browse through the following directories and identify five commonly used commands, tools, utilities, and daemons in them: **/bin**, **/sbin**, **/usr/bin**, **/usr/sbin**, **/usr/local/bin**, and **/usr/local/sbin**. Clearly state the purpose of each command, tool, utility, or daemon and the name of the directory in which it is found.

33. We say that Ext4 is backward compatible. What does it mean?

34. What size volumes and files does Ext4 support? How many directories may be created on the Ext4 file system?

35. Repeat Question 26 for ZFS and XFS.

36. How many file systems are mounted on your system? How many of these are of type Ext4, XFS, ZFS, and cgroups type each? Show the commands you used to obtain your answers.

37. How many file systems each are mounted under /sys and /dev on your system. Show the command you use to obtain your answer along with the output of the command.

38. How many files can a directory have under ZFS? Please give number in trillion.

39. Use the du command with and without option(s) to display disk usage of the following files and
    directories in terms of 1 K blocks and in terms of K, M, and G bytes:

    a. /etc/passwd

    b. /var/log/syslog

    c. Your home directory

    d. Root directory

    e. /usr/bin directory

Advanced Questions and Problems

40. How would you accomplish obtaining and displaying (on standard out) the following file attri-
    butes, for file objects in your current working directory?

    a. Sorted files based on modification time, showing all long-form attributes.

    b. A listing of all hidden files and subdirectories.

    c. A listing of all files and directories in reverse alphabetical order.

    d. A listing of all files and directories alphabetically, based on file extensions.

    e. A listing of subdirectories in a recursive display.

    f. A long-form attribute listing of filenames, including their inode numbers.

    g. A listing of only the names of file and directories, and their sizes.

    h. A listing of owner, group, and author in reverse alphabetical order.

    i. A listing of names, sizes, and type of files, using the special indicators.

41. If you insert a USB thumbdrive on a Linux computer that is not running a GUI desktop, how
    can you easily find out if it has automounted, and what (if it has automounted) the complete
    pathname to where it has automounted in the file system is? What commands do you use to
    navigate to that place? What command do you use to mount (if it hasn't been automounted!)
    and unmount the thumbdrive? What kind of partition table and file system type is present on
    the thumbdrive? What command(s) can you use to find out all of the earlier information?

42. What Linux command allows you to list, modify, or create partitions on a dispensible USB
    thumbdrive (one that you really don't care about the data on) already automounted on your
    system? Using that command, its subcommands, and any other necessary commands, do the
    following-

    a. Unmount the thumbdrive.

    b. Print a listing of partition information about that thumbdrive at stdout.

    c. Delete whatever partitions are on the thumbdrive.

    d. Create a primary partition of type Linux filesystem, and numbered 1, on it.

    e. Let that partition use the whole disk.

    f. Write the changes to the disk.

    Is it then possible to then simply mount the thumbdrive (if it is not already mounted), and
    immediately start putting files on it? What else would you have to do to accomplish that?

        The ext4 file system is called a journaling file system and ZFS is called a transactional file
    system. What is the difference between a journaling file system and a transactional file system?
    What is a journal?

43. How many times bigger a file can be on ZFS-based Linux system compared with those based
    on Ext4 and XFS? Show your work.

Projects

Project 1

Put a dispensable USB thumbdrive (one that you really don't care about the data on) into your computer, and in the primary partition, create an ext4 file system on it. It is assumed that the USB thumbdrive already has a primary partition on it, which is usually preformatted. Use two different approaches. First, use the mkfs command. Then change the volume label to "USER_DATA" with the proper mke2fs command and then verify that change with the e2label command.

For a GUI-installed Linux system, use the Gparted graphical application on your Linux system to delete the partition you created with mke2fs and recreate it in Gparted. In each approach, mount the ext4 partition at a convenient location in your Linux file system, create some files on it, and then unmount it. Which approach is more useful for you?

Looking for more? Visit our sites for additional readings, recommended resources, and exercises.

CRC Press e-Resource: https://www.crcpress.com/9781138710085

Authors' GitHub: https://github.com/bobk48/linuxthetextbook

5

File Security

OBJECTIVES

- To show the protection and security mechanisms that Linux provides
- To describe the types of users of a Linux file
- To discuss the basic operations that can be performed on a Linux file
- To explain the concept of file access permissions/privileges in Linux
- To discuss how a user can determine access privileges for a file
- To describe how a user can set and change permissions for a file
- To discuss special protection bits, set-user-ID, set-group-ID, and sticky bit, and describe their purpose
- To cover the following commands and primitives:

    ?, ~, *, chmod, mcrypt, groups, ls -l, ls -ld, openssl, umask, umask -S

5.1 Introduction

As we pointed out earlier, a time-sharing system offers great benefits. However, it poses the main challenge of protecting the hardware and software resources in it. These resources include the input/output devices, central processing unit (CPU), main memory, and the secondary storage devices that store user files. The CPU runs user and kernel processes, the main memory stores user processes and important operating system code and data structures while the system is running, and the secondary storage devices store user files and operating system code on a permanent basis. We limit this chapter to a discussion about the protection of a user's files from unauthorized access by other users. Linux provides three mechanisms to protect your files.

The most fundamental scheme to protect user files is to give every user a login name and a password, allowing a user to use a system (see Chapter 2). To prevent others from accessing your files, keep the password for your computer account strictly confidential. The second scheme protects individual files by converting them to a form that is completely different from the original version by means of encryption. This technique is used to protect your most important files, so that the contents of these files cannot be understood even if someone somehow gains access to them on the system. The third file protection scheme allows you to protect your files by associating access privileges with them, so that only a subset of users can access these files for a subset of file operations. In other words, the owner of the files can decide to whom to grant access to these files. All three mechanisms are described in this chapter, with emphasis on the third scheme.

5.2 Password-Based Protection

The first mechanism that allows you to protect your files from other users is the login password scheme. Every user of a Linux-based computer system is assigned a login name (a name by which the user is known to the Linux system) and a password. Both the login name and the password are assigned by the system administrator and are required for a user to enter and use a Linux system. All login names are public knowledge and can be found in the **/etc/passwd** file. A user's password, however, is given to that user only. This scheme prevents users from accessing each other's files. Users are encouraged to change their passwords frequently by using the passwd command (see Chapter 2).

On some networked systems, you may have to use the yppasswd or nispasswd command to change your password on all the network's computer systems. However, the execution of the command requires superuser privileges (see Section 5.4.1). If you don't find this command on your Linux system, you have to install the nis or yp-tools package. Consult your instructor about the command that you need to use on your particular system. Usually, only the system administrator can change login names. Under some Linux installations, users are also allowed to change their usernames.

The effectiveness of this protection scheme depends on how well protected a user's password is. If someone knows your password, that person can log on to the system and access your files. There are, primarily, the following methods of discovering a user's password:

1. You, as the owner of an account, inform others of your password.
2. A user guesses (or cracks) another user's password using several techniques such as the dictionary, brute force, rainbow table, and spidering attacks.
3. Using phishing to obtain your password.
4. Using social engineering to obtain your password.
5. Using malware or keylogger to obtain your password.

Never let anyone else know your password under any circumstances. As a safety measure, you should change your password regularly. Always choose passwords that would be difficult for others to guess. A good password is one that is a mixture of letters, digits, and punctuation marks—but it must be easy for you to memorize. Never write your password on a piece of paper, and never use birthdays or the names of relatives, friends, famous sportspersons, or favorite movie actors as passwords. Also, avoid using words as passwords. Cryptographers recommend that a good password is 20 characters long.

You may like to browse through the websites given in Table 5.6 (found at the book GitHub site) to know more about the details of the various techniques for hack or obtaining someone's password, including dictionary, brute force, rainbow table, spidering, phishing, malware, keylogger, and social engineering attacks. The table also contains a link to a website that describes ten most popular tools for cracking passwords. We briefly describe the brute force attack method of cracking someone's password. The dictionary, rainbow table, and spidering attacks are similar but less time consuming.

By using brute force, someone tries to learn your password by trying all possible combinations of characters until the user's password is found. Guessing someone's password is a time-consuming process and is commonly used by hackers. The brute force method can be made more time consuming for an infiltrator if the password is long and consists of letters, digits, and punctuation marks. To illustrate the significance of using a more complex password, consider a system in which the password is exactly eight characters consisting of decimal digits only. This would allow a maximum of 10^8 (100 million) passwords that the brute force method would have to go through, in the worst-case analysis. If the same system requires passwords to consist of a mixture of digits and uppercase letters (a total of 36 symbols: 10 digits and 26 uppercase letters), the password space would comprise 36^8 (about 2.8 trillion) passwords. If the system requires passwords that consist of a mixture of digits, uppercase letters, and lowercase letters, the password space would comprise 62^8 (about 218 trillion) passwords. Imagine the size of the password space if passwords could include punctuation marks too! Many systems force a short (e.g., 5 s) delay after an invalid password is entered before the next login prompt, to make the infiltrator's job even harder.

The following in-chapter exercise asks you to figure out how to change your password on your system.

Exercise 5.1

In some Linux systems you are not allowed to change your password. Does your system allow you to change your password? If so, change your password. What command did you use?

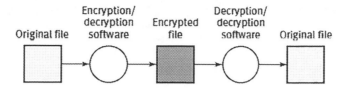

FIGURE 5.1 Process of encryption and decryption.

Note: Be sure to memorize your new password, because if you forget it you will have to request your system administrator to reset your password to a new value, unless your system allows you to change your password back to the previous password.

5.3 Encryption-Based Protection

In the second protection scheme, a software tool is used to convert a file to a form that is completely different from its original version. The transformed file is called an *encrypted file*, and the process of converting a file to an encrypted file is called *encryption*. The same tool is used to perform the reverse process of transforming the encrypted file to its original form, called *decryption*. You can use this technique to protect your most important files so that their contents cannot be understood even if someone else gains access to them. Figure 5.1 illustrates the encryption and decryption processes.

The Linux commands mcrypt and openssl can be used to encrypt and decrypt your files. You can learn more about these commands by running the man mcrypt and man openssl commands. We discuss these commands in detail in Chapter 7.

5.4 Protection Based on Access Permission

The third type of file protection mechanism prevents users from accessing each other's files when they are not logged on as a file's owner. As a file owner, you can attach certain access rights to your files that dictate who can and cannot access them for various types of file operations. This scheme is based on the types of users, the types of access permissions, and the types of operations allowed on a file under Linux. Without this protection scheme, users can access each other's files because the Linux file system structure (see Figure 4.2) has a single root from which all the files in the system hang. We use the terms access permissions, access privileges, and access rights synonymously throughout the book.

5.4.1 Types of Users

Each user in a Linux system belongs to a group of users, as assigned by the system administrator when a user is allocated an account on the system. A user can belong to multiple groups, but a typical Linux user belongs to a single group. All the groups in the system and their memberships are listed in the file **/etc/group** (see Figure 4.2). This file contains one line per group, with the last field of the line containing the login names of the group members. A user of a file can be the owner of the file, a user who belongs to the same group as the owner, or everyone else who has an account on the system. These, respectively, comprise the three types of users of a Linux file: *user*, *group*, and *others*. As the owner of a file, you can specify who can access it. The group name of a file is known as the *group owner* of the file.

Once a user of a system has logged in, he/she is known to the Linux system by an integer number, known as the *user ID* (UID), and not by the user's login name. Every Linux system has one special user who has access to all of the files on the system, regardless of the access privileges on the files. This user manages (administers) your Linux system and is commonly known as the *superuser* or *system adminis-trator*. The login name for the superuser is **root** and the user ID is **0**.

You can see the list of all user groups on your system by displaying the **/etc/group** file, shown as follows.

```
$ more /etc/group
root:x:0:
daemon:x:1:
bin:x:2:root,bin,daemon
sys:x:3:
adm:x:4:syslog,bob,ali,goldsmith,ravi
tty:x:5:
disk:x:6:
lp:x:7:
mail:x:8:
news:x:9:
uucp:x:10:
man:x:12:
proxy:x:13:
kmem:x:15:
dialout:x:20:davis,bob
fax:x:21:
voice:x:22:
cdrom:x:24:bob
floppy:x:25:
tape:x:26:
sudo:x:27:bob
audio:x:29:pulse
dip:x:30:bob
...
sambashare:x:130:bob,ali,ravi
lab:x:1001:davis,sam
faculty:x:1002:bob,sarwar,goldsmith,ravi,
ravi:x:1003:
goldsmith:x:1004:
bob:x:1005:
sarwar:x:1006:
$
```

There is one line in this file for every group on the system, each line having four colon-separated fields. The first field specifies the group name, the second specifies some information about the group, the third specifies the group ID as a number, and the last specifies a comma-separated list of users who are mem-bers of the group. For example, the **bin** group has group ID **2** and its members are users **root**, **bin**, and **daemon**. If the membership list in a line is missing (e.g., the **disk** group), this means that its membership is specified in the **/etc/passwd** file. The system administrator makes a user part of a group at the time of adding the user to the system. This group is known as a user's *default group*. The default group mem-bership of a user is specified in the user's entry in the **/etc/passwd** file. The system administrator can make a user part of another group (in addition to his/her default group) by placing his/her username in the comma-separated list of members for the group. You can use the groups command to display which groups on your system a user is a member of. The following session shows that **sarwar** is a member of the **faculty** group only; **bob** belongs to the groups **adm**, **dialout**, **sambashare**, and **courses**; and **davis** is a member of three groups: **dialout** and **lab**.

```
$ groups sarwar
faculty
$ groups bob
faculty courses
$ groups davis
dialout lab
$
```

TABLE 5.1

Summary of File Permissions in Linux

User Type	Permission Type			
	Read (r)	**Write (w)**	**Execute (x)**	**Meaning**
User (u)	X	X	X	Three bits for the user (owner) of a file
Group (g)	X	X	X	Three bits for the group (owner) of a file
Others (o)	X	X	X	Three bits for the rest of the user on the Linux machine

Note: X is 1 if the relevant permission is on and 0 if it is off.

TABLE 5.2

Possible Access Permission Values of a File for a User, Their Octal Equivalents, and Their Meanings

r	w	x	Octal Digit for Permission	Meaning
0	0	0	0	No permission
0	0	1	1	Execute-only permission
0	1	0	2	Write-only permission
0	1	1	3	Write and execute permissions
1	0	0	4	Read-only permission
1	0	1	5	Read and execute permissions
1	1	0	6	Read and write permissions
1	1	1	7	Read, write, and execute permissions

5.4.2 Types of File Operations/Access Permissions

In Linux, three types of access permissions/privileges can be associated with a file: read (r), write (w), and execute (x). The read permission on a file allows you to read the file, the write permission allows you to write to or remove the file, and the execute permission allows you to execute/run the file. The execute permission should be set for executable files only, i.e., files containing binary code (i.e., executable code generated by a compiler) or shell scripts, as setting it for any other type of file does not make any sense. We discuss the purpose of execute permission on a directory in Section 5.4.3.

With three categories of file users and three types of permissions for each user type, a Linux file has nine types of permissions associated with it, as shown in Table 5.1. Note that permissions are read across row by row. As stated in Chapter 4, access privileges are stored in a file's inode.

The value of X can be 1 (permission granted) or 0 (permission not granted). Thus, one bit is needed to represent a permission type and a total of three bits are needed to indicate file permissions for one type of user. In other words, a user of a file can have one of the eight (2^3) possible types of permissions for a file at a given time. Octal numbers can represent these eight three-bit access permission values from 0 to 7, as shown in Table 5.2. Access permissions 0 (binary 000) and 7 (binary 111) mean no access permissions and all permissions, respectively.

The nine bits needed to express permissions for the three types of file users result in the possible access permission values of three-digit octal numbers 000–777 for file permissions. The first octal digit specifies permissions for the owner of the file, the second digit specifies permissions for the group that the owner of the file belongs to, and the third digit specifies permissions for everyone else. In the output of the ls -l command, a bit value of 0 for a permission is displayed as a dash (-), and a value of 1 is displayed as r, w, or x, depending on the position of the bit according to the table. Thus, a permission value of 0 in octal (no permissions granted) for a user of a file can be written as --- and a permission of 7 (all three permissions granted) can be denoted as rwx. The outputs of the ls -l commands in the following session show that the **/etc/passwd** file is read-only for everyone on the system except root, who has read and write permissions. The **client.c** file has read and write permissions for the owner (**sarwar**) and group (**faculty**), and read-only permission for others.

```
$ ls -l /etc/passwd
-rw-r--r-- 1 root root 2862 Dec 14  2016 /etc/passwd
$ ls -l client.c
-rw-rw-r-- 1 sarwar faculty 1277 Dec 12  2016 client.c
$
```

Because a user can be in many groups, you can expand access to your files to accommodate different users through groups.

5.4.3 Access Permissions for Directories

Next, we will look at what the read, write, and execute permissions mean for directories. The read permission for a directory allows you to read the contents of the directory; recall that the contents of a directory are the names of files and directories in it. Thus, the ls command can be used to list its contents. The write permission for a directory allows you to create a new directory or a file in it or to remove an existing entry from it. The execute permission for a directory is permission to search the directory but not to read from or write to it. Thus, if you do not have execute permission for a directory, you cannot use the ls -l command to list its contents or use the cd command to make it your current directory. The same is true if any component in a directory's pathname does not contain execute permission. We demonstrate these aspects of the search permission on directories in Section 5.5.2.

5.5 Determining and Changing File Access Privileges

The following sections describe how you can determine the access privileges for files and directories and how you can change them to enhance or limit someone's access to your files.

5.5.1 Determining File Access Privileges

You can use the ls command with the -l or -ld option to display access permissions for a list of files and/ or directories. The following is a brief description of the ls command with the two options.

> **Syntax:**
>   ls -l [file-list]
>   ls -ld [directory-list]
>
> **Purpose:** First syntax: Display the long list of files and/or directories in the space-separated **file-list** on the display screen; in the case where **file-list** contains directories, display the long list of all the files in these directories
>
>   Second syntax: Display the long list of directories in **directory-list** on the display screen.

If no **file-list** is specified, the command gives the long list of all the files (except hidden files) in the present working directory. Add the -a option to the command line to include the hidden files in the display. Consider the following session.

```
$ ls -l
drwxr-x---             2      sarwar        faculty      512     Jul 29    17:35
courses
-rwxrwxrwx             1      sarwar        faculty       12     May 01    13:22    labs
-rwxr--r--             1      sarwar        faculty      163     May 05    23:13    temp

$
```

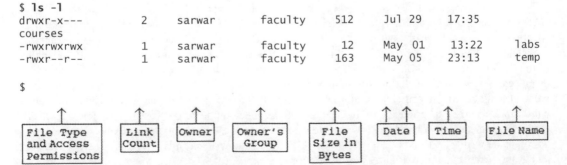

TABLE 5.3

Permissions for Access to the Files **courses**, **labs**, and **temp** for the Three Types of Users

File Name	User	Group	Other
courses	Read, write, search	Read and search	No permission
labs	Read, write, execute	Read, write, execute	Read, write, execute
temp	Read, write, execute	Read	Read

The leftmost character in the first field of the output indicates the file type (d for a directory and – for an ordinary file). The remaining nine characters in the first field show the file access privileges for user, group, and others, respectively. The second field indicates the number of hard links (discussed in Chapter 8) to the file. The third field shows the owner's login name. The fourth field shows the file's group owner. The fifth field displays the file's size (in bytes). The sixth, seventh, and eighth fields display the date and time of file's creation (or last update). The last field is the file's name. Table 5.3 shows who has what type of access privileges for the three files in this session: **courses**, **labs**, and **temp**.

If an argument of the ls -l command is a directory, the command displays the long list of all the files and directories in it. You can use the ls -ld command to display the long list of directories only. When executed without an argument, this command displays the long list of the current directory, as shown in the first command of the following session. The second and third commands show that when the ls -ld command is executed with a list of directories as its arguments, it displays the long list of those directories only. If an argument to the ls -ld command is a file, the command displays the long list of the file. The fourth command, ls -ld pvm/*, displays the long lists of all the files and directories in the **pvm** directory.

```
$ ls -ld
drwx--x--x      2   sarwar    faculty    11264   Jul  8 22:21  .
$ ls -ld ABET
drwx------      2   sarwar    faculty    512     Dec 18 1997   ABET
$ ls -ld ~/Images courses/ee446
drwx------      3   sarwar    faculty    512     Apr 30 09:52  courses/ee446
drwx--x--x      2   sarwar    faculty    2048    Dec 18 1997   /home/sarwar/Images
$ ls -ld pvm/*
drwx------      3   sarwar    faculty    512     Dec 18 1997   pvm/examples
drwx------      2   sarwar    faculty    1024    Oct 27 1998   pvm/qsort
-rw-------      1   sarwar    faculty    1606    Jun 19 1995   pvm/Book_PVM
-rw-------      1   sarwar    faculty    7639    Sep 11 1998   pvm/Jim_Davis
$
```

5.5.2 Changing File Access Privileges

You can use the chmod command to change access privileges for your files. The following is a brief description of the command.

Syntax:
```
chmod [options] octal-mode file-list
chmod [options] symbolic-mode file-list
```

Purpose: Change/set permissions for files in **file-list**

Commonly used options/features:

-R   Recursively descend through directories changing/setting permissions for all the files and subdirectories under each directory

-f   Force specified access permissions; no error messages are produced if you are the owner of the file

---

TABLE 5.4

Values for Symbolic Mode Components

Who	Operator	Privilege
u User	+ Add privilege	r Read bit
g Group	– Remove privilege	w Write bit
o Other	= Set privilege	x Execute/search bit
a All		u User's current privileges
ugo All		g Group's current privileges
		o Others' current privileges
		l Locking SGID privilege bit
		s Sets user or group ID mode bit
		t Sticky bit

TABLE 5.5

Examples of the chmod Command and Their Purposes

Command	Purpose
`chmod 700 *`	Sets access privileges for all files, including directories, in the current directory to read, write, and execute for the owner, and provides no access privilege to anyone else
`chmod 740 courses`	Sets access privileges for courses to read, write, and execute for the owner and read-only for the group, and provides no access for others
`chmod 751 ~/courses`	Sets access privileges for **~/courses** to read, write, and execute for the owner, read and execute for the group, and execute-only permission for others
`chmod 700 ~`	Sets access privileges for the home directory to read, write, and execute (i.e., search) for the owner, and no privileges for anyone else
`chmod u=rwx courses`	Sets owner's access privileges for **courses** to read, write, and execute and keeps the privileges of group and others at their present values
`chmod ugo-rw sample or chmod a-rw sample`	Does not let anyone read or write **sample**
`chmod a+x sample`	Gives everyone execute permission for **sample**
`chmod g=u sample`	Makes **sample**'s group privileges match its user (owner) privileges
`chmod go= sample`	Removes all access privileges to **sample** for group and others

The *symbolic mode*, also known as *mode control word*, has the form <who><operator> <privilege>, with possible values for "who," "operator," and "privilege" shown in Table 5.4. This table also shows the use of + and – operators in the chmod command to add and remove a permission. How and when the set-group-ID (SGID) privilege bit is set to S or l, is described in Section 5.6.2.

Note that u, g, or o can be used as a privilege with the = operator only. Multiple values can be used for "who" and "privilege," such as ug for the "who" field and rx for the "privilege" field. Some useful examples of the chmod command and their purposes are listed in Table 5.5.

The following session illustrates how access privileges for files can be determined and set. The chmod command is used to change (or set) access privileges, and the ls -l (or ls -ld) command is used to show the effect of the corresponding chmod command. After the chmod 700 courses command has been executed, the owner of the **courses** file has read, write, and execute access privileges for it, and nobody else has any privilege. The chmod g+rx courses command adds read and execute access privileges to the **courses** file for the group; the privileges of the owner and others remain intact. The chmod o+r courses command adds the read access privilege for the **courses** file for others. The chmod a-w * command takes away the write access privilege from all users for all the files in the current directory. The chmod go+x, o+r * command enables the execute permission for group and others, and read permission to others to all the files in the current directory. The chmod 700 [l-t]* command sets the access permissions to 700 for all the files that start with letters "l" through "t," as illustrated by the output of the last `ls -l` command, which shows access privileges for the files **labs** and **temp** changed to 700.

```
$ cd
$ ls -l
drwxr-x---    2        sarwar  faculty 512     Apr   23      09:37   courses
-rwxrwxrwx    1        sarwar  faculty 12      May   01      13:22   labs
-rwxr--r--    1        sarwar  faculty 163     May   05      23:13   temp
$ chmod 700 courses
$ ls -ld courses
drwx------    2        sarwar  faculty 512     Apr   23      09:37   courses
$ chmod g+rx courses
$ ls -ld courses
drwxr-x---    2        sarwar  faculty 512     Apr   23      09:37   courses
$ chmod o+r courses
$ ls -ld courses
drwxr-xr--    2        sarwar  faculty 512     Apr   23      09:37   courses
$ chmod a-w *
$ ls -l
dr-xr-x---    2        sarwar  faculty 512     Apr   23      09:37   courses
-r-xr-xr-x    1        sarwar  faculty 12      May   01      13:22   labs
-r-xr--r--    1        sarwar  faculty 163     May   05      23:13   temp
$ chmod go+x, o+r *
$ ls -l
dr-xr-xr--    2        sarwar  faculty 512     Apr   23      09:37   courses
-r-xr-xr-x    1        sarwar  faculty 12      May   01      13:22   labs
-r-xr-xr--    1        sarwar  faculty 163     May   05      23:13   temp
$ chmod 700 [l-t]*
$ ls -l
dr-xr-x---    2 sarwar         faculty 512     Apr   23      09:37   courses
-rwx------    1 sarwar         faculty 12      May   01      13:22   labs
-rwx------    1 sarwar         faculty 163     May   05      23:13   temp
$ chmod +x *
$ ls -l
dr-xr-x--x    2 sarwar         faculty 512     Apr   23      09:37   courses
-rwx--x--x    1 sarwar         faculty 12      May   01      13:22   labs
-rwx--x--x    1 sarwar         faculty 163     May   05      23:13   temp
$
```

The access permissions for all the files and directories under one or more directories can be set by using the chmod command with the -R option. In the following session, the first command sets access permissions for all the files and directories under the directory called **courses** to 711, recursively. The second command sets access permissions for all the files and directories under **~/personal/letters** to 700, recursively. "Recursively" means by traversing all subdirectories under the specified directories, i.e., **courses** and **~/personal/letters** in these examples.

```
$ chmod -R 711 courses
$ chmod -R 700 ~/personal/letters
$
```

If you specify access privileges with a single octal digit in a chmod command, it is used by the command to set the access privileges for "others"; the access privileges for "user" and "group" are both set to 0 (i.e., no access privileges). If you specify two octal digits in a chmod command, the command uses them to set access privileges for "group" and "others"; the access privileges for "user" are set to 0. In the following session, the first chmod command sets "others" access privileges for the **courses** directory to 7 (rwx) and 0 (---) for owner and group. The second chmod command sets "group" and "others" access privileges for the **personal** directory to 7 (rwx) and 0 (---), respectively, and no access rights for the file owner. The ls -l command shows the results of these commands.

```
$ chmod 7 courses
$ chmod 70 personal
$ ls -l
d------rwx   2       sarwar  faculty 512      Nov     10      09:43   courses
d---rwx---   2       sarwar  faculty 512      Nov     10      09:43   personal
drw-------   2       sarwar  faculty 512      Nov     10      09:43   sample
$
```

5.5.3 Access Privileges for Directories

As previously stated, the read permission on a directory allows you to read the contents of the directory (recall that the contents of a directory are the names of files and directories in it), the write permission allows you to create a file in the directory or remove an existing file or directory from it, and the execute permission for a directory is permission for searching the directory. It is important to note that read and write permissions on directories are not meaningful without the search permission. So, you must have both read and execute permissions on a directory to be able to list its contents. Similarly, you must have both write and execute permissions on a directory to be able to create a file in it.

In the following session, the write permission for the directory **courses** has been turned off. Thus, you cannot create a subdirectory **ee345** in this directory by using the mkdir command or copy a file **foo** into it. Similarly, as you do not have search permission for the directory **personal**, you cannot use the cd command to enter (change directory to) this directory. If the directory **sample** had a subdirectory, say **foobar**, for which the execute permission was turned on, you still could not change directory to **foobar**, because search permission for **sample** is turned off. Finally, as read permission for the directory **personal** is turned off, you cannot display the names of files and directories in it by using the ls command, even though search permission on it is turned on.

```
$ chmod 600 sample
$ chmod 500 courses
$ chmod 300 personal
$ ls -ld courses personal sample
dr-x------ 2 sarwar  faculty  512 Aug  4 06:36 courses
d-wx------ 2 sarwar  faculty  62 Aug  4 06:36 personal
drw------- 2 sarwar  faculty  88 Aug  4 06:36 sample
$ mkdir courses/ee345
mkdir: courses/ee345: Permission denied
$ cp foo courses
cp: courses/foo: Permission denied
$ cd sample
cd: sample: Permission denied
$ ls -l personal
total 0
ls: personal: Permission denied
$
```

The next session shows that simply having read or write permission on a directory is not sufficient to read its contents (e.g., display them with the ls command) or create a file or directory in it. For example, the directory **dir1** has write permission turned on, but you cannot copy the **prog1.cpp** file into it, because search permission on it is turned off. Similarly, you cannot remove the file **f1** from **dir2**. After you turn on its search permission with the chmod u+x dir2 command, you can remove the file **f1**.

```
$ ls -ld dir?
d-w------- 2 sarwar  faculty  2 Aug  4 06:59 dir1
d-w------- 2 sarwar  faculty  3 Aug  4 06:59 dir2
$ cp prog1.cpp dir1
cp: dir1/prog1.cpp: Permission denied
$ rm dir2/f1
rm: dir2/f1: Permission denied
```

```
$ chmod u+x dir2
$ ls -ld dir2
d-wx------  2 sarwar  faculty  3 Aug  4 06:59 dir2
$ rm dir2/fl
$
```

The following in-chapter exercises ask you to use the chmod and ls -ld commands to see how they work, and to enhance your understanding of Linux file access privileges.

Exercise 5.2

Create three directories called **courses**, **sample**, and **personal** by using the mkdir command. Set access permissions for the directory **sample** so that you have all three privileges, users in your group have read access only, and the other users of your system have no access privileges. Show your work.

Exercise 5.3

Use the chmod o+r sample command to allow others read access to the directory **sample**. Use the ls -ld sample command to confirm that read permission for **sample** has been enabled for others.

Exercise 5.4

Use the session preceding these exercises to understand fully how the read, write, and execute permissions work for directories. Run the session on your system and verify results.

5.5.4 Default File Access Privileges

When a new file or directory is created, Linux sets its access privileges based on the current *mask* value. On Linux systems, the default access privileges for the newly created files and directories are 777 for executable files and directories and 666 for text files. However, these default permissions may be different depending on the value of the mask set on your system. If a mask bit is 1, the corresponding permission will be disabled, and if it is 0, the permission will be determined by the system using a Boolean logic expression involving the current value of the mask and the default permissions. We describe this expression later in this section.

You can display the current mask value or set it to a new value by using the umask command. The following is a brief description of the command.

Syntax:
umask [-S] [mask]

Purpose: Set access permission bits for newly created files and directories to 0, if the corresponding bit in the mask is set to 1. Other permission bits are determined by using a Boolean logic expression involving the values of the mask and default permissions. Without any argument, the command displays the current value of the mask. With the -S option, the command displays the mask value in a symbolic form (see the following paragraphs).

When the command is executed without an argument, it displays the current value of the bit mask in octal, as is shown in the first command of the following session. The rightmost nine bits (i.e., the rightmost three octal digits) are for user, group, and others, and the leftmost three bits (i.e., the leftmost octal digit) are for special access bits described in Section 5.6. The umask -S command displays the symbolic value of the mask, showing the access privileges that will be set for a newly created directory or executable file for user, group, and others.

```
$ umask
0022
$ umask -S
u=rwx,g=rx,o=rx
$
```

When used with a mask as an argument, the umask command can be used to set the mask value. The mask value may be specified as a three-digit or four-digit octal number. In the following session, we show how the umask command can be used without and with the –S option.

```
$ umask 077
$ umask
0077
$ umask -S u=rwx,g=,o=
$ umask
0077
$
```

The umask command is normally placed in the system startup file ~/.**profile** in Linux, including Mint 18, so that it executes every time you log on to the system.

The argument of umask is a bit mask, specified in octal, that identifies the permission bits that are to be turned *off* when a new file is created. The values of other access permission bits are computed by the Boolean expression:

```
A = B AND C' = BC'
```

Here, A is the file access permission assigned to a newly created file or directory, B is the default access permission (777 for a directory or executable file and 666 for a text file), and C is the current mask value. C' (pronounced "NOT C" or "C prime") is called *negation*, or 1's complement of C. The 1's complement of a binary number is obtained by replacing 1s with 0s and vice versa. For example, the 1's complement of the four-bit binary number 1011 is 0100. The bitwise Boolean function AND of two binary variables returns 1 if and only if both the bits are 1; it returns 0 otherwise. While computing A in the earlier equation, the AND function compares the respective bits in B and C' and returns a 1 if and only if both bits are 1; otherwise it returns 0.

We now show a few examples how Linux assigns file access permissions to newly created files and directories for a given mask. We determine the access permissions for a newly created directory or executable file for the mask value 022 by using the earlier Boolean expression.

```
C = 022         = 000 010 010
C'              = 111 101 101
B = 777         = 111 111 111
A = B AND C'    = 111 101 101 = 755 (octal) = 111101101 (binary)
                                              rwxr-xr-x (symbolic)
```

Thus, file access permissions are 755 (read, write, execute for user, read and execute for group and others). For a text file, B is 666. Thus, access permissions for a newly created text file would be 644 as follows:

```
C'              = 111 101 101
B = 666         = 110 110 110
A = B AND C'    = 110 100 100 = 644 (octal) = 110100100 (binary)
                                              rw-r--r-- (symbolic)
```

The access permissions for a newly created text file for the mask value 077 would be 600, as follows:

```
C = 077         = 000 111 111
C'              = 111 000 000
B = 666         = 110 110 110
A = B AND C'    = 110 000 000 = 600 (octal) = 110000000 (binary)
                                              rw------- (symbolic)
```

We now do a Bourne shell session to verify the previous results. Note that with the mask set to 022, the permissions for the newly created directory (**labs**) and executable file (**hello**) are 755 (rwxl-xr-x), and are 644 (rw-r--r--) for the newly created text file, **tempfile**, as calculated earlier. With the mask set to 077, the permissions for the directory (**lectures**) and executable file (**greeting**) are 700 (rwx------), and are 600 (rw   ) for the text file (**textfile**).

```
$ umask 022
$ mkdir labs
$ cc hello.c -o hello
$ cat > tempfile
date
pwd
$ ls -ld labs hello tempfile
-rwxr-xr-x  1 sarwar  faculty  6945 Aug  5 22:27 hello
drwxr-xr-x  2 sarwar  faculty  2    Aug  5 22:26 labs
-rw-r--r--  1 sarwar  faculty  9    Aug  5 22:30 tempfile
$ umask 077
$ mkdir lectures
$ cc hello.c -o greeting
$ cat > testfile
date
echo "Hello, world!"
$ ls -ld lectures greeting testfile
-rwx------  1 sarwar  faculty  6945 Aug  5 22:28 greeting
drwx------  2 sarwar  faculty  2    Aug  5 22:27 lectures
-rw-------  1 sarwar  faculty  26   Aug  5 22:29 testfile
$
```

The authors prefer a mask value of 077 so that their new files are always created with full protection in place, that is, files have full access permissions for the owner and no permissions for anyone else. This mask allows you to have a completely private system, not allowing other users to read, write, or execute your files or read, write, or search your directories. Recall that you can change access privileges for files on an as-needed basis by using the chmod command. Another common umask value is 027, which gives default privileges to group members and no permissions to others.

Exercise 5.5 asks you to use the umask command to determine the current file protection mask.

Exercise 5.5

> Run all the shell sessions shown and discussed in this section to make sure that they work on your system too and to understand the use of the umask command works and how the Linux system decides about the access permissions on newly created files and directories.

5.6 Special Access Bits

In addition to the nine commonly used access permission bits described in this chapter, three additional bits are of special significance. These bits are known as the set-user-ID (SUID) bit, SGID bit, and sticky bit.

5.6.1 SUID Bit

We have previously shown that the external shell commands have corresponding files that contain binary executable codes or shell scripts. The programs contained in these files are not special in any way in terms of their ability to perform their tasks. Normally, when a command executes, it does so under the access privileges of the user who issues the command, which is how the access privileges system described in this chapter works. However, a number of Linux commands need to write to files that are otherwise protected from users who normally run these commands. An example of this file is **/etc/passwd**, the file that contains a user's login information (see Chapter 4). Only a superuser is allowed to write to this file to perform tasks such as adding a new login and changing a user's group ID. However, Linux users are normally allowed to execute the passwd command to change their passwords. Thus, when a user executes the passwd command, the command changes the user password in the **/etc/passwd** file on behalf of the user who runs this command. The problem is that we want users to be able to change their passwords, but at the same time, they must not have write access to the **/etc/passwd** file to keep information about other users in this file from being compromised. On Linux Mint, it is the nonreadable master password file **/etc/shadow** that changes.

As previously stated, when a command executes, it runs with the privileges of the user running the command. Another way of stating the same thing is that, when a command runs, it executes with the *effective user ID* of the user running the command. Linux has an elegant mechanism that solves the problem stated in the preceding paragraph—and many other similar security problems—by allowing commands to change their effective user ID and become privileged in some way. This mechanism allows commands such as passwd to perform their work, yet not compromise the integrity of the system. Every Linux file has an additional protection bit, called the SUID bit, associated with it. If this bit is set for a file containing an executable program, the program takes on the privileges of the owner of the file when it executes. Thus, if a file is owned by **root** and has its SUID bit set, it runs with superuser privileges. This bit is set, for example, for the passwd command. So, when you run the passwd command, it can write to the **/etc/passwd** file (replacing your existing password with the new password), even though you do not have access privileges to write to the file.

Several other Linux commands require **root** ownership and the SUID bit set, because they access and update operating system resources (files, data structures, etc.) that an average user must not have permissions for. Some of these commands are lp, mail, mkdir, mv, and ps. The authors of computer game software that maintains a scores file can make another use of the SUID bit. When the SUID bit is set for such software, it can update the scores file when a user plays the game, although the same user cannot update the scores file by explicitly writing to it.

The SUID bit is enabled if the execute bit for the owner is **s** (or **S**). If both SUID and execute bits are enabled, the bit is displayed as s in the output of the ls −l command. If SUID bit is enabled but execute bit is disabled, the bit is displayed as S by the ls −l command. The following session shows an example of each case.

```
$ ls -l cp.new foo
-rwSr--r-- 1 sarwar   faculty  14 Aug  4 23:38 cp.new
-r-sr-xr-x 1 sarwar   faculty  30 Aug  5 05:27 foo
$
```

The chmod command with the following syntax may be used to set the SUID bit.

Syntax:
    chmod 4xxx file-list
    chmod u+s file-list

Purpose: Change/set the SUID bit for files in **file-list**

Here, xxx is the octal number that specifies the read, write, and execute permissions for user, group, and others, and the octal digit 4 (binary 100) is used to set the SUID bit. When the SUID bit is set, the execute bit for the user is set to s (lowercase) if the execute permission is already set for the user; otherwise, it is set to S (uppercase). The following session illustrates the use of these command syntaxes. The first, the ls −l cp.new command, is used to show that the execute permission for the **cp.new** file is set. The chmod 4710 cp.new command is used to set the SUID bit and other nine bits of permission to octal 710. The second, the ls −l cp.new command, shows that the x bit value has changed to s. The two subsequent chmod commands are used to set the SUID and execute bits to 0. The ls −l cp.new command is used to show that execute permission has been taken away from the owner. The chmod u+s cp.new command is used to set the SUID bit again, and the last ls −l cp.new command shows that the bit value is S (uppercase) because the execute bit was not set before setting the SUID bit.

```
$ ls -l cp.new
-rwx--x--- 1 sarwar   faculty  14 Aug  4 23:26 cp.new
```

```
$ chmod 4710 cp.new
$ ls -l cp.new
-rws--x--- 1 sarwar  faculty  14 Aug  4 23:26 cp.new
$ chmod u-s cp.new
$ chmod u-x cp.new
$ ls -l cp.new
-rw---x--- 1 sarwar  faculty  14 Aug  4 23:26 cp.new
$ chmod u+s cp.new
$ ls -l cp.new
-rwS--x--- 1 sarwar  faculty  14 Aug  4 23:26 cp.new
$
```

Although the idea of the SUID bit is sound, it can compromise the security of the system if not implemented correctly. For example, if the permissions of any SUID program are set to allow write privileges to others, you can change the program in this file or overwrite the existing program with another program. Doing so would allow you to execute your (new) program with superuser privileges.

5.6.2 SGID Bit

The SGID bit works in the same manner in which the SUID bit does, but it causes the access permissions of the process to take the group identity of the group to which the owner of the file belongs. This feature is not as dangerous as the SGID feature, because most privileged operations require superuser identity, regardless of the current group ID. The SGID bit is enabled if the execute bit for the group is displayed as s (or S) by the ls -l command. If both SGID and execute bits are enabled, the bit is displayed as s in the output of the ls -l command. If SGID bit is enabled but the execute bit is disabled, the bit is displayed as S by the ls -l command. The following session shows an example of each case.

```
$ ls -l cp.new foo
-rw-r-Sr-- 1 sarwar  faculty  14 Aug  4 23:38 cp.new
-r-xr-sr-x 1 sarwar  faculty  30 Aug  5 05:27 foo
$
```

Using either of the following two command syntaxes can set the SGID bit.

Syntax:
    `chmod 2xxx file-list`
    `chmod g+s file-list`
Purpose: Change/set the SGID bit for files in **file-list**

Here, xxx is the octal number specifying the read, write, and execute permissions for the files in **file-list**, and the octal digit 2 (binary 010) specifies that the SGID bit is to be set for the same files. The following session on Linux Mint illustrates the use of these command syntaxes. The command chmod 2751 cp.new sets the SGID bit for the **cp.new** file and sets its access privileges to 751 (rwxr-x--x). The rest of the commands are similar to those in Section 5.6.1.

```
$ ls -l cp.new
-rwxr-x--x 1 sarwar  faculty  14 Aug  4 23:26 cp.new
$ chmod 2751 cp.new
$ ls -l cp.new
-rwxr-s--x 1 sarwar  faculty  14 Aug  4 23:26 cp.new
$ chmod g-s cp.new
$ chmod g-x cp.new
$ ls -l cp.new
-rwxr----x 1 sarwar  faculty  14 Aug  4 23:26 cp.new
$ chmod g+s cp.new
$ ls -l cp.new
-rwxr-S--x 1 sarwar  faculty  14 Aug  4 23:26 cp.new
$
```

You can set or reset the SUID and SGID bits by using a single chmod command. Thus, the command chmod ug+s cp.new can be used to perform this task on the **cp.new** file. You can also set the SUID and SGID bits along with the access permissions bits (read, write, and execute) by preceding the octal number for access privileges by 6 because the leftmost octal digit 6 (110) specifies that both the SUID and SGID bits be set. Thus, the command chmod 6754 cp.new may be used to set the SUID and SGID bits for the **cp.new** file and its access privileges to 754. If SGID is set on a directory, directories created in that directory will be owned not by the group of the user who created the directory, but by the group owner of the parent directory.

The ls -l command displays l in the seventh permission bit if your operating system and file system support mandatory file locking. The value l indicates that mandatory file locking is enabled for this file. Read the man page for the fcntl() system call and Chapter W21 for details on advisory (or discretionary) and mandatory file locks.

5.6.3 Sticky Bit

The last of the 12 access bits, the sticky bit, is on if the execute bit for others is t (or T), as in the case **/tmp** given as follows:

```
$ ls -l / | grep tmp
drwxrwxrwt  14 root root  4096 Jul  1 15:17 tmp
$
```

The sticky bit can be set for a directory to ensure that an unprivileged user cannot remove or rename files of other users in that directory. You must be the owner of a directory or have appropriate permissions to set the sticky bit for it. Some systems do not allow nonsuperusers to set the sticky bit. It is commonly set for shared directories that contain files owned by several users.

The output of the earlier command shows that the **/tmp** directory, owned by **root**, has the sticky bit on. It has read, write, and execute permissions for everyone. It means that any user can create a file or directory in **/tmp**. However, because the sticky bit is on, no user (other than the superuser) can remove a file or directory that he/she does not own. Thus, Linux allows you to remove any files in **/tmp** that you own, but no other file or directory, because you are not the owner of those files and directories.

Originally, this bit was designed to inform the kernel that the code segment of a program is to be shared or kept in the main memory or the *swap space* owing to the frequent use of the program. Thus, when this bit is set for a program, the system tries to keep the executable code for the program (the code segment only) in memory after it finishes execution—the processes literally "stick around" in the memory. If, for some reason, memory space occupied by this program is needed by the system for loading another program, the program with the sticky bit on is saved in the swap space (a special area on the disk used to save processes temporarily). That is, if the sticky bit is set for a program, the code segment of the program is either kept in memory or in the swap space after it finishes its execution. When this program is executed again, with the program in memory, its execution starts right away. If the program is in the swap space, the time needed for loading it is much shorter than if it were stored on disk as a Linux file. The advantage of this scheme, therefore, is that if a program with the sticky bit on is executed frequently, it is executed much more quickly.

This facility is useful for programs such as compilers, assemblers, editors, and commands such as ls and cat, which are frequently used in a typical computer system environment. However, care must be taken that not too many programs have this bit set. Otherwise, system performance will suffer because of the lack of free space, with more and more space being used by the programs whose sticky bit is set. This historical use of the sticky bit is no longer needed in newer Linux systems because virtual memory systems use page replacement algorithms that do not remove recently used program pages/segments. Linux Mint allows you to set the sticky bit on nondirectory files, as shown in the shell session at the end of this section.

Either of the following syntaxes may be used to set the sticky bit.

Syntax:
```
chmod 1xxx file-list
chmod +t file-list
```
Purpose: Change/set the sticky bit for files in **file-list**

Here, xxx is the octal number specifying the read, write, and execute permissions, and the octal digit 1 (binary 001) specifies that the sticky bit is to be set. When the sticky bit is set, the execute bit for others is set to t if others already has execute permission; otherwise, it is set to T. The following session on Linux Mint illustrates the use of these command syntaxes. The command chmod 1751 cp.new sets the sticky bit for the **cp.new** file and set its access privileges to 751. As the outputs of these commands show, Linux Mint allows you to enable the sticky bit for a nondirectory file. It also allows you to set this bit for **dir1**, a directory, with the command chmod +t dir1. Linux Mint allows you to set the sticky bit for nondirectory files as well as directories. The explanations for the lowercase t and uppercase T in the following session are the same as for the lowercase s and uppercase S for SUID and SGID bits.

```
$ ls -l cp.new
-rw-rw-rw- 1 sarwar  faculty  14 Aug  4 23:38 cp.new
$ ls -ld dir1
drwxr-xr-x 2 sarwar  faculty   2 Aug  5 07:59 dir1
$ chmod 1751 cp.new
$ ls -l cp.new
-rwxr-x--t 1 sarwar  faculty  14 Aug  4 23:38 cp.new
$ chmod +t dir1
$ ls -ld dir1
drwxr-xr-t 2 sarwar  faculty   2 Aug  5 07:59 dir1
$ chmod o-x dir1
$ ls -ld dir1
drwxr-xr-T 2 sarwar  faculty   2 Aug  5 07:59 dir1
$
```

Exercise 5.6

Run all the shell sessions shown and discussed in this section to make sure that they work on your system too and to understand how the chmod command is used to set and reset special access permission bits (SUID, SGID, and sticky) on your Linux system. Also, find out if the SGID and sticky bit may be set on your system for both files and directories.

Summary

A time-sharing system has to ensure protection of one user's files from unauthorized (accidental or malicious) access by other users of the system. Linux provides several mechanisms for this purpose, including one based on access permissions. Files can be protected by informing the system of the type of operations (read, write, and execute) that are permitted on the file by the owner, group (the users who are in the same group as the owner), and others (everyone else on the system). Linux allows a user to be part of multiple groups. Only the system administrator, also known as the superuser in Linux jargon, can add you to a group or remove you from a group. You can display the groups you (or any user) are a member of by using the groups command. These nine commonly used access permissions are represented by bits. This information is stored in the inode of the file. When a user tries to access a file, the system allows or disallows access based on the file's access privileges stored in the inode.

Access permissions for files can be viewed by using the ls -l command. When used with directories, this command displays attributes for all the files in the directories. The ls -ld command can be used to view access permissions for directories. The owner of a file can change access privileges on it by using the chmod command. The umask command, which is usually placed in the ~/.profile file in Linux, allows the user to specify a bit mask that informs the system of access permissions that are disabled

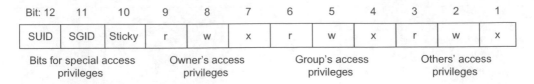

Bit: 12	11	10	9	8	7	6	5	4	3	2	1
SUID	SGID	Sticky	r	w	x	r	w	x	r	w	x

Bits for special access privileges	Owner's access privileges	Group's access privileges	Others' access privileges

FIGURE 5.2   Position of access privilege bits for Linux files as specified with the chmod command.

for the user, group, and others. When a file is created by the Linux system, it sets access permissions for the file according to the bit mask and the default access permissions for directories, executable files, and text files. File access permissions assigned to newly created files (A) are given by the expression A=B AND C', where B is the default privileges for the file, C is the value of the bit mask, and AND is the Boolean function for bitwise AND. In a typical system, the mask is set to 022 for default permissions on newly created files and directories. For a fully secure system, it should be set to 077. You can use the umask command to display the value of current flag in octal and umask –S command to display the flag in a symbolic form.

Linux also allows three additional bits—SUID, SGID, and sticky bit—to be set. The SUID and SGID bits allow the user to execute commands, including passwd, ls, mkdir, and ps, which access important system resources to which access is not allowed otherwise. The sticky bit can be set for a directory to ensure that an unprivileged user cannot remove or rename files of other users in that directory. Only the owner of a directory, or someone else having appropriate permissions, can set the sticky bit for the directory. It is commonly set for shared directories that contain files owned by several users, such as / tmp in the Linux file system structure. Historically, the sticky bit has served another purpose. It can be set for frequently used utilities so that Linux keeps them in the main memory or on a fixed area on the disk, called the swap space, after their use. This feature makes subsequent access to these files much faster than if they were to be loaded from the disk as normal files. However, due to the implementation of virtual memory systems using advanced algorithms of demand paging, demand segmentation, and paged segmentation on most modern multiuser time-sharing systems, the use of the sticky bit to keep a program memory or swap space resident is no longer required. Its use for directories remains useful and necessary. Therefore, all modern Linux systems support the sticky bit for directories, but support for the sticky bit for files is no longer maintained across the board. For example, Linux Mint does provide support for the sticky bit for nondirectory files.

The final format of the 12 access permissions bits, as used in the chmod command, is shown in Figure 5.2.

Questions and Problems

1. What are the three basic file protection schemes available in Linux?
2. List all possible two-letter passwords comprising digits and punctuation letters.
3. If a computer system allows six-character passwords comprising a random combination of decimal digits and punctuation marks, what is the maximum number of passwords that a user will have to try with the brute force method of breaking into a user's account? Why?
4. What is the maximum number of passwords that can be formed if a system allows digits, uppercase and lowercase letters, and seven punctuation marks (, ; : ! ' . and ?) to be used? Assume that passwords must be 12 characters long.
5. Suppose that a hacker is trying to guess a password—consisting of eight characters—using uppercase letters, lowercase letters, and digits. Further, suppose that the system forces a 5 s delay after each password guess. How long will it take the hacker to guess the password in the worst-case analysis? Repeat the exercise if we are allowed to use five punctuation marks as characters in a password. Why? Show all your work.

6. How does file protection based on access permissions work? Base your answer on various types of users of a file and the types of operations they can perform. How many permission bits are needed to implement this scheme? Why?

7. How do the read, write, and execute permissions work in Linux? Illustrate your answer with some examples.

8. How many user groups exist on your system? How did you get your answer? What groups are you a member of and what is your default group? How many groups is root a member of on your system? How did you obtain your answer? If you used any commands, show the commands and their outputs.

9. Create a file **test1** in your present working directory and set its access privileges to read and write for yourself, read for the users in your group, and none to everyone else. What command did you use to set privileges? Give another command that would accomplish the same thing.

10. The user **sarwar** sets access permissions to his home directory by using the command chmod 700 $HOME. If the file **cp.new** in his home directory has read permissions of 777, can anyone read this file? Why or why not? Explain your answer.

11. What is the effect of each command? Explain your answers.

   a. chmod 776 ~/lab5
   b. chmod 751 ~/lab?
   c. chmod 511 *.c
   d. chmod 711 ~/*
   e. ls -l
   f. ls -ld
   g. ls -l ~/personal
   h. ls -ld ~/personal

12. What is the effect of each command? Explain your answers.

   a. chmod u+rw,g-r,o-x ~/lab5
   b. chmod u+rw,g-r,o-x ~/lab?
   c. chmod u+x,g-wx *.c
   d. chmod ugo+x go-w ~/*
   e. chmod +x *

13. What does the execute permission mean for a directory, a file type for which execute operation makes no sense? Explain with an example.

14. Create a file **dir1** in your home directory and use cp /etc/passwd dir1/mypasswd command to copy the **/etc/passwd** file into it. Use the chmod command to have only the search permission on for it and execute the following commands. What is the result of executing these commands? Do the results make sense to you? Explain.

   a. cd dir1
   b. ls
   c. rm dir1/mypasswd
   d. cp /etc/passwd dir1

15. What umask command should be executed to set the permissions bit mask to 037? With this mask, what default access privileges are associated with any new file that you create on the system? Why? Where would you put this command so that every time you log on to the system this mask is effective?

16. Give a command line for setting the default access mode so that you have read, write, and execute privileges, your group has read and execute permissions, and all others have no permission for a newly created executable file or directory. How would you test it to be sure that it works correctly?

17. Give chmod command lines that perform the same tasks as the mesg n and mesg y commands. (*Hint*: Every hardware device, including your terminal, has an associated file in the **/dev** directory.)

18. What are the purposes of the SUID, SGID, and sticky bits?

19. Give one command line for setting all three special access bits (SUID, SGID, and sticky) for the file **cp.new**. (*Hint*: Use octal mode.)

20. In a Linux system, the cat command is owned by root and has its SUID bit set. Do you see any problems with this setup? Explain your answer.

21. Some Linux systems do not allow users to change their passwords with the passwd command. How is this restriction enforced? Is it a good or bad practice? Why?

22. Calculate the file access permissions assigned to newly created directories, executable files, and text files for bit mask 027. Show all your work.

23. Describe briefly the purpose of each of the following commands. Run these commands on your system after creating **~/prog1** (a file) and **~/dir1** (a directory) on it and show the outputs of the commands to verify your answers. If your system does not support any command on your system, explain why you think it does so.

    a.   chmod 4776 ~/prog1

    b.   chmod 1776 ~/prog1

    c.   chmod 6776 ~/prog1

    d.   chmod g+s ~/prog1

    e.   chmod +t ~/prog1

    f.   chmod +t ~/dir1

    g.   chmod ugo-r ~/prog1

    h.   chmod a-rw ~/prog1

    i.   chmod ug+x ~/dir1

    j.   chmod go= ~/dir1

    k.   chmod u=rwx ~/prog1

    l.   chmod g=o ~/dir1

    m.   chmod o-wx ~/dir1

Advanced Questions and Problems

24. How many discrete "fields" are there in a single entry in the /etc/passwd file on your Linux system, and what is the exact meaning of each field in that entry? What information do the fields in your user entry in the **/etc/passwd** file give you? What is the purpose of the "nobody" account that has an entry in the **/etc/passwd** file? Can you log into your Linux system as "nobody," and if you cannot, why?

25. When does the ls -l command display s, S, and l for SUID and SGID bits for files? Show a Bash session to illustrate your answers.

26. If SGID is set on a directory, D1, the group owner of the parent directory will own directories created in D1. Show a shell session to show that this statement is true.

27. What is the purpose of sticky bit for directories and files? When does the ls -l command display t and T for sticky bit for a directory or file? Show a Bash session to illustrate your answers.

    To extend your knowledge of discretionary access control beyond the traditional Linux permission bits, as have shown them in this chapter, work through the material presented in Chapter W26, Sections 9.3.1–9.3.3 at the book website. Then do the following problems

28. If you give a set of user permissions to a project directory using POSIX.1e ACLs, how can you ensure that subdirectories that are created by the project manager beneath that project directory provide the same access privileges to those users?

29. Create a project directory on your system and create a git repository in it for any number of local users on your computer system. Then, use POSIX.1e ACLs to give access to the project directory to the users that are collaborating in the project. This should allow those users to push to and pull from the git repository. Have your allowed users test the repository. Also test the security of the repository, i.e., can nonallowed users access it? See Chapter W24, Section 5.7, at the book website, to obtain complete information on creating a git repository.

Projects

Project 1

Mansoor is working with Bob on a project. He needs to read, write, create, and delete files related to the project, which are located in the Project directory in Bob's home directory. Bob and Mansoor are ordinary users without administrative privileges. They wish to do this project without contacting the system administrator to request new groups, group membership changes, sudo changes, etc. When the project is over, Bob will remove the modify permissions on his home and the Project directory for user "Mansoor" himself, instead of contacting the system administrator. They also must work solely on the same computer, using the default locally mounted disk drives.

On your own Linux system, possibly in conjunction with another user, use POSIX.1e ACLs to accomplish the following (substituting valid usernames on your system for Bob and Mansoor):

a. Create a project directory under Bob's home directory named "Project."

b. Set the POSIX.1e ACL on Bob's home directory, so Mansoor has read, write, and execute privileges on it.

c. Set the POSIX.1e ACL on the Project directory so that Mansoor has rwxo privileges on it.

d. Have Bob created some files in the Project directory.

e. Have Mansoor made Bob's home directory the current directory.

f. Have Mansoor tested whether he can

   i. delete files in Bob's home directory,

   ii. delete the Project directory from Bob's home directory,

   iii. list, create new files, or remove the files that Bob put in the Project directory.

g. Have Bob revoked Mansoor's x privileges on Bob's home directory and the directory Project.

h. Have Mansoor tested the revocation of modify privileges from step g.

i. Why can Mansoor still see the files in Bob's home directory, and the files in the **Project** directory, but not delete or modify any files in those directories after step g?

j. What chmod command(s) would Bob have to execute to deny Mansoor access to his home directory?

*Show verification of POSIX.1e ACL settings at as many steps as necessary to validate what you have done.

Project 2

In addition to setting traditional permission bits as shown in this chapter, *Access Control* security measures can be usefully grouped together into the general categories of discretionary, mandatory, or role-based.

Linux *Control Group (cgroup)* and *namespace* process isolation are system security techniques applied at the process level that can powerfully augment the traditional permission bit access controls. Various forms of cgroup and namespace techniques are employed effectively in virtualization methodologies, such as in LXC/LXD or Docker. But they can also be used in a much "lighter weight" way to "sandbox" your working environment without the overhead of creating the "heavier weight" forms of operating system virtualization implicit in those methodologies.

Create a verbose report on the forms of cgroup and namespace isolation in general, and more specifically, how the /mnt form of Linux namespaces can be used to securely isolate directories and files. Exactly why would you want to use this technique, instead of setting the traditional Linux permission bits and UID, GID, EUID, etc., granting sudo privileges, or manipulating POSIX.1e/ NFSv4 ACL's on directories and files? What exclusive Linux system call(s) do cgroups and namespaces rely upon? Which general category of access control does cgroups and namespaces belong to, if any? In your report, provide a working example of /mnt namespace isolation that can be executed from the Bash command line.

Looking for more? Visit our sites for additional readings, recommended resources, and exercises.

CRC Press e-Resource: https://www.crcpress.com/9781138710085

Authors' GitHub: https://github.com/bobk48/linuxthetextbook

6

Basic File Processing

OBJECTIVES

- To discuss how to display the contents of a file
- To explain copying, appending, moving/renaming, and removing/deleting files
- To describe how to determine the size of a file
- To discuss commands for comparing files
- To describe how to combine files
- To discuss printer control commands
- To cover the following commands and primitives:

  ```
  >, >>, ^, ~, [ ], *, ?, cancel, cat, cksum, cp, crc32, diff, diff3, head, lp,
  lpc, lpq, lpr, lprm, lpstat, lptest, less, ls, md5sum, more, mv, nl, patch, pg,
  pr, rm, sdiff, shasum, tail, uniq, vimdiff, wc
  ```

6.1 Introduction

This chapter describes how some basic file operations can be performed in Linux. These operations are primarily for nondirectory files, although some are applicable to directories as well; we previously discussed the most commonly used directory operations in Chapter 4. When discussing the file operations in this chapter, we also describe related commands and give examples to illustrate how these commands can be used to perform the needed operations. Remember, complete information on a particular command is available via the man command.

6.2 Viewing Contents of Text Files

You view files to identify their contents. You can use several Linux commands to display contents of text files on the display screen. These commands differ from each other in terms of the amount of the file content displayed, the portion of file contents displayed (initial, middle, or last part of the file), and whether the file's contents are displayed one screen or one page at a time. Recall that you can view only those files for which you have the read permission. In addition, you must have the execute (search) permission for all the directories involved in the pathname of the file to be displayed. Viewing does not mean edit, write, or update—just view.

6.2.1 Viewing Complete Files

You can display the complete contents of one or more files on screen by using the cat command. However, because the command does not display file contents one screen or one page at a time, you see only the last page of a file that is larger than one page (i.e., one screen) in size. The following is a brief description of the cat command.

Syntax:
```
cat [options] [file-list]
```

Purpose: Concatenate/display the files in `file-list` on standard output (screen by default), one after another

Output: Contents of the files in `file-list` displayed on the screen, one file at a time

Commonly used options/features:

-   **-e** Display $ at the end of each line; works in conjunction with the –v option
-   **-n** Put line numbers with the displayed lines
-   **-t** Display tabs as ∧ɪ and form feeds as ∧ʟ
-   **-v** Display nonprintable characters, except for the tab, form feed, and newline characters

---

Here, `file-list` is an optional argument that consists of the pathnames for one or more files, separated by spaces. For example, the following command displays the contents of the **student_records** file in the present working directory. If the file is larger than one page, the file contents quickly scroll off the display screen.

```
$ cat student_records
Jonh    Doe     ECE     3.54
Pam     Meyer   CS      3.61
Jim     DAVIS   CS      2.71
Jason   Kim     ECE     3.97
Amy     Nash    ECE     2.38
$
```

The following command displays the contents of files **lab1** and **lab2** in the directory **~/courses/ee446 /labs**. The command does not pause after displaying the contents of **lab1**.

```
$ cat ~/courses/ee446/labs/lab1 ~/courses/ee446/labs/lab2
[ contents of lab1 and lab2 ]
$
```

As discussed in Chapter 4, shell metacharacters can be used to specify file names. The contents of all the files in the current directory can be displayed by using the `cat *` command. The `cat exam?` command displays contents of all the files in the current directory starting with the string **exam** and followed by one character such as **exam1**. The contents of all the files in the current directory starting with the string **lab** can be displayed by using the `cat lab*` command.

As indicated by the command syntax, `file-list` is an optional argument. Thus, when the `cat` command is used without any arguments, it takes input from standard input, one line at a time, and sends it to standard output. Recall that, by default, standard input for a command is the keyboard, and standard output is the display screen. Therefore, when the `cat` command is executed without an argument, it takes input from the keyboard and displays it on the screen one line at a time. The command terminates when the user presses <Ctrl+D>, the Linux end-of-file, on a new line. As is the case throughout the book, the text typed by the user is shown in **bold**.

```
$ cat
This is a test.
This is a test.
In this example, the cat command takes input from stdin (keyboard)
In this example, the cat command takes input from stdin (keyboard)
and sends it to stdout (screen), one line at a time.
and sends it to stdout (screen), one line at a time.
However, this is not a typical use of this commend. It is normally
However, this is not a typical use of this commend. It is normally
```

```
used to display contents of a file, one line at a time, until it
used to display contents of a file, one line at a time, until it
encounters the end-of-file marker. When the cat command reads input from stdin,
encounters the end-of-file marker. When the cat command reads input from stdin,
<Ctrl+D> is the end-of-file marker, as shown below.
<Ctrl+D> is the end-of-file marker, as shown below.
<Ctrl+D>
$
```

At times, you need to view a text file with line numbers. You typically need to do so when, during the software-development phase, a compilation of your source code results in compiler errors having line numbers associated with them. The Linux utility nl allows you to display lines of text files with line numbers. Thus, the nl student_records command displays the lines in the **student_records** file with line numbers, as shown in the following session. The cat -n student_records command can also perform the same task.

```
$ nl student_records
     1  Jonh    Doe     ECE     3.54
     2  Pam     Meyer   CS      3.61
     3  Jim     DAVIS   CS      2.71
     4  Jason   Kim     ECE     3.97
     5  Amy     Nash    ECE     2.38
$
```

Also, if you need to display files with a time stamp and page numbers, you can use the pr utility. It displays file contents as the cat command does, but it also partitions the file into pages and inserts a header for each page. The page header contains today's date, current time, file name, and page number. The pr command, like the cat command, can display multiple files, one after the other. The following session illustrates a simple use of the pr command.

```
$ pr student_records


2017-07-09 12:02                student_records                        Page 1


Jonh    Doe     ECE     3.54
Pam     Meyer   CS      3.61
Jim     DAVIS   CS      2.71
Jason   Kim     ECE     3.97
Amy     Nash    ECE     2.38

<Blank lines until the end of page>

$
```

> **Exercise 6.1**
> Repeat the Bash sessions shown in this section on your system. Creates files and directories that you may have to for this purpose.

You can print files with line numbers and a page header by connecting the nl, pr, and lp (or lpr) commands. This method is discussed in Chapter 9.

6.2.2 Viewing Files One Page at a Time

If the file to be viewed is larger than one page, you can use the more command, also known as the Linux pager, to display the file a screenful at a time. The following is a brief description of the command.

Syntax:
  `more [options] [file-list]`
Purpose: Concatenate/display the files in `file-list` on standard output a screenful at a time
Output: Contents of the files in `file-list` displayed on the screen, one page at a time
Commonly used options/features:
  +/str   Start two lines before the first line containing **str**
  -nN   Display N line per screen/page
  +N   Start displaying the contents of the file at line number N

When run without **file-list**, the more command, like the cat command, takes input from the keyboard one line at a time and sends it to the display screen. If a **file-list** is given as an argument, the command displays the contents of the files in **file-list** one screen at a time. To display the next screen, press <Space>. To display the next line in the file, press <Enter>. At the bottom left of a screen, the command displays the percentage of the file that has been displayed up to that point. To display the next line, you press <Space>. To return to the shell, you press the q or Q key.

The following command displays the **sample** file in the present working directory a screenful at a time. Running this command is equivalent to running the cat sample | more command. We discuss the | operator, known as the *pipe* operator, in detail in Chapter 9.

```
$ more sample
[contents of sample]
$
```

The following command displays contents of the files **sample**, **letter**, and **memo** in the present working directory a screenful at a time. The files are displayed in the order they occur in the command.

```
$ more sample letter memo
[contents of sample, letter, and memo]
$
```

The following command displays the contents of the file **param.h** in the directory **/usr/ include/sys** one page at a time with ten lines per page after fully displaying the first page.

```
$ more -n10 /usr/include/sys/param.h
[contents of /usr/include/sys/param.h]
$
```

The following command displays, one page at a time, the contents of all the files in the present working directory that have the **.c** extension (i.e., files containing C source codes).

```
$ more ./*.c
[contents of all .c files in the current directory]
$
```

The less command can also be used to view a file page by page. It is similar to the more command but is more efficient and has many features that are not available in more. It has support for many of the vi and vim Command Mode commands. For example, it allows forward and backward movement of file contents one or more lines at a time, redisplaying the screen, and forward and backward string search. It also starts displaying a file without reading the whole file, which makes it more efficient than the more command or the vi or vim editor for large files.

Exercise 6.2
  Repeat the Bash sessions shown in this section on your system. Creates files and directories that you may have to for this purpose.

6.2.3 Viewing the Head or Tail of a File

Having the ability to view the head (some lines from the beginning) or tail (some lines from the end) of a file is useful in identifying the type of data stored in the file. For example, the head operation can be used to identify a *PostScript* file or a *uuencoded* file, which have special headers, and the tail information could be used to inspect status information at the end of a log file or error file. (We discuss encoding and decoding of files in Chapter 7.) The Linux commands for displaying the beginning lines or ending lines of a file are head and tail. The following is a brief description of the head command.

Syntax:
  head [option] [file-list]

Purpose: Display the initial portions (i.e., heads) files in **file-list**; the default head size is ten lines

Output: Heads of the files in **file-list** are displayed on the monitor screen

Commonly used options/features:

  -N Display first N lines

Without any option and the **file-list** argument, the command takes input from standard input (the keyboard by default). The following session illustrates the use of the head command. The cat sample command is used to display the contents of the **sample** file. The head sample command displays the first ten lines of the **sample** file. The head -5 sample command displays the first five lines of **sample**.

```
$ cat sample
Ann
Ben
Chen
David
Eto
Fahim
George
Hamid
Ira
Jamal
Ken
Lisa
Mike
Nadeem
Oram
Paul
Queen
Rashid
Srini
Tang
Ursula
Vinny
Wang
X Window System
Yen
Zen
$ head sample
Ann
Ben
Chen
David
Eto
Fahim
George
```

```
Hamid
Ira
Jamal
$ head -5 sample
Ann
Ben
Chen
David
Eto
$
```

You can display heads of multiple files by specifying them as arguments of the head command. For example, the head sample memo1 phones command displays the first ten lines each of the **sample**, **memo1**, and **phones** files. The head of each file is preceded by ==> filename <== at the top left.

   The following command, which displays the first ten lines of the file **otto**, shows that the file is a PostScript file. The output of the command gives additional information about the file, including the name of the software used to create it, the total number of pages in the file, and the page orientation. All of this information is important to know before the file is printed.

```
$ head otto
%!PS-Adobe-3.0
%%BoundingBox: 54 72 558 720
%%Creator: Mozilla (Firefox) HTML->PS
%%DocumentData: Clean7Bit
%%Orientation: Portrait
%%Pages: 1
%%PageOrder: Ascend
%%Title: Otto Doggie
%%EndComments
%%BeginProlog
$
```

Similarly, the following command shows that **data** is a uuencoded file and that, when uudecoded (see Chapter 7), the original file will be stored in the file **data.99**.

```
$ head -4 data
begin 600 data.99 M.OI$3T4L($IO92!#.B`@,#`#@,#`@P.3#P.3#H@OT4(@@`@(@`@`!34CH9&]E,4!S
M;6EL92YC;VTL,#$#,(C#R,J@;,#01DX("`@("`@@#6,<3#4^C<,#J@P
34Z144Z4U(Z<V%R=V%RO'5P+F5D=3HU,#,$Q+@@C$R
$
```

The tail command is used to display the last portion (tail) of one or more files. It is useful to ascertain, for example, that a PostScript file has a proper ending or that a uuencoded file has the required end on the last line. The following is a brief description of the command.

Syntax:
  tail [option] [file-list]

Purpose: Display the last portions (i.e., tails) of files in **file-list**; the default tail size is ten lines

Output: Tails of the files in **file-list** displayed on the monitor screen

Commonly used options/features:

-f	Follow growth of the file after displaying its tail and display lines, as they are appended to the file. The tail command run with this option is terminated with <Ctrl+C>
±n	Start **n** lines from the beginning of the file for **+n**, and **n** line (or **n** units) before the end of the file for **−n**; by default, **n** is 10
-n N	Display first **N** lines
-r	Display lines in the reverse order (last line first)

Like the head command, the `tail` command takes input from standard input if no **file-list** is given as an argument. The following session illustrates how the `tail` command can be used with and without options. We use the same **sample** file that we used for the head command. The `tail sample` command displays the last ten lines (the default tail size) of the **sample** file, and the `tail -5 sample` displays the last five lines of the **sample** file. The `tail +12 sample` command displays the tail of the file starting with line number 12 (not the last 12 lines). Finally, the `tail -5r sample` command displays the last five lines of the **sample** file in reverse order.

```
$ tail sample
Queen
Rashid
Srini
Tang
Ursula
Vinny
Wang
X Window System
Yen
Zen
$ tail -5 sample
Vinny
Wang
X Window System
Yen
Zen
$ tail +12 sample
Lisa
Mike
Nadeem
Oram
Paul
Queen
Rashid
Srini
Tang
Ursula
Vinny
Wang
X Window System
Yen
Zen
$ tail -5r sample
Zen
Yen
X Window System
Wang
Vinny
$
```

The following commands show that files **otto** and **data** have proper PostScript and uuencoded tails.

```
$ tail -5 otto
8 f3
( ) show
pagelevel restore
showpage
%%EOF
$ tail data
M;W4@:&%V90IN;WO@=')I960@;W5T(&9O<B!L;VYG('1I;64N("!(;;W=E=F5R
M+"!T;R!B92!S=6-C97-S9G5L+"!Y;W4@;;75S="!T<@I(96QL;;RP@5V]R;&0A
```

```
"(OH`
`

end
$
```

The -f option of the tail command is very useful if you need to see the tail of a file that is growing. This situation occurs quite often when you run a simulation program that takes a long time to finish (several minutes, hours, or days) and you want to see the data produced by the program as it is generated. It is convenient to do so if your Linux system runs the X Window System (see Chapter W28). In an X environment, you can run the tail command in an *xterm* (a console window) to monitor the newly generated data as it is generated and keep doing your other work concurrently. The following command displays the last ten lines of the **sim.data** file and displays new lines as they are appended to the file. You can terminate the command by pressing <Ctrl+C>.

```
$ tail -f sim.data
... last 10 lines of sim.data ...
... more data as it is appended to sim.data ...
```

Sometimes, while identifying problems in a Linux system, the system administrator needs to display files in the **/var** directory that keep growing because the kernel and applications keep appending new messages to them, including files in the **/var/spool**, **/var/mail**, and **/var/log** directories.

Traditionally, the system administrators are able to view these files as they are appended using the tail -f command, as in the following session to display the last ten lines in the **/var/log/messages** file and continue to show new messages from the kernel and applications as they are appended to this file.

```
$ tail -f /var/log/messages
... last 10 lines of /var/log/messages ...
... more data as it is appended to /var/log/messages ...
```

Using systemd, available in all major branches of Linux, the system administrators can also use the journalctl -f command to display messages being written to the system log in real time. It is possible, using the journalctl -r command, to view the last or most recent lines written to the system log. We show more uses of journalctl in Chapter 18, and Chapter W27 at the book website, on systemd.

In the following in-chapter exercises, you are asked to use the cat, head, more, pr, and tail commands for displaying different parts of text files, with and without page titles and numbers.

Exercise 6.3
Insert the **student_records** file used in Section 6.2.1 in your current directory. Add to it ten more students' records. Display the contents of this file by using the cat student_records and cat -n student_records commands. What is the difference between the outputs of the two commands?

Exercise 6.4
Display the **student_records** file by using the more and pr commands. What command lines did you use?

Exercise 6.5
Display the **/etc/passwd** file two lines before the line that contains your login name. What command line did you use?

Exercise 6.6
Give commands for displaying the first and last seven lines of the **student_records** file.

6.3 Copying, Moving, and Removing Files

In this section, we describe commands for performing copy, as well as move/rename and remove/delete operations on files in a file structure. The commands discussed are cp, mv, and rm.

6.3.1 Copying Files

The Linux command for copying files is `cp`. The following is a brief description of the command.

Syntax:
```
cp [options] file1 file2
```
Purpose: Copy **file1** to **file2**. If **file2** is a directory, make a copy of **file1** in this directory.

Commonly used options/features:

-**f** Force copying if there is no write permission on **file2**

 -**i** If **file2** exists, prompt before overwriting

-**p** Preserve file attributes such as owner ID, group ID, permissions, and modification times

-**r** Recursively copy files and subdirectories

You must have permission to read the source file (**file1**) and permission to execute (search) the directories that contain **file1** and **file2**. In addition, you must have the write permission for the directory that contains **file2** if it does not already exist. If **file2** exists, you don't need the write permission to the directory that contains it, but you must have the write permission to **file2**. If the destination file (**file2**) exists, by default, it will be overwritten without informing you if you have permission to write to the file. To be prompted before an existing file is overwritten, you need to use the –i option. If you do not have permission to write to the destination file, you will be informed of this. If you do not have permission to read the source file, an error message will appear on your screen.

The following `cp` command line makes a copy of **temp** in **temp.bak**. The `ls` commands show the state of the current directory before and after execution of the `cp` command. Figure 6.1 shows the same information in pictorial form.

```
$ ls
memo     sample temp
$ cp temp temp.bak
$ ls
memo     sample temp    temp.bak
$
```

The command returns an error message if **temp** does not exist or if it exists but you do not have permission to read its content. The command also returns an error message if **temp.bak** exists and you do not have permission to write to it. The following session illustrates these points. The first error message is reported because the **letter** file does not exist in the current directory. The second error message is reported because you do not have permission to read the **sample** file. The last command reports an error message because **temp.bak** exists and you do not have write permission for it. You can override the absence of write permission and force copying by using the –f option, as shown in the next command. The `ls -l memo temp.bak` command is used to show that the copying has actually taken place; that is,

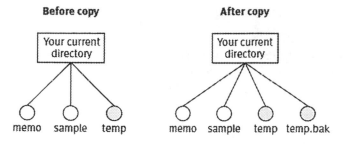

FIGURE 6.1   State of the current directory before and after the temp file has been copied to temp.bak.

the data has been copied, but the time stamp for the file is the current time. If you want to copy both the data and attributes of the source file, you need to use the cp command with –f and –p options, as in the last cp command that follows. The last ls –l memo temp.bak command is used to show that both data and file attributes, such as the time stamp, have been copied.

```
$ ls -l
total 3
-rwxr-----  1 sarwar  faculty  371 Aug 28 07:01 memo
--wxr-----  1 sarwar  faculty  164 Jul 25 12:35 sample
-r-xr-----  1 sarwar  faculty  792 Aug 28 07:01 temp
-r-xr-----  1 sarwar  faculty  792 Aug 28 07:05 temp.bak
$ cp letter letter.bak
cp: letter: No such file or directory
$ cp sample sample.new
cp: sample: Permission denied
$ cp memo temp.bak
cp: temp.bak: Permission denied
$ cp -f memo temp.bak
$ ls -l memo temp.bak
-rwxr-----  1 sarwar  faculty  0 Aug 28 07:01 memo
-rwxr-----  1 sarwar  faculty  0 Aug 28 07:22 temp.bak
$ cp -fp memo temp.bak
$ ls -l memo temp.bak
-rwxr-----  1 sarwar  faculty  0 Aug 28 07:01 memo
-rwxr-----  1 sarwar  faculty  0 Aug 28 07:01 temp.bak
$
```

The following command makes a copy of the **.profile** file in your home directory and puts it in the **.profile.old** file in the **sys.backups** subdirectory, also in your home directory. This command works regardless of the directory you are in when you run the command because the pathname starts with your home directory. You should execute this command before changing your runtime environment specified in the **~/.profile** file, so that you have a backup copy of the previous working environment in case something goes wrong when you set up the new environment. The command produces an error message if ~/.profile does not exist, if you do not have permission to read it, if the **~/sys.backups** directory does not exist or you do not have the execute (search) and write permissions for it, or if **.profile.bak** exists but you do not have permission to read it.

```
$ cp ~/.profile ~/sys.backups/.profile.bak
$
```

The following command copies all the files in the current directory, starting with the string lab to the directory **~/courses/ee446/backups**. The command also prompts you for overwriting if any of the source files already exist in the backups directory. In this case (in which multiple files are being copied), if **backups** is not a directory, or if it does not exist, an error message is displayed on the screen informing you that the target must be a directory.

```
$ cp -i lab* ~/courses/ee446/backups
$
```

If you want to copy a complete directory to another directory, you need to use the cp command with the –r option. This option recursively copies files and subdirectories from the source directory to the destination directory. It is a useful option that you can use to create backups of important directories periodically. Thus, the following command recursively copies the **~/courses** directory to the **~/backups** directory.

```
$ cp -r ~/courses ~/backups
$
```

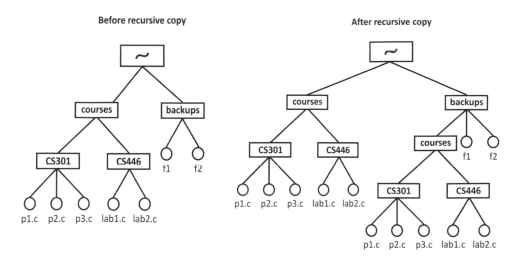

FIGURE 6.2 Current directory before and after the cp -r ~/courses ~/backups command.

This command creates copies of the **~/courses** directory, including all the files and directory hierarchies under the **~/courses** directory, and places it under the **~/backups** directory. Figure 6.2 shows the state of your home directory (~) before and after the execution of the command.

6.3.2 Moving Files

Files can be moved from one directory in a file structure to another. This operation in Linux can also result in simply renaming a file if it is on the same file system. The renaming operation is equivalent to creating a hard link (see Chapter 8) to the file, followed by removing/deleting (see Section 6.3.3) the original file. If the source and destination files are on different file systems, the move operation results in a physical copy of the source file to the destination, followed by removal of the source file. The command for moving files is mv. The following is a brief description of the command.

Syntax:
```
mv [options] file1 file2
mv [options] file-list directory
```

Purpose: First syntax: Move **file1** to **file2** or rename **file1** to **file2**
Second syntax: Move all the files in **file-list** to **directory**

Commonly used options/features:

    -f Force move regardless of the permissions of the destination file

    -i Prompt the user before overwriting the destination file

---

You must have the write and execute access permissions for the directory that contains the source file (**file1** in the description), but you do not need to have the read, write, or execute permissions for the file itself. Similarly, you must have the write and execute permissions for the directory that contains the target/destination file (**file2** in the description), execute permission for every directory in the pathname for the destination file, and write permission for the destination file if it already exists. If the destination file exists, by default, it is overwritten without informing you. If you use the –i option, you are prompted before the destination file is overwritten.

The following command moves **temp** to **temp.moved**. In this case, the **temp** file is renamed **temp. moved**. The mv command returns an error message if **temp** does not exist, or if you do not have the execute permission for the directory it is in. The command prompts you for moving the file if **temp.moved** already exists, but you do not have write permission for it.

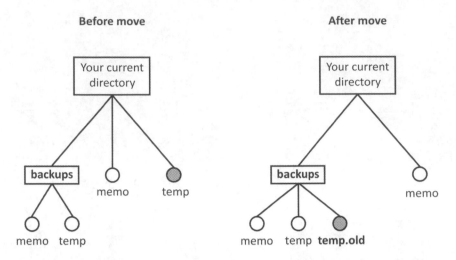

FIGURE 6.3   Current directory before and after the mv temp backups/temp.old command.

```
$ mv temp temp.moved
$
```

The following command moves **temp** to the backups directory as the **temp.old** file. Figure 6.3 shows the state of your current directory before and after the **temp** file is moved.

```
$ mv temp backups/temp.old
$
```

The following command is a sure move; you can use it to force the move, regardless of the permissions for the target file, **temp.moved**.

```
$ mv -f temp temp.moved
$
```

The following command moves all the files and directories (excluding hidden files) in **dir1** to the **dir2** directory. The command fails, and an error message appears on your screen if **dir2** is not a directory, if it does not exist, or if you do not have the write and execute permissions for it.

```
$ mv dir1/* dir2
$
```

   After the command is executed, **dir1** contains the hidden files only. You can use the ls -a command to confirm the status of **dir1**.

> **Exercise 6.7**
>   Create the directory structure discussed in this section under your home directory, including directories dir1 and dir2, and repeat the commands shown in this section on your system. Did command executions result as expected?

6.3.3 Removing/Deleting Files

When files are not needed anymore, they should be removed from a file structure to free up disk space to be reused for new files and directories. The Linux command for removing (deleting) files is rm. The following is a brief description of the command.

Syntax:
    `rm [options] file-list`

Purpose: Removes the files in `file-list` from the file structure (and disk)

Commonly used options/features:

    `-f` Force remove regardless of the permissions for `file-list`

    `-i` Prompt the user before removing the files in `file-list`

    `-r` Recursively remove the files in the directory, which is passed as an argument. *This removes everything under the directory, so be sure you want to do so before using this option.*

If files in `file-list` are pathnames, you need the read and execute permissions for all the directory components in the pathnames and the read, write, and execute permissions for the last directory (that contains the file or files to be deleted). You must also have write permission for the files themselves for their removal without prompting you. If you run the command from a terminal and do not have write permission for the file to be removed, the command displays your access permissions for the file and prompts you for your permission to remove it.

The following command lines illustrate the use of the `rm` command to remove one or more files from various directories.

```
$ rm temp
$ rm temp backups/temp.old
$ rm -f phones grades ~/letters/letter.john
$ rm ~/dir1/*
$
```

The first command removes **temp** from your current directory. The second command removes the **temp** file from your current directory and the **temp.old** file from the backups directory in your current directory. Figure 6.4 shows the semantics of this command. The third command removes the **phones**, **grades**, and **~/letters/letter.john** files, regardless of their access permissions. The fourth command removes all the files from the **~/dir1** directory; the directories are not removed.

Now, consider the following commands that use some shell metacharacter features (see Chapter 7).

```
$ rm [kK]*.prn
$ rm [a-kA-Z]*.prn
$
```

The first command removes all the files in current directory that have the **.prn** extension and names starting with k or K. The second command removes all the files in the current directory that have the **.prn** extension and names starting with a lowercase letter from a to k or an uppercase letter.

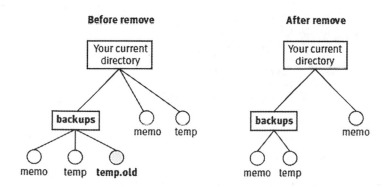

FIGURE 6.4 Current directory before and after execution of rm temp backups/temp.old command.

In Chapter 4, we talked about removing directories and showed that the rmdir command can be used to remove only the empty directories. The rm command with the -r option can be used to remove nonempty directories recursively. Thus, the following command recursively removes the **OldDirectory** in your home directory. This command prompts you if you do not have the permission to remove a file. If you do not want the system to prompt you and you want to force remove the **~/OldDirectory** recursively, then use the rm -rf ~/OldDirectory command. This command is one of the commands that you must never execute unless you really know its potentially catastrophic consequences: the loss of all the files and directories in a complete directory hierarchy. But the command is quite useful if you want to free up some disk space.

```
$ rm -r ~/OldDirectory
$
```

You should generally combine the -i and -r options to remove a directory (**~/OldDirectory** in this case) recursively, as shown in the following command. The -i option is for interactive removal, and when you use this option, the rm command prompts you before removing a file. This way you can ensure that you do not remove an important file by mistake.

```
$ rm -ir ~/OldDirectory
rm: examine files in directory /home/sarwar/OldDirectory (y/n)? y
rm: remove /home/sarwar/OldDirectory/John.11.14.2018 (y/n)? y
rm: remove /home/sarwar/OldDirectory/Tom.2.24.2018 (y/n)?
...
$
```

If you do not have write permission to a directory, you can neither delete its file or subdirectory, nor create a file or subdirectory in it. Similarly, if you do not have read and execute permissions to a directory, you can neither copy a file from it to the target directory nor delete a file from it. The reason being that you cannot search this directory with the execute permission off. For the same reason, you can also not display a listing of the contents of this directory.

In the following session, we show a few examples to illustrate these concepts. Since you do not have write permission to **d1**, the rm d1/f1 command fails to delete the **f1** file in it. Similarly, you are not able to create **f4** file and **d11** directory in **d1** using the touch d1/f4 and mkdir d1/d11 commands, respectively. After adding the write permission to d1 and taking away read and execute permissions from it using the chmod u+w, u-rx d1 command, you are allowed to copy the f1 file from it to the directory d2, remove f1, and list the contents of d1 using the cp d1/f1 d2, rm d1/f1, and ls -l d1 commands, respectively.

```
$ ls -ld *
dr-xr-xr-x 4 sarwar faculty 4096 Feb 12 22:35 d1
drwxr-xr-x 2 sarwar faculty 4096 Feb 12 22:37 d2
$ ls -l d1
total 356
-rw-rw-rw- 1 sarwar faculty     27 Feb 12 22:36 f1
-rw-rw-rw- 1 sarwar faculty   7878 Feb 12 22:36 f2
-rw-rw-rw- 1 sarwar faculty 340523 Feb 12 22:36 f3
$ rm d1/f1
rm: cannot remove 'd1/f1': Permission denied
$ touch d1/f4
touch: cannot touch 'd1/f4': Permission denied
$ mkdir d1/d11
mkdir: cannot create directory 'd1/d11': Permission denied
$ chmod u+w,u-rx d1
$ ls -ld d1
d-w-r-xr-x 4 sarwar faculty 4096 Feb 12 22:35 d1
$ cp d1/f1 d2
cp: cannot stat 'd1/f1': Permission denied
$ rm d1/f1
```

```
rm: cannot remove 'd1/f1': Permission denied
$ ls -l d1
ls: cannot open directory 'd1': Permission denied
$
```

Exercise 6.8

Repeat the rm, ls, cp, and touch commands discussed in this section on your system. Did the command executions result as expected?

6.3.4 Determining File Size

You can determine the size of a file by using one of several Linux commands. The two commands commonly used for this purpose that are available in all Linux versions are ls -l and wc. We described the ls -l command in Chapter 5, where we use it to determine the access permissions for files. We revisit this command here in the context of determining file size.

As mentioned earlier, the ls -l command displays a long list of the files and directories in the directory (or directories) specified as its argument. You must have the read and execute permissions for a directory to be able to run the ls command on it successfully; no permissions are needed on the files in the directory to be able to see the list. The command gives output for the current directory if none is specified as an argument. The output of this command has nine fields, and the fifth field gives file sizes in bytes (see Section 5.5). In the following command, the size of the **lab2** file is 709 bytes.

```
$ ls -l lab2
-rw-r--r--  1 sarwar  faculty  709 Apr  5 11:23 lab2
$
```

This command also displays the size of directory files. You can also use it to get the sizes of multiple files by specifying them in the command line and separating them by spaces. For example, the following command shows that sizes of the **lab1** and **lab2** files are 163 bytes and 709 bytes, respectively.

```
$ ls -l lab1 lab2
-rw-r--r--  1 sarwar  faculty  163 Jul  9 16:47 lab1
-rw-r--r--  1 sarwar  faculty  709 Apr  5 11:23 lab2
$
```

The following command uses the shell metacharacter * to display the long listing for all the files in the **~/courses/ee446** directory.

```
$ ls -l ~/courses/ee446/*
... output of the command ...
$
```

Whereas ls -l is a general-purpose command that can be used to determine many of the attributes of one or more files, including their sizes in bytes/characters, wc is a special-purpose command that displays only file sizes. The following is a brief description of the wc command.

Syntax:
  wc [options] file-list

Purpose: Display sizes of the files in **file-list** as number of lines, words, and characters

Commonly used options/features:

  -c Display only the number of characters
  -l Display only the number of lines
  -w Display only the number of words

```
$ wc sample
        6             44            227          sample
$
```

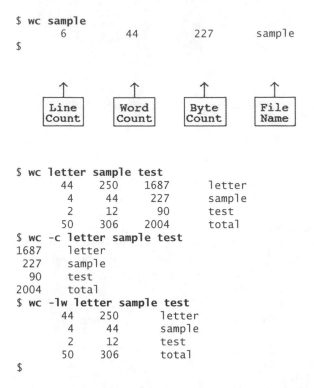

```
$ wc letter sample test
       44     250    1687        letter
        4      44     227        sample
        2      12      90        test
       50     306    2004        total
$ wc -c letter sample test
1687     letter
 227     sample
  90     test
2004     total
$ wc -lw letter sample test
       44     250        letter
        4      44        sample
        2      12        test
       50     306        total
$
```

The first command displays the number of lines, words, and characters in the **sample** file in the present working directory. The size of **sample** is 4 lines, 44 words, and 227 bytes. The second command displays the same information for the files **letter**, **sample**, and **test** in the present working directory. The last line in the output of this command also displays the total line count, word count, and byte count for all three files. The third command displays the number of characters in **letter**, **sample**, and **test**. The last command shows that multiple options can be used in a single command; in this case, the output is the number of words and letters for the three files in the command line.

On FreeBSD, the wc command with a directory argument returns three numbers along with the name of the directory: **0**, **1**, and the number of directory entries in the directory including the . (dot) and .. (dotdot) directories, as in

```
$ wc /etc
        0              1          116  /etc
$
```

This only applies to directories that contain regular files and directories as their contents. However, this style of output is not produced for directories that contain device (character special and block special) files and other types of files.

The wc command can be used with shell metacharacters such as * and ?. The following command displays sizes of all the files in the directory /usr/include/x86_64-linux-gnu/sys. The last line shows the total size of all the files in the directory.

```
$ wc /usr/include/x86_64-linux-gnu/sys/*
    108     501    3319 /usr/include/x86_64-linux-gnu/sys/acct.h
     37     196    1282 /usr/include/x86_64-linux-gnu/sys/auxv.h
      3      12      86 /usr/include/x86_64-linux-gnu/sys/bitypes.h
    444    2124   15412 /usr/include/x86_64-linux-gnu/sys/cdefs.h
     88     559    3575 /usr/include/x86_64-linux-gnu/sys/debugreg.h
     27     142     921 /usr/include/x86_64-linux-gnu/sys/dir.h
     29     159    1023 /usr/include/x86_64-linux-gnu/sys/elf.h
```

```
  142     640    4448 /usr/include/x86_64-linux-gnu/sys/epoll.h
    1       2      19 /usr/include/x86_64-linux-gnu/sys/errno.h
   44     208    1399 /usr/include/x86_64-linux-gnu/sys/eventfd.h
   38     187    1291 /usr/include/x86_64-linux-gnu/sys/fanotify.h
    1       2      19 /usr/include/x86_64-linux-gnu/sys/fcntl.h
   56     281    1721 /usr/include/x86_64-linux-gnu/sys/file.h
...
  175     934    6223 /usr/include/x86_64-linux-gnu/sys/wait.h
  105     674    4274 /usr/include/x86_64-linux-gnu/sys/xattr.h
 8165   38513  267732 total
$
```

Exercise 6.9

Repeat the ls and wc commands discussed in this section on your system. Create the requisite files and directories in order to replicate the Bash sessions on your system. Did the command executions result as expected?

6.4 Appending to Files

Appending to a file means putting new data at the end of the contents of the file. If the file does not exist, it is created to contain new data. The append operation is useful when an application or a user needs to augment a file by adding data to it. The following command syntax is used to append one or more files, or keyboard input, at the end of a file.

Syntax:
  cat [file-list] >> destination-file

Purpose: Append the contents of the files in **file-list**, in the order specified in the command
          line, at the end of **destination-file**

The >> operator is the Linux append operator. We discuss the >>, <, and > operators in detail in Chapter 9. That chapter describes how the input of your commands can be read as input from a file instead of the keyboard, and how the output and error messages of your commands can be redirected from the terminal (or console widow) to files. In this chapter, we use these operators only to describe how you can append new data at the end of the current contents of a file and how you can combine the contents of multiple files and put them in one file using the cat command.

The following session illustrates how the append operation works. The cat sample >> temp command appends the contents of the **sample** file at the end of the **temp** file. The cat commands before and after this command show the contents of the files involved. The command syntax can be used to append multiple files to a file, as shown in the command cat memo1 memo2 memo3 >> memos.record. This command appends the contents of the **memo1, memo2**, and **memo3** files at the end of the **memos.record** file.

```
$ cat temp
This is a simple file used to illustrate the working of append operation. The new
data will be appended right below this line.
$ cat sample
These are the new data that will be appended at the end of the test file.
$ cat sample >> temp
$ cat temp
This is a simple file used to illustrate the working of append operation. The new
data will be appended right below this line.
These are the new data that will be appended at the end of the test file.
$ cat memo1 memo2 memo3 >> memos.record
$
```

Without the optional **file-list** argument (see the command description), the command can be used to append keyboard input at the end of **destination-file**. The cat >> test.letter command takes input from the keyboard and appends it to a file called **test.letter**. The command terminates when you press <Ctrl+D> on a new line.

```
$ cat test.letter
John Doe
12345 First Lane
Second City, State 98765
$ cat >> test.letter
September 1, 2017
Dear John:
This is to inform you ...
...
<Ctrl+D>
$ cat test.letter
John Doe
12345 First Lane
Second City, State 98765
September 1, 2017
Dear John:
This is to inform you ...
...
$
```

Exercise 6.10

Repeat the Bash sessions shown in this section on your system. Create the requisite files on your system. Did the command executions result as expected?

6.5 Combining Files

The following command syntax can be used to combine multiple files into one file.

Syntax:
   cat [file-list] > destination-file

Purpose: Put the contents of the files in **file-list**, in the order specified in the command line, and put them in **destination-file**

The **destination-file** is overwritten if it already exists. If you do not have the write permission for the **destination-file**, the command displays an error message informing you that you do not have permission to write to the file. Without the optional **file-list** argument, you can use the command to put keyboard input in **destination-file**. Thus, this command syntax can be used to create a new file whose contents are what you enter from the keyboard until you press <Ctrl+D> on a new line, as is the case with the cat >> test.letter command in the previous session.

The following session illustrates how this command works with arguments. The ls -l command is used to view permissions for the files. The wc memo? command displays the sizes of all the files in the current directory that start with the string memo and have one character after this string. The third command combines the contents of the **memo1**, **memo2**, and **memo3** files and puts them in the **memos.y2k18** file in the order they appear in the command. The wc memos.y2k18 command is used to confirm that the **memos.y2k18** file has the same number of lines, words, and characters as the three memo files combined. Execution of the cat memo1 memo2 memo3 > memos.2018 command shows that you do not have permission to write to **memos.2018**.

```
$ ls -l
-r-xr--r--  1 sarwar  faculty  1687 Jan 10 19:15 memo1
```

```
-r-xr--r--  1 sarwar  faculty  1227 Feb 19 14:37 memo2
-r-xr--r--  1 sarwar  faculty   790 Sep  1 19:16 memo3
-r--------  1 sarwar  faculty  9765 Jan 15 22:11 memos.2018
$ wc memo?
      44     250    3352 memo1
      34     244    4083 memo2
      12     112     907 memo3
      90     606    3704 total
$ cat memo1 memo2 memo3 > memos.y2k18
$ wc memos.y2k18
      90     606    3704 memos.y2k18
$ cat memo1 memo2 memo3 > memos.2018
memos.2018: Permission denied.
$
```

You can also do the task of the `cat memo1 memo2 memo3 > memos.y2k18` by using the following command sequence.

```
$ cat memo1 > memos.y2k18
$ cat memo2 >> memos.y2k18
$ cat memo3 >> memos.y2k18
$
```

The following in-chapter exercises ask you to practice using the cp, mv, ls -l, wc, and cat commands and the operator for appending to a file.

Exercise 6.11
Copy the **.bashrc** file in your home directory to a file **.bashrc.old** in a directory called **backups**, also in your home directory. Assume that you are in your home directory. What command did you use?

Exercise 6.12
Create a directory called **new.backups** in your home directory and move all the files in the backups directory to **new.backups**. What commands did you use?

Exercise 6.13
Display the size in bytes of a file **lab3** in the **~/ece345** directory. What command did you use?

Exercise 6.14
Give a command for appending all the files in the **~/courses/ece446** directory to a file called **BigBackup.ece446** in the **~/courses** directory.

6.6 Comparing Files

At times, you will need to compare two versions of a program code or some other text document to find out where they differ from each other. You do this in order to synchronize these files so that they contain the same content. You may also need to compare nontext files to know if they are identical. There are several Linux tools that let you perform these tasks. We discuss them here, first for text files and then for nontext files.

6.6.1 Text Files

You can use the diff command to compare text files and identify their differences. The command compares two files and displays differences between them in terms of commands that can be used to convert one file to the other. You can then use another tool that synchronizes the files based on the commands generated by diff. The following is a brief description of the command.

Syntax:
  `diff [options] [file1] [file2]`

Purpose: Compare **file1** with **file2** line by line and display differences between them as a series of commands/instructions for the ed editor that can be used to convert **file1** to **file2** or vice versa; read from standard input if – is used for **file1** or **file2**

Commonly used options/features:

   -b Ignore trailing (at the end of lines) white spaces (blanks and tabs), and consider other strings of white spaces equal

   -e Generate and display a script for the ed editor that can be executed to change **file1** to **file2**

   -h Do fast comparison (the **-e** option may not be used in this case)

The **file1** and **file2** arguments can be directories. If **file1** is a directory, diff searches it to locate a file named **file2** and compares it with **file2** (the second argument). If **file2** is a directory, diff searches it to locate a file named **file1** and compares it with **file1** (the first argument). If both arguments are directories, the command compares all pairs of files in these directories that have the same names.

   The diff command does not produce any output if the files being compared are the same. When used without any options, the diff command produces a series of instructions for the ed editor that can be used to convert **file1** to **file2** if the files are different. The instructions are a (add), c (change), and d (delete) and are described in Table 6.1.

   The following session illustrates a simple use of the diff command.

```
$ cat Fall_OH
Office Hours for Fall 2017
Monday
9:00 - 10:00 A.M.
3:00 - 4:00 P.M.
Tuesday
10:00 - 11:00 A.M.
Wednesday
9:00 - 10:00 A.M.
3:00 - 4:00 P.M.
Thursday
11:00 A.M. - 12:00 P.M.
2:00 - 3:00 P.M.
4:00 - 4:30 P.M.
```

TABLE 6.1

File Conversion Instructions Produced by diff

Instruction	Description for Changing **file1** to **file2**
`L1aL2,L3` `> lines L2 through L3`	Append lines L2 through L3 from **file2** after line L1 in **file1**

`L1,L2cL3,L4`
`< lines L1 through L2 in file1`
`---`
`> lines L3 through L4 in file2`
	Change lines L1 through L2 in **file1** to lines L3 through L4 in **file2**

`L1,L2dL3` `< lines L1 through L2 in file1`	Delete lines L1 through L2 from **file1**

```
$ cat Spring_OH
Office Hours for Spring 2018
Monday
9:00 - 10:00 A.M.
3:00 - 4:00 P.M.
Tuesday
10:00 - 11:00 A.M.
1:00 - 2:00 P.M.
Wednesday
9:00 - 10:00 A.M.
Thursday
11:00 A.M. - 12:00 P.M.
$ diff Fall_OH Spring_OH
1c1
< Office Hours for Fall 2017
---
> Office Hours for Spring 2018
6a7
> 1:00 - 2:00 P.M.
9d9
< 3:00 - 4:00 P.M.
12,13d11
< 2:00 - 3:00 P.M.
< 4:00 - 4:30 P.M.
$
```

The instruction 1c1 asks the ed editor to change the first line in the **Fall_OH** file (Office Hours for Fall 2017) to the first line in the **Spring_OH** file (Office Hours for Spring 2018). The 6a7 instruction asks the ed editor to append line 7 in **Spring_OH** after line 6 in **Fall_OH**. The 12,13d11 instruction asks the ed editor to delete lines 12 and 13 from **Fall_OH**.

The following session illustrates use of the -e option with the diff command and how the output of this command can be given to the ed editor to make **Fall_OH** the same as **Spring_OH**. The command is used to show you what the output of the command looks like. The second diff command (with > diff. script) is used to save the command output (the ed script) in the **diff.script** file. The cat >> diff. script command is used to convert the **diff.script** file into a complete working script for the ed editor by adding two lines containing w and q. As previously stated, this command terminates with <Ctrl+D>. Finally, the ed command is run to change the contents of **Fall_OH**, according to the script produced by the diff -e command, and make it the same as **Spring_OH**. The numbers 209 and 177 show the sizes of the **Fall_OH** file before and after the execution of the ed command. The last command, diff Fall_OH Spring_OH, is run to confirm that the two files are the same.

```
$ diff -e Fall_OH Spring_OH
12,13d
9d
6a
1:00 - 2:00 P.M.
.
1c
Office Hours for Spring 2015
.
$ diff -e Fall_OH Spring_OH > diff.script
$ cat >> diff.script
w
q
<Ctrl+D>
$ ed Fall_OH < diff.script
209
177
$ diff Fall_OH Spring_OH
$
```

Most systems have a command called diff3 that can be used to do a three-way comparison; that is, three files can be composed. You can also use the vimdiff command to compare multiple files. The sdiff command allows you to compare two files side by side, optionally merge them manually, and output results. See Table 6.3, at the book GitHub site, for websites that describe these commands with examples.

You can use the patch command to produce a patched version of the two files. The command takes a patch file containing a difference listing produced by the diff command and applies those differences to one or more original files, producing patched versions. Here is a brief description of the patch command.

Syntax:
    patch [options] [originalfile [patchfile]]

Purpose: Takes patchfile containing the differences produced by the diff command and applies these differences to one or more original files, producing patched versions. By default, patched version overwrites the original files.

Commonly used options/features:

    -b Creates backup files before overwriting the originals

    -R Undo patch work, i.e., perform patching in reverse order and recover the original file(s)

In the following session, we use the diff command to generate the patch file for applying to **hello1.c** file using the patch command. The first diff command is used to generate the patch file, called **hello. patch**. The first and second lines (starting with --- and +++) show the names of first and second files, and the dates and times of their creation, respectively. The third line (starting and ending with @@) shows the total ranges of lines in the first and second files, respectively; 1–2 in the first file and 1–3 in the second file. The – character at the beginning of a line in the patch file means the line belongs to the first file, a + character means the line belongs to the second file, and a space implies the line is present in both files. The patch < hello.patch command is used to patch **hello1.c** file so that it becomes the same as **hello2.c**. The second diff command shows that **hello1.c** and **hello2.c** have no differences between them.

```
$ cat hello1
Hello, world!
Welcome to the patch command.
$ cat hello2.c
Hello, world!
Welcome to the diff and patch commands.
Goodbye!
$ diff hello1 hello2
2c2,3
< Welcome to the patch command.
---
> Welcome to the diff and patch commands.
> Goodbye!
$ diff -u hello1 hello2 > hello.patch
$ cat hello.patch
--- hello1      2018-02-20 22:44:18.538080809 +0500
+++ hello2      2018-02-20 22:40:21.646225787 +0500
@@ -1,2 +1,3 @@
 Hello, world!
-Welcome to the patch command.
+Welcome to the diff and patch commands.
+Goodbye!
$ patch < hello.patch
patching file hello1.c
$ diff hello1 hello2
$ cat hello1
```

```
Hello, world!
Welcome to the diff and patch commands.
Goodbye!
$
```

The execution of the patch < hello.patch command overwrites the exiting **hello1** file. If you want to keep a backup of **hello1**, run the patch command with –b option as in patch -b < hello.patch. The backup file is saved as hello1.orig. You can use the patch command with –V option to save the backup file in a numbered format. The patch -b -V numbered < hello.patch command saves the backup file as hello1.~1~.

Sometimes, we need to undo the patch operation, i.e., apply the patch process in the reverse order. You can do this by using the patch command with –R option, as shown below. The result is the recovery of the hello1 file to its original state, as shown below in the output of the cat hello1 command.

```
$ patch -R < hello.patch
patching file hello1
$ cat hello1
Hello, world!
Welcome to the patch command.
$
```

6.6.2 Nontext Files

Sometimes, you need to compress, encode, or encrypt your files for various reasons discussed in Chapter 7. These operations transform file data into a form that is completely different from the file's original content. These files have to be decompressed, decoded, or decrypted in order to recover their original content and use them for the purpose they were created. After having performed the recovery operations, you want to make sure the original content of a file has been recovered 100%. You usually deal with downloadable compressed files containing large datasets, images, tools, and operating system kernels such as that of Linux. You download such files, decompress them, and use them for their respective purposes.

However, before using a decompressed file, you want to make sure that it is the exact copy of the original file before it was compressed. However, the problem is that you don't have a copy of the original file to compare it with. The suppliers of the compressed files publish signatures of the respective original files along with the specific tools that were used to generate these signatures. This allows you to generate the signature for your decompressed file. If the published and your signature are the same, it means that the decompressed file is a true copy of the original.

There are several Linux tools that allow you to do so, including shasum, crc32, md5sum, and cksum. Browse the manual pages for these commands to know more about them. You cannot use the diff command (or its variants) in this case, because it does not generate any signature for a file. Without going into the details, the *signature* for a file is also called *hash value, checksum,* or *message digest* of the file. The shasum tool uses the Secure Hash Algorithm that produces a 160-bit hash value and md5sum uses MD5 hash function that produces a 128-bit hash value. Because researchers and practitioners have discovered several serious vulnerabilities in MD5, we therefore do not recommend the use of md5sum. You can run shasum with various algorithms with the –a option, such as sha1, sha224, sha256, etc. The default algorithm used by shasum is sha1. You can specify a specific algorithm with the –a option. See the manual page for the command for more details.

We show the practical use of shasum in Chapter 7. Here, we show the simple use of shasum and md5sum command to illustrate how they work. In the following session, we copy the **/bin/cp** file into our current directory as **my_cp** and produce message digests for them using the shasum and md5sum commands. In both cases, the files have identical message digests, guaranteeing that they are identical.

```
$ cp /bin/cp my_cp
$ shasum /bin/cp my_cp
```

```
ff3094b907d15cee91b8eecb0559011d2d1c175a  /bin/cp
ff3094b907d15cee91b8eecb0559011d2d1c175a  my_cp
$ md5sum /bin/cp my_cp
62eeb2e8a0073ab510bf75ec0876c7a6  /bin/cp
62eeb2e8a0073ab510bf75ec0876c7a6  my_cp
$
```

Exercise 6.15

Duplicate the interactive sessions given in this section to appreciate how the diff command works.

Exercise 6.16

Browse through the man pages of sdiff and vimdiff commands to see how you could use these commands to perform the task performed in Exercise 6.15.

Exercise 6.17

Duplicate the interactive sessions given in this section to appreciate how the diff and patch commands work.

Exercise 6.18

Copy **/bin/bash** to your current directory and name it **my_bash**. Calculate their message digests using the shasum and md5sum commands to show that they are identical files. Show your work.

6.7 Locating and Removing Repetition within Text Files

You can use the uniq command to remove all but one copy of the successive repeated lines in a text file. The command is intended for files of sorted content, although it can work on files without sorted content, as shown in the example. We discuss sorting in Chapter 7. The following is a brief summary of the command.

Syntax:
> uniq [options] [input-file] [output-file]

Purpose: Remove repetitious lines from the sorted **input-file** and send unique (nonrepeated) lines to **output-file**. The **input-file** does not change. If no **output-file** is specified, the output of the command is sent to standard output. If no **input-file** is specified, the command takes input from standard input.

Commonly used options/features:

-c   Precede each output line by the number of times it occurs

-d   Display the repeated lines

-f N Ignore the first N fields in input lines while doing comparisons

-i   Perform case-sensitive comparison of input lines

-s c Ignore the first c characters in input lines while doing comparisons

-u   Display the lines that are not repeated

The following session illustrates how the uniq command works. The cat command is used to show the contents of the **sample** file. The uniq sample command shows that only consecutive duplicate lines are considered duplicate. The uniq -c sample command shows the line count for every line in the file. The uniq -d sample command is used to output repeated lines only. Finally, the uniq -d sample out command sends the output of the command to the **out** file. The cat out command is used to show the contents of **out**. Note that the uniq command only works for unsorted files if repeated lines are adjacent.

```
$ cat sample
This is a test file for the uniq command.
It contains some repeated and some nonrepeated lines.
Some of the repeated lines are consecutive, like this.
Some of the repeated lines are consecutive, like this.
Some of the repeated lines are consecutive, like this.
And, some are not consecutive, like the following.
Some of the repeated lines are consecutive, like this.
The above line, therefore, will not be considered a repeated
line by the uniq command, but this will be considered repeated!
line by the uniq command, but this will be considered repeated!
$ uniq sample
This is a test file for the uniq command.
It contains some repeated and some nonrepeated lines.
Some of the repeated lines are consecutive, like this.
And, some are not consecutive, like the following.
Some of the repeated lines are consecutive, like this.
The above line, therefore, will not be considered a repeated
line by the uniq command, but this will be considered repeated!
$ uniq -c sample
   1 This is a test file for the uniq command.
   1 It contains some repeated and some nonrepeated lines.
   3 Some of the repeated lines are consecutive, like this.
   1 And, some are not consecutive, like the following.
   1 Some of the repeated lines are consecutive, like this.
   1 The above line, therefore, will not be considered a repeated
   2 line by the uniq command, but this will be considered repeated!
$ uniq -d sample
Some of the repeated lines are consecutive, like this.
line by the uniq command, but this will be considered repeated!
$ uniq -d sample out
$ cat out
Some of the repeated lines are consecutive, like this.
line by the uniq command, but this will be considered repeated!
$
```

The uniq command is commonly used with Linux pipes and filters (such as sort and grep commands) to perform more interesting tasks, as discussed in Chapter 9.

In the following in-chapter exercises, you will use the uniq command to appreciate the tasks they perform.

Exercise 6.19

Duplicate the interactive sessions given in this section to appreciate how the uniq command works. Create the requisite files on your system.

6.8 Printing Files and Controlling Print Jobs

We briefly discuss the Linux commands for printing files in Chapter 2. In this section, we cover file printing fully, including commands related to printing and printer control. These commands include commands for printing files, checking the status of print requests/jobs on a printer, and canceling print jobs. We describe two sets of commands for printing and controlling print jobs, based upon Berkeley Software Distribution (BSD) or System V UNIX, that are available in Linux Mint.

6.8.1 Linux Mechanism for Printing Files

The process of printing files is similar to the process of displaying files; in both cases, the contents of one or more files are sent to an output device. In the case of displaying output, the output device is a display

TABLE 6.2

List of Commands Related to Printing

System V Compatible Linux	BSD Compatible Linux	Purpose
lp	Lpr	Submits a file for printing
lpstat	Lpq	Shows the status of print jobs for one or more printers
cancel	lprm	Removes/purges one or more jobs from the print queue
	Lpc	Activates the printer control program
	lptest	Generates ripple pattern for testing the printer

screen, whereas in the case of printing output, the output device is a printer. Another key difference results primarily from the fact that every user has an individual display screen but that many users may share a single printer on a typical Linux (or any time-sharing) system. Thus, when you use the cat or more command to display a file, the contents of the file are immediately sent to the display screen by Linux. However, when you print a file, its contents are not immediately sent to the printer because the printer might be busy printing some other file (yours or some other users). To handle multiple requests, a *first-come first-serve* mechanism places a print request in a queue associated with the printer to which you have sent your print request and processes the request in its turn when the printer is available.

Linux maintains a queue of print requests, called the *print queue*, associated with every printer in the system. Each request is called a *job*. A job is assigned a number, called the *job ID*. When you use a command to print a file, the system makes a temporary copy of your file, assigns a job ID to your request, and puts the job in the print queue associated with the printer specified in the command line. When the printer finishes its current job, it is given the next job from the front of the print queue. Thus, your job is processed when the printer is available and your job is at the head of the print queue.

A Linux process called the *printer spooler* or *printer daemon* performs the work of maintaining the print queue and directing print jobs to the right printer. This process is called lpd. It starts execution in the background when the system boots up and waits for your print requests. We discuss daemons in Chapter 10, but for now, you can think of a daemon as a process that runs but you are not aware of its presence while it interacts with your terminal.

Linux systems have two different command sets for printing and controlling print jobs, based upon BSD or System V UNIX. Table 6.2 contains a list of the printing-related commands for both systems; all are available in Linux Mint. The superuser—the system administrator—normally uses the last two commands, lpc and lptest.

Additionally, you can utilize the facilities of the Common UNIX Printing System (CUPS) to accomplish everything shown in the sections below. We detail CUPS printing in the System Administration section entitled "CUPS Printing" in Chapter W26 at the book website.

6.8.2 Printing Files

As shown in Table 6.2, you can print files by using the lp command and the lpr command on a Linux system. It is very important to note that you should never try printing nontext files with the lp or lpr command, especially files with control characters (e.g., executable files such as **a.out**). Doing so will not print what you want printed and will waste many printer pages. Do not even try testing it. If by accident, you do send a print request for a nontext file, turn off the printer immediately and alert your system administrator that you need assistance.

The following is a brief description of the lp command.

Syntax:
    lp [options] file-list
Purpose: Submit a request to print the files in **file-list**

Commonly used options/features:

-P page-list	Print the pages specified in **page-list**
-d ptr	Send the print request to the **ptr** printer
-m	Send e-mail after printing is complete
-n N	Print N copies of the file(s) in **file-list**; default is one copy
-t title	Print **title** on a banner page
-w	Write to user's terminal after printing is complete

The following session shows how to use the lp command with and without options. The first command prints the **sample** file on the default printer. The system administrator sets the default printer for the users on a system. The job ID for the first print request is cpr-981, which tells you that the name of the printer is **cpr**. The second command uses the -d option to specify that the **sample** file should be printed on the **spr** printer. The third command is for printing three copies each of the **sample** and **phones** files on the **qpr** printer.

```
% lp sample
request id is cpr-981 (1 file(s))
% lp -d spr sample
request id is spr-983 (1 file(s))
% lp -d qpr -n 3 sample phones
request id is qpr-984 (2 file(s))
%
```

As mentioned earlier, the BSD counterpart of the lp command is the lpr command. The following is a brief description of this command.

Syntax:
    lpr [options] file-list

Purpose: Submit a request to print the files in **file-list**

Commonly used options/features:

-# N	Print N copies of the file(s) in **file-list**; default is one copy
-P ptr	Send the print request to the **ptr** printer
-T title	Print **title** on a banner page
-m	Send email after printing is complete
-p	Format output by using the pr command

The following session shows the BSD variant of the commands that perform the same print tasks as the lp command. Thus, the first lpr command sends the print request for printing the **sample** file on the default printer. The second command sends the request for printing the **sample** file on the **spr** printer. The third command prints three copies of the **sample** and **phones** files on the **qpr** printer.

```
% lpr sample
% lpr -P spr sample
% lpr -P qpr -# 3 sample
%
```

You can use the following command to print the **sample** file with the header information on every page produced by the pr command. The vertical bar (|) is called the *pipe* symbol, which we discuss in detail in Chapter 9.

```
% pr sample | lpr
%
```

You can perform the same task with the lpr -p sample command. You can print the **sample** file with line numbers and a pr header on each page using the following command. You can also perform the same task with the nl sample | lpr -p command.

```
$ nl sample | pr | lpr
$
```

You can enable the lpr command to print a nonstandard text file, such as a TEX file, by specifying an appropriate flag. For example, you can use the -t option to print a troff file and the -n option to print an nroff file.

6.8.3 Finding the Status of Your Print Requests

The lpstat command can be used to display the status of print jobs on a printer. The following is a brief description of the lpstat command.

Syntax:
  lpstat [options]

Purpose: Display the status of print jobs on a printer

Commonly used options/features:

-d	Display the status of print jobs sent to the default printer using the lp command
-o job-ID-list	Display the status of the print jobs in **job-ID-list**; separate job IDs with spaces and enclose the requests in double quotes for more than one job
-p printer-list	Display the status of print jobs on the printers specified in **printer-list**
-u user-list	Display the status of print jobs for the users in **user-list**

Without any option, the lpstat command displays the status of all your print jobs that are printing or waiting in the print queue of the default printer. The commands in the following session show some typical uses of the command. The lpstat -p command shows the status of all printers on the network. The lpstat -p qpr displays the status of print jobs on the **qpr** printer. The lpstat -u sarwar displays all print jobs for the user **sarwar**. The output of the command shows that there are three print jobs that **sarwar** has submitted: two to **qpr** (job IDs qpr-3998 and qpr-3999) and one to **tpr** (job ID tpr-203). Finally, the lpstat -a command displays all the printers that are up and accepting print jobs.

```
$ lpstat -p
printer cpr is idle. enabled since Tue Sep 2 10:43:48 GMT 2014. available. printer
mpr faulted. enabled since Mon Sep 1 10:48:29 GMT 2014. available. printer qpr now
printing qpr-53. enabled since Mon Sep 1 10:48:29 GMT 2014. available. printer spr
is idle. enabled since Mon Sep 1 10:48:29 GMT 2003. available.
$ lpstat -p qpr
printer qpr now printing qpr-53. enabled since Mon Sep 1 10:48:29 GMT 2014.
available.
$ lpstat -u sarwar
qpr-3998 sarwar          93874    Sep    2        22:05 on qpr
qpr-3999 sarwar          93874    Sep    2        22:05
tpr-203 sarwar 93874    Sep    2        22:05 on tpr
$ lpstat -a
cpr accepting requests since Tue Sep 2 10:43:48 GMT 2014
spr accepting requests since Mon Sep 1 10:48:29 GMT 2014
$
```

The following is a brief description of the BSD counterpart of the `lpstat` command, the `lpq` command.

Syntax:
   `lpq [options]`

Purpose: Display the status of print jobs on a printer

Commonly used options/features:

   `-P printer-list`   Display the status of print jobs on the printers specified in **printer-list**

   `-l`   Display the long format status of print jobs sent using the `lpr` command on the default printer

---

The most commonly used option is `-P`. In the following session, the first command is used to display the status of print jobs on the **mpr** printer. The output of the command shows that four jobs are in the printer queue: jobs 3991, 3992, 3993, and 3994. The active job is at the head of print queue. When the printer is ready for printing, it will print the active job first. The second command shows that the **qpr** printer does not have any jobs to print.

```
$ lpq -Pmpr
mpr is ready and printing
Rank    Owner   Job    Files          Total Size
active  sarwar  3991    mail.bob        1056 bytes
1st     sarwar  3992   csh.man         93874 bytes
2nd     davis   3993   proposal1.nsa 2708921 bytes
3rd     tom     3994   memo            8920 bytes
$ lpq -Pqpr
no entries
$
```

6.8.4 Canceling Your Print Jobs

If you realize that you have submitted the wrong file(s) for printing, you will want to cancel your print request(s). The command for performing this task is `cancel`. The following is a brief description of the command.

Syntax:
   `cancel [options] [printer]`

Purpose: Cancel the print requests sent through the `lp` command—that is, take these jobs out of the print queue

Commonly used options/features:

   `-jobID-list`   Cancel the print jobs specified in **jobID-lsit**

   `-ulogin`   Cancel all print requests issued by the user **login**

The following commands show how to cancel a print job. You can display the job IDs of the print jobs on a printer by using the `lpstat` or `lpq` command, as shown in Section 6.8.3. The first command cancels the print job mpr-3991. The second command cancels all print requests by the user **sarwar** on all printers. You can cancel your own print jobs only. The last command, therefore, works only when run by **sarwar** or the superuser.

```
$ cancel mpr-3991
request "mpr-3991" canceled
$ cancel -u sarwar mpr
request "mpr-3992" canceled
request "mpr-3995" canceled
$
```

The BSD counterpart of the cancel command is lprm. The following is a brief description of the command.

Syntax:
    lprm [options] [jobID-list] [user(s)]

Purpose: Cancel the print requests made by using the lpr command—i.e., remove these jobs from the print queue; the jobIDs in **jobID-list** are taken from the output of the lpq command

Commonly used options/features:

\-   Remove all the print jobs owned by the **user**

-P ptr   Specify the print queue for the **ptr** printer

The following lprm commands perform the same tasks as the cancel commands described in the previous session.

```
$ lprm -Pmpr 3991
mpr-3996 dequeued
$ lprm -Pmpr sarwar
mpr-3997 dequeued
mpr-3998 dequeued
$
```

When run without an argument, the lprm command removes the job that is currently active, provided it is one of your jobs.

The following in-chapter exercises will give you practice on using the printing-related commands.

Exercise 6.20
    How would you print five copies of the file **memo** on the printer **ece_hp1**? Give commands for both System V and BSD-style commands on a Linux system.

Exercise 6.21
    After submitting the two requests, you realize that you really wanted to print five copies of the file **letter**. How would you remove the print jobs from the print queue? Again, give commands for both System V and BSD-style commands on a Linux system.

Summary

The basic file operations involve displaying all or part of a file's contents, renaming a file, moving a file to another file, removing a file, determining a file's size, comparing files, combining files and storing them in another file, appending new contents (which can come from another disk file, keyboard, or output of a command) at the end of a file, and printing files. Linux provides several commands that can be used to perform these operations.

The cat and more commands can be used to display all the contents of a file on the display screen. The > symbol can be used to send outputs of these commands to other files, and the >> operator can be used to append new contents at the end of a file. The cat command sends a file's contents as continuous text, whereas the more command sends them in the form of pages. Furthermore, the more command has several useful features, such as the ability to display a page that contains a particular string. The less command supports even more features than the more command, including the vi-style forward and backward searching.

The head and tail commands can be used to display the initial or end portions (head or tail) of a file. These helpful commands are usually used to find out the type of data contained in a file, without using the file command (see Chapter 4). In addition, the file command cannot decipher contents of all the files.

A copy of a file can be made in another file or directory by using the cp command. Along with the > operator, the cat command can also be used to make a file copy, although there are differences between using the cp and cat > commands for copying files (see Chapter 9). A file can be moved to another file by using the mv command. However, depending on whether the source and destination files are on the same file system, its use might or might not result in actual movement of file data from one location to another. If the source and destination files are on the same file system, the file data is not moved and the source file is simply linked to the new place (destination) through a hard link (see Chapter 8) and the original/ source link is removed. If the two files are on different file systems, an actual copy of the source file is made at the new location and the source file is removed (unlinked) from the current directory. Files can be removed from a file structure by using the rm command. This command can also be used to remove directories recursively.

The size of a file can be determined by using the ls -l or wc commands; both give file sizes in bytes. In addition, the wc command gives the number of lines and words in the file. Both commands can be used to display the sizes of multiple files by using the shell metacharacters *, ?, [], and ∧.

The diff command can be used to display the differences between two files. The command, in addition to displaying the differences between the files, displays useful information in the form of a sequence of commands for the ed editor that can be used to make the two files the same. The diff3 and vimdiff commands allow multifile comparison. The sdiff command allows comparison of two files side by side and their manual merging. You can use the patch command instead of the ed command to merge files. The shasum and md5sum commands allow you to create unique signatures for files called message digests. These message digests are used to identify identical files. We do not recommend the use of md5sum command due to serious vulnerabilities discovered by researchers in the MD5 algorithm.

The uniq command can be used to remove all but one occurrence of successive repeated lines. With the -d option, the command can be used to display the repeated lines. The uniq command is commonly used using Linux pipes and filters (such as sort and grep commands) to perform more interesting tasks, as discussed in Chapter 7.

The lp (System V) or lpr (BSD) command can both be used for printing files on a Linux Mint printer. The lpstat (System V) or lpq (BSD) commands can be used for checking the status of all print jobs (requests) on a printer (waiting, printing, etc.). The cancel (System V) or lprm (BSD) commands can be used to remove a print job from a printer queue so that the requested file is not printed. All of these print commands are available in Linux Mint. Additionally, Chapter W26, Section 5, details CUPS printing.

Questions and Problems

1. List ten operations that you can perform on Linux files.
2. Give a command line for viewing the sizes (in lines and bytes) of all the files in your present working directory.
3. What does the tail -10r ../letter.John command do?
4. Give a command for viewing the size of your home directory. Give a command for displaying the sizes of all the files in your home directory.
5. Give a command for displaying all the lines in the **students** file, starting with line 25.
6. Give a command for copying all the files and directories under a directory **courses** in your home directory. Assume that you are in your home directory. Give another command to accomplish the same task, assuming that you are not in your home directory.
7. Repeat Problem 6, but give the command that preserves the modification times and permissions for the file.
8. What is the difference between the cp -r ~/courses ~/backups and cp -r ~/courses/ ~/backups commands?

9. Give an option for the `rm` command that could protect you from accidently removing a file, especially when you are using wild cards such as * and ? in the command.

10. What do the following commands do?

    a. `cp -f sample sample.bak`

    b. `cp -fp sample sample.bak`

    c. `rm -i ~/personal/memo*.doc`

    d. `rm -i ~/unixbook/final/ch??.prn`

    e. `rm -f ~/unixbook/final/*.o`

    f. `rm -f ~/courses/ece446/lab[1-6].[cC]`

    g. `rm -r ~/NotNeededDirectory`

    h. `rm -rf ~/NotNeededDirectory`

    i. `rm -ri ~/NotNeededDirectory`

11. Give a command for moving files **lab1**, **lab2**, and **lab3** from the **~/courses/ece345** directory to a **newlabs.ece345** directory in your home directory. If a file already exists in the destination directory, the command should prompt the user for confirmation.

12. Give a command to display the lines in the **~/personal/phones** file that are not repeated.

13. Refer to In-Chapter Exercise 15. Give a sequence of commands for the `ed` editor and use them to make **sample** and **example** the same files.

14. You have a file in your home directory called **tryit&**. Rename this file. What command did you use?

15. Give a command for displaying attributes of all the files starting with a string **prog**, followed by zero or more characters and ending with a string **.c** in the **courses/ece345** directory in your home directory.

16. Refer to Problem 15. Give a command for file names with two English letters between **prog** and **.c**. Can you give another command line to accomplish the same task?

17. Give a command for displaying files **got|cha** and **M*A*S*H** one screenful at a time.

18. Give a command for displaying the sizes of files that have the **.jpg** extension and names ending with a digit.

19. What does the `rm *[a-zA-Z]??[1,5,8].[^p]*` command do?

20. Give a command to compare the files **sample** and **example** in your present working directory. The output should generate a series of commands for the `ed` editor.

21. Use the `diff` command to generate the patch file for the Fall_OH file, called OH_patch. Display the contents of OH_patch file and identify differences between the two files. Use the `patch` command to produce the patched version of Fall_OH so that it becomes the same as Spring_OH as shown in Section 6.6.1.

22. What is message digest for a file? List five Linux commands that can be used to generate message digests for files.

23. What is a printer spooler? A print daemon? What is the name of printer daemon in Linux?

24. Give a command for producing ten copies of the report **file** on the **ece_hp3** printer. Each page should contain a page header produced by the `pr` command. Give commands for both System V and BSD-style printing on a Linux system.

25. Give the command to print the nroff file **Chapter 1** by using the `lpr` command. What command line would you use to print the troff file **sample** with the `lpr` command?

26. Give a command for checking the status of a print job with job ID ece_hp3-8971. How would you remove this print job from the print queue? Give commands for both System V and BSD-style printing on a Linux system.

27. What is the difference between the `tail -15 file1` and `tail +15 file1` commands? Which of the following command is equivalent to cat file1: `tail -$ file`, `tail -1 file1`, or `tail +1 file1`? Why?

28. What is the purpose of the more -n5 file1?

29. What are the differences between the more and less commands? Which of the two is more powerful and user friendly in your opinion and why?

30. Create a file in your current directory called **f1**. What is the inode number of the file? Move the file to your home directory and name it **f1.moved**. What is the inode number of the moved file? Move the moved file to the **/tmp** directory. What is the inode number of the **/tmp/f1.moved** file? Why are the inode numbers for **f1** and **f2** the same? Why is the inode number of **/tmp/ f1.moved** different from **f1** (or ~/**f1.moved**)?

Advanced Questions and Problems

31. Why can you not perform the following operations on a directory if you don't have read and execute permissions for it? Delete a file in it, copy a file from it to another directory, and list the contents of the directory? Clearly explain your answer in each case.

32. Use sdiff and vimdiff commands to perform the task performed in In-Chapter Exercise 6.15. Read through the man pages for these commands as well as the tutorials at the websites listed in Table 6.3 at the book GitHub site.

33. Browse the Web and briefly describe the SHA and MD5 algorithms. What are the lengths of message digests for crc32, md5sum, sha1sum, sha224sum, sha256sum, and sha512sum?

34. Using the procedures and techniques developed in Section G19.5 on CUPS printing, found at the GitHub site for this book, do the following exercise-

    You have two computers on an intranet LAN, named Proliant and Black_Dragon. You attach a USB-bus printer to Black_Dragon, and it is automatically detected and usable by CUPS facilities, but on Black_Dragon only. Use the steps shown in Chapter W26, Sections 5.2.1.1–5.2.6 at the book website to allow you to manage printers on two computers from your LAN. For example, from a Web browser on one computer, access the CUPS Web-based interface on the second computer. From the first computer, you can also use a Web browser and access the Web-based interface on the second computer to manage the printer on the first computer?

35. Enable the CUPS printing service using systemd on your Linux system, and use the systemctl commands shown in Chapter W26, Sections 5.1.1–5.1.5 at the book website, to manage that service. In particular, examine the log entries, using the systemd journalctl command to troubleshoot or gain an understanding of how viewing the log can be coupled with management of CUPS using systemd.

36. Use the CUPS Print Manager GUI on your Linux system, as illustrated in Chapter W26, Section 5.3 at the book website, to add, rename, and otherwise manage a new USB-attached printer.

37. Modify the 13 examples in Chapter 17, Section 6.3.1 on rsync to be able to execute them locally on your system, and remotely on your LAN, substituting file names, directories, account names, and IP addresses as necessary.

Projects

Project 1

Modify the Extended Python Script Example Using the rsync command to do a "Rolling" Backup, Chapter W19, Example 19.34 at the book website, to serve as an automated way of copying and backing up specific user directories and files on your Linux system. If you haven't already done so, do the prerequisites and Chapter W19, Examples 31–33, in Section 4.2.2 at the book website.

Project 2

If you haven't already done so, do the prerequisites and from Chapter W19, Examples 31–34, in Section 4.2.2 at the book website, but achieve the same results WITHOUT using Python. Just use the Linux commands shown embedded in the Python code to achieve these results. Most importantly, these examples illustrate the use of the rsync command to do file copying and backup.

Looking for more? Visit our sites for additional readings, recommended resources, and exercises.

CRC Press e-Resource: https://www.crcpress.com/9781138710085

Authors' GitHub: https://github.com/bobk48/linuxthetextbook

7

Advanced File Processing

OBJECTIVES

- To explain file compression and how it can be performed
- To explain the sorting process and how files can be sorted
- To discuss searching for commands and files in the Linux file structure
- To discuss the formation and use of regular expressions
- To describe searching files for expressions, strings, and patterns
- To describe how database-type operations of cutting and pasting fields in a file can be performed
- To discuss encoding and decoding of files
- To explain file encryption and decryption
- To cover the following commands and primitives:

  ^, *, /, \, |, +, >, ~, [,], md5sum, base64, bzip2, mcrypt, crypto, cut, egrep, fgrep, find, grep, md5sum, openssl, paste, pcat, shasum, sort, whereis, which, xz, zcat

7.1 Introduction

In this chapter, we describe some of the more advanced file processing operations on text and nontext files, and show how they can be performed in Linux. These operations include sorting files, searching for files and commands in the file system structure, searching files for certain strings or patterns, performing database-like operations of cutting fields from a table or pasting tables together, transforming non- American Standard Code for Information Interchange (ASCII) files to ASCII, compressing and decompressing file contents, and encrypting and decrypting files. Several tools are available in Linux for performing these tasks.

We discuss the important topic of *regular expressions*, which are a set of rules that can be used to specify one or more strings using a sequence of special text characters. While discussing the operations, we also describe the related shell commands and tools that make use of regular expressions. We also give examples to illustrate how these commands can be used to perform the required operations.

7.2 Sorting Files

Sorting means ordering a set of items according to some criteria. In computer jargon, it means ordering a set of items (e.g., integers, a character, or strings) in *ascending* (the next item is greater than or equal to the current item) or *descending* (the next item is less than or equal to the current item) order. So, for example, a set of integers {10, 103, 75, 22, 97, 52, 1} would become {1, 10, 22, 52, 75, 97, 103} if sorted in ascending order, and {103, 97, 75, 52, 22, 10, 1} if sorted in descending order. Similarly, words in a dictionary are listed in ascending order. Thus, the word apple appears before the word apply.

Sorting is a commonly used operation and is also performed in a variety of software systems. Systems in which sorting is used include

- Words in a dictionary
- Names of people in a telephone directory

- Airline reservation systems that display arrival and departure times for flights sorted according to flight numbers at airport terminals
- Names of people displayed in a pharmacy with ready prescriptions
- Names of students listed in class lists coming from the registrar's office

The sorting process is based on using a field, or portion of each item, known as the sort key. To determine the position of each item in the sorted list, you compare the items in a list (usually two at a time) by using their key fields. The choice of the field used as the key depends on the items to be sorted. If the items are personal records (e.g., student employee records), last name, student ID, and social security number are some of the commonly used keys. If the items are arrival and departure times for the flights at an airport, flight number and city name are commonly used keys.

The Linux `sort` utility can be used to sort items in text (ASCII) files. The following is a brief description of this utility.

Syntax:
  `sort [options] [file-list]`

Purpose: Sort lines in the ASCII files in **file-list**

Output: Sorted files to standard output

Commonly used options/features:

`-b`	Ignore leading blanks
`-d`	Sort according to usual alphabetical order: ignore all characters except letters, digits, and then blanks
`-c`	Check for sorted input; do not sort
`-f`	Consider lowercase and uppercase letters to be equivalent
`-m`	Merge already sorted files
`-n`	Compare according to string numerical value
`+n1[-n2]`	Specify a field as the sort key, starting with **+n1** and ending at **-n2** (or end of line if **-n2** is not specified); field numbers start with 0
`-o FILE`	Send sorted output to FILE instead of standard output
`-r`	Sort in reverse order

If no file is specified in **file-list**, `sort` takes input from standard input. The output of the `sort` command goes to standard output. By default, `sort` takes each line, starting with the first column, to be the key and performs case-insensitive sort in alphabetic order. In other words, it rearranges the lines of the file—that is, strings separated by the newline character (\n)—according to the contents of all the fields, going from left to right. Thus, in the output of the `sort` command, lines starting with digits (0–9) appear before lines starting with letters, lowercase letters appear before their uppercase counterparts as in a, A, b, B, c, etc., and 1,000 is treated smaller than 3 because 3 is greater than that of 1 (first letter in 1,000). The following session illustrates these points.

```
$ cat data
100
Hello
war
Zoo
20
apple
World
200
April
40
```

```
$ sort data
100
20
200
40
apple
April
Hello
war
World
Zoo
$
```

With the use of –n option, you can sort data in terms of their numerical values, placing letters before digits and sorting numbers according to their numerical values. The output of the following `sort` command illustrates this point.

```
$ sort -n data
apple
April
Hello
war
World
Zoo
20
40
100
200
$
```

In the output of the preceding command, lines starting with lowercase letters still appear before lines starting with corresponding uppercase letters. If you would like all the lines starting with uppercase letters appear before the lines starting with lowercase letters, you should set the environment variable **LC_ALL** to **C** and export it before using the `sort` command, as in the following session.

```
$ export LC_ALL=C
$ sort -n data
April
Hello
World
Zoo
apple
war
20
40
100
200
$
```

The following session illustrates the use of `sort` with and without some options using the **students** file, containing the items (student records, one per line) to be sorted. Each line contains four fields: first name, last name, e-mail address, and phone number. Each field is separated from the next by one or more space characters.

```
$ cat students
John Johnsen      john.johnsen@tp.com    503.555.1111
Hassaan Sarwar    hsarwar@k12.st.or      503.444.2132
David Kendall     d_kendall@msnbc.org    229.111.2013
John Johnsen      j.johnsen@psu.net      301.999.8888
Ibraheem Sarwar   ibraheem@abc.sci.com   222.123.4567
Kelly Kimberly    kellyk@umich.gov       555.123.9999
Maham Sarwar      smsarwar@k12.st.or     713.888.0000
```

```
Jamie Davidson     j.davidson@uet.edu      515.001.1212
Nabeel Sarwar      n.sarwar@xyz.net        434.555.1212
$ sort students
David Kendall      d_kendall@msnbc.org     229.111.2013
Hassaan Sarwar     hsarwar@k12.st.or       503.444.2132
Ibraheem Sarwar    ibraheem@abc.sci.com    222.123.4567
Jamie Davidson     j.davidson@uet.edu      515.001.1212
John Johnsen       j.johnsen@psu.net       301.999.8888
John Johnsen       john.johnsen@tp.com     503.555.1111
Kelly Kimberly     kellyk@umich.gov        555.123.9999
Maham Sarwar       smsarwar@k12.st.or      713.888.0000
Nabeel Sarwar      n.sarwar@xyz.net        434.555.1212
$
```

Note that the lines in the **students** file are sorted in ascending order by all characters, going from left to right (the whole line is used as the sort key). The following command sorts the file by using the whole line, starting with the last name—the second field (field number 1)—as the sort key.

```
$ sort +1 students
Jamie Davidson     j.davidson@uet.edu      515.001.1212
John Johnsen       j.johnsen@psu.net       301.999.8888
John Johnsen       john.johnsen@tp.com     503.555.1111
David Kendall      d_kendall@msnbc.org     229.111.2013
Kelly Kimberly     kellyk@umich.gov        555.123.9999
Hassaan Sarwar     hsarwar@k12.st.or       503.444.2132
Ibraheem Sarwar    ibraheem@abc.sci.com    222.123.4567
Nabeel Sarwar      n.sarwar@xyz.net        434.555.1212
Maham Sarwar       smsarwar@k12.st.or      713.888.0000
$
```

The following command sorts the file in reverse order by using the phone number as the sort key and ignoring leading blanks (spaces and tabs). The +3 option specifies the phone number to be the sort key (as phone number is the last field), the −r option informs sort to display the sorted output in reverse order, and the −b option asks the sort utility to ignore the leading white spaces between fields.

```
$ sort +3 -r -b students
Maham Sarwar       smsarwar@k12.st.or      713.888.0000
Kelly Kimberly     kellyk@umich.gov        555.123.9999
Jamie Davidson     j.davidson@uet.edu      515.001.1212
John Johnsen       john.johnsen@tp.com     503.555.1111
Hassaan Sarwar     hsarwar@k12.st.or       503.444.2132
Nabeel Sarwar      n.sarwar@xyz.net        434.555.1212
John Johnsen       j.johnsen@psu.net       301.999.8888
David Kendall      d_kendall@msnbc.org     229.111.2013
Ibraheem Sarwar    ibraheem@abc.sci.com    222.123.4567
$
```

The −b option is important if fields are separated by more than one space and the number of spaces differs from line to line, as is the case for the **students** file. The reason is that the space character is "smaller" (in terms of its ASCII value) than all letters and digits, and if we do not skip initial blanks, unexpected output will be generated. The sort keys can be combined, with one being the primary key and others being secondary keys, by specifying them in the order of preference (the primary key occurring first). The following command sorts the **students** file with the last name as the primary key and the phone number as the secondary key.

```
$ sort +1 -2 +3 -b students
Jamie Davidson     j.davidson@uet.edu      515.001.1212
John Johnsen       j.johnsen@psu.net       301.999.8888
John Johnsen       john.johnsen@tp.com     503.555.1111
David Kendall      d_kendall@msnbc.org     229.111.2013
```

```
Kelly Kimberly     kellyk@umich.gov        555.123.9999
Ibraheem Sarwar    ibraheem@abc.sci.com    222.123.4567
Nabeel Sarwar      n.sarwar@xyz.net        434.555.1212
Hassaan Sarwar     hsarwar@k12.st.or       503.444.2132
Maham Sarwar       smsarwar@k12.st.or      713.888.0000
$
```

The primary key is specified as +1 −2, meaning that the key starts with the last name (+1) and ends before the e-mail address field (−2) starts. The secondary key starts at the phone number field (+3) and ends at the end of line. As no field follows the phone number, it alone comprises the secondary key. For our file, however, the end result will be the same as for the command `sort +1 students`, because the first John Johnsen's e-mail address is "smaller" than the second's.

The −c option allows you to check whether data in your input file is already sorted. It returns with no output if file data is already in sorted order and at the first record that is not in sorted order, as shown in the following session. The `sort -c numbers` command returns with no output and the `sort -c names` command returns the first data value that is not in sorted order, i.e., Bill Gates.

```
$ cat numbers
10
20
30
40
$ sort -c numbers
$ cat names
Elon Musk
Bill Gates
Mark Zuckerberg
Jeff Bezos
Jack Ma
$ sort -c names
sort: names:2: disorder: Bill Gates
$
```

You can sort and merge data in multiple files by specifying files in the command line. In the following session, we show how to sort and merge data in files **numbers1** and **numbers2**. Finally, we show how to sort and merge data in both files and remove duplicates using the −u option.

```
$ cat numbers1
40
100
15
$ cat numbers2
15
200
90
40
$ sort -n numbers1 numbers2
15
15
40
40
90
100
200
$ sort -u -n numbers1 numbers2
15
40
90
100
200
$
```

Exercise 7.1

Repeat the abovementioned sessions on your system and verify that the sort command works on your system as expected.

Exercise 7.2

Consider the **numbers** file contains the following data, one per line: 10, 20, 30, 40, 100, 200, 300, 400. Sort the command with the `sort numbers`. Does the output make sense to you? Why or why not? Answer these questions for the output of the `sort -c numbers` command.

Exercise 7.3

What command would you use to sort data in the **numbers** file in the correct (expected) order, i.e., keep data in the given order?

7.3 Searching for Commands and Files

At times, you will need to find whether a particular command or file exists in your file structure. Or, if you have multiple versions of a command, you might want to find out which one executes when you run the command. We discuss three commands that can be used for this purpose: `find`, `whereis`, and `which`.

You can use the `find` command to search a list of directories that meet the criteria described by the expression (see the command description) passed to it as an argument. The command searches the list of directories recursively; that is, all subdirectories at all levels under the list of directories are searched. The following is a brief description of the command.

Syntax:
   `find directory-list expression`

Purpose: Search the directories in **directory-list** to locate files that meet the "criteria" described by the **expression** (the second argument); the expression comprises one or more "criteria" (see the examples)

Output: None unless it is explicitly requested in **expression**

Commonly used options/features:

`-L`	Follow symbolic links
`-exec` CMD	The file being searched meets the criteria if the command CMD returns 0 as its exit status (true value for commands that execute successfully); CMD must terminate with a quoted semicolon (\;)
`-group name`	Search for files belonging to group name or ID **name**
`-inum N`	Search for files with inode number N
`-links N`	Search for files with N links
`-mtime N`	Search for files whose data was modified within the last N days (for -N) and more than N days (for +N); 0 acts as −1
`-name pattern`	Search for files that are specified by the **pattern**
`-newer file`	Search for files that were modified after **file** (i.e., are newer than **file**)
`-ok` CMD	Like -**exec** except that the user is prompted first
`-perm octal`	Search for files if permission of the file is **octal**
`-print`	Display the pathnames of the files found by using the rest of the criteria; works even if file names contain white spaces such as newline characters
`-size ±N[c]`	Search for files of size N blocks; N followed by **c** can be used to measure size in characters; +N means size > N blocks, and −N means size < N blocks
`-type t`	File is of type **t**: b (block special), c (character special), d (directory), f (regular), l (symbolic link—never true with –L option), p (named pipe—FIFO), s (socket)

-user name	Search for files owned by the user name or ID **name**
\(expr \)	True if **expr** is true; used for grouping criteria combined with **OR** or **AND**
! expr	True if **expr** is false

You can use [-a] or a space to logically AND, and -o to logically OR two criteria. Note that at least one space is needed before and after an open bracket ([) or a close bracket (]), and before and after -o. A complex expression can be enclosed in parentheses, \(and \). We now discuss some illustrative examples.

The find (or find .) command lists pathname for all the files and directories under the current directory. The most common use of the find command is to search one or more directories for a file, as shown in the first example. Here, the command searches for the **USA.gif** and **Pakistan.gif** files in your home directory and displays the pathname of the directory that contains them. If the file(s) being searched for occurs in multiple directories, the pathnames of all the directories are displayed.

```
$ find ~ -name USA.gif -o -name Pakistan.gif
/home/sarwar/myweb/USA.html
/home/sarwar/myweb/Pakistan.html
$
```

The following command displays the absolute pathnames of all the files in your home directory that end in **.c** and **.C**.

```
$ find ~ -name '*.c' -o -name '*.C'
...
$
```

The next command searches the **/usr/include** directory recursively for a file named **socket.h** and prints the absolute pathname of the file.

```
$ find /usr/include -name socket.h
/usr/include/asm-generic/socket.h
/usr/include/x86_64-linux-gnu/sys/socket.h
/usr/include/x86_64-linux-gnu/asm/socket.h
/usr/include/x86_64-linux-gnu/bits/socket.h
/usr/include/linux/socket.h
$
```

You might want to know the pathnames for all the hard links (discussed in Chapter 8) to a file—that is, files that have the same inode number. The following command recursively searches the root directory for all the files that have inode number 36700164 and prints the absolute pathnames of all such files.

```
$ find /usr . -inum 36700164
/usr/include
$
```

You can use the -perm option to display files under a directory hierarchy with given permissions. For example, you can use the find . -perm 755 command to display a list of files in your current directory with octal permissions 755 (i.e., read, write, and execute for the file owner, and read and execute for other users on your system). You can use the ls -l command to verify the output of the find command.

```
$ find . -perm 755
.
./sandbox_dir
./my_cat
$ ls -l
total 1916
...
```

```
-rw-------  1 sarwar  faculty     502 Feb 27 01:42 memo1
-rwxr-xr-x  1 sarwar  faculty   52080 Feb 25 06:59 my_cat
-rw-r--r--  1 sarwar  faculty  151024 Feb 25 07:58 my_cp
-rw-r--r--  1 sarwar  faculty      56 Feb 28 04:05 names
-rw-r--r--  1 sarwar  faculty      28 Feb 28 04:14 numbers
drwxr-xr-x  2 sarwar  faculty    4096 Feb 27 18:32 sandbox_dir
-rw-r--r--  1 sarwar  faculty     457 Feb 24 13:19 students
$
```

You can use the –size option to display files of specific sizes or range of sizes. The following command displays regular files greater than 1 M bytes in your home directory.

```
$ find ~ -size +1M -type f
/home/sarwar/.linuxmint/mintMenu/apt.cache
/home/sarwar/linux2e/ch7/sandbox_dir/file1.nc
/home/sarwar/linux2e/ch7/sandbox_dir/file1
/home/sarwar/linux2e/ch7/bash_encoded
/home/sarwar/linux2e/newch14/bigdata.old
/home/sarwar/linux2e/newch14/bigdata1
/home/sarwar/linux2e/newch14/bigdata
$
```

You can use the –m option to display pathnames of the files in a directory hierarchy according to their modification time. The following command lists pathnames of the files in your current directory that have been modified within the last 2 days. The parameter –2 indicates within two days.

```
$ find . -mtime -2
./ch7
./ch7/secret_memo
./ch7/sandbox_dir
./ch7/sandbox_dir/secret_memo
./ch7/data
./ch7/names
./ch7/numbers
$
```

Similarly, you can identify files in your **~/linux2e/ch5**, **~/linux2e/ch6**, and **~/linux2e/ch7** directories that have been modified more than seven days ago with the find ~/linux2e/ch[5,6,7] -mtime +7.

The following command locates and removes file(s) named **foobar** in your home directory hierarchy. The curly braces, **{}**, identify the position where the name(s) of the matched file(s) are placed. The command ends with a semicolon. The curly braces must be quoted ('{}') and semicolon escaped (\;) to prevent the shell from interpreting them. If you want to be prompted for permission to remove the file, use the rm -i command or replace –exec with –ok, as shown below.

```
$ find ~ -name foobar -exec rm '{}' \;
$ find ~ -name foobar -ok rm '{}' \;
< rm ... /home/sarwar/linux2e/ch7/foobar > ? y
< rm ... /home/sarwar/linux2e/ch16/foobar > ? n
$ find ~ -name foo -exec rm -i '{}' \;
rm: remove regular file '/home/sarwar/linux2e/ch7/foo'? y
rm: remove regular file '/home/sarwar/linux2e/ch16/foo'? n
$
```

Another use of combining the use of find and exec is to search for text in one or more files. You can use the following command to locate all occurrences of text "int i;" in all C program files in your home directory along with line numbers.

```
$ find ~ -type f -name '*.c' -exec grep -n 'int i;'  {} \;
23:     int i;
33:     int i;
8:      int i;
21:     int i;
$
```

The following command searches the present working directory for files that have the name **core** or have extensions **.jpg** or **.o**, displays their absolute pathnames, and removes them from the file structure. Parentheses are used to enclose a complex criterion. Be sure that you use spaces before and after \(, \), and -o. The command does not prompt you for permission to remove ; in order to be prompted, replace –exec with –ok, or use the rm –i command.

```
$ find . \( -name core -o -name '*.jpg' -o -name '*.o' \) -print -exec rm {} \;
...
$
```

You can use the whereis command to find out whether your system has a particular command, and if it does, where it is in the file structure. You typically need to get such information when you are trying to execute a command that you know is valid but that your shell cannot locate because the directory containing the executable for the command is not in your search path (see Chapters 2 and 4). Under these circumstances, you can use the whereis command to find the location of the command and update your search path. Although whereis is a Berkeley Software Distribution (BSD) command, most Linux systems today have it because they have a BSD compatibility package. Depending on the system you are using, the command not only gives you the absolute pathname for the command that you are searching for but also gives you the absolute pathnames for its manual page and source files if they are available on your system. The following is a brief description of the command.

Syntax:
    whereis [options] [file-list]

Purpose: Locate binaries (executable files), source codes, and manual pages for the commands in file-list–a space-separated list of command names

Output: Absolute pathnames for the files containing binaries, source codes, and manual pages for the commands in file-list

Commonly used options/features:
    –b Search for binaries (executable files) only
    –s Search for source code only

The following examples illustrate the use of whereis command. The first command is used to locate the ftp command. The second command is used to locate the executable file for the cat command. The last command locates the information for the find and tar commands.

```
$ whereis ftp
ftp: /usr/bin/ftp /usr/share/man/man1/ftp.1.gz
$ whereis -b cat
cat: /bin/cat
$ whereis find tar
find: /usr/bin/find /usr/share/man/man1/find.1.gz /usr/share/info/find.info.gz
tar: /usr/lib/tar /bin/tar /usr/include/tar.h /usr/share/man/man1/tar.1.gz
$
```

In the outputs of these commands, the **/bin** directory contains the executable files (also called binaries) essential user commands, the directory **/usr/bin** contains the executable files for utilities, applications, and commands, the directory **/usr/share/man** contains several subdirectories that contain various sections of the Linux online manual, **/usr/lib** libraries for programming languages and packages, and the **/usr/include** directory contains header files.

In a system that has multiple versions of a command, the which utility can be used to determine the location (absolute pathname) of the version that is executed by the shell you are using when you type the command. The following sessions show sample runs of this command.

```
$ which cat
/bin/cat
$ which ftp
/usr/bin/ftp
$ which tar
/bin/tar
$
```

When a command does not work according to its specification, the which utility can be used to determine the absolute pathname of the command version that executes. A local version of the command may execute because of the way the search path is set up in the PATH variable (see Chapters 2 and 4). And, the local version has been broken due to a recent update in the code; perhaps it does not work properly with the new libraries that were installed on the system. The which command takes **command-list** (actually a **file-list** for the commands) as an argument and returns absolute pathnames for them to standard output.

In the following in-chapter exercises, you will get practice using the find, sort, and whereis commands, as well as appreciate the difference between the find and whereis commands.

Exercise 7.4
Give a command for sorting a file called **students** by using the whole line starting with the e-mail address.

Exercise 7.5
Give a command for finding out where the executable code for the traceroute command is on your system.

Exercise 7.6
You have a file called **phones** somewhere in your directory structure, but you do not remember the pathname of the directory it is in. What command would you use to locate it?

7.4 Regular Expressions

A *regular expression* is a sequence of constants and operator symbols (known as operators) that represents a set of strings, commonly known as *search patterns*. Most of the Linux utilities operate on text files a line at a time. Regular expressions are used for searching file contents for the desired patterns on a single line. They may not be used to search for patterns that start on one line and end on another.

Regular expressions allow you to search for strings of a certain size. You can search for strings with particular patterns of letters, numbers, and punctuation marks. You can search for a word that starts and ends with a vowel, and has two consecutive "l" letters. What you do with the strings identified by a regular expression that depends on the tool that you use to identify them. Text editor such as vim can replace them with new strings. Some simply display them on the screen.

Some of the shell metacharacters are similar to the operators used in regular expressions. Although metacharacters in a command line are processed and expanded by the shell before passing them to the command, regular expressions are handled by the command. Bourne shell, Bash, C shell, and utilities such as find and cpio use metacharacters and not regular expressions. On the other hand, utilities such as awk, egrep, emacs, expr, grep, more, perl, python, sed, and vim use regular expressions. To prevent the shell from processing them, operators in a regular expression must be quoted when passed as an option in a shell command.

Different utilities in Linux support different sets of operators, but the following operators are supported by almost all Linux tools that support regular expressions: (), [], ., ^, $, and *. Some use additional operators like |, ?, and +. Table 7.1 shows the purpose of these and other operators with examples.

The * operator may be used to specify zero or more occurrences of the preceding element. For example, a* represents an empty string, a, aa, aaa, and so on. The expression .com matches strings Acom, acom, Bcom, bcom, Ccom, ccom, and so on. Similarly, the expression A.B matches AAB, ABB, ACB, AOB, A1B, A2B, etc. It is important to note that regular expressions match the longest possible pattern. That is, the regular expression A.*B matches "AAB" as well as "AAAABBBBABCCCCAAABCDEFBBBAAAB." It is not the

TABLE 7.1

Regular Expression Operators and Their Support by Linux Tools

Name/Function	Operator	Example Usage	Sample Strings for Example	Supported by
Any character except newline	.	.com	Acom, acom, Bcom, bcom, Ccom, ccom, …	All
Beginning of line	^	^x	A line starting with x	All
End of line	$	x$	A line ending with x	All
Repetition: Zero or more occurrences of the preceding element	*	xy*	x, xy, xyy, xyyy, …	All
Grouping	( ) or \(\)	(xy)+	xy, xyxy, xyxyxy, …	All
Matches any character enclosed in brackets	[ ]	/[Hh]ello/	Hello, hello	All
Matches any character not enclosed in brackets	[^]	/[^A-KM- Z]ove/	Love	All
Concatenation (AND)	None	Xyx	xyz	All
Beginning of a word	\<			All
End of a word	\>			All
M to N duplicates of the preceding element	\{M,N\}			
Alternation (OR)	|	x|y|z	x, y, or z	awk, grep
Optional: Zero or one occurrence of the preceding element	?	xy?	x, xy	awk, egrep
Repetition: One or more occurrences of the preceding element	+	xy+	xy, xyy, xyyy, …	awk, egrep
Escape sequence: Cancels the special meaning of the metacharacter that follows it	\	*	*	ed, sed, vi
Delimiter: Marks the beginning or end of a regular expression	/	/L..e/	Love, Live, Lose, Lase, …	ed, sed, vi

cause of any problems when you are using searching tools such as grep, because such an expression will just match more lines than desired. However, if you use such patterns with an editor like vim or sed, you may end up deleting more than you wanted too.

The regular expression aa* represents a, aa, aaa, and so on. It is equivalent to the regular expression a+. The regular expression ^a represents letter a at the beginning of a line. Similarly, a$ matches lines that end with letter a.

The ? operator may be used to specify zero or one occurrence of the preceding element (a character or a pattern). For example, a? specifies a string with no character (an empty string) or a string with only a. The + operator may be used to specify one or more occurrence of the preceding element. For example, a+ represents a, aa, aaa, and so on. The | operator specifies alternatives. For example, a|b, pray|prey, or a|b*. The regular expression a|b* represents an empty string, a, b, bb, bbb, and so on.

The () operator is used to specify the scope and precedence of operators. For example, pr(a|e)y represents pray or prey, and is thus equivalent to pray|prey. You can use the [] operator to specify the exact characters you want to search for. For example, [AaBbAc] matches a single character that is A, a, B, b, C, or c. Similarly, [A-Za-z] matches a single uppercase or lowercase letter and [0-9]* matches zero or more digits.

At times you need to use an element (e.g., a letter) that you have already identified. You can enclose part of a pattern that you would like to refer in future between "\(" and "\)". Each occurrence of "\(" starts a new pattern. You can have nine different numbered patterns and can refer to a previous pattern

TABLE 7.2

Examples of Regular Expressions for vim and Their Meaning

Regular Expression	Meaning	Examples
/^Yes/	A line starting with the string **Yes**	**Yes**... **Yes**teryear... **Yes**terday... and so on
/th/	Occurrence of the string **th** anywhere in a word	**th**e, **th**ere, pa**th**, ba**th**ing, and so on
/:$/	A line ending with a colon	... following: ... below: ... follows: and so on
/[0-9]/	A single digit	0, 1, ..., 9
[a-z][0-9]/	A single lowercase English letter followed by a single digit	a0, a1, ..., a9, ..., b0, b1, ..., b9, ..., z0, ... z1, ..., z9
/\.c/	Any word that ends with **.c** (all C source code files)	**lab1.c**, **program1.c**, **client.c**, **server.c**, and so on
/[a-zA-Z]*/	Any string composed of letters (uppercase or lowercase) and spaces; no numbers and punctuation marks	All strings without numbers and punctuation marks such as He**ll**o world

TABLE 7.3

Some Commonly Used vim Commands Illustrating the Use of Regular Expressions

/ [0-9] /	Do a forward search for a single stand-alone digit character in the current file; digits that are part of strings are not identified.
?\.c[1-7] ?	Do a backward search for words or strings in words that end with **.c** followed by a single digit between 1 and 7.
:1,$s/:$/./	Search the whole file and substitute a colon (:) at the end of a line with a period (.).
:.,$s/^[Hh]ello / Greetings /	From the current line to the end of file, substitute the words "Hello" and "hello" starting a line with the word "Greetings."
:1,$s/^ *//	Eliminate one or more spaces at the beginning of all the lines in the file.

using "\" followed by a single digit starting with 1. Thus, the regular expression "\([0-9]\)\1" matches two identical digits such as 33 and 99.

Some of the commonly used Linux tools that allow the use of regular expressions are awk, ed, egrep, grep, sed, vi, and vim, but the level of support for regular expressions isn't the same for all these tools. Although awk and egrep have the best support for regular expressions, grep has the weakest.

Table 7.1 lists the regular expression operators, their names, example usage, meanings, and tools that support them. The regular expression operators overlap with shell metacharacters, but you can use single quotes around them to prevent the shell from interpreting them. The word "All" in the last column means that all Linux utilities support the corresponding operator. We do not use quotes for strings in the fourth column for brevity.

Table 7.2 lists some commonly used regular expressions in the vim editor and their meanings. Needless to say, regular expressions are used in the vim commands. We discuss examples for grep and egrep in Section 7.6. In the regular expression /\.c/, the backslash character (\) is used to escape the special meaning of dot (.) and take its literal meaning.

Table 7.3 lists some examples of the vim commands that use regular expressions and their meaning. Note that these commands are used when you are in vim's command mode.

In the following in-chapter exercises, you will use regular expressions in the vim editor to appreciate their power.

Exercise 7.7

List the strings that regular expression "\([0-9]\)\1\1" matches?

Exercise 7.8

Create a file that contains the words "Linux," "UNIX," "Windows," and "DOS." Be sure that some of the lines in this file end with those words. Replace the string Windows with Linux in the whole document as you edit it with the vim editor. What command(s) did you use?

Exercise 7.9

As you edit the document in Exercise 7.8, in a vim, run the command :1,$s/DOS\./LINUX\./ gp. What did the command do to the document?

Exercise 7.10

Create a file called **testregex** that has several (but not all) lines containing words express, expressing, expression, expressions, and expresses. What do you think would be the effect of executing the :1,$s/e....ss/DOODLING/gp command while editing **testregex**. Now edit the file with vim and run the command. What did the command do to the document? Does the result make sense? Why or why not?

7.5 Searching Files

Linux has powerful utilities for file searching that allow you to find lines in text files that contain a particular expression, string, or pattern. For example, if you have a large file that contains the records for a company's employees, one per line, you might want to search the file for line(s) containing information on John Johnsen. The utilities that allow file searching are grep, egrep, and fgrep. The following is a brief description of these utilities.

Syntax:
```
grep [options] pattern [file-list]
egrep [options] [string] [file-list]
fgrep [options] [expression] [file-list]
```
Purpose: Search the files in **file-list** for the given pattern, string, or expression; if no **file-list**, take input from standard input

Output: Lines containing the given pattern, string, or expression on standard output

Commonly used options/features:

-c Print the number of matching lines only

-i Ignore the case of letters during the matching process

-l Print only the names of files with matching lines

-n Print line numbers along with matched lines

-r Read files under a directory (home directory if none is specified) recursively; following symbolic links only if they are specified on command line

-s Useful for shell scripts; suppresses error messages (the *return status* is set to zero for success and nonzero for no success—see Chapter 10)

-v Print nonmatching lines

-w Search for the given pattern as a string

Of the three, the fgrep command is the fastest but most limited; egrep is the slowest but most flexible, allowing full use of regular expressions; and grep has reasonable speed and is fairly flexible in terms of its support of regular expressions. In the following sessions, we illustrate the use of these commands with some of the options shown in the description. We use the same **students** file in these sessions that we use in describing the sort utility in Section 7.3. We display the file by using the cat command.

```
$ cat students
John Johnsen        john.johnsen@tp.com     503.555.1111
Hassaan Sarwar      hsarwar@k12.st.or       503.444.2132
David Kendall       d_kendall@msnbc.org     229.111.2013
John Johnsen        j.johnsen@psu.net       301.999.8888
Ibraheem Sarwar     ibraheem@abc.sci.com    222.123.4567
Kelly Kimberly      kellyk@umich.gov        555.123.9999
Maham Sarwar        smsarwar@k12.st.or      713.888.0000
Jamie Davidson      j.davidson@uet.edu      515.001.1212
Sandy Khan          sandy.khan@isu.edu      515.101.9009
Nabeel Sarwar       n.sarwar@xyz.net        434.555.1212
$
```

The most common and simple use of the grep utility is to display the lines in a file containing a particular string, word, or pattern. In the following session, we display those lines in the **students** file that contain the string Sarwar. The lines are displayed in the order they occur in the file.

```
$ grep Sarwar students
Hassaan Sarwar      hsarwar@k12.st.or       503.444.2132
Ibraheem Sarwar     ibraheem@abc.sci.com    222.123.4567
Maham Sarwar        smsarwar@k12.st.or      713.888.0000
Nabeel Sarwar       n.sarwar@xyz.net        434.555.1212
$
```

The grep command can be used with the -n option to display the output lines with line numbers. In the following session, the lines in the **students** file containing the string John are displayed with line numbers.

```
$ grep -n John students
1: John Johnsen      john.johnsen@tp.com     503.555.1111
4: John Johnsen      j.johnsen@psu.net       301.999.8888
$
```

You can use the grep command to search a string in multiple files with regular expressions and shell metacharacters. In the following session, grep searches for the string "include" in all the files in the present working directory that end with **.c** (C source files). Note that the access permissions for **server.c** were set so that grep couldn't read it; the user running the command did not have the read permission for the **server.c** file.

```
$ grep -n include *.c
client.c: 21:   #include        <stdio.h>
client.c: 22:   #include        <ctype.h>
client.c: 23:   #include        <string.h>
lab1.c:   13:   #include        <stdio.h>
grep: can't open server.c
$
```

You can also use the grep command with the -l option to display the names of files in which the pattern occurs. However, it does not display the lines that contain the pattern. In the following session, the ~/**States** directory is assumed to contain one file for every US state, and this file is assumed to contain the names of all the cities in the state (e.g., Portland). The grep command, therefore, displays the names of files that contain the word "Portland"—that is, the names of states that have a city called Portland.

```
$ grep -l Portland ~/States
Maine
Oregon
$
```

Certain characters are treated specially by both shell and grep. Therefore, to make sure that shell passes the desired regular expression to grep, you need to enclose the regular expression in single or double quotes. You can pass quote a character by using backslash (\), unless the character is newline (\n). A single quote (') quotes every character except itself. A double quote (") quotes every character except ", $, |, or '. Thus, you may replace " with ' in any command that uses regular expressions enclosed in " and the command would work, but not vice versa. In the following sessions, we use single and double quotes interchangeably. Expressions enclosed in single and double quotes are also passed verbatim to grep and egrep.

The following command displays the lines in the **students** file that start with the letters A to H. In the command, ∧ specifies the beginning of a line.

```
$ grep '^[A-H]' students
Hassaan Sarwar       hsarwar@k12.st.or        503.444.2132
David Kendall        d_kendall@msnbc.org      229.111.2013
$
```

The following command displays the lines from the **students** file that contain eight consecutive lower-case letters.

```
$ grep '[a-z]\{8\}' students
Ibraheem Sarwar      ibraheem@abc.sci.com     222.123.4567
Maham Sarwar         smsarwar@k12.st.or       713.888.0000
Jamie Davidson       j.davidson@uet.edu       515.001.1212
$
```

The following command displays the lines from the **students** file that start five consecutive lowercase or upper case letters followed by a space character.

```
$ grep '^[A-Za-z]\{5\} ' students
David Kendall        d_kendall@msnbc.org      229.111.2013
Kelly Kimberly       kellyk@umich.gov         555.123.9999
Maham Sarwar         smsarwar@k12.st.or       713.888.0000
Jamie Davidson       j.davidson@uet.edu       515.001.1212
Sandy Khan           sandy.khan@isu.edu       515.101.9009
$
```

The character sequence \< is used to indicate the start of a word. Single (or double) quotes are used to ensure that the shell does not interpret any letter in the pattern as a shell metacharacter, as in '\<Ke' or in "\<Ke." Thus, the following command displays the lines that contain a word starting with the string "Ke."

```
$ grep "\<Ke" students
David Kendall        d_kendall@msnbc.org      229.111.2013
Kelly Kimberly       kellyk@umich.gov         555.123.9999
$
```

By using the regular expression "\<K," the output of the grep command displays lines that contain words starting with the letter K. Thus, the output of the command also includes the line for Sandy Khan, as follows.

```
$ grep "\<K" students
David Kendall        d_kendall@msnbc.org      229.111.2013
Kelly Kimberly       kellyk@umich.gov         555.123.9999
Sandy Khan           sandy.khan@isu.com       515.101.9009
$
```

The string \> is the end of the word anchor. Thus, the following command displays the lines that contain words that end with "net." If we replace the string net with the string war, what would be the output of the command?

```
$ grep 'net\>' students
John Johnsen        j.johnsen@psu.net        301.999.8888
Nabeel Sarwar       n.sarwar@xyz.net         434.555.1212
$
```

In the following command, the regular expression "Kimberly|Nabeel" is used to have egrep display the lines, and their numbers, that contain either "Kimberly" or "Nabeel." Note that the regular expression uses the pipe symbol (|) to logically OR the two strings.

```
$ egrep -n "Kimberly|Nabeel" students
6:Kelly Kimberly     kellyk@umich.gov         555.123.9999
10:Nabeel Sarwar     n.sarwar@xyz.net         434.555.1212
$
```

The egrep -v Kimberly\|Nabeel students command would also produce the same result because the pipe character has been escaped using \|.

You can use the -v option to display the lines that do not contain the string specified in the command. The following command produces all the lines not containing the words "Kimberly" and "Nabeel."

```
$ egrep -v Kimberly\|Nabeel students
John Johnsen        john.johnsen@tp.com      503.555.1111
Hassaan Sarwar      hsarwar@k12.st.or        503.444.2132
David Kendall       d_kendall@msnbc.org      229.111.2013
John Johnsen        j.johnsen@psu.net        301.999.8888
Ibraheem Sarwar     ibraheem@abc.sci.com     222.123.4567
Maham Sarwar        smsarwar@k12.st.or       713.888.0000
Jamie Davidson      j.davidson@uet.edu       515.001.1212
Sandy Khan          sandy.khan@isu.edu       515.101.9009
$
```

The following command displays the lines in the **students** file that start with the letter J. Note the use of ^ to indicate the beginning of a line.

```
$ egrep "^J" students
John Johnsen        john.johnsen@tp.com      503.555.1111
John Johnsen        j.johnsen@psu.net        301.999.8888
Jamie Davidson      j.davidson@uet.edu       515.001.1212
$
```

The following command displays the lines in the **students** file that start with the letters J or K. Note that ^J and ^K represent lines starting with the letters J and K.

```
$ egrep "^J|^K" students
John Johnsen        john.johnsen@tp.com 503.555.1111
John Johnsen        j.johnsen@psu.net   301.999.8888
Kelly Kimberly      kellyk@umich.gov    555.123.9999
Jamie Davidson      j.davidson@uet.edu  515.001.1212
$
```

The egrep \^J\|\^K students command could produce the same result. However, the egrep "^J|^K" students command uses simpler syntax. As a rule of thumb, if you need to escape multiple special characters in a regular expression, enclose the regular expression in single or double quotes, as the case may be.

In the following in-chapter exercises, you will use the commands of the grep family to understand their various characteristics.

Exercise 7.11

Give a command for displaying the lines in the ~/**Personal**/**Phones** file that contain the words starting with the string David.

Exercise 7.12

Give a command for displaying the lines in the **~/Personal/Phones** file that contain phone numbers with area code 212. Phone numbers are stored as xxx-xxx xxxx, where x is a digit from 0 to 9.

Exercise 7.13

Display the names of all the files in your home directory that contain the word "main" (without quotes). What command did you use?

Exercise 7.14

Display the names of all the files in your home directory that contain the word "main" (without quotes). What command did you use?

7.6 Cutting and Pasting

You can process files that store data in the form of tables in Linux by using the cut and paste commands. A table consists of lines, each line comprises a record, and each record has a fixed number of fields. Tabs or spaces usually separate fields, although any field separator can be used. The cut command allows you to cut one or more fields of a table in one or more files and send them to standard output. In other words, you can use the cut command to slice a table vertically in a file across field boundaries. The following is a brief description of the command.

Syntax:
```
cut -blist [-n] [file-list]
cut -clist [file-list]
cut -flist [-dchar] [-s] [file-list]
```
Purpose: Cut out fields of a table in a file

Output: Fields cut by the command

Commonly used options/features:

-b list	Treat each byte as a column and cut bytes specified in the **list**
-c list	Treat each character as a column and cut characters specified in the **list**
-d char	Use the character **char** instead of the <Tab> character as field separator
-f list	Cut fields specified in the **list**
-n	Do not split characters (used with **-b** option)
-s	Do not output lines that do not have the delimiter character

---

Here, **list** is a comma-separated list, with - used to specify a range of bytes, characters, or fields. The following sessions illustrate some of the commonly used options and features of the cut command. In this section, we use the file **student_addresses**, whose contents are displayed by the cat command.

```
$ cat student_addresses
John    Doe     jdoe@xyz.com     312.111.9999    312.999.1111
Pam     Meyer   meyer@uop.pk     666.222.1212    666.555.1212
Jim     Davis   jamesd@aol.org   713.999.5555    713.413.0000
Jason   Kim     j_kim@up.org     434.000.8888    434.555.2211
Amy     Nash    nash@state.gov   888.111.4444    888.827.3333
$
```

The file has five fields numbered 1–5, from left to right: first name, last name, e-mail address, home phone number, and work phone number. Although we could have used any character as the field separator, we chose the <Tab> character to give a "columnar" look to the table and the output of the following cut and paste commands. You can display a table of first and last names by using the -f option. Note that -f1,2 specifies the first and the second fields of the **student_addresses** file.

```
$ cut -f1,2 student_addresses
John    Doe
Pam     Meyer
Jim     Davis
Jason   Kim
Amy     Nash
$
```

We generate a table of names (first and last) and work phone numbers by slicing the first, second, and fifth fields of the table in the **student_addresses** file.

```
$ cut -f1,2,5 student_addresses
John    Doe     312.999.1111
Pam     Meyer   666.555.1212
Jim     Davis   713.413.0000
Jason   Kim     434.555.2211
Amy     Nash    888.827.3333
$
```

To generate a table of names and e-mail addresses, we use the following command. Here, –f1–3 specifies fields 1–3 of the **student_addresses** file.

```
$ cut -f1-3 student_addresses
John    Doe     jdoe@xyz.com
Pam     Meyer   meyer@uop.pk
Jim     Davis   jamesd@aol.org
Jason   Kim     j_kim@up.org
Amy     Nash    nash@state.gov
$
```

We recommend that you run this command on your machine to determine whether the desired output is produced. If the desired output is not produced, you have not used the <Tab> character as the field separator for some or all of the records. In such a case, correct the table and try the command again.

In the preceding sessions, we have used the default field separator, the <Tab> character. Depending on the format of your file, you can use any character as a field separator. For example, as we discussed in Chapter 4, the **/etc/passwd** file uses the colon character (:) as the field separator. You can therefore use the cut command to extract information, such as the login name, real name, group ID, and home directory for a user. Because the real name, login name, and home directory are the fifth, first, and sixth fields, respectively, the following command can be used to generate a table of names of all users, along with their login IDs and home directories. The first two lines of the output are for comments, as they start with #.

```
$ cut -d: -f5,1,6 /etc/passwd
# $FreeBSD$
#
root:Charlie &:/root
toor:Bourne-again Superuser:/root
daemon:Owner of many system processes:/root
...
sshd:Secure Shell Daemon:/var/empty
smmsp:Sendmail Submission User:/var/spool/clientmqueue
...
sarwar:Syed Mansoor Sarwar:/home/sarwar
...
$
```

Note that the –d option is used to specify : as the field separator, and it is also displayed as the field separator in the output of the command. For blank delimited files, use one or more space characters (blanks) after –d\, as shown in the following example. The cat sample command is used to display the blank delimited file, called **sample**, and the `cut -d\ -f1,6 sample` command is used to display fields 1 and 6 of this file.

```
$ cat sample
1 John CS Senior john@net2net.com 3.45
2 Jane CS Junior jane@net2net.com 3.76
3 Sara CS Senior sara@net3net.vom 3.33
$ cut -d\ -f1,6 sample
1 3.45
2 3.76
3 3.33
$
```

The `paste` command complements the `cut` command; it concatenates files horizontally (the `cat` command concatenates files vertically). Hence, this command can be used to paste tables in columns. The following is a brief description of the command.

Syntax:
   `paste [options] file-list`

Purpose: Horizontally concatenate files in **file-list**; use standard input if – (i.e., a hyphen) is used as a file

Output: Files in **file-list** pasted (horizontally concatenated)

Commonly used options/features:
   `-d list`   Use **list** characters as line separators; **<Tab>** is the default character

---

Consider the file **student_records**, which contains student names (first and last), major, and current Grade Point Average (GPA).

```
$ cat student_records
John    Doe     ECE     3.54
Pam     Meyer   CS      3.61
Jim     Davis   CS      2.71
Jason   Kim     ECE     3.97
Amy     Nash    ECE     2.38
$
```

We can combine the two tables, **student_records** and **student_addresses_shortened**, horizontally and generate another by using the `paste` command in the following session. To keep the resultant table small, we have used a shortened version of the original **student_addresses** file called **student_addresses_shortened**, which contains the first name, last name, and work phone number only. We generated this table by using the `cut -f1,2,5 student_addresses > student_addresses_short-ened` command. Note that the output of the `paste` command is displayed on the display screen and is not stored in a file. The resultant table has seven fields.

```
$ cat student_addresses_shortened
John    Doe     312.999.1111
Pam     Meyer   666.555.1212
Jim     Davis   713.413.0000
Jason   Kim     434.555.2211
Amy     Nash    888.827.3333

$ paste student_records student_addresses_shortened
John    Doe     ECE     3.54    John    Doe     312.999.1111
Pam     Meyer   CS      3.61    Pam     Meyer   666.555.1212
Jim     Davis   CS      2.71    Jim     Davis   713.413.0000
Jason   Kim     ECE     3.97    Jason   Kim     434.555.2211
Amy     Nash    ECE     2.38    Amy     Nash    888.827.3333
$
```

Suppose that you want to use the **student_addresses_shortened** and **student_records** tables to generate and display a table that has student names, majors, and home phone numbers. You may do so in one of two ways. When you use the first method, you cut the appropriate fields of the two tables, put them in separate files with the fields in the order you want to display them, paste the two tables in the correct order, and remove the tables. The following session illustrates this procedure and its result. Note that the new table is not saved as a file when the following commands are executed. If you want to save the new table in a file, use the paste table1 table2 > students_table command. The **students_table** contains the columns of **table1** and **table2** (in that order) pasted together.

```
$ cut -f1-3 student_records > table1
$ cut -f4 student_addresses > table2
$ paste table1 table2
John    Doe     ECE     312.111.9999
Pam     Meyer   CS      666.222.1212
Jim     Davis   CS      713.999.5555
Jason   Kim     ECE     434.000.8888
Amy     Nash    ECE     888.111.4444
$ rm table1 table2
$
```

The procedure just outlined is expensive in terms of space and time because you have to execute four commands, generate two temporary files (**table1** and **table2**) on disk, and remove these temporary files after the desired table has been displayed. You can use a different method to accomplish the same thing with the following command.

```
$ paste student_records student_addresses | cut -f1-3,8
John    Doe     ECE     312.111.9999
Pam     Meyer   CS      666.222.1212
Jim     Davis   CS      713.999.5555
Jason   Kim     ECE     434.000.8888
Amy     Nash    ECE     888.111.4444
$
```

Here, you first combine the tables in the two files into one table with nine columns by using the paste student_records student_addresses command and then displaying the desired table by using the cut -f1-3,8 command. Clearly, this second method is the preferred way to accomplish the task because no temporary files are created and only one command is needed. If you want to save the resultant table in the **students_table** file, use the command paste student_records student_addresses | cut -f1-3,8 > students_table.

Exercise 7.15

Create the **student_addresses** and **student_records** files used in Section 7.6. Then, run the cut and paste commands described in this section to see how these commands work.

7.7 Compressing Files

Reduction in the size of a file is known as file compression, which has both space and time advantages. A compressed file takes less disk space to store, less time to transmit from one computer to another in a network or Internet environment, and less time to copy. It takes time to compress a file, but if a file is to be copied or transmitted several times, the time spent compressing the file could be just a fraction of the total time saved. In addition, if the compressed file is to be stored on a secondary storage device (e.g., a disk) for a long time, the savings in disk space can be considerable. Another consequence of compression is that a compressed file reads as garbage. However, this is not a problem because the process is fully reversible and a compressed file can be converted back to its original form. If you can fully recover the original file from its compressed version, the compression is known as *lossless compression*. Lossless compression techniques are normally used for text files. There are *lossy compression* techniques too. These techniques are used to reduce the size of a file for storing, handling, and transmitting its content between computers on a network. Lossy compression techniques are normally used for image files such as Joint Photographic Experts Group (JPEG). In this chapter, we discuss tools for compressing and decompressing text and executable files. We also discuss commands for displaying and searching compressed files.

The Linux operating system has many commands for compressing and decompressing files and for performing various operations on compressed files. These commands include the Gnu tools gzexe (compress executable files), gzip (for compressing files), gunzip (for uncompressing files that were compressed with gzip), zcat (for displaying compressed files; gzcat does the same), gzcmp (for comparing compressed files), gzforce (for forcing the **.gz** extension onto compressed files so that gzip will not compress them twice), gzmore (for displaying compressed files one page at a time), and gzgrep (the grep command for compressed files; it searches possibly compressed files for a regular expression). This section will primarily discuss file compression and decompression.

7.7.1 The gzip Command

The gzip command is the Gnu tool for compressing files. The compressed file is saved as a file that has the same name as the original, with an extension **.gz** appended to it. As is the case with the compress command, the compressed files retain the access/modification times, ownership, and access privileges of the original files. The original file is removed from the file structure. With no file argument or with - as an argument, the gzip command takes input from standard input (keyboard by default), which allows you to use the command in a pipeline (see Chapter 9). We normally use the command with one or more files as its arguments. Here is a brief description of the command.

Syntax:
```
gzip [option] [file-list]
```
Purpose: Compress each file in **file-list** and store it in **filename.gz**, where **filename** is the name of the original file. If no file is specified in the command line or if – is specified, take input from standard input.

Output: The compressed **.gz** file or standard output if input is from standard input

Commonly used options/features:

- **-N** Control compression speed (and compression ratio) according to the value of **N**, with 1 being fastest and 9 being slowest; slow compression compresses more

- **-c** Send output to standard out; input files remain unchanged

- **-d** Uncompress a compressed (**.gz**) file

- **-f** Force compression of a file when its **.gz** version exists, or it has multiple links, or input file is **stdin**

- **-l** For the compressed files given as arguments, display sizes of the uncompressed and compressed versions, compression ratio, and uncompressed name

- **-r** Recursively compress files in the directory specified as arguments

- **-t** Test integrity of the compressed files specified as arguments

- **-v** Display compression percentage and the names of compressed files

---

7.7.2 The gunzip Command

The gunzip command can be used to perform the reverse operation and bring compressed files back to their original forms. The gzip -d command can also perform this task. With the gunzip command, the -N, -c, -f, -l, and -r options work just like they do with the gzip command.

The following session shows the use of the two commands with and without arguments. We use the man bash > bash.man and man tcsh > tcsh.man commands to save the manual page for the Bourne Again and TC shells in the **bash.man** and **tcsh.man** files, respectively. The gzip bash.man tcsh.man command is used to compress the **bash.man** and **tcsh.man** files, and the gzip -l bash.man.gz tcsh.man.gz command is used to display some information about the compressed and uncompressed versions of the **bash.man** and **tcsh.man** files. The output of the command shows, among other things, the percentage of compression achieved: 71.3% for **bash.man** and 67.6% for **tcsh.man**. The gzip bash.man.gz command is used to show that gzip does not compress an already compressed file that has a **.gz** extension. If a compressed file does not have the **.gz** extension, gzip will try to compress it again. The gunzip *.man.gz command is used to decompress the **bash.man.gz** and **tcsh.man.gz** files. The gzip -d *.man.gz command can be used to perform the same task. The ls -l commands have been used to show that the modification time, ownership, and access privileges of the original file are retained for the compressed file.

```
$ man bash > bash.man
$ man tcsh > tcsh.man
$ ls -l *.man
-rw-r--r--  1 sarwar  faculty  367470 Aug 30 21:39 bash.man
-rw-r--r--  1 sarwar  faculty  239996 Aug 30 21:39 tcsh.man
$ gzip bash.man
$ ls -l bash.man.gz
-rw-r--r--  1 sarwar  faculty  105342 Aug 30 21:39 bash.man.gz
$ gzip bash.man.gz
gzip: bash.man.gz already has .gz suffix - unchanged
$ gunzip bash.man.gz
$ gzip bash.man tcsh.man
$ gzip -l bash.man.gz tcsh.man.gz
  compressed uncompressed  ratio uncompressed_name
     105342        367470  71.3% bash.man
      77715        239996  67.6% tcsh.man
     183057        607466  69.8% (totals)
$ gunzip *.man.gz
$ ls -l *.man
-rw-r--r--  1 sarwar  faculty  367470 Aug 30 21:39 bash.man
-rw-r--r--  1 sarwar  faculty  239996 Aug 30 21:39 tcsh.man
$
```

7.7.3 The gzexe Command

The gzexe command can be used to compress executable files. An executable file compressed with the gzexe command remains executable and can be executed by using its name. This is not the case if an executable file is compressed with the gzip command. Therefore, an executable file is compressed with the gzexe command in order to save disk space and network bandwidth if the file is to be transmitted from one computer to another—for example, via e-mail over the Internet. The following is a brief description of this command.

Syntax:
  gzexe [options] [file-list]

Purpose: Compress the executable files given in **file-list**; backup files are created in **filename~** and should be removed after the compressed files have been successfully created

Commonly used options/features:

  -d Decompress compressed files

The following session illustrates the use of the gzexe command. Note that when you compress the executable file **sh.temp** with the gzexe command, it creates a backup of the original file in the **sh.temp~** file. After the **sh.temp** file has been compressed, it can be executed as an ordinary executable file. The gzexe -d sh.temp command is used to decompress the compressed file **sh.temp**. The backup of the compressed version is saved in the **sh.temp~** file.

```
$ cp /bin/sh sh.temp
$ file sh.temp
sh.temp: ELF 64-bit LSB executable, x86-64, version 1 (FreeBSD), dynamically
linked (uses shared libs), for FreeBSD 10.0 (1000510), stripped
$ gzexe sh.temp
sh.temp:            49.2%
4 ls -l sh*
-rwx------  1 sarwar  faculty   71133 Aug 30 21:59 sh.temp
-rwx------  1 sarwar  faculty  139264 Aug 30 21:59 sh.temp~
$ gzexe -d sh.temp
 49.2%
$ ls -l sh*
-rwx------  1 sarwar  faculty  139264 Aug 30 22:00 sh.temp
-rwx------  1 sarwar  faculty  139264 Aug 30 21:59 sh.temp~
$
```

7.7.4 The zcat and zmore Commands

Converting the compressed file back to the original and then displaying it is a time-consuming process because file creation requires disk input/output. If you only want to view the contents of the original file, you can use the Linux command zcat (the cat command for compressed files), which displays the contents of files compressed with gzip. The command decompresses a file before displaying it. The original file remains unchanged. The zmore command can be used to display the compressed files one screenful at a time. When no file or - is given as a parameter, these commands read input from **stdin**. Both commands allow you to specify one or more files as parameters. Here is a brief description of the zcat command.

Syntax:
  zcat [options] [file-list]

Purpose: Concatenate compressed files in their original form and send them to standard output; if no file is specified, take input from standard input

Commonly used options/features:

  -h Display help information

  -r Operate recursively on subdirectories

  -t Test integrity of compressed files

In the following session, the gzip command is used to compress the **bash.man** file and store it in the **bash.man.gz** file. When the more command is used to display a compressed file, garbage is displayed on the screen. The zmore command is used to display the contents of the original file. We did not use the zcat command because **bash.man** is a large, multipage file.

```
$ gzip bash.man
$ more bash.man.gz
<8B>^H^HA<FE>T<97><C3>
                      <DD><A0><CC>□<AE>j<82><B0>G<FE><EC>o<DC><99>U<DD><8F><AC><"
<E3><8E>_f<FA><E4><D9>o<8F><F9>O<E4><FF>fʮ<BE><FF><F9><BA><FA><FA>I
<F5><E4>w<D5><BE>xO<DE>><AF><FB><AB>x<BF><FA><FA><BB><F8>i<B7>ᶜ<9B><93>'<97>u
<BB><8E><CF>´<DB>,<97><EF><BD><F7><AC>z<F6><A7><EA>O<F0><C5>o<AA>o<BE><AD><BE>
<85>?<9F>VO<E1>_<FB>zu^Wu_<F5>W<D5>Uf<D1>m<86><B6>ÿ<F7>/<E3><8B>y<B7>Z<D5><EB>
<C5>Y?l<DB><F5>e<FC><CF>p<D1>.<9B><97><EF><BD><F7>Y<F5><8F><A3><FE><BE>
<FA>=
...
$ zmore bash.man.gz
BASH(1)                       BASH(1)

NAME
```

```
        bash - GNU Bourne-Again SHell

SYNOPSIS
        bash ÿoptions¦ ÿcommand_string f file¦

COPYRIGHT
        Bash is Copyright (C) 1989-2013 by the Free Software Foundation, Inc.

DESCRIPTION
        Bash  is  an  sh-compatible  command language interpreter that executes
        commands read from the standard input or from a file.  Bash also incor-
...
$
```

In the following session, the zcat command decompresses the **t2.gz** file and sends its output to standard output (the display screen in this case). The file **t2.gz** remains intact.

```
$ gzip t2
$ zcat t2.gz
This file will be used to test various Linux and Linux commands and tools.
Linux and Linux rule the networking world!
$
```

As given in this command description, multiple files can be displayed by the zcat command. For example, zcat t1.Z t2.Z t3.Z may be used to display the uncompressed forms of the three files **t1.Z**, **t2.Z**, and **t3.Z**. The zcat command may also be used to display the files compressed with the UNIX-only **compress** command—that is, files with the **.Z** extension.

7.7.5 The gzip, bzip2, and xz Commands

All three are famous tools for file compression and decompression. Their time and space performance tradeoff depends on the compression level used. Of the three, gzip has the lowest compression and decompression time, and xz provides the best compression ratio (i.e., smallest size compressed file) and second best decompression time. Thus, you should use gzip if you want to compress and decompress files as fast as possible, and xz if you want to save disk space and network bandwidth while downloading the compressed content. Bzip2 is a good middle of the road tool in terms of compression ratio and compression time. However, it takes the longest time to decompress. Thus, it is a good option if the compressed content will not be decompressed frequently.

In the final analysis, the choice really depends on whether you would like to save disk space or CPU time, and how frequently the compressed content will be decompressed. For content that will be compressed once and frequently downloaded and decompressed, you should use xz, as is done by kernel.org for compressing the Linux kernel. It saves disk space, download time, network bandwidth, and decompression time.

In the following in-chapter exercises, you will use the gzip, gunzip, gzmore, and zcat commands to appreciate their syntax and semantics.

Exercise 7.16
    Create the **t2** file used in this section. Use the gzip command to compress the file. What command line did you use?

Exercise 7.17
    Create the **bash.man** file used in this section. Use the gzip command to compress the file. What command line did you use?

Exercise 7.18
    Display the compressed version of the **t2** file on the display screen. What command line did you use?

Exercise 7.19

Give the command line for uncompressing the compressed files generated in Exercises 7.16 and 7.17. Where does the uncompressed (original) file go? Also, repeat the shell sessions shown in Sections 7.7.1–7.7.4.

7.8 Encoding and Decoding

E-mail messages are transported in clear (plain) text, and some e-mail systems are fussy about certain characters contained in the body of the message, such as the tilde character (~) in the first column for the `mail` and `mailx` utilities. This is a serious problem for mail systems, such as `mail`, that do not have convenient support for attachments when you need to attach items such as pictures, videos, or executable programs (binaries). You can use the `base64` utility to convert the contents of such a file into a format that contains printable ASCII characters only and then e-mail it as the body of your e-mail message. This conversion process is called *encoding*. The receiver can save the encoded contents in the e-mail body into a file and use the `base64` utility to convert the contents of this file to the original format. The process of converting the encoded content into original form is called *decoding*. In this section, we discuss the `base64` utility, starting with its brief description.

Syntax:
> `base64 [option] ... [source-file]`

Purpose: Encode/decode data using the Base64 method and send it to standard output

Commonly used options/features:

> **-d** Decode data

The `base64` command sends the encoded (ASCII) version of the **source-file** to standard output. The command takes input from standard input if no **source-file** is specified, or – is specified, in the command. The `base64 -d` command recreates the original file from the encoded file and sends the decoded (original) content to standard output. You can redirect the output of the `base64` command to a file by using the > symbol. The diagram shown in Figure 7.1 illustrates the process of encoding and decoding.

In the following session, we show the use of `base64` command when data to be encoded and decoded are supplied from standard input. The encoded version of `Hello, world!` is `SGVsbG8sIHdvcmxkIQo=`.

```
$ base64
Hello, world!
<Ctrl+D>
SGVsbG8sIHdvcmxkIQo=
$ base64 -d
SGVsbG8sIHdvcmxkIQo=
<Ctrl+D>
Hello, world!
$
```

In the following session, we show encoding of the binary file for the `cp` command in the **/bin** directory with the `base64` command and its conversion to original form with `base64 -d` command. The `shasum`

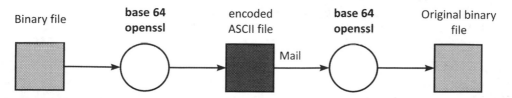

FIGURE 7.1   The process of encoding and decoding.

/bin/cp my_cp command is used to ascertain that **/bin/cp** and **my_cp** are identical files because the shasum command produces the same hash value for the data in both files.

```
$ cat /bin/cp
ELF>?=ÉÉ?FÉ8    ÉÉÉÉÉ?88É8ÉÉÉ?2?2 ?=?=b?=b?( ?=?=b?=TTÉTÉDDP?tdP?P?AP?A4    4
Q?tdR?td?=?=b?=b  /lib64/ld-linux-x86-64.so.2GNU GNU
                                        X:????Ä????2??qX???!?
?hL?#?????????????Pv?CÈ????2???????ö?????qX??ö?,crBE??9??bA????fUa?????Ka?
9???p?)????????6
??           BQ?.        ???ÜJ??p?F?????%??????U????cph????ÄK???????ÅVä?D???R?tu??m-
?i?+=6x?Xv?J???4???&?$?kdd?e.>?$H?
                        ä?D< å??w.Q??"CÅ6P ??jv,?????pBA?K_A??Eb?
8,É??,ÉP!?E?tEbTb?
...
$ base64 /bin/cp > cp_encoded
$ head cp_encoded
```

```
fOVMRgIBAQQAAAAAAAAAAAIAPgBAAAAwD1AAAAAAABAAAAAAAAAALBGAgAAAAAAAAAAEAAOAAJ
AEAAHQAcAAYAAAAFAAAAQAAAAAAAAAABAAEAAAAAAAEAAQAAAAAAA+AEAAAAAAAAD4AQAAAAAAAgA
AAAAAAAAwAAAAQAAAA4AgAAAAAAADgCQAAAAAAAOAJAAAAAAAAcAAAAAAAAABwAAAAAAAAQAA
AAAAAABAAAABQAAAAAAAAAAAAAAAABAAAAAAAAAAEAAAAAAAJwyAgAAAAAnDICAAAAAAAACAA
AAAAAAEAAAAGAAAA4D0CAAAAAADgPWIAAAAAOA9YgAAAAA1AcAAAAAAAAoFgAAAAAAAAAIAAA
AAAAgAAAAYAAAD4PQIAAAAAAPg9YgAAAAA+D1iAAAAAAAgAAAAAAAACAAAAAAACAAAAAAA
AAAEAAAABAAAAFQCAAAAAAAAAVAJAAAAAAAABUAkAAAAAAAEQAAAAAAAARAAAAAAAAAEAAAAAAA
UeVOZAYAAAAAAAAAAAAAAAAAAAAAAAAAAAAAAAAAAAAAAAAAAAAAAAAAAAAAAAAEAAAAAAAAABS
5XRkBAAAAOA9AgAAAAAA4D1iAAAAAADgPWIAAAAACACAAAAAAAAAIAIAAAAAAAABAAAAAAAAC9s
```

```
$ base64 -d cp_encoded > my_cp
$ ls -l my_cp /bin/cp cp_encoded
-rwxr-xr-x 1 root    root      151024 Feb 18  2016 /bin/cp
-rw-r--r-- 1 sarwar faculty 204018 Feb 25 07:56 cp_encoded
-rw-r--r-- 1 sarwar faculty 151024 Feb 25 07:58 my_cp
$ shasum /bin/cp my_cp
ff3094b907d15cee91b8eecb0559011d2d1c175a  /bin/cp
ff3094b907d15cee91b8eecb0559011d2d1c175a  my_cp
$
```

As the output of the last ls −l command in the abovementioned session shows, the size of encoded data generated by base64 (204,018 bytes) is about 35% *larger* than the data in the original file (151,024 bytes). Thus, for efficient use of the network bandwidth, you should compress binary files before encoding them and uncompress them after decoding them. Doing so is particularly important for large files containing multimedia data, such as videos.

You can also use the openssl utility to encode and decode data. In the following session, we encode and decode the **/bin/cp** file, and use the shasum command to confirm that the original and decoded files have identical content.

```
$ openssl enc -base64 < /bin/cp > cp.encoded
$ openssl enc -base64 -d < cp.encoded > my_cp
$ ls -l /bin/cp my_cp cp.encoded
-rwxr-xr-x 1 root    root      151024 Feb 18  2016 /bin/cp
-rw-r--r-- 1 sarwar faculty 204515 Feb 28 23:48 cp.encoded
-rw-r--r-- 1 sarwar faculty 151024 Feb 28 23:49 my_cp
$ shasum /bin/cp my_cp
ff3094b907d15cee91b8eecb0559011d2d1c175a  /bin/cp
ff3094b907d15cee91b8eecb0559011d2d1c175a  my_cp
$
```

Exercise 7.20

Copy the executable code for the grep command from the **/usr/bin** directory and encode it using base64. Run the ls -l command and report the sizes of original and encoded files. Then, decode the encoded file to convert it back to the original file.

7.9 File Encryption and Decryption

We briefly described *encryption* and *decryption* of files in Chapter 5. Here, we describe these processes in more detail with the help of the Linux command `crypt`.

Recall that encryption is a process by which a file is converted to a form completely different from its original version and that the transformed file is called an *encrypted* file; the reverse process of transforming the encrypted file to its original form is known as decryption. Figures 7.2 and 7.3 illustrate these processes.

You encrypt files to prevent others from reading them. You can also encrypt your e-mail messages to prevent hackers from understanding your message even if they are able to tap a network as your message travels through it. On a Linux system, you can use the `mcrypt` command to encrypt and decrypt your files. Ask your system administrator to install the command for you if it isn't already installed. The following is a brief description of the command.

Syntax:
```
mcrypt [options] [filename ...]
```
Purpose: Encrypt (decrypt) standard input and send it to standard output

Commonly used options/features:

-a ALGO	Algorithm used to perform encryption (and decryption)
-d	Decrypt
-g	Use openPGP file format for encrypted files so that OpenPGP compliant tools may access them
-z	Use gzip to compress files before encryption; decompresses files if the option is specified at decryption
-p	Use bzip2 to compress files before encryption; decompresses files if the option is specified at decryption
-u	Delete (unlink) the original file if the whole process of encryption/decryption completes

---

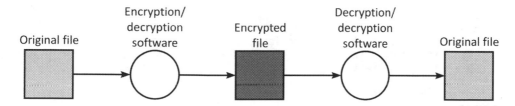

FIGURE 7.2 The process of encryption and decryption.

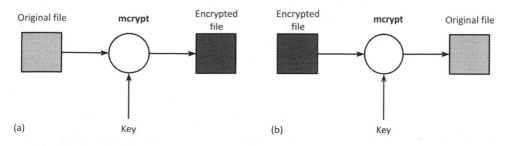

FIGURE 7.3 (a) Encryption and (b) decryption of a file by using the `mcrypt` command.

By default, the mcrypt command takes input from standard input and sends its output to standard output. The command is used mostly with actual files and not the keyboard input. The commonly used syntax for the mcrypt command is

```
mcrypt file
```

The encrypted contents of **file** are placed in **file.nc**. The new file keeps the modification date of the original and has protection mode 0600. The original file (the file to be encrypted) remains intact and must be explicitly removed from the system after the encrypted version has been generated. You can delete the original file successfully by specifying the -u option.

To decrypt an encrypted file, the process is reversed according to the syntax:

```
mcrypt -d file.nc
```

Remember that the key must be the same for both commands. Multiple files can be specified as source files and the command encrypts these files one-by-one, prompting for key for the encryption of each file. The semantics of these commands are shown in Figure 7.3.

The following session illustrates the use of mcrypt for encrypting and decrypting a file called **memo1**. The mcrypt memo1 command encrypts **memo1** (in the present working directory) by using the key (i.e., passphrase) that you are prompted to enter and puts the encrypted version in the file **memo1.nc**, also in the present working directory. The cat commands before and after the crypt command show the contents of the original and encrypted files. Note that the contents of the encrypted file, **memo1.nc**, are not readable, which is the objective. Also note that your shell prompt might get messy after the cat memo1.nc command has completed its execution. If this happens, you should logout and login again.

```
$ cat memo1
Dear Jim:
This is to inform you that the second quarter earnings do not
look good. This information will be made public on Monday next
week when The Wall Street Journal reports the company earnings.
Of course, the company stock will take a hit, but we need to keep
the morale of our employees high. I am calling a meeting of the
vice presidents tomorrow morning at 8:00 to talk about this
issue in the main conference room. See you then. Make sure this
information does not get out before time.
Nadeem
$ mcrypt memo1
Enter the passphrase (maximum of 512 characters)
Please use a combination of upper and lower case letters and numbers.
Enter passphrase: xxxxxxxxx
Enter passphrase: xxxxxxxxx

File memo1 was encrypted.
$ cat memo1.nc
m@rijndael-128 cbcmcrypt-sha1?@1???????<??^??^??sha1?=7ʂI?=??H@gnN???r?:p!}??J?o?B
}??&nI?1??ᵛ3Fδ??X??? ?e?|e!?Q
~??6U?:??aẽ"??
           "????g"?>f?????6;r`?,<?n?tFi?gjp?[??s!??K?t!qL4???.??|? 1??pEG5)
F)??{?n-?B?<???+:??b?m?'Py?<???Wj?w)?G?&~???/?2?VtgB??I?\D?=>o??]
{????Y$?L?{?????9???,a??屈?_?/o?å? ?E?TJ??g?cy?#@??K?hRh??B????{??!lW??_D*??1???aV??
????J???p???\??yxⱼ?B??,?(Z$

???X???TT$G?s?M?c6?n?'?@|?F??Hqⱼn???d??<t?rF?jK???*?zQ???t???y???
??      U??t-h??Zz)?????gH??????9";Y?b???:?9"??Q??Mtx>
~???'={J?T?J\??i護?o?Qu?P%?sarwar@Mint18 ~/linux2e/ch7 $
$
```

The command for decrypting the file **memo1.nc** and putting the original version in the file **memo1** is as follows. The cat command confirms that the original file has been restored. The output of the ls -l memo* command shows that the encrypted file is about 20% larger than the original file.

```
$ mcrypt -d memo1.nc
mcrypt: memo1 already exists; do you wish to overwrite (y or n)?y<Enter>
Enter passphrase: xxxxxxxxx
File memo1.nc was decrypted.
$ more memo1
Dear Jim:
This is to inform you that the second quarter earnings do not
look good. This information will be made public on Monday next
week when The Wall Street Journal reports the company earnings.
Of course, the company stock will take a hit, but we need to keep
the morale of our employees high. I am calling a meeting of the
vice presidents tomorrow morning at 8:00 to talk about this
issue in the main conference room. See you then. Make sure this
information does not get out before time.
Nadeem
$ ls -l memo*
-rw------- 1 sarwar faculty 502 Feb 27 01:42 memo1
-rw------- 1 sarwar faculty 605 Feb 27 01:42 memo1.nc
$
```

You can use the –k option to supply the key at the command line, as in the following session.

```
$ mcrypt memo1 -k $#123#$
Warning: It is insecure to specify keywords in the command line
File secret_memo was encrypted.
$ mcrypt -d memo1.nc -k $#123#$
Warning: It is insecure to specify keywords in the command line
File secret_memo.nc was decrypted.
$
```

You can use mcrypt with –z (or –p) option to compress files before encrypting. The encrypted file is placed in the **memo1.gz.nc** file. After this file is decrypted, the compressed file is placed in the **memo1.gz** file that has to be decompressed with gunzip, as in the following session.

```
$ mcrypt -k $#123#$ -z memo1
Warning: It is insecure to specify keywords in the command line
File memo1 was encrypted.
$ ls -l
total 8
-rw------- 1 sarwar faculty 502 Feb 27 09:03 memo1
-rw------- 1 sarwar faculty 445 Feb 27 09:03 memo1.gz.nc
$ mcrypt -k $#123#$ -d memo1.gz.nc
Warning: It is insecure to specify keywords in the command line
File memo1.gz.nc was decrypted.
$ ls -l
total 12
-rw------- 1 sarwar faculty 502 Feb 27 09:03 memo1
-rw------- 1 sarwar faculty 333 Feb 27 09:03 memo1.gz
-rw------- 1 sarwar faculty 445 Feb 27 09:03 memo1.gz.nc
$ gunzip -v memo1.gz
gzip: memo1 already exists; do you wish to overwrite (y or n)? y<Enter>
memo1.gz:        38.4% -- replaced with memo1
$
```

Again, the mcrypt command does not remove the file that it encrypts (or decrypts), and it is your responsibility to remove the original file after encrypting it. Note also that even a superuser cannot decrypt an encrypted file without having the correct key. You can use the md5sum command before and after encryption to verify that the source and decrypted files contain identical data.

The mcrypt command allows you to use any of the several encryption algorithms that it supports. You can run the mcrypt --list-hash command to display the names of these algorithms. You can use any of these algorithms to encrypt (and decrypt) files using the –h option, as in the following session.

```
$ mcrypt --list-hash
Supported Hash Algorithms:
crc32
md5
...
whirlpool
...
md2
$ mcrypt -h whirlpool memo1 -k $#123#$
Warning: It is insecure to specify keywords in the command line
File memo1 was encrypted.
$ ls -l memo1.nc
-rw------- 1 sarwar faculty 658 Feb 27 01:42 memo1.nc
$ mcrypt -d -h whirlpool memo1.nc -k $#123#$
Warning: It is insecure to specify keywords in the command line
mcrypt: memo1 already exists; do you wish to overwrite (y or n)?y<Enter>
File memo1.nc was decrypted.
$
```

If your work requires documentation and communication at higher levels of security, you should use the openssl tool. It supports the crypto library, which allows you to use several cryptographic functions, including the following:

- Encryption and decryption with ciphers
- Creation and management of public and private keys
- Public key cryptographic operations
- Handling of encrypted e-mail

The coverage of openssl and crypto is beyond the scope of this book. You can learn more about them by using the man openssl and man crypto commands.

Additionally, we show how to do various forms of disk encryption using the Linux gpg encryption tool in Chapter W26, Sections 9.7.3–9.7.5, at the book website. gpg is the OpenPGP-only version of the Gnu Privacy Guard (GnuPG). See the man page on your Linux system for gpg.

The following in-chapter exercises will give you practice using the mcrypt command and help you to understand its semantics with a hands-on session.

Exercise 7.21
Try the sessions for the mcrypt command given in this section on your system.

Summary

Several advanced operations have to be performed on text and nontext files from time to time. These operations include compressing and uncompressing file contents, sorting files, searching for files and commands in the file system structure, searching files for certain strings or patterns, performing database-like operations of cutting fields from a table or pasting tables together, transforming non-ASCII files to ASCII, and encrypting and decrypting files. Several tools are available in Linux for performing these tasks.

Some of these tools have the ability to specify a set of strings by using a single character string comprising of constants and operators called regular expressions. Commonly known as search patterns, regular expressions are used to search file contents for desired strings. The utilities that allow the use of regular expressions are awk, ed, egrep, grep, sed, and vim. In this chapter, we described regular expressions and their use in vim, egrep, and grep.

The sort command can be used to sort text files. Each line comprises a record with several fields, and the number of fields in all the lines in the file is the same. Text files can also be processed like tables by using the cut and paste commands that allow cutting of columns in a table and pasting of tables, respectively. The sort, cut, and paste commands can be combined via a pipeline (see Chapter 9) to generate tables based on different sets of criteria.

The find and whereis commands can be used to search the Linux file system structure to determine the locations (absolute pathnames) of files and commands. The find command, in particular, is very powerful and lets you search for files based on several criteria, such as file size. The which command can be used to determine which version a command executes, in case there are several versions available on a system.

Linux also provides a family of powerful utilities for searching for strings, expressions, and patterns in text files. These utilities are grep, egrep, and fgrep. Of the three, fgrep is the fastest but most limited; egrep is the most flexible but slowest of the three; and grep is the middle-of-the-road utility— reasonably fast and fairly flexible.

The gzip command can be used to compress and decompress files. The gunzip, and gzip -d commands, can be used to decompress files compressed with the gzip command. The gzexe command can be used to compress executable files and the gzexe -d command can be used to decompress them. Files compressed with gzexe can be executed without explicitly uncompressing them. The zcat and zmore commands can be used to display compressed files without explicitly uncompressing them. The bzip2 and xz can also be used for file compression and decompression. Of the three (gzip, bzip2, and xz), xz is the best in terms of savings in disk space, download time, network bandwidth, and decompression time.

The base64 utility is useful in situations when users want to e-mail non-ASCII files such as multimedia files, but the mailing system does not allow attachments. The base64 utility can be used to transform a non-ASCII file into an ASCII file, and base64 can transform the ASCII file back into the original non-ASCII version. Thus, the sender uses the base64 command before e-mailing a non-ASCII file as part of the e-mail body, and the receiver of a uuencoded file uses base64 to convert it back to its original form. The openssl tool can be used to encode and decode data.

In the Linux system, the mcrypt command can be used to encrypt and decrypt files that the user wants to keep secret. The techniques for converting an encrypted file back to its original are well known. However, the average user is not familiar with these techniques, so the use of mcrypt results in a fairly good scheme for protecting files. It also allows you to use any of the several encryption algorithms that it supports.

If your work requires higher-level secure documentation and communication, you should use the openssl and crypto tools. Additionally, Chapter W26 addresses the gpg encryption tool to accomplish various forms of disk and file encryption.

Questions and Problems

1. List five file-processing operations that you consider advanced.
2. What is sorting? Give an example to illustrate your answer. Name four applications of sorting. Name the Linux utility that can be used to perform sorting.
3. Go to http://cnn.com/weather and record the high and low temperatures for the following major cities in Asia: Kuala Lumpur, Karachi, Tokyo, Lahore, Manila, New Delhi, and Jakarta. In a file called **asiapac.temps**, construct an ASCII table comprising one line per city in the order: city name, high temperature, and low temperature. The following is a sample line:

Tokyo 78 72

Give commands to perform the following operations.
   a. Sort the table by city name.
   b. Sort the table by high temperature.
   c. Sort the table by using the city name as the primary key and the low temperature as the secondary key.

4. For the **students** file in Section 7.6, give a command to sort the lines in the file by using the last name only as the sort key.

5. What commands are available for file searching? State the purpose of each.

6. Give the command to add write permission for group on all C program files in your home directory. Give the command that searches your home directory and displays pathnames of all the files created after the file **/etc/passwd**.

7. On your Linux system, how long does it take to find all the files that are larger than 1000 bytes in size? What command(s) did you use?

8. Give a command to list all the regular files in your home directory with size greater than 2 M bytes.

9. What does the following command do?
   ```
   find ~ -size +2M -exec mv '{}' ~/bigfiles \;
   ```

10. Give commands to list all the files in your home directory that were modified within the last 24 hours and within the last one month. What commands did you use?

11. Give a command that searches your home directory and removes all PostScript and .gif files. The command must take your permission (prompt you) before removing a file.

12. Give a command to interactively remove all the files in your home directory that have not been modified for the last three months. How many such files existed in your system? What command did you use?

13. Repeat problem 12 for files that have been modified during the last three months.

14. What are regular expressions? What do the following regular expressions specify?

    a.   a|bc*

    b.   (a|b)*

    c.   a|(b*)c+

15. Give the vim command for replacing all occurrences of the string DOS with the string Linux in the whole document that is currently being edited. What are the commands for replacing all occurrences of the strings DOS and Windows with the string Linux from the lines that start or end with these strings in the document being edited?

16. Give the vim command for deleting all four-letter words starting with B, F, b, and f in the file being edited.

17. Give the vim command for renaming all C source files in a document to C++ source code files. *Note*: C source files end with **.c** and C++ source files end with **.cpp**.

18. What strings do the following regular expressions match?

    a.   ^A

    b.   A^

    c.   A$

    d. $A

    e.   ^^

    f. $$

19. What strings do the following regular expressions match?

    a.   ^A*

    b.   ^A*

    c.   ^AA*

    d.   ^AA*B

    e.   ^A\{2,4\}B

    f.   ^A\{3,\}B

    g.   \{2,4\}

    h.   A{2,4}

20. Give a regular expression for identifying blank lines.

21. What do regular expressions ^.$, ^.a*$, and ^.a+$ represent?

22. What do regular expressions [0123456789], [0-9], and ^[0-9]$ represent?

23. What does the regular expression [0-9_a-zA-Z] represent? What about [A-Z_a-z 0-9_] and ^[A-Z_a-z 0-9_]$?

24. What is the difference between regular expressions [a-z][a-z] and \([a-z]\)\1.

25. What does the command grep -n '^' student_addresses do? Assume that **student_ addresses** is the same file we use in Section 7.6.

26. Give the command that displays lines in **student_addresses** that start with the letter K or have the letter J in them. The output of the command should also have display line numbers.

27. What do the following commands do?

    a. grep [A-H] students

    b. grep [A,H] students

28. What would be the alternative of the grep ' \<Ke' students command that uses only back-slash (\) characters only (i.e., no quotes)?

29. What would be the alternative of the grep ' ^[A-H]' students command that uses backs-lashes (\) instead of single (or double) quotes?

30. What are the equivalent grep commands for the following commands: egrep, fgrep, zgrep, zegrep, and zfgrep?

31. What reaches the grep command in the following cases:

    a. grep '<K' students

    b. grep \<K students

    c. grep <K students

32. Give a command that displays names of all the files in your **~/courses/ece446** directory that contain the word "Linux."

33. Give a command that generates a table of user names for all users on your system along with their personal information. Extract this information from the **/etc/passwd** file.

34. Use the tables **student_addresses** and **student_records** to generate a table in which each row contains last name, work phone number, and GPA.

35. Imagine that you have a picture file **campus.bmp** that you would like to e-mail to a friend. Give the sequence of commands that are needed to convert the file to ASCII form, reduce its size, and encrypt it before e-mailing it.

36. What is the purpose of file encryption? Name the Linux command that you can use to encrypt and decrypt files. Give the command for encrypting a file called **~/personal/memo7** and store it in a **~/personal/memo_007** file. Be sure that, when the encrypted file is decrypted, it is put back in the **~/personal/memo7** file. Give the command to decrypt the encrypted file.

37. What is file compression? What do the terms *compressed files* and *decompressed files* mean? What commands are available for performing compression and decompression in Linux? Which are the preferred commands? Why?

38. Take three large files in your directory structure—a text file, a PostScript file, and a JPEG file—and compress them by using the gzip command. Which file was compressed the most? What was the percentage reduction in file size? Decompress the files by using the gunzip command. Show your work.

39. Which of gzip, bzip2, and xz is the best tool for file compression and decompression? Explain your answer with examples.

40. What is the difference between the following:

    a. encryption and encoding?

    b. encoding and compression?

41. Store the man pages for bash, sh, and csh in a file called **shell_man_pages**. Use the gzip command to compress with different levels of compression. Show your work and identify the command that does the maximum compression, along with the compression ratio.

42. Use the base64 and openssl commands to encode **/bin/bash** and place encoded content in the **my_bash** file in your current directory. Then, use base64 and openssl to decode the encoded file. Finally, use the shasum command to confirm that **/bin/bash** and **my_bash** are identical.

43. How many encryption algorithms does the mcrypt command support? What command did you use to obtain your answer? Create the **memo1** file discussed in Section 7.9 and encrypt it using the md5 and crc32 algorithms. Which of the two exhibits higher space efficiency? Show your work.

Advanced Questions and Problems

44. Consider the **persons** file contains the following data with each line containing a person's last name and his/her age. Give the command to sort persons based on age, numerically.

    ```
    $ cat persons
    Linus,47
    Judy,20
    Maham,25
    Adelson,85
    Amy,38
    Jacobson,25
    $
    ```

45. What is the difference between the outputs of the following commands:
    a. find ~
    b. find ~ -L

46. What is the difference between the outputs of the following commands:
    c. find ~ -name 'apple'
    d. find ~ -iname 'apple'
    e. find ~ -name 'apple' -type f
    f. find ~ -name 'apple' -type d
    g. find ~ | grep 'apple'

47. Give the command to display the names of files in your home directory that have been modified during the past 24 hours.

48. The executable files for the commands used normally by the superuser are found in /sbin and /usr/sbin directories. Give the command that displays pathnames for all the files in these directories that are executable but not readable.

49. Give the command that display the total number of files under your system's file system hierarchy owned by root, excluding files in the directories that you cannot open. How many such files exist on your system?

50. Give the command that displays the number of directories that you cannot search in your file system hierarchy. How many such directories do you have on your system? Show the command and its output on your Linux system.

51. Give the command to locate all C program files in or below your home directory, but limit subdirectory traversal to three levels beneath the home directory.

52. What do the regular expressions ^$, ^#*, and ^A* match?

53. What would be the regular expression to search for a string that
    a. starts with letter M,
    b. is the first word on a line,

  c. the second letter is a vowel,

  d. is exactly three letters long, and

  e. the third letter is c, d, m, n, or g.

54. Give the command that matches the lines containing strings "a hard question," "an easy question," and "a question" in the file called **questions**.

55. Give the regular expression that would match a three-letter palindrome such as "mom," "Mom," "MOM," or "Dad."

56. Give the regular expression that would match a five-letter palindrome such as "radar," "Radar," "RADAR," or "Maham."

57. Give the command that would display lines in the **questions** file that contain the following string:

  a. a hard question

  b. an easy question

  c. a question

58. Give the command that displays lines in the **students** file in Section 7.6 that start with four to six uppercase or lowercase letters followed by a space.

59. Use the gpg command on your Linux system to encrypt one or more important files in an ordinary user's account and then copy the encrypted file(s) to a USB thumbdrive, for transport to another Linux system. Then, on the other Linux system, use the gpg command to decrypt the important file(s). Refer to Chapter W26, Example 13 in Section 9.7.4, at the book website to assist in doing this problem.

60. Use the gpg and tar commands on your Linux system to encrypt an important directory in an ordinary user account, and then copy that directory to a USB thumbdrive for secure transport to another Linux system. Then, on the other Linux system, insert the thumbdrive, and use the appropriate commands to "untar" and decrypt the directory. Refer to Chapter W26, Example 12 in Section 9.7.3, at the book website to assist in doing this problem.

61. Use the cryptsetup utility on your Linux system to encrypt an entire USB thumbdrive. Then, insert the USB thumbdrive into another Linux system, and unlock it with the proper passphrase. Refer to Chapter W26, Example 14 in Section 9.7.5, at the book website, to assist in doing this problem.

Projects

Project 1

How can you encrypt an entire user account upon installation of your Linux system? Detail as carefully as you can, the installation procedures you would have to go through to achieve this. Test these procedures on a Linux system of your choice. What would be the major advantage of doing this, for the ordinary user, and for the security and integrity of the system itself?

 Looking for more? Visit our sites for additional readings, recommended resources, and exercises.

 CRC Press e-Resource: https://www.crcpress.com/9781138710085

 Authors' GitHub: https://github.com/bobk48/linuxthetextbook

8

File Sharing

OBJECTIVES

- To explain different ways of sharing files
- To discuss the Linux schemes and commands for implementing file sharing
- To describe Linux hard and soft (symbolic) links in detail and discuss their advantages and disadvantages
- To cover the following commands and primitives:

  *, ~, df, ln, ln -f, ln -s, ls -i, ls -l

8.1 Introduction

When a group of people works together on a project, they need to share information. If the information to be shared is on a computer system, group members have to share files and directories. For example, authors collaborating on a book or software engineers working on a software project need to share files and directories related to their project. In this chapter, we discuss several ways of implementing file sharing in a computer system. The discussion of file sharing in this chapter focuses on how a file can be accessed from various directories by various users in a Linux system. Under the topic of *version control* in Chapter 14 and in Chapter W25, Section 5.7 at the book website, we address how members of a team can work on one or more files simultaneously without losing their work.

Several methods can be used to allow a group of users to share files and directories. In this chapter, we describe duplicate shared files, common logins for members of a team, setting appropriate access permissions on shared files, common groups for members in a team, and sharing via links. All these methods can be used to allow a team of users to share files and directories in a Linux system. Although we describe each of these techniques, the chapter is dedicated primarily to a discussion of sharing via links in a Linux-based computer system.

8.2 Duplicate Shared Files

The simplest approach to files is to make copies of these files and give them to all team members. The members can put these copies anywhere in their own accounts (directory structures) and manipulate them in any way they desire. This scheme works well if members of the team are to work on the shared file(s) sequentially, but it has obvious problems if team members are to work on these files simultaneously. In the former case, team members work on one copy of the shared files one by one and complete the task at hand. In the latter case, because the members modify their own copies, the copies become inconsistent and no single copy of the shared files reflects the work done by all the team members. This outcome defeats the purpose of sharing.

8.3 Common Logins for Team Members

In this scheme, the system administrator creates a new user group comprising the members of a team and creates for them a new account to which they all have access; that is, they all know the login name and password for the account. The team owns all the files and directories created by any team member under this account and everyone has access to them.

It is a simple scheme that works quite well, particularly in situations in which the number of teams is small and teams are stable; that is, they stay together for long periods of time. Such is the case for teams of authors writing a book or programming teams working on large software projects that take several months to finish. However, this scheme also has a couple of drawbacks. First, the team's members have to use a separate account for their current project and cannot use their regular accounts to access shared files and directories. Second, the system administrator has to create a new account for every new team formed in the organization. Having to do so could create a considerable amount of extra work for the administrator if the duration of projects is short and new teams are formed for every new project. The scheme could be a real headache for the system administrator in a college-like environment where student teams are formed to work on class projects, resulting in a large number of teams every semester or quarter.

8.4 Setting Appropriate Access Permissions on Shared Files

In this scheme, the team members decide to put all shared files under one member's account, and the access permissions on these files are set so that all team members can access them. This scheme works well if *only* this team's members form the user group (recall the discussion of owner, group, and others in Chapter 5) because, if the group has other users in it, they will also have access to the shared files. For example, suppose that two university professors, Art Pohm and Jim Davis, belong to the user group **faculty**. They decide to put their shared files in Davis's account but set the group access permissions to read, write, and execute for all shared files. All the professors in the user group **faculty** then will have the same access permissions to these files, which will pose security problems. In particular, if the information to be shared is a small portion of the total amount of information residing in a member's account (say, two ordinary files out of tens of files and directories that the member owns), the risk of opening the door to all users in a group is too high, and a better technique must be used.

8.5 Common Groups for Team Members

This scheme works just like the preceding one, except that the system administrator creates a new user group consisting of the members of the team only. All team members get individual logins and set access permissions for their files so that they are accessible to other members of the team. This file-sharing scheme is effective and is used often, particularly in conjunction with a version control mechanism (see Chapter 14 and Chapter W25, Section 5.7 at the book website).

In Section 5.6.3, we show how you can set the sticky bit for a directory to allow a group of users to share files and directories in it, but to ensure that an unprivileged user cannot remove or rename files of other users in that directory.

8.6 File Sharing via Links

As described in Chapter 4, the attributes of a Linux file are stored in its inode on disk. When a file is opened, its inode is copied into the main memory, allowing speedy access to its contents. In this section, we describe how the use of an inode results in a mechanism that allows you to access a file by using multiple pathnames. System administrators commonly use this scheme to allow access to some files and directories through various other directories. Thus, for example, the home directories of all the users on the system may be accessed through **/home** or **/usr/home**. As discussed in Chapter 4, the commands, tools, and utilities for normal users are located in the **/bin**, **/usr/bin**, and **/usr/local/bin** directories, and commands, utilities, daemons (see Chapter 10), and tools for systems administration are located in the **/sbin**, **/usr/sbin**, and **/usr/local/sbin** directories.

A link is a way to establish a connection between the file to be shared and the directory entries of the users who want to have access to this file. Thus, when we say that a file has *N* links, we mean that the file has *N* directory entries somewhere in the file system hierarchy. The links therefore aid file sharing by providing different access paths to files to be shared. However, the appropriate setting of access permissions on these files controls the level of sharing. You can create links to files to which you do not have any access, but that gets you nowhere. Hence, file sharing via links is accomplished first by creating access paths to shared files by establishing links to them and then by setting appropriate access permissions on these files.

Linux supports two types of links: *hard links* and *soft/symbolic links*. The ln command may be used to create both types of links. The remainder of this chapter discusses methods of creating both types of links and their internal implementation in the Linux system.

8.6.1 Hard Links

A hard link is a *pointer* to the inode of a file. When a file is created in Linux, the system allocates a unique inode to the file and creates an entry in the directory in which the file is created. As we discussed in Chapter 4, the directory entry comprises an ordered pair (**inode number**, **filename**). The inode number for a file is used to access its attributes, including its contents on the disk for reading or writing (changing) them (see Chapter 4). Suppose that you create a file **Chapter3** in your present working directory and the system allocates inode number **53472** to this file; the directory entry for this file would be (**52473**, **Chapter3**).

If we assume that your present working directory previously contained files **Chapter1** and **Chapter2**, its logical structure is shown in Figure 8.1(a). The new file has been highlighted with a gray shade. Figure 8.1(b) shows the contents of the disk block that contains the present working directory. The connection between this directory entry and the file's contents is shown in Figure 8.1(c). The inode number in **Chapter3**'s directory entry is used to index the inode table in the main memory to access the file's inode. The inode contains the attributes of **Chapter3**, including its location on disk.

Here is a brief description of the ln command:

Syntax:
> ln [options] existing-file new-file
> ln [options] existing-file-list directory

Purpose: First syntax: Create a link to **existing-file** and name it **new-file**
> Second syntax: Create links to the ordinary files in **existing-file-list** in **directory**; links have the same names as the original file

Commonly used options/features:
> -f Force creation of link; don't prompt if **new-file** already exists
>
> -n Don't create the link if **new-file** already exists
>
> -s Create a symbolic link to **existing-file** and name it **new-file**

The ln command without any option creates a hard link to a file, provided the user has execute permission for all the directories in the path leading to the file. The following session illustrates how the ln command can be used to create a hard link in the same directory that contains **existing-file**. The only purpose of this example is to illustrate how the ln command is used; it isn't representative of how you would establish and use hard links in practice.

```
$ ls -il
total 52
16518415 -rw-r--r-- 1 sarwar faculty  1958 Mar  4 09:58 Chapter1
16518416 -rw-r--r-- 1 sarwar faculty  5188 Mar  4 09:58 Chapter2
16518417 -rw-r--r-- 1 sarwar faculty 39573 Mar  4 09:59 Chapter3
$ ln Chapter3 Chapter3.hard
$ ls -il
total 92
```

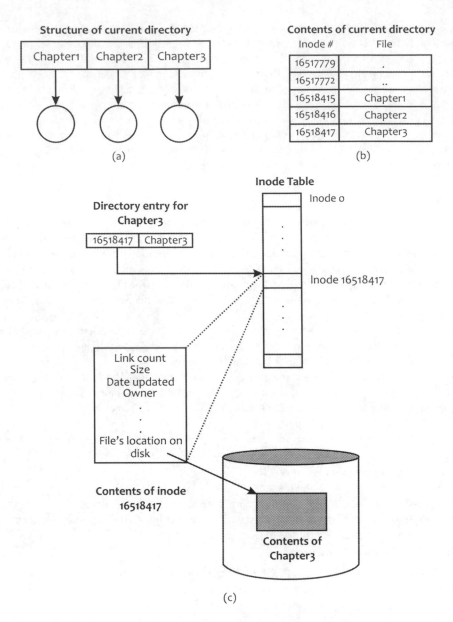

Structure of current directory

| Chapter1 | Chapter2 | Chapter3 |

(a)

Contents of current directory

Inode #	File
16517779	.
16517772	..
16518415	Chapter1
16518416	Chapter2
16518417	Chapter3

(b)

Inode Table

Inode 0

Directory entry for Chapter3

| 16518417 | Chapter3 |

Inode 16518417

Link count
Size
Date updated
Owner
.
.
.
File's location on disk

Contents of inode 16518417

Contents of Chapter3

(c)

FIGURE 8.1   (a) Logical structure of the current directory; (b) contents of the current directory; (c) relationship between directory entry, inode, and file contents.

```
16518415 -rw-r--r-- 1 sarwar faculty  1958 Mar  4 09:58 Chapter1
16518416 -rw-r--r-- 1 sarwar faculty  5188 Mar  4 09:58 Chapter2
16518417 -rw-r--r-- 2 sarwar faculty 39573 Mar  4 09:59 Chapter3
16518417 -rw-r--r-- 2 sarwar faculty 39573 Mar  4 09:59 Chapter3.hard
$
```

The `ls -il` command shows some of the attributes of all the files in the present working directory, including their inode numbers. The command `ln Chapter3 Chapter3.hard` creates a hard link to the file **Chapter3**; the name of the hard link is **Chapter3.hard**. The system creates a new directory entry (**16518417, Chapter3.hard**) for **Chapter3** in the present working directory. Thus, you can refer to **Chapter3** by accessing **Chapter3.hard** as well, because both names point to the same file on disk.

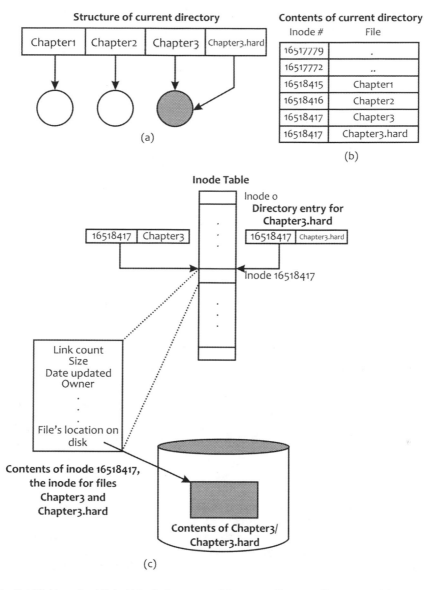

FIGURE 8.2 Establishing a hard link: (a) logical structure of the current directory; (b) contents of the current directory; (c) hard-link implementation by establishing a pointer to the inode of the file.

The second `ls -il` command is used to confirm that **Chapter3.hard** and **Chapter3** are two names for the same file, as both have the same inode number, **16518417**, and hence the same attributes. Therefore, when a hard link is created to **Chapter3**, a new pointer to its inode is established in the directory where the link (**Chapter3.hard**, in this case) resides, as illustrated in Figure 8.2.

Note that the output of the `ls -il` command also shows that both **Chapter3** and **Chapter3.hard** have link counts of 2 each. Thus, when a hard link is created to a file, the link count for the file increments by 1. That is, the same file exists in the file structure with two names (i.e., two pathnames). When you remove a file that has multiple hard links, the Linux system decrements by 1 the link count in the file's inode. If the resultant link count is 0, the system removes the directory entry for the file, releases the file's inode and all other kernel resources allocated to the file so they can be reused, and deallocates disk blocks allocated to the file so that they can be used to store other files and/or directories created in the future. If the new link count is not 0, only the directory entry for the removed file is deleted; the file contents and other directory entries for the file (hard links) remain intact. The following session illustrates this point.

```
$ rm Chapter3
$ ls -il
total 52
16518415 -rw-r--r-- 1 sarwar faculty  1958 Mar  4 09:58 Chapter1
16518416 -rw-r--r-- 1 sarwar faculty  5188 Mar  4 09:58 Chapter2
16518417 -rw-r--r-- 1 sarwar faculty 39573 Mar  4 09:59 Chapter3.hard
$
```

This session clearly shows that removing **Chapter3** results in the removal of the directory entry for this file but that the file still exists on disk and is accessible via **Chapter3.hard**. This link has the inode number and file attributes that **Chapter3** had, except that the link count, as expected, has been decremented from 2 to 1.

The following ln command can be used to create a hard link called **memo6.hard** in the present working directory to a file **~/memos/memo6**. The ls -il command is used to view attributes of the file before the hard link to it is created.

```
$ ls -il ~/memos/memo6
16518418 -rw-r--r-- 1 sarwar faculty 7878 Mar  4 10:05 /home/sarwar/memos/memo6
$ ln ~/memos/memo6 memo6.hard
$
```

After executing the ln command, you can run the ls -il command to confirm that both files (**~/memos/memo6** and **memo6.hard**) have the same inode number and attributes, as shown in the following session.

```
$ ls -il ~/memos/memo6
16518418 -rw-r--r-- 2 sarwar faculty 7878 Mar  4 10:05 /home/sarwar/memos/memo6
$ ls -il memo6.hard
16518418 -rw-r--r-- 2 sarwar faculty 7878 Mar  4 10:05 memo6.hard
$
```

The output shows two important things: first, the link count is up by 1; second, both files are represented by the same inode, **16518418**. Figure 8.3 shows the hard link pictorially.

In the following session, the ln command creates hard links to all nondirectory files in the directory called **~/linux2e/examples/dir1**. The hard links reside in the directory **~/linux2e/examples/dir2** and have the names of the original files in the **dir1** directory. The second argument, **dir2**, must be an existing directory, and you must have execute and write permissions to it. Note that the link counts for all the files in **dir1** and **dir2** are 2. The -f option is used to force creation of a hard link in case any of the files **f1**, **f2**, or **f3** already exist in the **~/linux2e/examples/dir2** directory.

```
$ cd linux2e/examples
$ more dir1/f1
Hello, World!
This is a test file.
$ ls -l dir1
-rw------- 1 sarwar faculty 35 Jun 22 22:21 f1
-rw------- 1 sarwar faculty 68 May 16 21:03 f2
-rw------- 1 sarwar faculty 94 Jul 11 11:39 f3
$ ln -f ~/linux2e/examples/dir1/* ~/linux2e/examples/dir2
$ ls -l dir1
-rw------- 2 sarwar faculty 35 Jun 22 22:21 f1
-rw------- 2 sarwar faculty 68 May 16 21:03 f2
-rw------- 2 sarwar faculty 94 Jul 11 11:39 f3
$ ls -l dir2
-rw------- 2 sarwar faculty 35 Jun 22 22:21 f1
-rw------- 2 sarwar faculty 68 May 16 21:03 f2
-rw------- 2 sarwar faculty 94 Jul 11 11:39 f3
```

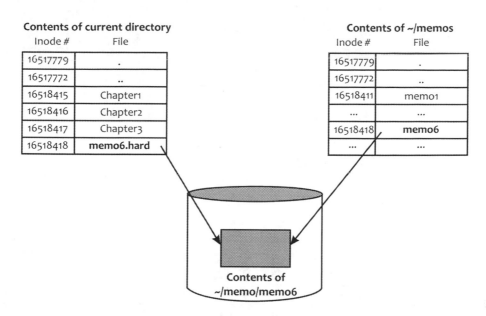

Contents of current directory

Inode #	File
16517779	.
16517772	..
16518415	Chapter1
16518416	Chapter2
16518417	Chapter3
16518418	**memo6.hard**

Contents of ~/memos

Inode #	File
16517779	.
16517772	..
16518411	memo1
...	...
16518418	**memo6**
...	...

Contents of
~/memo/memo6

FIGURE 8.3 Pictorial representation of the hard link between **~/memos/memo6** and **memo6.hard** in the current directory.

```
$ more dir2/f1
Hello, World!
This is a test file.
$
```

You can run the following command to create a hard link in your home directory to the file **/home/sarwar/linux2e/examples/demo1**. The hard link appears as a file **demo1** in your home directory. If **demo1** already exists in your home directory, you can overwrite it with the -f option. If **demo1** exists in the home directory and you don't use the -f option, an error message is displayed on the screen informing you that the **demo1** file exists. You must have the execute permission for the directories in the pathname **/home/sarwar/linux2e/examples/demo1**, and **demo1** must be a file.

```
$ ln -f /home/sarwar/linux2e/examples/demo1 ~
$
```

The user **sarwar** can run the following command to create a hard link **demo1** in a directory **dir1** in **bob**'s home directory that points to the file **/home/sarwar/linux2e/examples/demo1**. The name of the link in **bob**'s directory is **demo1**, the same as the original file. Figure 8.4 shows the establishment of the link.

```
$ ln -f /home/sarwar/linux2e/examples/demo1 /home/bob/dir1
$
```

The user **sarwar** must have execute permission for **bob**'s home directory and execute and write permissions for **dir1** (the directory in which the link is created). The user **bob** must have proper access permissions for **demo1** in **sarwar**'s directory structure to access this file. Thus, if **sarwar** and **bob** are in the same user group and **bob** needs to edit **demo1**, **sarwar** must set the group access privileges for the file to read and write. Then, **bob** is able to edit **demo1** by using, for example, the vim demo1 command from his home directory.

The following command accomplishes the same task. Remember that **sarwar** runs this command.

```
$ ln -f ~/linux2e/examples/demo1 /home/bob/dir1
$
```

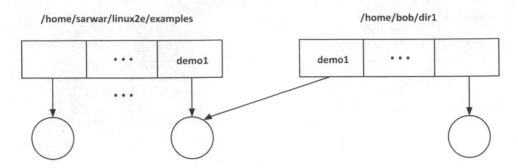

FIGURE 8.4   A hard link between **/home/sarwar/linux2e/examples** and **/home/bob/dir1**.

You can run the following command to create hard links to all nondirectory files in your ~/**linux2e/ examples** directory. The hard links reside in the **linux2e/examples** directory in user **john**'s home directory and have the names of the original files. The user **john** must first create the **linux2e** directory in his home directory and the **examples** directory in his **linux2e** directory. You must have the execute permission for **john**'s **linux2e** directory and execute and write permissions for his **examples** directory for the command to run successfully and accomplish the task.

```
$ ln -f ~/linux2e/examples/* /home/john/linux2e/examples
$
```

8.6.2 Drawbacks of Hard Links

Hard links are the traditional way of *gluing* the file system structure in Linux, which usually comprises several file systems. Hard links, however, have some problems and limitations that make them less attractive to the average user.

The first problem is that hard links cannot be established between files that are on different file systems. This inability is not an issue if you are establishing links between files in your own directory structure, with your home directory as the top-level directory, or with files in another user's directory structure that is on the same file system as yours.

However, if you want to create a hard link between a file (command) in the **/bin** directory and a file in your file structure, it most likely will not work, because on almost all systems, the **/bin** directory and your directory structure reside on different file systems. This problem also shows up when a file with multiple links is moved to another file system. The following session illustrates this point. The ls -il command shows that **Chapter3** and **Chapter3.hard** are hard links to the same file (note the same inode number). The mv command is used to move the file **Chapter3** to the **/dev/shm** directory, which is a different file system than the one that currently contains **Chapter3** (and **Chapter3.hard**). Note that, after the mv command has been executed, the link counts for **Chapter3.hard** and **/dev/shm/Chapter3** are 1 each and that the files have different inodes; **/dev/shm/Chapter3** has inode **4** and **Chapter3.hard** has the same old inode **16518417**. Note that although the ln command fails, the mv command is successful. The ln command cannot link **/dev/shm/Chapter3** to **Chapter3** because the two files are in different file systems.

```
$ ls -il
total 92
16518415 -rw-r--r-- 1 sarwar faculty  1958 Mar  4 14:07 Chapter1
16518416 -rw-r--r-- 1 sarwar faculty  5188 Mar  4 09:58 Chapter2
16518417 -rw-r--r-- 2 sarwar faculty 39573 Mar  4 14:10 Chapter3
16518417 -rw-r--r-- 2 sarwar faculty 39573 Mar  4 14:10 Chapter3.hard
$ mv Chapter3 /dev/shm
$ ls -il
total 52
```

```
16518415 -rw-r--r-- 1 sarwar faculty  1958 Mar  4 14:07 Chapter1
16518416 -rw-r--r-- 1 sarwar faculty  5188 Mar  4 09:58 Chapter2
16518417 -rw-r--r-- 1 sarwar faculty 39573 Mar  4 14:10 Chapter3.hard
$ ls -il /dev/shm/Chapter3
4 -rw-r--r-- 1 sarwar faculty 39573 Mar  4 14:10 /dev/shm/Chapter3
$ ln /dev/shm/Chapter3 Chapter3
ln: failed to create hard link 'Chapter3' => '/dev/shm/Chapter3':
Invalid cross-device link
$
```

The second problem is that only a superuser can create a hard link to a directory. The ln command gives an error message when a nonsuperuser tries to create a hard link to a directory **myweb**, as shown in the following session:

```
$ ln ~/myweb myweb.hard
ln: /home/sarwar/myweb: hard link not allowed for directory
$
```

The third problem is that some editors remove the existing version of the file you are editing and put new versions in new files. When that happens, any hard links to the removed file do not have access to the new file, thereby defeating the purpose of linking (file sharing). Fortunately, none of the commonly used editors do so. Thus, the text editors discussed in Chapter W25 (vi,vim, and emacs) at the book website are safe to use.

In the following in-chapter exercises, you will use the ln and ls -il commands to create and identify hard links, and to verify a serious limitation of hard links.

Exercise 8.1

Create a file **Ch8Ex1** in your home directory that contains this problem. Establish a hard link to this file, also in your home directory, and call the link **Ch8Ex1.hard**. Verify that the link has been established by using the ls -il command. What field in the output of this command did you use for verification?

Exercise 8.2

Execute the ln /dev/shm ~/tmp command on your Linux system. What is the purpose of the command? What happens when you execute the command? Does the result make sense? Why or why not?

8.6.3 Soft/Symbolic Links

Soft/symbolic links take care of all the problems inherent in hard links and are therefore used more often than hard links. They are different from hard links, both conceptually and in terms of how they are implemented. They do have a cost associated with them, which we discuss in Section 8.6.4, but they are extremely flexible and can be used to link files across machines and networks.

You can create soft links by using the ln command with the -s option. The following session illustrates the creation of a soft link.

```
$ ls -il
total 92
16518415 -rw-r--r-- 1 sarwar faculty  1958 Mar  4 14:07 Chapter1
16518416 -rw-r--r-- 1 sarwar faculty  5188 Mar  4 09:58 Chapter2
16518417 -rw-r--r-- 1 sarwar faculty 39573 Mar  4 14:17 Chapter3
16518417 -rw-r--r-- 1 sarwar faculty 39573 Mar  4 14:10 Chapter3.hard
$ ln -s Chapter3 Chapter3.soft
$ ls -il
total 92
16518415 -rw-r--r-- 1 sarwar faculty  1958 Mar  4 14:07 Chapter1
16518416 -rw-r--r-- 1 sarwar faculty  5188 Mar  4 09:58 Chapter2
16518417 -rw-r--r-- 1 sarwar faculty 39573 Mar  4 14:17 Chapter3
```

```
16518417 -rw-r--r-- 1 sarwar faculty 39573 Mar  4 14:10 Chapter3.hard
16518423 lrwxrwxrwx 1 sarwar faculty     8 Mar  4 14:17 Chapter3.soft -> Chapter3
$
```

The ln -s Chapter3 Chapter3.soft command is used to create a symbolic link to the file **Chapter3** in the present working directory, and the symbolic link is given the name **Chapter3.soft**. The output of the ls -il command shows a number of important items that reveal how symbolic links are implemented and how they are identified in the output. First, the original file (**Chapter3**) and the link file (**Chapter3. soft**) have different inode numbers: **16518417** for **Chapter3** and **16518423** for **Chapter3.soft**, which means that they are different files. Second, the original file (**Chapter3**) is of file type - (ordinary) and the link file (**Chapter3.soft**) is of type l (link). Third, the link count has not changed for **Chapter3** (and **Chapter3.hard**) and is 1 for **Chapter3.soft**, which further indicates that the two files are different. Fourth, the file sizes are different: 39,573 bytes for the original file (**Chapter3**) and 8 bytes file the link file (**Chapter3.soft**). Last, the name of the link file is followed by -> Chapter3, the pathname for the file that **Chapter3.soft** is a symbolic link to; the string after the -> sign is specified as the first argument in the ln -s command. The pathname of the existing file is the content of the link file, which also explains the size of the link file (eight characters in the word Chapter3). Figure 8.5 shows the logical file structure of the current directory, directory entries in the current directory, and a diagram that shows that **Chapter3** and **Chapter3.soft** are truly separate files and that the link file contains the pathname of the file to which it is a link.

In summary, when you create a symbolic link, a new file of type l is created. This file contains the pathname of the existing file as specified in the first argument of the ln -s command. When you make a reference to the link file, the Linux system sees that the type of the file is l and reads the link file to find the pathname for the actual file to which you are referring. For example, for the cat Chapter3. soft command, the system reads the contents of **Chapter3.soft** to get the name of the file to display (**Chapter3** in this case) and sends its contents to standard output. Hence, you see the contents of **Chapter3** displayed.

You can create soft links across file systems. In the following session, we create a symbolic link to the **/bin** directory. The name of the symbolic link is **symlinktobin,** and it is placed in the current directory. Note that the inode numbers of **/bin** and **symlinktobin** are different, as expected.

```
$ ln -s /dev symlinktodev
$ ls -ild symlinktodev /dev/
        2 drwxr-xr-x 19 root    root    4480 Feb 25 11:58 /dev/
16518424 lrwxrwxrwx  1 sarwar faculty     4 Mar  4 14:19 symlinktodev -> /dev
$
```

In the following session, we show an example to create a soft link to a file such that the file and its symbolic link reside in different file systems.

```
$ ln -s /dev/shm/Chapter3 temp.soft
$ ls -il /dev/shm/Chapter3 temp.soft
        4 -rw-r--r-- 1 sarwar faculty 39573 Mar  6 07:26 /dev/shm/Chapter3
16518443 lrwxrwxrwx 1 sarwar faculty    17 Mar  6 08:09 temp.soft -> /dev/shm/Chapter3
$
```

Here, the file **Chapter3** is copied from one file system (that contains this file) to another that contains the **/dev/shm** directory. Then, the ln -s /dev/shm/Chapter3 temp.soft command is used to create a symbolic link to the copied file. The command works without any problems, establishing a symbolic link to **/dev/shm/Chapter3** in **temp.soft**. Note that the inode numbers of the two files are different, indicating that the two files are distinct; **temp.soft** contains the pathname of the file for which it is a symbolic link, **/tmp/Chapter3**. Recall that in Section 8.6.1 a similar call to establish a hard link between **/tmp/ Chapter3** and **temp.hard** failed.

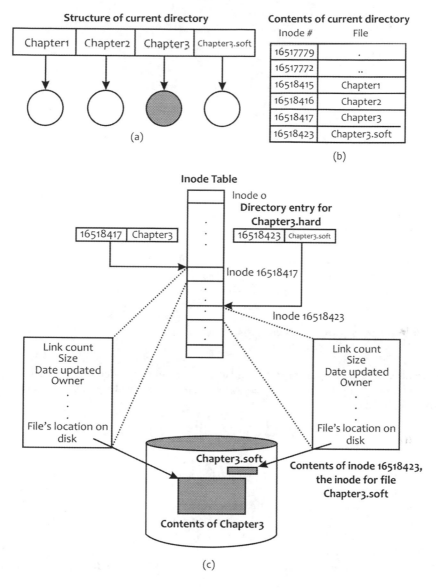

FIGURE 8.5 Establishing a soft link: (a) logical structure of the current directory; (b) contents of the current directory; (c) soft link implementation by establishing a pointer in the link file to (the pathname of) the existing file.

The following session shows how symbolic links can be created to all the files in a directory, including the directory files. The `ln -sf ~/linux2e/examples/dir1/* ~/linux2e/examples/dir2` command creates soft links to all the files in the directory called **~/linux2e/examples/dir1** and puts them in the directory **~/linux2e/examples/dir2**. You must have execute and write permissions for the **dir2** directory, and execute permission to all the directories in the pathname. The `-f` option is used to force creation of the soft link in case any of the files **f1**, **f2**, or **f3** already exist in **~/linux2e/examples/dir2**. On some systems, the `-f` and `-s` options may not work together, in which case you will use only the `-s` option.

```
$ cd ~/linux2e/examples
$ more dir1/f1
```

```
Hello, World!
This is a test file.
$ ls -l dir1
-rw------- 1 sarwar faculty  35 Jun 22 22:21 f1
-rw------- 1 sarwar faculty 168 Jun 22 22:33 f2
-rw------- 1 sarwar faculty 783 Jun 22 22:35 f3
$ ln -sf ~/linux2e/examples/dir1/* ~/linux2e/examples/dir2
$ ls -l dir2
lrwxr-xr-x 1 sarwar faculty 38 Jun 22 22:54 f1 -> /home/sarwar/linux2e/examples/dir1/f1
lrwxr-xr-x 1 sarwar faculty 38 Jun 22 22:54 f2 -> /home/sarwar/linux2e/examples/dir1/f2
lrwxr-xr-x 1 sarwar faculty 38 Jun 22 22:54 f3 -> /home/sarwar/linux2e/examples/dir1/f3
$ more dir2/f1
Hello, World!
This is a test file.
$
```

You can run the following command to create a symbolic link in your home directory to the file **/home/sarwar/linux2e/examples/demo1**. The soft link appears as a file called **demo1** in your home directory. If **demo1** already exists in your home directory, you can overwrite it with the -f option. If **demo1** exists in the home directory and you don't use the -f option, an error message is displayed on the screen informing you that the **demo1** file exists. You must have the execute permission for the directories in the pathname **/home/sarwar/linux2e/examples/demo1**, and **demo1** must be a file.

```
$ ln -sf /home/sarwar/linux2e/examples/demo1 ~
$
```

The user **sarwar** can run the following command to create a soft link called **demo1** in a directory **dir1** in **bob**'s home directory that points to the **/home/sarwar/linux2e/examples/demo1** file. Figure 8.6 shows how the soft link is established.

```
$ ln -sf /home/sarwar/linux2e/examples/demo1 /home/bob/dir1
$
```

The user **sarwar** must have execute permission for **bob**'s home directory, and execute and write permission for **dir1** (the directory in which the soft link is created). The user **bob** must have proper access permissions for **demo1** in **sarwar**'s directory structure to access this file. Thus, if **sarwar** and **bob** are in the same user group and **bob** has to edit **memo1**, then **sarwar** must set the group access privileges on the file to read and write. The user **bob** can then edit **demo1** by using, for example, the vi demo1 command from his home directory.

The following command accomplishes the same task. Remember that **sarwar** runs this command.

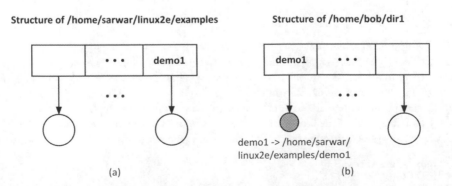

FIGURE 8.6   A soft link between (a) **/home/sarwar/linux2e/examples/demo1** and (b) **/home/bob/dir1**.

```
$ ln -sf ~/linux2e/examples/demo1 /home/bob/dir1
$
```

You can run the following `ln` command to create soft links to all the files, including directory files, in your **~/linux2e/examples** directory. These soft links reside in the directory called **linux2e/examples** in **john**'s home directory and have the names of the original files. The user **john** must create the **linux2e** directory in his **home** directory and the **examples** directory in his **linux2e** directory. You must have execute permission for **john**'s **linux2e** directory and execute and write permission for his **examples** directory in order for the command to run successfully.

```
$ ln -sf ~/linux2e/examples/* /home/john/linux2e/examples
$
```

8.6.4 Pros and Cons of Symbolic Links

As previously stated, symbolic links do not have the problems and limitations of hard links. Thus, symbolic links can be established to directories and between files across file systems. Also, files that symbolic links point to can be edited by any kind of editor without any ill effects, provided that the file's pathname doesn't get changed—that is, the original file is not moved.

Symbolic links do have a problem of their own that is not associated with hard links: if the file that the symbolic link points to is moved from one directory to another, it can no longer be accessed via the link. The reason is that the link file contains the pathname for the original location of the file in the file structure. When the file location is changed, the link becomes *dangling*; that is, it points to a file that does not exist at the specified (original) location. You also have a dangling pointer if the original file is deleted. The following session illustrates this point.

```
$ mv /dev/shm/Chapter3 .
$ cat temp.soft
cat: temp.soft: No such file or directory
$
```

Suppose that **temp.soft** is a symbolic link to the file **/dev/shm/Chapter3**. The mv command is used to move **/dev/shm/Chapter3** to the present working directory. The cat command fails because the soft link still points to the file with pathname **/dev/shm/Chapter3**. This result is quite logical but is still a drawback; in hard links, the cat command would not fail so long as the moved file stays within the same file system.

Some other drawbacks of the symbolic links are that Linux has to support an additional file type (the link type) and a new file has to be created for every link. Creation of the link file results in space overhead for an extra inode, disk space needed to store the pathname of the file to which it is a link, and other kernel data structures. Symbolic links also result in slower file operations because, for every reference to the file, the link file has to be opened and read in order for you to reach the actual file. The actual file is then processed for reading or writing, requiring an extra disk access to be performed if a file is referenced via a symbolic link to the file.

In the following in-chapter exercises, you will use the `ln -s` and `ls -il` commands to create and identify soft links, and to verify that you can create soft links across file systems.

Exercise 8.3
Establish a soft link to the file **Ch8Ex1** that you created in Exercise 8.1. Call the soft link **Ch8Ex1.soft**. Verify that the link has been established. What commands did you use to establish the link and verify its creation?

Exercise 8.4
Execute the `ln -s /dev/shm ~/tmp` command on your Linux system. What is the purpose of the command? What happens when you execute the command? Does the result make sense? Why or why not?

Summary

Any of the several techniques can be used to allow a team of users to share Linux files and directories. Some of the most commonly used methods of file sharing are duplicating the files to be shared and distributing them among team members, establishing a common account for team members, setting appropriate permissions on the files to be shared, setting up a Linux user group for the team members, and establishing links to the shared files in the directories of all team members. File sharing via hard and soft links is the main topic of this chapter. However, the issue of simultaneous access of shared files by team members is not discussed here (see Chapter 14 and Chapter W25, Section 5.7 at the book website).

Hard links allow you to refer to an existing file by another name. Although hard links are the primary mechanism used by Linux to glue the file system structure, they have several shortcomings. First, an existing file and its links must be in the same file system. Second, only a superuser can create hard links to directories. Third, moving a file to another file system breaks all links to it.

Soft links can be used to overcome the problems associated with hard links. When a soft link to a file is created, a new file is created that contains the pathname of the file to which it is a link. The type of this file is link. Soft links are expensive in terms of the time needed to access the file and the space overhead of the link file. The time overhead during file access occurs because the link file has to be opened in order for the pathname of the actual file to be read (or written to, whatever the case may be), and only then does the actual process of file opening and reading (or writing) take place. The link file that contains the pathname of the original file causes the space overhead.

Hard and soft links are established with the ln command. For creating soft links, the -s option is used with the command. The ls -il command is used to identify (or confirm establishment of) links. The first field of the output of this command identifies the inode numbers for the files in a directory, and all hard links to a file have the same inode number as the original file. The first letter of the second field represents file type (l for *soft link*) and the remaining letters specify file permissions. The third field identifies the number of hard links to a file. Every simple file has one hard link at the time it is created. The last field identifies file names; a soft link's name is followed by -> filename, where filename is the name of the original file. The -f option can be used to force the creation of a link—that is, to overwrite an existing file with the newly created link.

Questions and Problems

1. What are the five methods that can be used to allow a team of users to share files in Linux?
2. What is a link in Linux? Name the types of link that Linux supports. How do they differ from each other?
3. What are the problems with hard links?
4. Remove the file **Ch8Ex1** that you created in Exercise 8.1. Display the contents of **Ch8Ex1.hard** and **Ch8Ex1.soft**. What happens? What command did you use for displaying the files? Does the result make sense? Why or why not?
5. Search the **/usr/bin** directory on your system and identify three links in it. Write down the names of these links. Are these hard or soft links? How do you know?
6. While in your home directory, can you establish a hard and soft link to **/etc/passwd** on your system? Why? What commands did you use? Are you satisfied with the results of the command execution?
7. Every Linux directory has at least two hard links. Why?
8. Can you find the number of hard and soft links to a file? If so, what command(s) do you need to use?
9. Suppose that a file called **shared** in your present directory has five hard links to it. Give a sequence of commands to display the absolute pathnames of all these links. (*Hint*: Use the find command.)

10. Create a directory, **dir1**, in your home directory and three files, **f1**, **f2**, and **f3**, in it. Ask a friend to create a directory, **dir2**, in his or her home directory, with **dir1.hard** and **dir1.soft** as its subdirectories. Create hard and soft links to all the files in your **dir1** in your friend's **~/dir2/dir1.hard** and **~/dir2/dir1.soft** directories. Give the sequence of commands that you executed to do so.

11. For Problem 10, what are the inode numbers of the hard links and soft links? What command did you use to determine them? What are the contents of the link (both hard and soft) files? How did you get your answers?

12. What are the pros and cons of symbolic links?

13. Clearly describe how file sharing can be accomplished by using links (hard and soft) in Linux. In particular, do you need to do anything other than establish links to the files to be shared?

14. Suppose you have a collection of data files, say **file1.data**, ..., **file9.data**, that need to be shared (read only) among 100 programs in your group. Discuss the overhead involved for each of the following:
    a. Setting permissions
    b. Creating hard links
    c. Creating soft links
    d. Making individual private copies of each file

15. Browse through the root (**/**) directory and its subdirectories. Identify ten soft links and write them in a table, along with the name of each link and the directory in which it is found.

16. Symbolic links have the *dangling pointer* problem. What is it? Explain with an example.

17. Hard links may not be established between files across file systems. What is the technical reason for this limitation of hard links? (*Hint*: Think about inodes and file systems.)

Advanced Questions and Problems

18. You can use the df -h command to display information about all the file systems on your system. Can you create hard and soft links to a file under your home directory to a file under the following directories? Why or why not? Show your work and explain your answer.
    a. /tmp
    b. /dev/shm

19. What is the purpose of executing the ln command in the following session? Why does the command display an error message?

```
$ ln /bin/rm del
ln: failed to create hard link 'del' => '/bin/rm': Operation not permitted
$
```

20. Display the pathnames for the files in your home directory hierarchy with two or more hard links. What command did you use? Show the execution of the command on your system.

21. The presentation of the materials in Chapter W26, Sections 9.8.2 and 9.8.3, and particularly Example W26.27, all available at the book website, serve to further extend your idea of how file-sharing techniques can be used, by applying Linux "capabilities." Capabilities are permission checks implemented at the most fundamental level of system operation: the process level. Capabilities function to check permissions that a process can perform on file objects, in conjunction with and beyond what is presented in this chapter. A capability-enabled privileged process can operate with an effective User ID of 0 (root), as opposed to an unprivileged nonzero User ID. Capabilities give processes that they are applied to a more incisive, finer-grained, and limited scope privilege (or root privilege), rather than the all-or-nothing privileges exercised

with the role-based access control techniques of sudo (on Debian-family Linux systems) or superuser (on CentOS 7) assignment.

22. Take an executable program, one you have written yourself, and apply capabilities to it that give it a "limited" root set of privileges on files, in either the current working directory, or across the file system(s) available. Write the program by yourself in any language you are familiar with, such as C, C++, Python, and Bash scripting. More specifically, have the program interact with and be able to access a file or files that it would ordinarily be denied access to by the traditional permission bits set on those file objects. This is illustrated in Example W26.27. Although you can assign capabilities as a sudo user, as is shown in that example, execute your program as an ordinary, unprivileged user. What capability, from those illustrated in Table W26.14, would you use in your program to achieve the results asked for in this problem?

    Additionally, enhance your program to allow the setting and testing of as many other capabilities listed in Table G19.14 as you can, or as are assigned.

23. Can the file-sharing techniques shown in this chapter be applied to NFSv4-mounted network drives, as described in Chapter W26, Sections 9.3.3 and 9.3.4 at the book website, and the file system objects found on them? Which techniques would best apply to network drives? Fully explain your answer.

24. What do you believe are the major distinctions, differences, and relative advantages and disadvantages, of using all of the file-sharing techniques shown in this chapter, and those shown in Chapter W26, Section 9.3.4 "Setting NFSV4 ACLs on ZFS"? Is it possible for a single, stand-alone system to be both an NFSv4 server and client? How would you implement this scenario? Why would you want to implement this scenario, and what would its advantages and disadvantages be?

Projects

Project 1

Using the techniques that modify the traditional Linux access privileges developed in Section 5.5.2, and the git and NFSv4 material presented in Chapter W24, Section 5.7 and Chapter W26, Section 9.3.4, both found at the book website, create a git repository in an NFSv4 server directory, and share that directory with an NFSv4 client. Both server and client machines should be on the same intranet. Then, through the application of traditional Linux permissions, limit the access privileges of a nonprivileged user on the client, to the NFSv4-mounted git repository in a variety of ways. For example, only let the client user pull from the server-side git repository directory, or only let the client user push to that directory. The privileged user on the server should be able, via traditional permission control, to administrate the git repository directory. Test the git push and pull operations to the repository directory, from both the server (via a privileged account), and the client (via the nonprivileged account), given the variety of access modes you have set to it with the traditional permissions.

Project 2

Do the same thing as you did in Project 1, except have the server be a Virtualbox Virtual Machine (VM) instance that is any supported Linux flavor that you like, and let the client be the host system for that VirtualBox VM.

Looking for more? Visit our sites for additional readings, recommended resources, and exercises.

CRC Press e-Resource: https://www.crcpress.com/9781138710085

Authors' GitHub: https://github.com/bobk48/linuxthetextbook

9

Redirection and Piping

OBJECTIVES

- To describe the notion of standard files—standard input, standard output, and standard error files—and file descriptors
- To describe input and output redirection for standard files
- To discuss the concept of error redirection and appending to a file
- To explain the concept of pipes in Linux
- To describe how powerful operations can be performed by combining pipes, file descriptors, and redirection operators
- To discuss error redirection in the TC shell
- To explain the concept of FIFOs (also known as named pipes) and their command line use
- To cover the following commands and primitives:

  &, |, <, >, >>, cat, diff, find, grep, lp, mkfifo, more, pr, sort, stderr, stdin, stdout, tee, tr, uniq, wc, xargs

9.1 Introduction

All computer software (commands) performs one or more of the following operations: input, processing, and output; a typical command performs all three. The question for the operating system is: Where does a shell command (internal or external) take its input from, where does it send its output to, and where are the error messages sent to? If the input to a command is not part of the command code (i.e., data within the code in the form of constants and/or variables), it must come from an outside source. This outside source is usually a file, although it could be an input/output (I/O) device such as a keyboard or a network interface card. Command output and error messages can go to a file as well. For a command to read from or write to a file, it must first open the file.

There are default files where a command reads its input and sends its output and error messages, called *standard input*, *standard output*, and *standard error*. In Linux, these files are known as *standard files* for a command. The input, output, and errors of a command can be redirected to other files by using *file redirection facilities* in Linux. This allows you to connect several commands together to perform a complex task that cannot be performed by a single existing command. We discuss the notion of standard files and redirection of input, output, and error in Linux in this chapter.

9.2 Standard Files

In Linux, the kernel opens three files automatically for every command (or process) to read input from and send its output and error messages to. These files are known as standard files: standard input (**stdin**), standard output (**stdout**), and standard error (**stderr**). By default, these files are associated with the terminal on which the command executes. More specifically, the keyboard is standard input, and the display screen (or the console at which you are logged in) is standard output and standard error. Therefore, every command, by default, takes its input from the keyboard and sends its output and error messages

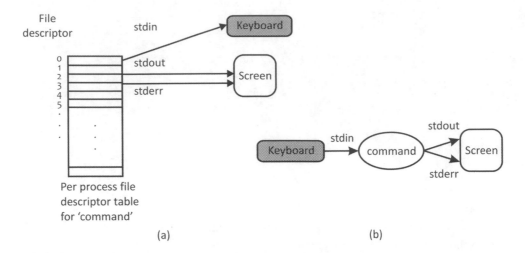

FIGURE 9.1   Standard files and file descriptors: (a) file descriptors; (b) semantics of a command execution.

to the display screen (or the console window), as shown in Figure 9.1. Recall our explanation of the per-process file descriptor table in Chapter 4. In the remainder of this chapter, we use the terms *monitor screen*, *display screen*, *console window*, and *display window* interchangeably.

9.3 Input Redirection

Input redirection is accomplished by using the less than symbol (<). The following syntax is used to detach the keyboard from the standard input of command and attach **input-file** to it. Thus, if command reads its input from standard input, this input will come from **input-file**, not the keyboard attached to the terminal on which the command is run. The semantics of the command syntax are shown in Figure 9.2. Note that the command input comes from **input-file**.

Syntax:
  `command < input-file`

Purpose: Input to command comes from **input-file** instead of from the keyboard

---

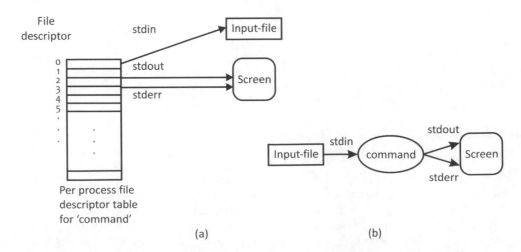

FIGURE 9.2   Input redirection: (a) file descriptors and standard files for command; (b) semantics of input redirection.

For example, the command `cat < tempfile` reads input from **tempfile** (as opposed to the keyboard, because the standard input for `cat` has been attached to **tempfile**) and sends its output to the display screen. So, effectively, the contents of **tempfile** are displayed on the monitor screen. This command is different from `cat tempfile`, in which **tempfile** is passed as a command line argument to the `cat` command; the standard input of `cat` does not change and is still the keyboard attached to the terminal on which the command is run.

Similarly, in the command `grep "John" < Phones`, the `grep` command reads its input from the **Phones** file in the current directory, not from the keyboard. The output and error messages of the command go to the display screen. Again, this command is different from `grep "John" Phones`, in which the **Phones** file is passed as an argument to `grep`; the standard input of `grep` does not change and is still the keyboard attached to the terminal on which the command executes. However, the net effect of the `grep` command is the same in both cases from a user's perspective. Similarly, the use of < is not needed in most cases because the command reads from a file in any case.

The `cat` and `grep` commands take input from standard input if they are not passed file arguments from the command line. The `tr` command takes input from standard input only and sends its output to standard output. The command does not work with a file as a command line argument. Thus, input redirection is often used with the `tr` command, as in `tr -s ''''< Bigfile`. When this command is executed, it substitutes multiple spaces in **Bigfile** with single spaces.

9.4 Output Redirection

Output redirection is accomplished by using the greater than symbol (>). The following syntax is used to detach the display screen from the standard output of `command` and attach **output-file** to it. Thus, if `command` writes/sends its output to standard output, the output goes to **output-file**, not the monitor screen attached to the terminal on which the command runs. The error messages still go to the display screen, as before. The semantics of the command syntax are shown in Figure 9.3.

Syntax:
   `command > output-file`

Purpose: Send output of `command` to the file **output-file** instead of to the monitor screen

---

Consider the `cat > newfile` command. Recall that the `cat` command sends its output to standard output, which is the display screen by default. This command syntax detaches the display screen from standard output of the `cat` command and attaches **newfile** to it. The standard input of `cat` remains attached

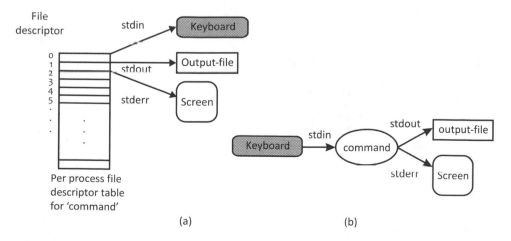

FIGURE 9.3 Output redirection: (a) file descriptors and standard files for `command`; (b) semantics of input redirection.

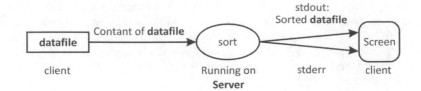

FIGURE 9.4   Semantics of the `ssh server sort < datafile` command run on **mymachine**.

to the keyboard. When this command is executed, it creates a file called **newfile** whose contents are whatever you type on the keyboard until you hit `<Ctrl-D>` in the first column of a new line. If **newfile** already exists, by default it is overwritten.

Similarly, the command `grep "John" Phones > Phone_John` sends its output (lines in the **Phones** file that contain the word "John") to a file called **Phone_John**, as opposed to displaying it on the monitor screen. The input for the command comes from the **Phones** file. The command terminates when `grep` encounters the end-of-file character in **Phones**.

In a network environment, the following command can be used to sort the file **datafile** residing on the computer that you are currently logged on to (the client computer), on the computer called **server**. The output of the command—that is, the sorted data—is sent to the display screen of the client computer. Figure 9.4 illustrates the semantics of this command.

```
$ ssh server sort < datafile
$
```

This command is a good example of how multiple computers can be used to perform various tasks concurrently in a network environment. It is a useful command if your computer (call it **client**) has a large file, **datafile**, to be sorted and you do not want to make multiple copies of the file on various computers on the network to prevent inconsistency in them. This command allows you to perform the task. Such commands are also useful if the server has specialized Linux tools that you are allowed to use but not allowed to make copies of on your machine. We discuss network-related Linux commands and utilities in Chapter 11. We have used this example to illustrate the power of the Linux I/O redirection feature, not to digress to computing in a network environment.

The following session shows the contents of the **Students** file on the local (client) machine and the output of the `sort` command executed on a remote (server) machine 198.102.10.20 under **sarwar**'s login. Since **sarwar**'s login is password protected, the system prompted him for a password before running the `sort` command on the remote computer and displaying its output on the local computer. Note that we have shown the Internet Protocol (IP) address of a fictitious `ssh` server, but the session was executed on a real remote machine.

```
$ cat Students
John Doe        ECE     3.54    A
Pam Meyer       CS      3.61    A
Jim Davis       CS      2.71    B
John Doe        ECE     3.54    A
Jason Kim       ECE     3.97    A
Amy Nash        ECE     2.38    C
$ ssh sarwar@198.102.10.20 sort < Students
Password for sarwar@pcbsd-srv:
Amy Nash        ECE     2.38    C
Jason Kim       ECE     3.97    A
Jim Davis       CS      2.71    B
John Doe        ECE     3.54    A
John Doe        ECE     3.54    A
Pam Meyer       CS      3.61    A
$
```

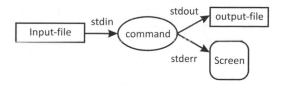

FIGURE 9.5   Combined use of input and output redirection.

9.5 Combining Input and Output Redirection

Input and output redirections can be used together, according to the syntax given in the following command description.

Syntax:
```
command < input-file > output-file
command > output-file < input-file
```
Purpose: Input to command comes from **input-file** instead of the keyboard and output of command goes to **output-file** instead of the display screen

---

When this syntax is used, command takes its input from **input-file** (not from the keyboard attached to the terminal) and sends its output to **output-file** (not to the display screen), as shown in Figure 9.5.

In the cat < lab1 > lab2 command, the cat command takes its input from the **lab1** file and sends its output to the **lab2** file. The net effect of this command is that a copy of **lab1** is created in **lab2**. Therefore, this command line is equivalent to cp lab1 lab2, provided that **lab2** does not exist. If **lab2** is an existing file, the two commands have different semantics. The cat < lab1 > lab2 command truncates **lab2** (sets its size to zero and the read/write pointer to the first byte position) and overwrites it with the contents of **lab1**. Because **lab2** is not recreated, its attributes (e.g., access permissions and link count) are not changed. In the case of the cp lab1 lab2 command, not only is the data in **lab1** copied into **lab2** but also its attributes from its inode are copied into the inode for **lab2**. Thus, the cp command results in a true copy (data and attributes) of **lab1** into **lab2**.

In the following in-chapter exercises, you will practice the use of input and output direction features of Linux.

Exercise 9.1
Write a shell command that counts the number of characters, words, and lines in a file called **memo** in your present working directory and shows these values on the display screen. Use input redirection.

Exercise 9.2
Repeat Exercise 9.1, but redirect output to a file called **counts.memo**. Use I/O redirection.

9.6 I/O Redirection with File Descriptors

As described in Section 4.6, the Linux kernel associates a small integer number with every open file, called the *file descriptor* for the file. The file descriptors for standard input, standard output, and standard error are 0, 1, and 2, respectively. The Bash and Portable Operating System Interface (POSIX) shells allow you to open files and associate file descriptors with them; the TC shell does not allow the use of file descriptors. The other descriptors, usually ranging from 3 onward, are used when a process opens files simultaneously. These descriptors are called *user-defined file descriptors*. The upper limit of these descriptors (and hence the number of files that a process may open simultaneously) varies from system to system and may be determined by running the uname -n command. Each descriptor is used to index a kernel table, called the *per-process file descriptor table*, as discussed briefly in Section 4.7. A more detailed discussion on this topic may be found in Chapters 15 and 16, and Chapters W20 and W21 at the book website. In the following sections, we describe I/O and error redirection under the Bash and POSIX shells. We discuss the TC shell syntaxes and give examples in Section 9.13.

By making use of file descriptors, standard input and standard output can be redirected in the Bash and POSIX shells by using the 0< and 1> operators, respectively. Therefore, cat 1> outfile, which is equivalent to cat > outfile, takes input from standard input and sends it to **outfile**; error messages go to the display screen. Similarly, ls -l foo 1> outfile is equivalent to ls -l foo > outfile. The output of this command (the long listing for **foo**) goes into a file called **outfile**, and error messages generated by it go to the display screen.

The file descriptor 0 can be used as a prefix with the < operator to explicitly specify input redirection from a file. In the command shown in the following session, the input to the grep command is the contents of **tempfile** in the present working directory.

```
$ grep "John" 0< tempfile
... command output ...
$
```

9.7 Redirecting Standard Error

The standard error of a command can be redirected by using the 2> operator (i.e., associating the file descriptor for standard error with the > operator) as follows:

Syntax:
    command 2> error-file

Purpose: Error messages generated by command and, by default, sent to **stderr** are redirected to error-file

---

With this syntax, command takes its input from the keyboard, sends its output to the monitor screen, and any error messages produced by command are sent to **error-file**. The semantics of the command syntax are shown in Figure 9.6. Command input may come from a file passed as a command line argument.

The command grep "John" Phones 2> error.log takes input from the **Phones** file, sends output to the display screen, and any error message produced by grep goes to a file called **error.log**. If **error.log**

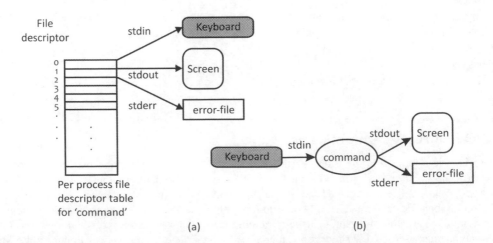

FIGURE 9.6   Error redirection: (a) standard descriptors and standard files for command; (b) semantics of error redirection.

already exists, it is overwritten; otherwise, it is created. The following example shows how the standard error of `ls -l` can be redirected to a file.

```
$ ls -l foo 2> error.log
... long listing for foo if no errors ...
$
```

The output of `ls -l foo` goes to the display screen, and error messages go to **error.log**. Thus, if **foo** does not exist, the error message `ls: foo: No such file or directory` goes into the **error.log** file, as shown in the following session. The actual wording of the message varies from system to system, but it basically informs you that **foo** does not exist.

```
$ ls -l foo 2> error.log
$ cat error.log
ls: foo: No such file or directory
$
```

Keeping standard error attached to the display screen and not redirecting it to a file is useful in many situations. For example, when the `cat lab1 lab2 lab3 > all` command is executed to concatenate files **lab1**, **lab2**, and **lab3** into a file called **all**, you would want to know whether any of the three input files are nonexistent or if you do not have permission to read any of them. In this case, redirecting the error message to a file does not make much sense because you want to see the immediate results of the command execution.

9.8 Redirecting **stdout** and **stderr** in One Command

Standard output and standard error can be redirected to the same file. One obvious way to do so is to redirect **stdout** and **stderr** to the same file by using file descriptors with the > symbol, as in the following command.

```
$ cat lab1 lab2 lab3 1> cat.output 2> cat.errors
$
```

In this case, the input of the cat command comes from the **lab1**, **lab2**, and **lab3** files, its output goes to the **cat.output** file, and any error message goes to the **cat.errors** file, as shown in Figure 9.7.

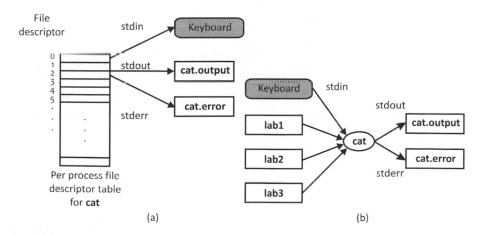

FIGURE 9.7   Error redirection: (a) file descriptors and standard files for `cat lab1 lab2 lab3 1> cat.output 2>` `cat.errors`; (b) semantics of the `cat` command.

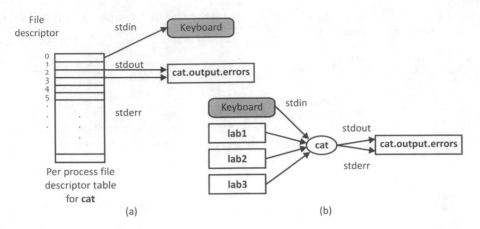

FIGURE 9.8 Error redirection: (a) file descriptors and standard files; (b) semantics of the cat lab1 lab2 lab3 1> cat.output.errors 2>&1 and cat lab1 lab2 lab3 2> cat.output.errors 1>&2 commands.

Note that, although not shown in Figure 9.7b, files **lab1**, **lab2**, and **lab3** have file descriptors assigned to them when they are opened for reading by the cat command. The command produces an error message if any one of the three "lab" files does not exist or if you do not have read permission for any of these files.

The following command redirects the **stdout** and **stderr** of the cat command to the **cat.output. errors** file. Thus, the same file (**cat.output.errors**) contains the output of the cat command, along with any error messages that may be produced by the command.

```
$ cat lab1 lab2 lab3 1> cat.output.errors 2>&1
$
```

In this command, the string 2>&1 tells the command shell to make descriptor 2 a duplicate of descriptor 1, resulting in the error messages going to the same place that the command output goes to. Similarly, the string 1>&2 can be used to tell the command shell to make descriptor 1 a duplicate of descriptor 2. Thus, the following command accomplishes the same task. Figure 9.8 shows the semantics of the two commands.

```
$ cat lab1 lab2 lab3 2> cat.output.errors 1>&2
$
```

The evaluation of the command line content for file redirection is left to right. Therefore, redirections must be specified in the left-to-right order if one notation is dependent on another. In the preceding command, first, **stderr** is changed to the file **cat.output.errors**, and then **stdout** becomes a duplicate of **stderr**. Thus, the output and errors for the command both go to the same file, **cat.output.errors**.

The following command therefore does *not* have the effect of the two commands just discussed. The reason is that, in this command, **stderr** is made a duplicate of **stdout** *before* output redirection. Therefore, **stderr** becomes a duplicate of **stdout** (the display screen at this time) first, and then **stdout** is changed to the file **cat.output.errors**. Thus, the output of the command goes to **cat.output.errors** and errors go to the display screen. The sequence shown in Figure 9.9 illustrates the semantics of this command.

```
$ cat lab1 lab2 lab3 2>&1 1> cat.output.errors
$
```

Note that Figure 9.9a and b are identical because the execution of cat lab1 lab2 lab3 2>&1 does not make any changes to **stdout** and **stderr**—they stay attached to the display screen before and after the command is executed.

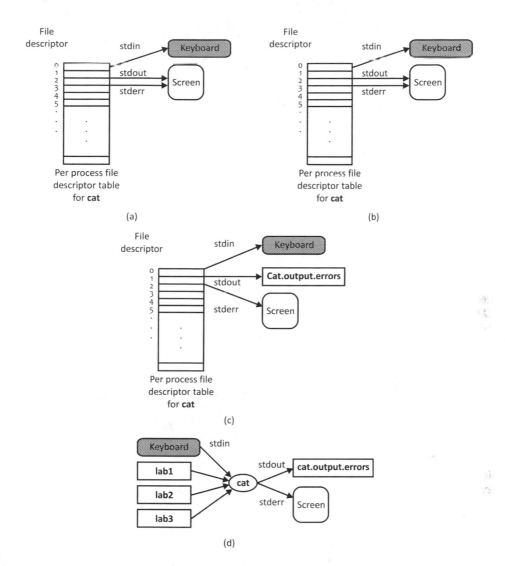

FIGURE 9.9 Output and error redirection: (a) file descriptors and standard files for the cat command; (b) standard files after `cat lab1 lab2 lab3 2>&1` with no change in **stdout** and **stderr**; (c) standard files after `cat lab1 lab2 lab3 2>&1 1> cat.output.errors`; and (d) command semantics.

9.9 Redirecting `stdin`, `stdout`, and `stderr` in One Command

Standard input, standard output, and standard error can be redirected in a single command according to the following syntax.

Syntax:
> `command 0< input-file 1> output-file 2> error-file`

Purpose: Input to command comes from **input-file** instead of the keyboard, output of command goes to **output-file** instead of the display screen, and error messages generated by command are sent to **error-file** instead of the display screen

---

The file descriptors 0 and 1 are not required because they are the default values for the < and > operators, respectively. The semantics of this command syntax are shown in Figure 9.10. Evaluation of the

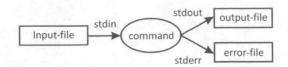

FIGURE 9.10   Redirecting **stdin**, **stdout**, and **stderr** in a single command.

command line content for file redirection is left to right, so the order of redirection is important. Consider the following command syntaxes. For the first command, if **input-file** is not found, the error message is sent to the display screen, because **stderr** has not been redirected yet. For the second command, if **input-file** is not found, the error message goes to **error-file** because **stderr** has been redirected to this file. If **error-file** exists, the outputs of the first and second commands go to **stdout** and **output-file**, respectively.

```
command 1> output-file 0< input-file 2> error-file command 2> error-file 1> output-file 0< input-file
```

The following sort command sorts lines in a file called **students** and stores the sorted file in **students.sorted**. If the sort command fails to start because the **students** file does not exist, the error message goes to the display screen as shown, not to the file **sort.error**. The reason is that, at the time the shell determines that the **students** file does not exist, **stderr** is still attached to the console.

```
$ sort 0< students 1> students.sorted 2> sort.error
cannot open students: No such file or directory
$
```

For the following command, the error message goes to the **sort.error** file if the sort command fails because the **students** file does not exist. The reason is that the shell processes error redirection before it determines that the **students** file is nonexistent.

```
$ sort 2> sort.error 0< students 1> students.sorted
$ cat sort.error
cannot open students: No such file or directory
$
```

9.10 Redirecting without Overwriting File Contents (Appending)

By default, output and error redirections overwrite contents of the destination file. To *append* output or errors generated by a command to the end of a file, replace the > operator with the >> operator. The default file descriptor with >> is 1, but file descriptor 2 can be used to append errors to a file. In the following command, the output of ls -l is appended to the **output.dat** file, and the error messages are appended to the **error.log** file.

```
$ ls -l 1>> output.dat 2>> error.log
$
```

The following command appends the contents of the files **memo** and **letter** to the end of the file **stuff**. If the command produces any error message, it goes to the **error.log** file. If **error.log** is an existing file, its contents are overwritten with the error message.

```
$ cat memo letter >> stuff 2> error.log
$
```

If you want to keep the existing contents of **error.log** and append new error messages to it, use the following command. For this command, the previous contents of **error.log** are appended with any error message produced by the cat command.

```
$ cat memo letter >> stuff 2>> error.log
$
```

The Bourne shell, by default, overwrites a file when the **stdout** or **stderr** of a command is redirected to it, but the Korn, TC, and Bash shells have a noclobber option that prevents you from overwriting important files accidentally, We discuss this option for the TC shell in Section 9.13, but discuss it for the Bash shell here.

You can set the noclobber option in the Bash shell by using the set command with the -o option as shown. Of course, if you want to set this option permanently, put the command in your ~/.**profile** file.

```
$ set -o noclobber
$
```

In the Bash shell, you can also set the option by using the set -C command. When you set the noclobber option, you can force overwriting of a file by using the >| operator.

In the following in-chapter exercises, you will practice the use of input, output, and error redirection features of Linux shells (excluding the TC shell) in a command line.

Exercise 9.3
Write a command that counts the number of characters, words, and lines in a file called **memo** in your present working directory and writes these values into a file **memo.size**. If the command fails, the error message should go to a file **error.log**. Use I/O and error redirections.

Exercise 9.4
Write a shell command to send the contents of a file **greetings** to **doe@domain.com** by using the mail command. If the mail command fails, the error message should go to a file **mail. errors**. Use input and error redirection.

Exercise 9.5
Repeat Exercise 9.2, but append error messages at the end of **mail.errors**.

9.11 Linux Pipes

The Linux system allows **stdout** of a command to be connected to the **stdin** of another command. You can make it do so by using the *pipe character* (|) according to the following syntax.

Syntax:
   command1 | command2 | command3 | ... | command

Purpose: The standard output of command1 is connected to the **stdin** of command2, the **stdout** of command2 is connected to the **stdin** of command3, ..., and the **stdout** of commandN-1 is connected to the **stdin** of commandN

Figure 9.11 illustrates the semantics of this command.

Thus, a pipe allows you to send output of a command as input to another command. The commands that are connected via a pipe are called *filters*. A filter belongs to a class of Linux commands that take input from **stdin**, manipulate it in some specific fashion, and send it to **stdout**. Pipes and filters are frequently used in Linux to perform complicated tasks that cannot be performed with a single command. Some commonly used filters are cat, compress, crypt, grep, less, lp, more, pr, sort, tr, uniq, and wc. If a command at the output of a pipe processes its input at a rate slower than the command connected to its input, the pipe stores the excess data and serves it to the command at the output on a first-in-first-out basis. Thus, if cmd2 in Figure 9.11 processes the incoming data at a slower rate than the rate at which cmd1 produces data, the excess data produced by cmd1 is stored in the pipe between cmd1 and cmd2, which serves it to cmd2 on a first-in-first-out basis. The maximum data that can be written to a pipe without interruption (i.e., atomically) in Linux, is dictated by the symbolic constant PIPE_BUF in **/usr/ include/linux/limits.h**.

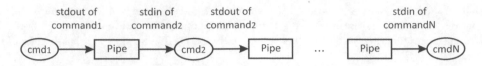

FIGURE 9.11   Semantics of a pipeline with *N* commands.

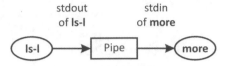

FIGURE 9.12   Semantics of the `ls -l | more` command.

For example, in `ls -l | more`, the `more` command takes the output of `ls -l` as its input. The net effect of this command is that the output of `ls -l` is displayed one screen at a time. The pipe really acts like a water pipe, taking the output of `ls -l` and giving it to `more` as its input, as shown in Figure 9.12.

This command does not use a disk to connect the standard output of `ls -l` to the standard input of `more`, because the pipe is implemented in the kernel area of the main memory. In terms of the I/O redirection operators, the command is equivalent to the following sequence of commands.

```
$ ls -l > temp
$ more < temp (or more temp)
[contents of temp]
$ rm temp
$
```

As you can see, not only do you need three commands to accomplish the same task, but the command sequence is also extremely slow, because file read and write operations are involved. Recall that files are stored on a secondary storage device, usually a disk. On a typical contemporary computer system, disk operations are about one million times slower than main memory (random access memory) operations. The actual performance gain in favor of pipes, however, is much smaller, owing to efficient caching of file blocks by the Linux kernel and the use of semiconductor disks.

You can use the `sort` utility discussed in Chapter 7 to sort lines in a file. Suppose that you have a file called **student_records** that you want to sort and that the file may have some repeated lines that you want to appear only once in the sorted file. The `sort -u student_records` command can accomplish this task. As discussed in Chapter 6, the `uniq` command can also do the task if it is given the sorted version of **student_records** with repeated lines in it. One way to perform the task is to use the following commands. The `more` command is used to show the contents of **student_records**.

```
$ more student_records
John Doe        ECE     3.54
Pam Meyer       CS      3.61
Jim Davis       CS      2.71
John Doe        ECE     3.54
Jason Kim       ECE     3.97
Amy Nash        ECE     2.38
Kim Coleman     CS      3.19
$ sort student_records > student_records_sorted
$ uniq student_records_sorted
Amy Nash        ECE     2.38
Jason Kim       ECE     3.97
Jim Davis       CS      2.71
John Doe        ECE     3.54
Kim Coleman     CS      3.19
Pam Meyer       CS      3.61
$
```

The same task can be accomplished in one command line by using a pipe, as follows:

```
$ sort student_records | uniq
Amy Nash          ECE      2.38
Jason Kim         ECE      3.97
Jim Davis         CS       2.71
John Doe          ECE      3.54
Kim Coleman       CS       3.19
Pam Meyer         CS       3.61
$
```

If you want to display the record for a student with first or last name Kim, you can use the grep student_ records Kim command to do so. However, if you do not know the location of the **student_records** file in your home directory, you can use the find ~ -name student_records | xargs grep Kim command, as shown in the following session. The xargs command executes the rest of the line as a shell command, i.e., grep Kim in this case.

```
$ find ~ -name student_records | xargs grep Kim
Jason Kim         ECE      3.97
Kim Coleman       CS       3.19
$
```

If you want to display the record for Kim Coleman, you can pass it as a string to grep, as in

```
$ find ~ -name student_records | xargs grep "Kim Coleman"
Kim Coleman       CS       3.19
$
```

At times, you may need to connect several commands. The following command line demonstrates the use of multiple pipes, forming a *pipeline* of commands. In this command line, we have used the grep and sort filters.

```
$ who | sort | grep "john" | mail -s "John's Terminal" doe@coldmail.com
$
```

This command sorts the output of who and sends the lines containing the string "john" (if any exists) as an e-mail message to **doe@coldmail.com**, with the subject line "John's Terminal." In terms of input and output redirection, this command line is equivalent to the following command sequence.

```
$ who > temp1
$ sort < temp1 > temp2
$ grep "john" temp2 > temp3
$ mail -s "John's Terminal" doe@coldmail.com < temp3
$ rm temp1 temp2 temp3
$
```

The command with pipes does not use any disk files, but the preceding command sequence needs three temporary disk files and six disk I/O (read and write) operations. The number of I/O operations may be a lot higher, depending on the sizes of these files, the system load in terms of the number of users currently using the system, and the run-time behavior of other processes running on the system.

A pipe, therefore, is a Linux feature that allows two Linux commands (processes) to communicate with each other. Hence, a pipe provides an interprocess communication (IPC) mechanism in Linux. More specifically, it can be used as a channel between two related processes on the same system to talk to each other. Typically, processes have a parent–child relationship (see Chapter 10) but processes with a common ancestor (parent, grandparent, etc.) can also communicate using a pipe. From a programmer's perspective, we discuss IPC using pipes in detail in Chapter W20 at the book website. The communication between processes is one-way only. For example, in ls | more, the output of ls is read by more as input. Thus, the one-way communication is from ls to more. For a two-way communication between processes, at least

FIGURE 9.13   Semantics of the grep "John" < Students | lpr –Pspr command.

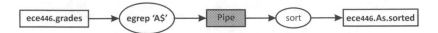

FIGURE 9.14   Semantics of the egrep 'A$' < ee446.grades | sort > ee446.As.sorted command.

two pipes must be used. This cannot be accomplished at the command shell level, but it can be done in C/C++ by using the pipe() system call. We explore this topic in Chapter W20 at the book website also.

I/O redirection and pipes can be used in a single command, as follows:

```
$ grep "John" < Students | lpr –Pspr
$
```

Here, the grep command searches the **Students** file for lines that contain the string "John" and sends those lines to the lpr command to be printed on a printer named **spr**. Figure 9.13 illustrates the semantics of this command.

In the following command, egrep takes its input from **ee446.grades** and sends its output (lines ending with the character A) to the sort utility, which sorts these lines and stores them in the file called **ee446.As.sorted** in the current directory. The end result is that the names, scores, and grades of those students who have A grades in the ECE446 course are stored in the **ee446.As.sorted** file in the current directory. Figure 9.14 illustrates the semantics of this command.

```
$ egrep 'A$' < ee446.grades | sort > ee446.As.sorted
$
```

Suppose that, before running the ssh server sort < datafile command in Section 9.4, you want to be sure that **datafile** on your local system is consistent with the updated copy on the server, called ~/**research/pvm/datafile.server**. You can copy **datafile.server** and compare it with your local copy, **datafile**. But, you will then have three copies, and if you are not careful you might remove the wrong copy. In this case, you can run the following command to see the differences between your local copy and the copy on the server without copying **datafile.server** on your (local) computer.

```
$ ssh server cat ~/research/pvm/datafile.server | diff datafile -
$
```

In this case, the cat command runs on the server side, and its output is fed as input to the diff command executed on the local machine. The output of the diff command also goes to the display screen on the local machine. Figure 9.15 illustrates the semantics of this command.

Exercise 9.6
Create the **student_records** file on your system and repeat all the commands in this section that were run using this file.

9.12 Redirection and Piping Combined

You cannot use the redirection operators and pipes alone to redirect the **stdout** of a command to a file as well as connect it to **stdin** of another command in a single command line. However, you can use the tee utility to do so. You can use this utility to tell the command shell to send the **stdout** of a command to one or more files specified as arguments of tee, as well as to another command. The following is a brief description of how the tee utility is normally used.

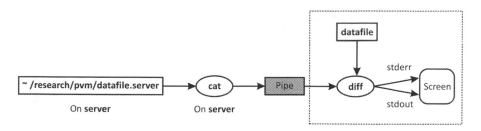

FIGURE 9.15 Semantics of the `ssh server cat ~/research/pvm/datafile.server | diff datafile –` command.

Syntax:
  `command1 | tee file1 file2 … fileN | command2`

Purpose: Standard output of `command1` is connected to the **stdin** of tee, and tee sends its input to files **file1** through **fileN** as well as the **stdin** of command2

The semantics of this command syntax are that `command1` is executed and its output is stored in files **file1** through **fileN** as well as sent to command2 as its input. An example use of the tee utility is given in the following command.

```
$ cat names students | grep "John Doe" | tee file1 file2 | wc –l
$
```

This command extracts the lines from the **names** and **students** files that contain the string "John Doe," pipes these lines to the tee utility, which puts copies of these lines into **file1** and **file2** as well as sending them to wc –l, which sends its output to the display screen. Thus, the lines in the **names** and **students** files that contain the string "John Doe" are saved in **file1** and **file2**, and the line count of such lines is displayed on the monitor screen. Figure 9.16 illustrates the semantics of this command line. Such commands are useful in a shell script where different operations have to be performed on **file1** and **file2** later in the script.

> **Exercise 9.7**
> Create the **names** and **students** files on your system and repeat all the commands in this section that were run using these files. Did the commands work as expected?

9.13 Output and Error Redirection in the TC Shell

The operators for performing the input, output, and append operations (<, >, >>) work in the TC shell as they do in other shells, as previously discussed. However, file descriptors cannot be used with these operators in the TC shell. Also, error redirection works differently in the TC shell than it does in other shells. In the TC shell, the operator for output and error redirection is >&.

Syntax:
  `command >& file`

Purpose: Redirect **stdout** and **stderr** of command to **file**

For example, the following command redirects output and errors of the `ls –l foo` command to the **error.log** file. The standard input of the command is still attached to the keyboard. Note that we have used the % sign as the shell prompt, which is the default for the TC shell.

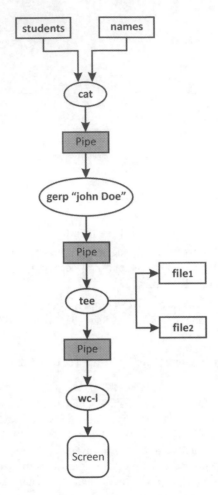

FIGURE 9.16   Semantics of the cat names students | grep "John Doe" | tee file1 file2 | wc -l command.

```
% ls -l foo >& error.log
%
```

The TC shell does not have an operator to redirect **stderr** alone. However, the **stdout** and **stderr** of a command can be attached to different files, if the command is executed in a subshell, by enclosing the command in parentheses. The following session illustrates this point.

```
% find ~ -name foo -print >& output.error.log
% (find ~ -name foo -print > foo.paths) >& error.log
%
```

The children of your current shell process, also known as subshells (see Chapter 10), execute all external shell commands. When the first command executes, the output and errors of the find command go to the **output.error.log** file. Because the subshell process is not created until the whole command line has been processed (interpreted), the **stdout** and **stderr** of the parent shell process are redirected to the **error.log** file because of the >& operator. Therefore, the subshell also has its **stdout** and **stderr** redirected to the **error.log** file.

In the second command line, the find command is executed under a subshell and inherits the standard files of the parent shell. When the find command in parentheses executes, it redirects the **stdout** of the

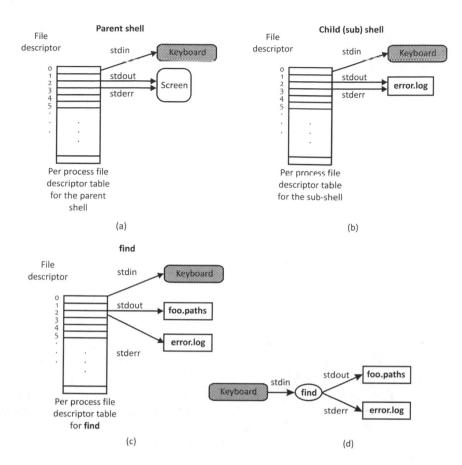

FIGURE 9.17 (a–d) Step-by-step semantics of the (find~-name foo -print > foo.paths) >& error.log command.

command to the **foo.paths** file; the **stderr** of the command remains attached to **error.log**. Thus, the output of the find command goes to the **foo.paths** file, and the errors generated by the command go to the **error.log** file. Figure 9.17 illustrates the semantics of the second find command.

You can use the >>& operator to redirect and append **stdout** and **stderr** to a file. For example, ls -l foo >>& output.error.log redirects the **stdout** and **stderr** of the ls command and appends them to the **error.log** file.

The TC shell also allows the **stdout** and **stderr** of a command to be attached to the **stdin** of another command with the |& operator. The following is a brief description of this operator.

Syntax:
   Command1 |& command2
Purpose: Redirect **stdout** and **stderr** of command1 to command2, that is, pipe **stdout** and **stderr** of command1 to command2

In the following command, the **stdout** and **stderr** of the cat command are attached to the **stdin** of the grep command. Thus, the output of the cat command, or any error produced by it (e.g., owing to the lack of read permission for **file1** or **file2**), is fed as input to the grep command.

```
% cat file1 file2 |& grep "John Doe"
%
```

The I/O redirection and piping operators (| and |&) can be used in a single command, as shown in the following session. This command is an extension of the previous command, in which the **stdout** of the grep command is attached to the **stdin** of the sort command. Furthermore, the **stdout** and **stderr** of the sort command are attached to the **stdin** of the wc -l command. Thus, if the command line completes successfully, it displays the number of lines in **file1** and **file2** that contain the string "John Doe."

```
% cat file1 file2 |& grep "John Doe" | sort |& wc -1
%
```

In the following in-chapter exercises, you will practice the use of Linux pipes, the tee command, and the error redirection feature of the TC shell.

Exercise 9.8

Write a shell command that sorts a file **students.records** and stores the lines containing "Davis" in a file called **Davis.record**. Use piping and I/O redirection.

Exercise 9.9

Write a command to copy a file **Scores** to **Scores.bak** and send a sorted version of **Scores** to **professor@university.edu** via the mail command.

Exercise 9.10

Write a TC shell command for copying a file **Phones** in your home directory to a file called **Phones.bak** (also in your home directory) by using the cat command and the >& operator.

The TC shell has a special built-in variable that allows you to protect your files from being overwritten with output redirection. This variable is called noclobber and, when set, prevents the overwriting of existing files with output redirection. You can set the variable by using the set command and unset it by using the unset command. Or, you can place the set noclobber command in your ~/.tcshrc or ~/.cshrc file (or some other startup file).

```
% set noclobber
[your interactive session]
...
% unset noclobber
%
```

If the noclobber variable is set, the command cat file1 > file2 generates an error message if **file2** already exists. If **file2** does not exist, it is created and the data from **file1** is copied into it. The command cat file1 >> file2 works if **file2** exists and noclobber is set, but an error message is generated if **file2** does not exist. You can use the >!, >>!, and >>&! operators to override the effect of the noclobber variable if it is set. Therefore, even if the noclobber variable is set and **file2** exists, the command cat file1 >! file2 copies data from **file1** to **file2**. For the cat file1 >>! file2 command, if the noclobber variable is set and **file2** does not exist, **file2** is created and the data from **file1** is copied into it. The >>&! operator works in a manner similar to the >>! operator.

9.14 Recap of I/O and Error Redirection

Table 9.1 summarizes the input, output, and error redirection operators in the Bash and TC shells. We did not discuss some of these operators in this chapter; we discuss them in detail in Chapters 12 and 13, and W29 and W30 at the book website under shell programming. We included these operators in this table because it seems to be the most appropriate place to show all of them together.

TABLE 9.1

Redirection Operators and Their Meaning in the Bash and TC Shells

Operator	Bash Shell	TC Shell
`< file`	Input redirection	Input redirection
`> file`	Output redirection	Output redirection
`>> file`	Append standard output	Append standard output
`0< file`	Input redirection	
`1> file`	Output redirection	
`2> file`	Error redirection	
`1>> file`	Append standard output to file	
`2>> file`	Append standard error to file	
`<&m`	Attach standard input to file descriptor m	
`>&m`	Attach standard input to file descriptor m	
`m>&n`	Attach file descriptor m to file descriptor n	
`<&-`	Close standard input	
`>&-`	Close standard output	
`m<&- or m>&-`	Close file descriptor m	
`>& file`	Output and error redirection to file	Output and error redirection to file
`>| file`	Ignore noclobber and assign standard output to file	
`>! file`		Ignore noclobber and assign standard output to file
`>>! file`		Ignore noclobber and assign standard output to file; if file does not exist, create it
`>>&! file`		Ignore noclobber and append standard output and standard error to file
`n>| file`	Ignore noclobber and force output from file descriptor n to file	
`<> file`	Assign standard input and standard output to file	
`n< file`	Set file as file descriptor n	
`n> file`	Direct file descriptor n to file	
`cmd1 | cmd2`	Connect standard output of command "cmd1" to standard input of command "cmd2"	Connect standard output of command "cmd1" to standard input of command "cmd2"
`cmd1 | &cmd2`		Connect standard output and standard error of command "cmd1" to standard input of command "cmd2"
`(cmd> file1)>&file2`		Redirect standard output of the command "cmd" to file1 and standard error to file2
`|&`		Allow stdout and stderr of a command to be attached to stdin of another command

9.15 FIFOs

FIFOs, also known as *named pipes,* can also be used for communication between two processes executing on a system. Although processes communicating with pipes must be related to each other through a common ancestor process that you execute, processes communicating with FIFOs do not have to have this kind of relationship—they can be independently executed programs on one system. For the command line use of pipes and FIFOs, this means that pipes can be used only for communication between commands connected via a pipeline, and FIFOs can be used for communication between separately run commands.

Another difference between pipes and FIFOs is that whereas a pipe is a main memory buffer maintained by the Linux kernel and has no name, a FIFO is created on disk and has a name like a filename. This means that, like files, FIFOs have to be created and opened before they can be used for communication between processes. Thus, accessing a FIFO requires an access to the secondary storage device where it resides. Pipes are *process persistent*, which means that they exist as long as the process that creates them exists. FIFOs on most Linux systems are *filesystem persistent*, meaning that they exist on the system until they are explicitly removed/deleted or the relevant file system is unmounted (i.e., removed from the system).

You can use the mknod() system call or the mkfifo() library call to create a FIFO in a program and the mkfifo command to create a file in a shell session. We discuss the command line use of FIFOs in this section. We discuss the use of FIFOs, pipes, and the related system calls and library calls in the chapters on system programming (Chapters 15 and 16, and W20 and W21 at the book website). Here is a brief description of the mkfifo command.

Syntax:
   mkfifo [option] file-list

Purpose: Create FIFOs with pathnames given in **file-list**

Output: FIFOs for the pathnames given in **file-list** are created in the relevant directories

Commonly used options/features:

  -m mode   Set access permissions for newly created FIFOs to "mode"; the access permissions are specified in "mode" as they are with the chmod command, such as 666, meaning read and write permissions for everyone and execute permission for nobody

---

In the following session, we use the first command to create a FIFO, called **myfifo1**, with default permissions based on the current value of umask (see discussion in Chapter 5). We use the second command to create a FIFO, called **myfifo2**, with read and write permissions for the owner and no permissions for all other users. We use the ls -al myfifo1 myfifo2 command to display the access permissions of the two FIFOs. Note that the first character in the long listing of the two FIFOs is "p," indicating that **myfifo1** and **myfifo2** are of the FIFO (named pipe) type.

```
$ mkfifo myfifo1
$ mkfifo -m 600 myfifo2
$ ls -l myfifo3 myfifo4
prw-r--r--  1 sarwar  faculty  0 Aug  9 08:30 myfifo3
prw-------  1 sarwar  faculty  0 Aug  9 08:31 myfifo4
$
```

The general method by which two commands, cmd1 and cmd2, can communicate with a FIFO, called **myfifo1**, is shown in the following command sequence.

```
cmd1 < myfifo1 &
cmd2 infile | tee myfifo1 | cmd3
```

Note that we run the first command in the background (see Chapter 10 for background processes) so that we could run cmd2. When we execute cmd1, it blocks, because **myfifo1** is empty. Note that cmd1 blocks and returns immediately if **myfifo1** is a file. When the output of cmd2 is sent to **myfifo1** and cmd3 via the tee utility, cmd1 unblocks and starts processing data in **myfifo1**. Outputs of commands cmd1 and cmd3 are sent to standard output (i.e., the display screen). Figure 9.18 shows these semantics with the help of a diagram.

In the following session, the command sequence displays the status of all the processes running on the system, the number of daemon processes, and the total number of processes running on the system. The two cat commands block until something is written into **myfifo1** and **myfifo2**. The ps -a command sends the status of all the processes running on the system to the tee command, which redirects this data to the two FIFOs as well as sending it to the grep command. The grep command extracts all the

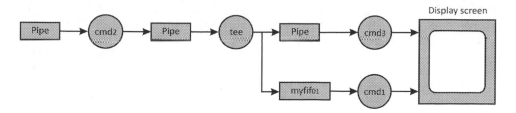

FIGURE 9.18 Semantics of execution of the command sequence `cmd1 < myfifo1 &` followed by `cmd2 infile | tee myfifo1 | cmd3`.

daemon processes from its input (i.e., the output of the `ps -a` command) and sends them to the `wc -l` command through a pipe. Thus, the first `cat` command displays the status of all the processes running on the system. The output of the second `cat` command is the number of processes running on the system and that of the third command (`ps -a`) is the number of daemons running on the system. As shown by the last two lines of the output, at the time of running this command sequence, the system was running 20 processes, of which two were daemons.

```
$ cat myfifo1 &
$ cat myfifo2 | wc -l &
$ ps -a | tee myfifo1 myfifo2 | grep 'd$' | wc -l
  PID TT  STAT    TIME COMMAND
 1533 v0- I     0:00.01 /bin/sh /usr/local/sbin/PCDMd
 1548 v0- IN    3:43.82 /bin/sh /usr/local/sbin/pbid
 1790 v0  Is+   0:00.00 /usr/libexec/getty Pc ttyv0
 1791 v1  Is+   0:00.00 /usr/libexec/getty Pc ttyv1
  ...
 1797 v7  Is+   0:00.00 /usr/libexec/getty Pc ttyv7
56478  1  Is    0:00.24 -csh (csh)
57216  1  S     0:00.12 /bin/sh
58002  1  S     0:00.01 cat myfifo1
58045  1  S     0:00.01 cat myfifo2
58046  1  I     0:00.01 wc -l
58089  1  R+    0:00.02 ps -a
58090  1  S+    0:00.01 tee myfifo1 myfifo2
58091  1  S+    0:00.02 grep d$
58092  1  S+    0:00.01 wc -l
       2
$      20
```

The sequence of the output in this session is dependent on the scheduling of the three commands; the output shown previously is what was produced by our system and is the most likely output. An interesting exercise would be to come up with a sequence of commands to ensure that the output is always produced in the same order as seen here.

When you no longer need to use a FIFO, you can remove it just like you remove an ordinary file. This means that you can use the `unlink()` system call (from within a process) or the `rm` command at the command line for removing a FIFO from your file system hierarchy. In the following session, we use the `rm` command for removing the **myfifo1** and **myfifo2** FIFOs. The output of the `ls` command before and after the `rm` command shows that the two FIFOs have in fact been removed.

```
$ ls
myfifo1 myfifo2
$ rm myfifo1 myfifo2
$ ls
$
```

The following in-chapter exercises are designed to give you practice using the `mkfifo` command and help you to understand its semantics with a hands-on session.

Exercise 9.11

Create three FIFOs, called **fifo1**, **fifo2**, and **fifo3**, with a single command. Write down the command that you used to perform the given task.

Exercise 9.12

Create a FIFO, called **fifo4**, with its access privileges set to read and write for owner and group, and no privileges for others. Show the command that you used to accomplish the given task.

Exercise 9.13

Try the shell session given in this section on your system. Does your system produce output in the same order as shown in our session? If not, show the output produced on your system.

Summary

Linux automatically opens three files for every command for it to read input from and send its output and error messages to. These files are called standard input (**stdin**), standard output (**stdout**), and standard error (**stderr**). By default, these files are attached to the terminal on which the command is executed. Thus, the shell makes the command input come from the keyboard and its output and error messages to go to the monitor screen. These default files can be changed to other files by using redirection operators: < for input redirection and > for output and error redirection.

The three standard files can be referred to by using the digits 0 (**stdin**), 1 (**stdout**), and 2 (**stderr**), called the file descriptors for the three standard files. All open files in Linux are referred to by similar integers that are used by the kernel to perform operations on these files. In the Bash and POSIX shells, the greater than symbol (>) is used in conjunction with descriptors 1 and 2 to redirect standard output and standard error, respectively.

The standard output of a command can be connected to the standard input of another command via a Linux pipe (|). Pipes are created in the main memory and are used to take the output of a command and give it to another command without creating a disk file, effectively making the two commands talk to each other. For this reason, a pipe is called an IPC channel, which allows related commands on the same machine to communicate with each other at the shell and application levels. The processes communicating through pipes must be related through a common ancestor; the relationship is usually parent–child or sibling.

The I/O and error redirection features and pipes can be used together to implement powerful command lines. However, redirection operators and pipes alone cannot be used to redirect the standard output of a command to a file as well as connecting it to standard input of another command. The tee utility can be used to accomplish this task, sending standard output of a command to one or more files as well as to another command. The commands and tee are connected through pipes.

The TC shell does not support I/O and error redirection with file descriptors. Also, redirecting standard output and standard error of a command to different files is specified differently in the TC shell than it is in the other shells.

FIFOs, also known as named pipes, allow related or unrelated processes on a system to communicate. Unlike a pipe, which is an in-memory buffer, a FIFO is a file created on a secondary storage device. For command line use, you can create a FIFO with the mkfifo command. The mknod() system call or mkfifo() library call may be used to create FIFOs within processes. When you no longer need a FIFO, you can remove it with the unlink() system call from within a running program (process) or the rm command at the command line. We discuss the Linux system calls and library calls related to pipes and sockets in Chapters 15 and 16, and W20 and W21 at the book website on Linux system programming.

Questions and Problems

1. What are standard files? Name them and state their purpose.
2. Briefly describe input, output, and error redirection. Write two commands of each to show simple and combined use of the redirection operators.

3. What are file descriptors in Linux? What are the file descriptors of standard files? How can the I/O and error redirection operators be combined with the file descriptors of standard files to perform redirection in the Bourne, Korn, Bash, and POSIX shells?

4. Sort a file **data1** and put the sorted file in a file called **data1.sorted**. Give the command that uses both input and output redirection.

5. Give the command to accomplish the task in Problem 4 by using a pipe and output redirection.

6. Give a set of commands equivalent to the command `ls -l | grep "sarwar" > output.p3` that use I/O redirection operators only. How does the performance of the given command compare with your command sequence? Explain.

7. What is the purpose of the `tee` command? Give a command equivalent to the command in Problem 6 that uses the `tee` command.

8. Write Linux shell commands to carry out the following tasks.
   a. Count the number of characters, words, and lines in a file called **data1** and display the output on the display screen.
   b. Count the number of characters, words, and lines in the output of the `ls -l` command and display the output on the display screen.
   c. Do the same as in part (b), but redirect the output to a file called **data1.stats**.

9. Give the command for searching a file **datafile** for the string "Internet," sending the output of the command to a file called **Internet.freq** and any error message to a file **error.log**. Draw a diagram for the command, similar to the ones shown in the chapter, to illustrate its semantics.

10. Give a command for accomplishing the task in Problem 9, except that both the output of the command and any error message go to a file called **datafile**.

11. Give a command to search for lines in **/etc/passwd** that contain the string "sarwar." Store the output of the command in a file called **passwd.sarwar** in your current directory. If the command fails, the error message must also go to the same file.

12. What is the Linux pipe? How is pipe different from output redirection? Give an example to illustrate your answer.

13. What do the following commands do under the Bash?
   a. `cat 1> letter 2> save 0< memo`
   b. `cat 2> save 0< memo 1> letter`
   c. `cat 1> letter 0< memo 2>&1`
   d. `cat 0< memo | sort 1> letter 2> /dev/null`
   e. `cat 2> save 0< memo | sort 1> letter 2> /dev/null`

14. Consider the following commands under the Bourne shell.
   a. `cat memo letter 2> communication 1>&2`
   b. `cat memo letter 1>&2 2> communication`

15. Where do output and error messages of the `cat` command go in each case if
   a. both files (memo and letter) exist in the present working directory, and
   b. one of the two files does not exist in the present working directory?

16. Send an e-mail message to **doe@domain.com**, using the mail command. Assume that the message is in a file called **greetings**. Give one command that uses input redirection and one that uses a pipe. Any error message should be appended to a file **mail.error**.

17. What happens when the following commands are executed on your Linux system? Why do these commands produce the results that they do?
   a. `cat letter >> letter`
   b. `cat letter > letter`

18. By using output redirection, send a greeting message "Hello, World!" to a friend's terminal.

19. Give a command for displaying the number of users currently logged on to a system.

20. Give a command for displaying the login name of the user who was the first to log on to a system.

21. What is the difference between the following commands?

    a. `grep "John Doe" Students > /dev/null 2>&1`

    b. `grep "John Doe" Students 2>&1 > dev/null`

22. Give a command for displaying the contents of (the files' names in) the current directory, with three files per line.

23. Give a command that reads its input from a file called **Phones**, removes unnecessary spaces from the file, sorts the file, and removes duplicate lines from it.

24. Repeat Problem 23 for a version of the file that has unnecessary spaces removed from the file but still has duplicate lines in it.

25. What do the following commands do?

    a. `uptime | cat - who.log >> system.log`

    b. `zcat secret_memo.Z | head -5`

26. Give a command that performs the task of the following command but with the `diff` command running on the machine called **server**: `ssh server cat ~/ research/pvm/datafile.server | diff datafile -`

27. Give a command for displaying the lines in a file called employees that are not repeated. What is the command for displaying repeated lines only?

28. Give a command that displays a long list for the most recently created directory.

29. Create a FIFO, called **myfifo1**. What are the default access privileges set for it? Create a FIFO, **myfifo2**, with read and write access privileges for everyone. Show your commands and their output for performing these tasks.

30. Give a set of commands for producing the output of the session given in Section 9.15. Your command sequence should ensure that the order of output is always the following: the status of all the processes running on the system, the number of daemons running on the system, and the total number of processes running on the system.

31. The number of files a process can open simultaneously on a Linux system is dependent on the size of the per-process file descriptor table on the system. Use a shell command with the features discussed in this chapter to display the value of the NR_OPEN variable in the **/usr/include/linux/limits.h** file. Show your command and its output and explain your answer.

    Hint: Look for a symbolic constant defined in the **/usr/include/linux/limits.h** file.

32. How much data can a pipe store on your system? Use a shell command to obtain your answer. What command did you use? Show the command and its output.

    Hint: Look for a symbolic constant defined in the **/usr/include/linux/limits.h** file.

33. What is the output of the following command? Give reasons for your answer.
    `cat file1 file2 |& grep "John Doe" | sort |& wc -1`

34. What is the purpose of the following command?
    `cat /usr/include/limits.h | nl | grep NR_MAX`

35. What do the following commands do?

    a. `mail mike@somewhere.org < to_do`

    b. `cat Phones | sort | uniq | pr | lpr`

36. The `ps -e` command displays information about all the processes running on your system. These processes include multiple instances of some processes such as systemd and sshd. Show a command that displays the number of unique processes running on the system.

Advanced Questions and Problems

37. What does the following command do? Clearly explain the purpose of every component and argument of the `find` and `xargs` commands.
    `find ~ 2> /dev/null -size +2G -print0 | xargs -0 -I '{}' mv '{}' ~/bigfiles`

38. What do the following commands do? Clearly explain the purpose of every component and argument of the `find` and `grep` commands.

   a. `find / 2>&1 | grep -v 'Permission denied'`

   b. `find / 2>&1 | grep 'Permission denied'`

39. Give a command to display the total number of files in your home director. The command should not display the "Permission denied" message.

40. Display the pathnames of the ten largest files under your home directory hierarchy in sorted order (largest first).

41. Give a command that displays the total number of files in your system with three hard links.

42. Write two programs in C that would obtain and print to stdout the values of the resource limits that you found in Problems 31 and 32.

43. Why would you want to find out the resource limits obtained in Problems 31 and 32? What kind of application, or user-written C program, would need to make use of not only those resource limits, but any of the others that the Single Unix Specification Version 3 (SUSv3) allows you to query, in either a determinate or indeterminate fashion?

44. Analyze the following line of Bash code, and in your own words, describe as exactly as you can what is being accomplished by each element on the line. The numbers 1, 3, etc. refer to file descriptors.
   `4>&1 >&3 3>&- | while read a; do echo "File_D4: $a"; done 1>&3 5>&- 6>&-`
   Exactly what does this structure `while read a; do echo "File_D4: $a"; done` accomplish?

Projects

Project 1

Familiarize yourself with the Bash Debugger, bashdb, which can be installed on Debian-family and CentOS 7 systems. Then do the following:

Run the script file found in the following session (which you can name "stdtest") from the command line, using the command `./stdtest`. Then, use bashdb to step through the script file, one command at a time. Exactly what does this script file accomplish? What are the times listed for bashdb-controlled execution? Why are they greater than execution of the script file without bashdb? How would you examine the assigned values of file descriptors during the bashdb debugging-controlled execution of this script?

```
#!/bin/bash
LOGFILE=my_logfile.txt
exec 6>&1
exec > $LOGFILE
time
cd /
df -hT
exec 1>&6 6>&-
ls -la
exit 0
```

Looking for more? Visit our sites for additional readings, recommended resources, and exercises.
   CRC Press e-Resource: https://www.crcpress.com/9781138710085
   Authors' GitHub: https://github.com/bobk48/linuxthetextbook

10

Processes

OBJECTIVES

- To describe the concept of a process, and execution of multiple processes on a computer system with a single central processing unit (CPU)
- To explain how a shell executes commands
- To discuss static and dynamic display of process attributes
- To discuss the main memory image of a Linux process
- To briefly describe the concept of CPU scheduling and scheduling classes in Linux
- To explain the concept of foreground and background processes, including a description of a daemon and its uses
- To describe sequential and parallel execution of commands
- To discuss process and job control in Linux: foreground and background processes, sequential and parallel processes, suspending processes, moving foreground processes into the background and vice versa, and terminating processes
- To describe the Linux process hierarchy
- To cover the following commands and primitives:

  ```
  <Ctrl+C>, <Ctrl+D>, <Ctrl+Z>, <Ctrl+\>, ;, &, ( ), bg, chrt, fg, jobs, kill,
  nice, nohup, ps, pstree, renice, sched, size, sleep, top, ulimit -s
  ```

10.1 Introduction

As we have mentioned in Chapter 1, a process is a program in execution. The program may be assembly language code, an executable code generated after compiling a source program written in a high-level language such as C++, or an interpreted code written in LISP, JavaScript, Perl, Python, Interpreted C (CINT), or a Linux shell. The Linux system creates a process every time you run an external command, and the process is removed from the system when the command finishes its execution. We use the terms *program* and *command* interchangeably.

Process creation and termination are the only mechanisms used by the Linux system to execute external commands. In a typical time-sharing system such as Linux, which allows multiple users to use a computer system and run multiple processes simultaneously, hundreds to thousands of processes are created and terminated every day. Remember that the central processing unit (CPU) in the computer executes processes. The question is, how does a system with a single CPU with one core execute multiple processes simultaneously? Even for systems with multiple CPUs or multiple cores in a CPU, the number of processes is greater than the number of CPUs or cores. How does a system with the number of processes larger than the number of CPUs or CPU cores execute these processes simultaneously? A detailed discussion of this topic is beyond the scope of this textbook, but we briefly address it in Section 10.2 and later in Section 10.5.1. You can get detailed information about CPU scheduling and related algorithms in a book on operating system concepts. Later in the chapter, we discuss viewing the static and dynamic state of processes, foreground and background processes, daemons, jobs, process and job attributes, and process and job control. We use the terms *time-sharing* and *multitasking* synonymously.

10.2 CPU Scheduling—Running Multiple Processes Simultaneously

On a computer system that contains a single CPU and runs a time-sharing operating system, multiple processes are simultaneously executed by quickly switching the CPU from one process to the next. That is, one process is executed for a short period of time, and then the CPU is taken away from it and given to another process. The new process executes for a short period of time and then the CPU is given to the next process. This procedure continues until the first process in the sequence gets to use the CPU again. The time a process is "in" the CPU before it is switched "out" of the CPU is called a *quantum* or *time slice*. The quantum is usually very short: one second or less for a typical Linux system.

On Linux Mint, the default kernel-set quantum, is 100 ms, as set by RR_TIMESLICE in the kernel include file /usr/src/linux-headers-4.4.0-21/include/linux/sched/rt.h. The actual quantum is *not* fixed for real-time (RT) processes and threads on Linux Mint, since the scheduler calculates it at run time, given the target level and how many processes are running at any given time.

When the CPU is free/idle (i.e., not used by any process) or when the current process has finished its quantum, the kernel uses an algorithm to decide which process gets to use the CPU next. The technique used to choose the process that gets to use the CPU is called *CPU scheduling*. The kernel code that performs this task is known as the *Short-term CPU Scheduler* or *CPU Scheduler*. On Linux Mint, this scheduler is known as the Completely Fair Scheduler (CFS). The details of CFS are beyond the scope of this book.

The process of taking the CPU away from the currently executing process and giving it to the newly scheduled process is called *context switching*. This task is performed by another part of the kernel, known as the *dispatcher*. In systems with multiple CPUs or a CPU with multiple cores, such as those multicore processors by Intel, Advanced Micro Devices (AMD), and other companies, if the number of processes on the system is more than the number of CPUs in the system (or the number of cores for a single CPU system), CPU scheduling and context switching still happen.

Thus, on a system with multiple users running multiple processes, the scheduler and dispatcher work in tandem to make you feel as if you are the only one using the system. This is the fundamental concept of concurrency as maintained and implemented by the kernel—an important higher-level abstraction issue we raise in Chapter 16, Section 16.1. It is also the foundation of the idea of autonomy, or the autonomous, insulated, self-contained, and independent user.

Although a focused discussion on CPU scheduling algorithms is beyond the scope of this book, we give a brief and simplistic view of how Linux schedulers work. In Section 10.5, we go into more detail about CPU scheduling for processes and threads in Linux. We do this by delineating a major aspect of scheduling: scheduler priorities and classes.

In a time-sharing system, a priority value is assigned to every process, and the process that has the highest priority gets to use the CPU next. Several methods can be used to assign a priority value to a process. One simple method is based on the time that it enters the system. In this scheme, typically, the process that enters the system first is assigned the highest priority and gets to use the CPU next; the result is called a *first-come, first-serve* (FCFS) scheduling algorithm. Another scheme is to assign a priority value based on the amount of time a process has used the CPU. Thus, a newly arriving process, or a process that spends most of its time performing input and/or output (I/O) operations, gets the highest priority. Processes that spend most of their time performing I/O are known as *I/O-bound processes*; they need less time in the CPU because I/O is sporadic. Processes that spend most of their time performing computations are known as *compute-bound or CPU-bound processes*; they need more time in the CPU because of the nature of their operations. An example of an I/O-bound process is a text editor such as vim. Classic examples of compute-bound processes are matrix calculations, database search algorithms, and forms of video processing.

In the *round robin* (RR) scheduling algorithm, the CPU is given to each process in the queue of processes for one time quantum, one after the other. This algorithm is a natural choice for time-sharing systems, wherein all users like to see progress by their processes.

If you are interested in other CPU scheduling algorithms and their details, we encourage you to read a book on operating system principles and concepts. The operating system code that implements the CPU scheduling algorithm is known as the *processor scheduler*. The processor scheduler code for most operating systems, including Linux, resides in the kernel.

You can assign a higher nice value to your processes by using the `nice` or `renice` command, but the nice value cannot be set to a negative number by a nonsuperuser. A higher nice value means a higher priority value and, hence, a lower priority. So, when you increase the nice value of your process, you are being nice to other user processes. The formula clearly indicates that the higher the recent CPU usage of a process, the higher its priority value and the lower its priority. Thus, Linux favors processes that have used less CPU time in the recent past. A text editor such as vim gets higher priority than a process that computes the value of pi (π) because vim spends most of the time waiting for I/O—that is, reading keyboard input, reading/writing to disk, and displaying file data or keyboard input on the screen. On the other hand, the process that computes π spends most of its time doing calculations—that is, using the CPU. Recalculating priority values of all the processes every second causes process priorities to change dynamically (up and down). In Section 10.5, we further explore the Linux scheduling concept. We also give a few more details about the use of the `nice` and `renice` commands in Section 10.5.2.

Exercise 10.1
Search the Web for context switch times for different CPUs. What is the range of these times for different CPUs. Which CPU has the shortest context time and how much is this time?

Exercise 10.2
Give three examples each of CPU-bound and I/O-bound processes.

10.3 Linux Process States

A Linux process can be in one of many states, moving from one state to another, eventually finishing its execution, normally or abnormally, and getting out of the system. A process terminates normally when it finishes its work and exits the system gracefully. A process terminates abnormally when it exits the system because of an *exception* (error condition) or *intervention* by its owner or the superuser. The owner of the process can intervene by using a command or a particular keystroke to terminate the process. We discuss these commands and keystrokes later in the chapter. The primary states that a process can be in are shown in the state diagram in Figure 10.1.

The *waiting state* encompasses several states; we use the term here to keep the diagram simple. Some of the states belonging to the waiting state are listed under the oval representing the state. Table 10.1

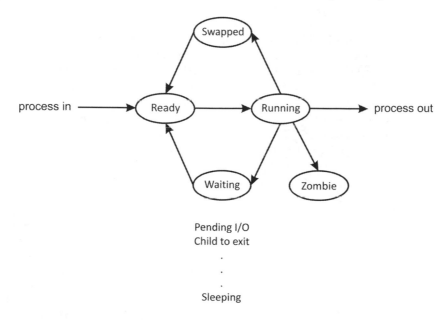

FIGURE 10.1 Linux process state diagram.

TABLE 10.1

A Brief Description of the Linux Process States

State	Description
Ready	The process is ready to run but does not have the CPU. Based on the scheduling algorithm, the scheduler decided to give the CPU to another process. Several processes can be in this state, but on a machine with a single CPU, only one can be executing/running (i.e., using the CPU).
Running	The process is actually running (i.e., using the CPU).
Waiting	The process is waiting for an event. Possible events are an I/O operation to complete (e.g., disk/terminal read or write), a child process to complete (the parent is waiting for one or more of its children to exit), or the process itself is waiting to be reawakened having been put to sleep.
Swapped	The process is ready to run, but it has been temporarily put on the disk (on the swap space); perhaps it needs more memory and there is not enough available at this time.
Zombie	A dying process is said to be in a zombie state. Usually, when the parent of a process terminates before it executes the exit call, it becomes a zombie process. The process finishes and finds that the parent is not waiting. The zombie processes are finished for all practical purposes and do not reside in the memory, but they still have some kernel resources allocated to them and cannot be taken out of the system. All zombies and their live children are eventually adopted by the granddaddy, the systemd process, which removes them from the system.

gives a brief description of these Linux process states. In the interest of brevity, and in keeping with the scope of this book, the other states that a Linux process can be in are not included in this discussion.

10.4 Execution of Shell Commands

A shell command can be internal (built in) or external. An *internal/built-in command* is one whose code is part of the shell process. Some of the commonly used internal commands are . (dot command), bg, cd, continue, echo, exec, exit, export, fg, jobs, pwd, read, readonly, return, set, shift, test, times, trap, umask, unset, and wait. An *external command* is one whose code is in a file; contents of the file can be binary code or a shell script. Some of the commonly used external commands are grep, more, cat, mkdir, rmdir, ls, sort, ftp, telnet, lp, and ps. A shell creates a new process to execute a command. While the command process executes, the shell waits for it to finish. In this section, we describe how a shell (or any process) creates another process and executes external commands. You can use the type command to determine whether your command is built in or external, as shown in the following session. Under Bash, you can also use the –a option of type to display all the locations of a command. Remember that if you created an alias for the type command in Chapter 2, Section 2.6, you will have to unalias it in order for the following session to work!

In the following session, you can see that the bg command is built in. The Bash shell may be invoked through an executable available in your file system structure as shown.

```
$ type bg
bg is a shell builtin
$ type -a bash
bash is /bin/bash
$
```

A Linux process can create another process by using the fork() system call, which creates an exact main memory copy of the original process (i.e., the process that calls fork()). Both processes continue execution, starting with the statement that follows the fork. The forking process is known as the *parent process*, and the created (forked) process is called the *child process*, as shown in Figure 10.2. Here, we show a Bash shell that has created a child process (another Bash). We discuss the use of fork() and other system calls needed for the creation of processes and *interprocess communication* in Chapters 15–16.

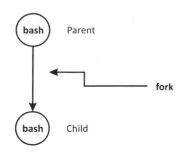

FIGURE 10.2 Process creation via the fork system call.

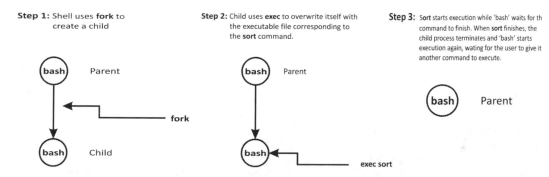

FIGURE 10.3 Steps for execution of a binary program sort by a Linux shell.

For executing an external binary command, a mechanism is needed that allows the child process to become the command to be executed. The Linux system call exec() can be used to do exactly that, allowing a process to overwrite itself with the executable code for another command. A shell uses the fork() and exec() calls in tandem to execute an external binary command. Figure 10.3 shows the sequence of events for the execution of an external command sort, whose code is in a binary file, **/usr/bin/sort**.

The execution of a shell script (a series of shell commands in a file; see Chapters 12 and 13) is slightly different from the execution of a binary command/file. In the case of a shell script, the current shell creates a child shell and lets the child shell execute commands in the script file, one by one. Each command in the script file is executed in the same way that commands from the keyboard are; that is, the child shell creates a child for every command that it executes. While the child shell is executing commands in the script file, the parent shell waits for the child to terminate. When the child shell hits the eof marker in the script file, it terminates. The only purpose of the child shell, like any other shell, is to execute commands, and eof means "no more commands." When the child shell terminates, the parent shell comes out of the waiting state and resumes execution. This sequence of events is shown in Figure 10.4, which also shows the execution of a find command in the script file.

Unless otherwise specified in the file containing the shell script, the child shell has the type of the parent shell. That is, if the parent is Bash, the child is also Bash. Thus, by default the shell script is executed by a "copy" of the parent shell. However, a shell script written for any shell (C, TC, Bourne, Bash, Korn, etc.) can be executed regardless of the type of the current shell. To do so, simply specify the type of the child shell under which the script should be executed in the first line of the file containing the shell script as #!full-path-name-of-the-shell. For example, the following line dictates that the child shell is Bash, so the script following this line is executed under the Bash.

```
#!/bin/bash
```

Also, you can execute commands in another shell by running that shell as a child of the current shell, executing commands under it, and terminating the shell. A child shell is also called a *subshell*. Recall that the commands to run various shells are sh for the Bourne shell, csh for the C shell, tcsh for the TC

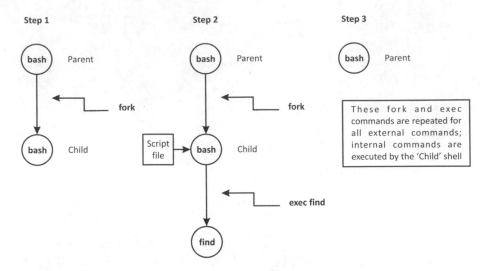

FIGURE 10.4 Steps for execution of a shell script by a Linux shell.

shell, ksh for the Korn shell, and bash for Bash. To start a new shell process, simply run the command corresponding to the shell you want to run. See Table 2.4 for a complete path to common Linux shells. If a shell is not installed on your Linux by default, ask your system administrator to install it for you or see Appendix A for instructions to install it on your system. Bash is the default shell on Linux Mint.

In the following session, we are running Bash as the default shell (see Chapter 2, Section 2.8.3, for information on how to set the default shell for the current session, or any subsequent sessions), and the Bourne shell runs as its child. The echo command is executed under the Bourne shell. Then another Bash shell is started, and the echo command is executed under it. The ps command shows the three shells running. Finally, both the TC and Bourne shells are terminated when <Ctrl+D> is pressed in succession, and control goes back to the original shell, Bash. The first <Ctrl+D> terminates the TC shell, giving control back to the Bourne shell. You can also exit a shell running the exit command. Figure 10.5 illustrates all the steps involved, showing the parent–child relationship between processes.

```
$ ps
  PID TTY          TIME CMD
14738 pts/0    00:00:00 bash
14760 pts/0    00:00:00 ps
$ /bin/sh
$ echo "This is Bourne shell."
This is Bourne shell.
$ bash
$ echo "This is Bourne Again SHell."
This is Bourne Again SHell.
$ ps
  PID TTY          TIME CMD
14738 pts/0    00:00:00 bash
14761 pts/0    00:00:00 sh
14762 pts/0    00:00:00 bash
14764 pts/0    00:00:00 ps
$ <Ctrl+D>
$ exit
$ <Ctrl+D>
$
```

Exercise 10.3

Repeat the previous session on your system to understand how shell processes are created and terminated. Draw the process hierarchy for the output of the last ps command.

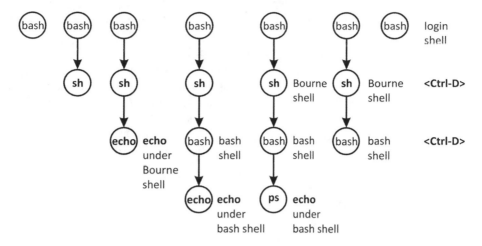

FIGURE 10.5 Execution of commands under the child shell (also called subshells).

10.5 Process Attributes

Every Linux process has several attributes, including owner ID (called user ID [UID] in Linux jargon), process ID (PID), PID of the parent process (PPID), process name, process state, command executed to start the process, priority of the process, process start time, percentage of the CPU time consumed by the process, percentage of the main memory consumed by the process, size of the process in virtual memory, size of the process currently in main memory, state of the process, the event a process is waiting for (in case it is not running), and the length of time the process has been running. From the user's and programmer's point of view, one of the most useful of these attributes is the PID, which is used as a parameter in several process control commands discussed later in this chapter.

Linux provides several tools that allow you to monitor the attributes of the processes currently running on your system, change the states of your processes, and perform various operations on them, including stopping/restarting them, sending them specific messages, having them communicate with each other, and terminating them. In this chapter, we will discuss some of the commands and tools that allow us to monitor the attributes of processes statically and dynamically, as well as perform various operations on processes just stated.

10.5.1 Static Display of Process Attributes

The ps command can be used to view a snapshot of the attributes of processes currently in the system. The following is a brief description of the ps command, with option/feature variants for both Berkeley Software Distribution (BSD)-style options (executed without a dash on the command line) and UNIX/POSIX-style options (executed with a dash on the command line).

Syntax:
 `ps [options]`

Purpose: Report static (one shot) information about process status/attributes

Output: A header line and a snapshot of the attributes of processes running on the system. Note: Gnu long options are preceded by two dashes.

Commonly used options/features (BSD variants without a preceding dash):

o — Display information about the comma- or space-separated keywords after the PID field in the default output. You can assign the header of your choice for a keyword by putting an = after the keyword, followed by the header value.

u	Display information about the processes for the users specified in the comma-separated list of usernames; no spaces before or after the commas
a	Lift the BSD-style "only yourself" restriction
e	Display the environmental information for each process
o	Similar to −o, except that it does not show the default fields and you can change header texts for multiple using multiple −o options. If no headers are specified with keywords, the header line is not displayed.
u	Display for each process information about the following keywords: user, pid, %cpu, %mem, vsz, rss, tty, state, start, time, and command.
x	Display information about processes that do not have controlling terminals (including daemons)

Commonly used options/features (UNIX/POSIX variants with a preceding dash):

−A	Select all processes
−e	Same as the −A option: display information about all processes, including session leaders
−f	Display a full format listing, including process owner ID (UID), PID, PPID, C, process start time (STIME), TTY, time executed for (TIME), full command line (CMD)
−l	Display a long listing: F, state (S), UID, PID, PPID, C, PRI, NI, ADDR, SZ, WCHAN, TTY, TIME, CMD
−u or −U	Display a list of processes belonging to UIDs or lognames listed in the comma-separated list

The output of the ps command is sorted first by the terminals associated with processes and then by their PIDs. You can change the default sort order by using different options. If multiple such options are specified, the command exercises the last option. The shell sessions in this section demonstrate the use of the ps command with and without options. We ran all of them on a Linux machine running the Bash shell.

The output of the ps command, as shown in the following session, displays four fields about processes, whose attributes are displayed one per line: PID, the terminal the process is attached to (TTY), the CPU time the process has consumed (TIME), and the command used by the user to run the process (CMD). The output shows that two processes are attached to terminal 0: bash (the login Bash) and ps. The PIDs of the processes attached to terminal 0, −bash and ps, are 27753 and 27783, and have run for 0 s each, respectively.

```
$ ps
  PID TTY          TIME CMD
27753 pts/0    00:00:00 bash
27783 pts/0    00:00:00 ps
$
```

You can also add columns of displayed output with the −o option. A comma-separated list must follow the option, with keywords designating which output format columns you want displayed. For example, the following command displays the default columns of output, plus a column showing the state of the processes shown.

```
$ ps -o pid,tty,stat,time,cmd
  PID TT       STAT     TIME CMD
27753 pts/0    Ss    00:00:00 -bash
27794 pts/0    R+    00:00:00 ps -o pid,tty,stat,time,cmd
$
```

The state of a process is displayed as a character string—for example, Ss, S+, and R+ in the previous session. Table 10.2 explains the various characters in the string listed under the process state (STAT) column. Thus, the process with the Ss state is a session leader process that is currently sleeping—that is, waiting for less than 20 s. Similarly, the process with the S+ state is a foreground process that has been

TABLE 10.2

A Brief Description of the Linux Process States Displayed by the ps Command

First Character	Description	Additional Character	Description
D	Process on a disk or other short-term interruptible wait	+	Foreground process
X	Dead process	<	Process has raised scheduling priority
R	Runnable process waiting to get CPU	L	Process locked in core—for example, for raw I/O
S	Sleeping process—waiting for < 20 s	N	Process has reduced CPU-scheduling priority
T	Stopped/suspended process	S	Session leader process—for example, login shell
W	Idle interrupt process/thread	L	Multithreaded
Z	Zombie process		

waiting for less than 20 s. Finally, the process with the R+ state is a foreground process that is ready to run and is waiting to be scheduled to use the CPU. We discuss foreground and background processes in detail in Section 10.6.

The ps -u command shows a long listing of all the processes belonging to the user running the command. VSZ is the virtual size of the process and RSS (resident set size) is the real size of the process in memory. Both sizes are in kilobytes. Note that **sarwar** has logged in on one terminal with Bash running.

```
$ ps -u
USER       PID %CPU %MEM    VSZ   RSS TTY      STAT START    TIME COMMAND
sarwar   27753  0.0  0.0  23192  5404 pts/0    Ss   07:23    0:00 -bash
sarwar   27797  0.0  0.0  37680  3356 pts/0    R+   07:39    0:00 ps -u
$
```

You can specify a comma-separated list of UIDs to display the same information about the processes belonging to them, as in ps -u 1004,1009,1020. As a reminder, UID is the third field of a line in the **/etc/passwd** file. Note that the state of the first three processes is Is+, which means that they are foreground idle session leader processes. In other words, they are foreground login shell processes that have not used the CPU for more than 20 s.

You can use the -U option to display the default status of all the processes belonging to the users whose comma-separated usernames are specified after the option. In the following example, all the processes belonging to the user **sarwar** are displayed. The ps -U john,bob,davis,sunil,goldman,ibraheem command would display the default status of all the processes belonging to the users **john**, **bob**, **davis**, **sunil**, **goldman**, and **ibraheem**.

```
$ ps -U sarwar
  PID TTY          TIME CMD
19919 ?        00:00:00 systemd
19923 ?        00:00:00 (sd-pam)
19937 ?        00:00:00 sshd
27753 pts/0    00:00:00 bash
27799 pts/0    00:00:00 ps
$
```

You can also use the comma-separated list of UIDs of the users to display the same information. Thus, the ps -U 1004 command displays the default status of all the processes belonging to the user with UID 1004. Similarly, the ps U 1004,1005,1001 command displays the default status of all the processes belonging to the users with UIDs 1001, 1004, and 1005.

You can use the -a and -x options to display information about your processes, processes belonging to other users, system processes, and processes not attached to any controlling terminals, such as daemons. You can determine the total number of processes, including all system processes and daemons, running on a system by using the ps -ax | wc -l command. The output contains a header line and a line each for the ps -ax and wc -1 processes. Thus, 176 processes are currently running on our Linux Mint system if we do not count the two processes created due to the command line itself.

```
$ ps -ax | wc -1
    177
$
```

You can also determine the number of processes running on a machine by using the ps -e | wc -1 command.

You can use the -H option to display the kernel-visible threads, known as *lightweight processes* (LWPs) in Solaris, in the processes running on your system. The output of the ps -axH | wc -1 command shows 283 as its output. Out of these 283 lines, one line is for the output header and one line each for the ps -axH and wc -1 commands each. Thus, 280 kernel-visible threads are running in the 176 processes that are currently running on the system.

```
$ ps -axH | wc -1
283
$
```

Although the -j, -1, -u, and -v options allow you to display the values of predetermined keywords in a known order and the standard header information, the -O and -o options allow you to display customized output: values of the keywords of your choice, in the order of your liking, and with the header of your taste. In the following session, we show the use of the ps command with these options. The keywords dsiz and tsiz are, respectively, the sizes (in kilobytes) of the data and text/code segments of the *memory image* of the process. wchan is the event a process is waiting for. The last command in the sessions again shows how you can customize the header of the output.

```
$ ps -o dsiz,tsiz,vsz,rss,wchan
DSIZ TSIZ     VSZ    RSS WCHAN
22217 974   23192   5472 wait
29114  89   29204   1604 -
$ ps -o dsiz,tsiz,lwp,nlwp,wchan,pcpu,pmem,flags,args
DSIZ TSIZ    LWP NLWP WCHAN  %CPU %MEM F COMMAND
22217 974  14738    1 wait    0.0  0.0 0 -bash
29142  89  14813    1 -       0.0  0.0 0 ps -o dsiz,tsiz,lwp,nlwp,wchan,pcpu,pmem
$ ps -o user,pid=SON -o ppid=MOM -o args
USER        SON   MOM COMMAND
sarwar    14738 14737 -bash
sarwar    14814 14738 ps -o user,pid=SON -o ppid=MOM -o args
$
```

Some of the fields and the corresponding keywords are obvious, and we have discussed a few of them earlier in this chapter. However, some keywords, which we will discuss now, are new and not obvious. You can display all keywords by using the ps L command, as shown:

```
$ ps L
%cpu          %CPU
%mem          %MEM
...
[Output truncated]
$
```

Table 10.3 shows some of the commonly used keywords and their meanings. Note that these fields are displayed in uppercase in the header of the output of the ps command (e.g., COMMAND for command and PID for pid).

The data, stack, and text/code segments are part of the memory image of a process. Three additional sections of the memory image of a Linux process are shared libraries, heap, and environment. Figure 10.6 shows the memory image of a Linux process. The environment consists of command line

TABLE 10.3

A Brief Description of the Commonly Displayed Fields/Keywords of the ps Command

Field	Description	Field	Description
command	Command executed to create process	Ppid	Parent's PID
dsiz	Data size in kilobytes	Pri	Scheduling priority
cpu	Short-term CPU usage for scheduling	Rss	Real process size (kilobytes) in memory
flags	Flags indicating process states/events	Sid	Session ID; PID of the session leader
Lwp	Lightweight process (thread) ID	Ssiz	Stack size in kilobytes
majflt	Total page faults (same as PAGEIN)	started	Time the process was started
mwchan	Event or lock of the locked/blocked process	Stat	Process state
nlwp	Number of threads tied to an LWP	Time	Time the process has executed for
pcpu	CPU utilization up to one previous minute	Tsiz	Text (code) size in kilobytes
pgid	Process group ID	Tt	Terminal the process is attached to
pid	Process ID	Vsz	Virtual size of the process in kilobytes
pmem	Percentage of memory used by process	wchan	Event (or address) on which a process waits

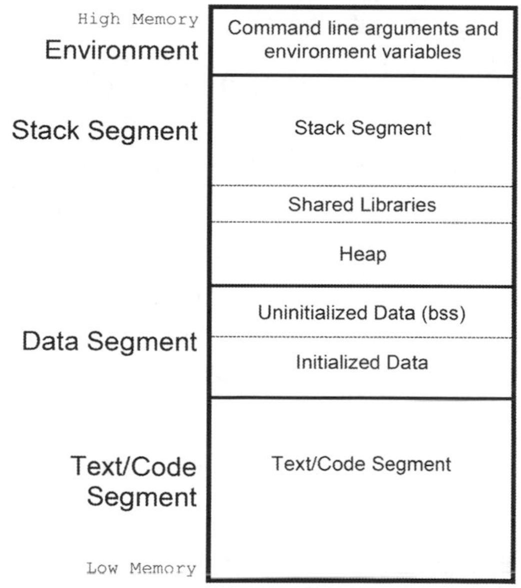

FIGURE 10.6   Memory image of a Linux process.

arguments and five shell variables (HOME, PATH, SHELL, USER, and LOGNAME), accessible through pointers to two arrays of pointers to null-terminated strings. The stack grows from high memory to low memory and the heap grows from low memory to high memory. Text and initialized data portions are read from the program file by the exec(2) system call, and the uninitialized data (bss) section is initialized to zero, also by exec(2).

You can also use the size command to display the sizes (in bytes) of text/code, initialized data, and uninitialized data (bss) for an executable program. The following is the sample run of the command on our Linux Mint system with **/usr/bin/sort** as the argument. The fourth number is the total size of the text and data sections (i.e., the sum of first three numbers) in decimal. The fifth number is the total size in hexadecimal.

```
$ size /usr/bin/sort
   text    data    bss     dec      hex filename
 104653    1880    1960  108493    1a7cd /usr/bin/sort
$
```

The ulimit -s command displays the virtual stack size of a Linux process to be 8,192 kilobytes (i.e., 8 MB).

```
$ ulimit -s
8192
$
```

The UNIX/POSIX options version of the ps command supports the following options of the BSD version of the command: S, a, e, r, v, w, and x. However, you invoke them without using a hyphen before them—for example, ps r. Similarly, the –o option also works under the UNIX/POSIX options, as it does under BSD, but not with all the keywords listed in Table 10.3 (or those displayed by the ps L command in the BSD version.). It works with the following keywords, most of which are common with the ps BSD version: user, ruser, group, rgroup, uid, ruid, gid, rgid, pid, ppid, pgid, sid, tasked, ctid, pri, opri, pcpu, pmem, vsz, rss, osz, nice, class, time, etime, stime, zone, zoneid, f, s, c, lwp, nlwp, psr, tty, addr, wchan, fname, comm, args, projid, project, pset, and lgrp.

The sample runs in the following session show the use of the ps command with various options. Note that the output of the ps –l command includes the obsolete fields F and ADDR, and the old-style (obsolete) PRI (known as opri in the list of keywords shown in the previous paragraph), where lower PRI value means higher priority. You can use the –y and –l options together to exclude the columns for the obsolete fields F and ADDR, include the RSS column, and display the PRI value in vogue.

```
$ ps -l
F S   UID   PID  PPID  C PRI  NI ADDR SZ WCHAN  TTY          TIME CMD
0 S  1004 14738 14737  0  80   0 -  5798 wait   pts/0    00:00:00 bash
0 R  1004 14832 14738  0  80   0 -  7308 -      pts/0    00:00:00 ps
$ ps -ly
S   UID   PID  PPID  C PRI  NI   RSS    SZ WCHAN  TTY          TIME CMD
S  1004 14738 14737  0  80   0  5472  5798 wait   pts/0    00:00:00 bash
R  1004 14839 14738  0  80   0  1476  7308 -      pts/0    00:00:00 ps
$
```

The second column, S, shows the state of the process. Process states in UNIX/POSIX are briefly described in Table 10.2, along with some fields that are different from the fields displayed by the ps command in the BSD variant. Note that the ps command is in the running state and bash is waiting for an event. The event in this case is the termination of its child, i.e., the ps command running in foreground. NI shows the nice value of the process; the higher the value, the lower the priority of the process. You can display the priorities of processes by using the –c option. The SZ field shows the size (in pages) of the process in virtual memory. We have discussed the rest of the fields in the output of the ps –ly command earlier in this section.

You can appreciate the difference between the old style and new style displays of priorities by running the following command. The first column for PRI is for the new style and the second column is for the obsolete PRI. Note the priority values displayed for the two processes in the obsolete and new styles.

```
$ ps -c -o pid,pri,opri,args
  PID PRI PRI COMMAND
14738  19  80 bash
14842  19  80 ps
$
```

You can use the –t option to display processes associated with a particular terminal, as shown in the following session. The ps –eLP command displays all the LWPs running in each process as well as the CPU core (shown under PSR) on which a thread runs. The first line of the output shows that systemd is running on core 4. The bottom three lines of the output of the command shows that our sshd, bash, and ps commands run on cores 1, 3, and 5, respectively. Note that systemd has six threads running, as shown by the output of the ps -eL | grep "systemd" | wc -l command. The output of the ps -eL | wc -l command shows that 282 LWPs are currently running on the system. The last command in the session shows how you can use the -o option to display various attributes about all the processes running on the system.

```
$ ps -t /dev/pts/0
  PID TTY              TIME CMD
 7074 pts/0        00:00:00 bash
 7224 pts/0        00:00:00 ps
$ ps -eLP
  PID    LWP PSR TTY           TIME CMD
    1      1   4 ?        00:01:39 systemd
    2      2   6 ?        00:00:00 kthreadd
    3      3   0 ?        00:00:13 ksoftirqd/0
    5      5   0 ?        00:00:00 kworker/0:0H
    7      7   4 ?        01:49:49 rcu_sched
    8      8   0 ?        00:00:00 rcu_bh
    9      9   0 ?        00:00:00 migration/0
   10     10   0 ?        00:00:20 watchdog/0
   11     11   1 ?        00:00:20 watchdog/1
   12     12   1 ?        00:00:00 migration/1
[Output Truncated]
26829 26829   3 ?        00:00:00 sshd
26842 26842   1 ?        00:00:00 sshd
26843 26843   3 pts/2    00:00:00 bash
26861 26861   5 pts/2    00:00:00 ps
$ ps -eL | grep "systemd" | wc -l
6
$ ps -eL | wc -l
282
$ ps -e -o user,uid,pid,ppid,pri,nice,vsz,rss,nlwp,args
USER       UID   PID  PPID PRI  NI    VSZ   RSS NLWP COMMAND
root         0     1     0  19   0 119836  4664    1 /sbin/init splash
root         0     2     0  19   0      0     0    1 [kthreadd]
root         0     3     2  19   0      0     0    1 [ksoftirqd/0]
root         0     5     2  39 -20      0     0    1 [kworker/0:0H]
root         0     7     2  19   0      0     0    1 [rcu_sched]
root         0     8     2  19   0      0     0    1 [rcu_bh]
root         0     9     2 139   -      0     0    1 [migration/0]
root         0    10     2 139   -      0     0    1 [watchdog/0]
root         0    11     2 139   -      0     0    1 [watchdog/1]
root         0    12     2 139   -      0     0    1 [migration/1]
[Output Truncated]
sarwar    1004 26842 26829  19   0 124556  4568    1 sshd: sarwar@pts/2
sarwar    1004 26843 26842  19   0  23160  5452    1 -bash
sarwar    1004 26860 26843  19   0  37680  3324    1 ps -e -o user,uid,pid,ppid,
$
```

Exercise 10.4

What command can you use to display the list of keywords that may be used with the O, -O, o, and –o options to display various attributes of running processes? Run the command on your system and show its output.

Exercise 10.5

Repeat the sessions discussed in Section 10.5 thus far on your system to understand how the ps command works with various options. Capture the shell commands and their outputs.

Exercise 10.6

Find the sizes of text, data, and bss for the executable code for bash and draw the process diagram for bash when it is in execution. Clearly show the sizes of as many sections as you can. State how you were able to find these sizes.

10.5.2 Linux Process and Thread Scheduling

There are two preliminary things you must realize about process and thread scheduling. First, Linux is predominantly a general-purpose operating system, not an RT operating system. That means that it is not intended to be used as the basis for embedded, RT control systems. This does not mean that it cannot or has not been applied very successfully that way. Examples of RT control systems are found in electromechanical embedded microprocessor-guided controllers in any industrial process application. Second, as Seymour Cray said, "Anyone can build a faster processor. The trick is to build a faster system." That's what the scheduler does; it gives you a faster system with higher throughput.

As can be seen in Chapter 16, Section 16.1, the steady-state maintenance of the Linux OS, and the hardware that it is running on, is achieved by the techniques of virtualization, concurrency, and persistence. The CPU scheduler provides the operating system virtualization and concurrency.

The CFS accomplishes virtualization in Linux. This means that the CPU appears as if it is more powerful than it really is in terms of system throughput and "simultaneous" execution of multiple processes, even on a single-core CPU. The CFS algorithm and its implementation by the scheduler are beyond the scope of what we cover in this book. It is sufficient to say that CFS tailors CPU usage for all executing processes and threads, to maximize CPU utilization and maximizes throughput.

10.5.2.1 Linux Scheduling Policies and Priorities

The scheduling of processes (or threads) on a CPU is dictated by a scheduling algorithm (or policy), which chooses the highest priority process from among the processes that are waiting to use the CPU and dictates the amount of time the chosen process can use the CPU for. Priorities are represented as integer numbers. We describe briefly the various scheduling policies used by the Linux scheduler and the mechanism used to assign priority values to processes. The technique used to calculate priorities is such that it allows kernel process to set their priorities higher than any user process. The CPU scheduling policies (or classes) supported by Linux are shown in Table 10.4, along with their purpose.

To gain a full appreciation of how scheduling priorities are assigned in Linux, browse through /usr/src/linux-headers-4.4.0-21/include/linux/sched/prio.h. A close reading of this header file will reveal that the priority of a process is dependent on its *nice value*, *nice width* (NICE_WIDTH), and *default priority*, DEFAULT_PRIO, of a process in Linux. The default priority is dependent on two symbolic constants, MAX_RT_PRIO and NICE_WIDTH, also defined in the same file. Table 10.5 shows the values of these and a few other parameters needed for the calculation of the priority of a process.

The priority of a process is calculated according to the following formula.

$$\text{Priority} = \left[(\text{nice value}) + \text{DEFAULT\_PRIO}\right]$$

This means that the smaller nice values result in smaller priority values and vice versa. Thus, if the nice value of a process is 0, its priority value will be 120. Similarly, the priority value of a process with a nice value 10 will be 130. In Linux, priority values are inverted, i.e., a smaller priority number equates to a higher CPU priority and vice versa. Thus, the process with priority value 120 will be given preference by the scheduling algorithm over the process with priority value 130.

TABLE 10.4

Scheduling Classes in Linux and Their Purpose

Policy Class	Purpose
SCHED_BATCH	Non-preemptive scheduling designed for CPU-bound processes/threads
SCHED_FIFO	Non-preemptive FIFO scheduling for time-critical (RT) processes/threads
SCHED_IDLE	Non-preemptive (?) scheduling for very low-priority processes/threads
SCHED_OTHER	The default Linux scheduling policy for most of the processes
SCHED_NORMAL	
SCHED_RR	Round Robin SCHED_FIFO scheduling

TABLE 10.5

Important Parameters for Computing the Priority of a Process in Linux

Parameter	Value
MAX_NICE	19 (least favorable to a process)
MIN_NICE	−20 (most favorable to a process)
NICE_WIDTH	(MAX_NICE − MIN_NICE + 1) = 40
MAX_RT_PRIO	100
MAX_PRIO	(MAX_RT_PRIO + NICE_WIDTH) = 100 + 40 = 140
DEFAULT_PRIO	(MAX_RT_PRIO + NICE_WIDTH / 2) = (100 + 40/2) = 120

The priority of a process can be 0–139. The priorities of RT processes can be 0–99. The user processes, i.e., SCHED_NORMAL/SCHED_OTHER (normal) and SCHED_BATCH (batch) processes have priorities of 100–139. For convenience, while working with various scheduler parameters, this priority range is translated to 0–39, achieved by simply subtracting MAX_RT_PRIO from the given priority value. This is the range of priorities that can be assigned to and affected by the nice value via the commands nice and renice. You can launch a program with your required priority using the general command nice -n nice_value program_name. Similarly, you can also change the priority of an already running process using the general command renice -n nice_value -p process_id.

Consider the following output of the ps command on our Linux Mint system, where the system is running at the multi-user.target level, and which requests an output format for scheduling priority class (CLS), scheduling priority (PRI), and nice value (NI), along with some standard identifying output format columns.

```
$ ps -ax --format uname,pid,ppid,tty,cmd,cls,pri,nice
USER       PID  PPID TT    CMD                        CLS PRI  NI
root         1     0 ?     /sbin/init splash          TS   19   0
root         2     0 ?     [kthreadd]                 TS   19   0
root         3     2 ?     [ksoftirqd/0]              TS   19   0
root         5     2 ?     [kworker/0:0H]             TS   39 -20
root         7     2 ?     [rcu_sched]                TS   19   0
root         8     2 ?     [rcu_bh]                   TS   19   0
root         9     2 ?     [migration/0]              FF  139   -
root        10     2 ?     [watchdog/0]               FF  139   -
...
[output truncated]
...
root     21936     2 ?     [kworker/u16:1]            TS   19   0
root     21942  1128 ?     sshd: sarwar [priv]        TS   19   0
sarwar   21944     1 ?     /lib/systemd/systemd --user TS  19   0
sarwar   21946 21944 ?     (sd-pam)                   TS   19   0
sarwar   21960 21942 ?     sshd: sarwar@pts/0         TS   19   0
sarwar   21961 21960 pts/0 -bash                      TS   19   0
sarwar   21982 21961 pts/0 ps -ax --format uname,pid,p TS  19   0
root     22407     1 tty1  /bin/login --              TS   19   0
root     22876     2 ?     [kworker/6:0]              TS   19   0
$
```

The /sbin/init splash command was run to create systemd, which has a priority of 19 and is in the class TS (SCHED_OTHER). Systemd is a collection of system management daemons, utilities, and libraries that serve as a replacement of the classical init daemon, initially designed for System V UNIX. It is the parent of all user-level processes and daemons. The same is true for the process kthreadd, the kernel thread daemon. Note that all the processes belonging to sarwar have priority value have the same class and priority as that of systemd. For more information on systemd, its service-starting capabilities, and how it allocates resources in unit files, see Chapter 20.

The chrt command sets or retrieves the RT scheduling attributes of a running process or runs a command with the specified attributes. You can set or retrieve both the scheduling policy and priority of a process. The default scheduling policy used by Linux is SCHED_OTHER (or SCHED_NORMAL). When run without any argument, chrt displays a brief summary of how it can be used, as shown in the following session.

```
$ chrt
Show or change the real-time scheduling attributes of a process.

Set policy:
 chrt [options] <priority> <command> [<arg>...]
 chrt [options] -p <priority> <pid>

Get policy:
 chrt [options] -p <pid>

Policy options:
 -b, --batch           set policy to SCHED_BATCH
 -f, --fifo            set policy to SCHED_FIFO
 -i, --idle            set policy to SCHED_IDLE
 -o, --other           set policy to SCHED_OTHER
 -r, --rr              set policy to SCHED_RR (default)

Scheduling flag:
 -R, --reset-on-fork  set SCHED_RESET_ON_FORK for FIFO or RR

Other options:
 -a, --all-tasks       operate on all the tasks (threads) for a given pid
 -m, --max             show min and max valid priorities
 -p, --pid             operate on existing given pid
 -v, --verbose         display status information

 -h, --help     display this help and exit
 -V, --version  output version information and exit

For more details see chrt(1).
$
```

The output of the chrt -m command shows that SCHED_OTHER/SCHED_NORMAL, SCHED_BATCH, SCHED_IDLE polices only allow for priority 0, while that of SCHED_FIFO and SCHED_RR can range from 1 to 99.

```
$ chrt -m
SCHED_OTHER min/max priority  : 0/0
SCHED_FIFO min/max priority   : 1/99
SCHED_RR min/max priority     : 1/99
SCHED_BATCH min/max priority  : 0/0
SCHED_IDLE min/max priority   : 0/0
$
```

In the following session, we use the ps command to display the status of current processes, including bash with PID 27223. The chrt -p 27223 command displays the RT scheduling attributes of bash. The attributes are current scheduling policy and current priority of the process.

```
$ ps
  PID TTY          TIME CMD
27223 pts/0    00:00:00 bash
27330 pts/0    00:00:00 ps
```

```
$ chrt -p 27223
pid 27223's current scheduling policy: SCHED_OTHER
pid 27223's current scheduling priority: 0
$
```

The first command in the following session is used to change the scheduling policy of bash from SCHED_OTHER to SCHED_BATCH. The second command displays the new scheduling policy as SCHED_BATCH and scheduling priority as 0.

```
$ chrt -b -p 0 27223
$ chrt -p 27223
pid 27223's current scheduling policy: SCHED_BATCH
pid 27223's current scheduling priority: 0
$
```

The man page for sched on your Linux system describes the API for CPU scheduling.

In the following in-chapter exercises, you will use the ps command with and without options to gain an appreciation of the command output.

Exercise 10.7

Use the ps command to display the status of processes that are running in your current session. Can you identify your login shell? What is it?

Exercise 10.8

Run the command to display the status of all the processes running on your system. What command did you run? What are their PIDs? What are the PIDs of the parents of all the processes?

Exercise 10.9

Replicate the earlier sessions involving the chrt command on your Linux system to verify that it works as expected.

10.5.3 Dynamic Display of Process Attributes

At times, you want to identify runaway processes or processes for a particular user that aren't advancing (i.e., stuck processes). The Linux top command allows you to perform these tasks by reporting the CPU activity and attributes of processes in real time. It displays the status of the most CPU-intensive processes on the system, by default, in terms of the following process attributes displayed from left to right:

PID USER PR NI VIRT RES SHR S %CPU %MEM TIME+ COMMAND

where PID is the process ID number, USER is the effective username of the owner, PR is the process priority, NI is the process nice value, VIRT is the size of the process in virtual memory, SHR is the amount of shared memory available to the process, RES is the resident part of the process (i.e., current size of the process in the main memory), S is the process status, %CPU is the percentage of CPU time used by a process, %MEM is the amount of memory used by the process, TIME+ is the amount of time a process has executed for, and COMMAND is the command name used to invoke the process. To see a more complete alphabetical list of process attributes that can be displayed, see the man page for top on your system.

The top command continues to run until you press <q> or <Q>, and keeps updating the status every second or so. The periodic update time can be specified through a command line option when the top command is executed or through an interactive command while top runs. The number of processes whose status is displayed depends on the size of the display screen or window size on a Graphical User Interface (GUI)-based system. On smart displays or windows, it is usually 15–30 lines, one per process.

The command also displays several other important system statistics in a short header, comprising five to eight lines, displayed before the statistics of the processes and/or threads. The header information

TABLE 10.6

Headers of the top Command for Linux Mint

top - 14:07:29 up 30 days, 21:30, 1 user, load average: 0.50, 0.36, 0.31
Tasks: 250 total, 1 running, 248 sleeping, 0 stopped, 1 zombie
%Cpu(s): 7.0 us, 3.9 sy, 0.0 ni, 88.9 id, 0.0 wa, 0.0 hi, 0.2 si, 0.0 st
KiB Mem : 5981672 total,   833348 free,   975496 used, 4172828 buff/cache
KiB Swap: 6157308 total, 6155404 free,   1904 used. 4641868 avail Mem

depends on the Linux variant on which the command runs. The information in the header can include load average, time (days+hours:minutes:seconds) the system has been up and running, current time, number of processes on the system, number of processes sleeping, number of zombie processes, CPU utilization (percentages of idle time, user time, kernel/system time, and interrupt-handling time), utilization of user area in the main memory (total and free), and swap space (total and free). Table 10.6 gives a listing of top command header information for Linux Mint.

Most of the features of the top command can be selected by an interactive keystroke command while top runs. The following syntax box gives common command line options, and Table 10.7 gives a brief description of the top command's interactive keystrokes.

Syntax:
    top [options]
Purpose: Display and periodically update information about processes currently running on the system
Output: The various attributes of processes currently running on the system, periodically updated
Commonly used options/features:

-H	Display statistics for each thread of a multithreaded process, one line per thread
-i	Do not display idle processes
-p	Monitor only processes with specified PIDs.
-s	Display/toggle system processes.
-u username	Display statistics for only those processes belonging to **username**
-o field	Sort display by **field**, a header of a column in output in lowercase—for example, cpu, pri, res, or time
-s	Starts top with secure mode forced, even for root.
-c	Display full command line for each process or just command name.

The options -H, -i, -p, -s, and -c are really toggles.

In the following text, we show example sessions of the top command output. The following is a run of the command without any options. The command output shows that the system has been up and running for 82 days without crashing. Further, 178 processes are currently in the system, with one running, 177 sleeping, and zero stopped or in the zombie state. The systemd process, owned by root, is PID 1. The rcu_sched process, PID 7, has used 112 h and 59.23 s of the CPU time that it and its dead children have used. You can appreciate other values in the header and attributes of the processes by using Table 10.8. If you observe the command output for a little while, you will notice that the processes move up and down as their priorities change. Similarly, the statistics in the header are also updated periodically. Also notice that pid 1 is listed as systemd, the program name of that process, not the path to the command. We can use the c interactive top command on our Linux Mint system to toggle the display of the COMMAND column to show the paths to the commands that execute the processes.

TABLE 10.7

Brief Description of Some of the Interactive Keystroke Commands of `top`

Command	Meaning
H	Display/toggle the threads statistics on separate lines
S	Display/toggle system processes
^L	Redraw screen
h or ?	Display summary of interactive keystrokes
k	Kill processes whose space-separated PID-list is specified
n or #	Change the number of processes to display
o or O	Other filtering—You will be prompted for the selection criteria which then determines which tasks will be shown in the "current" window
R	Change the nice value of a list of processes specified as space-separated PIDs
S	Change periodic update time (in seconds)
t	Toggle the display of `top` process
z	Switches the "current" window between your last used color scheme and the older form of black-on-white.

TABLE 10.8

Default Fields of the Output of the `top` Command with Brief Descriptions

Field	Description
%CPU	CPU usage—The task's share of the elapsed CPU time since the last screen update, expressed as a percentage of total CPU time
%MEM	Memory usage (RES)—A task's currently used share of available physical memory
COMMAND	Command name or command line—Display the command line used to start a task or the name of the associated program
NI	Nice value—The nice value of the task. A negative nice value means higher priority, whereas a positive nice value means lower priority
PID	Process id—The task's unique PID, which periodically wraps, though never restarting at zero
PR	Priority—The scheduling priority of the task
RES	Resident memory size (KiB)—The nonswapped physical memory a task is using
S	Process status—The status of the task which can be one of the following: D = uninterruptible sleep, R = running, S = sleeping, T = stopped by job control signal, t = stopped by debugger during trace, Z = zombie
SHR	The amount of VIRT memory that is shareable
TIME+	CPU time, hundredths—The same as TIME, but reflecting more granularity through hundredths of a second
USER	User name—The effective user name of the task owner
VIRT	Virtual memory size (KiB)—The total amount of virtual memory used by the task. It includes all code, data, and shared libraries plus pages that have been swapped out and pages that have been mapped but not used

```
$ top
top - 21:17:44 up 82 days,  2:25,  3 users,  load average: 0.37, 0.28, 0.29
Tasks: 178 total,   1 running, 177 sleeping,   0 stopped,   0 zombie
%Cpu(s):  7.7 us,  0.2 sy,  0.0 ni, 92.0 id,  0.0 wa,  0.0 hi,  0.0 si,  0.0 st
KiB Mem :  8079236 total,  2255528 free,   231152 used,  5592556 buff/cache
KiB Swap:  8292348 total,  8206680 free,    85668 used.  7063932 avail Mem

  PID USER      PR  NI    VIRT    RES    SHR S  %CPU %MEM     TIME+ COMMAND
24303 root      20   0  123204   7532   6392 S   0.7  0.1   0:00.06 sshd
24302 sarwar    20   0   42116   3780   3160 R   0.3  0.0   0:00.04 top
    1 root      20   0  119836   4664   2976 S   0.0  0.1   1:42.38 systemd
    2 root      20   0       0      0      0 S   0.0  0.0   0:00.81 kthreadd
    3 root      20   0       0      0      0 S   0.0  0.0   0:13.36 ksoftirqd/0
    5 root       0 -20       0      0      0 S   0.0  0.0   0:00.00 kworker/0:+
    7 root      20   0       0      0      0 S   0.0  0.0 113:00.38 rcu_sched
```

```
   8 root      20   0       0       0       0 S   0.0  0.0   0:00.00 rcu_bh
   9 root      rt   0       0       0       0 S   0.0  0.0   0:00.80 migration/0
  10 root      rt   0       0       0       0 S   0.0  0.0   0:21.12 watchdog/0
  11 root      rt   0       0       0       0 S   0.0  0.0   0:20.72 watchdog/1
  12 root      rt   0       0       0       0 S   0.0  0.0   0:00.71 migration/1
  13 root      20   0       0       0       0 S   0.0  0.0   0:32.22 ksoftirqd/1
  15 root       0 -20       0       0       0 S   0.0  0.0   0:00.00 kworker/1:+
  16 root      rt   0       0       0       0 S   0.0  0.0   0:35.27 watchdog/2
  17 root      rt   0       0       0       0 S   0.0  0.0   0:03.94 migration/2
  18 root      20   0       0       0       0 S   0.0  0.0   3:53.88 ksoftirqd/2
[Output truncated]
$
```

10.5.3.1 Top Interactive Keystroke Commands

You can interact with top while it runs by using various interactive keystrokes. You can press h to display the various keystrokes that allow you to interact with top to modify the information it displays. When you use an interactive command, top prompts you with one or more questions related to the chore that you want it to perform. For instance, when you press n, top prompts you for the number of processes to display. You input the number and hit the <Enter> key for top to start displaying information about the said number of processes. Similarly, if you want to terminate a process, press k and top prompts you for the PID of the process to be terminated. You input the PID of the process to be terminated and hit <Enter> for top to terminate the process. So, if you want to display the RT status of the processes owned by the user **sarwar**, you press u and enter the login name of the user. The following output shows the monitoring of **sarwar**'s processes.

```
  PID USER      PR  NI    VIRT    RES    SHR S  %CPU %MEM     TIME+ COMMAND
24302 sarwar    20   0   42116   3800   3160 R   0.3  0.0   0:00.60 top
23923 sarwar    20   0   45224   5172   4288 S   0.0  0.1   0:00.00 systemd
23927 sarwar    20   0  180020   2176     12 S   0.0  0.0   0:00.00 (sd-pam)
23941 sarwar    20   0  124552   4572   3368 S   0.0  0.1   0:00.08 sshd
23942 sarwar    20   0   23160   5376   3232 S   0.0  0.1   0:00.04 bash
```

You can use the interactive keystrokes listed in Table 10.7 to observe various effects on the command display. A more complete listing of interactive keystroke commands is given on the man page for top on your system.

Exercise 10.10

> Try the previous sessions for the top command on your system. How many processes are running on your system? What are the priority and nice values for the highest priority process? How would you toggle the display in the COMMAND column between showing the pathnames to the commands that execute the processes, and the actual program name? What is pid 1 on your system, and what does the display of the command that runs it, /sbin/init splash, mean?

10.6 Process and Job Control

Linux is responsible for several activities related to process and job management, including process creation, process termination, running processes in the foreground and background, suspending and resuming processes, and switching processes from foreground to background and vice versa. As a Linux user, you can request the process and job control tasks by using the shell commands discussed in this section.

10.6.1 Foreground and Background Processes and Related Commands

When you type a command and hit <Enter>, the shell executes the command and returns by displaying the shell prompt. While your command executes, you do not have access to your shell and therefore cannot execute any commands (i.e., continue work) until the current command finishes and the shell returns. When commands execute in this manner, we say that they execute in the *foreground*. By default, every process runs in the foreground, taking input from the keyboard and sending output to the display screen.

Linux allows you to run a command so that, while the command executes, you get the shell prompt back and can submit other commands. This capability is called running the command in the *background*. You can run a command in the background by ending the command with an ampersand (&). Of course, in a graphical environment, you can run a command in one terminal window in the foreground, open another terminal window, and use the shell in the new terminal window. However, this activity takes time and consumes additional system resources.

Background processes run at lower priorities compared with their foreground counterparts. Thus, they get to use the CPU only when no higher priority process needs it. When a background process generates output that is sent to the display screen, the screen looks garbled, but if you are simultaneously using another application, your work is not altered in any way. You can get out of the application and then get back into it to obtain a cleaner screen. Some applications such as vim allow you to redraw the screen without quitting it. In vim (see Chapter 3), pressing <Ctrl+L> in Command mode allows you to do so.

The syntaxes for executing commands in the foreground and background are as follows. Note that no space is needed between the command and & but that you can use space for clarity.

Syntax:
```
command (for foreground execution)
command & (for background execution)
```

Now consider the following command executed under the Bash shell. It searches the whole file structure for a file called **4plot** (which exists on our system) and stores the pathnames of the directories that contain this file in the file **4plot.paths**. The error messages are sent to the file **/dev/null**, which is the Linux black hole: whatever goes in never comes out. Note that, for the C shell, 2> should be replaced with >&. This command may take several minutes, perhaps hours, depending on the size of the file structure, system load (in terms of the number of users logged on), and the number of processes running on the system. So if you want to do some other work on the system while the command executes, you cannot do so because the command executes in the foreground.

```
$ find / -name 4plot -print > 4plot.paths 2> /dev/null
$
```

The find command is a perfect candidate for background execution because, while it runs, you have access to the shell and can do other work. Thus, the preceding command should be executed as follows:

```
$ find / -name 4plot -print > 4plot.paths 2> /dev/null &
[1] 26472
$
```

The number shown in brackets is returned by the shell and is the job number (also called job ID) for the process; the other number is the PID of the process. Here, the job number for the find command is 1 and the PID is 26,472. A job is a process that is not running in the foreground and is accessible only at the terminal with which it is associated. Such processes typically run in the back or are suspended processes.

The commands that perform tasks that do not involve user intervention and take a long time to finish are good candidates for background execution. Some examples are sorting large files (sort command), compiling large programs (cc, gcc, CC, make, etc.), computationally intensive programs such as one that determines whether a large integer number is prime and searching a large file structure for one or more files (find command). Commands that do terminal I/O, such as the vim editor, are, of course, not good candidates for background execution. The reason is that, when such a command executes in the background, it stops accepting input from the keyboard. The command needs to be brought back to the foreground before it can start running again. The fg command allows you to bring a background process to the foreground.

While running a command in the foreground, you might need to *suspend* it in order to go back to the shell, do something under the shell, and then return to the suspended process. For example, say that you are in the middle of editing a C program file with vim and need to compile the program to determine whether some errors have been corrected. You can save changes to the file, suspend vim, compile the program to view the results of the compilation, and return to vim. You can suspend a foreground process by pressing <Ctrl+Z>, move a suspended process to the foreground by using the fg command, and move a suspended process to the background by using the bg command. So you can suspend vim by pressing <Ctrl+Z>, compile the program to identify any other errors, and resume the suspended vim session by using the fg command.

Syntax:
```
fg [%jobid]
bg [%jobid-list]
```
Purpose: Syntax 1: Resume execution of the process with job number **jobid** in the foreground or move background processes to the foreground; a **jobid** starts with **%**

Syntax 2: Resume execution of suspended processes/jobs with job numbers in **jobid-list** in the background; a **jobid** starts with **%**

Commonly used options/features:

% or %+	Current job
%-	Previous job
%N	Job number N
%name	Job beginning with **name**
%?name	Command containing **name**

If there are multiple suspended processes, the fg command without an argument brings the current process into the foreground, and the bg command without an argument resumes execution of the current process in the background. The job using the CPU at any particular time is called the *current job*.

You can use the jobs command to display the job numbers of all suspended (stopped) and background processes and identify which one is the current process. The current process is identified by a + and the previous process by a - in the output of the jobs command. The following is a brief description of the command.

Syntax:
```
jobs [option] [%jobid-list]
```
Purpose: Display the status of the suspended and background processes specified in **jobid-list**; with no list, display the status of current job

Commonly used option/feature:

-l Also display PIDs of jobs

In the following sessions, we show the use of the fg, bg, <Ctrl+Z>, and jobs commands. It is assumed here that the example file bigdata is very large and takes a long time to sort and copy! Otherwise, your commands and the processes they generate don't live long enough to show up as running processes unless they comprise of infinite loops such as a C program like int main() { for(;;); }.

We run the sort and cp commands in the background. The jobID and PID pairs of these processes are [1] 60149 and [2] 60156 for the sort and cp commands, respectively. The sort bigdata command is in the currently running process/job, as shown in the output of the jobs and jobs -l commands, as well as the R state of the process in the output of the ps command. The cp bigdata bigdata1 command has been swapped on the disk, as indicated by the D state of the process in the output of the ps command.

```
$ sort bigdata > bigdata.sorted &
[1] 60149
$ cp bigdata bigdata1 &
```

```
[2] 60156
$ ps
  PID TT      TIME  COMMAND
56222  1   0:00.67  bash (bash)
60149  1   0:07.94  sort bigdata
60156  1   0:00.00  bash (bash)
60157  1   0:00.34  cp bigdata bigdata1
60164  1   0:00.00  ps
$ jobs
[1]  + Running                sort bigdata > bigdata.sorted
[2]  - Running                cp bigdata bigdata1
$ jobs -l
[1]  + 60149 Running              sort bigdata > bigdata.sorted
[2]  - 60156 Running              cp bigdata bigdata1
$
```

In the following session, the first fg command brings the current job into the foreground. The fg %2 command brings job number 2 into the foreground. A string that uniquely identifies a job can also be used in place of a job number. The string is enclosed in double quotes if it has spaces in it. The third fg command illustrates this convention. The jobs −l command, as expected, shows both jobs as suspended. The output of the ps command shows the state of the cp bigdata bigdata1 command as TW, which means that the process has been suspended and swapped out to disk temporarily. We use the bg %1 %2 command to start the background execution of the two suspended processes. We later confirm this status by using the jobs command, which shows both processes as running, with the sort process currently using the CPU. We also show that the bg command without argument puts the current (or the only suspended) process into the background. The ps −p 60149 command shows that the sort process is in the running state and has been in the system for 11 min and 3.78 s.

```
$ fg
sort bigdata > bigdata.sorted
<Ctrl+Z>
Suspended
$ fg %2
cp bigdata bigdata1
<Ctrl+Z>
Suspended
$ fg %"sort"
sort bigdata > bigdata.sorted
<Ctrl+Z>
Suspended
$ jobs -l
[1]  + 60149 Suspended           sort bigdata > bigdata.sorted
[2]  - 60156 Suspended           cp bigdata bigdata1
$ ps
  PID TT      TIME COMMAND
56222  1   0:00.96 bash (csh)
60149  1   0:29.29 sort bigdata
60156  1   0:00.00 bash (csh)
60157  1   0:00.00 cp bigdata bigdata1
61261  1   R+  0:00.02 ps
$ bg %1 %2
[1]    sort bigdata > bigdata.sorted &
[2]    cp bigdata bigdata1 &
$ jobs
[1]  + Running                sort bigdata > bigdata.sorted
[2]  - Running                cp bigdata bigdata1
$ fg %1
sort bigdata > bigdata.sorted
<Ctrl+Z>
Suspended
$ bg
```

```
[1]     sort bigdata > bigdata.sorted &
$ jobs
[1]  + Running                     sort bigdata > bigdata.sorted
[2]  - Running                     cp bigdata bigdata1
$ ps -p 60149
  PID TT      TIME COMMAND
60149  2  11:03.78 sort bigdata
$
```

As discussed earlier in this section, we discuss the following session as an additional example to explain the working of the fg and <Ctrl+Z> commands. The gcc -o lab8 lab8.c command is used to compile the C program in the **lab8.c** file and put the executable in a file called **lab8**. Understanding what compilation means is not the point here, and a fuller discussion of the syntax and semantics of the gcc command is presented in Chapter 15. Here, we are merely emphasizing that processes that take a long time to start or those that have executed for a considerable amount of time are usually good candidates for processes to be suspended. The example of suspending the vim command is presented only as an illustration. This sequence of events is shown in the following session. Note that the output of the ps command after vim has been suspended shows the status of vim as T, which means that this process has been suspended/stopped.

```
$ ps
PID TT      TIME COMMAND
587  1  0:00.17 bash (bash)
616  1  0:00.00 ps
$ vim lab8.c
#include <stdio.h>
#define SIZE 100
int main (int argc, char *argv[])
{
...
~
~
~
<Ctrl+Z>
Suspended
$ ps
 PID TT      TIME COMMAND
 587  1 0:00.41 bash (bash)
 812  1 0:00.13 vim lab8.c
1035  1 0:00.02 ps
$ gcc -o lab8 lab8.c
$ fg %1
#include <stdio.h>
#define SIZE 100
main (int argc, char *argv[])
{
...
~
~
~
:q!
$
```

In the following in-chapter exercise, you will practice the creation and management of foreground and background processes by using the bg, fg, and jobs commands.

Exercise 10.11

Run the sessions presented in this section on your system to practice foreground and background process creation and switching processes from the foreground to the background (with the bg command) and vice versa (with the fg command). Use the jobs command to display the job IDs of the active and suspended processes.

10.6.2 Linux Daemons

Although any process running in the background can be called a *daemon*, in Linux jargon a daemon is a system process running in the background. Daemons are frequently used in Linux to offer various types of services to users or running software and handle system administration tasks. For example, printing, logging, e-mail, Web browsing, remote login via Secure Shell, file transfer, interaction through social networking sites, and finger services are provided via daemons. Printing services are provided via systemd by the printer daemon cupsd. Finger services (see Chapter 11) are handled by the finger daemon fingerd, if that service has been started and enabled at system boot. The traditional inetd daemon, the Linux Internet *superserver*, has been replaced by the NetworkManager daemon running as a service under systemd. More information on systemd daemons and services can be found in Chapter 20.

10.6.3 Sequential and Parallel Execution of Commands

You can type multiple commands on one command line for the sequential and/or parallel execution of these commands. The following is a brief description of the syntax for sequential execution of commands specified in one command line.

Syntax:
    cmd1; cmd2; ...; cmdN
Purpose: Execute commands cmd1, cmd2, ..., cmdN sequentially as separate processes

---

Note that the semicolon is used as a command separator and, therefore, does not follow the last command. No spaces are needed before and after a semicolon, but you can use spaces for clarity. These commands execute one after the other, as though each were typed on a separate line. In the following session, the date and echo commands execute sequentially as separate processes.

```
$ date; echo Hello, World!
Wed Oct 11 22:16:05 PKT 2017
Hello, World!
$
```

You can specify parallel execution of commands in a command line by ending each command with an ampersand (&). The commands that end with an ampersand also execute in the background. No spaces are required before or after &, but you can use spaces for clarity. When you execute a command in the background, the shell displays the following pair as output: [jobID] PID, where the first job ID is 1, increasing linearly by adding one to the last-used job ID. Like PIDs, a job ID may be recycled if it has not been assigned to a process currently. The following is a brief description of the syntax for parallel execution of shell commands specified in one command line.

Syntax:
    cmd1& cmd2& ... cmdN&
Purpose: Execute commands cmd1, cmd2, ..., cmdN in parallel as separate processes

---

The following sessions were executed under Bash. The date and echo commands execute in parallel and in the background, followed by the sequential execution of the uname and who commands in the foreground. In general, since the who command executes at the end, its output is always displayed at the end. The outputs of the other three commands (date, echo, and uname) may be displayed in any order. This order is due to the scheduling of processes and the amount of time each takes to execute. Thus, the same output order may or may not be reproduced if you execute the command line again. In the following session, the outputs are displayed in the order echo, date, uname, and who. The job and PIDs for the date and echo commands are [1] 32429 and [2] 32430, respectively.

```
$ date& echo Hello, World\!& uname; who
[1] 32429
[2] 32430
Hello, World!
Fri Jul  7 20:16:28 PDT 2017
Linux
[1]-  Done                    date
[2]+  Done                    echo Hello, World\!
bob       tty8         2017-06-20 14:14 (:0)
$
```

The last & in a command line puts all the commands since the previous & in one process. In the following command line session, therefore, the date command executes as one process and all the commands in who; whoami; uname; echo Hello, World!& as another process.

As shown in the following session, the job ID and PID pair for the two processes are [1] 32445 and [1] 32449, respectively. Note that the job ID for each of the two jobs is 1. Further, the PIDs for the two processes are nonconsecutive. Finally, as shown in the shaded regions of the session, since the shell prompt returns after the two processes (jobs) have been created but before any or some output has been displayed on the screen, we have to hit the <Enter> key to display the shell prompt again.

```
$ date & who; whoami; uname; echo Hello, World\! &
[1] 32445
Fri Jul  7 20:22:28 PDT 2017
bob       tty8         2017-06-20 14:14 (:0)
[1]+  Done                    date
bob
Linux
[1] 32449
Hello, World!

$ <Enter>
[1]+  Done                    echo Hello, World\!
$
```

As will be discussed briefly in Chapter 12, Linux allows you to group commands and execute them as one process by separating commands using semicolons and enclosing them in parentheses. This is called *command grouping*. Because all the commands in a command group execute as a single process, they are executed by the same subshell. However, all the commands execute sequentially, one after the other. The following is a brief description of the syntax for command grouping.

Syntax:
   (cmd1; cmd2; ...; cmdN)

Purpose: Execute commands cmd1, cmd2, ..., cmdN sequentially, but as one process

---

In the following Bash session, therefore, the date and echo commands execute sequentially, but as one process.

```
$ (date; echo Hello, World\!)
Fri Jul  7 20:38:21 PDT 2017
Hello, World!
$
```

You can combine command grouping with sequential execution by separating command groups with other commands or command groups. In the following session, the date and echo commands execute as one process, followed by the who command executing as a separate process.

```
$ (date; echo Hello, World\!); who
Fri Jul  7 20:40:54 PDT 2017
Hello, World!
bob        tty8           2017-06-20 14:14 (:0)
$
```

Command groups can be nested. Hence, ((date; echo Hello, World!); who) and ((date; echo Hello, World!); (who; uname)) are valid commands and produce the expected results. Command grouping makes more sense when groups are executed as separate processes, as shown in the following session.

```
$ (date ; echo Hello, World\!)&
[1] 446
$ Fri Jul  7 20:42:28 PDT 2017
Hello, World!
<Enter>
[1]+ Done                  ( date; echo Hello, World\! )
$ (date; echo Hello, World)& (who; uname)& whoami
[1] 453
[2] 454
bob
bob@bob-sbc-flt1 ~ $ bob     tty8           2017-06-20 14:14 (:0)
Fri Jul  7 20:44:48 PDT 2017
Hello, World
Linux

[1]-  Done                 ( date; echo Hello, World )
[2]+  Done                 ( who; uname )
<Enter>
[2]    Done                        ( who; uname )
$
```

In the second group of commands, (date; echo Hello, World) and (who; uname) execute in the background and the whoami command executes in the foreground; all three commands execute in parallel. Again, the order of output is dependent on the scheduling of these commands.

In the following in-chapter exercises, you will practice sequential and parallel execution of Linux commands.

Exercise 10.12

Run the sessions presented in this section on your system to practice sequential and parallel execution of shell commands.

Exercise 10.13

Which of the following commands run sequentially and which run in parallel? How many of the processes run in parallel? (who; date) & (cat temp; uname & whoami)

10.6.4 Abnormal Termination of Commands and Processes

When you run a command, it terminates normally after successfully completing its task. A command (process) can terminate prematurely because of a bad argument that you passed to it, such as a directory argument as source file to the cp command, or because of a run-time error. At times, you might also need to terminate a process abnormally. The need for *abnormal termination* arises when you run a process with a legal but wrong argument (e.g., a wrong file name to the find command) or when a command is taking too long to finish, perhaps, due to an infinite loop. Here, we address abnormal termination in relation to both foreground and background processes.

You can terminate a foreground process by pressing <Ctrl+C> or using the kill command from another shell. You can terminate a background process in one of two ways: (1) by using the kill command or (2) by first bringing the process into the foreground by using the fg command and then pressing <Ctrl+C>. The primary purpose of the kill command is to send a *signal* (also known as a *software interrupt*) to a process. The Linux operating system uses a signal to get the attention of a process. You can send any one of the several signal types supported by your Linux system to a process that you own or you have the permission to do so. A process can take one of the three actions upon receiving a signal:

1. Accept the default action as determined by the Linux kernel
2. Ignore the signal
3. Intercept the signal and take a user-defined action

For most signals, the default action, in addition to some other events, always results in termination of the process. Ignoring a signal does not have any impact on the process. A user-defined action is specified as a program statement (usually a function call) that takes control to a specific piece of code in the process. In a shell script, you can specify these actions by using the `trap` command in the Bash shell. The C shell provides a limited handling of signals via the `onintr` command. In a C program, you can specify these actions by using the library call `signal`. We discuss the `trap` command in detail in Chapter 13. We describe the `signal(2)` system call in detail in Chapter 17. For more details, view its manual page by using the `man signal` or `man -S3 signal` commands.

Signals can be generated for various reasons. The processes themselves cause some of these reasons, whereas others are external to processes. A signal caused by an event internal to a process is known as an *internal signal* or a *trap* (not to be confused with the `trap` command in the Bash shell). For example, the execution of a divide-by-zero instruction in a process generates a trap. A signal caused by an event external to a process or a hardware device in the computer system is called an *external signal*. If an external signal is for a hardware device such as a CPU or disk controller, it is called a *hardware interrupt* or *interrupt*. An external event meant to get attention of one or more processes is known as a *software interrupt* or *signal* in the Linux jargon. For example, an internal signal is generated for a process when the process tries to execute a nonexisting instruction or access a memory region that it is not allowed to access, such as memory belonging to some other process or the Linux kernel. You can generate an external signal by pressing <Ctrl+C>, by logging out, or by using the `kill` command. The `kill` command can be used to send any type of signal to a process. The following is a brief description of the `kill` command.

Syntax:
```
kill [-s signal_name] proc-list
kill [-signal_name] proc-list
kill [-signal_number] proc-list
kill -l [exit_status]
```

Purpose: Syntaxes 1–3: Send the signal for **signal_number** or symbolic **signal_name** to processes whose PIDs or jobIDs are specified in space-separated **proc-list**; jobIDs must start with **%**.

Syntax 4: The command **kill -l** returns a list of all the signals along with their numbers and names. The operand **exit_status** specifies a signal number or the exit status of a terminated or completed process.

Commonly used options/features:

   1   HUP (Hang-up)
   2   INT (Interrupt: <Ctrl+C>)
   3   QUIT (Quit: <Ctrl+\>)
   6   ABRT (Abort)
   9   KILL (Sure kill: Nonignorable, noninterceptable)
   14 ALRM (Alarm clock)
   15 TERM (Software termination: the default signal number)

You can use the `kill -1` (number one) command to send the default signal to all of your processes. Only a superuser can send a signal to the processes belonging to other users. Thus, a superuser can use this command to send signals to all the processes running on the system. You can use the `kill -1` (lowercase L) command to display the signals supported by your Linux system. Linux Mint supports 64 signal types, as shown in the following session. You can use the man signal command to verify for yourself. The syntax box shows details of some of the more commonly used signals.

```
$ kill -1
 1) SIGHUP       2) SIGINT       3) SIGQUIT      4) SIGILL       5) SIGTRAP
 6) SIGABRT      7) SIGBUS       8) SIGFPE       9) SIGKILL     10) SIGUSR1
11) SIGSEGV     12) SIGUSR2     13) SIGPIPE     14) SIGALRM     15) SIGTERM
16) SIGSTKFLT   17) SIGCHLD     18) SIGCONT     19) SIGSTOP     20) SIGTSTP
21) SIGTTIN     22) SIGTTOU     23) SIGURG      24) SIGXCPU     25) SIGXFSZ
26) SIGVTALRM   27) SIGPROF     28) SIGWINCH    29) SIGIO       30) SIGPWR
31) SIGSYS      34) SIGRTMIN    35) SIGRTMIN+1  36) SIGRTMIN+2  37) SIGRTMIN+3
38) SIGRTMIN+4  39) SIGRTMIN+5  40) SIGRTMIN+6  41) SIGRTMIN+7  42) SIGRTMIN+8
43) SIGRTMIN+9  44) SIGRTMIN+10 45) SIGRTMIN+11 46) SIGRTMIN+12 47) SIGRTMIN+13
48) SIGRTMIN+14 49) SIGRTMIN+15 50) SIGRTMAX-14 51) SIGRTMAX-13 52) SIGRTMAX-12
53) SIGRTMAX-11 54) SIGRTMAX-10 55) SIGRTMAX-9  56) SIGRTMAX-8  57) SIGRTMAX-7
58) SIGRTMAX-6  59) SIGRTMAX-5  60) SIGRTMAX-4  61) SIGRTMAX-3  62) SIGRTMAX-2
63) SIGRTMAX-1  64) SIGRTMAX
$
```

Table 10.9 shows some examples of the kill command and their meanings.

The *hang-up* signal is generated when you log out, the *interrupt* signal is generated when you press <Ctrl+C>, and the *quit* signal is generated when you press <Ctrl+\>. The kill command sends signal number 15 to the process whose PID is specified as an argument. The default action for this signal is termination of the process that receives it. This signal can be intercepted and ignored by a process, as can most of the other signals. To terminate a process that ignores signal 15 or other signals, signal number 9, known as *sure kill*, has to be sent to it. The kill command terminates all the processes whose PIDs are given in the **proc-list**, provided that these processes belong to the user who is using kill. The following session presents some instances of how the kill command can be used. But don't kill the login shell process because it will log you out.

```
$ jobs -1
[1]   + 14338 Running           sort bigdata > bigdata.sorted
[2]   - 14667 Running           cp biggerdata biggerdata.bak
[3]     14668 Running           grep sh biggerdata > lines.sh
$ kill 14668
$ kill -2 14667
[3]     Terminated              grep sh biggerdata > lines.sh
$ kill -9 14338
[2]     Interrupt               cp biggerdata biggerdata.bak
$ jobs
[1]     Killed                  sort bigdata > bigdata.sorted
$ jobs
$
```

TABLE 10.9

Some Examples of the kill Command and Their Meanings

Command	Meaning
kill 1234	Send the default signal (SIGTERM) to the process with PID 1234
kill -9 1234	Send SIGKILL (guaranteed termination signal) to the process with PID 1234
kill -s kill 1234	
kill -s KILL 1234	
kill -9 1234 -1004	Send SIGKILL (guaranteed termination signal) to the process with PID 1234 and to all the processes with process group ID (PGID) 1004
kill -s kill 1234 -1004	
kill -s KILL 1234 -1004	
kill -TERM -1004	Send SIGTEM to all the processes with process group ID 1004
kill -- -1004	

In the first case, the kill command sends signal number 15 to a process with PID 14668. In the second case, signal number 2 (SIGINT) is sent to a process with PID 14667. In both cases, because the specified signal numbers are not intercepted, the processes are terminated. The kill command can be used to terminate a number of processes with one command line. For example, the command kill -9 13586 20581 terminates processes with PIDs 13586 and 20581.

Process ID 0 can be used to refer to all the processes created during the current login. Thus, the kill -9 0 command terminates all processes resulting from the current login, as shown in the following session. Note that the command has terminated all the processes resulting from **sarwar**'s current login session, including the process that maintains the Secure Shell login connection with the client. That is why you see the prompt of the local machine, a terminal window running Bash on a MacBook Pro under Mac OS X (Darwin), MacBook-Pro:~ syedsarwar$. This has serious consequences because the order in which kill terminates processes is not known in this case. Thus, if the login shell and the process that maintains the Secure Shell connection with the client machine are terminated first, then the processes executing under the login shell continue to run. You would notice this when you login again and realize that the ssh command is taking longer than usual to establish the connection with the remote machine, and once you are connected and have a login shell running, system response time is poor. The ps command would show that your background processes from the previous login are still running, as shown in the following session in the shaded portions. Note that the statuses of these "leftover" processes are D (swapped out) and DL (swapped out and locked), and they are not associated with any terminal (note the negative terminal numbers). You need to terminate these processes to improve the system response time, as shown in the last line of the session.

```
$ ps -U sarwar
PID TT  STAT    TIME COMMAND
15361  -  S     0:00.01 sshd: sarwar@pts/1 (sshd)
15368  1  Ss    0:00.07 -csh (csh)
15424  1  D     0:07.05 sort bigdata
15437  1  D     0:00.60 cp biggerdata biggerdata.bak
15456  1  D     0:01.31 grep sh biggerdata
15521  1  R+    0:00.00 ps -U sarwar
15158  2- DL    1:58.99 sort bigdata
$ kill -9 0
Connection to 192.102.169.10 closed.
MacBook-Pro:~ syedsarwar$ ssh sarwar@192.102.169.10
Password for sarwar@pcbsd-srv:
Last login: Sun Oct  5 10:17:45 2014 from 139.15.192.135
FreeBSD 10.0-RELEASE-p6 (GENERIC) #0 acf484b(releng/10.0): Mon Feb 24 15:14:38 EST
2014

Welcome to FreeBSD!
...
$ ps
  PID TT    TIME COMMAND
15424  1- 0:23.56 sort bigdata
15437  1- 0:09.15 cp biggerdata biggerdata.bak
15456  1- 0:21.06 grep sh biggerdata
15158  2- 1:59.38 sort bigdata
15866  3  0:00.06 bash (bash)
15913  3  0:00.00 ps
$ kill 15158 15424 15437 15456
$
```

The kill command also works with job numbers. Hence, the following command can be used to terminate a process with job number 1. You can terminate multiple processes by specifying their job numbers in the command line. For example, kill -9 %1 %3 can be used to terminate processes with job numbers 1 and 3.

```
$ kill -9 %1
[1] +  Killed find / -name foo -print > foo.paths &
$
```

When you log out, all the processes running in your session get a hang-up signal (signal number 1) and are terminated per the default action. If you want processes to continue to run even after you have logged out, you need to execute them so that they ignore the hang-up signal when they receive it. You can use the Linux command nohup to accomplish this task. The following is a brief description of the syntax for this command.

Syntax:
   `nohup command [args]`

Purpose: Execute **command** and ignore the hang-up signal

You need to use the nohup command for processes that take a long time to finish, such as a program sorting a large file containing hundreds of thousands of customer records. Obviously, you would run this type of program in the background so that it runs at a lower priority. The following is a simple illustration of the use of the nohup command. Here, the find command runs in the background and is not terminated when you log out or send it signal number 1 (hang-up) via the kill command. If output of the command is not redirected, it is appended to the **nohup.out** file by default.

```
$ nohup find / -name foo -print 1> foo.paths 2> /dev/null &
[1] 62808
$ kill -1 62808
$ jobs
[1]+  Running                 nohup find / -name foo -print > foo.paths 2> /dev/null &
$ kill 62808
$ <Enter>
[1]+  Terminated              nohup find / -name foo -print > foo.paths 2> /dev/null
$ jobs
$
```

If you separate commands with semicolons, you can run them with nohup. In the following session, GenData generates some data and puts it in a file called **employees**, and the sort command sorts the file and stores the sorted version in the **employees.sorted** file.

```
$ nohup GenData > employees ; sort employees > employees.sorted &
[2] 15931
$
```

In the following in-chapter exercises, you will use the kill command to practice abnormal termination of processes, and the nohup and ps -a commands to appreciate how you can run processes that do not terminate when you log out.

Exercise 10.14
> Give a command for terminating processes with PID 10974 and jobID 3.

Exercise 10.15
> Run the first of the nohup commands, use ps to verify that the command is executing, log out, log in again, and use the ps -a command to determine whether the find command is running.

10.7 Linux Start-Up, Login, and Process Hierarchy

When you turn on the power to your Linux computer system, there are basically two operations that are performed to make your computer behave in the way that you normally would expect it to. As detailed in Chapter 17, these two operations are booting and start-up. We are concerned here with the start-up operation, particularly with the creation of processes in user space, or user-side processes. We are also concerned with systemd as the controlling process during the start-up phase, and how processes are arranged and controlled in a hierarchical fashion by systemd during the entire time the computer is powered on.

TABLE 10.10

Some of the Key Linux System Processes

Process	Purpose
systemd	Granddaddy of all user processes; all user-level services, including all Internet services, and user processes run under the children of this process
kthreadd	Kernel thread daemon
agetty	Creates and controls tty's
NetworkManager	Manages your network devices and connections, attempting to keep active network connectivity when available
mdm	Mint Display Manager, controls desktop environment
cupsd	Linux print daemon, manages printers and printing
systemd-journal	Logging mechanism for systemd

As a part of the start-up operation, the Linux kernel is extracted from the boot medium and put into a working configuration. After performing some checks and other housekeeping tasks, the kernel then creates the first kernel space, or kernel side, process from scratch. In Linux, this process is called swapper or scheduler. This process, which has no parent, has PID 0. The PPID of this process is also 0. This first process then spawns several other kernel-side processes, which are intended to handle several important kernel tasks, including tasks related to virtual memory, file handling, interrupt handling, and performing various tasks in the continuing boot sequence, such as initializing hardware ports. To see a brief listing and description of key Linux processes, see Table 10.10.

As start-up continues, in all major branches of Linux, systemd is the first user-side process that is created. systemd is the granddaddy of all subsequent user processes that are created, so long as the system is up and running. The systemd process has a PID of 1 and runs with superuser privileges. Configurations in the **/etc/systemd/system** directory guides the initialization through various "target" states, depending upon whether the system is being used as a server or a desktop computer, for example. To see an illustration of the steps systemd goes through when it brings a Linux system up to nominal performance, see Section 18.2.

10.7.1 Linux Login and **agetty**

In terms of a user space interactive session, the process that is of critical importance during the start-up procedure is agetty, traditionally named getty in a legacy UNIX or Linux system. You can use any of the three methods to connect to a Linux system: through a GUI or Character User Interface (CUI), or via ssh (or a similar command for remote login). Whatever method you use to connect to a Linux machine, when you log on to the Linux system, the system creates the first process for you, and that is called your login shell. The login shell interprets/executes your commands by creating processes for all the commands that you execute (see Section 10.3 for details of command execution).

If you log in through a GUI, the display manager handles the details of agetty process generation. On Mint Linux, this is mdm. If you log in using a CUI method, agetty goes through the following steps to handle your login. At the login: prompt, when you type your login name and press <Enter>, the agetty process forks a child. The child process executes the exec system call to become a login process, with your login name as its parameter. The login process prompts you for your password and checks the validity of your login name and password. If it finds both to be correct, the login process execs to become your login shell. If the login process does not find your login name in the /etc/passwd file or finds that the password that you entered does not match the one in the /etc/passwd file, it displays an error message and terminates. Control goes back to the agetty process, which redisplays the login: prompt. Once in your login shell, you can do your work and terminate the shell by pressing <Ctrl+D>, typing logout, and then pressing <Enter>, or running the exit command. When you do so, the shell process terminates and control goes back to the agetty process, which displays the login: prompt, and life goes on.

When you log in to a remote server via an sshd daemon, you run the ssh command on your machine. The sshd daemon on the remote machine creates a child process, your private sshd, that handles communication with your client. The server sshd goes back and looks for more ssh connection requests from other clients. Your private sshd spawns another sshd process that overwrites itself with a pseudoterminal, which prompts you for the password. When you enter the correct password, it runs the login shell process for you. By default, the default login shell is Bash on a remote Linux machine. You run your commands under your login shell. When you press <Ctrl+D>, type logout, and then press <Enter>, or run the exit command, your login shell terminates along with your private sshd daemon.

10.7.2 Process Lifetime and pstree

Two Linux processes that exist throughout the running time of a powered-on system are scheduler and systemd. The agetty process, which monitors various forms of terminal lines, must live for as long as a terminal is attached to the system. Your login shell process lives for as long as you are logged on. All other processes are usually short lived, or are running for as long as a command, utility, or service must execute. Many of the services and their processes that are shown in the following sessions are essential for normal system operation. Of course, depending upon the use cases that your Linux system is designed for, many of the other services and their processes run continuously.

You can use the pstree command on a Linux system to display the process tree of the currently running processes on the system in a graphical form, showing the parent–child relationships. You can display the process tree for a process or for a user. For more information on the pstree command, see its man page on your system.

To view the process tree for a user named bob, the following session illustrates the hierarchical graphical form:

```
$ pstree -a bob
sshd
  └─bash
       └─pstree -a bob

systemd --user
  └─(sd-pam)
$
```

In the following sessions, we use the pstree command to show our system's process hierarchy for Linux Mint graphical.target (GUI) and also multi-user.target (CUI) systems. The outputs show that the granddaddy of all user processes is, as expected, the systemd process. Notice that four agetty processes are running on our system in the graphical.target state.

```
$ pstree
systemd─┬─ModemManager──┬─{qdbus}
        │               └─{gmain}
        ├─NetworkManager──┬─dnsmasq
        │                 ├─{gdbus}
        │                 └─{gmain}
        ├─accounts-daemon──┬─{gdbus}
        │                  └─{gmain}
        ├─acpid
        ├─4*[agetty]
        ├─avahi-daemon──avahi-daemon
        ├─cgmanager
        ├─console-kit-dae──┬─62*[{console-kit-dae}]
        │                  ├─{gdbus}
        │                  └─{gmain}
        ├─cron
        ├─cups-browsed──┬─{gdbus}
        │               └─{gmain}
```

```
  ├─2*[dbus-daemon]
  ├─dbus-launch
  ├─irqbalance
  ├─login───bash
  ├─lvmetad
  ├─mdm───mdm─┬─Xorg────3*[{Xorg}]
  │           └─mdmwebkit─┬─{dconf worker}
  │                       ├─{gdbus}
  │                       ├─{gmain}
  │                       ├─12*[{mdmwebkit}]
  │                       └─{pool}
  ├─ntpd
  ├─polkitd─┬─{gdbus}
  │         └─{gmain}
  ├─rsyslogd─┬─{in:imklog}
  │          ├─{in:imuxsock}
  │          └─{rs:main Q:Reg}
  ├─rtkit-daemon────2*[{rtkit-daemon}]
  ├─sshd───sshd───sshd───bash───pstree
  ├─2*[systemd───(sd-pam)]
  ├─systemd-journal
  ├─systemd-logind
  ├─systemd-udevd
  ├─thermald───{thermald}
  ├─udisksd─┬─{cleanup}
  │         ├─{gdbus}
  │         ├─{gmain}
  │         └─{probing-thread}
  └─upowerd─┬─{gdbus}
            └─{gmain}
$
```

The following pstree command shows our system's process hierarchy for Linux Mint multi-user.target output on a CUI system like a server.

```
$ pstree
systemd─┬─ModemManager─┬─{gdbus}
        │              └─{gmain}
        ├─NetworkManager─┬─dhclient
        │                ├─dnsmasq
        │                ├─{gdbus}
        │                └─{gmain}
        ├─agetty
        ├─avahi-daemon───avahi-daemon
        ├─cgmanager
        ├─cron
        ├─cups-browsed─┬─{gdbus}
        │              └─{gmain}
        ├─cupsd
        ├─dbus-daemon
        │
        ├─gpg-agent
        ├─irqbalance
        ├─mdadm
        ├─ntpd
        ├─polkitd─┬─{gdbus}
        │         ├─{gmain}
        │         └─{pool}
        ├─rsyslogd─┬─{in:imklog}
        │          ├─{in:imuxsock}
        │          └─{rs:main Q:Reg}
        ├─sd_cicero───sd_cicero
```

```
                        ├─{sd_cicero}
                        └─{threaded-ml}
        ├─sd_dummy──{sd_dummy}
        │              └─{threaded-ml}
        ├─sd_espeak──3*[{sd_espeak}]
        │              └─{threaded-ml}
        ├─sd_generic──{sd_generic}
        │               └─{threaded-ml}
        ├─speech-dispatch──{speech-dispatch}
        ├─sshd──sshd──sshd──bash──pstree
        ├─systemd-journal
        ├─systemd-logind
        ├─systemd-udevd
        └─vsftpd
$
```

Further analysis of the output of the two commands mentioned previously shows that the user is using the system through an Secure Shell (SSH) session and is currently using Bash to run the pstree command.

10.7.2.1 *ps* versus *pstree*

You can use the ps -auf command to display the parent–child relationship between processes. However, the output for this command is not as complete and fancy as it is for the pstree command, as can be seen in the following session. The ps -uf command displays the process hierarchy for the current user.

```
$ ps -auf
USER       PID %CPU %MEM    VSZ    RSS TTY     STAT START    TIME COMMAND
sarwar   29016  0.0  0.0  23160   5400 pts/0    Ss   07:38   0:00 -bash
sarwar   29127  0.0  0.0  37680   3340 pts/0    R+   07:47   0:00  \_ ps -auf
root     10096  5.9  6.0 888264 490852 tty7     Ssl+ Jul25 6817:12 /usr/lib/xorg/
root      5015  0.0  0.0  16256      0 tty2     Ss+  Jul24   0:00 /sbin/agetty --
root     22407  0.0  0.0 104912   3692 tty1     Ss   Jul24   0:00 /bin/login --
root     12694  0.0  0.0  23160   5152 tty1     S+   Jul25   0:00  \_ -bash
root      3191  0.0  0.0  16256      0 tty5     Ss+  Jul21   0:00 /sbin/agetty --
root      3187  0.0  0.0  16256      0 tty4     Ss+  Jul21   0:00 /sbin/agetty --
root      3180  0.0  0.0  16256      0 tty3     Ss+  Jul21   0:00 /sbin/agetty --
$ ps -uf
USER       PID %CPU %MEM    VSZ    RSS TTY     STAT START    TIME COMMAND
sarwar   29016  0.0  0.0  23160   5400 pts/0    Ss   07:38   0:00 -bash
sarwar   29130  0.0  0.0  37680   3388 pts/0    R+   07:48   0:00  \_ ps -uf
$
```

You can run the ps -aflx | more command to display information about all the processes running on your system and some of their important attributes, including PID, PPID, priority, nice value, virtual size, size currently in main memory, status, event waiting for, and full command name used to start the process.

Exercise 10.16

Run the pstree, ps -auf, and ps -uf commands on your system. Show outputs of these commands. Identify the process on your system that take the highest percentage of CPU time. Also identify the process that uses highest percentage of main memory. Clearly state the amount of virtual memory and physical memory this process is using.

10.8 The proc Filesystem

It is instructive to know how to look at, assess, and even change characteristics of the RT operation of the Linux kernel. For example, seeing what processes are running, and what characteristics these processes have, and then taking actions based upon these observations. A key component of this interaction is the *proc filesystem*, often referred to simply as proc. Here we give a brief overview of proc, and show how an understanding of it contributes to helping you in assessing the tasks that Linux manages on your computer. Those tasks are traditionally, and very commonly, referred to as processes and threads.

The proc filesystem was historically derived from the Plan 9 distributed operating system, developed at Bell Labs between the mid-1980s to 2002. This operating system was designed based on most of the principles of UNIX. It glues almost all computing resources, including files, network interfaces, and peripheral devices, through the file system. proc makes it easy for an ordinary user to gain access to kernel data structures, memory allocation, and information about processes on the system. The Linux version of the ps command directly accesses proc to show information about processes on the system. The proc filesystem is mounted at **/proc**, which gives you the ability to navigate through its hierarchy with the cd command.

The **/proc** directory structure on our Linux Mint system includes subdirectories for all running processes and other attendant facilities. For all processes on the system, there is a directory named **/proc/ PID**, where PID is the process ID number for any particular process. In that directory, there are numerous files and subdirectories with information about that process. For example, directory 1 is for systemd (PID 1). In the following session, we show commands that gave information about the systemd process.

```
$ cd /proc
$ ls
1       13      22659   29157   49      847     bus          misc
10      15      22662   29158   5       85      cgroups      modules
100     16      2280    3       50      86      cmdline      mounts
10083   17      22825   30      5015    87      consoles     mtrr
10096   177     23      31      51      88      cpuinfo      net
101     179     2306    315     52      884     crypto       pagetypeinfo
10110   18      25      317     53      89      devices      partitions
[Output truncated]
$ cd 1
$ ls
ls: cannot read symbolic link 'cwd': Permission denied
ls: cannot read symbolic link 'root': Permission denied
ls: cannot read symbolic link 'exe': Permission denied
attr              cpuset   limits     net            projid_map   stat
autogroup         cwd      loginuid   ns             root         statm
auxv              environ  map_files  numa_maps      sched        status
cgroup            exe      maps       oom_adj        schedstat    syscall
clear_refs        fd       mem        oom_score      sessionid    task
cmdline           fdinfo   mountinfo  oom_score_adj  setgroups    timers
comm              gid_map  mounts     pagemap        smaps        uid_map
coredump_filter   io       mountstats personality    stack        wchan
$ sudo ls -la exe
password for bob: xxxxx
lrwxrwxrwx 1 root root 0 Jul 24 09:03 exe -> /lib/systemd/systemd
$ ls -l /lib/systemd/systemd
-rwxr-xr-x 1 root root 1573136 Apr 12  2016 /lib/systemd/systemd
$ file /lib/systemd/systemd
/lib/systemd/systemd: ELF 64-bit LSB shared object, x86-64, version 1 (SYSV),
dynamically linked, interpreter /lib64/ld-linux-x86-64.so.2, for GNU/Linux 2.6.32,
BuildID[sha1]=80c38f5ef6fd5a367d6b043d91b43b11dcee91fb, stripped
$ more cmdline
/sbin/init
$ ls -al cmdline
-r--r--r-- 1 root root 0 Jul 25 02:15 cmdline
$
```

The ls command in the preceding session shows the subdirectories of proc for our purposes here, the running process "numbered" directories. Changing to the 1 directory (the systemd directory), we use the ls command to see the contents of this directory and find out that we have cwd, root, and exe files.

The exe file is a symbolic link to the **/lib/systemd/systemd** file that contains the executable program that creates systemd. If you have superuser privileges, you can use the su ls -la exe command to verify this information. We then use the ls -la command to find the command line file that originally starts the systemd process. Using the more command to examine the contents of the **cmdline** file, we see it as a symbolic link to /sbin/init splash, the command that begins the execution of the systemd program.

In addition, there are numerous other subdirectories and files in **/proc**, which can be queried by Python for example, to extract kernel-related and other values from. Some of those are seen in the outputs of the commands in the previous session. For more in-depth information on the **/proc** filesystem, and the files and directories in **/proc/PID**, for example, see the man page for procfs on your Linux system. For more information on systemd, see Chapter 18.

Exercise 10.17

Replicate the Bash sessions discussed in this section on your system. Do the commands produce expected outputs? How are outputs of the various commands different on your system?

Summary

A process is a program in execution. Being a time-sharing system, Linux allows execution of multiple processes simultaneously. On a computer system with one CPU, processes are executed concurrently by scheduling the CPU time and giving it to each process for a short time called a quantum. Each process is assigned a priority by the Linux system, and when the CPU is available, it is given to the process with the highest priority.

The shell executes commands by creating child processes using the fork and exec system calls. When a process uses the fork system call, the Linux kernel creates an exact main memory image of the process. The shell itself executes an internal command. An external binary command is executed by the child shell overwriting itself by the code of the command via an exec call. For an external command comprising a shell script, the child shell executes the commands in the script file one by one.

Every Linux process has several attributes, including PID, PPID, process name, process state (running, suspended, swapped, zombie, etc.), the terminal the process was executed on, the length of time the process has run, and process priority. The ps command may be used to view a static display of these attributes. The top command may be used to view a dynamic display of the various system statistics and attributes of the processes running on the system, and permit interactive commands via various keystrokes.

Linux processes can be run in the background or the foreground. A foreground process controls the terminal until it finishes, so the shell cannot be used for anything else while a foreground process runs. When a process runs in the background, it returns the shell prompt so that the user can do other work as the process executes. Because a background process runs at a lower priority, a command that takes a long time is a good candidate for background execution. The background system processes that provide various services are called daemons. A set of commands can be run in a group as separate processes or as one process. Multiple commands can be run from one command line as separate processes by using a semicolon (;) as the command separator; enclosed in parentheses, these commands can be executed as one process. Commands can be executed concurrently by using an ampersand (&) as the command separator.

Suspending processes, moving them from the foreground to the background and vice versa, having the ability to display their status, interrupting them via signals, and terminating them are all known as job control, and Linux has a set of commands that allow these actions. Foreground processes can be suspended and moved to the background pressing <Ctrl+Z> followed by executing the bg command. Suspended and background processes can be moved to the foreground by using the fg command. Commands that are suspended or run in the background are also known as jobs. The jobs command can be used to view the status of all your jobs. You can press <Ctrl+C> to terminate a foreground process.

The kill command can terminate a process with its PID or job ID. The command can be used to send various types of signals, or software interrupts, to processes. Upon receipt of any signal except one, a process can take the default (kernel-defined) action, take a user-defined action, or ignore it. No process can ignore the `kill -9` command, the sure kill signal, which was put in place by the Linux designers to make sure that every process running on a system could be terminated. Commands executed with the `nohup` command keep running even after a user logs out. The `kill -9 0` command is the sure kill for all the processes associated with the current login of a user.

When you power up your Linux system, the operations of booting and start-up begin the process of bringing your computer to a state that you would normally expect it to be in when you work with it. When the Linux kernel takes over during the start-up phase, the systemd superkernel is the first userland process that is started. The agetty program handles console or terminal connections that allow you to interact with the system.

You can use the `pstree` command on a system running Bash to display tree structures for your or some other user's processes. With the –a option, the output displayed contains a line for the granddaddy `systemd` process. The `ps -auxfd` command displays the complete hierarchy of all the processes on the system and the `ps -d` command displays the process hierarchy for a particular user.

A key component of the user interaction possible between kernel space and user space is the */proc filesystem*, often referred to simply as proc. We gave a brief overview of proc, and showed how an understanding of it contributes to helping you in assessing the tasks that the Linux kernel and its components manage on your computer.

We present some comparisons that can allow you to know what the difference is, if any, between a process and a thread, in the context of tasks in Linux. Several of these comparisons are cross-referenced to sections in this chapter, as well as to other chapters and sections in the book.

Questions and Problems

1. What is a process? What is the PID of a process?
2. What is CPU scheduling? How does a time-sharing system run multiple processes on a computer system with a single CPU? Be brief but precise.
3. What is a quantum? What is the default value of a quantum on your Linux machine? Where did you find this information? What is the quantum value on your system for RT processes?
4. Name three famous CPU-scheduling algorithms. Which are parts of the Linux scheduling algorithm?
5. What is a time-sharing system? What is a batch system?
6. What are the main states that a process can be in? What does each state indicate about the status of the process?
7. What is the difference between built-in (internal) and external shell commands?
8. How does a Linux shell execute built-in and external commands? Explain your answer with an example.
9. Name ten process attributes.
10. What is the purpose of the `nice` command in Linux?
11. What are foreground and background processes in Linux? How do you run shell commands as foreground and background processes? Give an example for each.
12. In Linux jargon, what is a daemon? Give examples of five daemons.
13. What are signals in Linux? Give three examples of signals. What are the possible actions that a process can take upon receiving a signal? Write commands for sending these signals to a process with PID 10289.
14. Give a command that displays the status of all running processes on your system.
15. Give a command that returns the total number of processes running on your system.
16. Compute the priority number of a Linux process with a recent CPU usage of 31, a threshold priority of 60, and a nice value of 20. Show your work.

17. What command line will you use to display all the processes running under the SCHED_FIFO priority policy class? Show the results of running the command on your system. What command will you use to display the total number of processes running under the SCHED_OTHER priority policy class?

18. What is the most important SCHED_OTHER process?

19. Give the sequence of steps (with commands) for terminating a background process.

20. Create a zombie process on your Linux system. Use the `ps` command to show the process and its state.

21. The `ps -auxw` or `ps auxw` command is one of the most useful commands. What does it display? Explain your answer.

22. Give two commands to run the `date` command after 10 s. Make use of the `sleep` command; read the relevant manual page to find out how to use it.

23. Run a command that would remind you to leave for lunch after an hour by displaying the message `Time for Lunch!`

24. Give a command for running the `find` and `sort` commands in parallel.

25. Give an example of a Linux process that does not terminate with <Ctrl+C>.

26. Run the following commands on one command line so that they do not terminate when you log out. What command did you use?
    ```
    find / -inum 23476 -print > all.links.hard 2> /dev/null
    find / -name foo -print > foo.paths 2> /dev/null
    ```

27. Run the following sequence of commands under your shell. What are the outputs of the three pwd commands? Why are the outputs of the last two pwd commands different?

    ```
    $ pwd
    $ sh
    $ cd /usr
    $ pwd
    ...
    $ <Ctrl+D>
    $ pwd
    ...
    $
    ```

28. What are the names of processes with PIDs 0, 1, 2, and 3 on your Linux system? How did you get the answer to the question? Show your work.

29. Suppose you are running various programs in a session—ssh, vim, etc.—and the terminal locks up or the remote login program crashes, causing you to be disconnected from the host. Or, perhaps your keyboard or mouse suddenly stops working. Explain how you could log in from another physical terminal and use a sequence of Linux commands to recover from the situation. Give the sequence of Linux commands you would use.

30. What command would you use to display the hierarchical structure of processes on your system? What is the name of the process with PID 0? How many children does this process have and what are their names? What is the pathname of the executable for the `systemd` process (not its symbolic link!)?

31. Write commands for displaying the number of threads in the `kthreadd` process (the grandmother of all kernel threads) and the `systemd` process (the granddaddy of all user processes). How long has the `kernel` process been running on your system? What command(s) did you use to find this out?

32. The `ps -U root,bin,goldman,ibraheem` command displays the default status of all the processes belonging to the users **root**, **bin**, **goldman**, and **ibraheem**. However, the output does not show the logname of the process owner. Write down the Linux command, along with a sample run on your machine that would display the username for each process.

33. Make a process "map" of your system using the `pstree` command. How does it differ from the two maps of multi-user.target and graphical.target Linux systems we present in Section 10.7.2? To what extent is your map the same as either of our maps? What daemons and services are running on your system, which are not running on ours, and vice versa?

34. In the output of the command `ps ax --format uname,pid,ppid,tty,cmd,cls,pri,nice`, you will notice that almost all processes have the same priority value (19 on our Linux Mint system). This includes all of the important one, such as systemd and kthreadd. Explain why they are all the same, and why they all have such a low priority.

35. Present a `ps` command with modifiers that would show the output columns for PID, Command Name, Thread Group, and Task ID, for any particular user's processes on your Linux system.

36. What do the following commands display?

    a. `ps H`

    b. `ps m`

    c. `ps j`

    d. `ps l`

    e. `ps av`

    f. `ps ax`

    g. `ps -L`

    h. `ps -P`

37. Give commands to

    a. Display processes with the number of LWPs (i.e., threads) in each process.

    b. Display the number of the processor on which an LWP is bound, i.e., executed on.

    c. Display processes, except session leaders, i.e., login shells for user processes and parent for the kernel processes.

    d. Display the tree structure for processes in ASCII

    e. Display a list of processes associated with a pst/1

    f. Display information about processes with a list of PIDs

38. Run the `top` command on your system. What are the priority and nice values of the highest priority process? Run commands to have `top` display information about the top-10 processes and refresh its output every 7 s. What commands did you use?

39. As you monitor the top session, display processes for the user **john**. What command did you use? Show your work.

40. Use the `pstree` command to display the tree structure for the processes running in your current session. What command did you use? Which process is the grandparent of all your processes and what is its PID? What command will you use to display the tree hierarchy for the processes that the user **kent** has run on your system?

41. What does **/proc** contain? What do directories **/proc/1** through **/proc/10** contain? Which command starts the systemd program? Where is it located?

Advanced Questions and Problems

42. How many processes are currently running on your system? What states are they in? Write the names and function of each of the processes on your system with PIDs 1–10. What is the priority of each of these processes? Which of them are RT processes? Show details of your work, including any commands that you ran along with their outputs.

43. Table 10.8 shows the default fields in the output of the `top` command with brief descriptions. There are around 30 more output fields for the `top` command that you can display with various options. List each of these fields and what they mean.

44. In Section 10.5.1, the `ulimit -s` command displays the virtual stack size for a process to be 8 MB. This seems an awfully large stack size for a process. Browse the Web and/or use other resources to explain how a Linux process uses stack.

45. Browse the Web and/or use other resources to report the stack size of a pthread and stack size of a kernel process thread. If these stack sizes are different from the stack size of a Linux process, explain why that is.

46. What is contained in the /proc subdirectory for the kthreadd process on your Linux system? Using the online chapter resource at this book's GitHub site for Python, write some simple Python script files that not only search for interesting strings in the kthreadd process's files in the subdirectory of /proc, but also query and search for interesting string information in the files contained in other subdirectories of /proc on your Linux system.

47. Explain the similarities and differences between a process capability and the namespace the process executes in. Formulate your answer in terms of an overall execution environment and security model, for both a bare-metal installation in a host machine environment, and as applied in a Virtual Machine (VM) environment. That VM environment could be either Virtualbox, LXD containers, or Docker. Which component(s) of the kernel-provided functions of concurrency, virtualization, and persistence do capabilities and namespaces implement and impact most strongly? How and why?

48. With respect to system initialization with systemd on your Linux system, what is contained in all the target.wants directories in /etc/systemd/system?

49. Referring to Chapter W27, Sections 5.1.8.2–5.1.8.10, at the book website, answer the following with respect to system initialization with systemd-:

   a. What systemctl subcommand lists all target unit statuses?

   b. What systemctl subcommand would allow you to see the targets that are currently active?

   c. How would you bring the system to any new target state?

Projects

Project 1

To extend your knowledge and understanding of Linux process state, execution, and security, read and do the exercises presented (as much as possible in the following order) in Chapter 10, Sections 5 and 6, and Chapter 17, Section 9.6. In addition, read and do the exercises presented in Chapter W26, Sections 9.8.2–9.9 (particularly Examples W26.27 and W26.28) at the book website.

Looking for more? Visit our sites for additional readings, recommended resources, and exercises.

CRC Press e-Resource: https://www.crcpress.com/9781138710085

Authors' GitHub: https://github.com/bobk48/linuxthetextbook

11

Networking and Internetworking

OBJECTIVES

- To describe networks and internetworks and explain why they are used
- To briefly discuss the Transmission Control Protocol/Internet Protocol (TCP/IP) suite, IP addresses, protocol ports, and Internet services and applications
- To explain what the client–server software model is and how it works
- To discuss various network software tools for electronic communication, remote login, file transfer, remote command execution, tracing a route in the Internet, and status reporting
- To describe in detail the Secure Shell and other secure commands
- To cover the following commands and primitives:

```
finger, ftp, ip, host, nslookup, ping, scp, sftp, ssh, talk, telnet, traceroute
```

11.1 Introduction

The history of computer networking and Internet dates back to the late 1960s, when the Department of Defense's Advanced Research Projects Agency (ARPA) started funding networking research. This research resulted in a wide area network, called Advanced Research Projects Agency Network (ARPANET), by the late 1970s, with five nodes—University of California at Los Angeles (UCLA), Stanford, UC Santa Barbara, University of Utah, and Bolt, Beranek, and Newman (BBN) Technologies. In 1982, a prototype Internet that used Transmission Control Protocol/Internet Protocol (TCP/IP) became operational and was utilized by a few academic institutions, industrial research organizations, and the US military. By early 1983, all US military sites connected to ARPANET were on the Internet, and computers on the Internet numbered 562. By 1986, this number had more than quadrupled to 2,308. From then on, the size of the Internet doubled every year for the next 10 years, until it served about 9.5 million computers by 1996. The first Web browser, called Mosaic, was developed at the National Center for Supercomputer Applications (NCSA) and launched in 1991. As a result, World Wide Web (shortened to the www, or just the Web) browsing surpassed File Transfer Protocol (FTP) as the major use of the Internet by 1995. The first website, info.cern.ch, was launched on August 6, 1991. Since the first edition of this book was written, the social networking sites have had a major impact on the use of the Internet. Facebook, Flickr, YouTube, Reddit, Twitter, Tumblr, Dropbox, Instagram, and Pinterest were born during this period of time. As of the writing of this book, Facebook and Twitter have over 2 billion and over 320 million active users, respectively, generating over 5 billion likes and 670 million tweets per day, and Google+ and Pinterest have over 550 million and 295 million active users.

The Internet has grown from less than 1% of world population in 1995 to over 51% of the world population. Today, the Internet serves around 3.9 billion users and has over 1.01 billion hosts, around 1.27 billion websites live on the Internet, over 1 yottabyte (10^{24} bytes) of data resides on the Internet, over 1.2 billion domain names are in use, over 220 billion e-mails are sent every day, over 5.4 billion Google searches are made daily, more than 6.1 billion YouTube videos are watched every day, about 70 million photos are uploaded on Instagram daily, over 240 million Skype calls are made every day, and over 259 of 263 countries, colonies or territories, and disputed territories in the world provide Internet access.

It is projected that, by 2020, over 60% (5 billion) of the planet's population will be connected by the Internet. Linux has a special place in the world of networking in general and internetworking in particular, because most of the networking protocols were initially implemented on UNIX and Linux platforms. Also, server processes running on Linux-based computers provide most of the Internet services.

11.2 Computer Networks and Internetworks

When two or more computer hardware resources (computers, printers, scanners, plotters, etc.) are connected, they form a computer *network*. A hardware resource on a network or an internetwork is usually referred to as a *host*. Figure 11.1a shows a schematic diagram of a network with six hosts, H1–H6.

Computer networks are categorized as *local area networks* (LANs), *metropolitan area networks* (MANs), and *wide area networks* (WANs), based on the maximum distance between two hosts on the network. Networks that connect hosts in a room, building, or buildings of a campus are called LANs. The distance between hosts on a LAN can be anywhere from a few meters to about 1 km. Networks that are used to connect hosts within a city, or between small cities, are known as MANs. The distance between hosts on a MAN is about 1–20 km. Networks that are used to connect hosts within a state or country are known as WANs. WANs are also known as *long-haul networks*. The distance between the hosts on a WAN is in the range of tens of kilometers to a few thousand kilometers.

An *internetwork* is a network of networks. Internetworks can be used to connect networks within a campus or networks that are thousands of kilometers apart. The networks in an internetwork are connected to each other via specialized devices called *routers* or *gateways*. The Internet is the ubiquitous internetwork of tens of thousands of networks throughout the world. Figure 11.1b shows an internetwork of four networks. The four networks, **Net1–Net4**, are connected via five routers, **R1–R5**. Not all of the networks are directly connected, and two networks can be connected to each other via multiple routers. In Figure 11.1b, for example, **Net2** and **Net4** are not directly connected and **Net3** and **Net4** are connected to each other directly via two routers, **R4** and **R5**. Note that the router **R4** also connects directly **Net3** and **Net1**. Routers such as **R4** that can connect more than two networks are known as *multiport* routers.

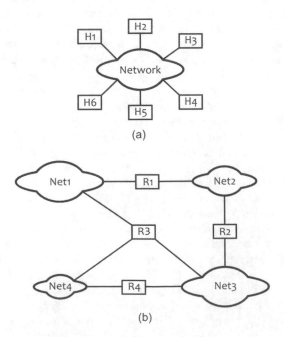

(a)

(b)

FIGURE 11.1 (a) A network of six hosts; (b) an internetwork of four networks.

11.3 Reasons for Computer Networks and Internetworks

There are numerous reasons for using networks of computers as opposed to stand-alone personal computers, powerful minicomputers, mainframe computers, or supercomputers. The main reasons include the following:

- *Sharing of computing resources*: Users of a computer network can share hardware resources including computers, printers, plotters, and scanners, and software resources such as files (data and software).
- *Network as a communication medium*: A network is an inexpensive, fast, and reliable communication medium between people who live far from each other.
- *Cost efficiency*: For the same price, you get more computing power with a network of workstations than with a minicomputer or mainframe computer.
- *Less performance degradation*: With a single powerful minicomputer, mainframe computer, or supercomputer, the work comes to a screeching halt if anything goes wrong with the computer, such as a bit in the main memory going bad. With a network of computers, if one computer crashes, the remaining computers on the network are still up and running, allowing continuation of work.

11.4 Network Models

Various questions arose in the design and implementation of networks, and these questions dictated the design of the two best-known network models:

1. The type of physical communication medium, or communication channel, used to connect hardware resources: It can be a simple RS-232 cable, telephone line, coaxial cable, fiber-optic cable, microwave link, or satellite link.
2. The topology of the network—that is, the physical arrangement of hosts on a network: Some commonly used topologies are bus, ring, mesh, and general graph.
3. The set of rules, called protocols, used to allow a host on a network to access the physical medium before initiating data transmission.
4. The protocols are used for routing application data (e.g., a Web page) from one host to another in a LAN or from a host in one network to a host in another network in an internetwork.
5. The protocols are used for transportation of data from a process on a host to a process on another host in a LAN or from a process on a host in one network to a process on a host in another network in an internetwork.
6. The protocols are used by network-based software to provide specific applications such as ftp.

The two best-known network models are the International Standards Organization's Open System Interconnect Reference Model (commonly known as the Open Systems Interconnection (OSI) Seven-Layer Reference Model) and the TCP/IP Five-Layer Model. The OSI model was proposed in 1981 and the TCP/IP model in the late 1970s. In March 1982, the U.S. Department of Defense adopted the TCP/IP model as the standard for all military networks. The TCP/IP model, which has its roots in ARPA, is the basis of the Internet and is, therefore, also known as the IP model. This model consists of five layers, each having a specific purpose and a set of protocols associated with it. The diagram in Figure 11.2 shows the two models, along with an approximate mapping between the two.

Because the TCP/IP model is used in the Internet, this will be our focus. In terms of the six issues previously listed, the first layer in the TCP/IP model deals with the first two issues, the second layer deals with the third issue, the third layer deals with the fourth issue, the fourth layer deals with the fifth issue, and the fifth layer deals with the sixth issue. In terms of their implementation, the first four layers

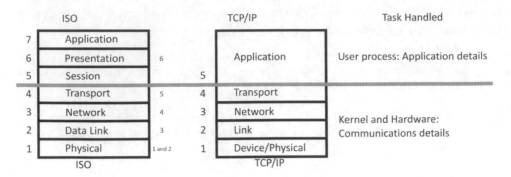

FIGURE 11.2 ISO and TCP/IP layered models, mapping between the two, and the general purpose of a group of layers.

deal with the details of communication between hosts, and the fifth layer deals with the details of the Internet services provided by various applications. Most of the first layer is handled by hardware (type of communication medium used, attachments of hosts to the medium, etc.). The *network interface card* (NIC) in a host handles the rest of the first layer and the second layer. Layer 2 consists of medium access control (MAC) addresses, network cards, drivers, and switches. Layers 3 and 4 are fully implemented in the operating system kernel on most existing systems. The first two layers are network hardware specific, whereas the remaining layers work independently of the physical network. On newer gigabit *Ethernet* interfaces where the processing overhead of the network stack becomes significant, the TCP offload engine (TOE) technology is used in NIC to offload processing of the TCP/IP stack to the network controller.

Exercise 11.1
Ask your system administrator: How many hosts are connected on your LAN? What type of computers are they (PCs or workstations)?

Exercise 11.2
What is the physical medium for your network (coaxial cable, twisted pair, or glass fiber)? Ask your instructor or system administrator about the topology of your network (bus, ring, etc.).

Exercise 11.3
Ethernet is the most commonly used link-level protocol for LANs. Does your LAN use Ethernet? If not, what does it use?

11.5 The TCP/IP Suite

Several protocols are associated with various layers in the TCP/IP model. These protocols result in what is commonly known as the *TCP/IP suite*, which is illustrated in Figure 11.3. The description of most of the protocols in the suite is beyond the scope of this textbook, but we briefly describe the purpose of those that are most relevant to our discussion. As a user, you see the application layer in the form of applications and utilities that can be executed to invoke various Internet services. Some of the commonly used applications are for electronic mail, Web browsing, file transfer, and remote login. We discuss some of the most useful applications in Section 11.8.

11.5.1 TCP and UDP

The purpose of the transport layer is to transport application data from your machine to a remote machine and vice versa. This delivery service can be a simple, best-effort service that does not guarantee reliable delivery of the application data or one that guarantees reliable and in-sequence delivery of the application data. The *User Datagram Protocol* (UDP) provides a best-effort delivery service and the TCP offers

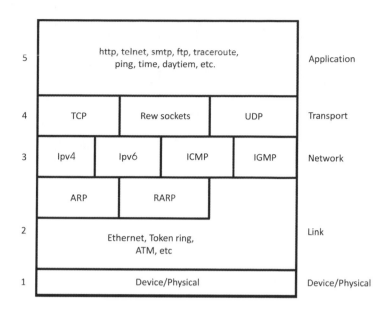

FIGURE 11.3 The TCP/IP suite.

completely reliable, in-sequence delivery. The UDP is a connectionless protocol; that is, it simply sends the application data to the destination without establishing a virtual connection with the destination before transmitting the data. Hence, the UDP software on the sender host does not "talk" to the UDP software on the receiver host before sending data. The TCP is a connection-oriented protocol that establishes a virtual connection between the sender and receiver hosts before transmitting application data, leading to reliable, error-free, and in-sequence delivery of data. Of course, the overhead for establishing the connection makes TCP more costly than UDP. Many well-known Internet applications such as ftp use TCP. Applications where efficiency of data delivery is more important than error-free, in-sequence delivery, such as video streaming, use UDP. In Internet jargon, a data packet transported by TCP is called a *segment* and a data packet transported by UDP is called a *datagram*.

Because multiple client and server processes might be using TCP and/or UDP at any one time, these protocols identify every process running on a host by 16-bit positive integers (0–65,535) called *port numbers*. Port numbers from 0 to 1,023 are called *well-known ports* and are controlled by the Internet-Assigned Numbers Authority (IANA). Well-known services such as http are assigned ports that fall in the well-known range (excluding 0). Most of these services allow the use of either TCP or UDP, and the IANA tries to assign the same port number to a given service for both TCP and UDP. For example, the ssh service is assigned port number 22 and the http (Web) server is assigned port number 80, for both TCP and UDP. Most clients can run on any port and are assigned a port by the operating system at the time the client process starts execution. Some well-known clients such as ssh require the use of a reserved port as part of the client–server authentication protocol. These clients are assigned ports in the range 513–1,023. Although server processes run forever, client processes for services that are not well known are assigned ports for as long as they run. Such ports, known as ephemeral ports, are in the range of 1,024–65,553.

11.5.2 Routing of Application Data: The IP

As mentioned earlier, the network layer is responsible for routing application data to the destination host. The protocol responsible for this is the IP, which transports TCP segments or UDP datagrams containing application data in its own packets called *IP datagrams*. The routing algorithm is connectionless, which means that IP routing is the best-effort routing and does not guarantee delivery of TCP segments and UDP datagrams. Applications that need guaranteed delivery use TCP as their transport-level protocol or have it built into the application itself. There are two versions of IP: the older version is IPv4 and the new version is IPv6 (commonly known as IPng or Internet Protocol: The Next Generation). In this textbook, we primarily discuss IPv4. The discussion on the actual routing algorithms used by IP is beyond our scope here. However, we describe a key component of routing on the Internet—the IP addressing (naming) scheme to uniquely identify a host on the Internet.

The key to routing is the IP assignment of a unique identification to every host on the Internet. IP does so by uniquely identifying the network the host is on and then uniquely identifying the host on that network. The ID, a 32-bit positive integer in IPv4 and a 128-bit positive integer in IPv6, is known as the host's *IP address*. Every IP datagram has the sender's and the receiver's IP address in it. The sender's IP address allows the receiver to identify and respond to the sender. Hosts and routers perform routing by examining the destination IP address on an IP datagram.

In IPv4, the IP address is divided into three fields: address class, network ID, and host ID. The address class field identifies the class of the address and dictates the number of bits used in the network ID and host ID fields. This scheme results in five address classes: A, B, C, D, and E, with classes A–C being the most common. Figure 11.4 shows the structures of the five address types. The IP addresses belonging to classes D and E have special use, and their discussion is beyond the scope of this textbook. A central authority, the Network Information Center (NIC; www.internic.net), assigns all IP addresses.

The maximum number of networks of classes A–C that can be connected to the Internet is given by the expression: $2^7 + 2^{14} + 2^{21}$. Here, 7, 14, and 21 are the number of bits used to specify network IDs in class A–C addresses, respectively. Thus, there are 2^7 class A networks, 2^{14} class B networks, and 2^{21} class C networks. The sum of these numbers gives a total of 2,113,664 networks. Similarly, the number of bits used to identify host IDs in the three classes of addresses can be used to get the maximum number of hosts that can be connected to the Internet. Thus, there are roughly 2^{24} hosts per class A network, 2^{16} hosts per class B network, and 2^8 hosts per class C network. The sum of all the hosts on the three types of networks is a total of 3,758,096,400 hosts. The actual numbers of class A–C networks and hosts are somewhat smaller than numbers shown, due to some special addresses (e.g., *broadcast* and *localhost* addresses). The broadcast addresses are used to address all hosts on a network. A host uses the localhost address to send a datagram to itself. Hence, an IP datagram with **localhost** as its destination address is never put on the network.

To slow down the use of IPv4 addresses and to reduce the growth of routing tables on Internet routers, the Internet Engineering Task Force (IETF; www.ietf.org) introduced Classless Internet-Domain Routing (CIDR) in 1993. Under CIDR, network address spaces in IPv4 are allocated on any address bit boundary, not necessarily on 8-bit sections. There are a lot of classless networks on the Internet.

The number of class A addresses is very small, so these addresses are assigned only to very large commercial organizations, educational institutions, and government agencies, such as US national

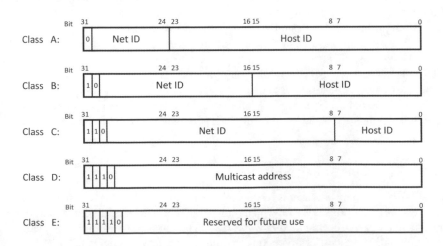

FIGURE 11.4 IPv4 address classes.

laboratories, the Massachusetts Institute of Technology (MIT), the University of California at Berkeley, Bell Labs, and National Aeronautics and Space Administration (NASA). The number of class B addresses is relatively large, and these addresses are assigned to large commercial organizations and educational institutions. Hence, corporations such as IBM and Oracle, educational institutions such as Iowa State University, and numerous other national and international universities have been assigned class B addresses. The total number of class C addresses is quite large, so these addresses are assigned to individuals and small- to medium-sized organizations, such as local Internet service providers, small consulting and software companies, community colleges, and universities.

Although the IPv4 addressing scheme can be used to identify a large number of networks and hosts, it has not been able to cope with the rapid growth of the Internet. Among the many advantages of IPv6 is the extremely large number of hosts that can be connected. With the 128-bit address, the maximum number of hosts on the Internet will increase to roughly 2^{128}, which is greater than 3.4×10^{38}. This number is roughly 6×2^{28} times the present world population. One disadvantage of IPing is that, as the address size is very large, remembering IPv6 addresses becomes very difficult. However, because most users prefer to use symbolic names, remembering IPv6 addresses should not present a problem. Also, some compact notations similar to dotted decimal notation (DDN) have been proposed for IPv6 addresses as well. Many large companies and academic institutions have been assigned IPv6 addresses, including Google, IBM, Intel, Microsoft, MIT, UC Berkeley, and Iowa State University. You can use the host command to find out if an organization has been assigned an IPv6 address, as follows:

```
$ host stanford.edu
stanford.edu has address 171.67.215.200
stanford.edu has IPv6 address 2607:f6d0:0:925a::ab43:d7c8
stanford.edu mail is handled by 10 mxa-00000d03.gslb.pphosted.com.
stanford.edu mail is handled by 10 mxb-00000d03.gslb.pphosted.com.
$
```

Note that IPv6 addresses are displayed in the *colon-hex notation*. In this notation, two hexadecimal digits are used to specify every byte of the address. A colon (:) is inserted between every two bytes—that is, four hex digits—of an address. If a number between two consecutive colons is not four digits, it represents the least significant nibbles in a two-byte sequence. For example, 82 in an address in fact is 0082, 0 is 0000, and 10 is 0010. The right most hex number for the two-byte sequence is for the least significant two bytes of the 128-bit address. Finally, two consecutive colons between the least significant hex number and previous hex number represent all zeros. The 128-bit IPv6 address of stanford.edu, consisting of 32 nibbles, is therefore represented as follows:

MSD																														LSD	
2	6	0	7	f	6	d	0	0	0	0	0	9	2	5	a	0	0	0	0	0	0	0	0	a	a	4	3	d	7	c	8

Most Linux commands and tools have been enhanced to handle IPv6 addresses in the colon-hex notation, as follows:

```
$ host 2607:f6d0:0:925a::ab43:d7c8
8.c.7.d.3.4.b.a.0.0.0.0.0.0.0.a.5.2.9.0.0.0.0.0.d.6.f.7.0.6.2.ip6.arpa domain
name pointer web.stanford.edu.
$
```

Here, the hex numbers are shown in the reverse order of significance—that is, least significant hex digit first. As you can see, the IPv6 address is for thumb.iastate.edu. A more detailed discussion on IPv6 addresses, including the purpose of each byte of the address, is beyond the scope of this book.

11.5.2.1 IPv4 Addresses in Dotted Decimal Notation

Although hosts and routers process IPv4 addresses as 32-bit binary numbers, they are difficult for people to remember. For this reason, the IPv4 addresses are given in *DDN*. In this notation, all four bytes of an IPv4 address are written in their decimal equivalents and are separated by dots. Thus, the 32-bit IP address

```
11000000 01100110 00001010 00010101
```

is written as

```
192.102.10.21
```

in DDN. The ranges of valid IP addresses belonging to the five address classes in DDN are shown in Table 11.1. Some of the addresses given in the table are special addresses.

The internetwork shown in Figure 11.5 connects four networks via four routers, **R1–R4**. **Net1** is a class A network, **Net3** is a class B network, and **Net2** and **Net4** are class C networks. The way to identify the class of a network is to look at the leftmost decimal number in the IP address of a host on the network—in this case, the IP addresses of the routers. Note that the routers are assigned as many IP addresses as the number of networks they connect. Here, for example, router **R1** connects **Net1** and **Net2** and has IP addresses 121.1.1.1 and 192.102.10.1. Similarly, **R3** is assigned three IP addresses, as it interconnects three networks **Net1**, **Net3**, and **Net4**.

Of the special addresses, 127.0.0.0 (or 127.x.x.x, where x can be any number between 0 and 255), also known as *localhost*, is used by a host to send a data packet to itself. It is also commonly used for testing new applications before they are used on the Internet. Another special address, in which the host ID field is all 1s, is the *directed broadcast address*. This address is used to send a data packet to all hosts on a network—that is, for broadcasting on a local network whose host is using the address as a destination address.

11.5.3 Symbolic Names

People prefer to use symbolic names rather than numeric addresses, because names are easier to remember, especially with the transition of the 128-bit-long numeric addresses in IPv6. Also, symbolic names

TABLE 11.1

IPv4 Address Classes and Valid IP Addresses

	Range of Valid IP Addresses	
Address Class	**Lowest**	**Highest**
A	0.0.0.0	127.255.255.255
B	128.0.0.0	191.255.255.255
C	192.0.0.0	223.255.255.255
D	224.0.0.0	239.255.255.255
E	240.0.0.0	247.255.255.255

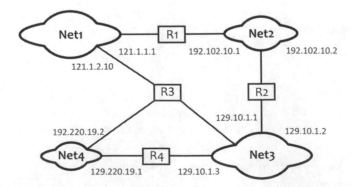

FIGURE 11.5   An internetwork of four networks with one class A, one class B, and two class C networks.

can remain the same even if numeric addresses change. Like its IP address, the symbolic name of a host on the Internet must be unique. The Internet allows the use of symbolic names by using a hierarchical naming scheme. The symbolic names have the format

```
hostname.domain_name
```

where `domain_name` is the symbolic name referring to the site and is assigned by various registrars whose list is maintained by the NIC. The `hostname` is assigned and controlled by the site that is allocated the `domain_name`. The `domain_name` consists of two (or more) strings separated by a period (.). The rightmost string in a domain name is called the *top-level domain* (TLD). The string to the left of the rightmost period identifies an organization and can be chosen by the organization and assigned to it by the NIC. If the string has already been assigned to another organization under the same TLD, another string is assigned to keep the domain names unique. There are three types of TLDs: *special TLDs*, *generic TLDs* (gTLDs), and *country code TLDs* (ccTLDs). According to the IANA's Root Domain Database (www.iana.org/domains/root/db), 1,776 TLDs have been registered at the time of writing. Details of these domains are given in Table 11.2.

For the domain names that consist of more than two strings, the remaining strings are assigned by the organization that owns the domain. Some example domain names are www.infinione.com, stanford.edu, intel.com, whitehouse.gov, uu.net, omsi.org, cs.berkeley.edu, www.cam.ac.uk, amazon.com.jp, pucit.edu.pk, www.beaverton.k12.org.us, www.abc.tv, www.nato.int, www.darpa.mil, example.info, and bbc.co.uk. The authorities in a country assign strings to the left of that country's domain. Figure 11.6 illustrates the domain name hierarchy.

Attaching the name of a host to a domain name with a period between them yields the *fully qualified domain name* (FQDN) for the host—for example, cs.stanford.edu, where cs is the name of a host in the department of computer science at Stanford University. However, FQDNs for the hosts on the Internet do not always have three parts. Most organizations allow various groups within the organization to choose the primary names for the hosts that they control and are responsible for. For example, the Department of Computer Science at Stanford, which uses the primary name cs.stanford.edu, uses www.cs.stanford.edu as the FQDN for its HTTP server. The School of Business Administration at Duke, which uses the primary name fuqua.duke.edu, can use the host name www.fuqua.duke.edu for its Web server.

11.5.4 Translating Names to IP Addresses: The Domain Name System

Because Internet software deals with IP addresses and people prefer to use symbolic names, application software translates symbolic names to equivalent IP addresses. This translation involves the use of a service provided by the Internet known as the *Domain Name System* (DNS). The DNS implements a distributed database of name-to-address mappings. A set of dedicated hosts run server processes called *name servers* that take requests from application software (also called the client software; see Section 11.7) and work together to map domain names to the corresponding IP addresses. Every organization runs at least one name server, often the *Berkeley Internet Name Domain* (BIND) program. The applications use resolver functions such as gethostbyname to invoke the DNS service. The gethostbyname resolver function maps a hostname (simple or fully qualified) to its IP address, and gethostbyaddr maps an IP address to its hostname.

An alternative, and old, scheme for using the DNS service is to use a static hosts file, usually **/etc/hosts**. This file contains the domain names and their IP addresses, one per line. The following command displays a sample **/etc/hosts** file.

```
$ more /etc/hosts
127.0.0.1      localhost
127.0.1.1      Mint18

# The following lines are desirable for IPv6 capable hosts
::1     ip6-localhost ip6-loopback
fe00::0 ip6-localnet
ff00::0 ip6-mcastprefix
```

TABLE 11.2

Top-Level Internet Domains

Domain Type	Top-Level Domain	Assigned To/For
Special	ARPA	Used exclusively for Internet; currently second-e164.arpa, in-addr.arpa, ip6.arpa, uri.arpa, urn.arpa
Generic	ACCOUNTANTS	Knob Town, LLC
	ACTOR	United TLD Holdco, Ltd.
	AERO	Reserved for members of air transport industry
	AIRFORCE	United TLD HoldCo, Ltd.
	ARMY	United TLD HoldCo, Ltd.
	ATTORNEY	United TLD HoldCo, Ltd.
	BEER	Top-Level Domain Holdings Limited
	BIKE	Grand Hollow, LLC
	BIZ	Reserved for businesses
	BUILDERS	Atomic Madison, LLC
	CAREERS	Wild Corner, LLC
	CHRISTMAS	Uniregistry, Corp.
	CHURCH	Holly Fields, LLC
	COM	Reserved for commercial organizations
	COOP	Reserved for cooperative associations
	EDU	Reserved for U.S. postsecondary educational institutions that are accredited by an agency on the U.S. Department of Education's list of Nationally Recognized Accrediting Agencies
	GOV	Reserved for the U.S. government
	INFO	First unrestricted TLD since .com, so it can be used by anyone—businesses, marketers, and so on
	INT	Reserved for organizations established by treaties between governments
	MIL	Reserved for the U.S. military
	MUSEUM	Reserved for museums
	NAME	Reserved for individuals
	NET	Intended for internet service providers (ISPs) and telephone service providers
	ORG	Intended for noncommercial communities but all are eligible to register
	PRO	Restricted to credentialed professionals (this domain is being established)
	...	
	ZIP	Charleston Road Registry, Inc.
	ZONE	Outer Falls, LLC
Country Code	AU	Australia
	DE	Germany (Deutschland)
	FI	Finland
	JP	Japan
	PK	Pakistan
	...	
	UK	United Kingdom
	US	United States

```
ff02::1 ip6-allnodes
ff02::2 ip6-allrouters
$
```

There are two problems with this scheme. First, its implementation depends on how the system administrator configures the system. Second, owing to the sheer size of the Internet and its current rate of growth, the static file can be extremely large.

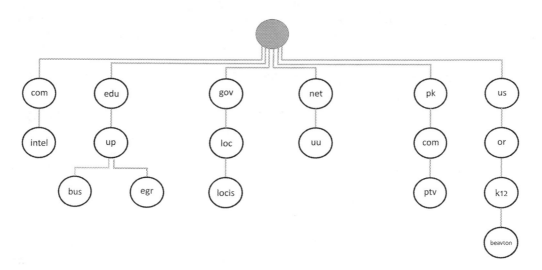

FIGURE 11.6 A portion of the Internet domain name hierarchy.

You can use the ifconfig command to view and set network interface parameters, including the IP address, localhost, *netmask*, *broadcast address*, and *maximum transmission unit* (MTU). However, this command is not maintained anymore and has been replaced with more contemporary ip command in most modern Linux systems. It can be used for doing post-install network configuration of network interface parameters as well as see their current settings. You can do with the ip command all that you can do with the ifconfig command and more. In this section, we show few sample runs of the ip command for viewing the various interface parameters. We give extensive information and examples of the use of the ip command for checking and setting network interface parameters in Chapter 17, Section 8.2.

Netmask, also known as subnetmask, is a bit mask used by TCP/IP to identify whether a host is on a remote network or on a local subnet. A broadcast address is used by the IP layer in a host to send a datagram to all the hosts on a subnet or to a remote network. The Ethernet broadcast address is all 1s. The MTU in a TCP/IP network is the maximum size of an IP datagram (packet) and is dependent on the technology used at the data link layer for connection to other hosts.

The following is an example run of ip on our Linux Mint machine. The output of the ip addr show (or ip a s) command shows that you are logged on to a host that has localhost interface and network interfaces enp2s0 and eno1, each having IPv4 and IPv6 addresses. For IPv4, localhost is 127.0.0.1, with MTU 65,536 bytes. For IPv6, localhost is ::1/128. The IPv4 network addresses are 202.147.169.195 and 172.16.0.101; each with MTU 1,500 bytes, and broadcast IP addresses 202.147.169.223 and 172.16.255.255, respectively. Further, the IP address 202.147.169.195 is associated with MAC address 00:10:18:30:ee:6e. For IPv6, network IP addresses for external traffic are fe80::67ab:4f1a:21fa:f0f5 and fe80::e352:b2f4:9833:ceb0, respectively, for enp2s0 and eno1.

```
$ ip addr show
1: lo: <LOOPBACK,UP,LOWER_UP> mtu 65536 qdisc noqueue state UNKNOWN group default
qlen 1
    link/loopback 00:00:00:00:00:00 brd 00:00:00:00:00:00
    inet 127.0.0.1/8 scope host lo
      valid_lft forever preferred_lft forever
    inet6 ::1/128 scope host
      valid_lft forever preferred_lft forever
2: enp2s0: <BROADCAST,MULTICAST,UP,LOWER_UP> mtu 1500 qdisc mq state UP group
default qlen 1000
    link/ether 00:10:18:30:ee:6e brd ff:ff:ff:ff:ff:ff
    inet 202.147.169.195/27 brd 202.147.169.223 scope global enp2s0
      valid_lft forever preferred_lft forever
    inet6 fe80::67ab:4f1a:21fa:f0f5/64 scope link
```

```
        valid_lft forever preferred_lft forever
3: eno1: <BROADCAST,MULTICAST,UP,LOWER_UP> mtu 1500 qdisc pfifo_fast state UP
group default qlen 1000
    link/ether 64:51:06:49:70:46 brd ff:ff:ff:ff:ff:ff
    inet 172.16.0.101/16 brd 172.16.255.255 scope global eno1
        valid_lft forever preferred_lft forever
    inet6 fe80::e352:b2f4:9833:ceb0/64 scope link
        valid_lft forever preferred_lft forever
$
```

You can also use the ip command to display information about a particular interface. For example, the ip a show eno1 command displays information about the interface eno1. The system administrator can enable or disable these interfaces or any of its parameters, as discussed in Chapter 17. Further discussion on the output of the command, and its various options for viewing and setting different network parameters is beyond the scope of this chapter.

The ip command is normally located in the **/bin** and/or **/sbin** directories on Linux Mint. If you get an error message that ip is not found, you should include the relevant directories in your search path and re-execute the command. You can run the cat /etc/hosts command to display the domain names and IP addresses of the hosts on your network.

You can use the host command to do the DNS lookup for a host whose domain name is passed as a command line argument to it. The command allows you to display IP address(es) for a domain name or vice versa. In the following session, we use the host command to display the IP addresses of the hosts iastate.edu, berkeley.edu, and facebook.com, and the domain names corresponding to two IPv6 addresses. The output of the host iastate.edu command displays the IPv4 and IPv6 addresses of iastate.edu (Iowa State University). It also displays that ten machines handle e-mail at Iowa State. The output of the second command shows that the given IPv6 address is for thumb.iastate.edu. The outputs of the fourth and sixth commands show the IPv4 and IPv6 addresses for stanford.edu and facebook.com, as well as the fact that ten hosts each handle e-mail for them. Finally, the output of the last command shows that the name server failed to map the IP address 198.175.96.33 to any domain name, which does not mean that the domain does not exist.

```
$ host iastate.edu
iastate.edu has address 129.186.235.2
iastate.edu has IPv6 address 2610:130:101:113::2
iastate.edu mail is handled by 10 mailin.iastate.edu.
$ host 2610:130:101:113::2
2.0.0.0.0.0.0.0.0.0.0.0.0.0.0.0.3.1.1.0.1.0.1.0.0.3.1.0.0.1.6.2.ip6.arpa domain
name pointer redirect.its.iastate.edu.
$ host redirect.its.iastate.edu
redirect.its.iastate.edu has address 129.186.235.2
redirect.its.iastate.edu has IPv6 address 2610:130:101:113::2
$ host stanford.edu
stanford.edu has address 171.67.215.200
stanford.edu has IPv6 address 2607:f6d0:0:925a::ab43:d7c8
stanford.edu mail is handled by 10 mxa-00000d03.gslb.pphosted.com.
stanford.edu mail is handled by 10 mxb-00000d03.gslb.pphosted.com.
$ host 2607:f140:0:82::10
0.1.0.0.0.0.0.0.0.0.0.0.0.0.0.0.2.8.0.0.0.0.0.0.0.4.1.f.7.0.6.2.ip6.arpa domain
name pointer calweb-farm-prod.ist.berkeley.edu.
$ host facebook.com
facebook.com has address 157.240.20.35
facebook.com has IPv6 address 2a03:2880:f11c:8183:face:b00c:0:25de
facebook.com mail is handled by 10 msgin.vvv.facebook.com.
$ host 198.175.96.33
Host 33.96.175.198.in-addr.arpa. not found: 3(NXDOMAIN)
$ host 52.6.169.253
253.169.6.52.in-addr.arpa domain name pointer ec2-52-6-169-253.compute-1.
amazonaws.com.
$
```

On Linux systems, you can use the `nslookup` command to do the DNS lookup. Here is a sample run of the command.

```
$ nslookup google.com
Server:    127.0.1.1
Address:   127.0.1.1#53

Nonauthoritative answer:
Name:      google.com
Address:   216.58.208.78

$
```

Similarly, you can use the `dig` command to do the DNS lookup on Linux. Here is a sample run of the command.

```
$ dig google.com

; <<>> DiG 9.10.3-P4-Ubuntu <<>> google.com
;; global options: +cmd
;; Got answer:
;; ->>HEADER<<- opcode: QUERY, status: NOERROR, id: 8082
;; flags: qr rd ra; QUERY: 1, ANSWER: 1, AUTHORITY: 0, ADDITIONAL: 1

;; OPT PSEUDOSECTION:
; EDNS: version: 0, flags:; udp: 512
;; QUESTION SECTION:
;google.com.                    IN      A

;; ANSWER SECTION:
google.com.            298      IN      A       216.58.210.78

;; Query time: 34 msec
;; SERVER: 127.0.1.1#53(127.0.1.1)
;; WHEN: Sat Mar 31 10:51:24 PKT 2018
;; MSG SIZE   rcvd: 55

$
```

11.5.5 Requests for Comments

The TCP/IP standards are described in a series of documents, known as *Requests for Comments* (RFCs). RFCs are first published as *Internet Drafts* and are made available to all Internet users for review and feedback by placing them in known RFC repositories. After the review process is complete, a draft can become a standard. But not all RFCs are *Internet Standards* documents; some are for information only and others are experimental.

An RFC citation has the following format:

```
####    Title. Authors (up to three). Issue date. (Format: TXT=size-in-bytes,
PS=size-in-bytes, PDF=size-in-bytes) (Obsoletes xxx) (Obsoleted by RFC####)
(Updates RFC####) (Updated by RFC####) (Also FYI ####) (Status: ssssss)
```

where `####` is a four-digit decimal number; `Format` can be TXT (ASCII), PS (PostScript), and PDF (Portable Document Format); and `Status` can be UNKNOWN, PROPOSED STANDARD, DRAFT STANDARD, STANDARD, INFORMATIONAL, EXPERIMENTAL, or HISTORIC. Here is an example citation:

```
          1180    TCP/IP tutorial. T.J. Socolofsky,
                  C.J. Kale. Jan-01-1991. (Format:
```

```
TXT=65494 bytes) (Status:
INFORMATIONAL)
```

You can view and download an RFC by accessing any of the repositories maintained on ftp or websites. The most common method of accessing an RFC is to browse the Web page at www.ietf.org/rfc.html. As of the writing of this chapter, there are 8,349 RFCs available in the RFC index maintained on this Web page, the last one being submitted in October 2017 as Internet Standards Track document. If you want to be notified of the announcement of a new RFC, you can subscribe to the following distribution list: http://mailman.rfc-editor.org/mailman/listinfo/rfc-dist.

To display the text version of an RFC in your browser, type www.ietf.org/rfc/rfcNNNN.txt into the location field of your browser, where NNNN is the RFC number. So, to display the text version of RFC 2020, type www.ietf.org/rfc/rfc2020.txt into the location field of your browser.

The following in-chapter exercises are designed to enhance your depth of understanding of your own network environment by way of learning the domain names and IP addresses of hosts on your network. You will also use the host command to translate domain names to IP addresses and vice versa.

Exercise 11.4

Give the domain names of some hosts on your LAN. Ask your instructor for help if you need any. How did you obtain your answer?

Exercise 11.5

List the IP addresses of the hosts identified in Exercise 11.4 in DDN. What is the class of your network (A, B, C, or classless)? How did you find out?

Exercise 11.6

Does your network have an IPv6 address? What is its value? Show the command that you used to obtain your answer to this question.

Exercise 11.7

Repeat the shell sessions given in this section demonstrating the host command on your system. Do you get the same results? If not, how do the outputs of your commands differ from those shown in this section?

Exercise 11.8

Browse the Web page at www.ietf.org/rfc.html, find the citation for RFC1118, and write it down.

11.6 Internet Services and Protocols

Most users do not understand the intricacies of the IPs and its architecture—nor do they need to. They access the Internet by using programs that implement the application-level protocols for various Internet services. Some of the most commonly used services and the corresponding protocols are listed in Table 11.3. The services are listed in alphabetic order and not according to their frequency of use. You can display the **/etc/services** file on your host to view the Internet services and their well-known port numbers.

The Linux operating system has some network-related services that are not necessarily available in other operating systems. They include services for displaying all the users logged on to the hosts in

TABLE 11.3

Popular Internet Services and Corresponding Protocols

Service	Protocol
Electronic mail	SMTP (Simple Mail Transfer Protocol)
File transfer	FTP (File Transfer Protocol)
Remote login	SSH (Secure Shell) and TELNET
Web browsing	HTTP (Hypertext Transfer Protocol) and HTTPS (HTTP Secure)

a LAN, remote execution of a command, real-time chat in a network, and remote copy. We discuss software for most of these services in Section 11.8.

11.7 The Client–Server Software Model

Internet services are implemented by using a paradigm in which the software for a service is partitioned into two parts. The part that runs on the host that the user is logged on to is called the *client software*. The part that handles client requests and usually starts running when a host boots up is called the *server software*. On the one hand, the server runs forever, waiting for a client request to come. Upon receipt of a request, it services the client request and waits for another request. On the other hand, a client starts running only when a user runs the program for a service that the client offers. It usually prompts the user for input (command and/or data), transfers the client's request to the server, receives the server's response, and forwards the response to the user. Most clients terminate with some sort of "quit" or "exit" command.

Many of the applications are connection-oriented client–server models, in which the client sends a connection request to the server and the server either accepts or rejects the request. If the server accepts the request, the client and server are connected through a *virtual connection*. From this point on, the client sends user commands to the server as requests. The server process serves client requests and sends responses to the client, which sends them to the user in a particular format. Communication between a client and a corresponding server—and the client's interaction with the user—is dictated by the protocol for the service offered by an application. Figure 11.7 shows an overview of the client–server software model.

Thus, when you run a program, such as Firefox, that allows you to surf the Web, an http client process starts running on your host. By default, most clients display the home page of the organization that owns the host on which the client runs, although it can be set to any page, including a blank page. When you want to view the Web page of a site, you give the site's *Uniform Resource Locator* (URL) to the client process. For displaying a home page, the URL has the format:

```
http://host/page
```

where `host` can be the FQDN or IP address (in DDN) of the computer that has the home page you want to display and `page` is the pathname for the file containing the page to be displayed—for example, http://cnn.com, www.stroustrup.com, http://mitadmissions.org/index.php, and http://profiles.stanford.edu/russ-altman. The client tries to establish a connection with the http server process on the site corresponding to the URL. If the site has the http server running and no security protections such as a password are in place, a connection is established between the client and server. The server then sends the Web page to the client, which displays it on the screen, with any audio or video components sent to appropriate devices. Note that `http` can be replaced with `ftp` if you want to access an ftp site through your browser, or with `ssh` if you want to remotely logon via the Secure Shell (SSH) protocol.

You can invoke the client programs for most Internet services by using the corresponding commands, such as `ssh` for the SSH service and `ftp` for the FTP service. Most of these commands permit a domain

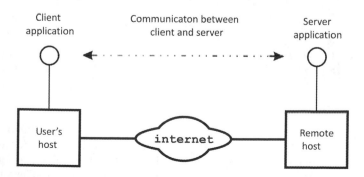

FIGURE 11.7 Depiction of the client–server software model.

name or IP address of the host on which the server runs as an argument in the command. Some commands also allow port number as an argument. Client software that has such flexibility built in is known as a *fully parameterized client*. Such clients are important in terms of the flexibility they offer. They also allow testing of updated server software by running it on a port that is not well known and contacting it with the client. A telnet client, discussed in Section 11.8, is a good example of a fully parameterized client.

11.8 Application Software

Numerous programs that implement the application-level protocols just discussed are available on networks of Linux hosts. Of the most commonly used applications described here, some are available on Linux- and Linux-based systems only, whereas others are available to all the hosts on the Internet.

11.8.1 Displaying the Host Name

Network-based applications use the **user@host** address format to identify a user on a network on the Internet. You can use the hostname and uname commands to display the name of the host you are logged on to. On some systems, the host name is shown in the short, simple name format, and on others it is displayed in the long, FQDN format. If you have to identify the host on the network that you are logged on to, you can use the hostname -s command to display the short format, which is simply the name of the host (the leftmost string in the FQDN format). You can use the uname -n command to display the host name of the computer that you are logged on to in the FQDN format. The uname -a command displays complete information about a host, including the operating system it is running and the name of the CPU. The following are some examples of the hostname and uname commands.

```
$ hostname
Mint18
$ uname -n
Mint18
$ uname -a
Linux Mint18 4.4.0-21-generic #37-Ubuntu SMP Mon Apr 18 18:33:37 UTC 2016 x86_64
x86_64 x86_64 GNU/Linux
$
```

You can use the arch command to print the processor architecture name. This command outputs the same information as does uname -m, as can be seen in the following session.

```
$ arch
x86_64
$ uname -m
x86_64
$
```

> **Exercise 11.9**
>
> Ask your friends and fellow students about the two network services they use most often. Which services are they? Which is the more popular of the two? How many people did you ask?

11.8.2 Testing a Network Connection

You can test the status of a network or a particular host on it by using the ping command. If the ping command does not work on your system, use the type ping command to find its location, update your search path, and try the command again. It is normally in the **/bin** directory on a Linux Mint system. You can also use the whereis ping command to find the location of command. The following is a brief description of the command.

Syntax:
```
ping [optins] hostname
```
Purpose: Send IP datagrams to **hostname** to test whether it is on the network (or Internet); if the host is alive, it simply echoes the received datagram

Output: Message(s) indicating whether the machine is alive

Commonly used options/features:

-c count	Send and receive count packets
-f	Send 100 packets per second or as many as can be handled by the network; only the superuser can use this option
-s packetsize	Send **packetsize** packets; the default is 56 bytes (plus an 8-byte header)

The following session illustrates the use of the ping command on Linux, with and without options. The output of the command is different for some systems, with the command displaying the echoed messages until you press <Ctrl+C>. We used the -c option in the following session to send and receive three messages. The -c and -s options are used to send and receive three 32-byte messages plus an 8-byte Internet Control Message Protocol (ICMP) header.

```
$ ping stanford.edu
PING stanford.edu (171.67.215.200) 56(84) bytes of data.
64 bytes from web.stanford.edu (171.67.215.200): icmp_seq=1 ttl=231 time=317 ms
64 bytes from web.stanford.edu (171.67.215.200): icmp_seq=2 ttl=231 time=464 ms
64 bytes from web.stanford.edu (171.67.215.200): icmp_seq=3 ttl=231 time=508 ms
64 bytes from web.stanford.edu (171.67.215.200): icmp_seq=4 ttl=231 time=402 ms
64 bytes from web.stanford.edu (171.67.215.200): icmp_seq=5 ttl=231 time=451 ms
64 bytes from web.stanford.edu (171.67.215.200): icmp_seq=6 ttl=231 time=410 ms
^C
--- stanford.edu ping statistics ---
9 packets transmitted, 8 received, 11% packet loss, time 8002ms
rtt min/avg/max/mdev = 317.866/440.723/531.807/62.377 ms
$ ping -c 3 stanford.edu
PING stanford.edu (171.67.215.200) 56(84) bytes of data.
64 bytes from web.stanford.edu (171.67.215.200): icmp_seq=1 ttl=231 time=406 ms
64 bytes from web.stanford.edu (171.67.215.200): icmp_seq=2 ttl=231 time=493 ms
64 bytes from web.stanford.edu (171.67.215.200): icmp_seq=3 ttl=231 time=470 ms

--- stanford.edu ping statistics ---
3 packets transmitted, 3 received, 0% packet loss, time 2000ms
rtt min/avg/max/mdev = 406.913/457.112/493.781/36.735 ms
$ ping -c 3 -s 32 stanford.edu
PING stanford.edu (171.67.215.200) 32(60) bytes of data.
40 bytes from web.stanford.edu (171.67.215.200): icmp_seq=1 ttl=231 time=568 ms
40 bytes from web.stanford.edu (171.67.215.200): icmp_seq=2 ttl=231 time=547 ms
40 bytes from web.stanford.edu (171.67.215.200): icmp_seq=3 ttl=231 time=317 ms

--- stanford.edu ping statistics ---
3 packets transmitted, 3 received, 0% packet loss, time 2170ms
rtt min/avg/max/mdev = 317.216/477.698/568.159/113.787 ms
$ ping -c 3 -s 128 stanford.edu
PING stanford.edu (171.67.215.200) 128(156) bytes of data.
136 bytes from web.stanford.edu (171.67.215.200): icmp_seq=1 ttl=231 time=340 ms
136 bytes from web.stanford.edu (171.67.215.200): icmp_seq=2 ttl=231 time=476 ms
136 bytes from web.stanford.edu (171.67.215.200): icmp_seq=3 ttl=231 time=325 ms

--- stanford.edu ping statistics ---
3 packets transmitted, 3 received, 0% packet loss, time 1999ms
rtt min/avg/max/mdev = 325.864/381.032/476.819/67.994 ms
$
```

You can use the IP address of a host in place of its hostname. For example, you can use ping 171.67.215.200 instead of ping stanford.edu, as in

```
$ ping 171.67.215.200
PING 171.67.215.200 (171.67.215.200) 56(84) bytes of data.
64 bytes from 171.67.215.200: icmp_seq=1 ttl=231 time=617 ms
64 bytes from 171.67.215.200: icmp_seq=2 ttl=231 time=312 ms
64 bytes from 171.67.215.200: icmp_seq=3 ttl=231 time=437 ms
64 bytes from 171.67.215.200: icmp_seq=4 ttl=231 time=398 ms
64 bytes from 171.67.215.200: icmp_seq=5 ttl=231 time=438 ms
^C
--- 171.67.215.200 ping statistics ---
5 packets transmitted, 5 received, 0% packet loss, time 4000ms
rtt min/avg/max/mdev = 312.828/441.068/617.665/99.423 ms
$
```

11.8.3 Displaying Information About Users

You can use the finger command to display information about users on a local or remote host. The information displayed is extracted from a user's ~/**.plan** and ~/**.project** files. By default, the finger command is not installed on our Linux Mint system. If your system does not have finger available on your Linux system, see Appendix A for instructions to install it on your system. The following is a brief description of the command.

Syntax:
  finger [options] [user-list]

Purpose: Display information about the users in **user_list**; without a **user_list**, the command displays a short status report about all the users currently logged on to the specified hosts

Output: User information extracted from the ~/**.project** and ~/**.plan** files

Commonly used options/features:

  -m Match **user_list** to login names only

  -s Display output in a short format

The following session shows the simplest use of the command in which information about a user, **bob**, on the host is displayed.

```
$ finger bob
Login: bob                       Name: Robert Koretsky
Directory: /home/bob             Shell: /bin/bash
On since Sun Oct 15 20:21 (PDT) on pts/1 from 103.255.5.99
    1 hour 39 minutes idle
On since Sun Oct 15 21:53 (PDT) on pts/0 from 103.255.5.99
    3 seconds idle
No mail.
No Plan.
$
```

You can use the finger command with the -s option to display the command's output in a short format and the -m option to match **user-list** to login names only. The finger -m tree command displays the same information as the finger Birch command if the login name of the user is **tree** (uppercase and lowercase letters are considered the same by the networking commands). However, if the login name of the user is **btree** and the login name **birch** does not exist in the system, the finger command displays the message informing you accordingly.

When run without any argument, the finger command returns the status of all the users who are currently logged on to your machine. The amount of information displayed varies somewhat, depending on the Linux system that your host runs. The following command runs on Linux Mint-

```
$ finger
Login      Name              Tty        Idle  Login Time    Office     Office Phone
bob        bob               tty8      12:37  Jul 25 17:53  (:0)
malik      Malik             tty8        86d  Jul 21 18:54  (:0)
root       root             *tty1         2d  Jul 25 08:12
sarwar     Mansoor Sarwar    pts/1      1:42  Oct 15 20:21  (103.255.5.99)
$
```

You can use the `finger` command to display information about a user on a host on the Internet, provided the host offers the finger service and has the finger server (fingerd; the finger daemon) running. The remote finger is disabled on most systems. The `finger Pohm@iastate.edu Ashfaq@iastate.edu` command can be used to display information about the users **Pohm** and **Ashfaq** at the **iastate.edu** if it runs the finger daemon. If a host does not run the finger server, the `finger` command displays the `Connection refused` message for you, as in

```
$ finger crenshaw@up.edu
finger: connect: Connection refused
$
```

If DNS cannot find a mapping for a domain name, the `finger` command returns an appropriate error message. When this happens, you can run the `host` command to find the IP address for the destination host and rerun the `finger` command by using the IP address instead of the domain name.

With `*@hostname` as its argument, the command displays the status of all the users who are currently logged on to **hostname**. Some sites put restrictions on the use of the wild card `*`—for example, requiring the use of at least two characters in all queries. Most sites today do not allow the use of the wild card `*`.

In the following in-chapter exercises, you will use the `ping` and `finger` commands to understand their syntax and various characteristics.

Exercises 11.10

Run the `ping` command to determine whether a remote host that you know about is up.

Exercises 11.11

Give the command for displaying information about yourself on your Linux host.

Exercises 11.12

Give the command for displaying information about a user on your host, with "John" as his first or last name.

Exercises 11.13

How many users with name Jack are found at iastate.edu? What command did you use to obtain the answer? Show the command and its output, and clearly identify your answer in the output.

11.8.4 File Transfer

There are several tools in Linux that you can use to transfer files to and from a remote host on the same network or another network. The classical of these tools is implemented using the well-known FTP protocol, with server side running the ftpd daemon and the client side invoking the service using the `ftp` command. You can use the `ftp` command used to transfer files to and from a remote host on the Internet. There is also the sftpd daemon that allows the use of client-side `sftp` command for secure transfer of files using the SSH protocol. We discuss this command in Section 11.8.9.

It is also possible, as we show in Appendix A, Section 2.3.4, and particularly in Example 19.1, to download, install, and start (using systemd) a very secure ftp daemon (vsftpd) on your Linux system. Example 19.1 shows the download, installation, and service starting using systemd on a Linux Mint system. vsftpd has many security advantages over the default installation of ftp on your system. The major security feature of vsftpd is that, unlike ftp that sends passwords in plain text, vsftpd encrypts passwords.

In this section, we show the use of the `ftp` command. The following is a brief description of the command.

Syntax:
  `ftp [options] [host]`

Purpose: To transfer files from or to a remote **host**

Commonly used options/features:

  `-d` Enable debugging

  `-i` Disable prompting during transfer of multiple files

  `-v` Show all remote responses

As mentioned earlier in the chapter, the FTP is a client–server protocol based on TCP. When you run the `ftp` command, an ftp client process starts running on your host and attempts to establish a connection with the ftp server process running on the remote host. If the ftp server process is not running on the remote host before the client initiates a connection request, the connection is not made and an `Unknown host` error message is displayed by the `ftp` command. A site running an ftp server process is called an *ftp site*. When an ftp connection has been established with the remote ftp site, you can run several `ftp` commands for effective use of this utility. However, you must have appropriate access permission to transfer files to the remote site. Table 11.4 presents some useful `ftp` commands.

Most ftp sites require that you have a valid login name and password on that site to transfer files to or from that site. A number of sites allow you to establish ftp sessions with them using **anonymous** as the login name and **guest** or your full e-mail address as the password. Such sites are said to allow anonymous ftp. In other words, *anonymous ftp* is a method of downloading public files using the ftp protocol.

TABLE 11.4

A Summary of Useful ftp Commands

Command	Meaning
`! [cmd]`	Runs `cmd` on the local machine; without the `cmd` argument, invokes an interactive shell
`Help [cmd]`	Displays a summary of `cmd`; without the `cmd` argument, displays a summary of all ftp commands
`ascii`	Puts the ftp channel into ASCII mode; used for transferring ASCII-type files such as text files
`binary`	Puts the ftp channel into binary mode; used for transferring non-ASCII files such as files containing executable codes or pictures
`cd`	Changes directory; similar to the Linux `cd` command
`close`	Closes the ftp connection with the remote host, but stays inside ftp
`dir remotedir localfile`	Saves the listing of **remotedir** into **localfile** on the local host; useful for long directory listings, as pipes cannot be used with the ftp commands
`get remotefile [localfile]`	Transfers **remotefile** to **localfile** in the present working directory on the local machine; if **localfile** is not specified, **remotefile** is used as the name of the local file
`ls [dname]`	Shows contents of the designated directory **dname**; current directory if none specified
`mget remotefiles`	Transfers multiple files from the remote host to the local host
`mput localfiles`	Transfers multiple files from the local host to the remote host
`open [hostname]`	Attempts to open a connection with the remote host; prompts if hostname not specified as parameter
`put localfile [remotefile]`	Transfers **localfile** to **remotefile** on the remote host; if **remotefile** is not specified, use **localfile** as name of remote file
`quit`	Terminates the ftp session
`user [login_name]`	If unable to log on, log on as a user on the remote host by specifying the **user_name** as the command argument; prompt appears if **user_name** is not specified

An anonymous log in is usually download only. If you have an account on the computer of the ftp site (i.e., it's not an anonymous login), then you are almost certainly allowed to download and upload files.

The following session illustrates the use of the ftp utility to do an anonymous ftp with the site ftp. FreeBSD.org and transfer the compressed kernel for release 10.1. In the process, we demonstrate the use of the ftp commands cd, get, ls, and pwd. We also demonstrate the execution of the ls command on the local host. Finally, we terminate the ftp session with the quit command. This site does not require any password for anonymous ftp. Thus, you hit <Enter> when prompted for a password. Note that at the time you execute the commands in the following sessions, more contemporary releases of the software will be available.

```
$ ftp ftp.FreeBSD.org
Trying 149.20.53.23:21 ...
Connected to ftp.geo.FreeBSD.org.
220 This is ftp0.isc.freebsd.org - hosted at ISC.org
Name (ftp.FreeBSD.org:sarwar): anonymous
331 Please specify the password.
Password: <Enter>
230-
230-This is ftp0.isc.FreeBSD.org, graciously hosted by
230-Internet Systems Consortium - ISC.org.
230-
230-FreeBSD files can be found in the /pub/FreeBSD directory.
230-
230 Login successful.
Remote system type is UNIX.
Using binary mode to transfer files.
ftp> cd pub/FreeBSD
250-ISO images of FreeBSD releases may be found in the releases/ISO-IMAGES
250-directory.  For independent files and tarballs, see individual
250-releases/${machine}/${machine_arch} directories.  For example,
250-releases/amd64/amd64 and releases/powerpc/powerpc64.
250 Directory successfully changed.
ftp> pwd
257 "/pub/FreeBSD" is the current directory
ftp> ls
200 PORT command successful. Consider using PASV.
150 Here comes the directory listing.
-rw-r--r--    1 ftp        ftp              4259 May 07  2015 README.TXT
-rw-r--r--    1 ftp        ftp                35 Apr 02 02:30 TIMESTAMP
drwxr-xr-x    9 ftp        ftp                10 Apr 02 02:30 development
-rw-r--r--    1 ftp        ftp              3665 Apr 01 10:00 dir.sizes
drwxr-xr-x   28 ftp        ftp                52 Nov 12 18:20 doc
drwxr-xr-x    5 ftp        ftp                 5 Nov 12 17:44 ports
drwxr-xr-x   11 ftp        ftp                13 Apr 02 02:30 releases
drwxr-xr-x   11 ftp        ftp                13 Nov 13 06:57 snapshots
226 Directory send OK.
ftp> cd releases/i386/i386/11.1-RELEASE
250 Directory successfully changed.
ftp> ls
200 PORT command successful. Consider using PASV.
150 Here comes the directory listing.
-rw-r--r--    1 ftp        ftp               894 Jul 21  2017 MANIFEST
-rw-r--r--    1 ftp        ftp          47846592 Jul 21  2017 base-dbg.txz
-rw-r--r--    1 ftp        ftp          84673592 Jul 21  2017 base.txz
-rw-r--r--    1 ftp        ftp           1428680 Jul 21  2017 doc.txz
-rw-r--r--    1 ftp        ftp          54097532 Jul 21  2017 kernel-dbg.txz
-rw-r--r--    1 ftp        ftp          32255748 Jul 21  2017 kernel.txz
-rw-r--r--    1 ftp        ftp          36753048 Jul 21  2017 ports.txz
-rw-r--r--    1 ftp        ftp         147661560 Jul 21  2017 src.txz
-rw-r--r--    1 ftp        ftp           4236044 Jul 21  2017 tests.txz
226 Directory send OK.
ftp> get kernel.txz
```

```
local: kernel.txz remote: kernel.txz
200 PORT command successful. Consider using PASV.
150 Opening BINARY mode data connection for kernel.txz (32255748 bytes).
226 Transfer complete.
32255748 bytes received in 295.89 secs (106.4592 kB/s)
ftp> !ls
kernel.txz
ftp> quit
421 Timeout.
$
```

Once you have established an ftp connection, most sites put you in binary mode so that you can transfer non-ASCII files, such as files containing audio and video clips. You can explicitly put the ftp session into binary mode by using the binary command, which ensures proper file transfer. You can revert to ASCII mode by using the ascii command.

In the following in-chapter exercise, you will use the ftp command to transfer a file from a remote host on the Internet.

> **Exercise 11.14**
>
> Establish an anonymous ftp session with the host **ftp.FreeBSD.org** and transfer all the files related to the latest release of the software, using the techniques of the earlier command line session. Ftp-transfer them into your system by using the mget command.

11.8.5 SSH and Related Commands

The BSD-originated r commands (rsh, rcp, rlogin, ruptime, etc. and not available in Linux) operate on *plaintext* (also called *cleartext*) and are insecure, as is telnet. SSH is a modern substitute for them that eliminates the need for the **.rhosts** and **hosts.equiv** files and can authenticate users by cryptographic key. It also increases the security of other TCP/IP-based applications, such as X Windows *forwarding*, which transparently "tunnels" them via SSH-encrypted connections. The program that handles the secure connection protocol is known as the SSH daemon, sshd.

The ssh protocol uses strong cryptography for transmitting data, including commands, command outputs, passwords, and files. Whenever your data is transmitted to your network, ssh automatically encrypts it and automatically decrypts it when the data reaches its intended recipient. The result is *transparent encryption*—that is, the sender and receiver are unaware of the encryption/decryption process.

For these reasons, ssh has become a de facto standard for secure terminal connections within a LAN or the Internet. The encrypted sessions used by ssh, scp, and sftp, for example, prevent anyone from making sense of the ongoing communication while *packet sniffing* it. Also, authentication methods used in ssh prevent any kind of *spoofing*, such as IP address spoofing.

SSH Version 2 is the default protocol. It has three components:

1. SSH Transfer Protocol: Handles server identification, encryption algorithm and key encryption, and manages the channel between client and server
2. SSH Authentication Protocol: Verifies the validity of the login role of user on the client machine
3. SSH Channel Protocol: Handles the encrypted channel and logical connection status for a remote shell session, port forwarding, or X11 forwarding

In Chapter 2, Section 2.3.3, we give details of how to log onto a Linux system using the ssh command, the client side of the SSH protocol. In that section, we also describe some of the possible ways of interacting with an sshd (SSH Daemon) service on a remote system. The following subsections assume that the ssh service is installed and enabled on your host (i.e., the client machine), and the sshd daemon is running on the remote server host machine.

11.8.5.1 Remote Login with ssh

Most Linux systems support two commands that allow you to log on to a remote host. They are telnet and ssh, both based on the Internet services for remote login. When you use telnet, commands and data travel in plaintext on the transmission channel and may be sniffed as it travels. For this reason, `telnet` is hardly ever used for remote login these days, and we do not discuss it in the book. In case of `ssh`, commands and data travel through encrypted channels.

The `ssh` command allows you to basically perform the same tasks that you can perform with the `telnet`, `rlogin`, and `rsh` commands, but in a more secure manner. You are allowed to log in to the remote host or execute a command on it if the ssh server (sshd) is running on the remote machine and your machine is listed in /etc/hosts.equiv or /etc/ssh/shosts.equiv on the server machine, and the user name is the same on the two machines. Alternatively, a user is considered for login if the files ~/.rhosts or ~/.shosts exist in the user's home directory on the server machine and include a line containing the name of the client machine and the name of the user on that machine. Additional security features are used by ssh, including public-key cryptography, to authenticate a user and its host before allowing the user to log in to the remote machine. Particularly, the server must be able to verify the client's host key in ~/.ssh/known_hosts before permitting login. This file contains server's domain name and/or IP address and its public key. You can run the `more ~/.ssh/known_hosts` command on your client machine to see what this file contains. It is created when you run the ssh user@domain_name or ssh user@IP_address for the first time.

We describe only some of the rather basic features of the `ssh` command that are needed for remote login, remote execution of commands, and file transfer between your client and the server machines. Here is a brief description of the `ssh` command:

Syntax:
`ssh [-option(s)] [-option argument(s)] [user@]hostname [command]`

Purpose: ssh (SSH client) is a program for logging into a remote machine and for executing commands on a remote machine. It is intended to replace rlogin and rsh, and provide secure encrypted communications between two untrusted hosts over an insecure network

Output: A remote connection over a secure channel to a host on a network or the Internet

Commonly used options/features:

`-D [bind_address:]port`	Specify a local "dynamic" application-level port forwarding. This works by allocating a socket to listen to the port on the local side, optionally bound to the specified **bind_address**.
`-L [bind_address:]port:host:hostport`	Specify that the given port on the local (client) host is to be forwarded to the given host and port on the remote side.
`-C`	Request on all data. The compression algorithm is the same as used by the gzip command.
`-l login_name`	Specify the user to log in as on the remote machine. This also may be specified on a per-host basis in the configuration file.
`-o option`	Specify options for which there is no explicit option/flag, using the format given in the configuration file.
`-p port`	Specify the ports to connect to if the sshd is not running on the well-known port (22).

In the following session, we start an ssh session on a remote host running the SSH server named **192.168.0.12**, using the default username (which in this case is the same username you have on the local client machine), execute shell commands, and log out of the remote host. The remote machine happens to be an OpenIndiana UNIX machine that has never had an ssh connection made to it from this client, i.e., your host.

```
$ ssh 192.168.0.12
The authenticity of host '192.168.0.12 (192.168.0.12)' can't be established.
RSA key fingerprint is 37:a2:db:ce:97:17:ce:37:26:bf:58:9d:a2:ed:1c:0f.
No matching host key fingerprint found in DNS.
Are you sure you want to continue connecting (yes/no)? yes
Warning: Permanently added '192.168.0.12' (RSA) to the list of known hosts.
Password: xxxxxx
Last login: Tue Sep 23 08:51:38 2014
OpenIndiana (powered by illumos)     SunOS 5.11    oi_151a9     November 2013
$
...
[ Execute command line commands ]
...
$ <Ctrl+D>
Connection to 192.168.0.12 closed.
$
```

Notice that since this session is the first ssh connection from the local client to the remote server host, an exchange of authentication keys is necessary. On every subsequent connection, the key exchange is not necessary if the keys have not been changed.

Using the following command, you log in to a host **202.147.169.195 (an ssh server running Linux Mint18 4.4.0)** with username **sarwar**, use the uname –a command to display the machine information of the server-side machine (which is running Linux Mint18 4.4.0), log out, and use the uname –a command on the client machine (which is running Darwin 10.8.0) to display information of the client-side machine. Note that there is no exchange of authentication keys in this particular session, because this is not the first ssh connection between the client and server machines.

```
$ ssh -l sarwar 202.147.169.195
sarwar@202.147.169.195's password: xxxxxxxx
Welcome to Linux Mint 18 Sarah (GNU/Linux 4.4.0-21-generic x86_64)

 * Documentation:  https://www.linuxmint.com
Last login: Fri Oct 27 13:56:47 2017 from 119.160.96.60
$ uname -a
Linux Mint18 4.4.0-21-generic #37-Ubuntu SMP Mon Apr 18 18:33:37 UTC 2016 x86_64
x86_64 x86_64 GNU/Linux
$
...
[ Your use of the remote Linux machine using the command line interface ]
...
$ <Ctrl+D>
Connection to 202.147.169.195 closed.
$ uname -a
Darwin Sarwars-MacBook-Pro.local 10.8.0 Darwin Kernel Version 10.8.0: Tue Jun  7
16:32:41 PDT 2011; root:xnu-1504.15.3~1/RELEASE_X86_64 x86_64
$
```

Following is a somewhat atypical, but nonetheless important and very instructive ssh log-in session, ended by a graceful logout and termination of the ssh channel. It is done between a Linux Mint client (192.168.0.13) and a Linux Mint host (192.168.0.8), where sshd has already been installed on the host system. This example illustrates the use of *known hosts* concept of ssh. Known hosts are established so that the client and server authenticate each other to make a secure connection between the two. When you connect to any host for the first time, a public counterpart of the host key is stored on the client. The known host concept prevents what is known as *man-in-the-middle attacks*.

After the first failed attempt to establish an ssh connection, an error message indicates that the authentication key has changed on the host. This could have happened for any number of reasons, including something you yourself did, rather than some attacker. Therefore, in our scenario that follows, a removal of the offending key in the file /home/bob/.ssh/known_hosts on the client machine is done by deleting that file. Then a new key is generated when you make another attempt to ssh into the host from the client, an exchange can take place, and a normal client-host login can proceed. What the user types-in is shown in bold text, and it is assumed that you are executing the commands from your home directory, in our case, /home/bob:

```
$ ssh 192.168.0.8
@@@@@@@@@@@@@@@@@@@@@@@@@@@@@@@@@@@@@@@@@@@
@@@@@@@@@@@@@@@@@@@@@@@@
@ WARNING: REMOTE HOST IDENTIFICATION HAS CHANGED! @
@@@@@@@@@@@@@@@@@@@@@@@@@@@@@@@@@@@@@@@@@@@
@@@@@@@@@@@@@@@@@@@@@@@@
IT IS POSSIBLE THAT SOMEONE IS DOING SOMETHING NASTY!
Someone could be eavesdropping on you right now (man-in-the-middle attack)!
It is also possible that a host key has just been changed.
The fingerprint for the ECDSA key sent by the remote host is
SHA256:uZpqi4U6uBN5SOBVFRbqbl5HspmV3eZAw/nUvPBTS5I.
Please contact your system administrator.
Add correct host key in /home/bob/.ssh/known_hosts to get rid of this message.
Offending ECDSA key in /home/bob/.ssh/known_hosts:2
ECDSA host key for 192.168.0.8 has changed and you have requested strict checking.
Host key verification failed.
$ cd .ssh
$ rm known_hosts
$ cd
$ ssh 192.168.0.8
The authenticity of host '192.168.0.8 (192.168.0.8)' can't be established.
ECDSA key fingerprint is SHA256:uZpqi4U6uBN5SOBVFRbqbl5HspmV3eZAw/nUvPBTS5I.
Are you sure you want to continue connecting (yes/no)? yes
Warning: Permanently added '192.168.0.8' (ECDSA) to the list of known hosts.
bob@192.168.0.8's password:
Welcome to Linux Mint 18.2 Sonya (GNU/Linux 4.4.0-45-generic x86_64)

 * Documentation:  https://www.linuxmint.com
Last login: Thu Jul 27 14:29:00 2017 from 192.168.0.25
$
...
[ Execute command line Linux commands ]
...
$ <Ctrl+D>
$
```

11.8.5.2 Remote Command Execution with ssh

You can use the ssh command to execute a shell command on a remote host on the Internet. Remote login is a relatively time-consuming process, but the ssh command gives you a faster way to execute commands on remote machines if the purpose of your remote login is to execute only a few commands.

The following session shows how you can run the ps command on behalf of the user sarwar on the remote host with IP address 202.147.169.195. The remote machine requires you to enter the password for the user sarwar before running the ps command on the remote machine and displaying the output of the command on the local machine. The ssh sarwar@202.147.169.195 ps command performs the same task. Figure 11.8 shows the semantics of the execution of the command.

```
$ ssh -1 sarwar 202.147.169.195 ps
sarwar@202.147.169.195's password: xxxxxxxx
  PID TTY          TIME CMD
32585 ?        00:00:00 systemd
32587 ?        00:00:00 (sd-pam)
32601 ?        00:00:00 sshd
32602 ?        00:00:00 ps
$
```

The following session shows that the third command line executes the sort students command on the remote host with IP address 202.147.169.195, taking input from the **students** file on remote machine,

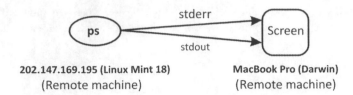

FIGURE 11.8   The semantics of the ssh -l sarwar 202.147.169.195 ps command.

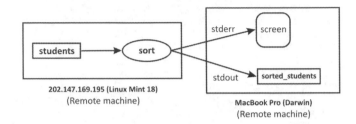

FIGURE 11.9   The semantics of the ssh sarwar@202.147.169.195 sort students > sorted_students command.

and sends the results back to the **sorted_students** file on the local machine. If the **students** file does not exist, the error message is also sent back to the local machine. The semantics of the command are illustrated in Figure 11.9. The session also shows the outputs of the ls -l students and cat students commands when executed on remote and local machines.

```
$ ssh sarwar@202.147.169.195 ls -l students
sarwar@202.147.169.195's password: xxxxxxxx
-rw-r--r-- 1 sarwar faculty 91 Oct 27 14:01 students
$ ssh sarwar@202.147.169.195 cat students
sarwar@202.147.169.195's password: xxxxxxxx
John   Doe    ECE    3.54
Pam    Meyer  CS     3.61
Jim    Davis  CS     2.71
Jason  Kim    ECE    3.97
Amy    Nash   ECE    2.38
$ ssh sarwar@202.147.169.195 sort students > sorted_students
sarwar@202.147.169.195's password: xxxxxxxx
$ ls -l sorted_students
-rw-r--r--  1 Mansoor  staff  91 Oct 27 14:06 sorted_students
$ cat sorted_students
Amy    Nash   ECE    2.38
Jason  Kim    ECE    3.97
Jim    Davis  CS     2.71
John   Doe    ECE    3.54
Pam    Meyer  CS     3.61
$
```

In the following command line, however, the sort command runs on the remote machine but takes input from the **students** file on the local machine. As with the previous command, the output is sent to the **sorted_students** file on the local machine.

```
$ ssh sarwar@202.147.169.195 sort < students > sorted_students
sarwar@202.147.169.195's password: xxxxxxxx
$
```

If you want the sort command to take input from a local file, called **students**, run the command on the remote machine, and store the sorted results in the **sorted_students** file on the remote machine, you must quote the remote command with output redirection, as in the following session.

```
$ cat students | ssh sarwar@202.147.169.195 'sort > sorted_students'
sarwar@202.147.169.195's password: xxxxxxxx
$
```

You can combine the I/O redirection operators with the pipe operator to create powerful command lines that take input from local and remote files, execute commands on local and remote machines, and send the final results to a file on the local or remote machine. The following command executes the ps -A | grep d$ command on the remote server host **192.168.0.6** (a Linux Mint machine) and displays (on the local client machine) the status of all the daemons on that remote host. As shown, you will be prompted for a password because of the default configuration of ssh.

```
$ ssh sarwar@202.147.169.195 ps -A | grep d$
sarwar@202.147.169.195's password: xxxxxxxx
    1 ?        00:00:06 systemd
    2 ?        00:00:00 kthreadd
    7 ?        00:22:06 rcu_sched
   49 ?        00:00:00 khungtaskd
...
[ Output truncated ]
...
$
```

In the following session, the ps -e | grep d$ command runs on the remote machine and its output is piped to the grep sshd$ command that runs on the local machine. The output of the command line is the status of the ssh daemons running on the remote host.

```
$ ssh sarwar@202.147.169.195 'ps -e | grep d$' | grep sshd$
sarwar@202.147.169.195's password: xxxxxxxx
 1128 ?        00:00:03 sshd
 5953 ?        00:00:00 sshd
 5955 ?        00:00:00 sshd
 5956 ?        00:00:00 sshd
 5973 ?        00:00:00 sshd
$
```

11.8.6 Remote Copy with scp

As discussed earlier, you can use the ftp command to transfer files to and from a remote host on another network, but doing so requires that you log on to the remote host. The scp (secure remote copy) command allows you to copy files to and from a remote machine on the same LAN, without logging on to the remote host. This command is not needed in a local area environment if you are using a network-based file system such as the Network File System (NFS). In this case, the storage of your files is completely transparent to you, and you can access them from any host on your network, without specifying the name of the host that contains them.

You can use the scp command to copy files between two hosts on a network using ssh. Both hosts may be remote. Copying takes place under encrypted sessions after proper authentication of the local host and user. Pathnames for files may contain a user and host specification to indicate that the file is to be copied to/from that user on the given host. Host names are followed by the colon (:) character. A user on a host is specified as user@host, where host can be a symbolic name or an IP address, as shown in the following examples. Here is a brief description of the scp command.

Syntax:
   scp [-option(s)] [[user@]host1:]file1 ... [[user@]host2:]file2

Purpose: To securely copy file(s) between hosts on a network using ssh

Commonly used options/features:

-3	Copying between two remote hosts is carried out through the local host.
-4	Use IPv4 addresses only.

-6	Use IPv6 addresses only.
-C	Enable compression by passing this flag to ssh.
-c cipher	This option is passed to ssh for selecting the cipher to encrypt data transfer.
-o ssh_option	Pass options to ssh in the format used in ssh_config file.
-r	Copy directories recursively.
-p	Preserve modification times, access times, and modes from the source file.

The following command copies the **students** file in your current directory into the user sarwar's **~/linux2e/ch11/** directory on 202.147.169.195. The outputs of the ls –l commands on the local and remote hosts confirm that the file has been copied from the local host to remote host.

```
$ scp students sarwar@202.147.169.195:~/linux2e/ch11/
sarwar@202.147.169.195's password: xxxxxxxx
students                               100%  153      0.2KB/s   00:00
$ ls -l students
-rw-r--r--  1 Mansoor  staff  153 Oct 28 00:52 students
$ ssh sarwar@202.147.169.195 'ls -l ~/linux2e/ch11/students'
sarwar@202.147.169.195's password: xxxxxxxx
-rw-r--r-- 1 sarwar faculty 153 Oct 28 00:54 /home/sarwar/linux2e/ch11/students
$
```

You can use a remote machine's symbolic name instead of its IP address.

```
$ scp students sarwar@linux18:~/linux2e/ch11/
sarwar@linux18's password: xxxxxx
...
$
```

The following command recursively copies sarwar's entire home directory on the 202.147.169.195 host to your current directory on the local machine. Don't forget to type the period (.) after the space character found after ~. Note that we terminate the command with <Ctrl+C>, as can be seen on the last line of the command output.

```
$ scp -r sarwar@202.147.169.195:~ .
Password: xxxxxx
temp                               100%    0      0.0KB/s   00:00
.profile                           100%   568     0.6KB/s   00:00
local.login                        100%   170     0.2KB/s   00:00
welcomesir                         100%   139     0.1KB/s   00:00
known_hosts                        100%   397     0.4KB/s   00:00
scp: /export/home/sarwar/.bash_history: Permission denied
local.cshrc                        100%   166     0.2KB/s   00:00
ps.man                             100%   28KB   27.9KB/s   00:00
bigdata                              3%   327MB  11.2MB/s   12:35 ETA
<Ctrl+C> Killed by signal 2.
$
```

11.8.7 Secure File Transfer with sftp

The sftp program is the secure version of ftp, which performs operations using ssh. It works just like the ftp command, except that stronger authentication takes place before file transfer starts, and command and data transfer takes place in encrypted sessions. The vsftpd and Filezilla applications that we illustrate in Chapter 17, and Appendix A, also achieve secure ftp connections. Here is a brief description of the sftp command.

Syntax:
   sftp [-option(s)] host
   sftp [user@]host[:file ...]
   sftp [user@]host[:dir[/]]
   sftp sftp -b batchfile [user@]host

Purpose: To log on to the specified host and enter the interactive command mode like ftp. The second syntax allows retrieval of files automatically under noninteractive mode or after interaction. The third syntax allows sftp to start in a remote directory. The last syntax allows automated file transfer sessions with commands taken from the batchfile instead of standard input. If batchfile is "-," then the commands are taken from the standard input

Commonly used options/features:

-C	Enable compression by passing this flag to ssh
-o ssh_option	Pass options to ssh in the format used in ssh_config file

The following sessions show sample commands executed in sftp. As the output of the ? command shows, the commands for the local host start with the letter "l" (except for ln) such as lls, lmkdir, lpw, and lumask. Alternatively, any Linux shell command preceded with the letter '!' executes on the local host such as !pwd.

    Connecting and getting help

```
$ sftp sarwar@202.147.169.195
Connecting to 202.147.169.195...
sarwar@202.147.169.195's password: xxxxxxxx
sftp> ?
Available commands:
bye                                Quit sftp
cd path                            Change remote directory to 'path'
chgrp grp path                     Change group of file 'path' to 'grp'
chmod mode path                    Change permissions of file 'path' to 'mode'
chown own path                     Change owner of file 'path' to 'own'
df [-hi] [path]                    Display statistics for current directory or
                                   filesystem containing 'path'

exit                               Quit sftp
get [-Ppr] remote [local]          Download file
reget remote [local]               Resume download file
help                               Display this help text
lcd path                           Change local directory to 'path'
lls [ls-options [path]]            Display local directory listing
lmkdir path                        Create local directory
ln [-s] oldpath newpath            Link remote file (-s for symlink)
lpwd                               Print local working directory
ls [-1afhlnrSt] [path]             Display remote directory listing
lumask umask                       Set local umask to 'umask'
mkdir path                         Create remote directory
progress                           Toggle display of progress meter
put [-Ppr] local [remote]          Upload file
pwd                                Display remote working directory
quit                               Quit sftp
rename oldpath newpath             Rename remote file
rm path                            Delete remote file
rmdir path                         Remove remote directory
symlink oldpath newpath            Symlink remote file
version                            Show SFTP version
!command                           Execute 'command' in local shell
!                                  Escape to local shell
?                                  Synonym for help
sftp>
```

Listing the local present working directory

```
sftp> lpwd
Local working directory: /Users/Mansoor
sftp>
```

Listing the remote present working directory

```
sftp> pwd
Remote working directory: /home/sarwar
sftp>
```

Listing the files in the remote present working directory

```
sftp> ls
Desktop              Documents         Downloads         Music
Pictures             Public            Servers           Templates
Videos               bin               courses           greeting
linux2e              personal          sorted_students   students
sftp>
```

List the files in the local present working directory

```
sftp> lls
Desktop              Movies                      Send Registration
Documents            Music                       Sites
Downloads            Pictures                    Students
Library              Public                      Terminal Saved Output.txt
sftp>
```

Uploading a single file named **alien** to the remote host

```
sftp> put students
Uploading students to /home/sarwar/students
students                                    100%   191      0.2KB/s    00:00
sftp>
```

Uploading multiple files that all end in **.txt** to the remote host

```
sftp> mput *.txt
Uploading emacs.txt to /home/bob/emacs.txt
emacs.txt                                   100%   20KB   20.5KB/s    00:00
Uploading fvwm.txt to /home/bob/fvwm.txt
fvwm.txt                                    100%  458KB  458.2KB/s    00:01
Uploading vi.txt to /home/bob/vi.txt
vi.txt                                      100%   50KB   50.3KB/s    00:00
sftp>
```

Getting (or downloading) single or multiple files from the remote to the local system

```
sftp> get testfile
Fetching /home/bob/testfile to testfile
/home/bob/testfile                          100%   22    0.0KB/s    00.00
sftp>
```

Switching from one directory to another directory in local and remote machines

> On the remote machine:
> ```
> sftp> cd test
> sftp>
> ```

On the local machine:
```
sftp> lcd Documents
sftp>
```

Creating new directories on local and remote machines

On the remote machine:
```
sftp> mkdir test
sftp>
```

On the local machine:
```
sftp> lmkdir Documents
sftp>
```

Removing a directory or file on the remote machine

```
sftp> rm Report.xls
sftp> rmdir Documents
sftp>
```

Note that to remove/delete any directory from the remote location, the directory must be empty.

Exiting from sftp:
```
sftp> quit
$
```

The following in-chapter exercise gives you practice for using the ssh, scp, and sftp commands in your environment.

Exercise 11.15

Use each of the ssh, scp, and sftp commands to access a remote host on your LAN, execute a command remotely, and copy a file from your machine to another machine on your network. What command did you use? What are their semantics? Write a short description of how you did this, and what the output from the local and remote systems was.

11.8.8 Packaged ssh Applications

In addition to the command line ssh operations shown earlier, which you can perform in a terminal or console window, there are three very powerful and expedient graphical front-end application programs readily available on Linux, Windows, iOS/OS X machines that accomplish the same things that ssh, scp, and sftp do.

The first application is called FileZilla, which allows you to do file and directory transfers over an SSH connection between different computer systems. We will consider some of its capabilities in Chapter 17 on system administration.

The second application, PuTTY, is the most popular SSH client. As shown in Chapter 2, it allows you to login and execute commands on a Linux machine from a Windows machine through an SSH connection.

The third application is termius, a free app available for iOS devices such as iPods and iPads at the App Store. It allows you to establish an ssh connection from your iPod or iPad to a remote server running Linux, and open a full-featured CUI terminal window into that server. Mac and Linux users can also use the terminal window for using the ssh command to establish a connection with the SSH server.

11.8.9 Interactive Chat

You can use the talk command for an interactive chat with a user on your host or on a remote host over a network. The following is a brief description of the command.

Syntax:
   `talk user [tty]`
Purpose: To initiate interactive communication with **user** who is logged in on a **tty** terminal

---

The **user** parameter is the login name of the person if he or she is on your host. If the person you want to talk to is on another host, use **login_name@host** for **user**. The **tty** parameter is needed if the person is logged on to the same host more than once.

When you use the `talk` command to initiate a communication request, the other user is interrupted with a message on his or her screen, informing that person of the request. The other user needs to execute the `talk` command to respond to you. That establishes a communication channel, and both users' display screens are divided into two halves. The upper half contains the text that you type and the lower half contains the other user's responses. Both you and the other user can type simultaneously. The `talk` command simply copies the characters that you type at your keyboard on the screen of the other user. The chat session can be terminated when either of you presses <Ctrl+C>. If you are using the vim editor and your screen is corrupted during the communication, you can use <Ctrl+L> to redraw the screen.

Suppose that user **sarwar** wants to talk to another user, **bob**, and that both are logged on to the same host. The following command from **sarwar** initiates a talk request to **bob**.

   `$ talk bob`

As soon as **sarwar** hits <Enter>, the following message is displayed at the top of **bob**'s screen.

```
[Waiting for your party to respond]
Message from Talk_Daemon@upibm7.egr.up.edu at 13:36 ...
talk: connection requested by sarwar@upibm7.egr.up.edu.
talk: respond with: talk sarwar@upibm7.egr.up.edu
```

When **bob** runs the `talk sarwar` command, both **bob**'s and **sarwar**'s screens are divided in half, with the upper halves containing the message `[Connection established]` and the cursor moved to the top of both screens. Both **bob** and **sarwar** are now ready to talk. If **bob** wants to ignore **sarwar**'s request while using a shell, he can simply press <Enter>.

If **bob** is logged in on another host—say, **Linux10**—**sarwar** needs to run the following command to initiate the talk request.

`$ talk bob@Linux10`

If **sarwar** is logged in once on **Linux20** and **bob** wants to communicate with **sarwar**, his response to the preceding request should be `talk sarwar@Linux20`. If **bob** is logged in on **Linux10** multiple times, the following command from **sarwar** initiates a talk request on terminal **ttyp2** (one of the terminals **bob** is logged on to).

`$ talk bob@Linux10 ttyp2`

If you want to block all talk requests because users keep bothering you with too many requests, execute

```
$ mesg n
$
```

This command works only for your current session. If you want to block all talk and write requests whenever you log on, put this command in your **~/.profile** file. Doing so simply takes away the write permission on your terminal file in the **/dev** directory for your group and others. Thus, you can accomplish the same by using the `chmod 600` command on your terminal file, as shown in the following session. Notice the change in the permissions of the **/dev/pts/2** file before the after the execution of the `mesg n`, `mesg y`, `chmod 600 /dev/pts/2`, and `chmod 620 /dev/pts/2` commands.

```
$ who
david               pts/0           Oct  9 08:17 (:0)
sarwar              pts/1           Oct 25 13:23 (182.185.251.51)
$ ls -l /dev/pts/2
crw--w----  1 sarwar   tty  0x98 Oct 25 15:46 /dev/pts/2
$ mesg n
% ls -l /dev/pts/2
crw-------  1 sarwar   tty  0x98 Oct 25 15:47 /dev/pts/2
$ mesg y
$ ls -l /dev/pts/2
crw--w----  1 sarwar   tty  0x84 Oct 25 13:45 /dev/pts/2
$ chmod 600 /dev/pts/2
$ ls -l /dev/pts/2
crw-------  1 sarwar   tty  0x84 Oct 25 13:45 /dev/pts/2
$ chmod 620 /dev/pts/2
$ ls -l /dev/pts/2
crw--w----  1 sarwar   tty  0x98 Oct 25 15:51 /dev/pts/2
$
```

Without any argument, the `mesg` command displays the current status.

In the following in-chapter exercises, you will use the `talk` command to establish a chat session with a friend on your network and appreciate the various characteristics of the command.

Exercise 11.16

Establish a chat session using the `talk` command with a friend who is currently logged on.

Exercise 11.17

Run the last shell session on your system to identify the pathname of your terminal file and verify the effect of the `mesg n` and `mesg y` commands on the permissions of your terminal file.

11.8.10 Tracing the Route from One Site to Another

The `traceroute` command uses IP's *time to live* (TTL) field to display the route (the names of the routers in the path) that your e-mail messages, `ssh` commands, and downloaded files from an ftp site can take from your host to the remote host and vice versa. It also gives you a feel for the speed of the route. Because this command poses some security threats, most system administrators disable its execution. The security threat stems from the fact that, by displaying a route to a host on the Internet, someone can figure out the internal structure of the network to which the host is connected and the IP addresses of some machines on the network.

The `traceroute` command is not installed by default on a Linux Mint system, so you have to ask your system administrator to install it for you. If you are using Linux Mint on your personal system and have root access, you can install it using the command `sudo apt install traceroute`.

The following is a simple execution of the command to show its output and demonstrate the inner workings of the Internet a bit more. The following command shows the route from our host to **mit.edu** on a Linux Mint machine. Since the route to a host is location dependent, the output of the command on your system may be different.

```
$ traceroute mit.edu
traceroute to mit.edu (104.74.143.40), 64 hops max
  1    202.147.169.193  0.693ms  0.442ms  0.372ms
  2    10.1.24.149  1.163ms  1.003ms  1.192ms
  3    119.30.106.21  1.224ms  1.059ms  1.055ms
  4    119.30.106.13  1.646ms  1.607ms  1.358ms
  5    119.30.106.5  1.689ms  1.404ms  1.787ms
  6    119.30.106.26  196.453ms  179.987ms  181.562ms
  7    117.20.31.5  17.532ms  17.353ms  17.377ms
  8    110.93.252.164  18.098ms  17.946ms  *
  9    213.144.176.232  139.907ms  136.842ms  137.637ms
```

```
10    195.22.210.195  151.871ms   153.840ms   158.610ms
11    129.250.8.205   216.851ms   215.512ms   212.654ms
12    129.250.6.13    218.558ms   218.324ms   219.499ms
13    129.250.2.26    218.508ms   221.273ms   220.868ms
14    129.250.3.13    236.175ms   239.262ms   239.244ms
15    129.250.3.218   232.616ms   *  232.316ms
16    129.250.4.43    229.882ms   229.536ms   234.037ms
17    212.119.27.186  341.867ms   434.138ms   446.160ms
18    104.74.143.40   224.929ms   224.949ms   225.059ms
$
```

The default probe datagram (packet) length is 64 bytes, but you can specify a larger (or smaller) length using the –m option, as in traceroute cs.berkeley.edu –m40. As shown in the first line of the output of our sample run, traceroute uses a 64-byte packet size. A line in the trace contains the times taken by the three 64-byte packets sent by traceroute as they go from one router (also known as *hop*) to the next. The output also contains the IP addresses of the various routers on the way from our host to the destination host. A total of 18 hops are traversed by anything that goes from our host to **mit.edu**. The MIT machine is on a class A network with network ID **104.74.143** and host ID **40**. If traceroute does not receive a response within a 5-second timeout interval, it prints an asterisk (*) for that probe. Some of the asterisks are unexplainable and may be the result of bugs in the Linux (or other relevant operating system) network code.

11.9 Important Internet Organizations

Table 11.5 lists the names of some of the important organizations that manage the Internet and formulate plans and policies for its growth.

TABLE 11.5

Important Organizations That Manage the Internet and Formulate Plans and Policies for Its Growth

Organization	Purpose
Internet Society (ISOC) www.isoc.org	An international, nonprofit organization that was established to encourage and promote the use of the Internet. ISOC is the host for Internet Architecture Board (IAB)
Internet Architecture Board (IAB) www.iab.org	A group of people responsible for setting policies and standards for the Internet and the TCP/IP suite
Internet Engineering Task Force (IETF) www.ietf.org	An open group of individuals (network designers, vendors, operators, and researchers) who are responsible for the evolution of the Internet architecture and the Internet's smooth operation. IETF has the responsibility to design and test new technologies for the Internet and the TCP/IP suite. IETF is the technical arm of IAB
Internet Research Task Force (IRTF) www.irtf.org	A group of individuals who are responsible for promoting research that is important for the evolution of the Internet in all relevant areas: protocols, applications, architecture, and technology. IETF is the research arm of IAB
Internet Assigned Numbers Authority (IANA) www.iana.org	Assignment of domain names and protocol port numbers for well-known Internet services, such as ftp
Internet's Network Information Center (InterNIC) www.internic.net	Maintains a list of the currently operating registrars of TLDs, information about new TLDs, problem reports about registrars, and information about registered domains
Internet Corporation for Assigned Names and Numbers (ICANN) www.icann.org	ICANN is a technical coordination body whose primary objective is to ensure the stability of the Internet's system of assigned names and numbers. Every business that wants to become a registrar with direct access to ICANN-designated top-level domains must be accredited by ICANN for this purpose

Summary

Computer networking began when the UCLA student Charley Kline sent the first successful message on ARPANET at 10:30 p.m. on October 29, 1969. At that time, ARPANET consisted of four hosts. Today, computing without networking is unthinkable because of the ubiquitous Internet. Web browsing, file transfer, interactive chat, electronic mail, and remote login are some of the well-known services commonly used by today's computer users. The e-commerce phenomenon has changed the way people do everyday chores and conduct business across the globe. Linux has a special place in the world of networking in general and internetworking in particular, because most of the networking protocols were initially implemented on Linux platforms. Today, Linux based computers run a majority of the server processes that provide most of the Internet services.

The core of internetworking software is based on the TCP/IP protocol suite. This suite includes, among several other protocols, the well known TCP and IP protocols for transportation and routing of application data. The key to routing in the Internet is 32-bit IP addresses (in IPv4) and 128-bit addresses (in IPv6). The most heavily used Internet services are for Web browsing (and all the services that it offers, such as e-commerce and social networking sites), electronic mail, file transfer, and remote login. Not only do Linux systems support all the Internet services, but they also have additional utilities to support local network activities.

The topics discussed in this chapter include the general structure of a network and an internetwork, networking models, the TCP/IP suite, IP addresses, the DNS, Internet protocols and services, and Linux utilities for performing networking- and internetworking-related tasks. These utilities are implemented by using the client–server software model. The utilities discussed in this chapter are `finger` (for finding information about users on a host), `ftp` and `sftp` (for file transfer), `ip`, `host`, and `nslookup` (for checking and setting network interface parameters, and for translation of domain names to IP addresses and vice vera), `ping` (to find the status of a host), `scp` (to remote copy on a Linux host), and `ssh` (for logging on to a remote host on a network and remote command execution), `talk` (for interactive chat), `telnet` (for remote login), and `traceroute` (for tracing the route of data from your host to a destination host).

Questions and Problems

1. What are computer networks and why are they important?
2. What is an internetwork? What is the Internet?
3. What are the key protocols that form the main pillars of the Internet? Where were they developed?
4. What is an IPv4 address? What is its size in bits and bytes? What is DDN? What is the size of an IPv6 address in bits and bytes?
5. What are the classes of IPv4 addresses? Given an IPv4 address in binary, how can you tell which class the address belongs to? How can you tell the class of the address when it is expressed in DDN?
6. What is the DNS? Name the Linux command that can be used to translate a host name to its IP address.
7. List two domain names each for sites that are in the following TLDs: edu, com, gov, int, mil, net, org, autos, beer, biz, careers, cancerresearch, church, museum, au, de, ir, kw, pk, and uk. How did you find them? Do not use examples given in this textbook.
8. Read the **ftp://ftp.isi.edu/in-notes/iana/assignments/port-numbers** file to identify port numbers for the following well-known services: ftp, http, time, daytime, echo, ping, ssh, and quote-of-the-day.
9. What is the timeout period for the finger protocol? How did you get your answer?

10. Give a command that accomplishes the same task that the following command does.
    `rsh upsun29 sort < students > sorted_students`

11. Show the semantics of the following command by drawing a diagram similar to the ones shown in Figures 11.8 and 11.9. Assume that the name of the local machine is **upsun10**.
    `cat students | ssh upsun29 sort | ssh upsun21 uniq > sorted_uniq_students`

12. Display the **/etc/services** file on your system and list the port numbers for well-known ports for the following services: daytime, time, quote-of-the-day (qotd), echo, smtp, and finger. Did you find all of them? Do the port numbers match those found in Problem 8?

13. Use the `telnet` command to get current time via the daytime service at **mit.edu**. Write down your command.

14. Fetch the files **history.netcount** and **history.hosts** from the directory **nsfnet/statistics** using anonymous ftp from the host **nic.merit.edu**. These files contain the number of domestic and foreign networks and hosts on the NSFNET infrastructure. What is the size of Internet in terms of the number of networks and hosts according to the statistics in these files? Although the statistics are somewhat dated, what is your prediction of its size a year from now? Why? Show your work.

15. You create the following entries in your **~/.rhosts** file on a host on your network.
    ```
    host1    john.doe
    host2    mike.brich
    ```
    What are the consequences if **john.doe** and **mike.birch** are users on hosts **host1** and **host2** in your network? Both users belong to your user group.

16. Give a command for displaying simple names of all the hosts on your network.

17. Use the `telnet` command to display information about all the users at **mit.edu** who have **Smith** as part of their name. Show the command that you used to obtain your answer.

18. Give the command for displaying page-by-page information about the users who have **Chen** as part of their name at **mit.edu**. How many such users exist?

19. Describe the semantics of the following command. Clearly state which commands are executed locally and which are executed on the remote host. What is the output of the command?
    `ssh cs00.syi.pcc.edu " ps -el | grep d$ " | grep '\<httpd'$ | wc -l`

20. Use the `traceroute` command to determine the route from your host to loc.gov. What is the approximate travel time for data from your host to locis.loc.gov? If their site is blocking `traceroute` to either skip this question or to go a website that does `traceroute` for you such as traceroute.org or ping.eu/traceroute.

21. Find a host that offers the quote-of-the-day (`qotd`) service. What is the quote of the day today?

22. What kind of network traffic is generated when an IP datagram is sent to **localhost**.

23. Which of the following domains do not have an IPv6 address: google.com, twitter.com, amazon.com, instagram.com, ibm.com? Show your commands and their outputs.

24. How many machines (hosts) do Google and IBM use to handle its mail service? How did you find out? Show the command(s) that you used to obtain your answer, along with the outputs of these commands.

25. Use the `ifconfig` command to give the following information about the local and network interfaces on your machine for IPv4 and IPv6 addresses: localhost addresses and network IP addresses along with their MTUs.

26. What is the domain name of the host with the following IPv6 address: **2a03:2880:2130:cf05: face:b00c:0:1**? Show the command(s) that you used to obtain your answer, along with the outputs of these commands.

27. What is the current RFC count? What is the last RFC about? What is its category (Standard, proposed standard, etc.)? How did you obtain your answer?

28. Explain the semantics of the `ssh sarwar@202.147.169.195 'pgrep inetd'` command. Clearly state where the `pgrep inetd` command is executed and where the output of the command goes.

29. Describe the semantics of the following commands. In particular, state clearly which commands in the command line execute on the local machine and which execute on the remote machine.

    a. `cat students | ssh 122.147.110.13 'sort | grep David'`

    b. `cat students | ssh 122.147.110.13 'sort' | grep David`

Advanced Questions and Problems

30. Install the vsftpd application on your Linux system, using the instructions given in Appendix A, Section 17.2.3.4. Then, do the ftp command line sessions shown in Section 11.8.4.

31. Install the Filezilla application in a GUI desktop environment on your Linux system. Then, do the ftp command line sessions in Section 11.8.4 using Filezilla. Which do you prefer, the typed-in command sessions or Filezilla?

32. Install the openssh-server on a Virtualbox Virtual Machine (VM) instance running on your Linux system, and then do all the things necessary to ssh to that VM instance from another machine on your intranet LAN, or from the Internet.

33. Install the openssh-server on your Linux system, and then do all the things necessary to ssh to your system from another machine on your intranet LAN, or from the Internet.

Projects

Project 1. For Debian 9.1, Ubuntu 16.04, or Linux Mint 18.2 and Later Releases

In preparation for this project, it will be necessary to complete Chapter W23, Sections W23.1–W23.2.6, particularly Examples W23.1 and W23.2, found at the book website. Then, on your Linux Debian-family system, following the example steps in Example W23.2 "Enabling an ssh server in an LXD container," install an openssh-server inside of an LXD container. Finally, ssh from and to the LXD container and the host system, and another machine on your intranet LAN.

Project 2. For CentOS 7.4 and Later

In preparation for this project, it will be necessary to install LXC on CentOS 7 as shown in Chapter W23, Section 2.7 at the book website. Then, on your CentOS 7 system, using the Debian 9 (Stretch) template available in CentOS LXC, do the following:

    a. Create the LXC container, start it, and use the **adduser** command in the container as root to add a user,

    b. ssh from the inside of the Debian container, from the user account you created in a., to the host CentOS 7 system,

    c. ssh from the host CentOS 7 system, from the user account on the host, into the user account you created on the Debian container.

Project 3. For CentOS 7.4 or Later

Repeat Project 2, but allow a user to ssh from another machine on your intranet LAN into the Debian LXC container you created in Project 2.

Looking for more? Visit our sites for additional readings, recommended resources, and exercises.
CRC Press e-Resource: https://www.crcpress.com/9781138710085
Authors' GitHub: https://github.com/bobk48/linuxthetextbook

12

Introductory Bash Programming

OBJECTIVES

- To introduce the concept of shell programming
- To discuss how shell programs are executed
- To describe the concept and use of shell variables
- To discuss how command line arguments are passed to shell programs
- To explain the concept of command substitution
- To describe some basic coding principles
- To discuss the various control structures for Bash scripting
- To explain the syntax and semantics of various types of operators in Bash and their precedence order
- To write and discuss some Bash scripts
- To describe how Bash commands may be grouped and executed sequentially
- To cover the following commands and primitives:

  ```
  *, =, ", ', `, &, &&, <, >, ^, ;, |, ||, \, /, [], :, ;, bash, break, case,
  continue, exit, export, env, for, if, ls, read, readonly, set, sh, shift, test,
  uname, while, until, unset
  ```

12.1 Introduction

The Bourne again shell (Bash) is more than a command interpreter. It has a programming language of its own that can be used to write shell programs for performing various tasks that cannot be performed by any existing command. A shell program, commonly known as a *shell script*, consists of shell commands to be executed in a shell, one command at a time, and is stored in an ordinary Linux file. The shell allows the use of a read/write storage place, known as a *shell variable*, for users and programmers to use as a scratch pad for completing a task. The shell also contains program control flow commands (also called *statements*) that allow nonsequential execution of the commands in a shell script and repeated execution of a block of commands—similar to high-level programming languages like C.

12.2 Running a Bash Script

There are three ways to run a Bash script. The first step for all three methods is to make the script file executable by adding the execute permission to the existing access permissions for the file. You can do so by running the `chmod u+x hello_world` command, where **hello_world** is the name of the file containing the shell script. Without making the script file executable, Bash displays the error message that the script in the **hello_world** file cannot be executed.

```
$ cat hello_world
echo "Hello, world!"
```

```
$ ./hello_world
-bash: ./hello_world: Permission denied
$ chmod u+x hello_world
$
```

Clearly, in this case, you make the script executable for yourself only. However, you can set appropriate access permissions for the file if you also want other users to be able to execute it. Once you have made the script file executable, you can type ./hello_world as a command to execute the shell script, as shown:

```
$ ./hello_world
Hello, world!
$
```

If your search path (the *PATH* variable) includes your current directory (.), you can simply use the hello_world command, instead of using the ./hello_world command. For the rest of this chapter, we assume that your *PATH* variable includes your current directory.

As described in Chapter 10, a child of the current shell process executes the script. Thus, with this method, the script executes properly if you are using Bash but not if you are using any other shell. If you are currently using some other shell, first execute the /bin/bash command to run Bash and then run the hello_world command, as shown in the following example. Here, we assume that your current shell is C (or TC), shell (with the % prompt). After the script has completed its execution, we press <Ctrl+D> to terminate bash and return to C shell.

```
% /bin/bash
$ hello_world
Hello, world!
$ <Ctrl+D>
%
```

The second method of executing a shell script is to run the /bin/bash command with the script file as its parameter. Thus, the following command executes the shell script in **hello_world**.

```
$ /bin/bash hello_world
Hello, world!
$
```

If your *PATH* variable includes the **/bin** directory, you can simply use the bash command, instead of using the /bin/bash command.

The third method, which is also the most commonly used method, is to force the current shell to execute a script in Bash, regardless of your current shell. You can do so by beginning a shell script with the line containing #!/bin/bash, as shown in the following session:

```
$ cat hello_world
#!/bin/bash
echo "Hello, world!"
$
```

When your current shell encounters the string #!, it takes the rest of the line as the absolute pathname for the shell to be executed, under which the script in the file is executed. If your current shell is the C shell, you can replace this line with a colon (:), which is known as the *null* command in Bash. When the C shell reads : as the first character, it runs a Bash process that executes the commands in the script. The : command returns true. We discuss the return values of commands later in the chapter.

Throughout this chapter, we would use the chmod u+x script_file command to make **script_file** executable by the owner of the file and run the script by using the ./script_file command.

12.3 Shell Variables and Related Commands

A *variable* is a main memory location with a name. It allows you to reference the memory location by using its name instead of its address. The name of a shell variable comprises digits, letters, and underscores, with the first character being a letter or underscore. Because main memory is read/write storage, you can read a variable's value or assign it a new value. For Bash, the value of a variable is always a string of characters, even if you store a number in it. There is no theoretical limit on the length of a variable's value.

Shell variables can be one of two types: *shell environment variables* and *user-defined variables*. Environment variables are used to customize the environment in which your shell runs and for proper execution of shell commands. A copy of these variables is passed to every command that executes in the shell as its child. Most of these variables are initialized when the **/etc/profile** file executes as you log on. This file is written by your system administrator to set up a common environment for all the users of the system. You can customize your environment by assigning different values to some or all of these variables, as well as define other variables, in your **~/.profile** startup file, which executes when you log on. You can also initialize an interactive Bash session by defining environment variables in your ~/.**bashrc** file that Bash runs when it is started interactively. Table 12.1 lists most of the environment variables whose values you can change. We described some of these variables in previous chapters.

These shell environment variables are *writable*, and you can assign any values to them. Other shell environment variables are *read only*, which means that you can use (read) the values of these variables but cannot change them. These variables are most useful for processing command line arguments (also known as *positional arguments*) or parameters passed to a shell script at the command line. Examples of command line arguments are the source and destination files in the cp command. Some other read-only shell variables are used to keep track of the process ID of the current process, the process ID of the most recent background process, and the exit status of the last command. Some important read-only shell environment variables are listed in Table 12.2. These read-only variables are established at the time the process is invoked rather than at the login time, as with other environment variables.

TABLE 12.1

Some Important Writable Bash Environment Variables

Environment Variable	Purpose of the Variable
CDPATH	Contains the names of the directories that are searched, one by one, by the cd command to find the location of the directory passed to it as a parameter; the cd command searches the current directory if this variable is not set
EDITOR	Contains the name of the default editor used in programs such as an e-mail program
ENV	Contains the path along which Linux looks to find configuration files
HOME	Contains the name of the directory where the login shell places you in the directory structure when you first log on
MAIL	Contains the name of user's system mailbox file
MAILCHECK	Contains a number that specifies how often (in seconds) the shell should check a user's mailbox for new mail and inform the user accordingly
PATH	Contains search path for Bash for the user—that is, the sequence of directories that Bash searches to locate an external command or program specified in a command line
PPID	Contains the process ID of the parent process
PS1	Contains the primary shell prompt that appears on the command line, usually set to $
PS2	Contains the secondary shell prompt displayed on second line of a command if the shell thinks that the command is not finished, typically when the command terminates with a backslash (\), the escape character. Usually set to >.
PWD	Contains the absolute pathname of the current working directory
TERM	Contains the type of user's console terminal

TABLE 12.2

Some Important Read-Only Bash Environment Variables

Environment Variable	Purpose of the Variable
$0	Name of program
$1–$9	Values of command line arguments 1–9
$*	Values of all command line arguments
$@	Values of all command line arguments; each argument individually quoted if $@ is enclosed in quotes, as in "$@"
$#	Total number of command line arguments
$$	PID of the current process
$?	Exit status of the most recent command
$!	PID of the most recent background process

User-defined variables are used within shell scripts as temporary storage places, whose values can be changed when the program executes. These variables can be made read only as well as passed to the commands that execute in the shell script in which they are defined. Unlike most other programming languages, in Bash programming language, you do not have to declare and initialize shell variables. An uninitialized shell variable is initialized to a null string by default.

You can display the names of all shell variables (including user-defined variables) and their current values by using the set command without any parameters. As described later in this chapter, the set command can also be used to change the values of some of the read-only shell environment variables. The following is a sample run of the set command on our machine:

```
$ set
BASH=/bin/bash
BASHOPTS=checkwinsize:cmdhist:complete_fullquote:expand_
aliases:extglob:extquote:force_fignore:histappend:interactive_comments:login_shell:
progcomp:promptvars:sourcepath
... [output truncated] ...
PPID=25922
PS1='\[\e]0;\u@\h \w\a\]${debian_chroot:+($debian_chroot)}\[\033[01;32m\]\u@\h\
[\033[00m\] \[\033[01;34m\]\w \$\[\033[00m\] '
PS2='> '
PS4='+ '
PWD=/home/sarwar/linux2e/ch12
SHELL=/bin/bash
SHELLOPTS=braceexpand:emacs:hashall:histexpand:history:
interactive-comments:monitor
SHLVL=1
SSH_CLIENT='103.255.5.80 5141 22'
SSH_CONNECTION='103.255.5.80 5141 202.147.169.195 22'
SSH_TTY=/dev/pts/0
TERM=xterm-color
UID=1004
USER=sarwar
... [output truncated] ...
$
```

You can also use the env and printenv commands to display the names of the environment variables and their values, but the list is not as complete as the one displayed by the set command. In particular, the output does not include any user-defined variables. The following is a sample output of the env command on the same system that we ran the set command on. The printenv command produces the same output.

```
$ env
LC_PAPER=ur_PK
```

```
LC_ADDRESS=ur_PK
XDG_SESSION_ID=1760
LC_MONETARY=ur_PK
TERM=xterm-color
SHELL=/bin/bash
XDG_SESSION_COOKIE=4603ef0b0eb640e2836974a04d12e5ab-1497116101.710678-1886286059
SSH_CLIENT=103.255.5.80 5141 22
LC_NUMERIC=ur_PK
SSH_TTY=/dev/pts/0
USER=sarwar
... [output truncated] ...
OLDPWD=/home/sarwar/linux2e/ch13
$
```

In the following in-chapter exercises, you will create a simple shell script and make it executable. Also, you will use the set, printenv, and env commands to display the names and values of the shell variables on your system.

Exercise 12.1

Display the names and values of all the shell variables on your Linux machine. What command(s) did you use?

Exercise 12.2

Create a file that contains a shell script comprising the date and pwd commands, one on each line. Make the file executable and run the shell script. List all the steps for completing this task.

12.3.1 Controlling the Prompt

Bash allows you to control your prompt in an easy manner. You can assign one or more of several special characters to your prompt variables (*PS1*, *PS2*, etc.) to display several prompts. Some of the commonly used special characters are described in Table 12.3.

In the following session, we show how various special characters change the primary prompt for your shell. Note that you can combine several special characters and assign them to your prompt variable.

```
$ PS1='\w$ '
~/linux2e/ch12$ date
Sun Jun 11 14:28:47 PKT 2017
~/linux2e/ch12$ PS1='\d$ '
Sun Jun 11$ PS1='\h$ '
Mint18$ PS1='\t$ '
14:30:03$ PS1='\s-\v$ '
 bash-4.3$
```

TABLE 12.3

Some Useful Prompt Characters and Their Descriptions

Special Character	Description
\H	Fully qualifies domain name of the host, such as alumni.harvard.edu
\T	Time in the 12-hour hh:mm:ss format
\d	The date in "Weekday month date" format
\h	Hostname of the computer up to the first dot, such as alumni
\s	The name of your shell
\t	Time in the 24-hour hh:mm:ss format
\u	User name of the current user
\v	Version of Bash such as 4.3
\w	Current working directory

12.3.2 Variable Declaration

Bash does not require you to declare variables and initialize them, but you can use the `declare` and `typeset` commands to declare variables, initialize them, and set their attributes. The attributes of a variable dictate the type of values that can be assigned to it and its scope (i.e., where it is accessible). A Bash variable is a string variable by default, but you can define a variable to be of integer type. You can also declare functions and arrays (see Chapter 13) by using these commands. You can mark a variable read only and make a variable's value available to a child process. The following is a brief description of these commands:

Syntax:
>   `declare`   `[options] [name[=value]]`
>   `typeset`   `[options] [name[=value]]`

Purpose: Declare variables, initialize them, and set their attributes. Inside functions, new copies of the variables are created. Using + instead of − turns attributes off

Output: Without name and options, display names of all shell variables and their values in the environment of the current shell. With options, display names of variables with the given attributes and their values

Commonly used options/features:

>   `-a`   each "name" is an array
>
>   `-f`   each "name" is a function
>
>   `-i`   "name" is an integer
>
>   `-r`   mark each "name" read-only (cannot be turned off by using **+x**)
>
>   `-x`   mark each "name" exported

An undeclared and uninitialized variable has an initial value of a null string. Bash also does not require you to specify a variable's attributes at the time of its declaration. You can initialize a variable and set its attributes at a later time by using the `declare` or `typeset` command, or by using other commands for setting specific attributes. Declaring an existing variable does not change its current value/attributes. In the following session, we demonstrate the use of the `declare` command; you can replace it with the `typeset` command if you so desire.

```
$ declare age=58
$ declare -rx OS=Linux
$ echo $age
58
$ echo $OS
Linux
$ declare OS
$ declare age
$ echo $OS
Linux
$ echo $age
58
$
```

When you use the `declare` and `typeset` commands without arguments, they display the names of all the shell variables in your environment and their values. You can use these commands to view the values of variables with particular attributes. The following session shows some examples. We have shown partial outputs for some commands to save space. Note that the `declare -ir` command displays all integer and read-only variables in your environment.

```
$ declare -i
declare -ir BASHPID
```

```
declare -ir EUID="1004"
declare -i HISTCMD
declare -i LINENO
declare -i MAILCHECK="60"
declare -i OPTIND="1"
declare -ir PPID="13147"
declare -i RANDOM
declare -ir UID="1004"
$ declare -x
declare -x HOME="/home/sarwar"
... [output truncated] ...
declare -x MAIL="/var/mail/sarwar"
declare -x OLDPWD="/home/sarwar"
declare -x PATH="/usr/local/sbin:/usr/local/bin:/usr/sbin:/usr/bin:/sbin:/bin:/
usr/games:/usr/local/games"
declare -x PWD="/home/sarwar/linux2e/ch12"
declare -x SHELL="/bin/bash"
declare -x SHLVL="1"
declare -x SSH_CLIENT="103.255.5.92 6489 22"
declare -x SSH_CONNECTION="103.255.5.92 6489 202.147.169.195 22"
declare -x SSH_TTY="/dev/pts/1"
declare -x TERM="xterm-color"
declare -x USER="sarwar"
... [output truncated] ...
$ declare -ir
declare -r BASHOPTS="checkwinsize:cmdhist:complete_fullquote:expand_aliases:
extglob:extquote:force_fignore:histappend:interactive_comments:login_shell:
progcomp:promptvars:sourcepath"
declare -ir BASHPID
declare -r BASH_COMPLETION_COMPAT_DIR="/etc/bash_completion.d"
declare -ar BASH_VERSINFO='([0]="4" [1]="3" [2]="42" [3]="1" [4]="release"
[5]="x86_64-pc-linux-gnu")'
declare -ir EUID="1004"
… [output truncated] ...
declare -ir UID="1004"
$ declare
BASH=/bin/bash
BASHOPTS=checkwinsize:cmdhist:complete_fullquote:expand_aliases:
extglob:extquote:force_fignore:histappend:interactive_comments:login_shell:
progcomp:promptvars:sourcepath
BASH_ALIASES=()
BASH_ARGC=()
BASH_ARGV=()
BASH_CMDS=()
BASH_COMPLETION_COMPAT_DIR=/etc/bash_completion.d
BASH_LINENO=()
BASH_SOURCE=()
BASH_VERSINFO=([0]="4" [1]="3" [2]="42" [3]="1" [4]="release"
[5]="x86_64-pc-linux-gnu")
BASH_VERSION='4.3.42(1)-release'
COLUMNS=80
DIRSTACK=()
EUID=1004
...
$
```

You can change the value of a variable by using the name=value syntax, discussed in detail in Section 12.3.3. You can use this syntax to declare generic variables that can be assigned a string value. An integer variable cannot be assigned a noninteger value; doing so results in the variable getting a value of zero. Noninteger variables can be assigned any value because every value is stored as a string. The following session shows some examples. Since *age* is declared to be an integer variable, assigning it a value "Fifty-eight" results in 0 being assigned to it. On the other hand, since *name* and *place* are generic variables, they can be assigned any type of value (integer or string).

```
$ declare -i age=58
$ echo $age
58
$ age="Fifty-eight"
$ echo $age
0
$ name=John
$ declare place=Portland
$ echo $name $place
John Portland
$ name=2017 place=007
$ echo $name $place
2017 007
$
```

12.3.3 Reading and Writing Shell Variables

You can use the following syntax to assign a value to (write) one or more shell variables. The command syntax `variable=value` comprises what is commonly known as the *assignment statement*, and its purpose is to assign `value` to `variable`. The evaluation of the assignment statement is right to left. Thus, `value` is evaluated first and then assigned to `variable`. If there is a problem in evaluating `value`, an error is reported. The following is the general syntax of the assignment statement:

Syntax:
   variable1=value1 [variable2=value2 … variableN=valueN]

Purpose: Assign values **value1**, …, **valueN** to variables **variable1**, …, **variableN**, respectively; no space allowed before or after the equal sign

---

Note that there is no space before and after the equals sign (=) in the syntax. If a value contains spaces, you must enclose the value in quotes. Single and double quotes work differently, as discussed later in this section. You can refer to (i.e., access) the current value of a variable by placing a dollar sign ($) before the variable name; there is no space between $ and the variable name. You can use the echo command to display the values of shell variables. The various syntaxes and substitution operators are described in Table 12.4.

   In the following session, we show how shell variables can be read (interpreted) and written (created). The first echo command displays a blank line for an uninitialized variable called *name*. Then the *name* variable is initialized to David and the subsequent echo command is used to display the value of *name*, which now contains the value David. The output of the echo `${name:-John} ${place:-Portland}` command is David Portland, because *name* has the value David and *place* has the value Portland. The echo `${place:?"Not defined"}` command displays the message -bash: place: Not defined, because the variable *place* is not defined. The echo `${name:+"Defined"}` command displays Defined because the *name* variable has been initialized to a nonnull value. The echo `${place:+"Not defined"}` command displays the null string because the *place* variable has not been defined anymore and is therefore null. The echo `${place:="San Francisco"}` command assigns the value San Francisco to the *place* variable and displays the same value because the *place* variable is undefined (was unset earlier in the same session). Finally, the echo `${name:-John} ${place:-Portland}` command displays David San Francisco, because the *name* variable has the value David and the *place* variable has the value San Francisco.

```
$ echo $name

$ name=David
$ echo $name
```

TABLE 12.4

Variable Substitution Operators and Their Descriptions

Operator	Purpose	Description
$variable	To get the value of a variable or null if it is not initialized	Returns the value of "variable" or null if it is not initialized
${variable}	To get the value of a variable or null if it is not initialized; used when something else is to be appended to the value	Return the value of "variable" or null if it is not initialized
${variable:-string}	To set a variable to a known value if it is undefined and return that value	If "variable" exists and is not null, return its value; otherwise, assign "string" to "variable" and return it
${variable:=string}	To set a variable to a known value if it is undefined and return that value	If "variable" exists and is not null, return its value; otherwise, assign "string" to "variable" and return it
${variable:?string}	To display a message if a variable is undefined	If "variable" exists and is not null, return its value; otherwise, display "variable": followed by "string" as a message
${variable:+string}	To test the existence of a variable	If "variable" exists and is not null, return "string"; otherwise, return null

```
David
$ echo $place

$ place=Portland
$ echo ${name:-John} ${place:-Portland}
David Portland
$ echo ${place:?"Not defined"}
Portland
$ unset place
$ echo ${place:?"Not defined"}
-bash: place: Not defined
$ echo ${name:+"Defined"}
Defined
$ echo ${place:+"Not defined"}

$ echo ${place:="San Francisco"}
San Francisco
$ echo ${name:-John} ${place:-Portland}
David San Francisco
$
```

Single quotes should be used to preserve the literal meanings of all characters, except single quotes. Double quotes preserve the literal meaning of all characters except single quotes, dollar signs, and backslashes. A backslash preserves the literal meaning of the character that follows, except n. The newline character (\n) has special meaning, and to preserve this meaning, it must be enclosed in single quotes, as in '\n'. A backslash in double quotes remains literal, unless it precedes the following characters: backslash, single quote, double quotes, dollar sign, and newline. Enclosing characters between $' and ' preserves the literal meaning of all characters, except single quotes and backslashes. \t and \b also have special meaning, and stand for tab and backspace, respectively.

The following session shows a few examples of quoting. $$ stands for the process ID (PID) of the process that executes the echo command, i.e., the current shell process. The output of the ps command verifies that 21914 is, in fact, the PID of your current bash process. Note that * represents the names of the files and directories in the current working directory.

```
$ echo '$name * ? \'
$name * ? \
```

```
$ echo "$ * \n ?"
$ * \n ?
$ echo "$$ * \n ?"
21914 * \n ?
$ ps
   PID TTY              TIME CMD
21914 pts/0     00:00:00 bash
21934 pts/0     00:00:00 ps
$ echo "\$ \' \" \\ \\n"
$ \' " \ \n
$ echo $'a d ? * " $'
a d ? * " $
$ echo $'a \ ? * " $'
a \ ? * " $
$ echo $'a ' ? * " $'"
a  ? Desktop Documents Downloads Music Pictures Public Templates Videos   $'
$
```

A command consisting of $variable only results in the value of *variable* executed as a shell command. If the value of *variable* comprises a valid command, the expected results are produced. If *variable* does not contain a valid command, the shell, as expected, displays an appropriate error message. The following session makes this point with some examples. The variable used in the session is *command*.

```
$ command=pwd
$ $command
/home/sarwar/linux2e/ch12
$ command=hello
$ $command
The program 'hello' can be found in the following packages:
 * hello
 * hello-traditional
Ask your administrator to install one of them
$
```

The following session shows how shell variables can be created and read, and how their values may be changed.

```
$ name=John
$ echo $name
John
$ name=Jimmy Doe
No command 'Doe' found, did you mean:
 Command 'joe' from package 'joe' (universe)
 Command 'joe' from package 'joe-jupp' (universe)
 Command 'toe' from package 'ncurses-bin' (main)
Doe: command not found
$ echo $name
John
$ name=James more ~/.profile | grep PATH
# set PATH so it includes user's private bin if it exists
    PATH="$HOME/bin:$PATH"
PATH=$PATH:.
$ echo $name
John
$ name=David date
Sat Jun 10 23:19:49 PKT 2017
$ echo $name
John
$ name="John Doe"
$ echo $name
John Doe
```

```
$ name=John*
$ echo $name
John.Bates.letter John.Johnsen.memo John.email
$ echo "$name"
John*
$ name=John*
Sat Jun 10 23:19:49 PKT 2017
$ echo $name
John.Bates John.email John.Johnsen
$ echo "The name $name sounds familiar!"
The name John* sounds familiar!
$ echo \$name
$name
$ echo '$name'
$name
$
```

If the right-hand side of an assignment statement is not enclosed in quotes and includes spaces, the shell assumes that the second and remaining words in the right-hand side form a command, and tries to execute it. The shell displays an error message if the assumed command does not correspond to a valid command, as shown in the statements name=Jimmy Doe. On the other hand, for the name=David date command, the output of the date command is displayed. Similarly, for name=James more ~/.profile | grep PATH the shell displays the output of the more ~/.profile | grep PATH command. In all these cases, regardless of whether the second and subsequent words form a valid command or not, the first command does not execute. Thus, the value of the *name* variable remains initialized to with the previous value, John, as assigned at the beginning of the session.

After the name=John* statement has been executed and $name is not quoted in the echo command, the shell lists the file names in your present working directory that match John*, with * considered as the shell metacharacter. However, if there is no file in your current directory that starts with the string John, the echo commands displays John*. However, if your current directory does not contain any file that starts with the string John, the echo John* command displays John*. The variable $name must be enclosed in quotes to refer to John*, as in echo "$name."

12.3.4 Command Substitution

Command substitution allows you to replace a command with its output. The following is a brief description of command substitution. Note the use of back quotes (also known as *grave accents*) in the first form of the syntax. In our example, shell sessions and scripts throughout the book, we will use the first syntax.

Syntax:
> `command`
> $(command)

Purpose: Execute command and substitute `command` (or $(command), if you use this syntax) with the output of the command

---

The following session illustrates this concept with a few examples. In the first assignment statement, the variable *command* is assigned a value pwd. In the second and third assignment statements, the output of the pwd command is assigned to the *command* variable.

```
$ command=pwd
$ echo "The value of command is: $command."
The value of command is: pwd.
$ command=`pwd`
$ echo "The value of command is: $command."
The value of command is: /home/sarwar/linux2e/ch12.
```

```
$ command=$(pwd)
$ echo "The value of command is: $command."
The value of command is: /home/sarwar/linux2e/ch12.
$
```

Command substitution can be specified in any command. For example, in the following session, the output of the date command is substituted for `date` before the echo command is executed.

```
$ echo "The date and time are `date`."
The date and time are Sat Jun 10 23:32:58 PKT 2017.
$
```

We demonstrate the real-world use of command substitution in various ways throughout this chapter and Chapter 13.

The following in-chapter exercises are designed to reinforce the creation and use of shell variables and the concept of command substitution.

> **Exercise 12.3**
>
> Assign your full name to a shell variable called *myname* and echo its value. How did you accomplish the task? Show your work.
>
> **Exercise 12.4**
>
> Assign the output of the echo "Hello, world!" command to the *myname* variable and then display the value of *myname*. List the commands that you executed to complete your work.

12.3.5 Exporting Environment

When a variable is created in a shell, subsequent shells do not have automatic access to it. The export, declare -x, or typeset -x command passes the *value* of a variable to subsequent shells. Thus, when a shell script is called and executed in another shell script, it does not get automatic access to the variables defined in the original (i.e., caller) script unless they are explicitly made available to it. These commands can be used to pass the value of one or more shell variables to any subsequent script. All read/write shell environment variables are available to every command, script, and subshell, so they are exported at the time they are initialized. The following is a brief description of the three commands:

> **Syntax:**
>   declare -x [name-list]
>   typeset -x [name-list]
>   export [name-list]
>
> **Purpose:** Export the names and copies of the current values in **name-list** to every command executed from this point on

The following session presents a simple use of the declare -x command. The *place* variable is initialized to **Notre Dame** and is exported to subsequent commands executed under the current shell and any subshells that run under the current shell.

```
$ declare -x place="Notre Dame"
$ export place
$
```

We now illustrate the concept of exporting shell variables via some simple shell scripts. However, it is important to know that unless your search path (i.e., *PATH* variable) is set to include your current directory (.), you would have to explicitly place ./ before the name of a script file on the command line or inside a script file. If you do not do so, the shell will display the following error message.

```
file_name: command not found
```

Now, consider the following session. Note that the two-line shell script in the **display_name** file displays a null string even though we initialized the *name* variable to John Doe just before executing this script. The reason is that the *name* variable is not exported before running the script, and the *name* variable used in the script is local to the script. Because this local variable *name* is uninitialized, the echo command displays the null string—the default value of every uninitialized variable. Note that if your *PATH* variable contains your current directory, you can run the script without using ./ before **display_name**.

```
$ cat display_name
echo $name
exit 0
$ name="John Doe"
$ ./display_name

$
```

You can use the exit command to transfer control to the calling process—the current shell process in the preceding session. The only argument of the exit command is an optional integer number, which is returned to the calling process as the exit status of the terminating process. All Linux commands return an exit status of zero upon *success* (i.e., if they successfully perform their tasks and terminate normally) and nonzero upon *failure*. The return status value of a command is stored in the read-only environment variable *$?* which can be checked and/or displayed by the calling process. In shell scripts, the status of a command is commonly checked and subsequent action taken. We show the use of *$?* in some shell scripts later in the chapter. When the exit command is executed without an argument, the Linux kernel sets the return status value for the script.

In the following session, the *name* variable is exported after it is initialized, thus making it available to the **display_name** script. Because the *name* variable is exported before the **display_name** script is executed, the script displays John Doe and not a null string, as was the result of its execution in the previous session. The session also shows that the return status of the **display_name** script is 0. Note that you can combine the initialization of a variable and exporting it in one command as in export name="John Doe."

```
$ declare -x name="John Doe"
$ ./display_name
John Doe
$ echo $?
0
$
```

We now show that a copy of an exported variable's value is passed to any subsequent command. In other words, a command has access to the value of the exported variable only; it cannot assign a new value to the variable. Consider the script in the **export_demo** file.

```
$ cat export_demo
#!/bin/bash
declare -x name="John Doe"
display_change_name
display_name
exit 0
$ cat display_change_name
#!/bin/bash
echo $name
name="Plain Jane"
echo $name
exit 0
```

```
$ ./export_demo
John Doe
Plain Jane
John Doe
$
```

When the **export_demo** script is invoked, the *name* variable is set to John Doe and exported so that it becomes part of the environment of all the commands that execute under **export_demo**. The first echo command in the **display_change_name** script displays John Doe as the value of the exported variable (nonlocal) *name*. It then initializes a local variable, *name*, to Plain Jane. The second echo command therefore echoes the current value of the local variable *name* and displays Plain Jane. When the **display_change_name** script has finished its execution, the **display_name** script executes and displays the value of the exported (nonlocal) *name*, thus displaying John Doe.

If your *PATH* variable does not contain your current directory, you must include it. Otherwise, you will see the following error messages on the screen:

```
$ ./export_demo
./export_demo: line 4: display_change_name: command not found
./export_demo: line 5: display_name: command not found
$
```

To avoid this issue, put the following line in your **~/.profile** file, logout, and login again. You should place your current directory as the last component of *PATH*, as is done in the following session, so that your current directory is not searched unnecessarily for commonly used commands.

```
PATH=$PATH:.
```

Including the abovementioned line in your **~/.profile** file and running the . ~/.profile command has the same effect. If you want your current directory to be included in your search path only for your current session, run the following command.

```
$ PATH=$PATH:.
$
```

12.3.6 Resetting Variables

A variable retains its value as long as the script in which it is initialized is running. You can reset the value of a variable to null (the default initial value of all variables) by either explicitly initializing it to null or using the unset command. The following is a brief description of this command:

Syntax:
> unset [name-list]

Purpose: Reset or remove from the shell environment the variables or functions corresponding to the names in **name-list**, where **name-list** is a list of names separated by spaces

---

We discuss functions in Bash in Chapter 13, so we limit the discussion of the unset command here to variables only. The following session shows a simple use of the command. The variables *name* and *place* are set to John and Corvallis, respectively, and the echo command displays the values of these variables. The unset command resets *name* to null. Thus, the echo "$name" command displays a null string (a blank line).

```
$ declare name=John place=Corvallis
$ echo $name $place
John Corvallis
$ unset name
$ echo $name
```

```
$ echo $place
Corvallis
$
```

The following command removes the variables *name* and *place* from the shell environment.

```
$ unset name place
$
```

Another way to reset a variable is to assign it explicitly a null value by assigning it no value and simply hitting <Enter> after the = sign, as in

```
$ country=
$ echo "$country"

$
```

12.3.7 Creating Read-Only Defined Variables

When programming, you sometimes need to use constants. You can use literal constants, but using symbolic constants is a good programming practice, primarily because it makes your code more readable. Another reason for using symbolic names is that a constant used at various places in code might need to be changed. With a symbolic constant, the change is made at one place only, but a literal constant must be changed every place it was used. A symbolic constant can be created in Bash by initializing a variable with the desired value and making it read only using the readonly, declare -r, or typeset -r command. This command is rarely used in shell scripts, but we discuss it briefly for the sake of complete coverage of shell variables. The following is a brief description of these commands:

Syntax:
    declare -r [name-list]
    typeset -r [name-list]
    readonly [name-list]

Purpose: Prevent assignment of new values to the variables in **name-list**

In the following session, the *name* and *place* variables are made read only after initializing them with John and Ames, respectively. Once they have become read only, assignment to either variable fails. In the case of last statement in the following session, Bash terminates the command when it finds out that the *name* variable is read only. Thus, *place* remains initialized to Ames.

```
$ declare -r name=John place=Ames
$ echo "$name $place"
John Ames
$ declare name=Art place="Ann Arbor"
-bash: name: is read only
-bash: place: is read only
$ name=Art place="Ann Arbor"
-bash: name: readonly variable
$
```

When the readonly, declare -r, or typeset -r command is executed without arguments, it displays all read-only variables, as in the following session. BASHPID, BASH _ VERSIONINFO, EUID, PPID, SHELLOPTS, and UID are environment variables. The **name** and **place** are user-defined read-only variables created in the preceding session.

```
$ declare -r
declare -r BASHOPTS="checkwinsize:cmdhist:complete_fullquote:expand_
aliases:extglob:extquote:force_fignore:histappend:interactive_comments:login_shell:
progcomp:promptvars:sourcepath"
declare -ir BASHPID
declare -r BASH_COMPLETION_COMPAT_DIR="/etc/bash_completion.d"
declare -ar BASH_VERSINFO='([0]="4" [1]="3" [2]="42" [3]="1" [4]="release"
[5]="x86_64-pc-linux-gnu")'
declare -ir EUID="1004"
declare -ir PPID="3347"
declare -r
SHELLOPTS="braceexpand:emacs:hashall:histexpand:history:interactive-
comments:monitor"
declare -ir UID="1004"
declare -r name="Jim"
declare -r place="Ames"
$
```

On some Linux systems, the output of the readonly command shows the list of read-only variables along with the values of these variables. You cannot reset the value of a read-only variable. The following session shows an example:

```
$ unset name
-bash: unset: name: cannot unset: readonly variable
$
```

12.3.8 Reading from Standard Input

So far, we have shown how you can assign values to shell variables statically at the command line level or by using the assignment statement in your programs. If you want to write an interactive shell script that prompts the user for keyboard input, you need to use the read command to store the user input in a shell variable. This command allows you to read one line of standard input. The following is a brief description of the command:

Syntax:
   read [options] [variable-list]

Purpose: Read one line from standard input and assign words in the line to variables in variable-list

Commonly used options/features:

   -a name   Read words into the "**name**" array

   -e   Read the whole line into the first variable; the rest of the variables are null

   -p prompt Display "**prompt**" if reading from a terminal

A line is read in the form of words separated by white spaces (<Space> or <Tab> characters, depending on the value of the shell environment variable IFS). The words are assigned to the variables in the order of their occurrence, from left to right. If the number of words in the line is greater than the number of variables in **variable-list**, the last variable is assigned the extra words. If the number of words in a line is less than the number of variables, the remaining variables are reset to null.

We illustrate the semantics of the read command by way of the following script in the **read_demo** file.

```
$ cat read_demo
#!/bin/bash
echo -n "Enter input: "
read line
echo "You entered: $line"
echo -n "Enter another line: "
read word1 word2 word3
```

```
echo "The first word is: $word1"
echo "The second word is: $word2"
echo "The rest of the line is: $word3"
exit 0
$
```

We now show how the input that you enter from the keyboard is read by the read command in that script. In the following run, you enter the same input: Linux rules the network computing world! The first read command takes the whole input and puts it in the shell variable *line* without the newline character. In the second read command, the first word of your input is assigned to the variable *word1*, the second word is assigned to the variable *word2*, and the rest of the line (without the newline character) is assigned to the variable *word3*. The outputs of the echo commands for displaying the values of these variables confirm this point.

```
$ ./read_demo
Enter input: Linux rules the network computing world!
You entered: Linux rules the network computing world!
Enter another line: Linux rules the network computing world!
The first word is: Linux
The second word is: rules
The rest of the line is: the network computing world!
$
```

The –n option used in the two echo commands is used to force the cursor to stay at the same line after the echo command has displayed the quoted text. If you do not use this character, the cursor moves to the next line, which is what you like to see happen while displaying information and error messages. However, when you prompt the user for keyboard input, you should keep the cursor in front of the prompt for a user-friendlier interface.

In Linux, several special characters can be used in the Bash echo -e command as control characters. These characters start with a backslash (\). Thus, for example, the echo -n "Enter input: " command in the **read_demo** script may be replaced with echo -e "Enter input: \c". Some of the control characters, along with their meanings, are listed in Table 12.3. Note that, on some Linux systems, you may have to use double backslash (\\) instead of single backslash for the special characters to work. For example, use \\c instead of \c to keep the cursor on the same line.

The lines

```
echo -n "Enter input: "
read line
```

can be replaced with read –p "Enter input: " line to prompt the user for input and store a line of input in the shell variable *line*.

In the following in-chapter exercises, you will use the read and export commands to practice reading from **stdin** in shell scripts and exporting variables to child processes.

Exercise 12.5

Give commands for reading a value into the *myname* variable from the keyboard and exporting it so that commands executed in any child shell have access to the variable.

Exercise 12.6

Copy the value *myname* variable to another variable, *anyname*. Make the *anyname* variable read only and unset both the *myname* and *anyname* variables. What happened? Show all your work.

12.4 Passing Arguments to Shell Scripts

In this section, we describe how command line arguments can be passed to shell scripts and manipulated by them. As discussed in Section 12.3, you can pass command line arguments, or *positional parameters*, to a shell script. The values of these arguments, starting with the first argument, are stored in variables $1, $2, $3, $4, and so on, respectively. The variable name $0 contains the name of the script file (i.e., the command name). On some Linux systems, up to only the first nine arguments are stored in these variables. You can use the names of these variables to refer to the values of these arguments. If the positional argument that you refer to is not passed as an argument, it has a value of null. The positional arguments beyond the first nine, such as $10 and $11, contain unknown (garbage) values. The environment variable $# contains the total number of arguments passed in an execution of a script. The variables $* and $@ both contain the values of all of the arguments, but $@ has each individual argument in quotes if it is used as "$@". The shell script in the **cmdargs_demo** file shows how you can use these variables. Note that values a0 and a1 are displayed for $10 and $11, respectively, in the first run of the **cmdargs_demo** program. In the second run, the values displayed for these variables are One0 and One1. When multiple command line arguments are enclosed in double quotes, they are treated as one argument. In the following session, the last cmdargs_demo command is passed only four command line arguments, One, Two, 3, and "Four 5 6," as is evident from the output of the command shows the total number of arguments as 4.

```
$ cat cmdargs_demo
#!/bin/bash
echo "The command name is: $0."
echo "The number of command line arguments passed as parameters is $#."
echo "The values of the command line arguments are: $1 $2 $3 $4 $5 $6 $7 $8 $9 $10
$11"
echo "Another way to display the values of all of the arguments is: $@."
echo "Yet another way is: $*."
exit 0

$ ./cmdargs_demo a b c d e f g h i j k l m n
The command name is: ./cmdargs_demo.
The number of command line arguments passed as parameters is 14.
The values of the command line arguments are: a b c d e f g h i a0 a1
Another way to display the values of all of the arguments is: a b c d e f g h i j
k l m n.
Yet another way is: a b c d e f g h i j k l m n.
$ ./cmdargs_demo One Two 3 Four 5 6
The command name is: ./cmdargs_demo.
The number of command line arguments passed as parameters is 6.
The values of the command line arguments are: One Two 3 Four 5 6    One0 One1
Another way to display the values of all of the arguments is: One Two 3 Four 5 6.
Yet another way is: One Two 3 Four 5 6.
$ cmdargs_demo One Two 3 "Four 5 6"
The command name is: ./cmdargs_demo.
The number of command line arguments passed as parameters is 4.
... [output truncated] ...
$
```

On systems where the shell maintains up to nine positional arguments at a time in the shell variables $1–$9, you can write scripts that can accept and process more than nine arguments. To do so, use the shift command. By default, the command shifts the command line arguments to the left by one position, making $2 become $1, $3 become $2, and so on. The first argument, $1, is shifted out. Once shifted, the arguments cannot be restored to their original values. The number of positions to be shifted can be more than one and specified as an argument to the command. The following is a brief description of the command:

Syntax:
 shift [N]
Purpose: Shift the command line arguments *N* positions to the left

---

The script in the **shift_demo** file shows the semantics of the shift command. The first shift command shifts the first argument out and the remaining arguments to the left by one position. The second shift command shifts the current arguments to the left by three positions. The three echo commands are used to display the current value of the program name (*$0*), the values of all positional parameters (*$@*), and the values of the first three positional parameters, respectively. The results of execution of the script are obvious.

```
$ cat shift_demo
#!/bin/bash
echo "The program name is $0."
echo "The arguments are: $@"
echo "The first three arguments are: $1 $2 $3"
shift
echo "The program name is $0."
echo "The arguments are: $@"
echo "The first three arguments are: $1 $2 $3"
shift 3
echo "The program name is $0."
echo "The arguments are: $@"
echo "The first three arguments are: $1 $2 $3"
exit 0
$ ./shift demo 1 2 3 4 5 6 7 8 9 10 11 12
The program name is ./shift_demo.
The arguments are: 1 2 3 4 5 6 7 8 9 10 11 12
The first three arguments are: 1 2 3
The program name is ./shift_demo.
The arguments are: 2 3 4 5 6 7 8 9 10 11 12
The first three arguments are: 2 3 4
The program name is ./shift_demo.
The arguments are: 5 6 7 8 9 10 11 12
The first three arguments are: 5 6 7
$
```

You can use the set command to set the values of shell flags, options, and positional arguments. The most effective use of this command is in conjunction with command substitution. The following is a brief description of the command:

Syntax:
  set [options] [argument-list]

Purpose: Allows you to set shell flags, options, and positional arguments; with no options, it displays the names and values of shell variables

Commonly used options/features:

`--`	Don't consider words starting with a "-" as options
`-C`	Set noclobber to on and force >| to overwrite existing files
`-a`	Automatically export variables upon assignment
`-o [option]`	Set options; display current settings without options. Some of the options are
`hashall`	Save command locations in an internal hash table (default setting); can also be set by using the –h option
`history`	Enable history; default option in interactive shell
`noclobber`	Don't allow overwriting of existing files by output redirection; same as –C option
`ignoreeof`	The interactive shell will not exit on reading end-of-file
`-v [option]`	Verbose mode: display shell input lines as they are read

As you can see, the set command has a variety of features. Here, we discuss its features for setting command line arguments.

The following session involves a simple interactive use of the `set` command. The `date` command is executed to show that the output has six fields. The `set \`date\`` command sets the positional parameters to the output of the `date` command. In particular, *$1* is set to Tue, *$2* to Jul, *$3* to 11, *$4* to 14:51:33, *$5* to PKT, and *$6* to 2017. The `set $(date)` command may be used to perform the same task. The `echo "$@"` command displays the values of all positional arguments. The third `echo` command displays the date in a commonly used form.

```
$ date
Tue Jul 11 14:51:33 PKT 2017
$ set `date`
$ echo "$@"
Tue Jul 11 14:51:33 PKT 2017
$ echo "$2 $3, $6"
Jul 11, 2017
$
```

An option commonly used with the `set` command is --. It is used to inform the `set` command that, if the first argument starts with a -, it should not be considered an option for the `set` command. The script in **set_demo** shows another use of the command. When the script is run with a file argument, it generates a line that contains the file name, the file's inode number, and the file size (in bytes). Note that, on some systems, the size variable needs to be set to *$5*. The `set` command is used to assign the output of the `ls -l` command as the new value of the positional arguments *$1–$9*. If you do not remember the format of the output of the `ls -l` command, we suggest you run this command on a file before studying the code. The first string in the output starts with a – for an ordinary file. Thus, with the use of the – option, the `set` command does not regard – (the first character in the permissions string) as the start of an option for the command and deals with it literally.

```
$ cat set_demo
#!/bin/bash
filename="$1"
set -- `ls -l $filename`
perms="$1"
size="$5"
set `ls -i $filename`
inode="$1"
echo "File Name:        $filename"
echo "Inode Number:     $inode"
echo "Permissions:      $perms"
echo "Size (bytes):     $size"
exit 0
$ ./set_demo set_demo
File Name:        set_demo
Inode Number:     16517945
Permissions:      -rwxr-xr-x
Size (bytes):     245
$
```

In the following in-chapter exercises, you will use the `set` and `shift` commands to reinforce the use of command line arguments and their processing.

Exercise 12.7

Write a shell script that displays all command line arguments, shifts them to the left by two positions, and redisplays them. Show the script along with a few sample runs.

Exercise 12.8

Update the shell script in Exercise 12.7 so that, after accomplishing the original task, it sets the positional arguments to the output of the `who | head -1` command and displays the positional arguments again.

12.5 Comments and Program Headers

You should develop the habit of putting comments in your programs to describe the purpose of a particular series of commands. At times, you should even briefly describe the purpose of a variable or assignment statement. Also, you should use a program header for every shell script that you write. These are simply good software engineering practices. A *program header* is a set of introductory comments used to explain the script. Program header and in-code comments help a programmer who has been assigned the task of maintaining (i.e., modifying or enhancing) your code to understand it quickly. They also help you understand your own code, in particular, if you reread it after some period of time. Long ago, putting comments in the program code or creating separate documentation for programs was not a common practice. Such programs, when inherited by a programmer or a team, are very difficult to understand and maintain, and are commonly known as *legacy code*. You may find different definitions for legacy code in the literature.

A good program header must contain at least the following items. In addition, you can insert any other items that you believe to be important or are commonly used in your organization or group as part of its coding rules. You can use the bash --version and uname -r commands to obtain the Bash version number and Linux release information, respectively.

1. Name of the file containing the script
2. Name of the author
3. Date written
4. Date last modified
5. Platform information (Bash and Linux versions)
5. Purpose of the script (in one or two lines)
6. A brief description of the algorithm used to implement the solution to the problem at hand

A comment line, including every line in the program header, must start with the number sign (#), as in the following session:

```
# This is a comment line.
```

However, a comment does not have to start at a new line; it can follow a command, as in the following session:

```
set -- `ls -l lab1`    # Assign new values to positional parameters and
                       # if the first argument starts with a -, do not
                       # consider it an option for the set command.
                       # This is to handle the output of the ls -l
                       # command if lab1 is an ordinary file.
```

The following is a sample header for the **set_demo** script:

```
# File Name:             ~/linux2e/chapter12/bash/set_demo
# Author:                Syed Mansoor Sarwar
# Date Written:          August 10, 1999
# Date Last Modified:    March 11, 2018 (by the original author)
# Platform Information:  Bash 4.3.42(1)on Linux 4.4.0-21-generic
# Purpose:               To illustrate how the set command works
# Brief Description:     The script runs with a filename as the only command
#                        line argument, saves the filename, runs the set command
#                        to assign output of ls -il command to positional
#                        arguments ($1-$9), and displays file name,its
#                        inode number, file permissions, and its size in bytes.
```

We do not show the program headers for all the sample scripts in this textbook for the sake of brevity.

12.6 Program Control Flow Commands

The program control flow commands/statements are used to determine the sequence in which statements in a shell script execute. There are three basic types of statements for controlling the flow of a script: two-way branching, multiway branching, and repetitive execution of one or more commands. The Bash statement for two-way branching is if, the statements for multiway branching are if and case, and the statements for repetitive execution of some code are for, while, and until.

12.6.1 The if-then-elif-else-fi Statement

The most basic form of the if statement is used for one-way branching, but the statement can also be used for two-way and multiway branching. The following is the syntax and a brief description of the multiway if statement. The words in monospace type are *keywords* and must be used as shown in the syntax. Everything in brackets is optional. All the command lists are designed to enable you to accomplish the task at hand.

Syntax:
```
if expression
  then
    [elif expression
      then
        then-command-list]
      ...
      [else
    else-command-list]
  fi
```
Purpose: To implement two-way or multiway branching

Here, **expression** is a list of commands. The execution of commands in **expression** returns a status of true (success) or false (failure). We discuss three versions of the **if** statement that together comprise the statement's complete syntax and semantics. The first version of the **if** statement is without any optional features, which results in the syntax for the statement that is commonly used for one-way branching.

Syntax:
```
if expression
  then
    then-commands
  fi
```
Purpose: To implement two-way branching

If **expression** is true, the **then-commands** are executed; otherwise, the command after **fi** is executed. The semantics of the statement are illustrated in Figure 12.1.

The expression can be evaluated with the test command. It evaluates an expression and returns true or false. The command has two syntaxes: One uses the keyword test and the other uses brackets. The following is a brief description of the command:

Syntax:
```
test [ expression ]
Or
[[ expression ]]
```
Purpose: To evaluate **expression** and return true or false status

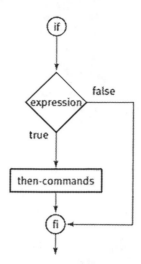

FIGURE 12.1 Semantics of the if-then-fi statement.

An important point about this second syntax is that the inside brackets indicate an optional expression and that the outside brackets are required because they comprise the test statement. Also, at least one space is required before and after an operator, a parenthesis, a bracket, or an operand. If you need to continue a test expression to the next line, you must use a backslash (\) before hitting <Enter> so that the shell does not treat the next line as a separate command. Recall that \ is a shell metacharacter. We demonstrate the use of the test command in the first session but then use the simpler syntax of [].

The test command supports many operators for testing files and integers, testing and comparing strings, and logically connecting two or more expressions to form complex expressions. Table 12.5 describes the meanings of the operators supported by the test command on most Linux systems.

We use the if statement to modify the script in the **set_demo** file so that it takes one command line argument only and checks on whether the argument is a file or a directory. The script returns an error message if the script is run with no or more than one command line argument, or if the command line argument is not an ordinary file. The name of the script file is **if_demo1**.

```
$ cat if_demo1
#!/bin/bash
if test $#  eq 0 then
        echo "Usage: $0 ordinary_file" exit 1
fi
if test $# -gt 1 then
        echo "Usage: $0 ordinary_file" exit 1
fi
if test -f "$1" then
          filename=$1
          set -- `ls -l $filename`
          perms="$1"
          size="$5"
          set `ls -i $filename`
          inode="$1"
          echo "File Name:        $filename"
          echo "Inode Number:     $inode"
          echo "Permissions:      $perms"
          echo "Size (bytes):     $size"
          exit 0
```

```
fi
echo "$0: argument must be an ordinary file" exit 1
$ ./if_demo1
Usage: if_demo1 ordinary_file
$ ./if_demo1 lab1 lab4
Usage: if_demo1 ordinary_file
$ ./if_demo1 dir1
if_demo1: argument must be an ordinary file
$ ./if_demo1 set_demo
File Name:        set_demo
Inode Number:     16517945
```

TABLE 12.5

Operators for the `test` Command

File Testing		Integer Testing		String Testing	
Expression	**Return Value**	**Expression**	**Return Value**	**Expression**	**Return Value**
`-d file`	True if **file** is a directory	`int1 -eq int2`	True if **int1** and **int2** are equal	`Str`	True if **str** is not an empty string
`-f file`	True if **file** is an ordinary file	`int1 -ge int2`	True if **int1** is greater than or equal to **int2**	`str1 = str2`	True if **str1** and **str2** are the same
`-r file`	True if **file** is readable	`int1 -gt int2`	True if **int1** is greater than **int2**	`str1 != str2`	True if **str1** and **str2** are not the same
`-s file`	True if length of the **file** is nonzero	`int1 -le int2`	True if **int1** is less than or equal to **int2**	`-n str`	True if the length of **str** is greater than zero
`-t [filedes]`	True if file descriptor **filedes** is associated with the terminal	`int1 -lt int2`	True if **int1** is less than **int2**	`-z str`	True if the length of **str** is zero
`-w file`	True if **file** is writable	`int1 -ne int2`	True if **int1** is not equal to **int2**		
`-x file`	True if **file** is executable				
`-b file`	True if **file** is a block file				
`-c file`	True if **file** is a character special file				
`-e file`	True if **file** exists				
`-L file`	True if **file** is a symbolic link				

Operators for Forming Complex Expressions

`!`	Logical NOT operator: true if the following expression is false	`( expression )`	Parentheses for grouping expressions; at least one space before and one after each parenthesis
`-a or &&`	Logical AND operator: true if the previous (left) and next (right) expressions are true	`-o or ||`	Logical OR operator: true if the previous (left) or next (right) expression is true

```
Permissions:        -rwxr-xr-x
Size (bytes):       245
$
```

In the preceding script, the first if statement displays an error message and exits the program if you run the script without any command line argument. The second if statement displays an error message and exits the program if you run the script with more than one argument. The third if statement is executed if conditions for the first two are false—that is, if you run the script with one argument only. This if statement produces the desired results if the command line argument is an ordinary file. If the passed argument is not an ordinary file, the condition for the third if statement is false and the error message if_demo1: argument must be an ordinary file is displayed. The exit command is used to take control out of the script. We normally use 1 as the argument (i.e., exit status) of exit when a script is to be terminated because of an erroneous condition. The exit status 0 is used for normal termination of the script.

We can rewrite the script in **if_demo1** using the alternative syntax for the test command and the logical AND operator, -a, as shown in Tables 12.5 and 12.6.

```
$ cat if_demo10
#!/bin/bash
if [ "$#" -eq 1  -a  -f "$1" ]
    then
        ...
fi
echo "$0:           Specify exactly one command line argument that"
echo "              must be an ordinary file."
exit 1
$
```

TABLE 12.6

Precedence of Bash Operators (High to Low)

Operator	Precedence
var++ var--	C-style postincrement, postdecrement
++var --var	C-style preincrement, predecrement
! ~	Negation logical/bitwise, inverts sense of following operator
**	Exponentiation
* / %	Multiplication, division, modulo
+ -	Addition, subtraction
<< >>	Bitwise shift left, shift right
-z -n	Unary comparison for a string is or is-not null
-e -f -t -x, etc.	Unary comparison for file-test
< -lt > -gt <= -le >= -ge	Compound comparison for strings and integers
-nt -ot -ef	Compound comparison for file test
== -eq != -ne	Equality/inequality test operators for strings and integers
&	Bitwise AND
^	Bitwise XOR (exclusive-OR)
|	Bitwise OR
&& -a	Logical AND
|| -o	Logical OR
?:	C-style trinary operator (a conditional statement)
=	Assignment
*= /= %= += -= <<= >>= &=	Combination assignment: times-equal, divide-equal, mod-equal
,	For linking a sequence of operations

Long expressions involving –a and –o (logical OR) operators can become confusing. You can write such expressions in a cleaner and more readable way by using the operators && (for logical AND) and || (for logical OR). The –o, &&, and || operators are also discussed in Tables 12.5 and 12.6. The expression in the **if_demo10** file can be written as follows.

```
$ cat if_demo10
#!/bin/bash
if [[ "$#" -eq 1 ]] && [[  -f "$1" ]]
    then
        ...
fi
...
$
```

The expression "A && B" returns true if both A and B are true. Further, while evaluating this expression, B is evaluated only if A succeeds (i.e., returns true). The expression "A || B" returns true if either A or B, or both are true. In case of this expression, B is evaluated only if A fails (i.e., returns false). We have discussed a couple of examples of these operators for expression evaluation in the scripts earlier. In the following session, we show a couple of examples of these operators for command execution. As expected, for the && operator, the echo command is not executed if the left-hand side of the operator is false. Also as expected, for the || operator, the echo command is no executed if the left-hand side of the operator is true.

```
$ true && echo "Hello, world!"
Hello, world!
$ false && echo "Hello, world!"
$ false || echo "Hello, world!"
Hello, world!
$ true || echo "Hello, world!"
$
```

As shown in Table 12.6, the operators &, |, and ∧ are bitwise AND, OR, and Exclusive-OR (XOR), respectively. When bitwise AND of two bits is performed, then the resultant bit is 1 only if both bits are 1; it is 0 otherwise. The bitwise OR of two bits results in 0 only if both bits are 0; it is 1 otherwise. Finally, the XOR of two bits is 1 only if one of the two bits is 1 and the other is 0; it is 0 otherwise.

An important practice in script writing is to correctly indent the commands/statements in it. Proper indentation of programs enhances their readability and makes them easier to understand and maintain. However, the white spacing neither impacts the syntactic correctness of the program nor its efficiency when executed. Note the indentation style used in our sample scripts and follow it when you write your own scripts.

We now discuss the second version of the if statement, which also allows two-way branching. The following is a brief description of the statement:

Syntax:

```
if expression
    then
            then-commands
    else
            else-commands
fi
```

Purpose: To implement two-way branching

If **expression** is true, the commands in **then-commands** are executed; otherwise, the commands in **else-commands** are executed, followed by the execution of the command after **fi**. The semantics of the statement are depicted in Figure 12.2.

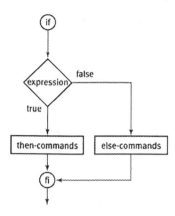

FIGURE 12.2 Semantics of the `if-then-else-fi` statement.

Next, we rewrite the **if_demo1** program, using the `if-then-else-fi` statement, and use the alternative syntax for the `test` command. The resulting script is in the **if_demo2** file, as shown in the following session. Note that the program looks cleaner and more readable.

```
$ cat if_demo2
#!/bin/bash
if [[ $# -eq 1 ]] && [[ -f "$1" ]]
    then
        filename=$1
        set -- `ls -l $filename`
        perms="$1"
        size="$5"
        set `ls -i $filename`
        inode="$1"
        echo "File Name:      $filename"
        echo "Inode Number:   $inode"
        echo "Permissions:    $perms"
        echo "Size (bytes):   $size"
        exit 0
    else
        echo "$0:     Specify exactly one command line argument "
        echo "        must be an ordinary file."
        exit 1
fi
exit 0
$
```

Finally, we discuss the third version of the `if` statement, which is used to implement multiway branching. The following is a brief description of the statement:

> **Syntax:**
>   **if expression1**
>   **then**
>   **then-commands**
>   **elif expression2**
>   **elif1-commands**
>   **elif expression3**
>   **elif2-commands**

```
...
     else
          else-commands
  fi
```

Purpose: To implement multiway branching

If **expression1** is true, the commands in **then-commands** are executed. If **expression1** is false, **expression2** is evaluated, and if it is true, the commands in **elif1-commands** are executed. If **expression2** is also false, **expression3** is evaluated. If **expression3** is true, the commands in **elif2-commands** are executed. If **expression3** is also false, the commands in **else-commands** are executed. The execution of any command list is followed by the execution of the command after **fi**. You can use any number of elifs in an if statement to implement multiway branching. The semantics of the statement are depicted in Figure 12.3.

We modify the script in the **if_demo** file so that, if the command line argument is a directory, the program displays the number of files and subdirectories in the directory, excluding the hidden files. In addition, the program ensures that the command line argument is an existing file or directory in the current directory before processing it. In addition to using two if statements, we also use the if-then-elif-else-fi statement in the implementation. The resulting script is in the **if_demo3** file, as shown in the following session. Note the use of the logical negation operator (!) in the condition for the if statement. This operator can also be used for bitwise negation, as described in Tables 12.5 and 12.6.

```
$ cat if_demo3
#!/bin/bash
if [[ ! $# -eq 1 ]]
   then
        echo "Usage: $0 file"
        exit 1
   elif [ -d "$1" ]
        then
            nfiles=`ls "$1" | wc -w`
            echo "The number of files in the $1 directory is $nfiles"
            exit 0
   else
        ls "$1" 2> /dev/null | grep "$1" 2> /dev/null 1>&2
        if [ $? -ne 0 ]
           then
                echo "$1: not found"
                exit 1
```

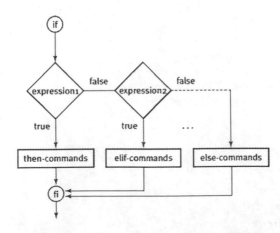

FIGURE 12.3 Semantics of the if-then-elif-else-fi statement.

```
        fi
        if [ -f "$1" ]
            then
                    filename=$1
                    set -- `ls -l $filename`
                    perms="$1"
                    size="$5"
                    set `ls -i $filename`
                    inode="$1"
                    echo "File Name:        $filename"
                    echo "Inode Number:     $inode"
                    echo "Permissions:      $perms"
                    echo "Size (bytes):     $size"
                    exit 0
            else
                    echo "$0: argument must be an ordinary file or directory"
                    exit 1
            fi
fi
$ ./if_demo3 lab2
lab2: not found
$ ./if_demo3 lab1 lab2
Usage: if_demo3 file
$ ./if_demo3 ~
The number of files in the /home/sarwar directory is 14
$ ./if_demo3 set_demo
File Name:        set_demo
Inode Number:     16517945
Permissions:      -rwxr-xr-x
Size (bytes):     245
$ ./if_demo3 /bin/ls
File Name:        /bin/ls
Inode Number:     10485856
Permissions:      -rwxr-xr-x
Size (bytes):     126584
$
```

If the argument is a directory, the number of files in it, excluding directories and hidden files, is saved in the nfiles variable. The command ls "$1" 2> /dev/null | grep "$1" 2>/dev/null 1>&2 is executed to check whether the file passed as the command line argument exists. The standard error is redirected to **/dev/null** (the Linux black hole), and the standard output is redirected to standard error by using 1>&2. Thus, the command does not produce any output or error messages; its only purpose is to set the command's return status value in *$?* If the command line argument exists, the ls command is successful and *$?* contains 0; otherwise, it contains a nonzero value. If the command line argument is a file, the required file-related data is displayed. Note the use of command substitution and pipe for setting the value of the *nfiles* variable.

In the following in-chapter exercises, you will practice the use of the if statement, command substitution, and manipulation of positional parameters.

Exercise 12.9

Create the **if_demo2** script file and run it with no argument, more than one argument, and one argument only. While running the script with one argument, use a directory as the argument. What happens? Does the output make sense?

Exercise 12.10

Write a shell script whose single command line argument is a file. If you run the program with an ordinary file, the program displays the owner's name and last update time for the file. If the program is run with more than one argument, it generates meaningful error messages.

12.6.2 The for Statement

The for statement is the first of the three statements that are available in Bash for repetitive execution of a block of commands in a shell script. These statements are commonly known as *loops*. The following is a brief description of the statement:

Syntax:

```
for variable [in argument-list]
do
    command-list
done
```

Purpose: To execute commands in `command-list` as many times as the number of items in the `argument-list`; without the optional part in `argument-list,` the arguments are supplied at the command line

---

The items in `argument-list` are assigned to `variable` one by one, and the commands in `command-list`, also known as the body of the loop, are executed for every assignment. This process allows the execution of commands in `command-list` as many times as the number of items in `argument-list`. Figure 12.4 illustrates the semantics of the for command. Items may be numbers of words.

The following script in the **for_demo1** file shows the use of the for command with optional arguments. The variable *people* is assigned the words in `argument-list` one by one, and each time, the value of the variable is echoed until no word remains in the list. At that time, control comes out of the for statement,

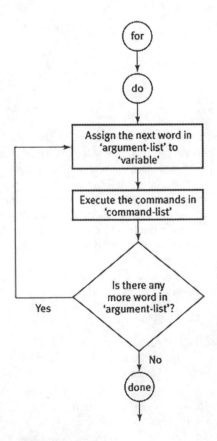

FIGURE 12.4   Semantics of the for statement.

and the command following done is executed. Then the code following the for statement, the exit 0 statement only in this case, is executed.

```
$ cat for_demo1
#!/bin/bash
for people in Debbie Jamie John Kitty Kuhn Shah
do
        echo "$people"
done
exit 0
$ ./for_demo1
Debbie
Jamie
John
Kitty
Kuhn
Shah
$
```

The following script in the **user_info** file takes a list of existing (i.e., valid) login names as command line arguments and displays each login name and the full name of the user who owns the login name, one per login. In the sample run, the first value of the user variable is **dheckman**. The echo command displays dheckman: followed by a <Tab>, and the cursor stays at the current line. The first grep command is used to check if **dheckman** has an entry in the **/etc/passwd** file. If the answer is no, the process is repeated for the second command line argument. If **dheckman** is found, the second grep command searches the **/etc/passwd** file for the login name **dheckman** and pipes it to the cut command, which displays the fifth field in the **/etc/passwd** line for **dheckman** (his full name). The process is repeated for the remaining two login names (**ghacker** and **msarwar**). No user is left in the list passed at the command line, so control comes out of the for statement and the exit 0 command is executed to transfer control back to shell. The command substitution "^"`echo $user":"` in the grep command can be replaced by "^"$user":".

```
$ cat user_info
#!/bin/bash
for user
do
# Don't display anything if a login name is not found in /etc/passwd
    grep "^"`echo $user":"` /etc/passwd 1> /dev/null 2>&1
    if [ $? -eq 0 ]
        then
            echo -n "$user:        "
            grep "^"`echo $user":"` /etc/passwd | cut -f5 -d':'
    fi
done
exit 0
$ ./user_info dheckman ghacker sarwar
dheckman: Dennis R. Heckman
ghacker:  George Hacker
sarwar:   Syed Mansoor Sarwar
$
```

Exercise 12.11

Verify the working of the scripts discussed in this section on your system.

Exercise 12.12

Write a shell script that takes an arbitrary number of command line arguments and displays them all, one by one, using a for loop and shift statement. Show the script along with a few example runs.

12.6.3 The while Statement

The while statement, also known as the while loop, allows repeated execution of a block of code based on the condition of an expression. The following is a brief description of the statement:

Syntax:
```
while expression
do
    command-list
done
```
Purpose: To execute commands in **command-list** as long as expression evaluates to true

---

The **expression** is evaluated and, if the result of this evaluation is true, the commands in **command-list** are executed and the **expression** is evaluated again. This sequence of expression evaluation and execution of **command-list**, known as one iteration, is repeated until the **expression** evaluates to false. At that time, control comes out of the **while** statement and the statement following **done** is executed. Figure 12.5 depicts the semantics of the while statement.

The variables and/or conditions in the expression that result in a true value must be correctly manipulated in the commands in **command-list** for well-behaved loops—that is, loops that eventually terminate and allow execution of the rest of the code in a script. Loops in which the expression always evaluates to true are known as *nonterminating*, or *infinite*, loops. Infinite loops, usually a result of poor design and/or programming, are undesirable because they continuously use CPU time without accomplishing any useful task. However, some applications do require infinite loops. For example, all the servers for Internet services, such as the Hypertext Transfer Protocol (HTTP) service (which allows us to browse Web pages on the Internet) are programs that run indefinitely, waiting for client requests. In case of the HTTP service, for example, the client requests come through Web browsers (i.e., HTTP clients) such as Mozilla Firefox. Once a server has received a client request, it processes it, sends a response to the client, and waits for another client request. The only way to terminate a process with an infinite loop is to

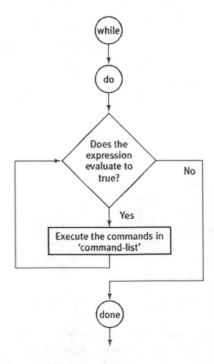

FIGURE 12.5 Semantics of the while statement.

kill it by using the `kill` command. Or, if the process is running in the foreground, pressing <Ctrl+C> would also do the trick, unless the process is designed to ignore <Ctrl+C>. In that case, you need to put the process in the background by pressing <Ctrl+Z> and use the `kill -9` command to terminate it. See Chapter 10 for details on processes.

The script in the **while_demo** file shows a simple use of the `while` loop. When you run this script, the *secretcode* variable is initialized to agent007 and you are prompted to make a guess. Your guess is stored in a local variable *yourguess*. If your guess is not agent007, the condition for the `while` loop is true and the commands between do and done are executed. This program displays a message tactfully informing you of your failure and prompts you for another guess. Your guess is again stored in the *your guess* variable, and the condition for the loop is tested. This process continues until you enter agent007 as your guess. At which time, the condition for the loop becomes false and the control comes out of the `while` statement. The echo command following done executes, congratulating you for being part of a great gene pool!

```
$ cat while_demo
#!/bin/bash
secretcode=agent007
echo "Guess the code!"
echo -e "Enter your guess: \c"
read yourguess
while [ "$secretcode" != "$yourguess" ]
do
        echo "Good guess but wrong. Try again!"
        echo -e "Enter your guess: \c"
        read yourguess
done
echo "Wow! You are a genius!!"
exit 0
$ ./while_demo
Guess the code!
Enter your guess: star wars
Good guess but wrong. Try again! Enter your guess: columbo
Good guess but wrong. Try again! Enter your guess: agent007
Wow! You are a genius!!
$
```

> **Exercise 12.13**
>
> Repeat Exercise 12.12 but replace the `for` loop with a `while` loop. Show the script along with a few example runs.

12.6.4 The `until` Statement

The syntax of the `until` statement is similar to that of the `while` statement, but its semantics are different. Although, in the `while` statement, the loop body executes as long as the expression evaluates to true, in the `until` statement, the loop body executes as long as the expression evaluates to false. The following is a brief description of the statement:

> **Syntax:**
>
>   until expression
>   do
>   command-list
>   done
>
> **Purpose:** To execute commands in **command-list** as long as expression evaluates to false

Figure 12.6 illustrates the semantics of the `until` statement. The code in the **until_demo** file performs the same task that the script in the **while_demo** file does (see Section 12.6.3), but it uses the `until` statement

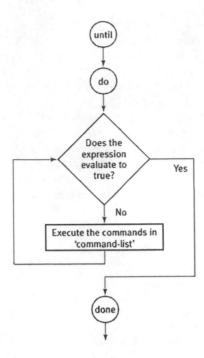

FIGURE 12.6 Semantics of the until statement.

instead of the while statement. Thus, the code between do and done (the loop body) is executed for as long as your guess is not agent007.

```
$ cat until_demo
#!/bin/bash
secretcode=agent007
echo "Guess the code!"
echo -n  "Enter your guess: "
read yourguess
until [ "$secretcode" = "$yourguess" ]
do
        echo "Good guess but wrong. Try again!"
        echo -n "Enter your guess: "
        read yourguess
done
echo "Wow! You are a genius!!"
exit 0
$ ./until_demo
Guess the code!
Enter your guess: Inspector Gadget Good guess but wrong. Try again! Enter your
guess: Peter Sellers Good guess but wrong. Try again! Enter your guess: agent007
Wow! You are a genius!!
$\
```

Exercise 12.14

Repeat Exercise 12.13 but replace the while loop with an until loop. Show the script along with a few example runs.

12.6.5 The break and continue Commands

The break and continue commands can be used to interrupt the sequential execution of the loop body. The break command transfers control to the command following done, thus terminating the loop

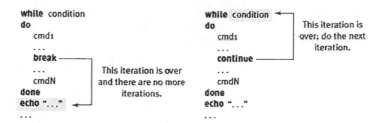

FIGURE 12.7   Semantics of the break and continue commands.

prematurely. The continue command transfers control to done, which results in the evaluation of the condition again and, hence, continuation of the loop. In both cases, the commands in the loop body following these statements are not executed. Thus, these statements are almost always used in an if or if-else setting. Figure 12.7 illustrates the semantics of these commands.

In the following in-chapter exercises, you will write shell scripts with loops by using the for, while, and until statements.

Exercise 12.15

Write a shell script that takes a list of host names on your network as command line arguments and displays whether the hosts are up or down. Use the ping command to display the status of a host and the for statement to process all host names.

Exercise 12.16

Rewrite the script in Exercise 12.15, using the while statement. Rewrite it again, using the until statement.

12.6.6 The case Statement

The case statement provides a mechanism for multiway branching similar to a nested if statement. However, the structure provided by the case statement is more readable. You would use the case statement when you can—that is, when you are testing a single variable to several distinct patterns. You would not use it when you want to test more than one variable. The following is a brief description of the statement:

Syntax:
```
case test-string in
    pattern1)        command-list1
                     ;;
    pattern2)        command-list2
                     ;;
    ...
    patternN)        command-listN
                     ;;
    esac
```
Purpose: To implement multiway branching like a nested if

The case statement compares the value in **test-string** with the values of all the patterns one by one until either a match is found or no more patterns with which to match **test-string** remain. If a match is found, the commands in the corresponding **command-list** are executed and control goes out of the case statement. If no match is found, control goes out of case. However, in a typical use of the case statement, a wild card pattern matches any value of **test-string**. Also known as the *default case*, it allows the execution of a set of commands to handle an exception (i.e., error) condition for situations in which the value in **test-string** does not match any pattern. Back-to-back semicolons (;;) are used to delimit a **command-list**. Without ;; the first command in the command list for the next pattern is executed, resulting in an unexpected behavior by the program. Figure 12.8 depicts the semantics of the case statement.

The following script in the **case_demo** file shows a simple but representative use of the case statement. It is a menu-driven program that displays a menu of options and prompts you to enter an option. Your option is read into a variable called *option*. The case statement then matches your option with one of the four available patterns (single characters in this case) one by one, and when a match is found, the corresponding **command-list** (a single command in this case) is executed. Thus, at the prompt, if you type d and hit <Enter>, the date command is executed and control goes out of case. Then, the program exits after the exit 0 command executes. A few sample runs of the script follow the code in this session.

```
$ cat case_demo
#!/bin/bash
echo "Use one of the following options:"
echo "  d:     To display today's date and present time"
echo "  l:     To see the listing of files in your present working directory"
echo "  w:     To see who's logged in"
echo "  q:     To quit this program"
echo -n "Enter your option and hit <Enter>: "
read option
case "$option" in
        d)      date
                ;;
        l)      ls
                ;;
        w)      who
                ;;
        q)      exit 0
                ;;
esac
exit 0
$ ./case_demo
Use one of the following options:
  d:     To display today's date and present time
  l:     To see the listing of files in your present working directory
```

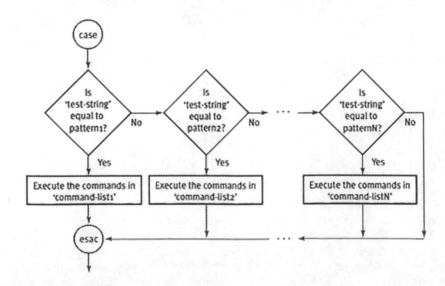

FIGURE 12.8 Semantics of the case statement.

```
  w:     To see who's logged in
  q:     To quit this program
Enter your option and hit <Enter>: d
Thu Jul 13 12:55:25 PKT 2017
$ ./case_demo
Use one of the following options:
  d:     To display today's date and present time
  l:     To see the listing of files in your present working directory
  w:     To see who's logged in
  q:     To quit this program
Enter your option and hit <Enter>: w
root       tty1          2017-04-06 18:57
sarwar     pts/0         2017-07-13 12:55 (119.160.96.110)
davis      tty8          2017-04-12 11:22 (:0)
jacob      tty2          2017-04-13 19:07
$ ./case_demo
Use one of the following options:
  d:     To display today's date and present time
  l:     To see the listing of files in your present working directory
  w:     To see who's logged in
  q:     To quit this program
Enter your option and hit <Enter>: l
case_demo               export_demo    if_demo2   shift_demo   while_demo
display_change_name     for_demo1      if_demo3   until_demo
display_name            if_demo1       set_demo   user_info
$ ./case_demo
Use one of the following options:
  d:     To display today's date and present time
  l:     To see the listing of files in your present working directory
  w:     To see who's logged in
  q:     To quit this program
Enter your option and hit <Enter>: q
$
```

From the output of the w option, it seems that **davis** is using the system through a remote login, most likely through an secure shell (ssh) session. Note that, when you enter a valid option, the expected output is displayed. However, when you enter input that is not a valid option (a in the preceding session), the program does not give you any feedback. The reason is that the case statement matches your input with all the patterns, one by one, and exits when there is no match. We need to modify the script slightly so that when you enter an invalid option, the script tells you so and then terminates. To do so, we add the following code.

```
*)       echo "Invalid option; try running the program again."
         exit 1
         ;;
```

We also enhance the script so that uppercase and lowercase inputs are considered to be the same. We use the pipe symbol (|) in the patterns to specify a logical OR operation. The enhanced code and some sample runs are shown in the following session.

```
$ cat case_demo
echo "Use one of the following options:"
echo "  d or D:    To display today's date and present time"
echo "  l or L:    To see the listing of files in your present working directory"
echo "  w or W:    To see who's logged in"
echo "  q or Q:    To quit this program"
echo -n "Enter your option and hit <Enter>: "
read option
case "$option" in
     d|D)    date
```

```
                ;;
    l|L)    ls
                ;;
    w|W)    who
                ;;
    q|Q)    exit 0
                ;;
    *)      echo "Invalid option; try running the program again."
            exit 1
                ;;
esac
exit 0
```

```
$ ./case_demo
Use one of the following options:
  d or D:    To display today's date and present time
  l or L:    To see the listing of files in your present working directory
  w or W:    To see who's logged in
  q or Q:    To quit this program
Enter your option and hit <Enter>: D
Thu Jul 13 13:00:57 PKT 2017
$ ./case_demo
Use one of the following options:
  d or D:    To display today's date and present time
  l or L:    To see the listing of files in your present working directory
  w or W:    To see who's logged in
  q or Q:    To quit this program
Enter your option and hit <Enter>: d
Thu Jul 13 13:01:34 PKT 2017
$ ./case_demo
Use one of the following options:
  d or D:    To display today's date and present time
  l or L:    To see the listing of files in your present working directory
  w or W:    To see who's logged in
  q or Q:    To quit this program
Enter your option and hit <Enter>: a
Invalid option; try running the program again.
$
```

Exercise 12.17

Update the previous script so that when the user enters u or U as input, the script invokes the **until_demo** script discussed in the previous section.

12.7 Command Grouping

A number of shell commands may be run as a group. The following is a brief description of command grouping:

Syntax:

(command-list)

Or

{ command-list; }

Purpose: Syntax 1—To execute commands in **command-list** under a child of the current Bash

Syntax 2—To execute commands in **command-list** as a group

The child shell inherits a copy of the parent shell's environment, except trapped but ignored signals. When commands are executed by using the second syntax, no child shell is created. It allows redirection of the outputs of the commands in **command-list**. In the second syntax, it is mandatory to have a semicolon at the end of the last command, a space after {, and a space before }. The following interactive session shows a few examples of command grouping using both syntaxes.

```
$ { date; }
Thu Jul 13 13:02:55 PKT 2017
$ { date; echo -n "Hello, "; echo "world!"; }
Sat Aug  2 06:35:27 PKT 2014
Hello, world!
$ { date; echo -n "Hello, "; echo "world!"; } > groupout
$ more groupout
Thu Jul 13 13:03:17 PKT 2017
Hello, world!
$ { date; echo; ps; echo; echo "Hello, world!"; }
Thu Jul 13 13:03:43 PKT 2017

  PID TTY          TIME CMD
14652 pts/0    00:00:00 bash
14802 pts/0    00:00:00 ps

Hello, world!
$ (date; pwd)
Thu Jul 13 13:04:10 PKT 2017
/home/sarwar/linux2e/ch12
$ (date; pwd) > groupout
$ cat groupout
Thu Jul 13 13:04:35 PKT 2017
/home/sarwar/linux2e/ch12
$ (date; echo; ps; echo; echo "Hello, world!")
Thu Jul 13 13:05:11 PKT 2017

  PID TTY          TIME CMD
14652 pts/0    00:00:00 bash
14836 pts/0    00:00:00 bash
14838 pts/0    00:00:00 ps

Hello, world!
$
```

The outputs of the commands in this session are self-explanatory. Note that no child shell is created to execute the commands in { date; echo; ps; echo; echo "Hello, world!"; }. However, the commands in (date; echo; ps; echo; echo "Hello, world!") are executed under a child shell and the PID of the child shell is 14836.

Exercise 12.18

Reproduce the previous shell session on your system.

Summary

Every Linux shell has a programming language that allows you to write programs for performing tasks that cannot be performed by existing commands. These programs are commonly known as shell scripts. In its simplest form, a shell script consists of a list of shell commands that are executed by a shell one by one, sequentially. More advanced scripts contain program control flow statements for implementing multiway branching and repetitive execution of a block of commands in a script. The shell programs that consist of Bash commands, statements, and features are called Bash scripts.

The shell variables are main memory locations that are given names and can be read from and written to. There are two types of shell variables: environment variables and user-defined variables. The environment variables are initialized by the shell at the time of user login and are maintained by the shell to provide a nice work environment. The user-defined variables are used as scratch pads in a script to accomplish the task at hand. Some environment variables such as positional parameters are read only in the sense that you cannot change their values without using the set command. User-defined variables can also be made read only by using the readonly command.

The Bash commands for processing shell variables are = (for assigning a value to a variable), set (for setting values of positional parameters and displaying values of all environment variables), env (for displaying values of all shell variables), export (for allowing subsequent commands to access shell variables), read (for assigning values to variables from the keyboard), readonly (for making user-defined variables read only), shift (for shifting command line arguments to the left by one or more positions), unset (to reset the value of a read/write variable to null), and test (to evaluate an expression and return true or false).

The program control flow statements if and case allow the programmer of a shell script to implement multiway branching; the for, until, and while statements allow the programmer to implement loops; and the break and continue statements allow the user to interrupt sequential execution of a loop in a script. Input/output redirection, command substitution, and other shell features can be used with control flow statements as with other shell commands (see Chapter 13).

Bash has a number of operators for file, integer, and string testing. It also has a number of logical and bitwise operators, as well as operators for sequential execution of commands individually and as groups using ;, (), and {}. Bash also has the & operator for background and parallel execution of commands. The operators && and || may be used for performing logical AND and OR operations, respectively. The use of these operators results in a cleaner and more compact code. The use of ;, (), and {} facilitates the execution of a group of commands sequentially, one after the other. The operators &, |, and ∧ may be used to perform bitwise AND, OR, and XOR operations.

Questions and Problems

1. What is a shell script? Describe three ways of executing a shell script.

2. What is a shell variable? What is a read-only variable? How can you make a user-defined variable read only? Give an example to illustrate your answer.

3. Which shell environment variable is used to store your search path? Change your search path interactively to include **~/bin** and your current directory (.). What would this change allow you to do? Why? If you want to make it a permanent change, what would you do? See Chapter 4 if you have forgotten how to change the search path of your shell.

4. What will be the output if the shell script **read_demo** in Section 12.3.6 is executed and you give * as input each time you are prompted?

5. Give the syntax for the read command that may be used to read a line from keyboard into the variable called line1. Run the command under Bash followed by running the echo line1. Capture and show the shell session, including the input entered from the keyboard and the output produced by the echo command.

6. Perform the task outlined in Problem 6 by using the three command sequence: echo, read, and echo.

7. Write a shell script that takes an ordinary file as an argument and removes the file if its size is zero. Otherwise, the script displays the following information about the file: name, size (in bytes), number of hard links, owner, and modify date (in this order) on one line. Your script must do the appropriate error checking.

8. Write a shell script that takes a directory as a required argument and displays the names of all zero-length files in it. Do the appropriate error checking.

9. Write a shell script that removes all zero-length ordinary files from the current directory. Do appropriate error checking.

10. Modify the script in Problem 6 so that it removes all zero-length ordinary files in the directory passed as an optional argument. If you don't specify the directory argument, the script uses the present working directory as the default argument. Do the appropriate error checking.

11. Run the script in **if_demo2** in Section 12.6.1 with if_demo2 as its argument. Does the output make sense to you? Why or why not?

12. Write a shell script that takes a list of login names on your computer system as command line arguments and displays these login names and full names of the users who own these logins (as contained in the **/etc/passwd** file), one per line. If a login name is invalid (i.e., not found in the **/etc/passwd** file), display the login name but nothing for the full name. The format of the output line is login name: username.

13. What happens when you run a stand-alone command enclosed in back quotes (grave accents), such as `date`? Why?

14. What happens when you type the following sequence of shell commands?

    a. name=date

    b. $name

    c. `$name`

15. Look at your ~/**.profile** and **/etc/profile** files and list the environment variables that are exported along with their values. What is the purpose of each variable?

    You have written three Bash scripts and placed them in files called **script1, script2,** and **script3**. The three scripts are placed in your ~/**linux2e/ch12** directory. In the body of script3, you use script1 and script2. Suppose you run script3 and get error message as shown in the following session.

    ```
    $ ./script3
    ./script3: line 4: script1: command not found
    ./script3: line 5: script2: command not found
    $
    ```

    Describe two methods that may be used to remove these error messages and enable script3 to work correctly by using script1 and script2.

16. Write a Bash script that takes a list of login names as its arguments and displays the number of terminals that each user is logged on to in a LAN environment.

17. Write a Bash script **domain2ip** that takes a list of domain names as command line arguments and displays their IP addresses. Use the nslookup command. The following is a sample run of this program:

    ```
    $ domain2ip usc.edu up.edu redhat.com
    Name: usc.edu
    Address: 128.125.253.136
    Name: up.edu
    Address: 64.251.254.23
    Name: redhat.com
    Address: 209.132.183.105
    $
    ```

18. Modify the script in the **case_demo** file in Section 12.6.6 so that it allows you to try any number of options and quits only when you use the q option.

19. Write a Bash script that displays the following menu and prompts you for one-character input to invoke a menu option, as follows:

    a. List all files in the present working directory

    b. Display today's date and time

    c. Invoke the shell script for Problem 14

d. Display whether a file is a *simple* file or a *directory*

e. Create a backup for a file

f. Start an ssh session

g. Start an ftp session

h. Exit

Option (c) requires that you ask for a list of login names; and for options (d) and (e), insert a prompt for file names before invoking a shell command/program. For options (f) and (g), insert a prompt for a domain name (or IP address) before initiating an ssh or ftp session. The program should allow you to try any option any number of times and should quit only when you give option x as input. A good programming practice for you to adopt is to build code incrementally—that is, write code for one option, test it, and then go to the next option.

20. Modify the Bash script for Problem 19 so that it executes code for each option under a child shell. Display the PID for the child shell whenever it is initiated to run the code for an option, before the code is executed.

21. Suppose the following expression is evaluated as a condition in an if statement. When does the expression evaluate to true?

```
[ "$#" -eq 1  -a  -f "$1" -o -d "$4" ]
```

Write the preceding expression using the && and || operators.

22. Clearly state the semantics of the following Bash commands, where A and B are commands:

a. A ; B

b. A && B

c. A || B

d. A&

23. Run the following commands on your system and explain their outputs.

a. echo Hello || echo "Hello, world!"

b. echo Hello && echo "Hello, world!"

24. Explain the tests in the following statements:

a. while [-n "$var1" -a "$var2" -gt 0]

b. if ["$v1" -gt "$v2" -o "$v1" -lt "$v2" -a -e "$filename"]

25. Transform expressions given in Problem 24 into equivalent forms using && and || operators.

26. What is the purpose of the echo * command? Run the command on your system and explain the output of the command.

27. Bash allows the following types of command groupings: (command-list) and { command-list; }. Run the following commands on your system and answer the questions that follow. If there are any errors, correct them and describe those corrections.

a. (date; pwd; who; echo "Hello, \c"; echo "world!")

b. {date; pwd; who; echo "Hello, world!"}

Make appropriate changes in the corrected versions of the two command groups so that, in each case, the outputs of the date and pwd commands go to **file1** and outputs of the who and echo commands are redirected to **file2**. Then, demonstrate that commands in the first group are executed under a child shell and those in the second group are executed under the current shell. For the first case, show two ways of displaying the PID of the child shell.

Advanced Questions and Problems

28. Give a one-line expression corresponding to the following script that uses the operators && and ||.

```
if [ -f Students ]
    then
          echo "file found"
    else
          echo "file not found"
fi
```

29. Following the Example 17.8 Bash script file presented in Chapter 17, Section 6.3.2, rewrite the code to produce three different script files that accomplish the following:

   a. The source directory is a subdirectory of your home directory on your Linux system, and the target directory is on a USB thumbdrive mounted on your machine.

   b. The source directory is a subdirectory of your home directory on your Linux system, and the target directory is a directory on another Linux machine on your intranet LAN.

   c. The source directory is a subdirectory of your home directory on your Linux system, and the target directory is a directory on another Linux machine in an accessible place on the Internet.

30. Repeat Problem 28 so that the Bash script files use the rsync command and no compression, instead of the **tar** command.

31. Design and implement a simple Bash script file that backs up what you consider critical directories and files from your home directory on your Linux system, using Linux commands of your choice inside the script file. Customize the script file, given your Linux system and its storage media configuration. That is, you may have access privileges to other locally mounted hard drives, NFSv4 network drives, or remote storage locations that you can use as backup targets.

32. Is it possible to use the capabilities of the Zettabyte File System (ZFS) to achieve exactly the same archiving results that Problems 28–30 accomplish? What functions and commands would you need to use from ZFS that would do this? Be explicit in detailing not only the capabilities but also the sequence of how they would be applied to do the same things as Problems 28–30.

Projects

Project 1: 1990s Linux

Design and implement an "automated" Bash script file that uses the rsync command and cron to do regularly scheduled backups of local user account home directories, given your local Linux system and its storage media configuration. That is, you may have access privileges to other locally mounted hard drives, NFSv4 network drives, or remote storage locations that you can use as backup targets. You may elect to only apply this script file to a small subset of all user accounts for testing. Implicit in the design and implementation of this project is that you have the requisite privileges on your system to accomplish what we are specifying here, and that the target media have adequate storage space that you can access. In your script file, how would you handle error conditions generated by inadequate space remaining on target storage media?

Project 2: 21st-Century Linux

Similar to Project 1, design and implement an "automated" Bash script file (or Python program) method that uses the rsync command and systemd (rather than **cron**) to do regularly scheduled backups of local user account home directories, given your local Linux system and its storage media configuration. That is, you may have access privileges to other locally mounted hard drives, NFSv4 network drives, or remote storage locations that you can use as backup targets.

You may elect to only apply this method to a small subset of all user accounts, for testing. Implicit in the design and implementation of this project is that you have the requisite privileges on your system to accomplish what we are specifying here, and that the target media have adequate storage space that you can access. In implementing your method, how would you handle error conditions generated by inadequate space remaining on target storage media?

Project 3

After providing a correct and complete answer to Problem 32, implement exactly what is asked for in that problem. Present your solution in a report format.

Looking for more? Visit our sites for additional readings, recommended resources, and exercises.

CRC Press e-Resource: https://www.crcpress.com/9781138710085

Authors' GitHub: https://github.com/bobk48/linuxthetextbook

13

Advanced Bash Programming

OBJECTIVES

- To discuss numeric data processing
- Arbitrary-precision mathematics using bc
- To describe how standard input of a command in a shell script can be redirected from data within the script
- To explain the signal/interrupt processing capability of Bash
- To explain functions in Bash including parameter passing and scoping of variables
- To describe how file input/output can be performed using file descriptors and how standard files can be redirected from within a shell script
- To discuss debugging of Bash scripts
- To cover the following commands and primitives:

  |, <, >, >>, <<<, bc, clear, declare, exec, expr, grep, kill, let, local, more, read, readonly, sort, stty, trap

13.1 Introduction

We discuss several important, advanced features of Bash in this chapter. They include processing of numeric data, the *here document*, signals and signal processing, and redirection of standard files from within a shell script. We also discuss Bash's support of functions that allow the programmer to write general-purpose and modular code. Finally, we describe how Bash scripts can be debugged.

13.2 Numeric Data Processing

The values of all Bash variables are stored as character strings. Although this feature makes symbolic data processing fun and easy, it does make numeric data processing a bit challenging. The reason is that integer data are actually stored in the form of character strings. To perform arithmetic and logic operations on them, you need to convert them to integers, perform the necessary operations on the integer data, and be sure the result is converted back to a character string for its proper storage in a shell variable. In this section, we discuss different ways of performing arithmetic operations, first by using features of Bash and then with bc, an arbitrary-precision calculator, that comes prepackaged with all Linux distributions.

13.2.1 Bash Features for Arithmetic Operations

There are three ways to perform arithmetic operations on numeric data in Bash without performing string to integer conversion and vice versa. These methods are

1. By using the let command
2. By using the shell expansion $((expression))
3. By using the expr command

The expression evaluation is performed in long integers and no overflow check is made. When shell variables are used in an expression, they are expanded (i.e., their values are substituted in the expression) and coerced to (i.e., their data type is treated as) the desired data type before the expression is evaluated. The arithmetic, logic, and relational operators supported by Bash are listed in Table 12.6 in decreasing order of precedence. The operators that are grouped together have equal precedence. The order of evaluation can be altered by enclosing expressions in parentheses. Constants with a leading 0 are treated as octal numbers; those with a leading 0x or 0X are treated as hexadecimal numbers. Otherwise, numbers are treated as decimal numbers.

The built-in Bash command `let` allows you to evaluate arithmetic expressions by specifying them as its arguments. If an expression contains spaces or other special characters, it must be enclosed in double quotes. The following is a brief description of the command:

Syntax:
   `let expression-list`

Purpose: Evaluate arithmetic expressions in **expression-list**

If the last expression evaluates to 0, `let` returns 1; otherwise it returns 0. The following session shows an interactive use of the `let` command. Note that $ is not needed to refer to the value of a shell variable and that quotes are used in the first *let* command because the expressions used in it contain spaces. These examples show variable declaration and arithmetic expression evaluation with the `let` command. The expression b**x in the last let command means *b* to the power *x*.

```
$ let a=8 b=13
$ let c=a+b
$ echo "The value of c is $c."
The value of c is 21.
$ let a*=b
$ echo "The new value of a is $a; the product of a and b."
The new value of a is 104; the product of a and b
$ b=2 x=10 y=20 z=30
$ let K=b**x M=b**y G=b**z
$ echo "In computer jargon 1K is $K, 1M is $M, and 1G is $G."
In computer jargon 1K is 1024, 1M is 1048576, and 1G is 1073741824.
$
```

The following Bash expression syntax can also be used to evaluate the arithmetic expression. The "expression" is evaluated and the result is returned.

Syntax:
   `$((expression))`

Purpose: Evaluate "expression" and return its value

In the following session, we use this syntax to perform the same task that we performed in the previous session.

```
$ b=2 x=10 y=20 z=30
$ echo "In computer jargon 1K is $((b**x)), 1M is $((b**y)), and 1G is $((b**z))."
In computer jargon 1K is 1,024, 1M is 1,048,576, and 1G is 1,073,741,824.
$
```

Note that you can store the values of $((b**x)), $((b**y)), and $((b**z)) in variables K, M, and G, as was done in the previous session. However, it is not needed unless you want to use these values later on.

The following is a brief description of the expr command:

Syntax:
   `expr args`

Purpose: Evaluate the expression arguments **args** and send the result to standard output

> **Commonly used options/features**:
> Arithmetic, relational, and logic operations are the most commonly used:
>
> | | |
> |---|---|
> | \| | Bitwise OR |
> | \& | Bitwise AND |
> | \^ | Bitwise exclusive-OR (XOR) |
> | =, \>, \>=, \<, \<=, != | Integer comparison operators: equal, greater than, greater than or equal to, less than, less than or equal to, not equal |
> | +, -, *, /, % | Integer arithmetic operators: add, subtract, multiply, integer divide (return quotient), remainder |
>
> **Other useful operator are**
>
> **arg1 : arg2**
>
> Search for the pattern "arg2" (a regular expression) in "arg1." If "arg2" is enclosed between \(and \), the portion of "arg1" that matches is returned; otherwise, the returned value is the number of characters that match. By default, the search starts with the first character of "arg1." Start the string with .* to match other parts of the string.
>
> **substr string start length**
>
> Search for and return a portion of "string" starting with "start" and "length" characters long. Return null if "start" or "length" is nonnumeric or negative.
>
> **index string character-list**
>
> Search "string" for the first possible character in "character-list" and return its position (first character in the string is at position 1). If no character is found, return 0.
>
> **length string**
>
> Return the length of "string" (i.e., the number of characters in it).

Shell metacharacters such as * must be escaped in an expression so that they are treated literally and not as shell metacharacters. In the following session, the first `expr` command increments the value of the shell variable *var1* by 1. The second `expr` command computes the square of *var1*. The last two `echo` commands show the use of the `expr` command to perform integer division and integer remainder operations on *var1*.

```
$ var1=10
$ var1=`expr $var1 + 1`
$ echo $var1
11
$ var1=`expr $var1 \* $var1`
$ echo $var1
121
$ echo `expr $var1 / 10`
12
$ echo `expr $var1 % 10`
1
$
```

The following examples demonstrate the use of the expr command for string processing. The first command displays the index of the location (the value 10) for the character "r" in the string "Hello, World." The second command displays 0, indicating that no character in the given set of characters ("i" and "p") is found in the string. The third command is used to display the length of the string, and the fourth command to display a substring of the string "Hello, World." Starting with the character at position 8 and having length 5. The fifth and sixth expr commands show that a string variable can be used in place of an actual string. The variable expansion must be enclosed in double quotes as shown; otherwise, an exception is caused.

```
$ expr index "Hello, World." arp
10
$ expr index "Hello, World." ip
0
```

```
$ expr length "Hello, World."
13
$ expr substr "Hello, World." 8 5
World
$ greeting="Hello, World."
$ expr length "$greeting"
13
$ expr substr "$greeting" 8 5
World
$
```

We now discuss a few scripts that use the expr command for arithmetic expression evaluation. The following countup script takes an integer as a command line argument and displays the range of numbers from 1 to the given number in one line, in ascending order. In the script, we use a simple while loop to display the current number (starting with 1) and then compute the next numbers, until the current number becomes greater than the number passed as the command line argument.

```
$ cat countup
#!/bin/sh
if [ $# != 1 ]
    then
        echo "Usage: $0 integer-argument"
        exit 1
fi
target="$1"     # Set target to the number passed at the command line
current=1       # The first number to be displayed
# Loop here until the current number becomes greater than the target
while [ $current -le $target ]
do
        echo -n "$current "
        current=`expr $current + 1`
done
echo    # Move cursor to the next line
exit 0
$ ./countup 5
1 2 3 4 5
$
```

The following script, addall, takes a list of integers as command line arguments and displays their sum. The while loop adds the next number in the argument list to the running sum (which is initialized to 0), updates the count of numbers that have been added, and shifts the command line arguments left by one position. The loop then repeats until all the numbers in the command line arguments have been added. The sample run following the code takes the list of the first eight perfect squares and returns their sum.

```
$ cat addall
#!/bin/bash

# File Name:    ~/linux2e/chapter13/bash/addall
# Author:       Syed Mansoor Sarwar
# Written:      August 18, 2004
# Modified:     August 18, 2004; July 28, 2014; July 9, 2017
# Purpose:      To demonstrate use of the expr command in processing
#               numeric data
# Brief Description:
#       Maintain the running sum of numbers in a numeric
#       variable called sum, initialized to 0. Read the next
#       integer and add it to sum. When all the integers
#       specified as command line arguments have been read,
#       display the answer, and terminate the program. If
```

```
#        the program is run with no arguments, inform the
#        user of the command syntax.
if [ $# = 0 ]
    then
        echo "Usage: $0 number-list"
        exit 1
fi
sum=0           # Running sum initialized to 0
count=0         # Count the count of numbers passed as arguments
while [ $# != 0 ]
do
    sum=`expr $sum + $1`        # Add the next number to the running sum
    count=`expr $count + 1`     # Update count of numbers added so far
    shift                       # Shift the counted number out
done
# Display final sum
echo "The sum of the given $count numbers is $sum."
exit 0
$ ./addall
Usage: ./addall number-list
$ ./addall 1 4 9 16 25 36 49 64
The sum of the given 8 numbers is 204.
$
```

Although this example neatly explains numeric data processing, it is nothing more than an integer addition machine. We now present a more useful example that uses the Linux file system. The **fs** (for file size) file contains a script that takes a directory as an optional argument and returns the size (in bytes) of all nondirectory files in it. On some Linux systems, running the fs command invokes xfs. If this happens on your system, change the name of this script to **files**, or whatever name you prefer to use.

When you run the program without a command line argument, the script assumes your current directory as the argument. If you run it with more than one command line argument, the script displays the command syntax and terminates. When you execute it with one nondirectory argument only, again the program displays the command syntax and exits. If the program is run with a nonexistent file as an argument, it displays an error message and terminates.

The gist of this script is the following code that runs when the script is run with a directory as a command line argument.

```
ls $directory | more |
while read file
do
    ...
done
```

This code generates a list of files in *directory* with the ls command, converts the list into one file name per line list with the more command, and reads each file name in the list, one by one, with the read command until no file remains in the list. The read command returns true if it reads a line and returns false when it reads the eof marker. The body of the loop—that is, the code between do and done—adds the file size to the running total if the file is an ordinary file. When no name is left in the directory list, the program displays the total space in bytes occupied by all nondirectory files in the directory and terminates.

If the value of the file variable is not an existing file, the [! -e "$file"] expression returns false and the error message Usage: fs [directory name], as shown in sample run, where **linux2e** is a nonexistent directory. On some systems, the message may display ./fs instead of fs, as in our session. The file="$directory"/"$file" statement is used to construct the relative path name of a file with respect to the directory specified as the command line argument. Without this, the set -- `ls -l "$file"` command will be successful only if the directory contains the name of the current directory.

```
$ cat fs
#!/bin/bash
# File Name:    ~/linux2e/chapter13/bash/fs
# Author:       Syed Mansoor Sarwar
# Written:      August 18, 2004
# Modified:     August 20, 2004; Jul 28, 2014; July 9, 2017
# Purpose:      To add the sizes of all the files in a directory passed as
#               command line argument
# Brief Description:
#       Maintain running sum of file sizes in a numeric variable
#       called sum, starting with 0. Read all the file names
#       by using the pipeline of the ls, more, and while commands.
#       Get the size of the next file and add it to the running
#       sum. Stop when all file names have been processed and
#       display the answer.
if [ $# = 0 ]           # If no command line argument, the
                        # set directory to current directory
    then
        directory="."
    elif [ $# != 1 ]    # If more then one command line argument
                        # then display command syntax
    then
        echo "Usage: $0 [directory name]"
        exit 1
    elif [ ! -e "$1" ]          # If one command line argument, but file
                                # does not exist, display error message
    then
        echo "$1: File does not exist"
        exit 1
    elif [ ! -d "$1" ]          # If one command line argument, but is
                                # not a directory, show command syntax
    then
        echo "Usage: $0 [directory name]"
        exit 1
    else
        directory="$1" # If one command line argument and it is a
                        # directory, prepare to perform the task
fi

# Get file count in the given directory; for empty directory, display a
# message and quit.
file_count=`ls $directory | wc -w`  # Get count of files in the directory
if [ $file_count -eq 0 ]        # If no files, display error message
    then
        echo "$directory: Empty directory."
        exit 0
fi
# For each file in the directory specified, add the file size to the running
# total. The more command is used to output file names one per line so can
# read command can be used to read file names.
sum=0  # Running sum initialized to 0.
ls "$directory" | more |
while read file
do
    file="$directory"/"$file"  # Store the relative path name for each file
    if [ -f "$file" ]           # If it is an ordinary file
        then                    # then
            set -- `ls -l "$file"` # set command line arguments
            sum=`expr $sum + $5`   # Add file size to the running total.
    fi
    # Code to decrement the file_count variable and display the final sum
    # if the last file has been processed.
```

```
        if [ "$file_count" -gt 1 ] # Are more files left? If so, continue.
            then
                file_count=`expr $file_count - 1`
            else
            # Spell out the current directory
            if [ "$directory" = "." ]
                then
                    directory="your current directory"
            fi
            echo "The size of all ordinary files in $directory is $sum bytes."
        fi
    done
    exit 0
$
$ pwd
/home/sarwar/linux2e/ch13
$ ./fs / /bin
Usage: ./fs [directory name]
$ ./fs linux2e
linux2e: File does not exist
$ ./fs ~/linux2e
The size of all ordinary files in /home/sarwar/linux2e is 0 bytes.
$ ./fs ..
The size of all ordinary files in .. is 0 bytes.
$ ./fs .
The size of all ordinary files in your current directory is 4060 bytes.
$ ./fs /
The size of all ordinary files in / is 16 bytes.
$ ./fs /bin
The size of all ordinary files in /bin is 16428442 bytes.
$ ./fs dir1
dir1: Empty directory.
$
```

Note that in the previous sample run of the fs script, the ./fs / command displays the total size of the two symbolic link files in the root directory.

In the following in-chapter exercise, you will create a Bash script that processes numeric data by using the expr command.

Exercise 13.1

Create the **addall** script in your directory and run it with the first ten numbers in the Fibonacci series. What is the result? Does the program produce the correct result? If you are not familiar with the Fibonacci series, browse through the following website: http://en.wikipedia.org/wiki/Fibonacci_number.

Exercise 13.2

Create the **fs** script in your directory and try it with sample runs given in the previous session.

Exercise 13.3

Try all of the shell sessions in the previous session related to the let command, expansion syntax, and the expr command.

13.2.2 Arithmetic with bc

bc is a language that supports arithmetic operations on arbitrary-precision numbers. It can be used as an interactive mathematical calculator or as a mathematical scripting language with C language like syntax. It supports the usual constructs and features of a scripting language to support arithmetic, assignment, comparison, increment/decrement, and logical/Boolean operations. It also supports conditional and iterative statements, and mathematical functions available in the Linux mathematics library, mathlib. You can also convert numbers from any number base to another, such as decimal to binary, binary to octal, octal to hexadecimal, and a number in base four to a number in base 21. Here is a brief syntax of bc:

Syntax:
  `bc [-hilwsqv] [long-options][file ...]`

Purpose: Perform arbitrary-precision arithmetic with interactive execution of statement as well as using scripts in a language with C-like syntax

Commonly used options/features:

  -i   Force interactive mode

  -l   Use functions in the Linux math library, mathlib

We show the interactive use of bc with a few examples, first by running bc as a shell and then under Bash. You can see how straightforward is expression evaluation. The second expression is evaluated without specifying precision, and thus the expression evaluates to 2. We then specify precision to three decimal digits (using `scale=3;`) and bc displays results as 2.254. In the final example, we specify precision up to six decimal digits and bc displays 27.048348 that is way different from 24, the result displayed with no precision.

```
$ bc
bc 1.06.95
Copyright 1991-1994, 1997, 1998, 2000, 2004, 2006 Free Software Foundation, Inc.
This is free software with ABSOLUTELY NO WARRANTY.
For details type 'warranty'.

10+13.7
23.7
29.37/13.03
2
scale=3; 29.37/13.03
2.254
29.37/13.03*12
24
scale=6; 29.37/13.03*12
27.048348
quit
$
```

The following session shows a few Bash commands that pass some of the mathematical expressions evaluated earlier to bs through a pipe.

```
$ echo '29.37/13.03' | bc
2
$ echo 'scale=4; 29.37/13.03' | bc
2.2540
$ echo '29.37/13.03*12' | bc
24
$ echo 'scale=6; 29.37/13.03*12' | bc
27.048348
$
```

You can also pass an expression to bc through by using the <<< operator, as shown in the following session:

```
$ bc <<< 'scale=6; 29.37/13.03*12'
27.048348
$
```

Finally, you can evaluate an expression with bc, save result in a shell variable, and display the result using the echo command, as in the following session:

```
$ x=`echo 'scale=6; 29.37/13.03*12' | bc`
$ echo $x
27.048348
$
```

You can use any of the mathlib functions by running bc with –l option. The following command shows the evaluation of the sine of an angle in radians using the sine function, s().

```
$ echo "s(2)" | bc -l
.90929742682568169539
$
```

The following session shows the use of if and for statements in bc. Note the similarities between these constructs and corresponding constructs of the C language.

```
$ echo 'a=10; b=20; if(a>b) print "a is greater" else print "b is greater\n" ' |
bc
b is greater
$ echo "for(i=1; i<=10; i++) {i;}" | bc
1
2
3
4
5
6
7
8
9
10
$
```

Exercise 13.4

Repeat the commands and scripts discussed in this section on your system. Were you able to reproduce the results? Log any differences.

13.3 Array Processing

Bash supports one-dimensional arrays. An *array* is a name collection of items of the same type stored in contiguous memory locations. Array items, known as members of the array, are numbered with the first one being number 0. There is no limit on the size of an array in Bash, and array elements are not required to be assigned contiguously. This means that once you have an array, you can assign a value to any array slot (see the following example). You can use any of the following syntaxes to declare array variables.

> **Syntax:**
> ```
> array_name[subscript]=value
> declare –a array_name
> declare –a array_name[subscript]
> local –a array_name
> readonly array_name
> ```
> **Purpose:** To declare "array_name" to be an array variable. The "subscript" is considered an arithmetic expression that must evaluate to a number greater than or equal to 0. The second and third syntaxes are equivalent as "subscript" in the third declaration is ignored. The fourth syntax can be used to declare array variables whose scope is local to a function. The fifth syntax can be used to declare read-only array variables.

Arrays can be initialized at the time of their declaration by using the syntax

```
array_name=(value1 ... valueN)
```

where "value1" is of the form [[subscript]=]string. Note that the subscript is optional but, if supplied, the corresponding value is assigned to the given array slot; otherwise, the slot in the array to which the value is assigned to is the last slot assigned plus one.

You can reference an array item by using the ${name[subscript]} syntax. The process is known as *array indexing*. If "subscript" is @ or *, all array elements are referenced. The difference between @ and * is that "${name[*]}" expands to a single word comprising all elements of the array *name* and "${name[@]}" expands each element of the array *name* to a separate word.

In the following example, the array variable *movies* is assigned three values; the first two values are assigned to locations 0 and 1, and the third is assigned to slot number 65. The echo commands are used to display the values of the three variables.

```
$ movies=("Silence of the Lambs" "Malcolm X" [65]="The Birds")
$ echo ${movies[0]}
Silence of the Lambs
$ echo ${movies[1]}
Malcolm X
$ echo ${movies[2]}

$ echo ${movies[65]}
The Birds
$ echo ${movies[*]}
Silence of the Lambs Malcolm X The Birds
$ echo ${movies[@]}
Silence of the Lambs Malcolm X The Birds
$
```

The readonly command with the −a option can be used to declare read-only array variables. In the following session, the *us_presidents* array is initialized to three items. The echo commands are used to display the first, fortieth, forty-fourth, and forty-fifth element of this array. The command for declaring the Fibonacci array fails, as the local command can only be used to declare variables within the scope of a function. The declare command for declaring the Fibonacci is successful.

```
$ readonly -a us_presidents=([1]="George Washington" [40]="Ronald Reagan" "Bush"
[42]="Bill Clinton" [44]="Barack Obama" "Donald Trump")
$ echo ${us_presidents[1]}
George Washington
$ echo ${us_presidents[40]}
Ronald Reagan
$ echo ${us_presidents[44]}
Barack Obama
$ echo ${us_presidents[45]}
Donald Trump
$ local -a Fibonacci=(0 1 1 2 3 5 8 13 21 34)
-bash: local: can only be used in a function
$ declare -a Fibonacci=(0 1 1 2 3 5 8 13 21 34)
$ echo ${Fibonacci[7]}
13
$
```

The size (in bytes) of an array item can be displayed by using the ${#[subscript]} syntax. If no subscript is used, the size of the first array element is displayed. If * is used as subscript, the number of array elements is displayed. In the following session, we display the sizes of items in the *movies* array. The first of the two assignment statements is used to insert a new item ("Forest Gump") in the array at slot 16, and the second assignment statement is used to replace the current value ("The Birds") in slot 65 with a new value ("Sleepless in Seattle"). Figure 13.1 shows the *movies* array before and after making these changes.

	Before			**After**	
	movies			movies	
movies[0]:	Silence of the Lambs		movies[0]:	Silence of the Lambs	
movies[1]:	Malcolm X		movies[1]:	Malcolm X	
...	...		...	...	
movies[16]:			movies[16]:	Forest Gump	
...	...		...	...	
movies[65]:	The Birds		movies[65]:	Sleepless in Seattle	

FIGURE 13.1   The movies array before and after changing its contents.

```
$ echo ${#movies}
20
$ echo ${#movies[65]}
9
$ echo ${#movies[*]}
3
$ movies[16]="Forest Gump"
$ movies[65]="Sleepless in Seattle"
$ echo ${#movies[*]}
4
$ echo ${movies[0]}; echo ${movies[1]}; echo ${movies[16]}; echo ${movies[65]}
Silence of the Lambs
Malcolm X
Forest Gump
Sleepless in Seattle
$
```

It should be obvious by now that any shell variable that is assigned multiple values by using the assignment statement becomes an array variable. Thus, when a variable is assigned a multiword output of a command as a value, it becomes an array variable and contains each field of the output in a separate array slot. In the following example, *files* is an array variable whose elements are the names of all the files in the current directory. The *numfiles* variable contains the number of files in the current directory. The number of files can also be displayed using the echo ${#files[*]} command, as shown. The echo ${files[*]} command displays the names of all the files in the current directory, and the echo ${files[2]} command displays the third element (i.e., element at slot number 2).

```
$ files=(`ls`) numfiles=`ls | wc -w`
$ echo ${files[*]}
addall canleave countup debug_demo dext dext_functions diff2 exec.demo1 find_sys_
const fs heredoc_demo trap_demo
$ echo $numfiles
12
$ echo ${#files[*]}
12
$ echo ${files[2]}
countup
$
```

In the following example, the num_array_demo file contains a script that uses an array of integers, called *Fibonacci*, computes the sum of integers in the array, and displays the sum on the screen. The *Fibonacci* array in this example contains the first ten numbers of the Fibonacci series. For those who are not familiar with the Fibonacci series, the first two numbers are 0 and 1, and the numbers are calculated by adding the previous two numbers. Mathematically, the series is represented as follows.

$$F_1 = 0$$

$$F_2 = 1$$

$$F_n = F_{n-1} + F_{n-2}, \text{ for } n \geq 3$$

Thus, the first ten numbers in the Fibonacci series are 0, 1, 1, 2, 3, 5, 8, 13, 21, and 34.

The script in the num_array_demo file is well documented and fairly easy to understand. It displays the sum of the first ten Fibonacci numbers. A sample run of the script follows the code.

```
$ cat num_array_demo
# File Name:             ~/linux2e/ch13/bash/num_array_demo
# Author:                Syed Mansoor Sarwar
# Date Written:          August 16, 1999
# Last Modified:         July 16, 2017
# Purpose:               To demonstrate working with numeric arrays in Bash
# Brief Description:      Maintain a running sum of the first 10 Fibinacci
#                         number in a numeric variable 'sum', starting with 0.
#                         Read the next number and add it to sum. When the
#                         last number has been added to sum, display answer.
#!/bin/bash -xv

# Initialize Fibonacci array to any number of Fibonacci numbers; first 15
# in this case.
declare -a Fibonacci=(0 1 1 2 3 5 8 13 21 34 55 89 144 233 377)
size=${#Fibonacci[*]}    # Size of the Fibonacci array as string
index=1                  # Index points to the second element
sum=0                    # Running sum initialized to 0
next=0                   # For storing the next array item
while [ $index -le $size ]
do
    next=$((${Fibonacci[$index]}))    # Get the next value as integer
    sum=$((sum + next))               # Update the running sum
    index=$((index + 1))              # Update array index by 1
done
echo "The sum of first ${#Fibonacci[*]} Fibonacci numbers is $((sum))."
exit 0
$ ./num_array_demo
The sum of first 15 Fibonacci numbers is 986.
$
```

Exercise 13.5
Write a Bash script that contains two numeric arrays, *array1* and *array2*, initialized to the values in the sets {1, 2, 3, 4, 5} and {1, 4, 9, 16, 25}, respectively. The script should produce and display an array whose elements are the sum of the corresponding elements in the two arrays. Thus, the first element of the resultant array is $1 + 1 = 2$, the second is $2 + 4 = 6$, and so on.

Exercise 13.6
Replicate all Bash sessions and scripts discussed in this section on your system to verify that they produce expected results.

13.4 The Here Document

The *here document* feature of Bash allows you to redirect standard input of a command in a script and attach it to data within the script, wrapped in a particular format, as will be explained. Obviously, this feature works with commands that take input from standard input. The feature is used mainly to display menus, although there are some other important uses of this feature. The advantage of maintaining data within the script is that it eliminates extra file operations such as open and read that would be required if the data was maintained in a separate file. The result is a much faster program. The following is a brief description of the here document:

Syntax:
```
command << [-] input_marker
... input data ...
input_marker
```
Purpose: To execute **command** with its input coming from the here document—data between the input start and end markers **input_marker**

The `input_marker` is a string that you choose to wrap the input data in for `command`. The closing marker must be on a line by itself and cannot be surrounded by any spaces. The command and variable substitutions are performed before the here document data is directed to **stdin** of the command. Quotes can be used to prevent these substitutions or to enclose any quotes in the here document. `input_marker` can be enclosed in quotes to prevent any substitutions in the entire document, as in

```
command <<'Marker'
...
'Marker'
```

A hyphen (-) after << can be used to remove leading tabs (not spaces) from the lines in the here document and the marker that ends the here document. This feature allows the here document and the delimiting marker to conform to the indentation of the script. The following session illustrates this point:

```
while [ ... ]
do
  grep ... <<- DIRECTORY
          John Doe ...
          ...
          Art Pohm ... DIRECTORY
  ...
done
```

One last, but very important point: output and error redirections of the command that uses the here document must be specified in the command line, not following the marker that ends the here document. The same is true of connecting the standard output of the command with other commands via a pipeline, as shown in the following session. Note that the grep ... <<- DIRECTORY 2> errorfile | sort command can be replaced by (grep ... 2> errorfile | sort) <<- DIRECTORY.

```
while [ ... ]
do
  grep ... <<- DIRECTORY 2> errorfile | sort John Doe ...
          ...
      Art Pohm ... DIRECTORY
          ...
done
```

We can illustrate the use of the here document feature with a simple instance of redirecting **stdin** of the `cat` command from the here document. The script in the **heredoc_demo** file is used to display a message for the user and then send a message to the person whose e-mail address is passed as a command line argument. In the following session, we use two here documents: one begins with << DataTag and ends with DataTag; and the other begins with << WRAPPER and ends with WRAPPER. Note that the Linux mail utility must have been installed before you can use this script.

```
$ cat heredoc_demo
#!/bin/bash

cat << DataTag
This is a simple use of the here document. This data is the
input to the cat command.
```

DataTag

```
# Second example
mail -s "Weekly Meeting Reminder" $1 << WRAPPER

Hello,

This is a reminder for the weekly faculty meeting tomorrow.

Mansoor

WRAPPER

echo "Sending mail to $1 ... done."
exit 0
$ ./heredoc_demo eecsfaculty
This is a simple use of the here document. These data are the input to the cat
command.
Sending mail to eecsfaculty ... done.
$
```

The following script is more useful and makes a better utilization of the here document feature. The dext (directory expert) script maintains a directory of names, phone numbers, and e-mail addresses. The script is run with the name as a command line argument and uses the grep command to display the directory entry corresponding to the name. The –i option is used with the grep command to ignore the case of letters.

```
$ more dext
#!/bin/bash
if [ $# = 0 ]
    then
        echo "Usage: $0 name"
        exit 1
fi
user_input="$1"
grep -i "$user_input" << DIRECTORY

        John Doe       555.232.0000    johnd@somedomain.com
        Jenny Great    444.6565.1111   jg@new.somecollege.edu
        David Nice     999.111.3333    david_nice@xyz.org
        Don Carr       555.111.3333    dcarr@old.hoggie.edu
        Masood Shah    666.010.9820    shah@Garments.com.pk
        Jim Davis      777.000.9999    davis@great.adviser.edu
        Art Pohm       333.000.8888    art.pohm@great.professor.edu
        David Carr     777.999.2222    dcarr@net.net.gov

DIRECTORY
exit 0
$ ./dext
Usage: ./dext name
$ ./dext Pohm
        Art Pohm       333.000.8888    art.pohm@great.professor.edu
$ ./dext Carr
        Don Carr       555.111.3333    dcarr@old.hoggie.edu
        David Carr     777.999.2222    dcarr@net.net.gov
$
```

If there are multiple entries for a name, the grep command displays all the entries. You can display the entries in sorted order by piping the output of the grep command to the sort command and enclosing them in parentheses, as in (grep -i "$user_input" | sort). We enhance the dext script in Section 13.7 to include this feature, as well as take multiple names from the command line.

The following in-chapter exercise is designed to reinforce your understanding of the here document feature of Bash.

Exercise 13.7

Create the `dext` script on your system and run it. Try it with as many different inputs as you can think of. Does the script work correctly?

13.5 Interrupt (Signal) Processing

We discussed the basic concept of signals in Chapter 10, where we defined them as software interrupts that can be sent to a process. We also stated that the process receiving a signal can take one of the three possible actions:

1. Take the default action as defined by the Linux kernel
2. Ignore the signal
3. Take a programmer-defined action

In Linux, several types of signals can be sent to a running program. Some of these signals can be sent via hardware devices such as the keyboard, but all can be sent via the `kill` command, as discussed in Chapter 10. The most common event that causes a hardware interrupt (and a signal) is generated when you press <Ctrl+C> and is known as the keyboard interrupt. The default kernel-defined action of this event is that the foreground process terminates. Other events that cause a process to receive a signal include termination of a child process, a process accessing a main memory location that is not part of its address space, and a software termination signal caused by execution of the `kill` command without any signal number. The address space of a process is the main memory area that the process owns legally and is allowed to access. Table 13.1 presents a list of some important signals, their numbers, and their purpose. The signal numbers, or the corresponding symbolic names, can be used to generate the respective signals with the `kill` command.

The interrupt processing feature of Bash allows you to write programs that can ignore signals, take actions as defined by the Linux kernel for those signals, or execute a specific sequence of commands when signals of particular types are sent to them. This feature is much more powerful than that of the

TABLE 13.1

Some Important Signals, Their Numbers, and Their Purpose

Signal Name	Signal #	Purpose
SIGHUP (hang up)	1	Informs the process when the user who ran the process logs out, and the process terminates
SIGINT (keyboard interrupt)	2	Informs the process when the user presses <Ctrl+C> and the process terminates
SIGQUIT (quit signal)	3	Informs the process when the user presses <Ctrl+|> or <Ctrl+\> and the process terminates
SIGKILL (sure kill)	9	Terminates the process when the user sends this signal to it with the kill -9 command
SIGSEGV (segmentation violation)	11	Terminates the process upon memory fault when a process tries to access memory space that does not belong to it
SIGTERM (software termination)	15	Terminates the process when the kill command is used without any signal number
SIGTSTP (suspend/stop signal)	18	Suspends the process; usually <Ctrl+Z>
SIGCHLD (child finishes execution)	20	Informs the process of termination of one of its children

C shell, which allows programs to ignore a keyboard interrupt (<Ctrl+C>) only. The trap command can be used to intercept signals. The following is a brief description of the command:

Syntax:
    `trap ['command-list'] [signal-list]`

Purpose: To intercept signals specified in **signal-list** and take default kernel-defined action, ignore the signals, or execute the commands in **command-list**; note that quotes around **command-list** are required.

When you use the trap command in a script without any argument (i.e., no **command-list** and no **signal-list**), the script takes default actions when it receives signals. Thus, using the trap command without any argument is redundant. When the trap command is used without any commands in single quotes, the script ignores the signals in **signal-list**. When both a **command-list** and a **signal-list** are specified, the commands in **command-list** execute when a signal specified in **signal-list** is received by the script.

Next, we enhance the script in the **while_demo** file from Chapter 12 so that you cannot terminate execution of this program with <Ctrl+C> (signal number 2), the kill command without any argument (signal number 15), or the kill -1 command (to generate the SIGHUP signal). The enhanced version is in the **trap_demo** file, as shown in the following session. Note that the trap command is used to ignore signals 1, 2, 3, 15, and 18. A sample run illustrates this point.

```
$ cat trap_demo
#!/bin/bash

# Intercept signals 1, 2, 3, 15, and 18 and ignore them
trap '' 1 2 3 15 18

# Set the secret code
secretcode=agent007

# Get user input
echo "Guess the code!"
echo -e "Enter your guess: \c"
read yourguess

# As long as the user input is the secret code (agent007 in this case),
# loop here: display a message and take user input again. When the user
# input matches the secret code, terminate the loop and execute the
# echo command that follows.
while [ "$secretcode" != "$yourguess" ]
do
    echo "Good guess but wrong. Try again!"
    echo -e "Enter your guess: \c"
    read yourguess
done
echo "Wow! You are a genius!"
exit 0
$ ./trap_demo
Guess the code!
Enter your guess: codecracker
Good guess but wrong. Try again!
Enter your guess: <Ctrl+C>
Good guess but wrong. Try again!
Enter your guess: agent007
Wow! You are a genius!
$
```

To terminate programs that ignore terminal interrupts, you have to use the kill command. You can do so by suspending the process by pressing <Ctrl+Z>, using the ps command to get the PID of the process, and terminating it with the `kill` command. Alternatively, you can login from another terminal and use the ps and kill commands as stated in the previous sentence.

You can modify the script in the **trap_demo** file so that it ignores signals 1, 2, 3, 15, and 18, clears the display screen, and turns off the echo. When echo has been turned off, whatever input you enter from the keyboard, is not displayed. Next, it prompts you for the code word and saves it. It then prompts you again to enter the same code word to make sure that you remember the word that you have entered. It gives you two chances for this purpose. If you do not enter the same code word both times, it reminds you of your bad short-term memory and quits. If you enter the same code word, it clears the display screen and prompts you to guess the code word again. If you do not enter the original code word, the display screen is cleared and you are prompted to guess again. The program does not terminate until you have entered the original code word. When you do enter it, the display screen is cleared, a message is displayed at the top left of the screen, and the echo is turned on. Because the terminal interrupt is ignored, you cannot terminate the program by pressing <Ctrl+C>. The stty -echo command turns off the echo. Thus, when you type the original code word (or any guesses), it is not displayed on the screen. The stty echo turns on the echo. The clear command clears the display screen and positions the cursor at the top-left corner. The resulting script is in the **canleave** file, as shown in the following session.

```
$ cat canleave
#!/bin/bash
# File Name:     ~/linux2e/chapter13/bash/canleave
# Author:        Syed Mansoor Sarwar
# Written:       August 18, 2004
# Modified:      May 8, 2004; Jul 29, 2014; July 9, 2017
# Purpose:       To allow a user to leave his/her terminal for a short
#                duration of time by locking the terminal after taking a code
#                from the user. Terminal is unlocked only when the user
#                re-enters the same code. Ignores command line arguments.
# Brief Description:
#       Clear screen and turn off echo (i.e., do not display what the user
#       types at the keyboard). Take user code, save it, and ask the user
#       to re-enter his/her code just to make sure that the user remembers
#       the code that he/she has entered. It is done twice. If the user does
#       not enter the same code, the program terminates after displaying a
#       message for the user. The user is prompted to enter the original
#       code. If the user enters the wrong code, the program keeps on
#       prompting the user until he/she enters the original code. The
#       keyboard is then unlocked, echo is turned on, and program exits.

# Ignore signals 1, 2, 3, 15, and 18
trap '' 1 2 3 15 18

# Clear the screen, locate the cursor at the top-left corner,
# and turn off echo
clear
stty -echo

# Set the secret code
echo -n "Enter your code word: "
read secretcode
echo " "

# To make sure that the user remembers the code word,
# ask the user to enter the secret code again.
echo -n "Enter your code word again: "
read same
if [ $secretcode != $same ]
   then
       clear
       echo "Wrong code. Try again."
fi
```

```
echo -n "Enter your code word again: "
read same
if [ $secretcode != $same ]
    then
        clear
        echo "Work on your short-term memory before using this code!"
        echo "Goodbye!"
        exit 1
fi

# Keyboard locked. Hit <Enter> to continue.
clear
echo -n "Keyboard locked. Hit <Enter> to continue."
read ignore
clear

# Get user guess to unlock the terminal
clear
echo -n "Enter the code word: "
read yourguess
echo " "

# As long as the user input is not the original code word, loop here: display
# a message and take user input again. When the user input matches the secret
# code word, terminate the loop and execute the following echo command.
while [ "$secretcode" != "$yourguess" ]
do
    clear
    echo "Wrong code. Try again."
    echo -n "Enter the code word: "
    read yourguess
done
# Set terminal to echo mode clear
clear
echo "Back again!"
stty echo
exit 0
$
```

You can use this script to lock your terminal before you leave it for a short period of time—for example, to pick up a printout or get a can of soda; hence, the name **canleave** (can leave). Using it saves you the time otherwise required for the logout and login procedures.

The following in-chapter exercise is designed to reinforce your understanding of the signal-handling feature of Bash.

Exercise 13.8

Test the scripts in the **trap_demo** and **canleave** files on your Linux system. Do they work as expected? Be sure that you understand them.

13.6 Functions in Bash

Bash allows you to write functions. Functions consist of a series of commands, called the *function body*, that are given a name. You can invoke the commands in the function body by using the function name.

13.6.1 Reasons for Using Functions

Functions are normally used if a piece of code is repeated at various places in a script. By making a function of this code, you save typing time. Thus, if a block of code is used at, say, nine different places in a script, you can create a function of it and invoke it where it is to be inserted by using the name of the function. Another advantage of functions is that any changes to the code, which would otherwise be needed at nine places, would now be needed at one place only—that is, in the function body. The tradeoff is that the mechanism of transferring control to the function code and returning it to the calling code (from where the function is invoked/called) takes time, which slightly increases the running time of the script.

Another way of saving typing time is to create another script file for the block of code and invoke this code by calling the script as a command. The disadvantage of using this technique is that, as the script file is on a secondary storage device such as a hard disk, the invocation of the script requires loading the script from the disk into main memory once it has been defined, which is an expensive operation. Whether they are located in **~/.profile**, defined interactively in a shell, or defined in the script, function definitions are always in the main memory. Thus, invocation of functions is several times faster than invoking shell scripts, which are on the disk.

13.6.2 Function Definition

Before you can use a function, you have to define it. For often-used functions, you should put their definitions in your **~/.profile** file. This way, the shell records them in its environment when you log on and allows you to invoke them while you use the system. The definitions for functions that are specific to a script are usually put in the file that contains the script. You must execute the **~/.profile** file with the . (dot) command after defining a function in it and, before using it, unless you want to log off and then log back on. You can also define functions while interactively using the shell. These definitions are valid for as long as you remain in the session that you were in when you defined these functions.

The syntax of a function definition is given as follows:

```
name () command
```

where name is the name (also called the label) of the function and command is the body of the function. For a function with multiple commands in the body, the command list is enclosed in curly brace, { and }. This allows you to format a function definition in the following ways:

```
function_name () command
```

```
function_name () { command-list; }
```

```
function_name () {
    command-list
}
```

```
function_name ( )
{
    command-list
}
```

The function_name is the name of the function that you choose, and the commands in **command-list** comprise the function body. The following session shows example function definitions corresponding to the four ways of defining functions and sample runs of the first three functions. The output of functions function4 and function5 would be the same as that of function3.

```
$ function1 () { date; }
$ function2 () { uname -s; }
$ function3 () { date; echo "Hello, world!"; ps; }
$ function4 () {
> date
> echo "Hello, world!"
> ps
> }
$ function5 ()
```

```
>  {
>      date
>      echo "Hello, world!"
>      ps
>  }
$ function1
Sun Apr  9 19:43:18 PKT 2017
$ function2
Linux
$ function3
Sun Apr  9 19:47:57 PKT 2017
Hello, world!
  PID TTY          TIME CMD
  874 pts/0    00:00:00 bash
 1497 pts/0    00:00:00 ps
$
```

Note that there is no export command for functions. Thus, function definitions are visible only to the shell in which they are defined.

13.6.3 Function Invocation/Call

The commands in a function body are not executed until the function is invoked—that is, called. You can invoke a function by using its name as a command. When you call a function, its body is executed and control comes back to the command following the function call. If you invoke a function at the command line, control returns to the shell after the function finishes its execution.

Variables declared within a function are visible outside the function and may be accessed accordingly. To make a variable *local* to a function, the keyword local may be used, as in the function f1 in the following example. A local variable inherits the value and *exported* and *read-only* flags of a variable with the same name defined in the surrounding scope. If there is no such variable in the surrounding scope, the local variable is initialized to null. The Bash uses *dynamic scoping*. This means that a local variable defined in the caller function is accessible to the called function. It is also demonstrated in the following session, where the variable x is defined as local within f1 (the caller function) and is accessible to f2 (the called function). However, x is not accessible outside f1. On the other hand, the variable y is not defined as local and is accessible outside f1. The outputs of the echo $x, echo $y, and f2 commands verify these scoping rules.

```
$ f1 () { local x=10; y=20; f2; }
$ f2 () { echo "x is $x"; echo "y is $y"; }
$ f1
x is 10
y is 20
$ echo $x

$ echo $y
20
$ f2
x is
y is 20
$
```

13.6.4 A Few More Examples of Functions

The execution of the following function, called machines, returns the names of all the computers on your local network.

```
$ machines ()
>  {
>      date
```

```
>       echo "These are the machines on the network: "
>       ruptime | cut -f1 -d' ' | more
> }
$ ./machines
Thu Jul 31 20:23:28 PKT 2014
These are the machines on the network: upibm0
...
upsun1
...
upsun29
$
```

We now enhance the dext script described in Section 13.3 so that it can take multiple names at the command line. The enhanced version also uses a function OutputData to display one or more output records (i.e., lines in the directory) for every name passed as the command line argument. In the case of multiple lines for a name, this function displays them in sorted order. A few sample runs are shown following the script.

```
$ cat dext
#!/bin/bash
# File Name:    ~/linux2e/chaper13/bash/dext
# Author:       Syed Mansoor Sarwar
# Written:      August 1999
# Modified:     August 28, 2004; July 29, 2014; April 9, 2017

if [ $# = 0 ]
   then
        echo "Usage: $0 name"
        exit 1
fi
OutputData()
{
    echo "Infomation about $user_input"
    (grep -i "$user_input"| sort) << DIRECTORY

        John Doe      555.232.0000    johnd@somedomain.com
        Jenny Great   444.6565.1111   jg@new.somecollege.edu
        David Nice    999.111.3333    david_nice@xyz.org
        Don Carr      555.111.3333    dcarr@old.hoggie.edu
        Masood Shah   666.010.9820    shah@Garments.com.pk
        Jim Davis     777.000.9999    davis@great.adviser.edu
        Art Pohm      333.000.8888    art.pohm@great.professor.edu
        David Carr    777.999.2222    dcarr@net.net.gov

DIRECTORY
    echo            # A blank line between two records
}

# As long as there is at least one command line argument (name), take the
# first name, call the OutputData function to search the DIRECTORY and
# display the line(s) containing the name, shift this name left by one
# position, and repeat the process.

while [ $# != 0 ]
do
    user_input="$1"     # Get the next command line argument (name)
    OutputData          # Display info about the next name
    shift               # Get the following name
done
exit 0
$ ./dext john
```

```
Infomation about john
        John Doe        555.232.0000    johnd@somedomain.com

$ ./dext jim
Infomation about jim
        Jim Davis       777.000.9999    davis@great.adviser.edu

$ ./dext pohm masood carr
Infomation about pohm
        Art Pohm        333.000.8888    art.pohm@great.professor.edu

Infomation about masood
        Masood Shah     666.010.9820    shah@Garments.com.pk

Infomation about carr
        David Carr      777.999.2222    dcarr@net.net.gov
        Don Carr        555.111.3333    dcarr@old.hoggie.edu

$
```

13.6.5 Parameter Passing in Functions

Parameters are passed to functions just like command line arguments are passed to Bash scripts, i.e., parameters are specified right after the function name. Within a script, these parameters are accessed via *$1*, *$2*, etc. A function may return a value, which may be accessed via *$?* within the script.

In the following example,

```
$ cat function_parameters
#!/bin/bash
#Passing parameters to functions

display_params () {
    if [ $# = 0 ]
        then
            exit 1
    fi

    while [ $# != 0 ]
    do
        echo $1
        shift
    done
}
# Function call
display_params "New York" London Paris
exit 0
$ function_parameters
New York
London
Paris
$
```

In a Bash script, variables are global by default. If you want to use a local variable in a function, you can declare it as local inside a function by using the keyword local before the variable name, as in local var1 or local var1=5. Local variables are not visible outside the function in which they are declared. If a local variable and a global variable have the same name, the local variable will be accessed within the function.

In the variable_scope script, we explain the concept of scoping of variable with an example. Global variables *var1* and *var2* are initialized to 100 and 200, respectively. The local variable *var1* inside change_vars function is initialized to 10. Thus, inside change_vars, *var1* refers to local *var1* and *var2* refers to global *var2*. After a call to change_vars, local *var1* vanishes and global *var1* becomes visible again. Since *var2* is global, any changes made to it are visible inside and outside the function. The execution of **variable_scope** in the following session shows that any changes made to *var1* inside **change_vars** are not visible outside the function but changes made to *var2* are visible inside as well as outside function. Thus, the initial value of global *var1* remains 100 before and after call to **change_vars**, but change in the value of global *var2* from 200 to 300 made inside **change_vars** is visible after the function call returns.

```
$ cat variable_scope
#!/bin/bash
#All variables are global unless they are explicitly specified as
#local with the keyword local, as in local var1 or local var1=5

var1=100
var2=200

change_vars () {
    local var1=10

    echo "Inside function: var1 is $var1 and var2 is $var2"
    var1=50
    var2=300
    echo "Inside function: var1 is $var1 and var2 is $var2"
}
echo "Before call to change_vars: var1 is $var1 and  var2 is $var2"
change_vars
echo "After call to change_vars: var1 is $var1 and var2 is $var2"
$ variable_scope
Before call to change_vars: var1 is 100 and  var2 is 200
Inside function: var1 is 10 and var2 is 200
Inside function: var1 is 50 and var2 is 300
After call to change_vars: var1 is 100 and var2 is 300
$
```

In the following in-chapter exercise, you will write a simple function and a Bash script that uses it.

Exercise 13.9

Replicate the previous Bash sessions about parameter passing and scoping of variables on your system. Do they reproduce the results shown here?

Exercise 13.10

Write a function called menu that displays the following menu. Then write a shell script that uses this function.

```
Select an item from the following menu:
d. to display today's date and current time,
f. to start an ftp session,
t. to start a telnet session, and
q. to quit.
```

13.7 The exec Command and File I/O

The exec command is the command-level version of the Linux loader. Normally, the exec command is used to replace the current process with a new process. Thus, when executed under a shell, exec cmd overwrites the current shell with cmd. The exec command may also be used to open and close *file descriptors*.

When the `exec` command is used in conjunction with the redirection operators, it allows commands and shell scripts to read/write any type of files, including devices. In this section, we describe both uses of this command, but focus primarily on the second use.

13.7.1 Execution of a Command (or Script) in Place of Its Parent Process

The `exec` command can be used to run a command (or a script) instead of the process, usually the shell that executes this command. It works with all shells. The following is a brief description of the command:

Syntax:
   `exec command`

Purpose: Overwrite the code for **command** on top of the process that executes the **exec** command (the calling process), which makes **command** run in place of the calling process without creating a new process

---

After you have run this command, the control cannot return to the calling process. When the `exec` command has finished, control goes back to the parent of the calling process. If the calling process is your login shell, control goes back to the `getty` process, which displays the `Login:` prompt after the `exec` command finishes execution, as in

```
% exec date
Sun Mar 19 18:03:35 PKT 2017
Login:
```

When `exec date` finishes, control does not go back to the shell process but to the `getty` process—that is, the parent of the login shell process. The semantics of this command execution are shown in Figure 13.2.

If the command is run under a subshell of the login shell, control goes back to the login shell, as clarified in the following session. Here, a C shell is run as a child of the login shell, also a C shell in this case, and `exec date` is run under the child C shell. When the `exec date` command finishes execution, control goes back to the login C shell. The sequence of three diagrams from left to right shown in Figure 13.3 depicts the semantics of these steps.

```
$ ps
  PID TTY          TIME CMD
 6990 pts/2    00:00:00 bash
 7152 pts/2    00:00:00 ps
$ /bin/csh
% ps
  PID TTY          TIME CMD
 6990 pts/2    00:00:00 bash
 7158 pts/2    00:00:00 csh
 7160 pts/2    00:00:00 ps
```

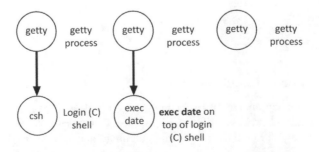

FIGURE 13.2 Execution of the `exec date` command under the login shell.

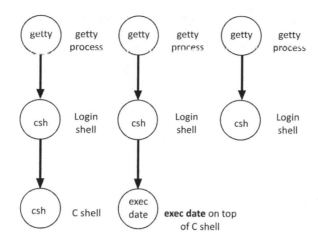

FIGURE 13.3 Execution of the `exec date` command under a subshell of the login shell.

```
% exec date
Sun Mar 19 18:04:24 PKT 2017
$ ps
  PID TTY          TIME CMD
 6990 pts/2    00:00:00 bash
 7163 pts/2    00:00:00 ps
$
```

13.7.2 File Input/Output via the exec Command

Different shells allow the use of different numbers of file descriptors at a time. As stated in Chapter 7, three of these descriptors are set aside for standard input (0), standard output (1), and standard error (2). Using the redirection operators with the `exec` command can use all of these descriptors for I/O. Table 13.2 describes the syntax of the `exec` command for file I/O.

TABLE 13.2

Syntax of the `exec` Command for File I/O

Syntax		Meaning
exec	< file	Opens file for reading and attaches standard input of the process to file
exec	> file	Opens file for writing and attaches standard output of the process to file
exec	>> file	Opens file for writing, attaches standard output of the process to file, and appends standard output to file
exec	n< file	Opens file for reading and assigns it the file descriptor n
exec	n> file	Opens file for writing and assigns it the file descriptor n
exec ... tag	n<< tag	Opens a *here document* (data between << tag and tag) for reading; the opened file is assigned a descriptor n
exec	n>> file	Opens file for writing, assigns it file descriptor n, and appends data to the end of file
exec	n>&m	Duplicates m into n; whatever goes into file with file descriptor n will also go into file with file descriptor m
exec	<&-	Closes standard input
exec	>&-	Closes standard output
exec	n<&-	Closes file with descriptor n attached to **stdin**
exec	n >&-	Closes file with descriptor n attached to **stdout**

When executed from the command line, the exec < sample command causes each line in the **sample** file to be treated as a command and executed by the current shell. That happens because the shell process, whose only purpose is to read commands from **stdin** and execute them, executes the exec command; as the **sample** file is attached to **stdin**, the shell reads its commands from this file. The shell terminates after executing the last line in **sample**. When executed from within a shell script, this command causes the **stdin** of the remainder of the script to be attached to **sample**. The following session illustrates the semantics of this command when it is executed at the command line. As shown, the **sample** file contains two commands: date and echo. A Bash is run under the login shell, which is a C shell, via the /bin/sh command. When the exec < sample command is executed, the commands in the **sample** file are executed, Bash (the child process of the login C shell) terminates after finishing execution of the last command in **sample** (the output of the third ps command shows that only the login shell runs after the exec < sample command has completed execution), and control returns to the login shell.

```
% cat sample
date
echo "Hello, world!"
% ps
  PID TTY          TIME CMD
  874 pts/0    00:00:00 bash
  933 pts/0    00:00:00 ps
% /bin/sh
% ps
  PID TTY          TIME CMD
  874 pts/0    00:00:00 bash
  942 pts/0    00:00:00 sh
  949 pts/0    00:00:00 ps
% exec < sample
$ Sun Apr  9 19:23:37 PKT 2017
$ Hello, world!
$
% ps
  PID TTY          TIME CMD
  874 pts/0    00:00:00 bash
  979 pts/0    00:00:00 ps
%
```

So, effectively, when the exec < sample command is executed from the command line, it attaches **stdin** of the current shell to the sample file. When this command is executed from a shell script, it attaches **stdin** of the shell script to the sample file. In either case, the exec < /dev/tty command must be executed to reattach **stdin** to the terminal. Here, **/dev/tty** is the pseudoterminal that represents the terminal on which the shell is executed. The following session illustrates the use of this command from the command line. The semantics of these steps are shown in Figure 13.4.

Similarly, when the exec > data command is executed from the command line, it attaches **stdout** of the current shell to the data file. Thus, it causes outputs of all subsequent commands executed under the shell to go to the data file. Thus, you do not see the output of any command on the screen. In order to see the output on the screen again, you need to execute the exec > /dev/tty command. After doing so, you can view the contents of the data file to see the outputs of all the commands executed prior to this command. When the exec > data command is executed from a shell script, it causes the outputs of all subsequent commands to go to the data file until the exec > /dev/tty command is executed from within the shell script.

The following session illustrates the use of this command from the command line. Note that, after the exec > data command has completed its execution, the outputs of all subsequent commands (date, echo, and more) go to the data file. To redirect the output of commands to the screen, the exec > /dev/tty command must be executed, as shown in the following session. Note that our system has an Intel x86 (or compatible) 64-bit CPU and, as expected, is running Linux Mint18.

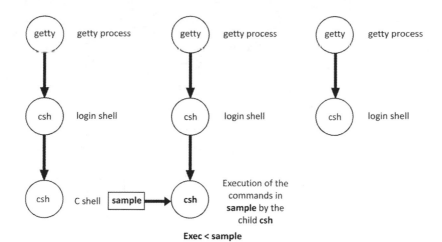

Exec < sample

FIGURE 13.4 Execution of the csh and exec < sample commands under the login shell.

```
$ exec > data
$ date
$ echo "Hello, world!"
$ uname -sm
$ exec > /dev/tty
$ date
Sun Jul  9 13:06:51 PKT 2017
$ more data
Sun Jul  9 13:06:25 PKT 2017
Hello, world!
Linux Mint18 4.4.0-21-generic #37-Ubuntu SMP Mon Apr 18 18:33:37 UTC 2016 x86_64
x86_64 x86_64 GNU/Linux
$
```

Similarly, you can redirect standard output and standard error for a segment of a shell script by using the following command:

```
exec > outfile 2> errorfile
```

In this case, output and error messages from the shell script following this line are directed to **outfile** and **errorfile**, respectively. (Obviously, file descriptor 1 can be used with > to redirect output.) If output needs to be reattached to the terminal, you can do so by using

```
exec > /dev/tty
```

Once this command has executed, all subsequent output goes to the monitor screen. Similarly, you can use the exec 2> /dev/tty command to send errors back to the display screen.

Consider the following shell session. When exec.demo1 is executed, **file1** gets the line containing Hello, world!, **file3** gets the contents of **file2**, and **file4** gets the line This is great!. The shell script between the commands exec < file2 and exec < /dev/tty takes its input from **file2**. Therefore, the command cat > file3 is really cat < file2 > file3. The cat > file4 command takes input from the keyboard as it is executed after the exec < /dev/tty command has been executed (which reattaches the **stdin** of the script to the keyboard). Figure 13.5 illustrates the semantics of the three cat commands in the shell script.

```
$ more exec.demo1
cat > file1
exec < file2
```

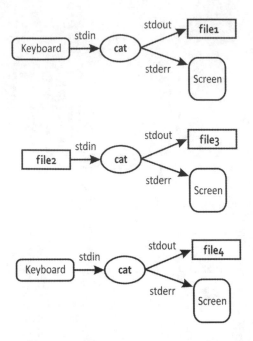

FIGURE 13.5 Detachment and reattachment of **stdin** and **stdout** inside a shell script.

```
cat > file3
exec < /dev/tty
cat > file4
$ ./exec.demo1
Hello, world!
<Ctrl+D>
This is great!
<Ctrl+D>
$ more file1
Hello, world!
$ more file2
Contents of file2.
$ more file3
Contents of file2.
$ more file4
This is great!
$
```

Now, we develop a shell script, diff2, which uses the file input/output (I/O) features of Bash. It takes two files as command line arguments and compares them line by line. If the files are the same, it displays a message that says so, and the program terminates. If one file is smaller than the other, it displays a message informing you of that and exits. As soon as the program finds the lines at which the two files differ, it displays an error message informing you of the lines from both files that are different and terminates. The following is the script and a few sample runs:

```
$ cat diff2
#!/bin/bash
# File Name: ~/linux2e/chaper13/bash/diff2
# Author:    Syed Mansoor Sarwar
# Written:   August 28, 2004
# Modified:  July 29, 2014; April 9, 2017
# Purpose:   To see if the two files passed as command line arguments
#        are same or different
```

```
# Brief Description:
#        Read a line from each file and compare them. If
#        the lines are the same, continue. If they are
#        different, display the two lines and exit. If one
#        of the files finishes before the other, display a
#        message and exit. Otherwise, the files are the
#        same; display an appropriate message and exit

if [ $# != 2 ]
    then
        echo "Usage: $0 file1 file2"
        exit 1
    elif [ ! -f "$1" ]
    then
        echo "$1 is not an ordinary file"
        exit 1
    elif [ ! -f "$2" ]
    then
        echo "$2 is not an ordinary file"
        exit 1
    else
        :
fi
file1="$1"
file2="$2"

# Open files for reading and assign them file descriptors 3 and 4

exec 3< "$file1"
exec 4< "$file2"

# Read a line each from both files and compare. If both reach EOF, then files
# are the same. Otherwise, they are different. 0<&3 is used to attach standard
# input of the read line1 command to file descriptor 3, 0<&4 is used to attach
# standard input of the read line2 command to file descriptor 4.

while read line1 0<&3
do
    if read line2 0<&4
        then
        # if lines are different, the two files are not the same
            if [ "$line1" != "$line2" ]
                then
                    echo "$1 and $2 are different."
                    echo " $1: $line1"
                    echo " $2: $line2"
                    exit 1
            fi
        else
            # if EOF for file2 reached, file1 is bigger than file2
            echo "$1 and $2 are different and $1 is bigger than $2."
            exit 1
        fi
done
# if EOF for file1 reached, file2 is bigger than file1. Otherwise, the two
# files are the same. 0<&4 is used to attach standard input of read to file
# descriptor 4
if read line2 0<&4
    then
        echo "$1 and $2 are different and $2 is bigger than $1."
        exit 1
    else
```

```
        echo "$1 and $2 are the same!"
        exit 0
fi

# Close files corresponding to descriptors 3 and 4
exec 3<&-
exec 4<&-
$ cat test1
Hello, world!
Not the same!
Another line.
$ cat test2
Hello, world!
$ cat test3
Hello, world!
Not the same!
$ cat test4
This is different file.
Hello, world!
$ ./diff2
Usage: ./diff2 file1 file2
$ ./diff2 test1 test2 test3
Usage: ./diff2 file1 file2
$ ./diff2 test1 test1
test1 and test1 are the same!
$ ./diff2 test1 test3
test1 and test3 are different and test1 is bigger than test3.
$ ./diff2 test1 test2
test1 and test2 are different and test1 is bigger than test2.
$ ./diff2 test2 test3
test2 and test3 are different and test3 is bigger than test2.
$ ./diff2 test1 test4
test1 and test4 are different.
 test1: Hello, world!
 test4: This is different file.
$
```

The exec command is used to open and close files. The exec 3< "$file1" and exec 4< "$file2" commands open the files passed as command line arguments for reading and assigns them file descriptors 3 and 4. From this point on, you can read the two files by using these descriptors. The commands read line1 0<&3 and read line2 0<&4 read the next lines from the files with files descriptors 3 (for **file1**) and 4 (for **file2**), respectively. The commands exec 3<&- and exec 4<&- close the two files. The colon sign (:) in the else part of the first if statement is a null statement that simply returns true. You may use the Bash command true for the same purpose. Incidentally, the false commands, as one would expect, returns false.

In the following in-chapter exercises, you will use the exec command to redirect the I/O of your shell to ordinary files. The concept of I/O redirection from within a shell script and file I/O by using file descriptors is also reinforced.

Exercise 13.11

Write a command for changing **stdin** of your shell to a file called **data** and **stdout** to a file called **out**, both in your present working directory. If the data file contains the following lines, what happens after the commands are executed?

```
echo -n "The time now is: "
date
echo -n "The users presently logged on are: "
who
```

Exercise 13.12

After finishing the steps in Exercise 13.10, what happens when you type commands at the shell prompt? Does the result make sense to you? Write the command needed to bring your environment back to normal.

Exercise 13.13

Create a file that contains the `diff2` script and try it with different inputs.

13.8 Debugging Shell Scripts

You can debug your Bash scripts by using the -x (echo) option of the sh command. This option displays each line of the script after variable substitution but before its execution. You can combine the -x option with the -v (verbose) option to display each line of the script, as it appears in the script file, before execution. You can also invoke the sh command from the command line to run the script, or you can make it part of the script, as in #!/bin/sh -xv. In the latter case, remove the -xv options after debugging is complete.

In the following session, we show how a shell script can be debugged. The script in the **debug_demo** file prompts you for a digit. If you enter a value between 1 and 9, it displays Good input! and quits. If you enter any other value, it simply exits. When the script is executed and you enter 4, it displays the message ./debug_demo: line 9: syntax error: unexpected end of file. This message usually means that you have not completely closed out a control construct such as missing fi in an if statement or missing done in a while statement.

```
$ cat debug_demo
#!/bin/bash

echo -e "Enter a digit: \c"
read var1
if ["$var1" -ge 1 -a "$var1" -le 9 ]
    then
        echo "Good input!"
exit 0
$ ./debug_demo
Enter a digit: 4
./debug_demo: line 9: syntax error: unexpected end of file
$
```

We debug the program by using the sh -xv debug_demo command. The error message clearly states that fi in the if statement is missing. We insert fi at the rightful place and rerun the program. We now get the error message ./debug_demo: 5: debug_demo: [4: not found. The shaded portion of the run-time trace shows the problem area. In this case, the error is generated because of a problem in the condition for the if statement. A closer examination of the shaded area reveals that a missing space between [and 4 is the problem. In other words, the comparison between [4 and 1 is the problem; it should be between 4 and 1. After we take care of this problem by changing ["$var1" to ["$var1", the script works properly.

```
$ sh -xv debug_demo
#!/bin/bash

echo -e "Enter a digit: \c"
+ echo -e Enter a digit: \c
-e Enter a digit: read var1
+ read var1
4
if ["$var1" -ge 1 -a "$var1" -le 9 ]
```

```
        then
              echo "Good input!"
exit 0
./debug_demo: 9: debug_demo: Syntax error: end of file unexpected (expecting "fi")
$ sh -xv debug_demo
#!/bin/bash

echo -e "Enter a digit: \c"
+ echo -e Enter a digit: \c
-e Enter a digit: read var1
+ read var1
4
if ["$var1" -ge 1 -a "$var1" -le 9 ]
        then
              echo "Good input!"
fi
+ [ 4 -ge 1 -a 4 -le 9 ]
./debug_demo: 5: debug_demo: [4: not found
exit 0
+ exit 0
$ ./debug_demo
Enter a digit: 4
Good input!
$
```

You can also use Bash Debugger Project (*bashdb*) to debug Bash scripts. Like the Gnu debugger for C/C++ programs, gdb (discussed in Chapter 14), bashdb allows you to single step through your Bash script line by line, set breakpoints, perform a backtrace, inspect values of variables, list program statements, etc. It does not come packaged with all distributions of Linux. If it doesn't come preinstalled on your system, ask your system administrator to install it for you.

The following in-chapter exercise is designed to enhance your understanding of interrupt processing and the debugging features of Bash.

Exercise 13.14

Test the scripts in the **trap_demo** and **canleave** files on your Linux system. Do they work as expected? Make sure you understand them. If your versions do not work properly, use the sh -xv command to debug them.

Summary

Bash does not have the built-in capability for numeric integer data processing in terms of arithmetic, logic, and shift operations. To perform arithmetic and logic operations on integer data, you can use the let or expr command. You can perform numeric expression evaluation using the shell expansion $((expression)). You can also evaluate arbitrary-precision mathematical expressions using the bc tool. It allows evaluation of expressions interactively as well as with scripts written in the language it supports, whose features and constructs are quite similar to those of the C language.

Bash supports one-dimensional arrays. It allows you to declare local and global arrays by using declare -a, local –a, and readonly –a commands, or the array_name[subscript]=value syntax. You can initialize an array by using the syntax name=(value1 ... valueN), where value1 is of the form [[subscript]=]string, such as Fibonacci=(0 1 1 2 3 5 8 13 21) or names=("Xi Jinping" "Tom Nelson" "Muhammad Yousaf" "Ravi Shankar" "Zach Goldberg" "Akihito"). You can index an array by using the ${name[subscript]} syntax.

The here document feature of Bash allows standard input of a command in a script to be attached to the data within the script. The use of this feature results in more efficient programs, because no extra file-related operations, such as file open and read, are needed, as the data is within the script file and should have been loaded into the main memory when the script was loaded.

Bash also allows the user to write programs that ignore signals such as keyboard interrupt (<Ctrl+C>). This useful feature can be used, among other things, to disable program termination when it is in the middle of updating a file. The trap command can be used to invoke this feature.

Bash allows the use of functions, including parameter passing and access to a function's return value. Scoping rules for local and global variables are clearly defined.

Bash has powerful I/O features that allow explicit processing of files. The exec command can be used to open a file for reading or writing and to associate a small integer, called a file descriptor, with it. The command exec n< file opens a file for reading and assigns it a file descriptor n. The command line exec n> file opens a file for writing and assigns it a file descriptor n. This feature allows writing scripts for processing files. The command line exec n<&- can be used to close a file with descriptor n. The exec command provides various other file-related features, including opening a here document and assigning it a file descriptor, which allows the use of a here document anywhere in the script.

Bash programs can be debugged using the -x and -v options of the sh command. This technique allows viewing the commands in the user's script after variable substitution but before execution. You can also debug Bash scripts using bashdb that has features similar to that of Gnu debugger for C/C++, gdb, discussed in Chapter 14.

Questions and Problems

1. Why is the expr command needed?

2. Modify the num_array_demo script to display the sum of the square of the given Fibonacci numbers.

3. Write a Bash script, Fibonacci_series, that takes an integer *N* as its only command line argument and displays the first N Fibonacci numbers.

4. What is the here document? Why is it useful?

5. Write a Bash script cv that takes the side of a cube as a command line argument and displays the volume of the cube.

6. Modify the countup script in Section 13.2 so that it takes two integer command line arguments. The script displays the numbers between the two integers (including the two numbers) in ascending order if the first number is smaller than the second, and in descending order if the first number is greater than the second. Name the script count_up_down.

7. Give two Bash commands each to evaluate the following expressions without precision and precision up to five decimal digits. Show commands along with their outputs.

   a. 97.03*57.71/23.97

   b. 23.4+93.17-61.721/13.97*13

8. Perform the following tasks by using bc. Show commands along with their outputs under Bash.

   a. Convert 10111011 in base 2 to equivalent number in decimal and hexadecimal

   b. Convert 10111011 in base 10 to equivalent numbers in binary, octal, and hexadecimal

   c. Convert 10111011 in base 20 to equivalent numbers in decimal, binary, octal, and hexadecimal

9. What are the outputs of the following commands? Explain each output.

   a. echo "var=25;++var" | bc

   b. echo "var=25;var++" | bc

   c. echo "var=700;var%=9;var" | bc

   d. echo "var=25;var^=3;var" | bc

      e.   echo "5 && 50" | bc

      f.   echo "5 || 50" | bc

      g.   echo "0 || 0" | bc

      h.   echo "1 || 1" | bc

      i.   echo "! 0" | bc

      j.   echo "obase=2;15" | bc -l

      k.   echo "ibase=2;1111" | bc -l

      l.   echo "ibase=2;obase=8;1110" | bc -l

10. Write a Bash script that prompts you for a user ID and displays your login name, your name, and your home directory.

11. Write a Bash script that takes a list of integers as the command line argument and displays a list of their squares and the sum of the numbers in the list of squares.

12. Write a Bash script that takes a machine name as an argument and displays a message informing you whether the host is on the local network.

13. What are signals in Linux? What types of signals can be intercepted in Bash scripts?

14. Write a Bash script that takes a file name and a directory name as command line arguments and removes the file if it is found under the given directory and is a simple file. If the file (the first argument) is a directory, it is removed (including all the files and subdirectories under it).

15. Write a Bash script that takes a directory as an argument and removes all the ordinary files under it that have **.o**, **.ps**, and **.jpg** extensions. If no argument is specified, the current directory is used.

16. Enhance the diff2 script in Section 13.6 so that it displays the line numbers where the two files differ.

17. Enhance the diff2 script of Problem 11 so that, if only one file is passed to it as a parameter, it uses standard input as the second file.

18. Write a Bash script for Problem 16 in Chapter 12, but use functions to implement the service code for various options.

19. Suppose that a function, f, is defined first, followed by defining f as a variable. Would f refer to the function or the variable after you have provided the second definition? Why? Show a shell session to support your answer.

20. How are parameters passed to a function in Bash?

21. Write a Bash script that implements the following menu options:

      a.   Display the CPU used by your system

      b.   Display the name of the operating system used by your computer

      c.   Display a and b on the screen separated by a vertical tab

      d.   Display the full path names for the commands that have been executed on your system

      e.   Display the maximum number of files a process may open

      f.   Display the maximum number of simultaneous processes a user may have on the system
         Your program should not terminate on signals 1, 2, 3, 15, and 18. Make use of a function to display the program menu.
         Hint: Review the man pages for the following commands: hash, ulimit, uname.

Advanced Questions and Problems

22. Give while loop equivalent of the for loop given in Section 13.2.2 to display the first ten decimal digits, one per line.

23. Change the while loop in Problem 20 such that the digits are displayed in one line with a space between them.

24. Browse the Web to locate a good Youtube video for a tutorial on bashdb. Learn bashdb and run all Bash script discussed in it. Fix any problems that it report, log the problems for every script and their fixes.

25. What are the scoping rules for functions in Bash? Show an example (other than the one discussed in the book) to illustrate how these rules work. Your example should make at least one call to a function from within another function.

26. How can the return value of a function be accessed when a function call return? Illustrate your answer with an example script.

27. Create a Bash script file named "mother," that contains three or more function definitions in it. What those functions accomplish is irrelevant to this problem, as long as they produce some clear and definite output at stdout when executed. If you execute the script file mother by calling it inside of a second script file, named "daughter," are the functions you put in mother available to be executed in the environment of the script file daughter? If not, why not? How can you make sure that those functions in mother are available in the environment of the script file daughter? Explain your answers to these questions fully, and provide the code for the script file mother and the script file daughter that calls it.

28. Define a Bash function on the command line that creates a new directory in the current working directory you execute the function in, and any number of subdirectories, and sub-subdirectories under those, recursively. It should take an argument list to the function, when it is actually executed, named specifically as dir1/dir2/dir3. It should be able to handle erroneous arguments, such as zero arguments supplied, or illegal arguments of any kind. Additionally, the function should create a single file in each of the directories, and that single file should be copied from a "master" template that exists in the current working directory in which you execute the function in. What command did you use to actually execute the function?

29. All iterative programming constructs (whether they are determinate or indeterminate) can be formulated as recursive functions. Take the enhanced version of the dext Bash script, first shown in Section 13.4 with an overall iterative construct, and then refined in Section 13.6.4, and convert it so that the iterative construct in it uses a recursive Bash function call to achieve *exactly* the same results.

Projects

Project 1

Simpson's rule is a very accurate method of numerically integrating a continuous function over some closed interval. The integral of a function $f(x)$ between a and b is approximated as

$$h/3[y0+4y1+2y2+4y3+2y4+...+2yn-2+4yn-1+yn]$$

where $h = (b - a)/n$, for some even integer n, and yk = f(a + kh). Larger values of n increase the accuracy of the approximation.

Using the Gnu calculator command **bc** (which comes preinstalled on our Linux Mint 18.2 and CentOS 7.4 systems) and the here document, define a mathematical function in **bc** that takes as arguments $f(x^2)$, a, b, and n, and returns the value of the Simpson's Rule numerical integral accurate to at least ten significant figures to the right of the decimal point. Integrate between 0 and 1 (with sample n values of $n = 100$ and $n = 1,000$).

Invocation of the function, which you can name "sintegrate," on the **bc** command line, would be done using something like (sintegrate square 0 1 100.0). In addition, extend your system in **bc** so that other valid definitions of the mathematical function can be entered upon successive invocations of the **bc** function.

Project 2

Repeat Project 1 using only Python, instead of **bc** and the here document. As in Project 1, allow for extension of the valid provisioning of the math function. For reference, see Chapter W19 at the book website, which covers the basics of Python programming. This reference primarily illustrates its objectives using Python 2.7.X, but you can use Python 3.X if you prefer for this project.

Project 3

How would you proceed to design and implement your own basic shell program as a software system in Linux? What would you use as a model, and in what programming language would you code the shell? Would you use a compiled or interpreted language, such as C++, or Python 2.X/3.X? A very critical determination you have to make first is exactly why and how would implementing the shell in Python, for example, be advantageous? The specifications of your shell, for example, should include implementing at least commands such as cat or more, ls, mkdir, cd, rm, rmdir, with as many options to those commands as you think are necessary.

And more extensively, you can construct shell commands that you find useful given your particular use case(s) for the hardware you are working on. That does not necessarily mean you are constrained to the X86 architecture, or its instruction set, but your software system should be implemented on a Linux, or Linux-based operating system that "rides" on Reduced Instruction Set Computer (RISC) or Advanced RISC Machine (ARM) processor instruction sets.

Once you have answered the previous questions, and designed and specified your own personal shell program as a software system, implement it in the language you've chosen, and then test it in a two-person team, as rigorously as necessary.

Looking for more? Visit our sites for additional readings, recommended resources, and exercises.

CRC Press e-Resource: https://www.crcpress.com/9781138710085

Authors' GitHub: https://github.com/bobk48/linuxthetextbook

14

Linux Tools for Software Development

OBJECTIVES

- To summarize computer programming languages at different levels
- To discuss interpreted and compiled languages and the compilation process
- To briefly describe the software engineering life cycle
- To discuss Linux program generation tools for C to perform the following tasks: editing, indenting, compiling (of C, C++, and Java programs), handling module-based software, creating libraries, source code management, and revision control
- To describe Linux tools for static analysis of C programs: verifying code for portability and profiling
- To discuss Linux tools for dynamic analysis of C programs: debugging, tracing, and monitoring performance
- To cover the following commands and primitives:

  ar, cppcheck, g++, gcc, gdb, gprof, grep, help, indent, javac, make, memcheck, nm, perf, perl, python, ranlib, rlog, ruby, strip, time, valgrind

---

14.1 Introduction

A typical Linux system supports several high-level languages (HLLs), both interpreted and compiled. These languages include C, C++, Pascal, Java, LISP, and FORTRAN. However, most of the application software for the Linux platform is developed in the C language, the language in which the Linux operating system is written. Thus, a range of software engineering tools is available for use in developing software in this language. Many of these tools can also be used for developing software in other programming languages, C++ in particular.

The Linux operating system has a wealth of software engineering tools for program generation and static and dynamic analysis of programs. They include tools for editing source code, indenting source code, compiling and linking, handling module based software, creating libraries, profiling, verifying source code for portability, source code management, debugging, tracing, and performance monitoring. In this chapter, we describe some of the commonly used tools in the development of C-based software. The extent of discussion of these tools varies from brief to detailed, depending on their usefulness and how often they are used in practice. Before discussing these tools, however, we briefly describe various types of languages that can be used to write computer software. In doing so, we also discuss both interpreted and compiled languages.

14.2 Computer Programming Languages

Computer programs can be written in a wide variety of programming languages. The native language of a computer is known as its *machine language*, the language comprising the *instruction set* of the CPU inside the computer. Recall that the instruction set of a CPU consists of instructions that the CPU understands. These instructions enable the performance of various types of operations on data, such as arithmetic, logic, shift, and input/output operations. Today's CPUs are made of bistate devices (devices that operate in *on* or *off* states), so CPU instructions are in the form of 0s and 1s (0 for the off state and 1 for the on state). The total number of instructions for a CPU and the maximum length (in bytes) of an instruction is CPU dependent. Although *reduced instruction set computer* (RISC)-based CPUs have several hundred simple instructions, *complex instruction set computer* (CISC)-based CPUs have a much smaller number of complex instructions. A program written in a CPU's machine language is known as *machine program*, commonly known as *machine code*. The machine language programs are the most efficient because they are written in a CPU's native language. However, they are the most difficult to write because the machine language is very different from any spoken language; the programmer has to write these programs in 1s and 0s, and a change in one bit can cause major problems. Debugging machine language programs is a very challenging and time-consuming task. For these reasons, programs today are rarely written in machine languages.

In *assembly language programming*, machine instructions are written in English-like words called *mnemonics*. Because programs written in assembly language are closer to the English language, they are relatively easier to write and debug. However, these programs must be translated into the machine language of the CPU used in your computer before you can execute them. This process of translation is carried out by a program called an *assembler*. You have to execute a command to run an assembler, with the file containing an assembly language program as its argument. Although assembly languages are becoming less popular, they are still used to write time-critical programs for controlling real-time systems (e.g., the controllers in drilling machines for oil wells) that have limited amounts of main storage.

In an effort to bring programming languages closer to the English language—and make programming and debugging tasks easier—HLLs were developed. Commonly used HLLs are Ada, C, C++, Java, JavaScript, BASIC, FORTRAN, LISP, and Prolog. Some of these languages are *interpreted* (e.g., JavaScript, LISP, and all shell scripts), whereas others are *compiled* (e.g., C, C++, and Java). On the one hand, programs written in an interpreted language are executed one instruction at a time by a program called an *interpreter*, without translating them into the machine code for the CPU used in the computer. On the other hand, programs written in compiled languages must be translated into the machine code for the underlying CPU before they are executed. A program, called a *compiler*, which generates the assembly version of the HLL program, carries out this translation. The assembly version has to go through further translation before the executable code is generated. The compiled languages run many times faster than the interpreted languages, because compiled languages are directly executed by the CPU, whereas the interpreted languages are executed by a piece of software (an interpreter).

However, the Java language is not compiled in the traditional sense. Java programs are translated into a form known as the Java *bytecode*, which is then interpreted by an interpreter.

To simplify the task of writing computer programs even more, languages at a higher level even than the HLLs were developed. They include scripting and visual languages such as Linux shell programming, Perl, Matlab, Ruby-on-Rails, Python, Visual Basic, and Visual C++. Some of these languages are interpreted; others are compiled. Figure 17.1 shows the proximity of various types of programming languages to the computer hardware, ease of their use, and relative speed at which programs are executed.

As the level of programming languages increases, the task of writing programs becomes easier, programs become more readable, and code more portable. The tradeoff is that programs written in HLLs take longer to run. For interpreted programs, the increase in program running time is due to the fact that another program (the interpreter) is running the program. For compiled languages, the compilation process takes longer, and the resulting machine code is usually much bigger than it would be if written in assembly language by hand. However, time is saved because the ease of programming in HLLs far outweighs the increase in code size. Figure 14.1 also shows some language statement examples to demonstrate the increased readability of programs as the level of programming languages increases.

14.3 The Compilation Process

Because our focus in this chapter is on Linux tools—primarily for the C programming language (a compiled language)—we need to describe briefly the compilation process before moving on. As we stated in Section 14.2, computer programs written in compiled languages must be translated to the

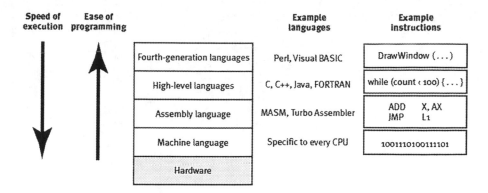

Speed of execution	Ease of programming		Example languages	Example instructions
		Fourth-generation languages	Perl, Visual BASIC	DrawWindow (. . .)
		High-level languages	C, C++, Java, FORTRAN	while (count < 100) { . . . }
		Assembly language	MASM, Turbo Assembler	ADD X, AX JMP L1
		Machine language	Specific to every CPU	1001110100111101
		Hardware		

FIGURE 14.1 Levels of programming languages, with examples, ease of programming, and speed of execution.

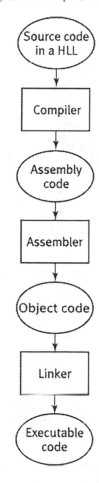

FIGURE 14.2 The process of translating a high-level language program to an executable code.

machine code of the CPU used in the computer system on which they are to execute. This translation is usually a three-step process consisting of *compilation*, *assembly*, and *linking*. The compilation process translates the source code (e.g., a C program) to the corresponding assembly code for the CPU used in the computer system. The assembly code is then translated to the corresponding machine code, known as *object code*. Finally, the object code is translated to the *executable code*. Figure 14.2 outlines the translation process. In a large project, you may have multiple source files. In such a case, you compile them into separate object files and pass them to the linker to generate one executable file.

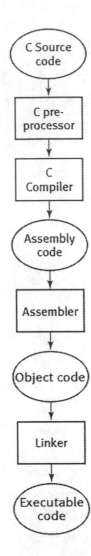

FIGURE 14.3   The process of translating C programs to executable code.

The object code consists of machine instructions, but it is not executable because the source program might have used some library functions that the assembler cannot resolve references to, because the code for these functions is not in the source file(s). The linker performs the task of linking (connecting) the object code for a program and the object code in a library, and generates the executable binary code.

The translation of C programs goes through a preprocessing stage before it is compiled. The C preprocessor translates program statements that start with the # sign. Figure 14.3 outlines the compilation process for C programs. The entire translation process is carried out by a single compiler command. We discuss various Linux compilers later in this chapter. As stated earlier, the linker may be given multiple object files for generating an executable file.

14.4 The Software Engineering Life Cycle

A software product is developed in a sequence of phases, collectively known as the *software development life cycle*. Several life cycle models are available in the literature and used in practice. The life cycle used for a specific product depends on its size, the nature of the software to be developed (scientific, business, etc.), and the design methodology used (object oriented or classical). Some of the commonly used life cycle models are *build-and-fix*, *waterfall*, *spiral*, *agile software development*, and *rapid application development*. The common phases in most classical life cycle models are *requirement analysis*, *specifications*, *planning*, *design*, *coding*, *testing*, *installation/deployment*, and *maintenance*. A full discussion of life cycle models and their phases is outside the scope of this textbook, but we discuss the coding phase in detail—in particular, the Linux program development, analysis, and debugging tools that can be used in this phase.

The program development process consists of three steps: *program generation*, *static analysis of the source code*, and *dynamic analysis of the executable code*. The purpose of the program generation phase is to create source code and generate the executable code for the source code. Hence, it involves tools for editing text files, indenting the source code properly, compiling the source code, handling module-based software, creating libraries, managing the source code, and controlling revisions. The static analysis phase consists of verifying the source code for portability and measuring metrics related to the source code (e.g., the number of calls to a function and the time taken by each function). The dynamic analysis phase comprises debugging, tracing, and monitoring the performance of the software, including testing it against product requirements. In the rest of this chapter, we describe Linux tools for all three steps. The depth of discussion on each tool depends on its usefulness for an average software developer, the frequency of its use, and how widely it is available on various Linux platforms.

14.5 Program Generation Tools

The program generation phase consists of creating source code and generating the executable code for it. Hence, it involves tools for editing text files, indenting the source code properly, compiling the source code, handling module-based software, creating libraries, managing the source code, and controlling revisions. We now discuss the Linux tools for supporting these tasks.

14.5.1 Generating C Source Files

Any text editor can be used to generate C program source files. We discussed the most frequently used Linux editors, including nano in Chapter 3 and emacs in Chapter W25 at the book website.

14.5.2 Indenting C Source Code

Proper indentation of source code is an important part of good coding practice, primarily because it enhances the readability of the code, and readable code is easier to maintain (debug, correct, and enhance). The best known indentation style for C programs was proposed by Brian Kernighan and Dennis Ritchie in *The C Programming Language* (1978), the first book on the C language. It is commonly known as the *K&R* (Kernighan and Ritchie) indentation style. Most non-C programmers are not familiar with this style unless they have read the book. The Linux utility `indent` can be used to indent a C program properly. It can also be used to convert a C program written in one style to another. The current version of `indent` defaults to the Gnu coding style. The following is a brief description of the `indent` command.

Syntax:
```
indent [options] [input-files]
indent [options] [single-input-files] [-o output-file]
```
Purpose: This command reads a syntactically correct program specified as input, indents it according to some commonly accepted C program structure as specified in the options, saves the formatted program in the input file, and saves the original program in a backup file

Commonly used options/features:

`/*INDENT OFF*/`	The source code between these two
`/*INDENT ON*/`	comment is not formatted by indent

The source code between these two comments is not formatted by indent.

-bad	Force blank lines after declarations
-bap	Force blank lines after function bodies
-br	Format according to the more commonly used K&R-like syntax, the default setup
-linux	Format according to Linux style coding
-kr	Format according to K&R style coding
-orig	Format according to the original Berkeley style coding
-st	Send formatted program to standard output

Several other options allow you to format your code in various styles. You can specify these options before or after the file names. The indent command makes sure that the names of **input-file** and **output-file** are different; if they are the same, it gives an error message and quits. By default, the backup (i.e., original) file is saved in a file that has the name of the original file with "~" appended to it, and the original file contains the formatted file. We show a simple use of the command in the following session using the C program file called **second.c**. The indent command saves the original contents of the source file **second.c** in the **second.c~** file in the current directory. The **second.c** file contains the newly formatted code.

```
$ more second.c
#include <stdio.h>
main()
{
        int i, j;

        for (i=0,j=10; i < j; i++)
        {
                printf("Linux Rules the Networking World!\n");
        }
}
$ indent second.c
$ cat second.c
#include <stdio.h>
main()
{
        int i, j;

        for (i = 0, j = 10; i < j; i++) {
                printf("Linux Rules the Networking World!\n");
        }
}
$
```

If you don't specify any input file, or specify "-" as input file, input is read from standard input. With the -o option, you explicitly specify the name of the formatted file, as in indent `first.c -o first_formatted.c`. With the -st option, indent sends the formatted code to standard output. Thus indent -st `first.c > first_gnu.c` saves the Gnu-style formatted file in first_gnu.c.

In the following in-chapter exercise, you will use indent to practice indentation of C programs.

Exercise 14.1

Create the **second.c** file just described and indent it according to the K&R style by using the indent command. What command line(s) did you use?

14.5.3 Running C, C++, Java, Perl, Python, and Ruby Programs

Linux supports compliers and interpreters for a large number of contemporary languages, including C, C++, Java, Perl, Python, and Ruby. In the following subsections, we show how you can compile and execute programs in C, C++, and Java, and run scripts in Perl, Python, and Ruby. We also show how you can handle multimodule programs in C and link libraries with your object modules.

14.5.3.1 Compiling and Executing C Programs

Several C compilers are available on Linux, including cc and gcc. The most commonly used C compiler for Linux is the Gnu C/C++ compiler, gcc. This compiler is written for American National Standards Institute's specification for C, called ANSI C, the most recent standard for C language. All C++ compilers, such as the Gnu compiler for C++, g++, can also be used to compile C programs. The g++ compiler invokes gcc with options necessary to make it recognize C++ source code. We primarily discuss the gcc compiler in this chapter.

The gcc command can be used with or without options. We describe some basic options here and some in later sections of this chapter. One of the commonly used options, even by the beginners, is -o. You can use this option to inform gcc that it should store the executable code in a particular file instead of the default **a.out** file. In the following session, we show compilation of the C program in the **first.c** file, with and without the -o option. The gcc first.c command produces the executable code in the **a.out** file and the gcc -o slogan first.c command produces the executable code in the slogan file. The ls command is used to show the names of the executable files generated by the two gcc commands.

Syntax:
```
gcc [options] file-list
```
Purpose: This command can be used to invoke the C compilation system. When executed, it preprocesses, compiles, assembles, and links to generate executable code. The executable code is placed in the **a.out** file by default. The command accepts several types of files and processes them according to the options specified in the command line. The files can be archive files (**.a** extension), C source files (**.c** extension), C++ source files (**.C**, **.cc**, or **.cxx** extension), assembler files (**.s** extension), preprocessed files (**.i** extension), or object files (**.o** extension). When a file extension is not recognizable, the command assumes the file to be an object or library/archive file. The files are specified in **file-list**

Commonly used options/features:

-ansi	Enforce full ANSI conformance
-c	Suppress the linking phase and keep object files (with the **.o** extension)
-g	Create symbol table, profiling, and debugging information for use with gdb (Gnu debugger)
-llib	Link to the **lib** library
-mconfig	Optimize code for **config** CPU (**config** can specify a wide variety of CPUs, including Intel 80386, 80486, Motorola 68K series, RS6000, AMD 29K series, and MIPS processors)
-o file	Create executable in **file**, instead of the default file **a.out**
-O[level]	Optimize. You can specify 0–3 as **level**; generally, the higher the number for **level**, the higher the level of optimization. No optimization is done if **level** is 0
-pg	Provide profile information to be used with the profiling tool gprof
-S	Do not assemble or link .c files, and leave assembly versions in corresponding files with the **.s** extension
-v	Verbose mode: Display commands as they are invoked
-w	Suppress warnings

```
$ cat first.c
#include <stdio.h>

int main ()
{
        printf("Linux Rules the Networking World!\n");
        return (0);
}
```

```
$ ls
first.c  second.c
$ gcc first.c
$ ls
a.out    first.c  second.c
$ a.out
Linux Rules the Networking World!
$ gcc -o slogan first.c
$ ls
a.out    first.c  second.c slogan
$ slogan
Linux Rules the Networking World!
$
```

If your shell's search path does not include your current directory (.), you will get the message a.out: Command not found., as shown in the following session. If this happens, then you have two options: you can either run the command as ./a.out (that is, explicitly inform the shell that it should run the **a.out** file in your current directory) or include your current directory in your shell's search path and rerun the command as a.out. The following session illustrates both options.

```
$ a.out
a.out: Command not found.
$ ./a.out
Linux Rules the Networking World!
$ PATH=$PATH:.
$ a.out
Linux Rules the Networking World!
$
```

Note that the change in your search path is effective for your current session only. For a permanent change in the search path, you need to change the value of the *PATH* variable in your ~/.**profile** file. See Chapter 2 for details.

14.5.3.2 Dealing with Multiple Source Files

You can use the gcc command to compile and link multiple C source files and create an executable file, all in a single command line. For example, you can use the following command line to create the executable file called **polish** for the C source files **driver.c**, **stack.c**, and **misc.c**.

```
$ gcc driver.c stack.c misc.c -o polish
$
```

If one of the three source files is modified, you need to retype the entire command line, which creates two problems. First, all three files are compiled into their respective object modules, although only one needs recompilation. This results in longer compilation time, particularly if the files are large. Second, retyping the entire line may not be a big problem when you are dealing with three files (as here), but you certainly will not like having to do it with a much larger number of files. To avoid these problems, you should create object modules for all source files and then recompile only those modules that are updated. All the object modules are then linked together to create a single executable file.

You can use the gcc command with the -c option to create object modules for the C source files. When you compile a program with the -c option, the compiler leaves an object file in your current directory. The object file has the name of the source file and an **.o** extension. You can link multiple object files by using another gcc command. In the following session, we compile three source modules—**driver.c**, **stack.c**, and **misc.c**—separately to create their object files, and then use another gcc command to link them and create a single executable file, **polish**.

```
$ gcc -c driver.c
$ gcc -c stack.c
```

```
$ gcc -c misc.c
$ gcc misc.o stack.o driver.o -o polish
$ polish
[output of the program]
$
```

You can also compile multiple files with the -c option. In the first of the following command lines, we compile all three source files with a single command to generate the object files. The compiler shows the names of the files as it compiles them. The order in which files are listed in the command line is not important. The second command line links the three object files and generates one executable file, **polish**.

```
$ gcc -c driver.c stack.c misc.c
$ gcc misc.o stack.o driver.o -o polish
$
```

Now if you update one of the source files, you need to generate only the object file for that source file by using the gcc -c command. Then you link all the object files again using the second of the gcc command lines to generate the executable file.

14.5.3.3 Linking Libraries

The C compilers on Linux systems link appropriate libraries with your program when you compile it. Sometimes, however, you have to tell the compiler explicitly to link the required libraries. You can do so by using the gcc command with the -l option, immediately followed by the letters in the library name that follow the string lib and before the extension. Most libraries are in the **/lib** directory. You have to use a separate -l option for each library that you need to link. In the following session, we link the math library (**/lib/i386-linux-gnu/libm.so.6** or **/lib/x86_64-linux-gnu/libm.so.6** depending on whether your machine is 32-bit or 64-bit) to the object code for the program in the **power.c** file. We used the first gcc command line to show the error message generated by the compiler if the math library is not linked. The message says that the symbol pow is not found in the **power.o** file, the file in which it is used. The name of the math library is **libm.so.6**, so we use the letter m, which follows the string lib and precedes the extension, with the -l option.

```
$ cat power.c
#include <stdio.h>
#include <math.h>

int main()
{
        float   x,y;

        printf ("The program takes x and y from stdin and displays x^y.\n");
        printf ("Enter integer x: ");
        scanf   ("%f", &x);
        printf ("Enter integer y: ");
        scanf   ("%f", &y);
        printf ("x^y is: %6.3f\n", pow((double)x,(double)y));
        return(0);
}
$ gcc power.c
/tmp/ccoxeTLL.o: In function `main':
power.c:(.text+0x7e): undefined reference to `pow'
collect2: error: ld returned 1 exit status
$ gcc power.c -lm -o power
$ power
The program takes x and y from stdin and displays x^y.
Enter integer x: 9.82
```

```
Enter integer y: 2.3
x^y is: 191.362
$
```

14.5.3.4 Compiling and Executing C++ and Java Programs

You can use the g++ compiler to compile C++ programs. However, it is not preinstalled in Linux Mint 18. You can ask your system administrator to install it for you. If you have superuser privileges, you can install it yourself as described in Appendix A.

The file containing C++ source must have one of the following extensions: **.C**, **.CPP**, **.cpp**, **.c++**, or **.cc**. In the following session, we show a small C++ program, its compilation, and execution.

```
$ more Hello.cpp
// My first C++ program on Mint 18 Linux
#include <iostream>

int main()
{
    std::cout << "Hello, World!\n";
    return 0;
}
$ g++ Hello.cpp -o Hello
$ Hello
Hello, World!
$
```

Java source code is compiled (translated) into Java *bytecode* and is interpreted by the *Java Virtual Machine* (JVM), also known as the *Java Interpreter*. The Java compiler on our Linux system is called javac, and the JVM is java. Thus, to run a Java program in a file, say **Hello.java**, we use the javac compiler to compile it. It produces the Java bytecode and stores it in the **Hello.class** file, which is interpreted with the java command, as shown in the following session.

```
$ more Hello.java
public class Hello {
    public static void main(String[] args) {
        System.out.println("Hello, World!");
        System.exit(0);
    }
}
$ javac Hello.java
$ java Hello
Hello, world!
$
```

In the following in-chapter exercises, you will use the gcc, g++, and javac compiler commands to compile simple C, C++, and Java programs on your Linux system and run them.

Exercise 14.2
    Replicate all of the sessions discussed in Sections 14.5.3.1–14.5.3.4 on your system to verify that they work as expected.

Exercise 14.3
    Create simple C, C++, and Java programs on your Linux system. Compile and run them to appreciate the basic working of the two compilers (gcc and javac) and the Java Virtual Machine, java.

14.5.3.5 Running Perl, Python, and Ruby Scripts

You can run Perl, Python, and Ruby scripts on Linux by using their interpreters. The interpreters for Perl, Python, and Ruby are **/usr/bin/perl**, **/usr/bin/python**, and **/usr/bin/python**, which may be invoked by using the command perl, python, and ruby. If in doubt, use the which command to find the locations of the interpreters in the filesystem, as shown in the following session.

```
$ which perl
/usr/bin/perl
$ which python
/usr/bin/python
$ which ruby
/usr/bin/ruby
$
```

Ruby is not installed by default on Linux Mint. You can ask your system administrator to install it for you. You can install it yourself on your personal Linux machine by using the instructions given in Appendix A.

The extensions for the script files for Perl, Python, and Ruby are **.pl**, **.py**, and **.rb**, respectively. As is the case with all script files in Linux, the first line specifies the full pathname for the interpreter, followed by the script. The Perl, Python, and Ruby scripts are executed in the same manner that shell scripts are executed: create the script file, make the file executable, and run the script by either using the name of the script file as command name or by specifying the script file as an argument to the command for the relevant interpreter (perl, python, or ruby). The following session shows how the "Hello, World!" Perl script looks like and how it can be executed.

```
$ cat Hello.pl
#!/usr/bin/perl
# The Hello World program in Perl

print "Hello, World!\n"
$ chmod +x Hello.pl
$ Hello.pl
Hello, World!
$ perl Hello.pl
Hello, World!
$
```

The following session shows how the "Hello, World!" Python script look like and how it can be executed.

```
$ cat Hello.py
#!/usr/bin/python
# The Hello World program in Python

print "Hello, World!"
$ chmod +x Hello.py
$ Hello.py
Hello, World!
$ python Hello.py
Hello, World!
$
```

Finally, the following session shows how the "Hello, World!" Ruby script look like and how it can be executed.

```
$ cat Hello.rb
#!/usr/bin/ruby
# The Hello World program in Ruby

puts "Hello, World!"
$ chmod +x Hello.rb
$ Hello.rb
Hello, World!
```

```
$ ruby Hello.rb
Hello, World!
$
```

You can also run Perl and Python from the command line, as shown in the following session.

```
$ perl -e 'print "Hello, World!\n"'
Hello, World!
$ python -c 'print "Hello, World!"'
Hello, World!
$ ruby -e 'puts "Hello World!\n"'
Hello World!
$
```

> **Exercise 14.4**
> Repeat the earlier sessions on your Linux system to appreciate how to run Perl, Python, and Ruby from the command line and how to execute their scripts.

14.5.4 Handling Module-Based C Software with make

Most of the useful C software is divided into multiple source (**.c** and **.h**) files. This software structure has several advantages over a *monolithic* program stored in a single file. First, it leads to a more modular software, which results in smaller program files that are less time-consuming to edit, compile, test, and debug. It also allows recompilation of only those source files that are modified, rather than the entire software system. Furthermore, the multimodule structure supports information hiding, the key feature of object-oriented design and programming.

However, the multimodule implementation also has its disadvantages. First, you must know the files that comprise the entire system, the interdependencies of these files, and the files that have been modified since you created the last executable system. Also, when you are dealing with multimodule C software, compiling multiple files to create an executable one sometimes becomes a nuisance because two long command lines have to be typed: one to create object files for all C source files, and the other to link the object files to create one executable file. An easy way out of this inconvenience is to create a simple shell script that does this work. The disadvantage of this technique is that, even if a single source file (or header file) is modified, all object files are recreated, most of them unnecessarily.

Linux has a much more powerful tool called make, which allows you to manage the compilation of multiple modules into an executable. The make utility reads a specification file called the *makefile* that describes how the modules of a software system depend on each other. The make utility uses this dependency specification in the makefile and the times when various components were modified, to minimize the amount of recompilation. This utility is very useful when your software system consists of tens of files and several executable programs. In such a system, remembering and keeping track of all header, source, object, and executable files can be a nightmare. The following is a brief description of the make utility.

> **Syntax:**
>   `make [options] [targets] [macro definitions]`
>
> **Purpose:** This utility updates a file based on the dependency relationship stored in a file called **makefile** or **Makefile**; the dependency relationship is specified in **makefile** in a particular format. "options," "targets," and "macro definitions" can be specified in any order
>
> **Commonly used options/features:**
>   `-d`   Display debugging information
>   `-f file`   This option allows you to instruct make to read interdependency specification from "file"; without this option, the file name is treated as **makefile** or **Makefile**. The file name "-" implies standard input
>   `-h`   Display a brief description of all options

-n	Do not run any makefile commands; just display them
-s	Run in silent mode, without displaying any messages

The make utility is based on interdependencies of files, target files that need to be built (e.g., executable or object file[s]), and commands that are to be executed to build the target files. These interdependencies, targets, and commands are specified in the makefile as *make rules*. The following is the syntax of a make rule.

Syntax:
```
target-list: dependency-list
<Tab> command-list
```
Purpose: The syntax of a make rule

Here, **target-list** is a list of target files separated by one or more spaces, **dependency-list** is a list of files (object, header, source code, etc.) separated by one or more spaces that the target files depend on, and **command-list** is a list of commands—separated by the newline character—that have to be executed to create the target files. Each command in the **command-list** starts with the <Tab> character. The comment lines start with the # character. Files in the **target-list** and **dependency-list** may use shell wildcards ?, *, [], and {}. An alternative syntax for a make rule is target-list: dependency-list; command-list where "command-list" is a series of commands separated by semicolons. No tabs should precede targets.

With the : operator, a target is considered out of date if its modification time is less than any of the files in the **dependency-list**. If the operator is !, the target is always recreated after examining and recreating the sources in the **dependency-list**. With the :: operator, the targets are always recreated if no sources are specified in **dependency-list**. In case of the : and ! operator, the target is removed if make is interrupted, but not in case of the :: operator. In this book, we use only the most commonly used : operator.

The makefile consists of a list of make rules that describe the dependency relationships between files that are used to create an executable file. The make utility uses the rules in the makefile to determine which of the files that comprise your program need to be recompiled and relinked to recreate the executable. Thus, for example, if you modify a header (**.h**) file, the make utility recompiles all those files that include this header file. The files that contain this header file must be specified in the corresponding makefile. The directory that contains the source files and the makefile is commonly called the build directory.

The following makefile can be used for the power program discussed in Section 14.5.3.

```
$ cat makefile
# Sample makefile for the power program
# Remember: each command line starts with a <TAB>
power: power.c
        gcc power.c -o power -lm
$
```

If the executable file **power** exists and the source file **power.c** has not been modified since the executable file was created, running make will give the message that the executable file is up to date for **power.c**. Therefore, make has no need to recompile and relink **power.c**. At times, you will need to force the remaking of an executable because, for example, one of the system header files included in your source has changed. To force recreation of the executable, you will need to change the last update time. One commonly used method for doing so is to use the touch command and rerun make. The following session illustrates these points.

```
$ make
make: 'power' is up to date.
$ touch power.c
$ make
gcc power.c -o power -lm
$
```

When you use the touch command with one or more existing files as its arguments, it sets their last update time to the current time. When used with a nonexistent file as an argument, it creates a zero-length (i.e., empty) file with that name.

In the following in-chapter exercise, you will use the make command to create an executable for a single source file.

Exercise 14.5

Create the executable code for the C program in the **power.c** file and place it in a file called **XpowerY**. Use the make utility to perform this task by using the makefile given previously. Run **XpowerY** to confirm that the program works properly.

To show a next-level use of the make utility, we partition the C program in the **power.c** file into two files: **power.c** and **compute.c**. The following session shows the contents of these files. The **compute.c** file contains the compute function, which is called from the main function in **power.c**. To generate the executable in the **power** file, we need to compile the two source files independently and then link them, as shown in the two cc command lines at the end of the session.

```
$ cat power.c
#include <stdio.h>

double compute(double x, double y);
int main()
{
        float    x,y;

        printf   ("The program takes x and y from stdin and displays x^y.\n");
        printf   ("Enter integer x: ");
        scanf    ("%f", &x);
        printf   ("Enter integer y: ");
        scanf    ("%f", &y);
        printf   ("x^y is: %6.3f\n", compute(x,y));
}
$ cat compute.c
#include <math.h>
double compute (double x, double y)
{
        return (pow ((double) x, (double) y));
}
$ gcc -c compute.c power.c
$ ls
compute.c compute.o power.c   power.o
$ gcc compute.o power.o -o power -lm
$
```

The dependency relationship of the two source files is quite simple in this case. To create the executable file **power**, we need two object modules: **power.o** and **compute.o**. If either of the two files **power.c** or **compute.c** is updated, the executable needs to be recreated. Figure 14.4 shows this first cut on the dependency relationship.

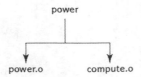

FIGURE 14.4   First cut on the make dependency tree.

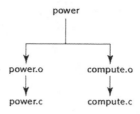

FIGURE 14.5   Second cut on the make dependency tree.

The make rule corresponding to this dependency relationship is, therefore, the following. Note that the math library has to be linked because the `compute` function in the **compute.c** file uses the pow function in this library.

```
power:  power.o compute.o
        gcc power.o compute.o -o power -lm
```

We also know that the object file **power.o** is built from the source file **power.c** and that the object file **compute.o** is built from the source file **compute.c**. Figure 14.5 shows the second cut on the dependency relationship.

Thus, the make rules for creating the two object files are

```
power.o:        power.c
gcc -c power.c
compute.o:      compute.c
gcc -c compute.c
```

The final makefile is shown as

```
$ cat makefile
power:          power.o compute.o
                gcc power.o compute.o -o power -lm
power.o:        power.c
                gcc -c power.c
compute.o:      compute.c
                gcc -c compute.c
$
```

We then execute the make utility with the preceding makefile:

```
$ make
gcc -c power.c
gcc -c compute.c
gcc power.o compute.o -o power -lm
$
```

In the following in-chapter exercise, you will use the make utility to create the executable code for a C source code that is partitioned into two files.

Exercise 14.6

Create the two source files **power.c** and **compute.c** and follow the steps just discussed to create the executable file **power** by using the make utility.

We now change the structure of this software and divide it into six files called **main.c**, **compute.c**, **input.c**, **compute.h**, **input.h**, and **main.h**. The contents of these files are shown in the following session. Note that the **compute.h** and **input.h** files contain declarations (prototypes) of the compute and

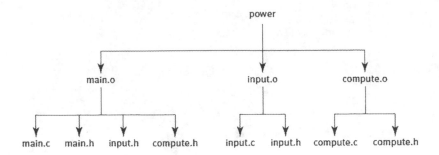

FIGURE 14.6   The make dependency tree for the sample C software.

input functions but not their definitions; the definitions are in the **compute.c** and **input.c** files. The **main.h** file contains two prompts to be displayed to the user. Figure 14.6 shows the new dependency tree.

```
$ cat compute.h
/* Declaration/Prototype of the "compute" function */
double compute(double, double);
$ cat input.h
/* Declaration/Prototype of the "input" function */
double input (char *);
$ cat main.h
/* Declaration of prompts to users */
#define PROMPT1 "Enter the value of x: "
#define PROMPT2 "Enter the value of y: "
$ cat compute.c
#include <math.h>
#include "compute.h"
double compute (double x, double y)
{
        return (pow ((double) x, (double) y));
}
$ cat input.c
#include <stdio.h>
#include "input.h"
double input(char *s)
{
        float x;

        printf ("%s", s);
        scanf ("%f", &x);
        return (x);
}
$ cat main.c
#include <stdio.h>
#include "main.h"
#include "compute.h"
#include "input.h"

int main()
{
        double x, y;

        printf ("The program takes x and y from stdin and displays x^y.\n");
        x = input(PROMPT1);
        y = input(PROMPT2);
        printf ("x^y is: %6.3f\n", compute(x,y));
}
$
```

To generate the executable for the software, you need to generate the object files for the three source files and link them into a single executable. The following commands are needed to accomplish this task. Note that, as before, you need to link the math library while linking the **compute.o** file to generate the executable in the power file.

```
$ gcc -c main.c input.c compute.c
$ gcc main.o input.o compute.o -o power -lm
$
```

The makefile corresponding to this dependency relationship is

```
$ cat makefile
power:          main.o input.o compute.o
                gcc main.o input.o compute.o -o power -lm
main.o:         main.c main.h input.h compute.h
                gcc -c main.c
input.o:        input.c input.h
                gcc -c input.c
compute.o:      compute.c compute.h
                gcc -c compute.c
$
```

The execution of the make command results in the execution of the rules associated with all targets in the makefile.

```
$ make
gcc main.o input.o compute.o -o power -lm
$
```

If you run make and get the message make: 'power' is up to date., remove the **power** file and rerun make. If the object files have not been created already, the make command will first create the object files for all the modules and then create the executable file, as shown in the following session.

```
$ make
gcc -c main.c
gcc -c input.c
gcc -c compute.c
gcc main.o input.o compute.o -o power -lm
$
```

In the following in-chapter exercise, you will use the make utility to create an executable for a multimodule C source.

Exercise 14.7

Create the three source and header files just discussed, and then use the make command to create the executable in the file **power**. Use the preceding makefile to perform your task.

If the make rules are in a file other than **makefile** (or **Makefile**), you need to run the make command with the -f option, as in make -f my.makefile.

The make rules as shown in the preceding makefile contain some redundant commands that can be removed. The make utility has a predefined rule that invokes the gcc -c xxx.c -o xxx.o command for every rule, as in

```
xxx.o: xxx.c zzz.h
gcc -c xxx.c
```

Furthermore, the make utility recognizes that the name of an object file is usually the name of the source file. This capability is known as a *standard dependency*, and because of it, you can leave **xxx.c** from the dependency list corresponding to the target **xxx.o**. The following makefile, therefore, works as well as the one given previously.

```
$ cat makefile
power:          main.o input.o compute.o
                gcc main.o input.o compute.o -o power -lm
main.o:         main.h input.h compute.h
input.o:        input.h
compute.o:      compute.c compute.h
                gcc -c compute.c
$
```

Running the make command with this makefile produces the following result:

```
$ make
gcc main.o input.o compute.o -o power -lm
$
```

The make utility supports simple macros that allow simple text substitution. You must define the macros before using them; they are usually placed at the top of the makefile. A macrodefinition has the following syntax.

Syntax:
    macro_name = text

Purpose: Substitute **text** for every occurrence of **$(macro_name)**

With this rule in place, text is substituted for every occurrence of $(macro_name) in the rest of the makefile. In addition, the make utility has some built-in macros, such as CFLAGS, that are set to default values and are used by the built-in rules, such as execution of the cc $(CFLAGS) -c xxx.c -o xxx.o command for a predefined rule, as previously described.

The default value of the CFLAGS macro is usually -O (for optimization), but it can be changed to any other flag(s) for the cc compiler. On our system, CFLAGS is set to null; that is, there are no default options. The make utility uses several built-in macros for the built-in rules.

The following makefile shows the use of user-defined macros and some useful make rules that can be invoked at the command line. It also shows that the commands for make rules are not always compiler or linker commands; they can be any shell commands.

```
$ cat makefile
CC = gcc
OPTIONS = -O4 -o
OBJECTS = main.o input.o compute.o
SOURCES = main.c input.c compute.c
HEADERS = main.h input.h compute.h

power:          main.c $(OBJECTS)
                $(CC) $(OPTIONS) power $(OBJECTS) -lm
main.o:         main.c main.h input.h compute.h
input.o:        input.c input.h
compute.o:      compute.c compute.h

all.tar:        $(SOURCES) $(HEADERS) makefile
                tar cvf - $(SOURCES) $(HEADERS) makefile > all.tar

clean:
                rm *.o
$
```

When the make command is executed, the commands for the last two targets (all.tar and clean) are not executed, as these targets do not depend on anything and nothing depends on them. You can invoke the commands associated with these targets by passing the targets as parameters to make. The advantage of putting these rules in the **makefile** is that you do not have to remember which files to archive (by using the tar command in this case) and which to remove once the final executable has been created. The make clean command invokes the rm *.o command to remove all object files that are created in the process of creating the executable for the software. The following session shows the output of make when executed with two targets as command line arguments. The tar archive is placed in the **all.tar** file. The output of the file all.tar commands shows that indeed the Gnu tar archive has been created in the **all.tar** file.

```
$ make all.tar clean
tar cvf - main.c input.c compute.c main.h input.h compute.h makefile > all.tar
main.c
input.c
compute.c
main.h
input.h
compute.h
makefile
rm *.o
$ file all.tar
all.tar: POSIX tar archive (GNU)
$
```

In the following in-chapter exercise, you will run the previous sessions on your system to further enhance your understanding of the **make** utility.

Exercise 14.8
Use the preceding **makefile** to create the executable in the file power.

14.5.5 Building Object Files into a Library Using ar and MRI Librarian

The Linux operating system allows you to archive (bundle) object files (also called modules or members) into a single file, called a *library*. This process lets you to use the name of one file instead of a number of object files in a **makefile** and allows function-level software reuse of C programs. The ar tool, also called the *librarian*, allows you to perform this task. The following is a brief description of this utility.

Syntax:
  ar key archive-name [file-list]

Purpose: Allows the creation and manipulation of archives; for example, to create an archive of the object files in **file-list** and store it in the file called **archive-file**

Commonly used options/features:

  d   Delete one or more files (modules/members) from an archive

  p   Print the specified modules or all modules (if none is specified) of the archive

  q   Append a module to an existing archive

  r   Create a new archive or overwrite (i.e., replace) existing modules with those specified

  s   Force generation of the archive symbol table

  t   Display the table of contents of an archive

  u   Update (when used with the **r** key) or extract (when used with the **x** key) modules only if they are newer than the existing ones

  v   Generate a verbose output

  x   Extract one or more modules from an archive and store them in the current working directory

---

The **archive-name** must end with the **.a** extension. Once an archive has been created for a set of object modules, the C compiler and the Linux loader (ld) can access these modules by specifying the archive file as an argument. The compiler or the loader automatically links the object modules needed from the archive. The ld command can be used to explicitly link object files and libraries.

A *key* is like an option for a command. However, unlike with most Linux commands, you do not have to insert a hyphen (-) before a key for the ar command, but you can use it if you want to. In the following examples of the ar command, we do not use a hyphen before a key.

14.5.5.1 Creating an Archive

You can create an archive by using the ar command with the r key. The following command line creates an archive of the **input.o** and **compute.o** files in **mathlib.a**.

```
$ ar r mathlib.a input.o compute.o
ar: creating mathlib.a
$
```

Note that if **mathlib.a** does not exist, it is created. If it already exists and has the **input.o** and **compute.o** modules in it, they are replaced with the ones specified in the command line. Once the archive has been created in your current directory, you can link it to the **main.c** file by using the compiler command such as cc, as in

```
$ gcc main.c mathlib.a -o power -lm
$
```

You can use the q key to append the object modules at the end of an existing archive. If the archive specified in the command does not exist, it is created. In the following example, the object module **main.o** is appended at the end of the existing archive **mathlib.a**. After **main.o** has been appended to **mathlib.a**, the gcc command is used to create the executable in **power**.

```
$ ar q mathlib.a main.o
$ gcc mathlib.a -o power -lm
$
```

Once you have created an archive of some object modules, you can remove the original modules to save disk space, as in the following session:

```
$ rm compute.o input.o main.o
$
```

14.5.5.2 Displaying the Table of Contents

You can display the table of contents of an archive by using the ar command with the t key. The command displays the table of contents of the **mathlib.a** archive.

```
$ ar t mathlib.a
input.o
compute.o
main.o
$
```

14.5.5.3 Deleting Object Modules from an Archive

You can delete one or more object modules from an archive by using the ar command with the d key. In the following session, the first ar command deletes the object module **main.o** from the **mathlib.a** archive, and the second displays the new table of contents confirming the removal of the **main.o** object module from the archive.

```
$ ar d mathlib.a main.o
$ ar t mathlib.a
input.o
compute.o
$
```

Note that creating a brand new archive from scratch is more efficient than modifying an existing archive by using the d, q, and r keys.

14.5.5.4 Extracting Object Modules from an Archive

You can extract one or more object modules from an archive by using the ar command with the x key. The extracted module remains in the archive. The following command line can be used to extract the object module **cpstr.o** from the **stringlib.a** archive and put it in your current directory.

```
$ ar x stringlib.a cpstr.o
$
```

You can run the ls -l cpstr.o command to see that the **cpstr.o** object file has been extracted, and the ar t mathlib.a command to see that this object file remains a part of the archive.

Although we have shown the use of the ar command from the command line, you can also run the command as part of a makefile so that an archive of the object files of a software product is created after the executable file has been created. Doing so allows future use of any general-purpose object modules (one or more functions in these modules) created as part of the software. It is done at the end of a makefile with an explicit make rule, as in

```
mathlib.a:      input.o compute.o
                ar rv mathlib.a input.o compute.o
                rm input.o compute.o
```

The following makefile is an enhancement of the makefile from the previous section that can be used to create an archive of **input.o** and **compute.o**, called **mathlib.a**. It then removes the **input.o** and **compute.o** files before creating the executable power by using the archive **mathlib.a**.

```
$ cat makefile
CC = gcc
OPTIONS = -O4 -o
OBJECTS = main.o input.o compute.o
SOURCES = main.c input.c compute.c
HEADERS = main.h input.h compute.h

power:          main.o mathlib.a
                $(CC) $(OPTIONS) power main.o mathlib.a -lm
main.o:         main.h input.h compute.h

mathlib.a:      input.o compute.o
                ar rv mathlib.a input.o compute.o
                rm input.o compute.o

all.tar:        $(SOURCES) $(HEADERS) makefile
                tar cvf - $(SOURCES) $(HEADERS) makefile > all.tar

clean:
                rm *.o
$
```

For each of these rules, the make utility executes a sequence of built-in commands that generate the object module by using the cc command and archives this object module by using the ar command. The following is a sample run of the preceding makefile. Make sure to remove compute.o, input.o, main.o, and mathlib.a files, if they exist in your current directory, before running the make command.

```
$ make
gcc     -c -o main.o main.c
gcc     -c -o input.o input.c
```

```
gcc    -c -o compute.o compute.c
ar rv mathlib.a input.o compute.o
ar: creating mathlib.a
a - input.o
a - compute.o
rm input.o compute.o
gcc -04 -o power main.o mathlib.a -lm
$
```

In the following in-chapter exercise, you will use the ar command with different options to appreciate its various characteristics in dealing with the libraries of object files.

Exercise 14.9
Use the commands just discussed to create an archive, delete an object file from the archive, display the table of contents for an archive, and extract an object file from the archive. Show your work.

14.5.5.5 *Working with the MRI Librarian*

You can run the ar command with the –M option to invoke the MRI librarian, which allows you to manage libraries with commands from standard input. In the following session, we show how you can use this interface of the ar command to create an archive, add a module to the archive, list modules currently in the archive, delete an archive, extract an archive, save changes in the archive, and close this interface of ar. Note that the default prompt for the MRI librarian is AR >.

```
$ ar -M
AR >create math2lib.a
AR >addmod input.o compute.o
AR >list
Current open archive is tmp-math2lib.a
input.o
compute.o
AR >delete compute.o
AR >list
Current open archive is tmp-math2lib.a
input.o
AR >addmod compute.o
AR >list
Current open archive is tmp-math2lib.a
compute.o
input.o
AR >save
AR >end
$ ar t math2lib.a
compute.o
input.o
s
```

In this session, the create command is used to create an archive, addmod to add one or more object modules to the newly created archive, list to display the name of the modules that are currently in the archive, delete to delete one or more modules from the archive, save to save the archive on the disk, and end to quit the MRI librarian. You can use the extract to extract one or more modules from the archive.

You can work with an existing archive by opening it with the open command. In the following session, we show how you can work with math2lib.a using the open command.

```
$ ar -M
AR >open math2lib.a
AR >list
```

```
Current open archive is tmp-math2lib.a
rw-r--r-- 0/0    1416 Jan  1 05:00 1970 compute.o
rw-r--r-- 0/0    1736 Jan  1 05:00 1970 input.o
...
AR >save
AR >end
$
```

You can also create an archive, or work with an existing archive, by using a script that you supply to the MRI librarian via standard input. In the following session, we supply the script in the **ar_script** file to the MRI librarian. The script creates the math3lib.a archive and performs the same operations on it that were performed on math2lib.a using the command line session at the beginning of this section.

```
$ more ar_script
create math3lib.a
addmod input.o compute.o
delete compute.o
list
addmod compute.o
list
save
end
$ ar -M < ar_script
Current open archive is tmp-math3lib.a
input.o
Current open archive is tmp-math3lib.a
compute.o
input.o
$ ar t math3lib.a
compute.o
input.o
$
```

You can browse through the manual page of ar for more information about the additional features of the MRI librarian. Note that instead of modifying an existing archive (i.e., deleting and adding existing objects from the archive), it is more efficient to remove the existing archive and create a brand new one from scratch.

Exercise 14.10

Repeat the previous session on your system to understand how the MRI librarian works.

14.5.6 Working with Libraries Using `ranlib` and `nm`

A library is an archive of object modules. Working with libraries, therefore, involves creating libraries, ordering modules in a library, and displaying library information. We discussed library creation and manipulation in several ways in the previous section. In this section, we discuss the remaining two operations: ordering archives and displaying library information.

14.5.6.1 Ordering Archives

Object files are not maintained in any particular order in an archive file created by the ar command. On some Linux systems, the caller function must occur before the called function, regardless of whether they are in the same or different modules. This condition is a problem because the gcc and ld commands cannot locate object modules unless they are properly ordered. When they cannot locate object modules, these commands display an undefined symbol error message when they encounter a call to a function in an object module in an archive. The easiest way to handle this problem is to use the ranlib utility, which adds a table of contents to one or more archives that are passed as its parameters. This utility performs the same task as the ar command with the s key. The following is a brief description of the ranlib utility.

> **Syntax**:
>   ranlib [archive-list]
>
> **Purpose**: Adds a table of contents to each archive in **archive-list**

The following ranlib command adds a table of contents to the **mathlib.a** archive. The ar s mathlib.a command can also be used to perform the same task.

```
$ ranlib mathlib.a
$
```

14.5.6.2 Displaying Library Information

The nm utility can be used to display the symbol table (names, types, sizes, entry points, etc.) of libraries, object files, and executable files. The command displays one line for each object (function and global variable) in the specified files. This output informs you about the functions available in a library and the functions that these functions depend on. Each output line includes the size (in bytes) of an object, the type of the object (data object, function, file, etc.), scope of the object, and the name of the object. This information is quite useful for debugging libraries. The following is a brief description of the utility.

> **Syntax**:
>   nm key archive-name [file-list]
>
> **Purpose**: Allows display of the symbol table of the library and object files specified in **file-list**
>
> **Commonly used options/features**:
>   -A   Display before each symbol the name of the input file (or archive) in which it was found
>   -v   Display the version number of the command
>   -n   Display symbols according to their address and not alphabetically
>   -s   Display names of the modules that contain the definitions of the symbols

The following characters specify symbol types:

A/a	Global/local absolute symbol
B/b	Global/local symbol in the uninitialized data segment (also known as bss)
D/d	Global/local symbol in the initialized data segment
f	Symbol name in the source file
L/l	Global/static thread-local symbol
R/r	Global/local symbol in a read-only data segment
T/t	Global/local symbol in the text (code) segment
U	Undefined symbol

In the following session, the nm -v command is used to display the version of the nm command, and the nm mathlib.a command is used to display the information about the **mathlib.a** library that we created in Section 14.5.5.1.

```
$ nm -v
GNU nm (GNU Binutils for Ubuntu) 2.26.1
Copyright (C) 2015 Free Software Foundation, Inc.
This program is free software; you may redistribute it under the terms of
the GNU General Public License version 3 or (at your option) any later version.
```

```
This program has absolutely no warranty.
$ nm mathlib.a
compute.o:
0000000000000000 T compute
                 U pow

input.o:
0000000000000000 T input
                 U __isoc99_scanf
                 U printf
                 U __stack_chk_fail
$
```

The output of the nm command shows that it contains two object modules: **input.o** and **compute.o**. Further, the printf, scanf, and pow symbols are undefined (U) and the symbols input and compute are in the text (code) sections of the relevant object modules in the library.

The nm and grep commands are often run together to retrieve information about a specific object. The following command is used to display information about the pow symbol in all archive files in the current directory. The output shows that the symbol is undefined and found in two archives, math2lib.a and math3lib.a.

```
$ nm -A *.a | grep pow
math2lib.a:compute1.o:               U pow
math3lib.a:compute2.o:               U pow
$
```

You can use the nm command with –u option to display all undefined symbols in an object, executable, or archive file. The following command is used to display all undefined symbols in **math2lib.a**. Note that all of the undefined symbols are in the various libraries and are resolved by the linker.

```
$ nm -u *.a

math2lib.a:

compute.o:
                 U pow

input.o:
                 U __isoc99_scanf
                 U printf
                 U __stack_chk_fail
$
```

Exercise 14.11
Repeat the commands in the previous section on your system to verify that they generate the same outputs.

14.5.7 Version Control with git

See Appendix A for instructions on how to install git on our four representative Linux systems. It can be used very effectively for version control on your Linux system. We present a complete description of the git command Chapter W24, Section 5.7 at the book website. Instructions for obtaining the materials at the book GitHub site, via a Web browser interface, are given in the Preface. To use the git command to "pull" the entire repository for this book at https://github.com/bobk48/linuxthetextbook, onto your own Linux system, do the following steps.

1. Begin by setting up, on your own Linux system, a new local repository Working Directory and initializing it as a git repository.

```
$ mkdir linuxthetextbook3
$ cd linuxthetextbook3
$ git init
$ git remote add origin https://github.com/bobk48/linuxthetextbook
$
```

This has initialized an empty git repository in /usr/home/your_username/linuxthetextbook3/.git/, where your_username is your username on your Linux system. It has also set the origin for pulling from a remote GitHub site.

2. Put a file in the new repository.

```
$ touch Readme.txt
$
```

3. Examine the status of the new repository.

```
$ git status
On branch master
Initial commit
Untracked files:
(use "git add <file>..." to include in what will be committed)
Readme.txt
nothing added to commit but untracked files present (use "git add" to track)
$
```

4. Stage the Readme.txt file, and make your initial commit into the new repository.

```
$ git add Readme.txt
$ git commit -m "first commit"
[master (root-commit) 57e0400] first commit
1 file changed, 0 insertions(+), 0 deletions(-)
create mode 100644 Readme.txt
$
```

5. Use the git pull command to pull and merge the entire book GitHub repository from the branch named master.

```
$ git pull https://github.com/bobk48/linuxthetextbook master

From https://github.com/bobk48/linuxthetextbook
$
```

You now have the entire contents of the book GitHub site on your own Linux system.

14.6 Static Analysis Tools

Static analysis of a program involves analyzing the structure and properties of your program without executing it. These analyses are usually meant to determine the level of portability of your code for multiple platforms, the number of *lines of code* (LOC), the number of *function calls/points* (FPs) in your program, and the percentage of time taken by each function in the code. During the planning phase of a software project, parameters such as LOC and FPs are commonly used in software cost models that are used to estimate the number of person months needed to complete a software project and, hence, the software cost.

Static analysis tools allow you to measure these parameters as well as report bugs that the compiler does not catch such as finding memory leaks, invalid pointers, and other memory problems. Some of the commonly used Linux tools for this purpose are cppcheck, valgrind, and memcheck. Some of the valgrind tools can profile your programs in detail in addition to automatically detecting memory-related bugs. gprof also allows you to profile your program comprehensively. All of these tools supplement a source code debugger (discussed in Section 14.7).

In the following sections, we discuss briefly cppcheck and gprof that allow you to perform these analyses.

14.6.1 Verifying Code for Portability with cppcheck

Most C compilers do fairly well at checking for type mismatches, but few handle portability. You can use cppcheck, which is a static analysis tool that comes preinstalled on Linux Mint 18. It is also integrated in many popular program development environments including Code::Blocks, Eclipse, git, and Visual Studio.

You can use it to identify bugs in your C/C++ software that compilers normally do not detect. It is one of the most useful tools in Linux for developing high-quality, clean, and portable C software. It detects program features that are likely to be bugs, nonportable, or wasteful of system resources. However, it reports real bugs and no false warnings. It also does not report structural problems, such variables declared but not used and unreachable code. cppcheck can detect various types of bugs in your code and reports accordingly, including the following:

- Out of bounds checking
- Memory leaks
- Null pointer dereferences
- Use of obsolete or unsafe functions

You can use the cppcheck --doc command to display the list of checks performed by cppcheck. Although cppcheck can handle various compiler extensions, inline assembly code, and preprocessor files, the discussion here is limited to its use with C program files. The following is a brief description of the cppcheck command.

Syntax:
  cppcheck [options] [file(s) or path(s)]

Purpose: Analyze the given source file(s), or files in case of directory, for errors that a compiler may not catch

Commonly used options/features:

--doc	Display the list of checks performed by cppcheck
-inconclusive	Report error even if analysis is inconclusive; false positives are possible in this case
-v	Report more detailed errors

If the specified path is a directory, cppcheck recursively checks for all source files and analyzes them. We demonstrate the use of cppcheck with simple programs shown in the following session to illustrate that it reports out-of-bounds problem, null pointer dereferencing, and memory leaks. We have used the nl command with -ba option to number all the source lines (including blank lines), because the line numbers reported in cppcheck's error messages include blank lines.

```
$ nl -ba buggy1.c
     1  /* Out of bounds check */
     2  #define SIZE 10
     3
     4  int main()
```

```
   5 {
   6      char str[SIZE];
   7
   8      str[SIZE] = '\0';
   9      return 0;
  10 }
$ gcc buggy1.c -o buggy1
$ cppcheck buggy1.c
Checking buggy1.c...
[buggy1.c:8]: (error) Array 'str[10]' accessed at index 10, which is out of
bounds.
$
```

In the following buggy2.c program, we show how cppcheck reports the null pointer dereferencing error that gcc couldn't detect.

```
$ cat buggy2.c
/* Null pointer dereferencing error */
int main()
{
    char *str;

    *str = '\0';
}
$ gcc buggy2.c
$ cppcheck buggy2.c
Checking buggy2.c...
[buggy2.c:6]: (error) Uninitialized variable: str
$
```

The following example illustrates the memory leak that wasn't reported by gcc but caught by cppcheck.

```
$ cat buggy3.c
/* Memory leak bug */
#include <stdlib.h>

struct student {
    char name[100];
    int ID;
};

int main()
{
    struct student s;
    char *s1, *s2;
    int size;

    size = sizeof(s);
    s1 = malloc(size);
    s2 = malloc(size);
    free(s1);

    return 0;
}
$ gcc buggy3.c
$ cppcheck buggy3.c
Checking buggy3.c...
[buggy3.c:20]: (error) Memory leak: s2
$
```

You can specify multiple files in the command line, as shown in the following session.

```
$ cppcheck buggy1.c buggy2.c buggy3.c
Checking buggy1.c...
[buggy1.c:8]: (error) Array 'str[10]' accessed at index 10, which is out of
bounds.
1/3 files checked 25% done
Checking buggy2.c...
[buggy2.c:6]: (error) Uninitialized variable: str
2/3 files checked 43% done
Checking buggy3.c...
[buggy3.c:20]: (error) Memory leak: s2
3/3 files checked 100% done
$
```

The following in-chapter exercise is designed to give you an appreciation of the use of the cppcheck utility and to help you understand some of the error messages that it produces.

Exercise 14.12
  Go through all the sessions presented in this section to appreciate how cppcheck works. Does cppcheck produce the same error messages on your system for all the buggy programs discussed in this section?

Exercise 14.13
  Run the cppcheck --doc command to find out the types of errors that cppcheck reports.

We strongly recommend that you create a make rule for running the cppcheck utility on your modules before compiling them. The following is an example make rule and its execution:

```
$ cat makefile
SOURCES = compute.c input.c main.c
CPPCHECKFLAGS =
...
cppcheck:       $(SOURCES)
cppcheck $(CPPCHECKFLAGS) $(SOURCES)
...
$ make cppcheck
cppcheck compute.c input.c main.c
...
$
```

14.6.2 Source Code Metrics with gprof

Software profiling is important because it allows you to identify function call sequences and portions of your software that are time consuming. This information allows you to focus more closely on those code portions or functions that are causing bottlenecks in the software. You can rewrite such pieces of code to make them more time efficient and make your programs execute faster. With profiling, you can also identify functions in your program that are being called with higher or lower frequency than expected. This information may help you identify bugs in your program that might have gone unnoticed otherwise.

  The profiler builds its data based on the actual execution of a program. Thus, while profiling a complex or a large program, it is important that you run it with input that invokes all of its features during execution. If the program input does not invoke a set of program features, no profile information will be collected for them.

  To obtain a profile of a program and analyze it, you need to perform the following steps:

  1. Compile and link your software with profiling enabled
  2. Run the software to collect profile information in a file
  3. Analyze the profile information by running a profiling tool with the profile file as input

With the profiling option enabled, the compiler inserts appropriate code in the object module to count the number of times each function is executed, construct function call sequences, and gather the time spent in each function. The profiling tool may generate output in the form of a *flat profile*, *call graph*, or *annotated source code file*.

The flat file output shows the amount of time each function took when the program executed. The call graph shows the function call sequence and the amount of time taken by a function and children (i.e., the called functions). This allows to identify function call sequences that may cause bottlenecks in the program. Lastly, in the annotated source code file each function is labeled with its call frequency (i.e., the number of times it was called).

Several profiling tools are available for Linux, however, gprof and perf are most famous. Whereas perf is mostly used to profile the use of hardware components, commands, and processes, gprof is meant for source code profiling. Although gprof is a powerful tool with a whole array of options for performing various types of analyses on the profile data file, we briefly describe its use with a small example. Here is a brief description of gprof:

Syntax:
   `gprof [options]`

Purpose: Produces execution profile of programs written in C and a few other languages

Commonly used options/features:

   `-A` Display profile of the program as annotated source code

   `-c` Display the function names and the number of times each was called

   `-p` Display flat profile

   `-q` Display profile as a call graph

We use a small C program in the **gprof_test.c** file to illustrate the use of gprof. First, we compile the program with the `-pg` option to enable profiling in the resultant executable file. We also use the `-Wall` option so that the compiler reports all sorts of warnings. Second, we run the program to generate the profile data file, **gmon.out**. Finally, we run the gprof command to analyze the data in **gmon.out** and display the profile information in one of the three forms discussed earlier. In order to generate the annotated source file, you must compile the source code with `-g` option, in addition to the `-pg` option.

The following session shows an example C program and the steps required to generate the profile information. In this case, we save the profile information in the file called **prof.out**. Note that without an option, gprof generates the flat profile followed by call graph, as shown by the cat command at the end of the session. The terminology used in the output is also explained in the output (that we have truncated to save space). The flat profile shows that functions f1, f2, and f3 were called 1, 1025, and 1074790401 times, respectively. It further shows that the total time taken by the program to complete was 0.87 s, out of which 0.47 s (i.e., 54.12% of the total time—see column of the output) was taken by f3.

```
$ cat gprof_test.c
/* Example for gprof */
#include <stdio.h>

#define COUNT1 1024
#define COUNT2 1048576 /* 1024 x 1024 */

void f1(int);
void f2(int);
void f3(void);

int main(void)
{
    f1(COUNT1);
    f2(COUNT2);
    f3();
```

```
        printf("Hello, World!\n");

        return 0;
}

void f1(int count)
{
    int i;

    for (i=0; i < count; i++)
        f2(COUNT2);

    return;
}

void f2(int count)
{
    int i;

    for (i=0; i < count; i++)
        f3();

    return;
}

void f3(void)
{ }
$ gcc -Wall -pg gprof_test.c -o gprof_test
$ gprof_test
Hello, World!
$ gprof gprof_test gmon.out > prof.out
$ cat prof.out
Flat profile:

Each sample counts as 0.01 seconds.
  %   cumulative   self              self     total
 time   seconds   seconds    calls  ms/call  ms/call  name
54.12      0.47      0.47 1074790401   0.00     0.00  f3
47.06      0.87      0.40     1025     0.39     0.85  f2
 0.00      0.87      0.00        1     0.00   869.27  f1
```

```
 %          the percentage of the total running time of the
time        program used by this function.

cumulative a running sum of the number of seconds accounted
 seconds    for by this function and those listed above it.

...
[ Output truncated ]
...
                 Call graph (explanation follows)


granularity: each sample hit covers 2 byte(s) for 1.15% of 0.87 seconds

index % time    self  children    called     name
                                                 <spontaneous>
[1]    100.0    0.00    0.87                 main [1]
                0.00    0.87       1/1           f1 [3]
                0.00    0.00       1/1025        f2 [2]
                0.00    0.00       1/1074790401      f3 [4]
-----------------------------------------------
```

```
              0.00      0.00        1/1025            main [1]
              0.40      0.46     1024/1025            f1 [3]
[2]   100.0   0.40      0.47     1025             f2 [2]
              0.47      0.00 1074790400/1074790401       f3 [4]
----------------------------------------------
              0.00      0.87        1/1            main [1]
[3]    99.9   0.00      0.87        1          f1 [3]
              0.40      0.46     1024/1025          f2 [2]
----------------------------------------------
              0.00      0.00        1/1074790401     main [1]
              0.47      0.00 1074790400/1074790401      f2 [2]
[4]    53.5   0.47      0.00 1074790401          f3 [4]
----------------------------------------------
```

```
This table describes the call tree of the program, and was sorted by
the total amount of time spent in each function and its children.

Each entry in this table consists of several lines.  The line with the
index number at the left-hand margin lists the current function.
The lines above it list the functions that called this function,
and the lines below it list the functions this one called.
This line lists:
    index     A unique number given to each element of the table.
              Index numbers are sorted numerically.
              The index number is printed next to every function name so
              it is easier to look up where the function is in the table.

    % time    This is the percentage of the `total' time that was spent
              in this function and its children.  Note that due to
              different viewpoints, functions excluded by options, etc.,
              these numbers will NOT add up to 100%.
...
[output truncated]
$
```

Note that gprof generates the profile data file, **gmon.out**, only if your program exists normally via exit() or return. No profile file is generated if the program terminates abnormally or via _exit(). Since the profile file is created in the current working directory, your program must have appropriate privileges to create the file.

> **Exercise 14.14**
> Repeat the previous session on your Linux system. Does it produce the same results?
>
> **Exercise 14.15**
> Show the command sequence to generate the annotated source file for the program in the gprof_test.c file?

14.7 Dynamic Analysis Tools

Dynamic analysis of a program involves its analysis during run time. As mentioned earlier, this phase comprises debugging, tracing, and performance monitoring of the software, including testing it against product requirements. In this section, we discuss the two useful Linux tools for tracing the execution of a program and debugging (gdb), and measuring the running time of a program in actual time units (time).

14.7.1 Source Code Debugging with gdb

The task of debugging software is time-consuming and difficult. It consists of monitoring the internal working of your code while it executes, examining values of program variables and values returned by functions, and executing functions with specific input parameters. As stated earlier, many C programmers tend to use the `printf` calls (`cout` for C++) at various places in their programs to locate the origin of a bug and then remove it. This technique is simple and works quite well for small programs. However, for large-size software, where an error may be hidden deep in a function call hierarchy, this technique ends up taking a lot of editing time for adding and removing `printf` (or `cout`) calls in the source file. A more efficient debugging method under such circumstances is to use a symbolic debugger. A typical *symbolic debugger* offers several facilities for observing the run-time behavior of a program, including the following:

- Running programs
- Setting breakpoints
- Single stepping
- Listing source code
- Editing source code
- Accessing and modifying variables
- Tracing program execution
- Searching for functions and variables
- Identifying what a program was doing when it crashed

Several symbolic debuggers are available on Linux platforms, the most common being the freeware Gnu debugger, gdb. The default source code debuggers on Solaris and PC-BSD are adb and gdb, respectively. They offer similar facilities. adb is a link to mdb, the *modular debugger*, which allows you to examine processes, user process core dumps, as well as live operating system and operating system dumps. We primarily describe gdb, as it is the standard debugger on most Linux systems. Although gdb has several features for debugging C++ classes as well, we discuss its features for debugging C programs only. The following is a brief description of the utility.

Syntax:
  `gdb [options] [executable-file [core-file or PID]]`

Purpose: Allows source-level debugging and execution of the program in **executable-file**, which was generated using a C or C++ source file; or **core-file**, created due to a C/C++ program crash; or process ID (**PID**) of a running program

Commonly used options/features:

-c file	Examine the file **file** as the core dump (i.e., file created by Linux when a program crashes)
-h	List all options along with their brief explanations
-n	Do not run commands from any **.gdbinit** initialization files
-x file	Execute gdb commands from the file **file** (or from the **.gdbinit** initialization file(s) if **file** is not specified)

During startup, gdb searches for **.gdbinit**. The search order is: your current directory (.) and then your home directory (~). The -x option allows you to use any file as a startup file. Table 14.1 gives a brief description of some of the commonly used gdb commands.

14.7.1.1 Using gdb

Before debugging a program with gdb (or any other debugger), you must compile it with the -g compiler option to include the symbol table and relocation, debugging, and profiling information in the executable. This information is used by the debugging and profiling tools. We use the program in the **bugged.c** file to show various features of gdb. The program prompts you for keyboard input, displays the input, and exits. We use several functions to demonstrate the features of gdb for setting breakpoints and displaying

TABLE 14.1

Commonly Used gdb Commands

Command	Command Syntax	Brief Description
break	break <line_num>:	Set breakpoint at line number line_num
	break <function_name>:	Set breakpoint at function function_name
continue	Continue:	Continue execution after breakpoint
clear	clear <line>:	Delete breakpoint set at line number line
	clear <function>:	Delete breakpoint set at function function
delete	delete <num>:	Delete breakpoint number num
	delete:	Delete all breakpoint numbers
frame	frame:	Show all stack frames
	frame <num>:	Set current stack frame to frame number num
help	help <num>:	List a brief description of the command classes
	help command:	Display a brief description of a command or command class command
info	info break:	Show information about current breakpoints
	info frame:	Show information about current stack frame
	info locals:	Show contents of local variables on the current stack frame
	info registers:	Display values of CPU registers
list	list:	List next few (ten by default) lines of the program
	list <line>:	List ten lines around line number line
	list <start>,<end>:	List ten lines from lines start through end
	list <function>:	List ten lines of the function
next	next:	Like the step command, except that it treats a function
	next <count>:	call as one instruction
print	print <expr>:	Display the value of the expression expr
	print identifier:	Display the current value of identifier
quit	quit:	Quit gdb
set	set <var> = <expr>:	Set variable var to expression expr
step	step:	Execute the next program instruction, stepping into a function (i.e., not treating a function call as one instruction)
	step <count>:	Execute next count lines of program code
run	run [command-line-args]:	Execute the program that was an argument of the gdb command
whatis	whatis identifier:	Display the type of identifier

the stack trace at function boundaries, tracing program execution by executing program statements one by one, viewing types and values of variable, and so on. The following session shows the program code, its compilation without the -g option, and its execution.

```
$ nl -ba bugged.c
    1 /*
    2 *  Sample C program bugged with a simple, yet nasty error
    3 */
    4
    5
    6 #include <stdio.h>
    7
    8 #define PROMPT "Enter a string: "
    9
   10 void get_input(char *, char *);
   11 void null_function1 ();
   12 void null_function2 ();
```

```
13
14  int main ()
15  {
16      char *s_val;
17
18      null_function1 ();
19      null_function2 ();
20      get_input(PROMPT, s_val);
21      (void) printf("You entered: %s\n", s_val);
22      (void) printf("The end of buggy code!\n");
23      return (0);
24  }
25
26  void get_input(char *prompt, char *str)
27  {
28      (void) printf("%s", prompt);
29      for (*str = getchar(); *str != '\n'; *str = getchar())
30          str++;
31          *str = '\0'; /* string terminator */
32      }
33
34  void null_function1 ()
35  { }
36
37  void null_function2 ()
38  { }
$ gcc bugged.c -o bugged
% bugged
Enter a string: Hello!
Segmentation Fault
$
```

Note that the program prompts you for input and faults without echoing what you enter from the keyboard. This happens frequently in C programming, particularly with programmers who are new to C or are not careful about initializing variables in their programs and rely on the compiler. It is time to use gdb!

14.7.1.2 Entering the gdb Environment

As stated earlier, in order to enter the gdb environment, you must compile your C program with the -g compiler option. This option creates an executable file that contains the symbol table and debugging, relocation, and profiling information for your program. After the source program compiles successfully, you can then use the gdb command to debug your code, as in the following session. We ran the gdb -q command to prevent the introductory messages from being displayed. Note that (gdb) is the prompt for the gdb debugger.

```
$ gcc -g bugged.c -o bugged
$ gdb -q bugged
Reading symbols from bugged...done.
(gdb)
```

Once inside the gdb environment, you can run many commands to monitor the execution of your code. You can use the help command to get information about the gdb commands. Without any argument, the help command displays the names of all of the gdb *command classes*. A command class specifies the type of operations you can perform on your executable code, a core file, or a process. Under a command class, you can run the commands available for that class. You can get information about any gdb command class by passing the command name as an argument to the help command. In the following session, the help command shows the names of all gdb commands and the help tracepoints command displays a brief description of the command class tracepoints and commands supported by gdb under this class.

```
(gdb) help
List of classes of commands:

aliases -- Aliases of other commands
breakpoints -- Making program stop at certain points
data -- Examining data
files -- Specifying and examining files
[ ... output truncated ... ]
(gdb) help tracepoints
Tracing of program execution without stopping the program.

List of commands:

actions -- Specify the actions to be taken at a tracepoint
collect -- Specify one or more data items to be collected at a tracepoint
end -- Ends a list of commands or actions
[ ... output truncated ... ]
(gdb) help trace
Set a tracepoint at specified location.

trace [PROBE_MODIFIER] [LOCATION] [thread THREADNUM] [if CONDITION]
PROBE_MODIFIER shall be present if the command is to be placed in a
probe point.  Accepted values are `-probe' (for a generic, automatically
guessed probe type), `-probe-stap' (for a SystemTap probe) or
`-probe-dtrace' (for a DTrace probe).
LOCATION may be a linespec, address, or explicit location as described
below.
[ ... output truncated ... ]
(gdb)
```

In addition to the gdb-specific commands, gdb also allows you to execute all shell commands.

14.7.1.3 Executing a Program

You can run your program inside the gdb environment by using the run command. The following command executes the program called **bugged**.

```
(gdb) run
Starting program: /home/sarwar/linux2e/ch14/bugged
Enter a string: Hello!

Program received signal SIGSEGV, Segmentation fault.
0x000000000040063b in get_input (prompt=0x400704 "Enter a string: ", str=0x0)
    at bugged.c:29
29          for (*str = getchar(); *str != '\n'; *str = getchar())
(gdb)
```

The program (now a process) prompts you for input. When you enter the input (Need gdb! in this case) and hit <Enter>, the process fails when it tries to execute the command at line 30. The error message is quite cryptic for beginners and those who are not familiar with Linux jargon. The error message says that the process received a signal of type SIGSEGV (segmentation violation, i.e., address space violation) when it was executing the statement at line 29. The message explains that SIGSEGV was sent to the program because of Segmentation fault. A segmentation fault is generated by the Linux kernel when a process tries to access a memory region (a segment) that it is not allowed to access. In other words, your process tried to access a memory location that did not belong to its *address space*.

14.7.1.4 Listing Program Code

You can use the list command to display all or part of a source program. By default, it displays ten lines around the line (or function) that you specify as its argument. You can display LOCs in a particular function or on a range of lines. In the following example, the list get_input command is used to display ten lines in the code around the get_input function, and the list 14,24 command is used to display the source program at lines 14 24. Note that the list get_input command displays five lines before the start of the function code and five lines after it. To display the first ten lines of the function code, we use the list get_input command. Since get_input has less than ten lines, a few extra lines after the function are also displayed. Use the help list command to get more information about the list command.

```
(gdb) list get_input
22              (void) printf("The end of buggy code!\n");
23              return (0);
24      }
25
26      void get_input(char *prompt, char *str)
27      {
28              (void) printf("%s", prompt);
29              for (*str = getchar(); *str != '\n'; *str = getchar())
30                  str++;
31              *str = '\0'; /* string terminator */
(gdb) list get_input,
27      {
28              (void) printf("%s", prompt);
29              for (*str = getchar(); *str != '\n'; *str = getchar())
30                  str++;
31              *str = '\0'; /* string terminator */
32      }
33
34      void null_function1 ()
35      { }
36
(gdb) list 14,24
14      int main ()
15      {
16          char *s_val;
17
18          null_function1 ();
19          null_function2 ();
20          get_input(PROMPT, s_val);
21          (void) printf("You entered: %s\n", s_val);
22          (void) printf("The end of buggy code!\n");
23          return (0);
24      }
(gdb)
```

14.7.1.5 Tracing Program Execution

To find out what went wrong in our process, we need to identify the part of the code that may be problematic. There are several ways of doing so, including line-by-line tracing of the whole program, tracing statements of a function, calls to a particular function, and changes to a variable. We do so by backtracking the program using the where or backtrace command of gdb, as shown in the following session. The output of both commands is the same: the location of the program where the fault occurred and how this location was reached. The how part is identified by the function call sequence identified in the *stack trace* (also known as *stack frame* or *activation record* in the programming language jargon) that the command displays. Line #0 shows the top of stack with the get_input() function's statement where the crash occurred, and #1 shows that the get_input() function was called at line 20 in the main() function. The main() function is called the *caller* and the get_input() function is called the *callee*.

```
(gdb) where
#0  0x000000000040063b in get_input (prompt=0x400704 "Enter a string: ",
```

```
        str=0x0) at bugged.c:29
#1   0x00000000004005e3 in main () at bugged.c:20
(gdb) backtrace
#0   0x000000000040063b in get_input (prompt=0x400704 "Enter a string: ",
        str=0x0) at bugged.c:29
#1   0x00000000004005e3 in main () at bugged.c:20
(gdb)
```

The output of the command shows that the program was at location (memory address) 0x000000000040063b at line 30 in the get_input() function when it received the SIGSEGV signal. The stack frame shows that the call sequence is main() => get_input() and the program was executing the *str=getchar() statement at line 29 when it crashed (see the program listing in the previous section). Note that the code on this line has two assignment statements and one comparison statement. In all of these statements, we dereference the string pointer str. Two statements use the C library function getchar(). This library function is well tested and has been in use for many years. Thus, the problem must be with the pointer variable str. We pursue this issue in a later section.

14.7.1.6 Setting Breakpoints

Viewing execution of all or part of your code statement by statement is called *program/code tracing*. In order to trace code, you need to set breakpoints in your program. You can trace a program up to a particular statement or function by using the break command, as described in Table 14.1. It allows you to run a program without interruption until the control reaches the line or function that you want to study more closely. The process of stopping a program in this way is known as setting *breakpoints*. As shown in the previous section, the main function starts at line 14 but its first executable statement is at line 18. We set the breakpoint at the first executable statement in main() and run the program. The program stops execution at line 18, the only breakpoint we have set, having statement null_function1();

```
(gdb) break main
Breakpoint 1 at 0x4005be: file bugged.c, line 18.
(gdb) run
The program being debugged has been started already.
Start it from the beginning? (y or n) y
Starting program: /home/sarwar/linux2e/ch14/bugged

Breakpoint 1, main () at bugged.c:18
18            null_function1 ();
(gdb)
```

14.7.1.7 Single-Stepping through Your Program

Always set breakpoints in your program to be able to view execution of all or part of your code statement by statement. The process of tracing program execution statement by statement is known as *single-stepping* through your program. Single-stepping, combined with tracing of variables, allows you to study program execution closely. Single-stepping can be done with the step command, which executes the next program statement, stepping into a function if the statement is a function call. You can use the next command to single-step through your code, but it executes a function into its entirety. If you are tracing a variable, it shows you the value of the variable when a statement within the scope of the variable executes. Run the help scope command to get more information about the scoping rules.

After setting the breakpoint at main and running the program in the previous session, we then run each statement of the program one by one by using the next command. After the third next statement, the control reaches the call to the get_input (PROMPT, temp); function at line 20 in the main() function. We then use the step command to step into and execute each statement in the get_input() function. The first step command takes control to the first executable statement in the get_input() function, the for loop. The second step statement prompts you for keyboard input. As soon as you hit <Enter> after typing Hello World!, the system displays the error message Program received signal SIGSEGV, Segmentation fault along with the problematic source code statement and line number (29 in this case) and the value of the str variable. The hexadecimal number at the beginning of the second line of the error message is the memory address (**0x000000000040063b**) of the getchar() function that caused the exception. We can use the x command to display the contents of this memory location. The output of the command shows that the error occurred at the 49th byte in the get_input() function. We still do not know the reason why the program faulted.

```
(gdb) next
19          null_function2 ();
(gdb) next
20          get_input(PROMPT, s_val);
(gdb) next
Enter a string: Hello!

Program received signal SIGSEGV, Segmentation fault.
0x000000000040063b in get_input (prompt=0x400704 "Enter a string: ", str=0x0)
    at bugged.c:29
29              for (*str = getchar(); *str != '\n'; *str = getchar())
(gdb) x 0x000000000040063b
0x40063b <get_input+49>:        0x12eb1088
(gdb)
```

14.7.1.8 Accessing Identifiers (Variables and Functions)

You can access the location of an *identifier* (variable or function) in the program source and view its type, value, and places of use by using various gdb commands. In the following session, we illustrate the use of these commands with examples. The outputs of the commands are fairly self-explanatory. The print str command displays the value of the str variable as zero and the whatis str command displays the type declaration of the str variable: char *, as expected. The print get_input command displays the types of input parameters of the get_input() function, the type of its return value (i.e., void), and the starting address of the function (i.e., 0x40060a).

```
(gdb) delete
Delete all breakpoints? (y or n) y
(gdb) run
The program being debugged has been started already.
Start it from the beginning? (y or n) y

Starting program: /home/sarwar/linux2e/ch14/bugged
Enter a string: Hello!

Program received signal SIGSEGV, Segmentation fault.
0x000000000040063b in get input (prompt=0x400704 "Enter a string: ", str=0x0)
    at bugged.c:29
29              for (*str = getchar(); *str != '\n'; *str - getchar())
(gdb) whatis str
type = char *
(gdb) print str
$1 = 0x0
(gdb) print get_input
$2 - {void (char *, char *)} 0x40060a <get_input>
(gdb)
```

14.7.1.9 Fixing the Bug

After finding out that the program faults at line 29, the first thing you should do is determine the variables involved in the statement that caused the fault. Then you should display the values of these variables.

The session immediately reveals that the value of the actual parameter to the get_input() function, s_val, is nil (0). This causes the formal parameter in the get_input() function, str, to have a starting value of nil. When we dereference the str variable to store user input, we try to access memory

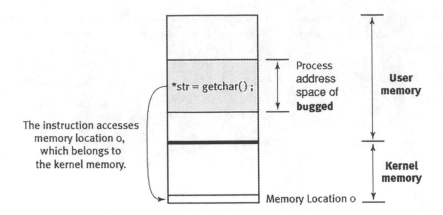

FIGURE 14.7   The memory (segmentation) access violation causing program failure.

location with address **0**. This location belongs to the Linux kernel space and is used to store the resident part of the operating system. Therefore, the process tries to write to a location that is outside its address space—that is, does not belong to it. This attempt is a clear violation that results in the SIGSEGV signal sent to the running program, causing its termination—the default action on this signal. Hence, you see the error message Segmentation Fault when you run the program from the command line. Figure 14.7 illustrates segmentation violation.

To fix the bug in the program, all you need do is initialize the s_val pointer to a memory space that has been allocated to the program, statically or dynamically. We use a character array called user_input[SIZE] and set the s_val pointer to point to the first byte of the array. The revised main() function is shown in the following session, along with its compilation and proper execution. The changes in the code are the additions of lines 8, 17, 19, and 20 in the program, as follows:

```
$ nl -ba working.c
    1  /*
    2   *  Sample C program bugged with a simple, yet nasty error
    3   */
    4
    5  #include <stdio.h>
    6
    7  #define PROMPT "Enter a string: "
    8  #define SIZE    255
    9
   10  void get_input(char *, char *);
   11  void null_function1 ();
   12  void null_function2 ();
   13
   14  int main ()
   15  {
   16          char *s_val, *temp;
   17          char user_input[SIZE];
   18
   19          s_val = user_input; /* Initialize s_val to an array */
   20          temp = s_val;
   21          null_function1 ();
   22          null_function2 ();
   23          get_input(PROMPT, temp);
   24          (void) printf("You entered: %s\n", s_val);
   25          (void) printf("The end of buggy code!\n");
   26          return (0);
   27  }
   28
```

```
29  void get_input(char *prompt, char *str)
30  {
31          (void) printf("%s", prompt);
32          for (*str = getchar(); *str != '\n'; *str = getchar())
33                  str++;
34          *str = '\0'; /* string terminator */
35  }
36
37  void null_function1 ()
38  { }
39
40  void null_function2 ()
41  { }
42
$ gcc working.c -o working
$ working
Enter a string: Hello!
You entered: Hello!
The end of working code!
$
```

14.7.1.10 Leaving gdb and Wrapping Up

You can use the quit command to leave gdb and return to your shell.

```
(gdb) quit
$
```

Once your code has been debugged, you can decrease the size of the binary file, releasing some disk space, by removing from it the information generated by the -g option of the C compiler to be used by debugging and profiling utilities. You can do so by using the strip command. The information stripped from the file contains the symbol table and relocation, debugging, and profiling information. In the following session, we show the long list for the working file before and after the execution of the strip command. Note that the size of the file has decreased from 8880 bytes to 6336 bytes, resulting in a saving of about 29% disk space. Alternatively, you can recompile the source to generate an optimized executable by using various options.

```
$ ls -l working
-rwxr-xr-x 1 sarwar faculty 8880 Oct  3 10:41 working
$ strip working
$ ls -l working
-rwxr-xr-x 1 sarwar faculty 6336 Oct  3 10:42 working
$
```

In the following in-chapter exercise, you will make extensive use of gdb to understand its various features.

Exercise 14.16
Go through all gdb commands discussed in this section to appreciate how gdb works. If some of the commands used in this section do not work on your system, use the help command to list the gdb commands and use those that are available in your version of gdb.

14.7.2 Run-Time Performance with time

The run-time performance of a program or any shell command can be measured and displayed by using the time command. This command reports three times: *real time*, *system time*, and *user time*. Real time is the actual time taken by the program to finish running, system time is the time taken by system (kernel) activities while the program was executing, such as handling the clock interrupt, and user time is the time taken by execution of the program code. Because Linux is a time-sharing system, real time is not always equal to the sum of system and user time, as many other user processes may be running while your program executes. The following is a brief description of the command.

Syntax:
    time [command]

Purpose: Report the run-time performance of command in terms of its execution time. It reports three times: real time (actual time taken by command execution), system time (time spent on system activities while the command was executing), and user time (time taken by the command code itself).

There are two versions of the time command: the built-in command for the C shell and the /usr/bin/time command. The output of the built-in time command is quite cryptic, whereas the output of the /usr/bin/time command is readable. When the C shell version of the time command is executed without a command argument, it reports the length of time the current C shell has been running. The reported time includes the time taken by all its children—that is, all the commands that have run under the shell. The other version of the command does not have this feature. The time command sends its output to **stderr**. So, if you want to redirect the output of the time command to a disk file, you must redirect its **stderr** (not its **stdout**) to the file. If you are not interested in the output of the program (or command) that you want to time but only want to know how long the program took to complete, you can redirect standard output to **/dev/null**.

The following command reports the time taken by the cp command that copies a 10.3 GB file. Note how neat the output looks.

```
$ ls -l bigdata
-rw-r--r-- 1 sarwar faculty 10302182842 Oct  3 10:52 bigdata
$ time cp bigdata bigdata.old

real    1m47.155s
user    0m0.040s
sys     0m6.424s
$
```

As stated earlier, the sum of the user and system times does not always equal real time, especially if a program is idle and does not use the CPU for some time. In the output of the time command, the real time is 1 min 47.155 s, which clearly does not equal the sum of the user and system times (0.040 and 6.424 s, respectively).

Because the time command can be used to measure the running time of any program, you can use it with an executable of your own—a binary image or a shell script. The following session shows the running time of the find command. Since we are only interested to know how long the find command takes to complete and not its output, we redirect its output to /dev/null.

```
$ time find /usr -name socket.h 1> /dev/null

real    0m0.313s
user    0m0.152s
sys     0m0.160s
$
```

There are other ways of measuring the running time of a program that give you better precision. But, using the time command to perform this task is the easiest way, and we certainly recommend it for beginners.

Exercise 14.17
    Replicate the earlier sessions on your system to verify their working as expected.

Summary

Linux supports all contemporary HLLs (both interpreted and compiled), including C, C++, Java, Javascript, FORTRAN, BASIC, and LISP. We described the translation process that a program in a compiled language such as C has to go through before it can be executed. We also described briefly a typical software engineering life cycle and discussed in detail the program development process and the tools available in Linux for this phase of the life cycle. The discussion of tools focused on their use for developing production-quality C software.

The program development process comprises three phases: code generation, static analysis, and dynamic analysis. The Linux code generation tools include text editors (emacs, pico, and vi), C-language enhancers (cb and indent), compilers (cc, gcc, xlc, CC, cpp, and g++), tools for handling module-based software (make), tools for creating libraries (ar, nm, and ranlib), and the most commonly used version control tool, git/GitHub, and its related commands.

The purpose of the static analysis phase is to identify features of the software that might be bugs or nonportable, and to measure metrics such as LOCs, FPs, and repetition count for functions. The Linux tools that can be used for this purpose include cppcheck, memcheck, prof, gprof, and valgrind.

The purpose of the dynamic analysis phase is to analyze programs during their execution. The tools used during this phase are meant to trace program execution in order to debug them and measure their run-time performance in terms of their execution time. The commonly used Linux tools for this phase of the software life cycle are debuggers (gdb, etc.) and tools for measuring running times of programs (time).

Linux has several tools for other phases of a software life cycle, but a discussion of them is outside the scope of this textbook.

Questions and Problems

1. What are the differences between compiled and interpreted languages? Give three examples of each.

2. Give one application area each for assembly and HLLs.

3. Write the steps that a compiler performs on a source program to produce an executable file. State the purpose of each step. Be precise.

4. What is the purpose of the cat sieve.c | indent -o sieve.out command? In particular, what coding style is the program in the sieve.c file converted to, which file contains the newly formatted program, and where is the backup of the original program saved?

5. Give the indent command line that performs the function of the command given in Problem 4, but uses output redirection (>) instead of a pipe (|).

6. Give an equivalent of the indent command in Problem 4 that does not use a pipe or output redirection. Then, give an equivalent of this command so that instead of saving the newly formatted code in the Gnu style, it is saved it in the K&R style.

7. What are the -o, -1, and -x0 options of the cc command used for? Give an example command line for each and describe what it does.

8. Give the compiler commands to create an executable called **prog** from C source files **myprog.c** and **misc.c**. Assume that **misc.c** uses some functions in the math library. What is the purpose of each command?

9. What are the three steps of the program development process? What are the main tasks performed at each step? Write the names of Linux tools that can be used for these tasks.

10. Give a shell command that can be used to determine the LOC in the program stored in the **scheduler.c** file.

11. Write advantages and disadvantages of automating the recompilation and relinking process by using the make utility, as opposed to manually doing this task.

12. Consider the following makefile and answer the questions that follow:

```
CC = cc

OPTIONS = -xO4 -o
OBJECTS = main.o stack.o misc.o
SOURCES = main.c stack.c misc.c
HEADERS = main.h stack.h misc.h
polish: main.c $(OBJECTS)

$(CC) $(OPTIONS) power $(OBJECTS) -lm
main.o: main.c main.h misc.h
stack.o: stack.c stack.h misc.h misc.o: misc.c misc.h
```

List the following:

a. Names of macros

b. Names of targets

c. Files that each target is dependent on

d. Commands for constructing the targets named in part (b)

13. For the makefile in Problem 12, give the dependency tree for the software.

14. What commands are executed for the **main.o**, **stack.o**, and **misc.o** targets? How do you know?

15. For the makefile in Problem 12, what happens if you run the make command on your system? Show the output of the command.

16. Give the command line to display the object files or library archives in your current directory that contain the object x.

17. Give the command line to display the undefined symbols in the a.out file in your current directory.

18. Give the command line to display all the symbols in the a.out file in your current directory.

19. The following is the output of an nm command. What does each field in each line represent? What do T, R, and B tell us about the relevant symbols?

20. Write a C program to copy standard input to standard output. Compile the program with the executable program saved in a.out. Give the command line to display all dynamic symbols in a.out along with the command output.

21. Give the command line to display all external symbols in a.out along with the command output.

22. Write a C program that contains the following bugs: out-of-bound, memory leak, and null pointer deferencing. Compile the following program. Did gcc report the bugs in the program? Use cppcheck to see if catches the bug? Show your program code, compilations using gcc, and bug catching using cppcheck.

23. Generate the flat profile, call graph, and annotated source code file for the program in the **gprof_test.c** file in Section 14.6.2. Show all your work and explain the contents of each file.

24. The flat profile for the program in the **gprof_test.c** file shows that functions f2 and f3 are called/executed 1,025 and 1,074,790,401 times, respectively. Explain these numbers.

25. The flat profile for the program in the **gprof_test.c** file shows that functions f2 and f3 take 0.85 ms and 0.00 ms total time per call, respectively. Explain these numbers. What is the difference between total ms/call and self ms/call in the flat profile?

26. The following code is meant to prompt you for integer input from the keyboard, read your input, and display it on the screen. The program compiles with one warning, but it doesn't work properly. Use gdb to find the bugs in the program. What are they? Fix the bugs, recompile the program, and execute it to be sure that the corrected version works. Show the working version of the program.

```
#include <stdio.h>

#define PROMPT   "Enter an integer: " void get_input(char *, int *);
void main ()
{
    int user_input;
    get_input(PROMPT, user_input);
    (void) printf("You entered: %d.\n", user_input);
}

void get_input(char *prompt, int *ival)
{
    (void) printf("%s", prompt);

    scanf ("%d", &ival);
}
```

27. What does the `time sh` command line do when executed under the C shell?

28. Give the command line to redirect the output of the `/usr/bin/time` polish command to a file called **polish_output**. Assume that you are using the Bourne shell.

29. How long does it take to complete the search to locate stat.h file on your system if searh start with **/**? What is the answer if search starts with **/usr**? Show commands that you used to obtain your answer.

30. What command will you use to display the output of the time command for performing the earlier task without displaying command output or error message, except the time taken by the command. You may not use output and error redirection. *Hint*: think about the tail of the output shown on the screen.

Advanced Questions and Problems

31. How can you view the assembly version of source code under gdb? How can you view the source code and assembly version side by side. Explain your answer with a sample C program.

    To prepare for the completion of the following questions, problems, and projects, complete everything in Chapter W24, Section 5.7, entitled "Version Control," at the book website.

32. What do you think the role of an integrator of a project would be, in terms of what a revision control system, such as git, accomplishes for a software development and maintenance program?

33. What do you think would be the quickest and easiest way to completely delete a local git repository on your Linux system?

34. Why do you think git would not be effective, or even work at all, in tracking content changes in C program executable image files, or in files like Word .docx files?

35. What git command can you use to see the abbreviated list of commits in the current branch of a repository, and their commit comments? Is this possible in GitHub, and how?

36. As an alternative to using git pull in Example W24.5 to obtain the source code for this book from the listed GitHub repository (**https://github.com/bobk48/linuxthetextbook**), use the git clone command, as shown in Example W24.4, from your home directory on your Linux system. What will be the name of the repository directory created by git on your local machine that contains the source code files?

Projects

Project 1

Create a three-branch repository of commits exactly like Chapter W24, Figure W24.16 at the book website. Use any number of text files that you modify between commits on the three branches. Keep the default name for the branch **master**, but name the other two branches **test** and **dev**, as seen in Chapter W24, Figure W24.16 at the book website.

Project 2

Create a three-branch repository of commits exactly like Figure W24.14. Use any number of text files that you modify between commits on the three branches. Keep the default name for the branch master, but name the other two branches **test** and **dev**, as seen in Figure W24.14.

Project 3

Before doing this project, go through all of the material in Chapter W22 on the Zettabyte File System (ZFS), including the in-chapter exercises, and the questions and problems presented at the end of that chapter. Then, in your own words, write a report that describes the differences and similarities between git, GitHub, and ZFS. For example, the commands `zfs snapshot` and `git commit` are very similar. In what way do they differ? In your opinion, is it possible to use the commands `zpool` and `zfs` to achieve the same or similar results as using git and GitHub? Which suite of commands offers more to achieve the goals of distributed version control, and exactly why? Also, specify what other Linux commands would be needed to augment the less-capable system of commands (of course in terms of distributed version control) to attain results that are similar to the other more-capable system's functions, commands, and capabilities?

Looking for more? Visit our sites for additional readings, recommended resources, and exercises.
CRC Press e-Resource: https://www.crcpress.com/9781138710085
Authors' GitHub: https://github.com/bobk48/linuxthetextbook

15

System Programming I: File System Management

OBJECTIVES

- To explain the concept of system programming
- To briefly describe the concept of system calls
- To discuss the execution details of a system call in Linux
- To briefly describe the types of system calls in Linux
- To explain the concept of file and file descriptors in Linux
- To discuss the concept of per-process file descriptor tables, system wide file tables, and inode tables
- To briefly discuss standard input/output (I/O) and low-level I/O
- To describe in detail the system calls for I/O of file data and attributes
- To briefly describe the purpose of system calls related to directories and setting file attributes
- To discuss the concept of file holes in Linux
- To discuss several small programs to illustrate the use of various system calls for the I/O and management of file data and attributes
- To discuss the concept of blocking I/O and restarting a system call
- To briefly explain several system calls and C library functions for manipulating directories and file attributes
- To cover the following system calls, library calls, and primitives:

    access(), chdir(), chmod(), chown(), close(), closedir(), creat(), fclose(), fopen(), fread(), fstat(), fwrite(), lseek(), lstat(), mkdir(), opendir(), printf(), read(), readdir(), rename(), rewinddir(), seekdir(), stat(), telldir(), truncate(), umask(), unlink(), utime(), write()

15.1 Introduction

The kernel does the steady-state maintenance of a Linux operating system, and the kernel provides services to user programs through the *system call interface* (SCI). This maintenance is achieved by using three techniques: *virtualization, concurrency,* and *persistence.* Virtualization allows devices, such as a single or multicore CPU, to act as if it were many CPUs acting simultaneously in a time-sharing manner. Concurrency allows multithreaded or multiprocess programs to access the virtualized resources of the hardware. Persistence allows data to be retained over time via mechanisms of I/O onto devices such as a *solid-state drive* (SSD).

In this and the subsequent three chapters, we describe how various components of the SCI perform their roles in the realization of virtualization, concurrency, and persistence. This chapter deals with persistence. Chapter 16 deals with virtualization and concurrency. Chapter W20 deals with the communication aspect of virtualization and concurrency. Chapter W21 deals with the various practical issues related to virtualization, concurrency, and persistence.

In this chapter, we discuss the use of the API provided by the SCI for file handling. We assume that the reader is familiar with file handling through the use of the standard I/O library that is built around the ISO C standard specification and is available on multiple operating system platforms, including Linux and Microsoft Windows. Our coverage of file handling is focused on performing data I/O with regular files. We also investigate how you can access and display a file's attributes by reading certain kernel data structures. We do not describe in detail the system calls for changing file attributes, but explain them briefly so you can explore them on your own. Also, we do not cover the details of directory handling. However, we provide a summary of the relevant system calls and library functions for performing such chores. Our coverage of the various topics is closely linked with the underlying kernel data structures and operating system concepts where appropriate. For the compilation of our sample C programs, we use the Gnu C compiler, gcc.

15.2 What Is System Programming?

Application programming is the skill of writing programs to provide services to users, including word processing, text and graphics editing, video processing, voice streaming, and Internet services such as Web browsing. *System programming* is the ability of writing the kernel code to manage system hardware, including main memory and disk space management, disk formatting and defragmentation, CPU scheduling, and management of I/O devices through device drivers. It also includes the development of utilities outside the kernel that directly interface with the operating system through the use of system calls. Writing compilers or any part of the operating system code also falls in the realm of system programming. Although application programming requires users to be savvy in using language libraries, system programming requires a high degree of understanding of the computer hardware, assembly language, and a language like C.

In Linux jargon, system programming uses the SCI to access hardware resources such as the CPU, disk and files, main memory, and the status of processes and files. Such programs include, for instance, a shell, a language assembler or compiler, tools for providing information about a computer's hardware resources, including the usage of CPU, disk, and main memory, and programs that provide the status of software resources such as processes and files.

15.3 Entry Points into the Operating System Kernel

There are several reasons for control to transfer from a user process to the operating system kernel. These reasons collectively fall into four categories, known as four *entry points* into the kernel, as shown in Figure 15.1. Because the purpose of these entry points is to invoke the relevant pieces of codes in the kernel to provide different services, they are also known as *service points* into the kernel.

Two of these entry points, *trap* and *interrupt*, are caused by computer hardware. An *interrupt* is a "signal" that a peripheral hardware device sends to the CPU in order to get its attention. For example, after a disk controller has finished reading a disk block (or cluster), it needs to inform the CPU about it. For this purpose, it generates an interrupt, and the kernel takes over control in order to execute the code

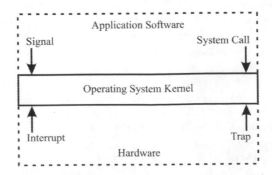

FIGURE 15.1   Entry points into the operating system kernel.

to service this interrupt, called the *interrupt service routine* for this interrupt. A *software interrupt* is a form of trap generated by an application by calling a special assembly code instruction such as int 21h for an x86 CPU.

Although a peripheral device generates an interrupt, the CPU itself generates a *trap* in order to handle an exception in the code being executed. There are several reasons for the CPU to do so, including execution of an illegal instruction (instruction not in the instruction set of the CPU), a potential run-time error like a divide-by-zero situation, and an instruction trying to access a main memory area outside the process address space.

A *signal* in Linux jargon is a mechanism that allows the interruption of a process. A signal is a high-level abstraction provided by the operating system that allows communication between the kernel and a user process or between two processes (usually for process control). In Section 13.4, we discussed this topic in detail. However, the discussion was focused on shell-level handling of signals. Table 13.1 shows some of the commonly used signals and their purpose. Note that some of the signals are generated through keystrokes but most are software generated. We discuss signals and signal handling from a system programmer's point of view in detail in Chapter 16.

The fourth entry point, *system call*, is the topic of discussion in the remainder of this and next three chapters.

15.4 Fundamentals of System Calls

The SCI is a mechanism that allows a user process to execute a piece of code in the operating system kernel. Processes use system calls to invoke services that allow access to hardware devices and kernel resources (code and data) that processes are not allowed to access otherwise, particularly in a multiuser, time-sharing system. This limitation is necessary in order to protect the resources (data and code) belonging to the kernel and other users from accidental or judicious access by a user. The SCI ensures such protection and security.

15.4.1 What Is a System Call?

A user process is not allowed to have direct access to computer hardware, such as a disk drive and kernel data structures, in order to ensure that resources belonging to other users and the kernel remain protected. However, a user does need to access resources that he/she owns on the system such as files, as well as information about various system resources (hardware and software) such as CPU utilization or any number of processes running on the system. A system call is a mechanism that allows a process to perform privileged tasks that it is not allowed to perform by directly accessing (reading or writing) an I/O device or executing a piece of kernel code.

Thus, a system call is an entry point into the Linux kernel code. In other words, it is a way for a process to execute a piece of code in the kernel, ensuring protection of resources that do not belong to the process. Linux offers this mechanism to provide several types of services, including the following:

- Opening a file
- Reading and writing file data and attributes
- Accessing file attributes
- Setting file attributes
- Obtaining information about system hardware and operating system
- Creating a process
- Creating a channel for communication between processes
- Getting attention of a process and having it perform a particular task
- Obtaining information about the processes currently running on the system
- Creating and terminating processes

- Stopping processes
- Making processes wait for different events
- Creating communication channels for processes to communicate with each other
- Accessing process attributes
- Accessing utilizations of hardware resources such as memory and CPU utilization

These service points are provided through *wrapper* (library) functions, one for each system call. System programmers use these library functions to invoke relevant services.

15.4.2 Types of System Calls

There are several types of system calls dealing with the various aspects of the Linux services. We categorize them as follows:

- Process control
- File management/manipulation
- Device management
- Information maintenance
- Communications

In this and the subsequent chapters on system programming in this book, we will primarily cover some commonly used system calls related to process control, file management/manipulation, and interprocess communication. We discuss system calls related to file management/manipulation in this chapter. In Chapters W20 and W21, we describe system calls for process management and interprocess communication.

> **Exercise 15.1**
> Browse the Web for manual pages on Linux system calls and list two system calls each for the five types listed in Section 15.4.2 along with their purpose.

15.4.3 Execution of a System Call

Although not necessary to the learning of system programming and becoming good at it, knowing the low-level details of system call execution enhances your understanding of the process an operating system goes through to execute a system call. These steps are hardware dependent and vary from system to system. Here is an example of a sequence of steps:

1. The user program makes a call to a library function that acts as a wrapper for the system call.
2. The library function:
   i. Puts the parameters of the system call on the process stack.
   ii. Passes a *call number* (N) that uniquely identifies the system call via a register or stack, or as a parameter to the `trap` function (see item iii).
   iii. Executes a CPU instruction, called the `trap` instruction on some CPUs (such as the Motorola microprocessors), to switch the CPU mode from *user* to *kernel/system* and transfer control to the `syscall()` function in the Linux kernel.
3. The `syscall()` function is the system call handler and performs the following tasks:
   i. Identifies the system call invoked based on the call number.
   ii. Copies the appropriate number of parameters from the user stack to the kernel stack.
   iii. Uses the call number to index the *dispatch table* (which contains pointers to service routines for system calls) to execute the code for the requested service.

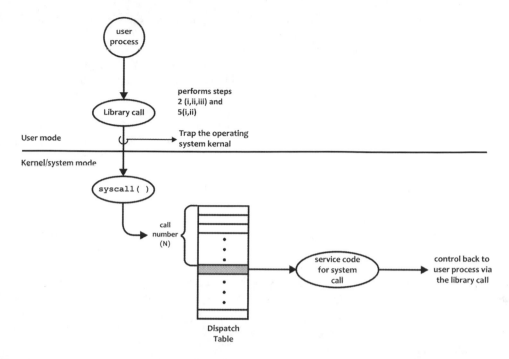

FIGURE 15.2 The sequence of steps required for the execution of a system call.

4. The code for the respective service executes and
   i.   The return value is placed in one or two registers for transferring it to the caller process (the return value is −1 in case of failure).
   ii. In case of failure, the appropriate error code is placed in a register.
5. Control transfers back to the library code, which performs the following tasks:
   i.   In case of error, the error code, placed in a register in 4(ii), is saved at a known location (for errno).
   ii. The instruction following the trap instruction executes.
6. Execution of the user program continues.

Figure 15.2 gives a pictorial view of these steps.

15.5 Files: The Big Picture

Recall that nearly everything in Linux is a sequence of bytes: files, directories, I/O devices, network cards, and so on. Further, Linux treats all files as streams of bytes. Thus, it treats all files consistently. We discussed the structure of a directory entry in Linux, comprising of filename and inode number, and the concept of file descriptors in Chapter 4. We can access a file or its attributes without opening it through the file's inode. After opening a file using the corresponding system call, its contents are accessed through its file descriptor.

15.5.1 File Descriptors, File Descriptor Tables, File Tables, and Inode Tables

We discussed the concept of file descriptors in detail in Chapters 4 and 9. However, that discussion was focused on the use of standard file descriptors for I/O and error redirection for shell commands. In this chapter, we focus on the concept and use of file descriptors, including the standard file descriptors, from a system programmer's point of view.

File descriptors are nonnegative integers starting with 0 and are used to index the *per-process file descriptor tables* (*PPFDTs*). We discussed this concept in Sections 4.7–4.9 and Chapter 9, starting with Section 9.6. Figure 4.6 shows the relationship between PPFDTs, system-wide file tables (SFTs), inode tables, and a file's contents on a secondary storage.

Several Linux system calls, including open(), creat(), pipe(), and socket(), return file descriptors for the file, pipe, and socket that they open or create. Communication with these objects using the read() and write() system calls is carried out through these descriptors. We discuss the open() and creat() system calls later in this chapter and describe the pipe() and socket() calls in Chapters 16, W20, and W21, respectively.

Processes perform data I/O and error output also through file descriptors for the three files that the Linux kernel opens automatically for every process. As discussed in Chapters 4 and 9, these files are known as standard files: *standard input*, *standard output*, and *standard error*. Standard input is the default location from where a process takes its input. Standard output is the default location where a process's output goes to, and standard error is the file where errors generated by a process are sent. The integer values for standard input, standard output, and standard error are 0, 1, and 2, respectively. Thus, the first descriptor returned by a system call is 3, by default. However, if you close one of the standard files, execution of the open(), creat(), pipe(), or socket() system call will return the first available descriptor, starting with 0.

15.5.2 Why Two Tables?

The Linux PPFDT keeps track of all the files that a process has opened. The SFT keeps track of all the files open in a Linux system at any given time and links the PPFDTs and the inode table, as shown in Figure 15.3.

Why did the designers of the Linux kernel need to have the SFT? Why could they not have an entry in the PPFDT point directly to an entry in the inode table? The answer is that since Linux allows a file to be opened multiple times simultaneously, it needs to keep track of multiple read/write file pointers (See Section 15.7.5). Thus, an inode cannot be used to store all of the attributes of a file, and a separate data structure is needed, where file pointers for files that have been opened multiple times simultaneously

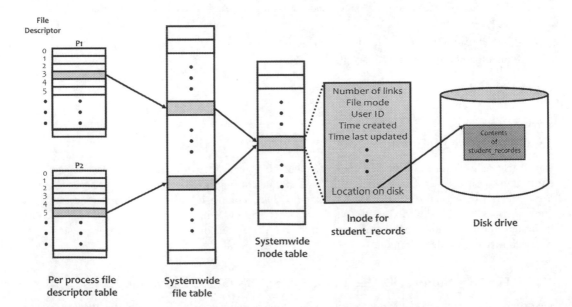

FIGURE 15.3 The relationship between PPFDT, SFT, inode table, and secondary storage.

may be maintained. The SFT is an array (i.e., table) of such a data structure that comprises a pointer to file's inode and read/write file pointer.

When multiple processes open a file simultaneously, multiple SFT slots are allocated, one for each process. The SFT slot for a process P_i contains the current position of the read/write file pointer for P_i and pointer to file's inode. The diagram in Figure 15.3 shows that the **student_records** file has been opened in two processes, P1 and P2. In P1, the file descriptor for the file is 3 and in P2 the descriptor is 5. Because the file has been opened twice, it has two entries in the SFT, each pointing to the inode of **student_records** and containing the position of the read/write file pointer for the relevant process.

Exercise 15.2

Browse through the relevant header files on your Linux system to determine the size of the PPFDT on your system. What is it? Clearly write down the exact definition as given in the header file.

Exercise 15.3

What is the size of the SFT on your system? How did you obtain your answer?

15.6 Fundamental File I/O Paradigm

The basic file I/O paradigm in Linux is *open*, *read*, *write*, and *close*. It means that a process must open a file before performing a data I/O (i.e., read/write) operation on it. If a file does not exist, it should be created and opened before data I/O can be performed on it. If a file is created using the open() system call (see Section 15.7.2 and Table 15.2 for details on system calls for file I/O), it can be created and opened according to the access permission specified as a parameter. If, however, the file is created using the creat() system call, it must be explicitly opened with the open() system call for I/O using the read() and write() system calls. As I/O is being performed on a file, the position of the read/write file pointer may be changed to carry out the read/write operations at the desired byte position in the file. We discuss this issue in Section 15.7.5. A file does not have to be opened for reading or writing its attributes. We discuss this topic in Section 15.8.

15.7 Standard I/O versus Low-Level I/O

The Linux API provides two interfaces for file I/O, one via the Standard I/O library and the second through the SCI. Since the SCI resides immediately above the kernel and the language libraries are built on top of the SCI, I/O via the SCI is called *low-level I/O*. Some of the functions in the C Standard Library (CSL) interface are standalone and some are built on top of the SCI. For example, strlen() is standalone, and fopen() is built on top of the open() system call. I/O via functions in the SCL interface is known as *standard I/O*.

Programmers normally prefer standard I/O for the following reasons:

a. Library functions are portable across all systems.

b. The low-level I/O interface via the SCI may be more time efficient because of its direct interface with the Linux kernel. However, it is cumbersome because the programmer needs to handle buffer issues such as buffer allocation/deallocation and choosing the right size block for I/O.

c. The library functions provide buffered I/O services for files (see Section 15.7.2) while system calls provide nonbuffered I/O services. This means that you could read file data using the fread() library function that provides buffered I/O, which means that not every call to fread() would result in a call to system call the read() system call. Thus, you should use standard I/O function fread() instead of read() if your application requires reading data from a file very frequently.

The real advantage of the I/O interface via the SCI is that open(), read(), write(), and close() calls can be used not just with files, but with network sockets and other file-like objects that don't interact well with buffers.

15.7.1 The CSL

The CSL implementation is based on the ISO C standard specification that has been implemented on several operating systems, including UNIX, Linux, and Microsoft Windows. Thus, code written using this library is portable across operating systems. The library consists of a set of predefined and well-tested functions, constants, and header files. It provides for you the mechanisms to handle tasks such as character I/O to file I/O, from time-related functions to complex math functions, and from string operations to signal handling. The header files in this library contain hundreds of variable types, functions, and macros. Some of the definitions are duplicated in multiple files for the sake of completion. For example, NULL is defined in multiple header files. Table 15.1 lists the header files that comprise the CSL.

15.7.2 File Data I/O Using the CSL

The Standard I/O library API is simple to use and handles the issues of buffer allocation and using appropriate size blocks for optimal I/O in terms of time (both CPU and clock times). The functions in this library use the file I/O calls available in the SCI. The purpose of using the right size buffers and block sizes is to minimize the number of disk I/O operations by minimizing the number of read() and write() system calls (see Sections 15.7.4 and 15.7.5).

TABLE 15.1

Header Files for CSL and Their Purpose

Header File	Purpose
<assert.h>	It contains a macro, called assert, that can be used to diagnose whether the assumptions made in your program are correct.
<ctype.h>	It contains a set of functions to test the attributes of characters (i.e., whether a character is a decimal digit, a control character, etc.). It also contains two functions to convert a lowercase character to uppercase and vice versa.
<errno.h>	It contains the definition of the errno variable that is set by system calls and some library functions. It also contains a few macros that indicate different error codes.
<float.h>	It contains a set of constants and macros that deal with floating point values in a machine-dependent way.
<limits.h>	It contains the minimum and maximum values of different variable types, including signed char, unsigned char, int, unsigned int, short int, unsigned short int, long, and unsigned long.
<locale.h>	It contains currency symbols and date formats.
<math.h>	It contains math functions, including sin, cos, tan, sqrt, log, ceil, and floor.
<setjmp.h>	It contains a macro, a function, and a variable used to bypass the normal function call/return mechanism.
<signal.h>	It contains functions and macros for signal handling (discussed in Chapters 13 and 15).
<stdarg.h>	It contains a functions and macro to handle variable number of arguments in a function.
<stddef.h>	It contains variable types and macros, some of which appear in other header files too.
<stdio.h>	It contains variable types, functions, and macros for I/O.
<stdlib.h>	It contains variable types, functions, and macros for general-purpose tasks, including conversion from ASCII to float, random number generation, and dynamic memory allocation/deallocation.
<string.h>	It contains variable types, functions, and macros for handling character arrays (i.e., character strings).
<time.h>	It contains variable types, functions, and macros for handling date and time, including conversion from one format to another.

The Standard I/O library supports three types of *buffered I/O*: *fully buffered*, *line buffered*, and *unbuffered*. The ISO C standard requires that standard input and standard output be fully buffered for noninteractive devices. On the other hand, standard error is never fully buffered; it is normally unbuffered so that error messages are displayed at the earliest possible time.

When we open a file using the Standard I/O interface, we get back a pointer to an object FILE, called the *file pointer*. We perform all subsequent file operations using this pointer and the standard I/O library functions. When we open a file using the SCI, we get back an object of type int, called the *file descriptor*. We perform all subsequent I/O operations on the file using this descriptor and the relevant system calls. Another term used for both of these objects is *file handle*. In this book, we primarily deal with the low-level I/O interface.

When we open a file using the Standard I/O library interface, a *stream* is associated with the file. Depending on the character set used, a character can be represented using one or multiple bytes. Thus, stream I/O depends on the "width" of a character in terms of bytes and determines the "orientation" of the stream. When a stream is created it has no orientation, and its orientation is set by the type of functions used for performing I/O with the stream. When we open a file with the open() system call, a *sequence of bytes* is associated with the file, and all subsequent I/O take place in terms of byte sequence.

We assume that the reader is familiar with file I/O using the Standard I/O library and do not discuss it any further.

15.7.3 Low-Level I/O in Linux via System Calls

Table 15.2 shows Linux system calls for file I/O along with their purpose. We will discuss most of these system calls in this chapter and the remaining in Chapter 16. As discussed in Section 15.3.5, these are in fact wrapper functions that eventually transfer control to the system call, syscall(), along with the respective call number and call parameters so that appropriate kernel function may be executed to serve the relevant system call. You need to include the **<fcntl.h>** file to be able to use the open() and creat() system calls. For the close() and lseek() system calls, you need to use the **<unistd.h>** file. Finally, for using read() and write() you need to include the **<sys/types.h>** and **<unistd.h>** files.

In this chapter, we discuss the system calls for performing the following operations on files: creating, opening, closing, reading, writing, preparing for random access, deleting a hard link to a file, and getting file attributes. You will get to practice some of the remaining calls in some of the programming exercises given at the end of the chapter.

TABLE 15.2

The Linux System Calls for File I/O

Creating, Opening for I/O, and Closing	
creat()	Creates a file
open()	Opens or creates and opens a file
close()	Closes a file
Data I/O	
read()	Reads from an open file
write()	Writes to an open file
Creating and Removing Files (Hard Links)	
link()	Creates a hard link to a file
unlink()	Removes a hard link to a file and, if the resultant link count becomes 0, deletes the file
remove()	Identical to unlink()
Setting Up for Random Access	
lseek()	Sets file for random access

15.7.4 System Call Failure and Error Handling

A Linux system call returns −1 on failure. Thus, while using a system call in your code, you must first check whether the call has failed and use appropriate code to handle it. In most cases, you would like to display an adequate error message to the user by using the CSL function `perror()` and terminate the program execution by using the `exit()` system call.

> **Exercise 15.4**
> Browse through **<stdio.h>** and identify the number of functions and macros defined in it.
>
> **Exercise 15.5**
> List the Standard I/O library calls for character and file processing. Give a one-sentence description of each call.

15.8 File Manipulation

In this section, we discuss the Linux system calls to access and manipulate file data and attributes. Our focus will be on the following operations:

- Opening a file
- Creating a file
- Reading file data
- Writing data to a file
- Positioning the read/write file pointer for random access
- Truncating a file
- Closing a file
- Checking the existence of a file
- Removing a file
- Obtaining and displaying file attributes

We will primarily focus on manipulating data in regular files. However, we discuss the access, display, and modification of file attributes for files of all types. We do not discuss the manipulation of directories, but briefly describe the SCI for doing so. The system calls for handling sockets and FIFOs are described in detail in Chapters W20 and W21.

15.8.1 Opening and Creating a File

You can use the `open()` system call to open an existing file for reading, writing, or performing both operations. If a file does not exist, you can create it and open it for writing by using the `creat()` system call. Obviously, reading from a newly created (empty) file does not make sense. You can also create a file by specifying appropriate flags in the `open()` call to request the kernel to create the file if it does not exist, and set it for reading, writing, or both. The following are the brief descriptions of the two system calls.

```
#include <fcntl.h>
int open(const char path, int flags, ... /* mode_t mode */ );
```

Success: Nonnegative integer, called a file descriptor

Failure: −1 and kernel variable `errno` set to indicate the type of error

```
#include <fcntl.h>
int creat(const char path, mode_t mode);
```

Success: Nonnegative integer, called a file descriptor

Failure: −1 and kernel variable `errno` set to indicate the type of error

The third argument (...) in the prototype of the open() call means that the number and type of remaining arguments may vary. The mode argument in the creat() call is used to specify the access permissions (see Chapter 5) for the newly created file. Both calls return a nonnegative integer called the file descriptor of the opened file, which can be used to perform the desired I/O operations by using the read() and write() system calls (discussed later), change the position of the read/write file pointer for random access, and close the file using the close() system call. The returned descriptor is the first unused descriptor in the PPFDT, starting with descriptor 0. If the file opens successfully, the read/write file pointer is set to the beginning of the file. It means that the first read or write operation starts at the first byte of the file. Both calls return −1 on failure.

Bitwise ORing two or more constants listed in the **<fcntl.h>** file forms the flag argument. Table 15.3 shows some of the commonly used flags used for file I/O and their meanings.

The open() call may fail for several reasons. Some of the commonly occurring reasons for the call to fail for regular files are given in Table 15.4, along with the corresponding symbolic values saved in the errno variable.

Since the creat() system call creates a file and, by default, opens it for writing only, the following calls are equivalent to each other:

```
creat(pathname, mode);
open(pathname, O_WRONLY | O_CREAT | O_TRUNC, mode);
```

TABLE 15.3

Commonly Used Flags for File I/O

Flag	Meaning
O_RDONLY	Open the file for reading only
O_WRONLY	Open the file for writing only
O_RDWR	Open the file for reading and writing
O_CREAT	Create the file if it does not exist. If you use this option, you must specify the access permissions for the newly created file as the third argument, mode
O_APPEND	Each write operation appends at the end of file
O_TRUNC	If the file exists, its length is set to 0. Previous file contents are not accessible anymore

TABLE 15.4

A Few Common Reasons for the Failure of the open() System Call

Reason for Failure	Value of errno
A component in the **path** does not exist	ENOENT
A component in the **path** is not a directory	ENOTDTR
The named file is a directory	EISDIR
Search permission is not set on a component in the **path**	EACCESS
The process has already opened the maximum number of files allowed by the system, i.e., the PPFDT is already full	EMFILE
The system has already reached the limit of the maximum number of files that may be opened on it simultaneously (i.e., the SFT is full)	EMFILE
The file opening operation was interrupted by a signal	EINTR
A component in **path** exceeds the file name size limit (255 characters) or the entire path exceeds the 1,023 characters	ENAMETOOLONG
The named file is a special file, but the device associated with the file does not exist	ENXIO
O_CREAT flag is specified to create a file but the file system is read only	EROFS
O_CREAT flag is specified and error occurred while creating an inode for the new file or making the directory entry for the file to be created	EIO
O_CREAT and O_EXCL flags are specified and the file specified in path exists	EEXIST

The open() system call follows a symbolic link and accesses the data or attributes of the file that the link points to. If you want to read the data or attributes of a symbolic link, you must use the readlink() system call. Similarly, the creat() system call can create only an ordinary file. If you want to create a symbolic link, you should use the symlink() system call. You should browse through the man pages of these calls to learn more about them.

Exercise 15.6

Give the prototypes of the symlink() and readlink() system call. Briefly describe each parameter for these calls.

15.8.2 Closing a File

You can close a file by using the close() system call, which takes a file's descriptor as an argument and deallocates the PPFDT slot corresponding to the given descriptor. The following is a brief description of the call:

```
#include <unistd.h>
int close(int fd);
```

Success: 0

Failure: −1 and kernel variable errno set to indicate the type of error

The close() system call may fail for several reasons. Two commonly occurring reasons are given in Table 15.5, along with the symbolic values of the errno variable corresponding to these errors.

If an open file is being referenced by more than one process—that is, if entries in multiple PPFDTs are pointing to an entry in the SFT—then closing the file would only deallocate the entry in the relevant PPFDT and decrement the reference count by 1. A file is closed only when this reference count becomes 0. When a process terminates, the kernel automatically closes all of its open files.

In the following example, we open the file passed as the only command line argument, display the file descriptor of the file, and close the file. The compilation and execution of the example program shows the program output. We use the compiler command with the −w option in order to suppress the warning messages.

```
$ cat open_close.c
#include <fcntl.h>
#include <unistd.h>

int main(int argc, char *argv[])
{
        int fd;

        /* Open file */
        if (argc == 1) {
                printf("No file specified as command line argument.\n");
                exit(1);
        }
        if ((fd = open(argv[1], O_RDWR)) == -1) {
                perror("File opening");
```

TABLE 15.5

Commonly Occurring Reasons for the Failure of the close() System Call

Reason for Failure	Value of errno
fd is not an active descriptor, i.e., does not correspond to an open file	EBADF
The file-closing operation was interrupted by a signal	EINTR

```
                        exit(1);
                } else
                        printf("The file descriptor is %d.\n",fd);
                /* Close file */
                if (close(fd) == -1) {
                        perror("File closing");
                        exit(1);
                }
                else
                        printf("File closed successfully.\n");
                exit(0);
}
$ gcc open_close.c -w
$ man talk > foobar
$ a.out foobar
The file descriptor is 3.
File closed successfully.
$
```

Note that this is the first (and the only) file that the process has opened. Since the kernel has already opened the standard files for this process, the kernel allocated file descriptor 3 when the program opened **foobar**.

We now modify the program slightly so that it first closes standard input and then opens the file passed to it as a command line argument. In order to do so, we only need to insert the following line before the open() system call:

```
close(0);
```

Here is the updated program:

```
$ cat open_close_new.c
#include <fcntl.h>
#include <unistd.h>

int main(int argc, char *argv[])
{
        int fd;

        /* Open file */
        if (argc == 1) {
                printf("No file specified as command line argument.\n");
                exit(1);
        }
        close(0); /* Close standard input */
        if ((fd = open(argv[1], O_RDWR)) == -1) {
                perror("File opening");
                exit(1);
        } else
                printf("The file descriptor is %d.\n",fd);
        /* Close file */
        if (close(fd) == -1) {
                perror("File closing");
                exit(1);
        }
        else
                printf("File closed successfully.\n");
        exit(0);
}
$ gcc open_close_new.c -w
$ a.out foobar
The file descriptor is 0.
```

```
File closed successfully.
$
```

As you can see, as expected, the kernel allocated file descriptor 0 to the newly opened file **foobar**.

Exercise 15.7

Compile and run the programs given in Sections 15.8.1 and 15.8.2. Do they work as intended?

15.8.3 Reading from a File

Once a file has been opened using the open() system call, you can read its contents by using the read() system call. Here is a brief description of the read() call.

```
#include <sys/types.h>
#include <unistd.h>
ssize_t read(int fd, void *buf, size_t nbytes);
```

Success: 0

Failure: −1 and kernel variable errno set to indicate the type of error

This call tries to read nbytes bytes from the file with the file descriptor fd into the main memory area pointed to by buf. The data is read starting with the current location of the read/write file pointer. Upon completion, the read() call returns the number of bytes actually read, which may be less than nbytes. The file pointer advances by the number of bytes actually read. For a regular file, the call guarantees reading nbytes if the file has these many bytes left before the end-of-file (EOF). However, with other files, such as pipes or sockets (see Chapters 16, W20, and W21), this is not guaranteed. The read() system call returns 0 when it encounters EOF.

On failure, read() returns −1 and errno is set to indicate the reason for failure. Table 15.6 shows some common reasons for the read() call to fail for regular files.

15.8.4 Writing to a File

Once a file has been opened using the open() or creat() system call, you can write to it using the write() system call. Here is a brief description of the write() call:

```
#include <sys/types.h>
#include <unistd.h>
ssize_t read(int fd, const void *buf, size_t nbytes);
```

Success: 0

Failure: −1 and kernel variable errno set to indicate the type of error

TABLE 15.6

Some Common Reasons for the read() System Call to Fail

Reason for Failure	Value of errno
The fd argument is not a valid descriptor for reading	EBADF
The buf argument points to a memory location outside the process address space	EFAULT
An I/O error occurred while reading from the file system	EIO
The file reading operation was interrupted	EINTR
The pointer associated with the fd argument is negative	EINVAL
The nbytes value is greater than INT_MAX, i.e., the maximum value of an integer	EINVAL

TABLE 15.7

Some Common Reasons for the `write()` System Call to Fail

Reason for Failure	Value of `errno`
The `fd` argument is not a valid descriptor for writing	EBADF
The `fd` argument is associated with a negative pointer	EINVAL
File size limit for the process or the maximum file size limit has reached	EFBIG
The file system containing the file is full	ENOSPC
The user's quota of disk blocks on the file system containing the file has been exhausted	EDQUOT
An I/O error occurred while reading from the file system	EIO
The file reading operation was interrupted	EINTR
The nbytes value is greater than INT_MAX, i.e., the maximum value of an integer	EINVAL

This call tries to write `nbytes` bytes from the main memory area pointed to by `buf` to the file with the file descriptor `fd`. The data is written starting with the current location of the read/write file pointer. Upon completion, the `write()` call returns the number of bytes actually written, which may be less than `nbytes`. The file pointer advances by the number of bytes actually written. For a regular file, the call guarantees writing `nbytes` if the disk is not full or the file size has not exceeded the maximum file size supported by the Linux system. However, with other files, such as sockets (see Chapter W20), this is not guaranteed.

On failure, `write()` returns −1 and `errno` is set to indicate the reason for failure. Table 15.7 shows some common reasons for the `read()` call to fail for regular files.

We now enhance the `open_close.c` program and convert it into a file copy program, `cpy`, with the following syntax:

```
cpy source target
```

The program copies the **source** file, an existing file, to the **target** file, a nonexistent file. The compilation and execution of the example program show the program output.

```
$ cat cpy.c
#include <fcntl.h>
#include <unistd.h>

#define SIZE 512

void closefd(int);

int main(int argc, char *argv[])
{
        int rfd, wfd, nr, nw;
        char buf[SIZE];

        if (argc != 3) {
                printf("Inappropriate number of command line arguments.\n");
                exit(0);
        }
        /* Open the source file for reading */
        if ((rfd = open(argv[1], O_RDONLY)) == -1) {
                perror("Source file opening");
                exit(1);
        }
        /* Open the destination file for writing */
        if ((wfd = open(argv[2], O_CREAT | O_WRONLY | O_TRUNC, 0666)) == -1) {
                perror("Creation and opening of destination file");
                exit(1);
        }
```

```
        while ((nr = read(rfd, buf, sizeof(buf))) != 0) {
                if (nr == -1) {
                        perror("File reading");
                        exit(1);
                }
                nw = write(wfd, buf, nr);
        }
        closefd(rfd);
        closefd(wfd);
        return(0);
}
void closefd(int fd)
{
        if (close(fd) == -1) {
                perror("File closing");
                exit(1);
        }
}
$ gcc cpy.c -w
$ a.out cpy.c cpy_bak.c
$ ls -l
total 32
-rwxr-xr-x 1 sarwar faculty 9000 Oct 22 21:14 a.out
-rw-r--r-- 1 sarwar faculty  843 Oct 22 21:15 cpy_bak.c
-rw-r--r-- 1 sarwar faculty  843 Oct 22 21:14 cpy.c
-rw-r--r-- 1 sarwar faculty 3647 Oct 22 21:11 foobar
-rw-r--r-- 1 sarwar faculty  467 Oct 22 21:09 open_close.c
-rw-r--r-- 1 sarwar faculty  505 Oct 22 21:12 open_close_new.c
$
```

Exercise 15.8

Compile and run the programs given in Section 15.8.4. Do they work as intended?

Exercise 15.9

Remove the O_TRUNC flag in the second open() system call. What is its effect? Explain your answer.

15.8.5 Positioning the File Pointer: Random Access

You can use lseek() to change the position of the read/write file pointer. This service allows you to access file data randomly. Here is a brief description for the call.

```
#include <unistd.h>
off_t lseek(int fd, off_t offset, int whence);
```

Success: 0
Failure: −1 and kernel variable errno set to indicate the type of error

The call sets the position of the read/write file pointer for the file with descriptor fd to offset according to the value of the whence argument. The repositioning of the descriptor is done according to Table 15.8.

On successful completion, lseek() returns the resulting offset location in bytes from the beginning of the file. It returns −1 on failure and errno is set to indicate the reason for error. Thus, if you want to append data to a file, you should set the file pointer to the end of the file by using the following statement:

```
n = lseek(fd, 0L, SEEK_END);
```

Note that this call sets the file pointer to 0 bytes away from the current EOF and returns the new position (in bytes) of the file pointer. The new position also indicates the size of the file in bytes. Thus, when lseek() returns, n contains the size of the file (in bytes).

TABLE 15.8

Repositioning of the Read/Write File Pointer for Random Access of File Data

The Value of whence	Position of the Read/Write Pointer
SEEK_SET	The position/offset is set to offset bytes from the beginning of the file
SEEK_CUR	The position/offset is set to offset bytes from the current location of the pointer
SEEK_END	The position/offset is set to offset bytes from the end of the file (i.e., to the size of the file plus offset)
SEEK_HOLE	The position/offset is set to the start of the next hole greater than or equal to offset
SEEK_DATA	The position/offset is set to the start of the next nonhole (i.e., data region) greater than or equal to offset

TABLE 15.9

Some Common Reasons for the lseek() System Call to Fail

Reason for Failure	Value of errno
The fd argument is not an open file descriptor	EBADF
The fd argument is associated with a file that it is not capable of seeking: pipe, FIFO, or socket	ESPIPE
The whence argument or the resulting offset would be negative for a noncharacter special file	EINVAL
The resulting offset cannot be represented in off_t	EOVERFLOW
There is no data region (SEEK_DATA) or hole region (SEEK_HOLE) beyond offset	ENXIO

You can determine the current position of the file pointer by using the following statement. When the call returns, n contains the current position (in bytes) of the file pointer from the beginning of the file.

```
n = lseek(fd, 0L, SEEK_CUR);
```

This call only works for file objects that are capable of seeking. The use of lseek() for other types of files fails and errno is set to indicate this reason. Three such file types are pipe, FIFO, and socket, and device files not capable of seeking, such as keyboard. Table 15.9 shows some common reasons for the lseek() call to fail.

The lseek() call allows you to position the file pointer beyond the EOF. If data is written at this point later on, a *hole* is created between the previous EOF and the position of the file pointer after lseek(). The hole contains null bytes—that is, bytes containing zeros. Figure 15.4 shows the state of a file with holes.

The create_holes.c program creates three holes of 512 bytes each in the file passed to it as a command line argument. After the sample run of the program, we use the ls -l data command to display the size of the data file with holes. We use the tail data | od commands to show the contents of the original file and after the holes have been inserted in it as octal dumps (using the od command). Note that asterisks (^) are used in the octal dump to show a series of zeros. The count of zeros in one hole will come out to be, as expected, 512. Note that you cannot view the holes with the tail data command because it does not display null bytes.

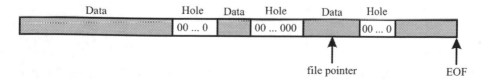

FIGURE 15.4 A Linux file with three holes of different sizes.

```
$ cat create_holes.c
#include <fcntl.h>
#include <unistd.h>
#include <string.h>

#define HSIZE   512L

int main(int argc, char *argv[])
{
        int fd, i, n, nholes, nw;
        char *buf="Random data for creating holes.";

        /* Open the given file */
        if (argc != 3) {
                printf("Inappropriate number of command line arguments.\n");
                exit(0);
        }
        nholes = atoi(argv[2]);
        if ((fd = open(argv[1], O_WRONLY)) == -1) {
                perror("File opening");
                exit(1);
        }
        for (i=0; i<nholes; i++) {
                n = lseek(fd, HSIZE, SEEK_END);
                nw = write(fd, buf, strlen(buf));
        }
         if (close(fd) == -1) {
                 perror("File closing");
                 exit(1);
         }
        return(0);
}
$ man ls > data
$ ls -l data
-rw-r--r-- 1 sarwar faculty 7878 Oct 22 21:18 data
$ tail --lines=5 data | od
0000000 042523 020105 046101 047523 020012 020040 020040 020040
0000020 072506 066154 062040 061557 066565 067145 060564 064564
0000040 067157 060440 035164 036040 072150 070164 027472 073457
0000060 073567 063456 072556 067456 063562 071457 063157 073564
0000100 071141 027545 067543 062562 072165 066151 027563 071554
0000120 005076 020040 020040 020040 067440 020162 073141 064541
0000140 060554 066142 020145 067554 060543 066154 020171 064566
0000160 035141 064046 063156 020157 024047 067543 062562 072165
0000200 066151 024563 066040 020163 067151 067566 060543 064564
0000220 067157 005047 043412 052516 061440 071157 072545 064564
0000240 071554 034040 031056 020065 020040 020040 020040 020040
0000260 020040 020040 020040 062506 071142 060565 074562 031040
0000300 030460 020066 020040 020040 020040 020040 020040 020040
0000320 020040 020040 020040 020040 020040 020040 020040 051514
0000340 030450 005051
0000344
$ gcc create_holes.c -w
$ a.out data 3
$ ls -l data
-rw-r--r-- 1 sarwar faculty 9507 Oct 22 21:26 data
$ tail data | od
0000000 020040 020040 020040 043440 046120 073040 071145 064563
0000020 067157 031440 067440 020162 060554 062564 020162 064074
0000040 072164 035160 027457 067147 027165 071157 027547 064554
0000060 062543 071556 071545 063457 066160 064056 066564 037154
0000100 005056 020040 020040 020040 052040 064550 020163 064440
0000120 020163 063040 062562 020145 071440 063157 073564 071141
```

```
0000140 035145 020040 067571 020165 060440 062562 063040 062562
0000160 020145 067564 061440 060550 063556 020145 067141 020144
0000200 062562 064514 072163 064562 072542 062564 064440 027164
0000220 020012 020040 020040 020040 064124 071145 020145 071551
0000240 047040 020117 040527 051122 047101 054524 020054 067564
0000260 072040 062550 062440 072170 067145 020164 062560 066562
0000300 072151 062564 020144 074542 066040 073541 005056 051412
0000320 042505 040440 051514 005117 020040 020040 020040 043040
0000340 066165 020154 067544 072543 062555 072156 072141 067551
0000360 020156 072141 020072 064074 072164 035160 027457 073567
0000400 027167 067147 027165 071157 027547 067563 072146 060567
0000420 062562 061457 071157 072545 064564 071554 066057 037163
0000440 020012 020040 020040 020040 071157 060440 060566 066151
0000460 061141 062554 066040 061557 066141 074554 073040 060551
0000500 020072 067151 067546 023440 061450 071157 072545 064564
0000520 071554 020051 071554 064440 073156 061557 072141 067551
0000540 023556 005012 047107 020125 067543 062562 072165 066151
0000560 020163 027070 032462 020040 020040 020040 020040 020040
0000600 020040 020040 043040 061145 072562 071141 020171 030062
0000620 033061 020040 020040 020040 020040 020040 020040 020040
0000640 020040 020040 020040 020040 020040 020040 046040 024123
0000660 024461 000012 000000 000000 000000 000000 000000 000000
0000700 000000 000000 000000 000000 000000 000000 000000 000000
*
0001660 000000 051000 067141 067544 020155 060544 060564 063040
0001700 071157 061440 062562 072141 067151 020147 067550 062554
0001720 027163 000000 000000 000000 000000 000000 000000 000000
0001740 000000 000000 000000 000000 000000 000000 000000 000000
*
0002720 000000 060522 062156 066557 062040 072141 020141 067546
0002740 020162 071143 060545 064564 063556 064040 066157 071545
0002760 000056 000000 000000 000000 000000 000000 000000 000000
0003000 000000 000000 000000 000000 000000 000000 000000 000000
*
0003760 051000 067141 067544 020155 060544 060564 063040 071157
0004000 061440 062562 072141 067151 020147 067550 062554 027163
0004020
$ tail data
        GPL version 3 or later <http://gnu.org/licenses/gpl.html>.
        This  is  free  software:  you  are free to change and redistribute it.
        There is NO WARRANTY, to the extent permitted by law.

SEE ALSO
        Full documentation at: <http://www.gnu.org/software/coreutils/ls>
        or available locally via: info '(coreutils) ls invocation'

GNU coreutils 8.25             February 2016                        LS(1)
Random data for creating holes.Random data for creating holes.Random data for cr
eating holes.
$
```

Exercise 15.10

Compile and run the program given in Section 15.8.5. Does it work as intended?

Exercise 15.11

Change the hole size to 2K. Compile and run the program. Verify that it works correctly.

15.8.6 Truncating a File

Truncating a file means chopping off contents from the tail of the file. The use of the open() system call with the O_TRUNC flag, as shown in Section 15.8.4, is a special case of truncation where the file size is reduced to 0 and the read/write file pointer and EOF are set to 0. You can use the truncate() or ftruncate() system calls to truncate a file. Here are the brief descriptions of these commands.

```
#include <unistd.h>
int truncate(const char *path, off_t length);
int ftruncate(int fd, off_t length);
```

Success: 0

Failure: −1 and kernel variable errno set to indicate the type of error

These functions truncate an existing file to length bytes. The difference between the functions is that truncate() works with a pathname and ftruncate() with a file descriptor. If length is smaller than the existing length of the file, the contents of the file beyond length bytes are not accessible anymore. If length is greater than the current file size, the file size is increased to length and the space between the previous EOF and new EOF is filled with zeros and becomes a hole (see Section 15.8.5).

The **truncate.c** program takes two command line arguments: a file and a number. It displays the size of the file by using the lseek() system call, truncates the file to the size specified as the second argument by using the truncate() system call, and displays the size of the truncated file. The sample run shows the sizes of the file, called data, before and after truncation. Note that the size of the input file is also confirmed with the ls −l data command. The tail commands are used to show the tails of the **data** file before and after truncation.

```
$ cat truncate.c
#include <fcntl.h>
#include <unistd.h>

#define HSIZE   512L

int main(int argc, char *argv[])
{
        int fd, newfilesize, size_o, size_t;

        /* Open the given file */
        if (argc != 3) {
                printf("Inappropriate number of command line arguments.\n");
                exit(0);
        }
        newfilesize = atoi(argv[2]);
        if ((fd = open(argv[1], O_WRONLY)) == -1) {
                perror("File opening");
                exit(1);
        }
        size_o = lseek(fd, 0, SEEK_END); /* Size of original file */
        if (truncate(argv[1], newfilesize) == -1) {
                perror("Truncating file");
                exit(1);
        }
        size_t = lseek(fd, 0, SEEK_END); /* Size of truncated file */
        printf("The size of the original file is %d bytes.\n", size_o);
        printf("The size of the truncated file is %d bytes.\n", size_t);
        if (close(fd) == -1) {
                perror("File closing");
                exit(1);
        }
        return(0);
}
$ man cat > data
$ ls -l data
-rw-r--r-- 1 sarwar faculty 1958 Oct 22 21:38 data
```

```
$ tail data
        This is free software: you are free  to  change  and  redistribute  it.
        There is NO WARRANTY, to the extent permitted by law.

SEE ALSO
        tac(1)

        Full documentation at: <http://www.gnu.org/software/coreutils/cat>
        or available locally via: info '(coreutils) cat invocation'

GNU coreutils 8.25            February 2016                         CAT(1)
$ gcc truncate.c -w
$ a.out data 1024
The size of the original file is 1958 bytes.
The size of the truncated file is 1024 bytes.
$ tail data

        -v, --show-nonprinting
                use ^ and M- notation, except for LFD and TAB

        --help display this help and exit

        --version
                output version information and exit

$
```

Exercise 15.12

Compile and run **truncate.c**. Does it work as intended?

15.8.7 Removing a File

You can use the unlink() system call to remove a file from the file system structure. The following is a brief description of the call:

```
#include <unistd.h>
int unlink(const char *path);
```

Success: 0

Failure: −1 and kernel variable errno set to indicate the type of error

This system call essentially decrements by 1, the hard link count for the file stored in its inode. If the resultant link count becomes 0 and no process has the file open, the file is removed from the file system. If link count becomes 0 but one or more processes have the file open, the removal of the file is delayed until the reference count for the file becomes 0. If the link count is not 0, the file remains in the file system structure and its directory entry is removed. Once a file has been removed from the file system, its directory entry, inode, and all other kernel resources associated with the file are returned to the system for recycling.

On successful completion, unlink() returns 0. It returns −1 on failure and errno is set to indicate the reason for failure. Table 15.10 shows some of the common reasons for the unlink() call to fail for regular files.

The **create_holes_delete.c** program shown in the following is an updated version of the **create_holes.c** program discussed in Section 15.8.5. The program saves the size of the original file, creates holes in it, displays the sizes of the original file and the file with holes, and then removes the file with holes. The program determines the sizes of the original and new file (with holes) by using the lseek() system call. The sample run shows the sizes of original file and the file with holes. The ls -l data command shows that the program did remove the file with holes.

TABLE 15.10

Some Common Reasons for the unlink() System Call to Fail

Reason for Failure	Value of **errno**
The named file is a directory	EISDIR
The file specified in **path** does not exist	ENOENT
A component in **path** is not a directory	ENOTDIR
Search permission is denied on a component of **path**	EACCES
Write permission does not exist on the directory containing the file/link to be removed	EACCES
A component in **path** exceeds the file name size limit (255 characters) and the entire path exceeds the 1,023 characters	ENAMETOOLONG
An I/O error occurred while deleting the directory entry for the file or deallocating file's inode	EIO
The file to be removed resides on a read-only file system or device	EROFS

```
$ cat create_holes_delete.c
#include <fcntl.h>
#include <unistd.h>
#include <string.h>

#define HSIZE   512L

int main(int argc, char *argv[])
{
        int fd, i, n, nholes, nw, size_o, size_h;
        char *buf="Random data for creating holes.";

        /* Open the given file */
        if (argc != 3) {
                printf("Inappropriate number of command line arguments.\n");
                exit(0);
        }
        nholes = atoi(argv[2]);
        if ((fd = open(argv[1], O_WRONLY)) == -1) {
                perror("File opening");
                exit(1);
        }
        size_o = lseek(fd, 0, SEEK_END); /* Size of original file */
        for (i=0; i<nholes; i++) {
                n = lseek(fd, HSIZE, SEEK_END);
                nw = write(fd, buf, strlen(buf));
        }
        size_h = lseek(fd, 0, SEEK_END); /* Size of file with holes */
        if (close(fd) == -1) {
                perror("File closing");
                exit(1);
        }
        printf("The size of the original file is %d bytes.\n", size_o);
        printf("The size of the file with holes is %d bytes.\n", size_h);
        unlink(argv[1]);
        return(0);
}
$ man cat > data
$ ls -l data
-rw-r--r-- 1 sarwar faculty 1958 Oct 22 21:43 data
$ gcc create_holes_delete.c -w
$ a.out data 3
The size of the original file is 1958 bytes.
```

```
The size of the file with holes is 3587 bytes.
$ ls -l data
ls: cannot access 'data': No such file or directory
$
```

The remove() system call is equivalent to the unlink() system call for files.

Exercise 15.13
Compile and run **create_holes_delete.c**. Does it work as expected?

15.9 Getting File Attributes from a File Inode

Recall that the Linux inode contains most of the attributes of a file, including the following:

- Size (in bytes)
- User (Owner) ID
- Group ID (of the owner)
- File mode and access permissions
- Hard link count
- Times for the creation, last access, and last modification
- Pointers to file blocks on secondary storage

Note that the file's name and the current position of the file's read/write pointer are not stored in the inode of the file. The name of the file, as discussed in Chapter 4 and in Section 15.11, is stored in the directory entry for the file and, as discussed Section 15.5.2, the file pointer is stored in the SFT. You can obtain a copy of a file's attributes stored in its inode by using the stat(), lstat(), and fstat() system calls. These calls fill in the stat structure (struct stat), defined in **<sys/stat.h>**, with the values of various attributes. We discuss the stat structure and the three calls in Sections 15.9.1 and 15.9.2, respectively.

15.9.1 The stat Structure

Most, but not all, of the attributes of a file are stored in the file's inode. For example, as discussed in Chapter 4, a file's inode number resides in the directory entry of the file and not in file's inode. The stat structure describes part of the inode. The following definition is found in the **/usr/include/asm-generic/stat.h** file.

```
struct stat {
        unsigned long  st_dev;       /* Device.  */
        unsigned long  st_ino;       /* File serial number; inode number  */
        unsigned int   st_mode;      /* File mode.  */
        unsigned int   st_nlink;     /* Link count.  */
        unsigned int   st_uid;       /* User ID of the file's owner.  */
        unsigned int   st_gid;       /* Group ID of the file's group. */
        unsigned long  st_rdev;      /* Device number, if device.  */
        unsigned long  __pad1;
        long           st_size;      /* Size of file, in bytes.  */
        int            st_blksize;   /* Optimal block size for I/O.  */
        int            __pad2;
        long           st_blocks;    /* Number 512-byte blocks allocated. */
        long           st_atime;     /* Time of last access.  */
        unsigned long  st_atime_nsec;
        long           st_mtime;     /* Time of last modification.  */
        unsigned long  st_mtime_nsec;
```

TABLE 15.11

Commonly Used Fields of `struct stat` Defined in **<sys/stat.h> >**

Field	Meaning
st_mode	File's access permissions
	Type of file
	• Status of special permission bits: SUID, SGID, and sticky (see Chapter 5)
st_ctime	Time when file's attributes were changed such as file access permissions through the `chmod` command
st_blocks	The number of 512-byte blocks allocated to the file

TABLE 15.12

Macros Defined in **<sys/stat.h>** for Checking the Type of a File

Macro	Test
S_ISBLK(m)	Test for block special file
S_ISCHR(m)	Test for character special file
S_ISDIR(m)	Test for directory
S_ISFIFO(m)	Test for named pipe (FIFO)
S_ISLN(m)	Test for symbolic link
S_ISREG(m)	Test for regular file
S_ISSOCK(m)	Test for socket
S_ISWHT(m)	Test for whiteout (not implemented)

```
        long          st_ctime;        /* Time of last status change.  */
        unsigned long st_ctime_nsec;
        unsigned int  __unused4;
        unsigned int  __unused5;
};
```

Most of the fields of the structure are self-explanatory, but not all. Some of the fields that need a bit more explanation are described briefly in Table 15.11.

The **/usr/include/linux/stat.h** file defines 23 distinct 24-bit flags for the st_mode field for file types, access permissions, special permission bits, and so on. A programmer does not need to know what these values are, but you should browse through the header file to see these fields. The **/usr/include/ linux/stat.h** file also contains several macros to check the type of file. Each macro takes st_mode value as an argument corresponding to a specified type and evaluates to a nonnegative if the test is true. If it evaluates to 0, then the test is false. Table 15.12 contains these macros and their purpose.

15.9.2 Populating the **stat** Structure with System Calls

You can use any of the three system calls to populate the stat structure with the attributes of a file: stat(), lstat(), and fstat(). Here are brief descriptions of these calls.

```
#include <sys/types.h>
#include <sys/stat/h>
int stat(const char *path, struct stat *sb);
int lstat(const char *path, struct stat *sb);
int fstat(int fd, struct stat *sb);
```

Success: 0
Failure: −1 and kernel variable errno set to indicate the type of error

The stat() system call reads information about the file specified in the path parameter and stores it in the stat structure (struct stat) pointed to by sb. All of the components in the path variable must be searchable (i.e., must have the execute permission on). Access permissions on the named file are irrelevant because this call does not deal with the contents of the file. If the named file is a symbolic link, the stat() system call returns the information about the file that the link points to.

The lstat() system call works like the stat() system call, except that if the named file is a symbolic link, the call returns information about the link file and not where it points to. The fstat() system call returns information about an open file.

15.9.3 Displaying File Attributes

Once you have obtained file attributes by using one of the preceding system calls and stored them in a variable of struct stat, you can display them by using the Standard I/O functions for output, such as printf() and sprint(). For example, the **stat.c** program that follows reads the sv variable (of struct stat type) the attributes of the file passed as a command line argument. It then displays the size of the file (in bytes) if it is a regular file. Otherwise, it displays an error message, stat: Success, using the library call perror().

```
$ cat stat.c
#include <sys/stat.h>
#include <sys/types.h>

int main(int argc, char *argv[])
{
        struct stat sv;

        if (argc != 2) {
                printf("Inappropriate number of command line arguments.\n");
                exit(0);
        }
        if (stat(argv[1], &sv) == 0 && S_ISREG(sv.st_mode))
                printf("File size is %d bytes\n", sv.st_size);
        else
                perror("stat");
}
$ gcc stat.c -w
$ a.out stat.c
File size is 368 bytes
$ ls -l stat.c
-rw-r--r-- 1 sarwar faculty 368 Oct 22 21:56 stat.c
$ a.out ~
stat: Success
$ a.out /usr
stat: Success
$
```

15.9.4 Accessing and Manipulating File Attributes

Linux provides several system calls for modifying file attributes, including file access permissions, file owner, and file access times. We do not discuss these system calls in detail. Table 15.13 contains brief descriptions of some system calls related to the manipulation of file attributes and their purpose.

Exercise 15.14
Browse through the stat structure definition on your Linux system. Are there any differences between the definitions of this structure on your system and those displayed in this section?

Exercise 15.15
Compile and run **stat.c**, discussed in Section 15.9.3. Does it work as intended?

Exercise 15.16
Browse through the man pages for the system calls given in Table 15.13 on your Linux system. Write small programs to practice them.

TABLE 15.13

Additional System Calls for Accessing and Changing File Attributes

Accessing and Setting File Attributes	
access()	Checks file access permissions
chmod()	Sets file access permissions
chown()	Changes file owner
rename()	Renames a file
umask()	Checks and sets umask
utime()	Changes access and medication times

15.10 Restarting System Calls

We should write code that has the ability to restart system calls that are interrupted during their execution. Such situations arise when you perform *blocking I/O* using the read() and write() system calls. We will also discuss blocking I/O in detail in Chapters 16 and W20 while discussing the concept of a communication channel, called a *Linux pipe*, used for communication between related processes. We discussed the command line use of pipes in detail in Chapter 9.

Simply saying, blocking input means that the read() system call waits as there is nothing to read because the pipe is empty. Another example of blocking is when the read() system call reads input interactively from a keyboard. The read() call blocks until the user enters keyboard input. Similarly, in the context of a pipe, a write() system call blocks if the pipe is full.

The read() system call may be interrupted when it is performing a blocking read from a slow device. When the read() call is used to read input interactively from a keyboard, it may be interrupted while waiting for user input. Most modern Linux implementations restart such system calls automatically. If you are not sure whether your code would be run on such a system, you need to write code to explicitly handle the restarting of an interrupted system call. The following code snippet may be used for this purpose.

```
Repeat:
    if ((nr = read(fd, buf, SIZE)) == -1)) {
        switch(errno) {
            case EINTR:  goto repeat;
                         break;
            /* handle other errors */
            default:
        }
    }
```

We address this issue for network-related system calls, particularly, the select() system call, in Chapter W20.

15.11 System Calls for Manipulating Directories

Linux provides a set of system calls for creating, deleting, and changing directories. Table 15.14 lists these calls along with their purpose. We will not discuss these calls any further. However, you will get to practice them in a programming exercise given at the end of the chapter.

As a system programmer, you do not need to know the details of the kernel data structures for directories. However, such knowledge enhances your understanding of the Linux operating system and your skills as a programmer. The most important directory data structure is defined in the **/usr/src/linux-headers-4.4.0-21/include/linux/dirent.h** file. Here is the definition of the directory entry in Linux Mint.

TABLE 15.14

Linux System Calls Related to Directories

System Call	Purpose
chdir()	Changes directory
mkdir()	Creates directory
rmdir()	Removes directory
remove()	Removes an empty directory; equivalent to rmdir()
rename()	Renames a directory

```
struct linux_dirent64 {
        u64             d_ino;          /* "inode number" of entry */
        s64             d_off;          /* offset of disk directory entry */
        unsigned short  d_reclen;       /* length of this record */
        unsigned char   d_type;         /* file type, see below */
        char            d_name[0];      /* name of file */
};
```

The definition has five fields: the inode number for the file, offset of the disk directory entry, length of the directory entry, file type, and file name.

You can use several C library functions to manipulate directories, including opendir(), closedir(), readdir(), telldir(), seekdir(), and rewinddir().

Exercise 15.17

Browse through the man pages for the previously stated calls to manipulate directories to understand their syntax and semantics. Then, write small programs to use them in order to enhance your understanding of these calls.

Summary

Linux provides two APIs for software development: the language libraries and the SCI. Application programmers typically use the interface provided by the language libraries and system programmers primarily use the SCI. We described the Linux system calls and their use in small programs for the I/O of file data and attributes. Our focus was on regular files.

Before discussing the system calls, we explained that four types of events cause the control to be transferred to the kernel code. They are interrupt, trap, signal, and system call, with the first two caused by the system hardware and the last two primarily caused by software. We also discussed the details of events that happen in order to execute a system call. We discussed briefly the CSL and the difference between Standard I/O via fopen(), fread(), fwrite(), and fclose() and low-level I/O through system calls using open(), read(), write(), and close().

We discussed the system calls for data I/O for regular files and I/O of file attributes using the open(), creat(), read(), write(), close(), truncate(), and stat() system calls. For random I/O, we used the lseek() system call to set the file pointer to the byte where the next read or write operation should be performed. We showed the use of the lseek() and write() system calls to illustrate how holes can be created in a regular file under Linux. We showed the holes created in a file by displaying its octal dump using the od command.

We briefly discussed several system calls and C library functions for accessing and manipulating file attributes and directories. These calls are access(), chdir(), chmod(), chown(), closedir(), mkdir(), opendir(), read(), readdir(), rename(), rewinddir(), seekdir(), telldir(), umask(), and utime().

We showed the kernel data structures for the stat structure and how the contents of the data structure may be read to know about the various file attributes.

We described a small piece of code that may be used to restart a system call if it is interrupted in the middle of execution and the underlying operating system kernel does not automatically restart it. Such interruption may occur when blocking I/O is done on slow devices, keyboards, and files such as pipes, FIFOs, and sockets.

Finally, we show the contents of a directory entry in Linux, as described in the various header files, and mention a few system calls that may be used to manipulate directories: chdir(), mkdir(), rmdir(), remove(), and rename().

Questions and Problems

1. What is a system call? List names of three system calls in each of the following categories:
   a. Process control
   b. File management/manipulation
   c. Device management
   d. Information maintenance
   e. Communications

2. List a few software interrupts, along with their purpose, generated by the INT instruction in computer with Intel x86 CPU.

3. Clearly describe the difference between file I/O using the CSL and the Linux SCI.

4. Why do programmers prefer to use the standard I/O library functions instead of the SCI interface for applications where files data is ready very frequently?

5. What is a stream in the context of standard I/O? Explain with an example.

6. What is the FILE object? What does it contain? Create a file called **foo**, open it using a standard C library function, and display the file descriptor of the file.

7. Describe clearly the differences between the PPFDTs and the SFT. Illustrate your answer with an example.

8. Describe in words the situations under which the same entry in multiple PPFDTs would point to the same entry in the SFT.

9. What happens when a process executes the close() system call on a file descriptor when pointers from more than one PPFDT point to the entry for this file in the SFT?

10. What is the size of the PPFDT on your Linux system? How did you obtain the answer?

11. What does the lseek() system call return after completing successful execution?

12. Write a piece of code that opens a file whose name is passed as a command line argument and sets the read/write file pointer to the beginning of the file.

13. Write a program that takes the following command line arguments: file, position, nbytes. It displays nbytes bytes from the file starting with the position byte of the file. Do the appropriate error handling.

14. Enhance the **create_holes.c** program in Section 15.7.5 so that it takes the hole size from the command line and appends after the hole data taken from the file specified as the second command line argument. Compile and show a few sample runs of the program.

15. What is the effect of executing the unlink() system call if there are multiple hard links to a file?

16. Give the syntax for the open() system call for creating a new file for reading and writing. open(pathname, O_RDWR | O_CREAT | O_TRUNC, mode);

17. Give the sequence of system calls for performing this operation by using the creat(), open(), and close() system calls.

18. Write a C program to verify that the Linux kernel allocates file descriptors sequentially starting with descriptor 0. Show the source code and execution of your program.

19. Create a large file running the man csh >> bigdata command five times. Then, create three holes in it of sizes 4 K, 8 K, and 16 K bytes. Display the file size by using the ls -l bigdata command. Then use the du bigdata command to determine the file size. Read the man page

for the du command to determine the block size used so you can perform this task. Now verify that the **bigdata** file contains holes equal to 28 K bytes.

20. What is the purpose of the following program? Explain your answer.

```
int main(void)
{
                    int nr, nw;

                    while ((nr = read(0, &c, 1)) !== 0)
                            nw = write(1, &c, 1);
                    return(0);
}
```

21. Write a C program that takes a file as a command line argument and outputs Yes if the file is empty and No if it isn't. Show the source code, compilation of the source code, and a few sample runs of the program.

22. What would happen if we did not use the O_TRUNC flag in the cpy.c program in Section 15.7.4?

23. Modify the truncate.c program discussed in Section 15.7.6 so that it takes any number of files as command line arguments and truncates them to the argument specified as the last argument. It displays the names and sizes of the original and truncated files in the form of a table and removes the truncated files from the file system. Show a few sample runs of the program.

24. Change the truncate.c program discussed in Section 15.7.6 so that it takes three command line arguments. The first two arguments are file names, with the first being the name of an existing file. The third argument specifies the new file length. The program creates the file with the name given as the second argument, copies the first file into it, truncates the file according to the third argument, displays the sizes of the original and truncated file, removes the truncated file from the file system, and terminates. Show a few sample runs of the program.

25. Browse through the <stdio.h> file and identify the values (or macros) for the following:
    a. Maximum number of files that the system allows to be opened concurrently
    b. Macros for standard files
    c. Macros that define SEEK_CUR, SEEK_END, and SEEK_SET

26. Browse through the **<stdio.h>** file and identify all of the functions and their purposes.

27. Write a program that opens a file with one hard link and removes the file with the unlink() system call. Read data from the opened file and send it to standard output using the write() system call. What happens? Is your program able to read file data? Explain your answer.

28. Clearly state the difference between the stat(), fstat(), and lstat() system calls. Which of these calls would you use if you need to display the attributes of a link-type file? Why?

29. Write a C program that uses the stat() system call to display the file size and last modification time for the file whose path is passed as a command line argument. The program then opens the file, appends data to it, and uses the fstat() system call to read and display the file size and modification time for the file.

30. Enhance the program from Problem 27 so that it takes the file as a command line argument and, in the output, first displays the type of the file followed by the file size in bytes, the file's inode number, the file's access permissions, the number of hard links to the file, the file size in 512-byte blocks, and the last modification time. Compile the program and show its execution for an ordinary file, a directory, a link file, a character-special file, and a block-special file.

Advanced Questions and Problems

31. Browse the Web for Linux kernel code and identify
    a. Call numbers for the following system calls: open(), creat(), read(), write(), close()

b.   Dispatch table for the system call

c.   Memory addresses for the code executed for the calls stated in a.

32. Draw the diagram shown in Figure 15.2 for the open() system call.

33. Read the manual page for the strace command and explain the output for strace ls and strace ps commands. How many unique system calls are used by ls and ps command each? List those commands in tables.

34. If a file is opened in multiple processes simultaneously, can the same file descriptor (number) be assigned to the file in these processes? Explain your answer.

35. Suppose a process creates a child, and both the parent and child open a file, called **student_grades**, for reading. Draw a figure showing connections between the PPFDTs for the two processes, the SFT, and the inode table.

36. Repeat Problem 32, except that the parent process opens the **student_grades** file for writing first and then forks a child. Show the figure if parent forks two children.

37. Explain the behavior of the close() system call in case of a fork(). Illustrate your answer with a diagram showing the PPFDT, the SFT, and the inode table.

38. Modify the **stat.c** program discussed in Section 15.9.3 so that if the command line argument is a link file, it displays the information about the link file and not where the link points. Do the appropriate error handling.

39. Write a program that takes a directory name as a command line argument and displays the names of all the files in the directory with the following information for each:

a.   Type

b.   Inode number

40. Write a program that takes a file name as a command line argument. If file type of the argument is nondirectory, show file's attributes in the ls –l style output. If file type is directory, display the ls –ld style output for the directory, followed by the ls –l style output for all the files and directories in the directory. Do appropriate error handling.

Projects

Project 1

Write a simple Linux shell that:

a.   Displays a prompt and waits for the user to enter a command terminated with <Enter>

b.   Executes the shell commands and prompts the user for input again

c.   Terminates when the user presses <Ctrl+D>

You should implement the following commands with your own code; that is, they are the built-in (intrinsic) shell commands:

Command	Purpose
finfo file	Display the following attributes of **file**: type, number of hard links, user ID, access permissions
truncate file size	Truncate size of **file** to size bytes
holes file number	Create number holes in **file** if it is a regular file
mkdir dname	Create the directory **dname**
rmdir dname	Remove the directory **dname**
ls [options] [file-list]	Generate output similar to the Bash ls command; support –l and –d options
cd [dname]	Change directory to **dname**

For the execution of all other shell commands, use the library function system(). Do the appropriate error handling. You will need to use the following system and library calls in addition to the Standard I/O library calls: chdir(), mkdir(), rmdir(), truncate(), lseek(), stat(), and system().

Looking for more? Visit our sites for additional readings, recommended resources, and exercises.

CRC Press e-Resource: https://www.crcpress.com/9781138710085

Authors' GitHub: https://github.com/bobk48/linuxthetextbook

16

System Programming II: Process
Management and Signal Processing

OBJECTIVES

- To explain the concepts of processes and threads
- To describe the general concept of the process control block and its implementation in Linux
- To discuss the details of main memory and disk images of Linux processes
- To describe threads in detail, including user- and kernel-level threads
- To explain the differences and commonalities between threads
- To discuss fundamental process management concepts in Linux: process creation, process termination, waiting for a child process to terminate and obtain its termination status, overwriting a process image with another, creating zombie processes, sharing open files between a process and its children, and duplicating file descriptors
- To explain the concept of signals in Linux, signal generation, and signal handling
- To explain process management concepts using various system and library calls using small programs
- To introduce thread programming using pthreads
- To cover the following system calls, library calls, commands, and primitives:

  alarm(), dup(), dup2(), execl(), execve(), execle(), exit(), file, fork(), gcc48,
  getpid(), getppid(), getuid(), getguid(), kill(), ls, signal(), size, wait(),
  waitpid(), wait3(), wait4(), wait6(), various pthread library calls

16.1 Introduction

In this chapter, we discuss the use of the API provided by the Linux SCI and libraries for process creation, termination, and management. We also describe the generation and handling of software interrupts, commonly known as *signals* in Linux lexicon. Our coverage starts with the concepts of processes and threads, and moves to the coverage of the Linux process control block (PCB), the structure of the memory and disk images of Linux processes, static and dynamic linking, single- and multithreaded processes, user- and kernel-level threads, and the differences between processes and threads. We also touch on the issues of the *critical section* and *critical section problems*.

Our treatment of Linux process management encompasses the following: system and library calls for process creation, process termination and status reporting to the parent process, getting the process ID (PID) and parent process ID (PPID), creating and handling zombie processes, a process overwriting itself with another executable image, duplicating file descriptors, file sharing between processes, sending different types of interrupts (Linux signals) to processes, handling signals, and setting up alarms in programs. As usual, our coverage of the various topics is closely linked with the underlying kernel data structures and operating system concepts where appropriate. As was the case for the programs in Chapter 15, for the compilation of our sample C programs, we use the default Gnu C compiler on Mint 18. The default compiler on our Mint 18 system is gcc.

16.2 Processes and Threads

In this section, we discuss the concepts of processes and threads. Our discussion focuses on the conceptualization of the two, how they are created, the system resources that they need for their execution, their address spaces, similarities between two, and how they differ from each other. After the discussion on generic processes, we focus primarily on Linux processes. The discussion of threads covers both user- and kernel-level thread libraries.

16.2.1 What Is a Process?

A simplistic and high-level view of a process is that it is a program in execution. Process execution starts at the entry point into the process (usually the first statement of the main function) and continues sequentially, one program statement at a time. Thus, a program counter (PC) is associated with a process that controls the sequence of execution of the program statements, including function calls. On a function call, the control transfers to the called function and the first statement in the function is executed, followed by the execution of the remaining body of code in the function sequentially, statement by statement. On return from the function, the statement following the function call in the caller function is executed. Process execution continues like this until the execution of the last statement in the process.

A deeper look into a process reveals that it in fact consists of the following three entities:

1. The address space
2. The CPU state
3. The PCB

The *address space* of the process is dictated by the main *memory image* of the process, as discussed in Section 10.5.1. The values of the CPU registers at any given time, including the value of the PC, comprise the *CPU state* of the process. We describe the enhanced version of the main memory image of a process, the address space of a process, and the *PCB* of a process in the next subsection.

16.2.2 Process Control Block

The PCB is a kernel data structure that keeps track of the run-time attributes of the process. In Linux, the PCB of a process consists of two parts: the *proc structure* and the *u area*. Although the proc structure contains the scheduling-related information of a process, the u area contains information about signal handling, resource allocation, and a reference to the proc structure. The proc structure of a process always remains in the main memory regardless of the state of the process. However, the u area is in the main memory only when the process is in the running state.

The proc structure contains the PID and PPID, its priority, scheduling and waiting queues, information about memory management (paging and segmentation), the process state, and the signal-handling mask. The u area contains a pointer to the proc structure, the CPU state (i.e., context) of a blocked process, the user ID (UID) and Group UID (GUID), the current directory, CPU usage of the process, the terminal the process is attached to, signal handling information, and the per-process file descriptor table (PPFDT).

A process may not access its PCB directly; the kernel updates the proc structure fields of a process as and when needed. For example, the kernel code updates the scheduling priority field in the PCB of a process when the process priority changes. Although a part of a process image, the u area is only accessible to a process through the kernel code that executes on behalf of the process, such as the code for a system call.

16.2.3 Process Memory Image (Process Address Space)

We discussed the memory image of a Linux process in detail in Chapter 10. As stated earlier, the process memory image, also known as the *process image*, consists of several sections, as shown in Figure 10.6.

Figure 16.1 shows the complete version of the process image, including the u area at the top of the image. The memory image of a process delineates the main memory region(s) that a process may access legally, known as the process address space.

Some of the regions of the process image come into being only while the process remains in the system (running or waiting for an event); other regions are an integral part of the process image, whether it is a process in the main memory or an executable image on disk. For example, the environment, stack, shared (dynamically linked) libraries, heap, and uninitialized data portions of the process are only required for as long as the process remains in the system. These regions have been labeled as "run-time areas" in Figure 16.1. We have discussed most of the sections of the address space of a process in this and/or other chapters, except shared libraries.

By default, all modern C compilers, including gcc, generate executable code by linking all library calls to the relevant library code at run time. When a process calls a library function, the code for the relevant library is loaded into the memory if it is not already in the memory because of a previous reference to this library by this or another process. Once a library's code has been loaded into the memory, it remains memory resident for future references to it by any process. Such linking of library code to an executable code is called *dynamic linking* and the libraries used in this fashion are called *shared libraries*. Thus, dynamically linked libraries are always shared.

Dynamic linking is preferred over static linking, primarily, for the following reasons:

1. The resultant executable code is smaller in size. It means that it requires less disk space to save it, a shorter time to load it into the main memory, and potentially less main memory to execute it (in case the library function is not called during execution).
2. Library code is loaded into the memory only once and is shared by multiple processes.
3. New executable code does not need to be generated (relinked) again if a library is updated, provided the prototypes for the library calls do not change.

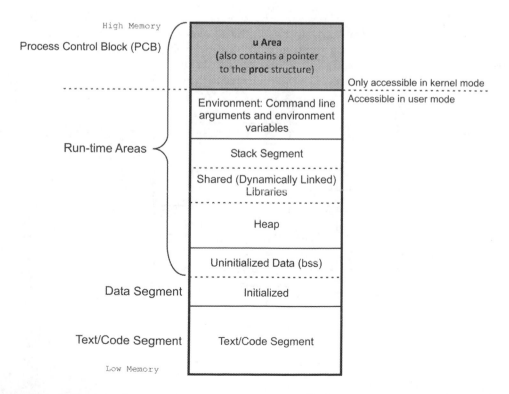

FIGURE 16.1 Main memory image (process address space) of a Linux process.

Some of the main disadvantages of dynamic linking are as follows:

1. The execution of a program is slower if it refers to a library that has not been referenced by any process previously and, hence, its code has not been loaded into the main memory previously.
2. If an executable program is moved to a system with different libraries, it fails to run.
3. If libraries are removed or upgraded, executable programs linked against them can break.

16.2.4 Process Disk Image

The disk image of an executable file in Linux has five sections, as shown in Figure 16.2. These sections are header, text/code, data, relocation information, and symbol table. The header section contains the following information:

- Magic number
- Size of text/code area
- Size of data area
- Size of initialized data area (bss)
- Size of the symbol table
- Information about the entry point into the text/code area and flags

The *magic number* of an executable file describes the type of the executable code in the file. A few types are as follows: binary executable generated by a compiler, shell script, Perl script, and Python code. Although commonly used in the computer literature now to describe file formats and protocols, the term *magic number* was first used in the seventh edition of Linux for identifying the type of an executable file. The Linux command `file`, discussed in Chapter 4, uses the magic number in a file to decipher the type of the file.

The relocation information describes whether the program is *relocatable*. An executable code is relocatable if it will run regardless of where it is loaded into the memory. Programs that are not relocatable must be loaded in a specific area in the main memory for it to execute properly.

As discussed in Chapter 10, you can use the `size` command to display the sizes of the text/code, data, and bss segments of an executable file. The following session shows the use of the `file`, `ls -l`, and `size`

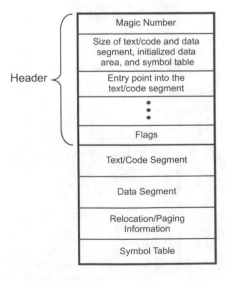

FIGURE 16.2 Disk image of an executable file.

commands to display the types of the executable files **client** and **createpipe**, their sizes in bytes, and the sizes in bytes of their code, data, and bss segments, as well as the sum of their sizes in bytes shown in decimal and hexadecimal notations.

```
$ file client createpipe
client:    ELF 64-bit LSB executable, x86-64, version 1 (SYSV), dynamically
linked, interpreter /lib64/ld-linux-x86-64.so.2, for GNU/Linux 2.6.32,
BuildID[sha1]=becc6bdae9989b9bda647121091fffa660e95741, not stripped
createpipe: ELF 64-bit LSB executable, x86-64, version 1 (SYSV), dynamically
linked, interpreter /lib64/ld-linux-x86-64.so.2, for GNU/Linux 2.6.32,
BuildID[sha1]=457bd89c9d371b63fb7601757b1f2b37f012174c, not stripped
$ ls -l client createpipe
-rwxr-xr-x 1 sarwar faculty 9048 Oct 21 18:42 client
-rwxr-xr-x 1 sarwar faculty 8816 Oct 21 17:19 createpipe
$ size client createpipe
   text    data    bss     dec     hex      filename
   2341     624      8    2973     b9d      client
   1641     584      8    2233     8b9      createpipe
$
```

The output of the `file` command shows that **client** and **createpipe** are both executable files in the ELF 64-bit format. The output of the `ls  -l` command shows that their sizes are 9,048 and 8,816 bytes, respectively. Finally, the output of the `size` command shows the sizes in bytes of their text, data, and bss segments, as well as the sum of their sizes.

Exercise 16.1

What is the difference between static linking and dynamic linking?

Exercise 16.2

Use the `size` command to determine the sizes of the text, data, and bss segments of the executable files for the Linux commands `find` and `sort`. Show the command runs and their outputs.

16.2.5 What Is a Thread?

When a process calls a function from its "main" function, the control of execution transfers from the main function to the code of the *called function*. On return from the function, the statement following the function call is executed in the *caller function*. If a called function calls another function, the control transfers to the newly called function, returns to its caller, and, eventually, returns to the main function. Thus, there is a single flow of control as the program executes, also known as the *thread of execution* (commonly called a *thread*) that, usually, starts with the main function and moves to the called functions one after another, eventually returning to the main function. When the main function finishes its execution, the only thread of execution in the process and the process itself terminate. Such processes are known as *single-threaded processes* and the only thread in them is known as the *main thread*. We can also say that single-threaded processes have only one *PC* associated with them and the value of the PC at any given time determines the address of the next instruction to be executed in the thread.

All conventional programs result in single-threaded processes. Figure 16.3 shows a typical single-threaded C program in which program execution starts with the main function and moves through different functions when they are called. The single execution path followed by the process from main to other functions and back—that is, the thread of execution through the program—is shown in Figure 16.4. Note how the only thread of execution moves from one function to another and back in the following order:

main → *f1* → *f2* → *f1* → *main* → *f3* → *main* → *f1* → *main*

You can create another thread of execution in your program that executes in parallel with the main thread by executing a piece of code in the program, usually a function, such that the function code executes

```
Code
#include <stdio.h>

int main(int argc, char* argv[])
{
        int i;
        char c;

        printf("Hello, world!\n");
        ...
        f1(...);
        ...
        f3(...);
        ...
        f1(...);
        ...
        exit(0);
}
void f1(int i, int* ip)
{
        ...
        if (...)
        then
                        f2(...);
        else
                ...
}
void f2(char c)
{
        ...
}
void f3(char* s)
{
        ...
}
```

FIGURE 16.3   Single-threaded process with only the main thread.

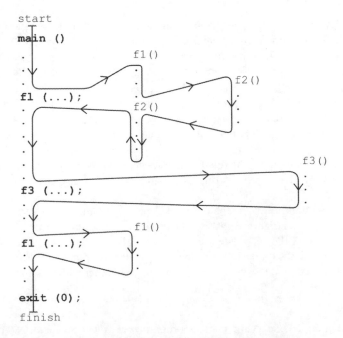

FIGURE 16.4   The flow of control (thread) of the program: a single-threaded process.

like an independent program but without a main function and within the address space of the process in which the function resides. This means that this thread of execution has its own PC and stack. However, because it executes within the address space of a process, it shares with all other threads of execution, including the main thread, data, code, and other segments of the program previously discussed.

You may use the API for one of the several thread libraries to create and manage threads. The threads created by user programs using these libraries, which are not known to the kernel and are managed solely by the user-level thread libraries, are known as *user-level threads*. The code and data structures for these libraries are maintained in the user space. The kernel does not know about and provides no support for such threads. It means that the kernel only manages processes and the relevant thread library manages threads, including their scheduling.

Some programming languages such as Java provide direct support for user-level threads. Such threads are created and managed by the program itself through the use of the API of a thread library.

If a thread library is implemented in the kernel, the operating system maintains the code and data structures for the library. Calls to functions in these libraries for creating and managing threads result in system calls. The threads created by the user programs using such libraries are known as *kernel-level threads*. The kernel handles both processes and threads, including their scheduling.

Several libraries are available for the implementation of user- and kernel-level threads. The POSIX standard *Pthreads* may be provided as user- or kernel-level libraries.

Kernels of almost all modern general-purpose operating systems are multithreaded. This means that they can serve multiple system calls simultaneously. This is done by running the code corresponding to a system call as a thread instead of a function call. Thus, for example, multiple invocations of the `read()` system call from threads within the same process or different processes may be served simultaneously. The operating system kernels that offer such multithreaded services are known as *multithreaded kernels*.

We use library calls to create, terminate, and manage threads. If the threads library were for kernel threads, each library call would eventually invoke a system call. The kernel would have the knowledge of threads and would be responsible for managing them, including their scheduling. Otherwise, the user-level library manages threads.

In Figure 16.5, we show a multithreaded process with four threads: the main thread, two threads of the function `f1()`, and one thread for the function `f3()`. A function that is designed to become a thread must terminate with a library function that terminates the thread, such as `pthread_exit(...)`. We use the names of the Pthreads library functions for the creation and termination of threads. Note that you need to include the header file **/usr/include/pthread.h** in your program and link the Pthreads library with the executable code using the –pthread option with the gcc compiler, as in

```
gcc –w –pthread prog.c –o prog
```

The typical graphical representations of single-threaded and multithreaded processes are shown in Figure 16.6. Note that after tid3 has started running and tid1 has not finished, four threads run simultaneously: main, tid1, tid2, and tid3.

16.2.6 Commonalities and Differences between Processes and Threads

Processes and threads have several things in common. For example, each has its own ID, stack, and PC.

There are several differences between processes and threads. For example, whereas processes operate within their own address spaces, threads within a process operate within the address space of the process. Similarly, processes have their own data and text (code) areas but threads share the text and data areas in the process.

16.2.7 Data Sharing among Threads and the Critical Section Problem

Data sharing among threads is a mixed blessing. Although this saves memory space by preventing the duplication of data, it does require that threads access data on a mutually exclusive basis by locking data before accessing it and unlocking it afterwards. If this were not done, the results produced by the process

```
Code
#include <stdio.h>
#incldue <pthreads.h>

int main(int argc, char* argv[])
{
          int i;
          char c;

          printf("Hello, world!\n");
          ...
          pthread_create(tid1, ..., f1, ...);
          ...
          pthread_create(tid2, ..., f3, ...);
          ...
          pthread_create(tid3, ..., f1, ...);
          ...
          exit(0);
}
void f1(int i, int* ip)
{
          ...
          if (...)
          then
                    f2(...);
          else
                    ...
}
void f2(char c)
{
          ...
}
void f3(char* s)
{
          ...
}
```

FIGURE 16.5   Multithreaded process with four threads including the main thread.

would be dependent on the sequence in which instructions within threads access shared data, and the results may or may not be correct. This is called the *race condition* and the final result produced by the process is dependent on the order in which instructions in different threads access the shared data.

The piece of code in a thread that accesses shared data is known as the *critical section*. Thus, multiple critical sections that access shared data must be executed mutually exclusively, one after the other. In other words, when there are simultaneous requests for the execution of multiple critical sections, the execution of these critical sections must be serialized. How such mutual exclusion is achieved is known as the *critical section problem*. Several solutions are available in the literature for solving this problem. However, further discussion on this topic is beyond the scope of this book. If you are interested in knowing more about this subject, you may read a textbook on operating system principles or operating system concepts.

Exercise 16.3

Browse the Web and write down the names of three thread libraries, in addition to those we have discussed in Section 16.2.5, that may be used for user-level threads.

Exercise 16.4

The Bakery Algorithm is an elegant solution for the N-process critical section problem. Browse the Web or see a book on operating system principles and concepts to find out the name of the author of this algorithm. Who is the author?

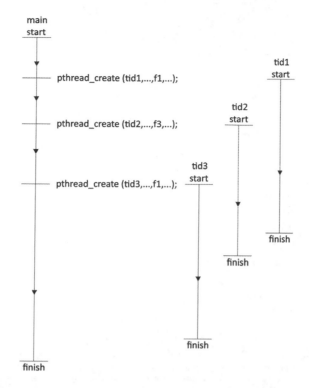

FIGURE 16.6 Typical graphical representations of a single-threaded process (left) and a multithreaded process (right).

16.3 Process Management Concepts

Process management entails several things, from process creation to termination and everything in between, including suspending a process, having a process wait for an event, such as the termination of a child process, sending a signal (software interrupt) to a process, handling signals, setting up an alarm in a process, and duplicating a file descriptor in the PPFDT. Table 16.1 shows a few system calls for process management.

TABLE 16.1

The Linux System Calls for Process Management

fork()	Create a clone (i.e., a child) of the calling process
execve()	Overwrite the main memory image of the caller process
wait()	Suspend the calling process and wait for a child process to terminate
waitpid()	Wait for the termination of the process with the given PID
exit()	Terminate the caller process and return status to the parent of the process
getpid()	Get the PID of the caller process
getppid()	Get the PPID of the parent of the caller process
getuid()	Get the UID of the owner of the caller process
getgid()	Get the GID of the owner of the caller process
signal()	Handle a signal by ignoring it, taking the default (kernel-defined) action, or taking the programmer-defined action specified for the relevant signal
kill()	Send a signal of a particular type to a process; normally used to terminate a process
alarm()	Set an alarm signal for the given number of seconds
dup(), dup2()	Duplicate a file descriptor in the PPFDT

Exercise 16.5

Browse the man page for the `execve()` system call and list the names of all the variants of this call, both system calls and library calls.

Exercise 16.6

Browse the man page for the `wait()` system call and list the names of all the variants of this call, both system calls and library calls.

16.3.1 Getting the PID and the PPID

You can use the `getpid()` and `getppid()` calls to display the ID of a process (PID) and the ID of its parent (PPID). Here are the brief descriptions of these calls.

```
#include <sys/types.h>
#include <unistd.h>
pid_t getpid();
```

Success: The ID of the process (PID)
Failure: −1 and kernel variable `errno` set to indicate the type of error

```
#include <sys/types.h>
#include <unistd.h>
pid_t getppid();
```

Success: The ID of the parent process (PPID)
Failure: −1 and kernel variable `errno` set to indicate the type of error

The **pids.c** program in the following session shows the use of these calls. The compilation and running of this program displays the PID of the process (14519) and the PID of its parent (14483). Note that the parent of the process is the login shell (`bash`) with PID 14483, as shown in the output of the `ps` command.

```
$ cat pids.c
void main(void)
{
        printf("Child's PID = %d\n", getpid());
        printf("Parent's PID = %d\n", getppid());
}
$ gcc -w pids.c -o pids
$ pids
Child's PID = 14519
Parent's PID = 14483
$ ps
  PID TTY          TIME CMD
14483 pts/0    00:00:00 bash
14520 pts/0    00:00:00 ps
$
```

Exercise 16.7

Compile and run the **pids.c** program to make sure it works on your system.

Exercise 16.8

Browse the man page for the `getpgrp()` and `getpgid()` system calls. What is the purpose of each of these calls?

16.3.2 Creating a Clone of a Process

The only way to create a process in Linux is to have a process create a clone of itself—that is, a replica of the main memory image of the process and most of the associated kernel data structures (discussed later in this section). A process can use the `fork()` system call for this purpose. The clone is called the *child process* and the caller of fork() is known as the *parent process*. Both processes have their unique PIDs. Here is a brief description of the fork() system call.

```
#include <sys/types.h>
#include <unistd.h>
pid_t fork(void);
```

Success: 0 to the child process and the PID of child to the parent process

Failure: −1 and kernel variable errno set to indicate the type of error

As you can see, the fork() call returns a value of type pid_t. A combined reading of **/usr/include/x86_64-linux-gnu/bits/types.h** and **/usr/include/x86_64-linux-gnu/bits/typesizes.h** reveals that pid_t is a signed 32-bit integer. This means that you would use fork() in an assignment statement, as in

```
pid_t   pid;
...
pid = fork();
...
```

As soon as fork() has completed the creation of the child process and before it returns, the parent and child start running concurrently as independent processes. Both execute the statement following fork(), which is the assignment statement meant to save the return value of the fork() call. The fork() call returns 0 to the child process and the PID of the child process to the parent process. It is not easy to comprehend how a system call could return two values because a function may return only one value. Note that it is possible for fork() to return two values because as soon as the child process has been created, fork() has to complete its execution in both processes by returning a value in each of the two processes.

The child process inherits several attributes and characteristics of the parent process. Here is a partial list:

- A copy of the parent's PPFDT
- The current working directory
- The environment
- The root directory
- The set-UID (SUID) and set-group-ID (SGID) status
- The signal settings
- The time left before an alarm goes off
- The value of the file creation mask, umask

Although the child process is an exact copy of the memory image of the parent process, the two differ from each other in many ways:

- Both processes have their own PIDs.
- Both have different parent processes.
- Both have their own copies of the PPFDT.
- The resource utilization for the child process is set to 0.
- The child process has only one thread of execution, and if the parent process is multithreaded, the other threads do not release the resources held by them.

Since the parent and child processes have copies of the same PPFDT, both can do I/O with files that the parent process had opened before creating the child process, and this I/O is visible in both processes. In other words, the change is the position of the file pointer due to read(), write(), and lseek() calls by

TABLE 16.2

Reasons for the fork() System Call to Fail

Reason for Failure	Value of errno
The limit on the maximum number of processes that may run on the system simultaneously would be exceeded	EAGAIN
The limit on the maximum number of processes that may run under a user would be exceeded	EAGAIN
The resource limit set for the process has been reached	EAGAIN
Insufficient swap space for the new process	ENOMEM

either process is visible in the other process. So, if either process reads N bytes from an already open file, the file pointer is incremented by N and the next read or write operation by parent or child is performed at the new position of the file pointer.

The fork() call may fail for the reasons listed in Table 16.2.

The vfork() system call is a lighter version of fork(). It creates a child process exactly like fork() but does not copy the address space of the parent; it copies the address space on demand—that is, when the child's memory image is needed. The vfork() system call is used instead of fork(), particularly if the child process uses the execve() system call (or a variant) or the execl() library call (or a variant) right after its creation to overwrite itself with another executable code. This is done in applications like a shell program that executes external commands by using and exec() call (execl(), execve(), fexecve(), etc.) immediately after fork(). The exec() call is covered later in the chapter.

Exercise 16.9

If you browse the man page for the fork() system call, you will see the names of the vfork() and rfork() calls. What is the purpose of each of these calls?

16.3.3 Reporting Status to the Parent Process

A Linux process reports its termination status to its parent via the _exit() system call or the exit() library call. Since most programmers use the exit() call to terminate a process, we will discuss it throughout the book. A process passes its termination status to its parent through the only parameter of the exit() call, usually, 0 or 1. The value 0 is used to indicate successful completion of the process and 1 means its unsuccessful termination.

The parent process accepts this status through the use of a system call in the wait() class (see Section 16.3.4). In reality, the status of the terminating process is saved in the proc structure of the process. All the data structures and resources allocated to the process are deallocated, except its proc structure. The proc structure is released back to the kernel after the parent has read the exit status of the child process. If a process does not use the exit() call to terminate itself, the kernel automatically sends the termination status of the process to its parent process.

Note that a child process and its parent rendezvous through the exit() and wait() system calls, with the parent receiving the exit status of the child and the reason for its termination through this meeting. This is the simplest form of interprocess communication available in Linux. By associating different connotations to the exit() parameter, the child can communicate meaningful messages to its parent. We discuss the issue of communication between Linux processes in detail in Chapter W20.

16.3.4 Collecting the Status of a Child Process

A process can use the wait() system call to wait for the termination of a child process and to know the reason of its termination. The process can use the waitpid() system call to wait for the termination of a particular child process. Here are brief descriptions of these system calls:

```
#include <sys/types.h>
#include <sys/wait.h>
pid_t wait(int *status);
pid_t waitpid(pid_t wpid, int *status, int options);
```

Success: PID of the child process
Failure: −1 and kernel variable errno set to indicate the type of error

The wait() call may fail for several reasons. Some of these reasons are listed in Table 16.3.

The programmer can decipher the termination status of a child process and additional information related to its termination, such as whether the process was terminated due to a signal, by examining the value of the status parameter. As shown in Figure 16.7, the exit status and the reason for the process to terminate are stored in status. The lower byte contains the reason for the process termination and the upper byte contains the exit status. If a process terminated on receiving a signal, the signal number is stored in the lower byte of the status variable. Thus, for example, if a process terminated due to <Ctrl+C> (keyboard interrupt), the lower byte contains SIGINT (i.e., value 2).

With the passage of time, Linux designers have added a number of variants of the wait() call. The wait() system call is the oldest, and waitpid() is the most restricted of these calls. The newest and broadest interface is the wait6() system call. Detailed discussions on wait3(), wait4(), and wait6() are beyond the scope of this book. Here are brief descriptions of the wait3() and wait4() calls:

```
#include <sys/types.h>
#include <sys/wait.h>
#include <sys/time.h>
#include <sys/resource.h>
pid_t wait3(int *status, int options, struct rusage *rusage);
pid_t wait4(pid_t wpid, int *status, int options, struct rusage *rusage);
```

Success: PID of the child process
Failure: −1 and kernel variable errno set to indicate the type of error

The wait4() system call instructs the process to wait for specific children processes and to retrieve the statistics about their resource usage. As shown in Table 16.4, the wpid parameter in the wait4() and waitpid() calls determines what child (or children) the parent process waits for.

The options argument in wait3() and wait4() contains a bitwise OR of any of the half a dozen options. You can see these options and their meaning by viewing the man pages for these calls and by browsing through the header file **/usr/include/linux/wait.h** or **/usr/src/linux-headers-4.4.0-21/**

TABLE 16.3

Some of the Reasons for the wait() System Call to Fail

Reason for Failure	Value of errno
The caller process has no children to wait for	ECHILD
The statuses from the terminated children are not available because the caller process is ignoring SIGCHLD and the system is discarding such statuses	ECHILD
The call was interrupted by a signal	EINTR
An invalid option was specified in the system call	EINVAL

Low byte		High byte	
System's understanding of the reason for child's termination: zero for normal and nonzero for abnormal		Argument of the exit ( ) call	
0	7	8	15

FIGURE 16.7   The meaning of the values in the two bytes of the value in the status parameter.

TABLE 16.4

Values of the wpid Parameter for the wait4() and waitpid() System Calls and
Children Processes Waited for

wpid	The Caller Waits for
−1	Any child process
0	Any child process in the process group of the caller
> 0	The child with PID wpid
< −1	Any child process whose process GID equals the absolute value of wpid

include/uapi/linux/wait.h. The man pages also describe several macros that you may use to determine the status of a terminated child process. For example, you can find out if a child process terminated
due to the arrival of a signal and, if so, the number of the signal that caused the process to terminate.
These macros are defined in the files **/usr/include/x86_64-linux-gnu/sys/wait.h** and **/usr/include/
x86_64-linux-gnu/bits/waitstatus.h**.

When the WNOHANG option is specified and no processes are there to report the status, wait4() and
wait3() do not block, and return 0 as the PID. Table 16.5 shows a few equivalences of the wait4() call
and other calls in the wait() class.

The **fork.c** program in the following session uses the fork() system call to create a child process.
The child process terminates after displaying its PID and its parent PID. The parent process waits for the
child process and terminates after displaying its PID and the child's PID. Note that the PID displayed by
the parent is, as expected, the same as the parent PID displayed by the child process.

```
$ cat fork.c
#include <unistd.h>

extern int errno;

int main(void)
{
        int pid, status;

        pid = fork();
        if (pid == -1) {
            perror("Fork failed.");
            exit(1);
        }
        /* Child process */
        if (pid == 0) {
            printf("\nCHILD: Child here with PID = %d.\n", getpid());
            printf("CHILD: My Mom has PID = %d.\n", getppid());
            exit(0);
        }
        /* Parent process */
        wait(&status);
        printf("\nPARENT: Mom here with PID = %d.\n", getpid());
        printf("PARENT: Well done, child!\n\n");
        exit(0);
}
$ gcc -w fork.c -o fork
```

TABLE 16.5

The Equivalences of the wait4() System Calls

wait4(..., ..., ..., 0);	waitpid(..., ..., ...);
wait4(-1, ..., ..., ...);	wait3(..., ..., ...);
wait4(-1, ..., 0, 0);	wait(...);

```
$ fork

CHILD: Child here with PID = 15736.
CHILD: My Mom has PID = 15735.

PARENT: Mom here with PID = 15735.
PARENT: Well done, child!

$
```

16.3.5 Overwriting a Process Image

A process may overwrite itself with another executable image. The execve() and fexecve() system calls may be used to do so. These calls differ from each other in the manner in which parameters are passed to the caller process. Here are brief descriptions of these calls:

```
#include <unistd.h>
int execve(const char *path, char *const argv[], char *const envp[]);
int fexecve(int fd, char *const argv[], char *const envp[]);
```

Success: Control does not return to the caller process because it has been overwritten with a new executable

Failure: −1 and kernel variable errno set to indicate the type of error

In the execve() call, path is the pathname of the ordinary file that contains the executable image of the new process. In the case of the fexecve() system call, fd is the file descriptor of the file that contains the image of the new process. The file may contain the binary executable code or a script to be executed by an interpreter, such as a shell or Perl script. A script file begins with a line in the following format:

```
#! interpreter [arg]
```

The execve() system call actually overlays the caller process with interpreter and passes the script file to it as an argument. If the arg parameter is not specified on the first line, the script file is specified as the first argument to interpreter; otherwise, it is specified as the second argument. The argv argument is a pointer to a null-terminated array of pointers to null-terminated strings that become the command line arguments of the script or the binary executable. The envp argument is structurally similar to argv and contains the values of different environment variables.

If the file that is to overlay the caller process is a binary executable, it is executed just like a main program in C with command line arguments is executed, as in the following session:

```
int main(int argc, char **argv, char **envp)
```

where argc is the number of command line arguments including the program name, and argv and envp are as described in the previous paragraph.

File descriptors open in the caller process remain open in the new process, except for those descriptors for which the close-on-exec flag is set. Signals set to be ignored in the caller process remain ignored in the new process and signals set to be caught in the caller process are set to default action in the new process. The new process also inherits the following identities and attributes of the caller process: PID, PPID, PGID, working directory, root directory, control terminal, access groups, resource usages, timers, resource limits, signal mask, and umask.

The execve() and fexecve() system calls may fail for different reasons, some of which are listed in Table 16.6.

The system call interface for overwriting a process with an executable image specified by the execve() and fexecve() calls is quite cumbersome. Linux provides several library functions with easier interfaces for performing the same task. The simplest of these is the execl() library call, which is the front

TABLE 16.6

Some of the Reasons for the execve() or fexecve() System Call to Fail

Reason for Failure	Value of errno
A component in the path argument, except the last component, is not a directory	ENOTDIR
The ordinary file specified as the last component in the path argument is not found	ENOENT
Search permission for a directory in path has not been given	EACCES
The last component in path is not an ordinary file	
The last component in path does not have the execute permission on	
The ordinary file does not have a valid magic number in its header	ENOEXEC
The process requires more virtual memory than the system-imposed limit	ENOMEM
One or more of the three arguments of the call point to an illegal address	EFAULT
An error occurred while reading the file	EIO
The fd argument is not a valid file descriptor	EBADF

end of the execve() system call. We use this call in the examples that we discuss in this book. Here is a brief description of the call:

```
#include <unistd.h>
int execl(const char *path, const char *arg, ... /*, (char *)0 */);
```

Success: Control does not return to the caller process because it has been overwritten with a new executable
Failure: −1 and kernel variable errno set to indicate the type of error

The execl() call may fail for the same reasons that the execve() system call fails.

The **fork_exec_1.c** program shown in the following session creates a child process that overwrites itself with the executable in the **/bin/date** file to display today's date and the current time. Thus, the child becomes the date process. The parent process displays the PID of the child process, waits for the child process to terminate, and displays the PID of the terminated child process. The sample run of the program shows that, as expected, the PID displayed by the parent for the child process and that of the terminated child process are the same.

```
$ cat fork_exec_1.c
#include <unistd.h>

extern int errno;

int main(void)
{
        int pid, status;

        pid = fork();
        if (pid == -1) {
            perror("Fork failed");
            exit(1);
        }
        /* Child process */
        if (pid == 0) {
            execl("/bin/date", "date", (char *) NULL);
            perror("execl failed");
            exit(1);
        }
        /* Parent process */
        printf("Child PID = %d.\n", (int) pid);
        pid = wait(&status);
        printf("Child with PID %d has terminated.\n", (int) pid);
```

```
        exit(0);
}
$ gcc -w fork_exec_1.c -o forke1
$ forke1
Child PID = 15754.
Sun Oct 22 09:31:34 PKT 2017
Child with PID 15754 has terminated.
$
```

The **fork_exec_2.c** program creates a child process that creates a directory, called **dir1**, in the current directory using the execl() call. The parent process creates a file called **foo** in **dir1**, opens this file, writes the Hello, world! message into **foo**, resets the file pointer to the beginning of the file, reads back the Hello, world! message from **foo**, displays this message on standard output, removes **foo** from **dir1**, and removes the now empty **dir1**. The compilation and a sample run of the program are shown in the following session.

```
$ cat fork_exec_2.c
#include <fcntl.h>
#include <unistd.h>
#include <errno.h>

#define SIZE 512
#define MODE 0644
#define Message "Hello, world!\n"

extern int errno;

int main(void)
{
        int n, nr, nw, fd, pid, status;
        char buf[SIZE];

        pid = fork();
        if (pid == -1) {
            perror("Fork failed");
            exit(1);
        }
        /* Child process */
        if (pid == 0) {
            execl("/bin/mkdir", "mkdir", "dir1", (char *) NULL);
            perror("execve failed.\n");
            exit(1);
        }
        /* Parent process */
        while (((pid = wait(&status)) -- -1) && errno -= EINTR)
            ;
        /* Write to foo */
        if (pid == -1) {
            perror("Wait failed");
            exit(1);
        }
        /* Open or create dir1/foo file */
        if ((fd = open("dir1/foo", O_RDWR|O_CREAT, MODE)) == -1) {
            perror("Open failed");
            exit(1);
        }
        /* Write to foo */
        if ((nw = write(fd, Message, strlen(Message))) == -1 ) {
            perror("Write failed");
            exit(1);
        }
```

```
        /* Rewind file pointer */
        n = lseek(fd, OL, SEEK_SET);

        /* Read back from foo */
        if ((nr = read(fd, buf, nw)) == -1) {
            perror("Read failed");
            exit(1);
        }
        /* Throw on standard output */
        write(1, buf, nr);
        if (close(fd) == -1) {
            perror("File closing");
            exit(1);
        }
        /* Remove dir1/foo to make dir1 an emptry directory */
        unlink("dir1/foo");
        /* Remove the now empty directory dir1 */
        rmdir("dir1");
        exit(0);
}
$ gcc -w fork_exec_2.c -o forke2
$ forke2
Hello, world!
$
```

Exercise 16.10

Compile and run the preceding program on your system. Does it produce the expected output?

16.3.6 Creating a Zombie Process

A process whose parent is not waiting (i.e., is sleeping or has finished execution) when it terminates cannot report its termination status to its parent and is called a *zombie process*. It has completed its work, but some of the resources allocated to it may not be returned to the system. Eventually, systemd, the grandparent of all user processes, takes over the parenthood of the zombie process, receives its status, and releases the remaining system resources.

The following piece of code spawns a child process and puts the parent process to sleep for 10 s. The child process displays its PID and exits. Because the parent process is not waiting when the child terminates, the child process becomes a zombie. In the sample run, we suspend the parent process by pressing <Ctrl+Z> and use the ps command to display the status as zombie. Note that the child process is marked as <defunct> and its status is Z (for zombie). We bring the zombie process into the foreground by using the fg program.

```
$ cat create_zombie.c
int main(void)
{
        int pid;

        pid = fork();
        if (pid == -1) {
            perror("Fork failed");
            exit(1);
        }
        /* Child process */
        if (pid == 0) {
            printf("Child's PID  = %d\n", getpid());
            exit(0);
        }
        /* Parent process */
        printf("Parent's PID = %d\n", getpid());
```

```
                sleep(10);
                /* Parent process does not wait for the child */
                /* process and child becomes a zombie process */
                exit(0);
        }
$ gcc -w create_zombie.c -o create_zombie
$ create_zombie
Parent's PID = 15831
Child's PID  = 15832
^Z
[1]+  Stopped                    create_zombie
$ ps -o pid,tt,stat,cmd
  PID TT          STAT CMD
15691 pts/2       Ss   -bash
15831 pts/2       T    create_zombie
15832 pts/2       Z    [create_zombie] <defunct>
15836 pts/2       R+   ps -o pid,tt,stat,cmd
$ fg
create_zombie
$
```

You can also use the ps -al command to display the status of the zombie process.

When a child of the systemd process terminates, systemd calls one of the wait() system calls to collect the exit status of the child. Thus, none of the children of systemd ever becomes a zombie. The child may be a process that was directly created by systemd or was inherited by systemd because the parent of the process had died (or was not waiting) when the process terminated.

Exercise 16.11

Replicate the preceding session on your system. Does the **create_zombie.c** program work as expected?

16.3.7 Terminating a Process

A process can use the kill() system call to terminate a process for which it has such permission. A process owned by the superuser may terminate any process. We discuss this system call in detail in Section 16.5.4.

16.4 Processes and the File Descriptor Table

We now discuss the importance of PPFDT in relation to file sharing between processes, I/O redirection, and communication between processes using different channels including pipes, FIFOs, and sockets. We discuss the concept of *file sharing* in Section 16.4.1 and *I/O redirection* in Section 16.4.2. Chapter W20 describes Linux interprocess communication in detail.

16.4.1 File Sharing between Processes

As discussed earlier, a child process gets a copy of the PPFDT of its parent. It means that if a process opens a file before creating a child process, the parent and child processes have access to the file descriptor of the open file. Consequently, the I/O performed on the open file by either process is visible to the other process.

The program **file_sharing.c** illustrates how parent and child processes share files that are open before the child process is created. We demonstrate the concept with only one file, but it extends to multiple files. The program takes filename and nbytes (the I/O size in bytes) as command line arguments and opens for reading and writing the file specified as the command line argument. It then creates a child process that inherits the PPFDT and, therefore, the file descriptor of the opened file. Both the parent and child processes do I/O with the same file. The program displays the data read by both processes and the position of the file

pointer after I/O in both processes, verifying that both processes do I/O with the same file. The nbytes
argument may vary from 0 to 512 because the size of the read/write buffer has been set to 512 bytes. We use
the sleep(1) call in the parent process to make sure that the child process performs I/O before the parent
process. It is done to make sure that the values of the file pointer are displayed correctly in both processes. A
few sample runs follow the listing of the program to verify that the file-sharing concept works as discussed.
Note that **foo** is the name of an existing file. We created this file using the man bash > foo command.

```
$ cat file_sharing.c
/* file_sharing.c: fileshare filename nbytes */

#include <fcntl.h>
#include <unistd.h>
#include <sys/wait.h>

#define SIZE 512

int main(int argc, char *argv[])
{
        int fd, n, nr, nbytes, pid, status;
        char buf[SIZE];

        /* Open file */
        if (argc != 3) {
            printf("%s filename nbytes\n", argv[1]);
            exit(1);
        }
        nbytes = atoi(argv[2]);
        if ((fd = open(argv[1], O_RDWR)) == -1) {
            perror("File opening");
            exit(1);
        }
        /* Create a child */
        pid = fork();
        /* Child process */
        if (pid == 0) {
            /* Read and display on stdout */
            if ((nr = read(fd, buf, nbytes)) == -1) {
                perror("Read failed");
                exit(1);
            }
            write(1, buf, nr); /* Throw on standard output */
            n = lseek(fd, 0L, SEEK_CUR);
            printf("\n\nCHILD: position of the file pointer after I/O is %d\n\n", n);
            close(fd);
            exit(0);
        }
        /* Parent process */
        sleep(1); /* Allow the child process to do I/O first. */
        /* Read and display on stdout */
        if ((nr = read(fd, buf, nbytes)) == -1) {
            perror("Read failed");
            exit(1);
        }
        write(1, buf, nr); /* Throw on standard output */
        n = lseek(fd, 0L, SEEK_CUR);
        printf("\n\nPARENT: position of the file pointer after I/O is %d\n\n", n);
        close(fd);
        while (wait3(&status, WNOHANG, 0) >= 0)
            ;
        exit(0);
}
```

```
$ gcc -w file_sharing.c -o fileshare
$ fileshare foo 32
BASH(1)                         Gene

CHILD: position of the file pointer after I/O is 32

ral Commands Manual

PARENT: position of the file pointer after I/O is 64

$ fileshare foo 64
BASH(1)                         General Commands Manual

CHILD: position of the file pointer after I/O is 64

        BASH(1)

NAME
        bash - GNU Bourne-Again SHell

SYNOP

PARENT: position of the file pointer after I/O is 128

$ fileshare foo 256
BASH(1)                         General Commands Manual                    BASH(1)

NAME
        bash - GNU Bourne-Again SHell

SYNOPSIS
        bash [options] [command_string | file]

COPYRIGHT
        Bash is Copyright (C) 1989-2013 by the Free Software Foundat

CHILD: position of the file pointer after I/O is 256

ion, Inc.

DESCRIPTION
        Bash  is  an  sh-compatible  command language interpreter that executes
        commands read from the standard input or from a file.  Bash also incor-
        porates useful features from the Korn and C shells (ksh and csh).


PARENT: position of the file pointer after I/O is 512

$
```

Exercise 16.12

Replicate the preceding session on your system. Does the **file_sharing.c** program work as expected?

Exercise 16.13

Modify the **file_sharing.c** program so that it takes the number of bytes to be read as the first command line argument and the filename as the second argument. Compile and run the program with different files and bytes-to-read as command line arguments.

TABLE 16.7

Reasons for the dup() and dup2() System Calls to Fail

Reason for Failure	Value of `errno`
The descriptor specified in `olddes` is not a valid active descriptor (i.e., has not been allocated using a system call such as `open()`) or the descriptor specified in `newdes` is greater than the size of PPFDT or is a negative number	EBADF
The process has already used the maximum number of descriptors (i.e., the PPFDT is being used to full capacity)	EMFILE

16.4.2 Duplicating File Descriptor

You can use the dup() and dup2() system calls to duplicate a file descriptor. A Linux shell uses these system calls to implement I/O redirection. Here are brief descriptions of these calls:

```
#include <unistd.h>
int dup(int olddes);
int dup2(int olddes, int newdes);
```

Success: A newly allocated file descriptor
Failure: −1 and kernel variable `errno` set to indicate the type of error

The dup() call duplicates the existing file descriptor `olddes` in the PPFDT and returns the duplicated file descriptor. As discussed earlier, the new file descriptor is the smallest unused descriptor in the PPFDT. The old and new file descriptors point to the same entry in the SFT. Thus, changes in the read/write file pointer, file contents, and file attributes are visible through both descriptors. This is true regardless of the file type that the descriptor points to: ordinary file, pipe, socket, or FIFO.

The dup2() call is similar to dup(), except that the newly allocated descriptor is explicitly specified as the second argument. If `olddes` is a valid descriptor and is equal to `newdes`, the call is successful and performs a null operation; that is, it does not do anything. If `oldfd` is a valid descriptor and is not equal to `newdes`, the dup2() call first deallocates `newdes` and then performs the duplicate operation. If `olddes` is not a valid descriptor, the call fails and no duplication operation is performed.

These calls may fail for the reasons specified in Table 16.7.

These calls are normally used to implement input, output, and error redirection for a process (see Chapter 9). You can do input redirection by closing standard input for the process, opening the file to which input has to be redirected, and attaching standard input to the file by duplicating the file descriptor to the PPFDT slot for standard input. Similarly, you can do output and error redirection for a process.

The following program shows sample uses of the dup() system call. The program takes a file as a command line argument, opens the file for writing, and saves the file descriptor for the file in `fd`. If the file does not exist, the program creates it and sets its access permissions to read and write. The program then closes the file descriptor for standard output by using the close(1) system call. After the standard file descriptor has been closed, slot number 1 of the PPFDT becomes free and the dup(fd) system call copies the entry for `fd` in the PPFDT into slot 1. After this has been done, anything directed to standard output is sent to the opened file. Thus, the output of the printf() call is redirected to the file passed as the command line argument. In the following session, we pass **foo** as the command line argument to the program. The output of the cat foo command verifies that the standard output has been redirected to **foo**.

```
$ cat dup.c
    #include <fcntl.h>
    #include <sys/stat.h>

    int main(int argc, char *argv[])
    {
        int fd;
```

```
         /* Open file */
         if (argc == 1) {
                 printf("No file specified as command line argument.\n");
                 exit(1);
         }
         if ((fd = open(argv[1], O_WRONLY|O_CREAT |O_TRUNC, S_IREAD|S_IWRITE)) == -1) {
                 perror("File opening");
                 exit(1);
         }
         /* Close standard output */
         close(1);
         /* Duplicate fd into file descriptor 1, i.e., stdout */
         if (dup(fd) == -1) {
                 perror("Duplicating file descriptor");
                 exit(1);
         }
         /* Close fd in order to release the extra slot in the PPFDT */
         if (close(fd) == -1) {
                 perror("File closing");
                 exit(1);
         }
         /* Stdout redirected to the file passed as command line argument */
         printf("Hello, world!\n");
         exit(0);
    }
$ gcc -w dup.c -o dup
$ dup foo
$ cat foo
Hello, world!
$
```

In the following session, we show the use of dup2() in the **dup2.c** program to perform the same task as performed by the **dup.c** program—that is, redirect standard output to the file passed to the program as a command line argument.

```
$ cat dup2.c
    #include <fcntl.h>
    #include <sys/stat.h>

    int main(int argc, char *argv[])
    {
         int fd;

         /* Open file */
         if (argc == 1) {
                 printf("No file specified as command line argument.\n");
                 exit(1);
         }
         if ((fd = open(argv[1], O_WRONLY|O_CREAT|O_TRUNC, S_IREAD|S_IWRITE)) == -1) {
                 perror("File opening");
                 exit(1);
         }
         /* Duplicate fd into file descriptor 1, i.e., stdout */
         if (dup2(fd, 1) == -1) {
                 perror("Duplicating file descriptor");
                 exit(1);
         }
         /* Close fd in order to release the extra slot in the PPFDT */
         if (close(fd) == -1) {
                 perror("File closing");
                 exit(1);
         }
    }
```

```
        /* Stdout redirected to the file passed as command line argument */
        printf("Hello, world!\n");
        exit(0);
    }
```

```
$ gcc -w dup2.c -o dup2
$ dup2 foo
$ cat foo
Hello, world!
$
```

You can redirect standard input (or standard error) in the same manner as discussed. All three standard files may be redirected to one file or multiple files, one each for standard input, standard output, and standard error.

> **Exercise 16.14**
>
> Replicate the preceding sessions on your system. Do the **dup.c** and **dup2.c** programs work as expected?

16.5 Getting the Attention of a Process: Linux Signals

Similar to how a peripheral device may use an interrupt to get the attention of the CPU and be served, you may use a signal to get the attention of a process. We discussed this topic in detail in Chapter 12 to explain signal/interrupt handling in shell scripts. Here, we discuss signal handling in a C program.

16.5.1 What Is a Signal?

As stated earlier, a signal in Linux vocabulary is an event that interrupts the execution of a process. We discussed the events that cause signals in Section 12.4 while discussing advanced Bourne shell programming. Some of the events that cause signals are listed in Table 12.1. We reproduce that table here as Table 16.8.

You can view the complete list of events that cause signals by viewing the man pages for the kill command or kill() system call. You can also use the kill -l command to display the list of signals.

TABLE 16.8

Commonly Used Signals, Their Number, and Their Purpose

Signal	Number	Purpose
SIGHUP (hang up)	1	Informs the process when the user who ran the process logs out and the process terminates
SIGINT (keyboard interrupt)	2	Informs the process when the user presses <Ctrl+C> and the process terminates
SIGQUIT (quit signal)	3	Informs the process when the user presses <Ctrl+|> or <Ctrl+\> and the process terminates
SIGKILL (sure kill)	9	Terminates the process with no further processing (e.g., exception handling) when the user sends this signal to it with the kill -9 command
SIGSEGV (segmentation violation)	11	Terminates the process upon memory fault when a process tries to access memory space that does not belong to it
SIGTERM (software termination)	15	Terminates the process when the kill command is used without any signal number
SIGTSTP (suspend/ stop signal)	18	Suspends the process; usually <Ctrl+Z>
SIGCHLD (child finishes execution)	20	Informs the process of termination of one of its children

A similar list is also found in the **/usr/include/x86_64-linux-gnu/asm/signal.h**, **/usr/include/asm-generic/signal.h**, and **/usr/src/linux-headers-4.4.0-21/include/uapi/asm-generic/signal.h** files.

You can send a signal to a process by using the kill command or the kill() system call. Both take two arguments, a signal number and the PID of the process to receive the signal. For example, the kill -16 12345 command sends signal number 16 to the process with PID 12345.

16.5.2 Intercepting Signals

A process may use the sigaction() system call or the signal() library call to intercept a signal and take one of the following three possible actions:

1. Ignore the signal
2. Take the default action as defined by the kernel
3. Take the programmer-defined action

The default actions are listed in the **/usr/src/linux-headers-4.4.0-21/include/linux/signal.h** file.

The interface for the sigaction() system call is rather complex. The library call signal() has a much simpler interface. For this reason, programmers normally use signal() for handling signals. Here is a brief description of this call:

```
#include <signal.h>
sign_t signal(int sig, sig_t func);
```

Success: 0
Failure: −1 and kernel variable errno set to indicate the type of error

The func argument is SIG_IGN for ignoring a signal and SIG_DFL for the default, kernel-defined action. A call to the signal() function may fail for two reasons, as shown in Table 16.9.

16.5.3 Setting Up an Alarm

The library call alarm() sends the SIGALRM signal to the calling process after the specified number of seconds. Thus, the call alarm(10) sends the SIGALRM signal to the caller process after 10 s. Here is a brief description of the call:

```
#include <unistd.h>
unsigned int alarm(unsigned int seconds);
```

Success: 0 if no alarm is currently set and the number of seconds left on the timer of the previous alarm()
Failure: −1 and kernel variable errno set to indicate the type of error

The maximum number of seconds allowed is 100,000,000. If an alarm has already been set but has not gone off (i.e., the signal has not been sent to the process), another call to alarm supersedes the previous call. The alarm(0) call cancels the current alarm and SIGALRM is never delivered to the calling process.

TABLE 16.9

Reasons for the Library Call signal() to Fail

Reason for Failure	Value of errno
The signal specified in sig is not a valid signal number	EINVAL
The process tried to ignore or specify a handler for the SIGKILL or SIGSTOP signal	EINVAL

The **signals.c** program shown in the following session ignores the SIGHUP signal and takes programmer-defined actions for SIGINT (signal number 2, <Ctrl+C>) and SIGALARM (signal number 14). For all other signals, the process takes the system-defined default action.

```
$ cat signals.c
#include <sys/signal.h>

#define TRUE 1

void nicetry(void);
void onalarm(int);

int main(void)
{
        signal(SIGHUP, SIG_IGN);
        signal(SIGINT, nicetry);
        signal(SIGALRM, onalarm);
        alarm(10);
        while (TRUE) {
            printf("Waiting for alarm.\n");
            sleep(9);
        }
}

void nicetry(void)
{
        printf("Nice try! Sorry you cannot terminate me like this.\n");
}

void onalarm(int signal)
{
        printf("Caught signal number %d. Going back to work.\n", signal);
}
$
```

In the following session, we show the compilation and working of the code by running the program. Note that when we press <Ctrl+C> (which appears as <Ctrl+C> in the shell session), the program responds with the message displayed by the nicetry() function. To test that the response to SIGHUP also works as expected, we put the signals process (PID 19156) in the background using <Ctrl+Z> (which appears as <Ctrl+Z> in the shell session) and send the SIGHUP signal to the process using the kill -1 19156 command. The output of the ps command shows that the signal process ignored SIGHUP and continued to run. The handling of SIGALRM also works as expected when the program displays the message Caught signal number 14. Going back to work. Eventually, we stop the signals process again with <Ctrl+Z> and use the sure kill signal to terminate it using the kill -9 16647 command.

```
$ gcc -w signals.c -o signals
$ signals
Waiting for alarm.
Waiting for alarm.
Caught signal number 14. Going back to work.
Waiting for alarm.
^CNice try! Sorry you cannot terminate me like this.
Waiting for alarm.
^CNice try! Sorry you cannot terminate me like this.
Waiting for alarm.
^Z
[1]+  Stopped                 signals
$ ps -U sarwar
  PID TTY          TIME CMD
14466 ?        00:00:00 systemd
```

```
[Output truncated]
16626 ?         00:00:00 sshd
16627 pts/0     00:00:00 bash
16647 pts/3     00:00:00 signals
16657 pts/0     00:00:00 ps
$ kill -1 16647
$ fg
signals
Waiting for alarm.
^CNice try! Sorry you cannot terminate me like this.
Waiting for alarm.
^Z
[1]+  Stopped                 signals
$ kill -9 16647
$ ps -U sarwar
  PID TTY           TIME CMD
14466 ?         00:00:00 systemd
[Output truncated]
16626 ?         00:00:00 sshd
16627 pts/0     00:00:00 bash
16659 pts/3     00:00:00 ps
[1]+  Killed                  signals
$
```

On some Linux systems, the settings for intercepting signals using the signal() call is effective only once. When a signal has been intercepted and handled according to signal settings, the signal settings have to be reestablished in order for signal handling to work correctly. For this purpose, the relevant signal handlers include the code for resetting signals for future signal handling. For the **signals.c** program to work correctly on such systems, you need to modify the nicetry() function, the handler for SIGINT, as follows:

```
void nicetry(void)
{
        signal(SIGINT, nicetry);
        printf("Nice try! Sorry you cannot terminate me like this.\n");
}
```

Exercise 16.15

Replicate the preceding sessions on your system. Does the **signals.c** program work as expected?

Several Internet services such as ftp are offered via server processes that provide services to client processes through multiple *slave processes*, one for each client. Slaves are precreated and/or created dynamically by the main server when needed. The main server is also known as the *master server*. Such servers are known as *multiprocess servers* or *concurrent servers*. A concurrent server runs in an infinite loop with the following code structure:

```
while (1) {
    wait for a client request
    create a slave process when a client request arrives
    slave handles the client request
}
```

Because concurrent servers keep creating slave processes as client requests arrive, it is important to terminate a slave process properly and remove it from the system after it has provided its service to a client process. This is done by using the exit() and wait() calls in tandem, in the slave and master processes, respectively. If the master server process does not remove the slave processes after they have provided their services, a large number of zombie processes will be created in the system. We would want the server to spawn a child, wait, and then when returned to, kill the child, but we can't do that with a concurrent server since it has to continue to process further client requests.

Recall that when a process terminates, the Linux kernel sends the SIGCHLD signal to its parent. A concurrent server process uses this feature to intercept all SIGCHLD signals to remove the terminating slave processes from the system by using the following code structure:

```
...
int main(...)
{
    ...
    signal (SIGCHLD, zombie_gatherer);
    while(1) {
        wait for a client request
        create a slave process when a client request arrives
        slave handles the client request
    }
    ...
}

void zombie_gatherer(int signal)
{
    int status;

    while (wait3(&status, WNOHANG, 0) >= 0)
        ;
}
```

Recall that the WNOHANG option for the wait3() system call makes it a nonblocking call, in the sense that if it does not find a child process that has performed exit(), it returns −1. When wait3() returns −1, the control returns from the zombie_gatherer() function to the line of code in the main function that was interrupted by SIGCHLD.

We discuss the algorithms for the various types of Internet servers in Chapter W20.

16.5.4 Sending Signals

A process can send a signal to another process by using the kill() system call. Here is a brief description of the call:

```
#include <sys/types.h>
#include <signal.h>
int kill(pid_t pid, int sig);
```

Success: 0

Failure: −1 and kernel variable errno set to indicate the type of error

If pid is 0, the signal is sent to all the processes whose group ID (GID) is equal to the sender process and for which the sender process has the permission to do so. If the sender process has superuser privileges and pid is −1, the specified signal is sent to all the processes having the same UID as the sender, the system processes, and the init process (PID=1). The kill() call may fail for the reasons listed in Table 16.10.

The **killer.c** program in the following session takes a PID and a signal number as command line arguments and uses the kill() system call to send the specified signal to the process with the given PID. After compiling the program and saving the executable code in the file called killer, we start a C shell process, and use the ps command to see the PID of the new C shell process. We then use the killer program to send signal number 2 (SIGINT) to the new C shell process (PID 16693). The output of the ps command shows that, as expected, the C shell does not terminate. When we send signal number

TABLE 16.10

Reasons for the kill() System Call to Fail

Reason for Failure	Value of errno
The signal specified in sig is not valid	EINVAL
The process specified in pid does not exist	ESRCH
The process using kill() does not have the permission to send the signal to the process specified in pid	EPERM

9 (SIGKILL; sure kill) to the new C shell process, the shell terminates. The SIGHUP signal would also terminate the shell process.

```
$ cat killer.c
/* killer.c: killer signal pid */

#include <sys/types.h>

int main(int argc, char *argv[])
{
        pid_t pid;
        int signal;

        if (argc != 3) {
            printf("Inappropriate number of command line arguments.\n");
            exit(0);
        }
        pid = atoi(argv[1]);
        signal = atoi(argv[2]);
        if (kill(pid, signal) == -1) {
            perror("Kill failed");
            exit(1);
        }
        exit(0);
}
$ gcc -w killer.c -o killer
$ csh
% ps
  PID TTY          TIME CMD
15995 pts/3    00:00:00 bash
16693 pts/3    00:00:00 csh
16695 pts/3    00:00:00 ps
% killer 16693 2

% ps
  PID TTY          TIME CMD
15995 pts/3    00:00:00 bash
16693 pts/3    00:00:00 csh
16702 pts/3    00:00:00 ps
% killer 16693 9
Killed
$ ps
  PID TTY          TIME CMD
15995 pts/3    00:00:00 bash
16704 pts/3    00:00:00 ps
$
```

Exercise 16.16

Replicate the preceding sessions on your system. Make sure that the **killer.c** program works as expected.

16.6 Thread Programming with pthreads

We discussed the general concept of a thread, differences between threads and processes, resources that threads share with each other, resources that threads share with the process that creates them, and the issues related to the sharing of data amongst threads in Sections 16.2.5–16.2.7. In this section, we introduce the Gnu/Linux implementation of the POSIX standard thread API, called *pthreads*.

16.6.1 The pthread API

The pthread functions and data types are defined in the header file /usr/include/pthreads.h. The pthreads functions are in the libpthread library and are not part of the standard C library. Thus, you should link this library with the –lpthread option in the command line for gcc. Table 16.11 shows a listing of the basic pthread API functions and their brief descriptions.

16.6.2 Thread Creation

A thread can be created using the pthread_create() function. Here is a brief description of the function:

```
#include <pthread.h>
int pthread_create(pthread_t *restrict tid, const pthread_attr_t *restrict attr,
                   void *(*tfn) (void *), void *restrict arg);
```

Success: 0
Failure: Error number

On successful creation of the thread, tid contains the thread ID (TID) of the newly created thread. The attr argument contains thread attributes, tfn is the address of the function where thread starts execution, and arg is the function argument. If you set attr to NULL, the thread is created with default attributes. With arg set to NULL, the thread function is executed without any argument.

Unlike a PID that is unique to a process on a system, a TID is unique to a thread within a process. A thread can obtain its ID by calling pthread_self() function. A TID is of pthread_t type, which is a structure type. This function always succeeds. You can use the pthread_equal() function to compare two TIDs. Here are the brief descriptions of these functions:

```
#include <pthread.h>
int pthread_self(void);
```

Success: TID of the caller thread

TABLE 16.11

Some Commonly Used pthread API Functions

Creation and Termination of Threads	
pthread_create()	Create a thread
pthread_join()	Join a thread making sure that it has finished
pthread_self()	Return the TID of the caller thread
pthread_equal()	Compare two TIDs
pthread_exit()	Terminate a thread
Initializing and Destroying Thread Attributes	
pthread_attr_init()	Initialize thread attributes to their default values
pthread_attr_setdetachstate()	Set the detach state in a thread attribute object
pthread_detach()	Set a joinable thread to detached thread
pthread_attr_destroy()	Release the attribute object

```
#include <pthread.h>
int pthread_equal(pthread_t tid1, pthread_t tid2);
```

Success: Nonzero if equal, 0 otherwise
Failure: Error number

Every process has at least one thread that normally starts at the main function. Thus, every process has a PID and a TID. The **process.c** program in the following session displays its PID as well as its TID. We compile the program and save its executable in the **proc** file. The sample run of **proc** shows that PID of the process is 17401 and its TID is 2877273856.

```
$ cat process.c
#include <pthread.h>
#include <stdio.h>

int main (void)
{
    pid_t pid;
    pthread_t tid;

    pid = getpid();
    tid = pthread_self();
    printf ("pid %u tid %u\n", (unsigned int) pid, (unsigned int) tid);
    exit(0);
}
$ gcc -w process.c  -o proc -lpthread
$ proc
pid 17401 tid 2877273856
$
```

We modify this program so that it creates a thread, and the thread displays its PID and TID. The main thread does not display its PID or TID. The **process_thread.c** program in the following session performs the task. We compile the program and save its executable in the **proc_t** file. The sample run of **proc_t** shows that the thread is in the process with PID 17741 and TID of the thread is 71923456. Note that we use the sleep(1); call to make the parent process sleep for 1 s. We discuss its purpose in Section 16.6.3.

```
$ cat process_thread.c
#include <pthread.h>
#include <stdio.h>

void thread_fn (void *);

int main (void)
{
    pthread_t tid;
    int rval;

    rval = pthread_create(&tid, NULL, thread_fn, NULL);
    if (rval != 0) {
        printf("Thread creation failed.\n");
        exit(1);
    }
    sleep(1);
    exit(0);
}

void thread_fn (void *arg)
{
    pthread_t tid;
```

```
    pid_t pid;

    pid = getpid();
    tid = pthread_self();
    printf ("pid %u tid %u\n", (unsigned int) pid, (unsigned int) tid);
}
$ gcc -w process_thread.c -o proc_t -lpthread
$ proc_t
pid 17741 tid 71923456
$
```

The **thread_ids.c** program in the following session creates a thread that displays its ID as well as the ID of the process that creates it. The main thread (i.e., the process) displays its PID and its TID.

```
$ cat thread_ids.c
#include <pthread.h>
#include <stdio.h>

void thread_fn(void *);
void displayids (const char *);

int main (void)
{
    pthread_t tid;
    int rval;

    rval = pthread_create(&tid, NULL, thread_fn, NULL);
    if (rval != 0) {
        printf("Thread creation failed.\n");
        exit(1);
    }
    displayids("Parent/Main thread:            ");
    sleep(1);
    exit(0);
}

void thread_fn (void *arg)
{
    displayids("Child thread:         ");
    return NULL;
}

void displayids (const char *s)
{
    pthread_t tid;
    pid_t pid;

    tid = pthread_self();
    pid = getpid();
    printf ("%s pid %u tid %u\n", s, (unsigned int) pid, (unsigned int) tid);
}
$ gcc -w thread_ids.c -o tids -lpthread
$ tids
Parent/Main thread:      pid 10249 tid 309786368
Child thread:            pid 10249 tid 301491968
$
```

Exercise 16.17

Replicate the preceding sessions on your system. Do they work as expected on your Linux system?

16.6.3 Thread Termination

A thread can terminate for any of the following reasons:

1. It returns from its thread function.
2. The thread calls the `pthread_exit()` function.
3. Another thread in the same process terminates it using the `pthread_cancel()` function.

In the **process_thread.c** and **thread_ids.c** programs, we use the `sleep(1)` call to make the process wait for 1 s. We do so to give time for the thread to complete its execution and return from the thread function, `thread_fn()`.

Threads created using `pthread_create()` can be *joinable* or *detached*. By default, they are joinable. The thread (or process) that creates a thread can use `pthread_join()` to wait for it to terminate. This is like the `wait()` system call (or any of its variants) used by a process to wait for a child process to terminate. Here is a brief description of the call:

```
#include <pthread.h>
int pthread_join(pthread_t tid, void **retval);
```

Success: 0

Failure: Error number

The first argument is the ID of the thread to wait for and `retval` is the location where the return value of the thread is placed. If the second argument is NULL, the return value of the thread is ignored. If the given thread has already terminated, `pthread_join()` returns immediately.

The following is a brief description of the `pthread_exit()` call.

```
#include <pthread.h>
int pthread_exit(void *retval);
```

Success: 0

Failure: Error number

Other threads in the process may access the `retval` pointer by calling `pthread_join()`. If a thread terminates by returning from the thread function, `retval` in `pthread_join()` will contain the return code. If a thread terminates because it was canceled by another thread, the memory location pointed to by `retval` contains PTHREAD_CANCELED.

If a joinable thread (the default) is not waited for with `pthread_join()`, it is not automatically cleaned up when it terminates. It hangs around like a zombie process until another thread calls `pthread_join()` to obtain its return value. A joinable thread may be turned into a detached thread by calling `pthread_detach()`, whose only parameter is the TID for the thread to be detached. On the other hand, a detached thread cannot be changed to joinable. You do not wait for a detached thread using `pthread_join()`; it is cleaned up when it terminates.

Exercise 16.18

Replace the `sleep(1)` call in the `main()` functions of the earlier programs, compile, and rerun them to verify that the programs work as expected with `pthread_join()`.

We modify the **thread_ids.c** program so that it creates two threads and passes thread number (i.e., 1 or 2) to each thread via the fourth parameter of `pthread_create()`. Changes are made to `main()` and `thread_fn()`; the `displayids()` function remains unchanged. Here is the listing of the new program:

```
$ cat thread_ids.c
#include <pthread.h>
#include <stdio.h>
```

```
/* Structure for passing parameters to thread_fn */
struct thread_params {
    int count;
};

void thread_fn(void *);
void displayids (const char *);

int main (void)
{
    pthread_t tid1, tid2;
    struct thread_params thread1_params, thread2_params;
    int rval;

    thread1_params.count = 1;
    thread2_params.count = 2;
    rval = pthread_create(&tid1, NULL, thread_fn, &thread1_params);
    if (rval != 0) {
        printf("Thread creation failed.\n");
        exit(1);
    }
    rval = pthread_create(&tid2, NULL, thread_fn, &thread2_params);
    if (rval != 0) {
        printf("Thread creation failed.\n");
        exit(1);
    }
    displayids("Parent/Main thread:   ");
    pthread_join(tid1, NULL);
    pthread_join(tid2, NULL);
    exit(0);
}

void thread_fn (void *args)
{
    /* Cast pointer to the correct type */
    struct thread_params* p = (struct thread_params *) args;

    if (p->count == 1)
        displayids("Thread1:            ");
    else
        displayids("Thread2:            ");
    return NULL;
}

void displayids (const char *s)
{
/* Same as before */
}
$
```

In the following session, we show compilation and a sample run of the program. Note that, in the sample run, the PIDs are displayed before the IDs of the two threads. In general, however, you cannot rely on the order of execution of threads. The order of execution of the threads is dependent on thread scheduling used by pthread library. If you run the program enough number of times, you will see another order of thread execution and outputs produced by the threads.

```
$ gcc -w thread_ids.c -o tids -lpthread
$ tids
Parent/Main thread:    pid 32167 tid 1123837696
Thread1:               pid 32167 tid 1115543296
Thread2:               pid 32167 tid 1107150592
$
```

Exercise 16.19

Replicate the earlier session on your system and run tids enough number of times so that a different order of outputs is produced. Show the different sequences of outputs that were produced.

16.6.4 Thread Attributes

Thread attributes dictate the behavior of threads. With the second argument to pthread_create() set to a null pointer, the thread is created with default attributes. The following are the attributes of a thread:

detachstate: the detached thread attribute

stackaddr: the lowest address of thread stack

stacksize: the size in bytes of thread stack

guardsize: the size in bytes of guard buffer at the end of thread stack

Guard buffer provides protection against overflow of the stack pointer.

You can use the pthread_attr_t structure to store the attributes of a thread and associate them with the threads that we create. You can initialize this structure to default attributes using the pthread_attr_init(). You can use other functions to change individual attributes stored in the pthread_attr_t structure. For example, you can use the pthread_attr_setdetachstate() function to set detach state to PTHREAD_CREATE_DETACHED to create a thread in detached state or PTHREAD_CREATE_JOINABLE to create a thread joinable state.

You can use pthread_attr_destroy() to set attributes in pthread_attr_t to invalid values. If pthread_attr_t was allocated dynamically, a call to pthread_attr_destroy() frees up the space.

In summary, you can create a thread with customized attributes using the following steps:

1. Create an object of pthread_attr_t type
2. Initialize pthread_attr_t to default attributes with pthread_attr_init()
3. Modify the attributes object to desired values using relevant functions
4. Pass a pointer to the attributes object to pthread_create() to create a thread with desired attributes
5. Release the attributes object by calling pthread_attr_destroy()

Here is a code structure for these steps:

```
...
int retval;
pthread_t tid;
pthread_attr_t attr;

retval = pthread_attr_init(&attr);
if (retval != 0)
    return(retval);
retval = pthread_attr_setdetachstate(&attr, PTHREAD_CREATE_DETACHED);
if (retval == 0)
    retval == pthread_create(&tid, &attr, thread_fn, arg);
pthread_attr_destroy(&attr);
...
```

Exercise 16.20

Write a program that creates thread in detached state and displays its PID and TID.

16.6.5 Thread-Specific Data

Since threads in a process share its address space, they have shared access to its data area. This means that a change made to a data item is visible to other threads. Sometimes, you would like each thread in a process to have its own copy of a data item, such as a variable. You may store a copy of a variable in a *thread-specific data area*. A change made to this variable is invisible to other threads.

You can create as many thread-specific data items as you want, each of type void*. Each such item is referenced by a key that you create using `pthread_key_create()`. The first parameter of this function is a pointer to an object of `pthread_key_t` type and the second is a cleanup function. The cleanup function is called automatically to deallocate a thread-specific data item when each thread terminates. This function is called even if the thread is canceled during its execution. Cleanup functions are used to prevent memory leaks. If you don't need a cleanup function, you set the second parameter to null.

Once a thread has created a key, it can use `pthread_setspecific()` to set the thread-specific value corresponding to the key. The first argument of the function is the key and the second is the thread-specific value of type void* to store, as shown in the following code snippet.

```
...
pthread_key_t thread_key;
...
void * thread_fn (void *args)
{
    ...
    int thread_item;
    ...
    pthread_setspecific(thread_key, thread_item);
    ...
}
```

You can access a thread-specific data item using the `pthread_getspecific()` function that takes the key as its only argument and returns the data item associated with the key. Since thread-specific values are stored as void* type, you must typecast them into the desired type, as in

```
...
int fd;
...
fd = (int) pthread_getspecific(thread_key);
...
```

16.6.6 Thread Synchronization

The pthread library provides support for a wide range of synchronization constructs (also known as primitives), including *mutexes*, *reader-write-locks*, *semaphores*, and *condition variables*. You can use these primitives to solve a range of problems in concurrent programming related to race conditions and the critical section problem. Table 16.12 shows a set of pthread library functions related to the abovementioned synchronization construct.

Further discussion on the synchronization topics is beyond the scope of this book. The interested reader may read more about them in a book on operating system concepts and principles, concurrent programming, or multithreaded programming.

Summary

We described the Linux system calls and their use in small programs for process creation, process termination, and process management. Before discussing the system calls, we explained in detail the concept of processes and threads. We then discussed how they differ from each other and the concept of user- and kernel-level threads. We also discussed the following concepts and kernel data structures in Linux: the PCB, memory and disk images of processes, zombie processes, signals, and signal handling.

TABLE 16.12

The pthread API Functions for Synchronization Constructs

Mutexes	
pthread_mutex_init()	Initialize a mutex with the attributes passed as an argument
pthread_mutex_lock()	Lock an unlocked mutex; wait until a locked mutex becomes unlocked
pthread_mutex_unlock()	Unlock a mutex
pthread_mutexattr_init()	Initialize a mutex to default attributes
pthread_mutexattr_setkind_np()	Set the mutex kind: fast, recursive, or error-checking
pthread_mutexattr_destroy()	Release the attributes object for a mutex
pthread_mutex_trylock()	Unlock a mutext but don't block if it is already locked
Reader-Writer Locks	
pthread_rwlock_init()	Initialize a reader-writer lock
pthread_rwlock_destroy()	Destroy a reader-writer lock
pthread_rwlock_rdlock()	Lock a reader-writer lock for reading
pthread_rwlock_wrlock()	Lock a reader-writer lock for writing
pthread_rwlock_unlock()	Unlock a reader-writer lock
Condition Variables	
pthread_cond_init()	Initialize a condition variable
pthread_cond_destroy()	Destroy a condition variable
pthread_cond_wait()	Block the caller thread until a thread signals the condition variable
pthread_cond_timedwait()	Same as pthread_cond_wait() with timeout in seconds and nanoseconds
pthread_cond_signal()	Signal a condition variable and unblock a thread blocked on it
pthread_cond_broadcast()	Signal a condition variable and unblock all threads blocked on it

We discussed process management using the following system calls and library functions: fork(), vfork(), exit(), _exit(), wait(), and its different variants, the getpid(), getppid(), signal(), dup(), dup2(), kill(), execve(), fexecve(), and execl() calls. We discussed the concept of zombie processes and used the exit() and wait() calls to illustrate how a zombie process can be created. We also discussed the concept of file sharing between the parent and children processes.

We discussed the concept of signal and signal handling in Linux as well as multithreaded programming using the pthread library using the related system calls and library functions.

Throughout the chapter, we showed the use of various system calls and library functions in small C programs to illustrate various process management concepts.

Questions and Problems

1. What is a thread? What are user threads? What are kernel threads?
2. List the differences between processes and threads.
3. Suppose a system supports multithreaded processes with user-level threads. Where is the scheduling of threads carried out?
4. Explain the difference between single- and multithreaded kernels by using an example.
5. Why is a separate function needed to terminate a thread? Why can we not terminate a thread with the _exit() system call or the exit() library function?
6. What is *atomic execution* of a piece of code?
7. What is a critical section? What is the critical section problem?

8. The Bakery Algorithm is an elegant solution for the N-process critical section problem. Browse the Web or read a book on operating system principles and concepts to find out the name of the author of this algorithm. Who is the author?

9. A process is said to be a program in execution. What really comprises a process in terms of the system resources that a process utilizes?

10. What is the PCB? What are the names of the two parts of a process's PCB in Linux? What kind of information do they contain about the process?

11. The PCB of a process is not accessible in user mode. Why?

12. Which part of the PCB of a Linux process always remains in the main memory? What information about the process does it contain?

13. Why is bss not part of the disk image of an executable file but is part of the process that is created when the file is executed?

14. The Linux command `size` may be used to display the size in bytes of the code, data, and bss segments of an executable file. Show an example to illustrate your answer.

15. What is the relationship of shared libraries with dynamic linking? What are the advantages of dynamic linking over static linking? What are the disadvantages of dynamic linking?

16. By default, the `gcc` compiler uses dynamic linking. What is the compiler option for creating an executable program using static linking? Which linking generates smaller executable code? Why?

17. Compile a program using static and dynamic linking and then display the program size using the `ls -l` command. Show your shell session.

18. Display the sizes of the code, data, and bss sections of the executable program generated in Problem 7. Show the shell session. Why is the sum size of the code, data, and bss sections generated by the `size` command not equal to the size of the same executable generated by using the `ls -l` command?

19. Consider the following C programs and answer the questions that follow:

```
$ cat bss_size_1.c
int BigData[1000000];

void main(void)
{
   BigData[0] = 0;
}
$ cat bss_size_2.c
int BigData[1000000] = {1, 2, 3};

void main(void)
{
   BigData[0] = 0;
}
$
```

    a. Which of these two programs takes more disk space? Why?

    b. If executable codes are generated for the two programs, which would take more disk space? Which would require more main memory to execute? Explain your answers.

    c. What are the sizes of the bss and data segments of the executable files for these programs? What commands did you use to obtain these sizes? Show your shell session.

20. What is the purpose of the magic number of a file in Linux? What command uses the magic number of a file to display its output? Show an example to illustrate your answer.

21. Write a small program that creates three children processes, P1, P2, and P3. The parent and each child display their PIDs and their parent PPIDs. The parent should display its PID and PPID only when all three of its children have terminated.

22. How many processes does the following C program create? Assume that all three fork() calls are successful. Draw the process tree for the program after the third fork() call has executed successfully. Explain your answer.

```c
#include <stdio.h>

int main(void)
{
    int pid1, pid2, pid3;

    pid1 = fork();
    pid2 = fork();
    pid3 = fork();
}
```

23. Browse the man pages for the vfork() and rfork() system calls. What is the purpose of each of these calls? How does vfork() differ from fork()?

24. Write a program that has one file opened by the parent process and one by the child. The child writes Long live Linux. to both files, reads and displays the contents of its own file, closes its file, and returns (i.e., exits). The parent reads and displays the contents of its file, closes the file, and terminates. Does the output of the program make sense? Explain your answer.

25. Write a program in which a parent opens two files (**file1** and **file2**), writes Long live Linux. to **file1**, and spawns a child. The child copies the contents of **file1** to **file2**, closes the two files, and returns (i.e., exits). The parent reads and displays the contents of the two files, closes them, and terminates. Does the output of the program make sense? Explain your answer.

26. When a process uses the execve() or fexecve() call to overwrite itself, signals set to be caught in the caller process are set to default action in the new process. Why are signals set to default action—why not the actions as specified in the caller process?

27. What are the differences between the _exit() and exit() calls?

28. What would happen if you terminated a multithreaded process with the exit() call? Explain your answer.

29. What are the wait4() equivalents of the following calls? The three dots (...) represent the value of the corresponding argument for the calls:
    - waitpid(..., ..., ...);
    - wait3(..., ..., ...);
    - wait(...);

30. Write a program that creates a child process that displays its PPID and sleeps for 60 s. The parent displays its PID, the termination status of child, and the reason for the child's termination (i.e., the signal that caused its termination). Run the program in the background, use the ps command to identify the PID of the child process, and then terminate the process by sending it a signal using the kill command. Show your code and a few sample runs. Make sure that the program displays the correct values for the return code of the child as well as the signal number that caused its termination.

31. What is the purpose of the kill() system call? What happens when 0 or −1 is specified as the PID to this call?

32. What is the effect of the kill(getpid(), SIGKILL) system call?

33. Suppose you run a program under a C shell. What will be the effect of the kill(getppid(), SIGKILL) system call in the program?

34. What is the effect of executing the dup() system call? What happens if the specified descriptor is invalid?

35. What happens if dup2() is used and the old and new descriptors specified in the call are the same, provided that the old descriptor is valid? What happens if the old descriptor is not valid?

36. The **dup2.c** program discussed in Section 16.4.2 illustrates how a shell implements output redirection. Modify this program so that it takes two files as arguments and implements both input redirection and output redirection.

37. What is the difference between fork() and vfork()? As a system programmer, when should you prefer to use vfork() over fork()?

38. What is the output of the following program? Explain your answer.

```
int main(void)
{
  pid_t pid;
  int i=100, status;

  if ((pid = vfork()) == -1) {
    perror ("vfork failed");
    exit(1);
  }
  if (pid == 0) {
    i++;
    exit(0);
  }
  wait(&status);
  printf("%d\n",++i);
  exit(0);
}
```

39. What will be the output of the program listed in Problem 38 if the following statements are swapped? Explain your answer.

```
wait(&status);
printf("%d\n",++i);
```

Will the program always produce the same output every time it is run? Explain your answer.

40. Write a C program that creates a zombie process and then runs the ps command from within the process to verify that the process status is zombie. Show your program and its sample run.

41. Remove the sleep(1); statement in the **file_sharing.c** program discussed in Section 16.4.1. Compile the program and run it a few times. You will notice that, for some sample runs, the file pointer value displayed by the parent and child processes is the same. Explain why this is so.

42. Identify the values of the following by browsing through the relevant header files on your Linux system. Write down the names of the header files that contain these values.

    a. The maximum number of processes that can run on your system concurrently

    b. The maximum number of children processes that a process can have at a given time

43. Use the limit command under the C shell or the ulimit command under Bash to determine the maximum number of processes that can run on your system concurrently and the maximum number of files that can be opened on your system concurrently.

44. Browse through the relevant header file(s) and identify all of the signals along with their numbers and their purpose. Write down the names of the file(s) in which you found these items?

45. Give two reasons each for the following system calls to fail: dup(), fork(), execve(), _exit(), wait(), kill(), and signal().

46. Browse through the /usr/include/pthread.h file and add missing functions along with their brief descriptions to Table 16.11.

47. Can multiple processes running on a system have the same ID? Can multiple threads running on a system have the same ID? Explain your answer.

48. How will the program behave if we remove the `pthread_join()` calls in the `main()` function in the **thread_ids.c** program discussed in Section 16.6.3? With this done, what are the issues if the `main()` function terminates before the two threads start running?

49. Can a detached thread be converted to a joinable thread? Is vice versa possible?

Advanced Questions and Problems

50. What do we mean by *race condition* in the context of multithreaded processes? What does a programmer need to do to handle the issue of race condition? Explain your answer with an example.

51. Browse through the literature on operating systems and distributed systems and define clearly the following terms:

    a. *Semaphore*

    b. *Spin lock*

    c. *Monitor*

    d. *Condition variable*

52. The following are a few classic problems in concurrency:

    a. First readers–writers problem

    b. Second readers–writers problem

    c. Dining philosophers problem

    d. Cigarette smokers problem

    e. Sleeping barbers problem

    Browse through literature on operating systems and distributing systems, and clearly state each problem.

53. What piece of code would you add to your code in the simple shell program that you wrote for Project 1 in Chapter 15 so that the shell only terminates when the user presses <Ctrl+D>? It should not terminate due to a keyboard interrupt—that is, when you press <Ctrl+C>.

54. Add to Table 16.11 the pthread API functions for

    a. Mutexes

    b. Condition variable

    c. Setting up and accessing thread-specific data

    d. Cancelation of synchronous and asynchronous threads

55. Unlike the UNIX implementation of pthreads, older versions of Gnu/Linux pthreads were implemented as "processes" managed by a *thread manager* process that is created on the first call to `pthread_create()`. Are pthreads implemented as processes on your system? Explain your answer with a small C program. Show the source code, its compilation, and execution to verify your answer.

56. Suppose that threads A, B, and C are reading from the same file at the same time using the following sequences of instructions:

Thread A	Thread B	Thread C
`lseek(fd, 200, SEEK_SET);`	`lseek(fd, 500, SEEK_SET);`	`lseek(fd, 800, SEEK_SET);`
`read(fd, bufA, 50);`	`read(fd, bufB, 50);`	`read(fd, bufB, 50);`

What would each thread read? Will the concurrent execution of these code snippets always produce the same result? Explain your answer. How can you make sure that the three threads would read the desired data from the desired location as stated in the code every time you run them?

Projects

Project 1

Implement the classic problems in concurrency stated in Problem 52 using *mutexes* and *condition variables* in the pthread library. Show the source codes for your solutions along with their compilations and sample runs.

Looking for more? Visit our sites for additional readings, recommended resources, and exercises.

CRC Press e-Resource: https://www.crcpress.com/9781138710085

Authors' GitHub: https://github.com/bobk48/linuxthetextbook

17

Linux System Administration Fundamentals

OBJECTIVES

- To recommend doing a fresh install of a 64-bit, X86 architecture version of Linux from ISO DVD media using a Graphical User Interface (GUI) installer, using a default installation with a GUI desktop environment
- To detail basic configuration considerations before and after installation of that system
- To detail the steps of the boot and startup processes, in the context of systemd
- To describe how to gracefully bring the system down, traditionally and with systemd
- To provide references on how to use systemd to manage system services
- To illustrate how to manage additional users and groups on the system
- To show how to design and maintain user accounts
- To show postinstallation of persistent media, such as hard disks, Solid State Drives (SSDs), and USB thumb drives
- To provide strategies for the backup and archiving of system and user files
- To show how to add/update/remove user application package repository software
- To explain and provide references for Linux system upgrades
- To show how to monitor and enhance the performance of the system
- To provide strategies for system security, particularly the system firewalls ufw and firewalld
- To describe basic virtualization techniques for Linux, including LXC/LXD and VirtualBox
- To cover the commands and primitives

```
ACL, addgroup, adduser, APT, apt, apt-get, Baobob, BIOS, capabilities, cgroups,
chgrp, chmod, chown, cipher text, clone, Clonezilla, cpio, CUI, DAC, dd,
delgroup, deluser, df, Docker, du, ext4, fail2ban, fdisk, Filezilla, find,
firewalld, free, GID, git, GitHub, Gparted, GRUB2, GUI, gunzip, gzip halt, IDS,
ifdown, ifup, inxi, ip, IPS, iSCSI, journalctl, LIO, ls, lsblk, LVM, LXC, LXD,
MAC, MBR, mdadm, mirror, mkfs, newusers, NFSv4, nice, NIDS, openssh, passwd,
passwd, persistent cgroup, pgrep, plain text, POSIX1.e ACL, ps, RAID, RBAC,
renice, repositories, rsync, SAN, shutdown, SSD, ssh, su, sudo, system.slice,
systemctl, systemd, systemd-cgls, systemd-run, tar, top, transient cgroup,
UEFI, ufw, UID, umount, untar, update, upgrade, usermod, vdev, VirtualBox,
Webmin, YaST, YUM, yum, ZFS
```

17.1 Introduction

As seen in previous chapters, two of the three major components of this book are as follows: basic Linux operations and commands, and Linux system programming.

This chapter is central to the third major component: Linux system administration.

In accordance with the "Distribution Information and Additions" section in the Preface, all of our command line session illustrations in this chapter are executed on Linux Mint by default. This is the representative distribution, or "flavor", we have chosen from the Debian branch of the three major families of Linux, which are Debian, Slackware (openSUSE), and Redhat (CentOS)). But particularly in this chapter, we also illustrate some important tasks, such as package management of software applications, by showing examples executed on Debian, Ubuntu, and CentOS. As indicated in the Preface, whenever the command, its output, or an operation in one of these systems is significantly different from the default, we prefix an appropriate command line prompt to help you differentiate the system on which the command, or operation, has been executed.

To install, maintain, and effectively use a Linux system composed of both hardware and software components, it is often necessary to perform a set of common tasks shown in the following sections. In this chapter, we target an individual, novice user, who performs these tasks exclusively for their own use, on their own personal desktop/laptop/tablet computer. The common tasks may also be performed by an appointed administrator, for a more complex computer system used by many people. It is possible to divide these common tasks into those performed by an administrator, and those performed by a single user, or a group of ordinary, autonomous users.

Even though we show the basics of those common tasks, it is possible to extrapolate from what is presented to the wider context of much larger-scaled computer systems, run by a system administrator.

Additionally, at the book website, we provide many more detailed descriptions of each task, and present other advanced system administration topics as well. We specifically reference these detailed descriptions and advanced topics at the appropriate place. For example, when illustrating system booting and startup in Section 17.2.3.1, we reference Chapter W26, Section W26.2.3.1.1, titled "systemd Bootup," at the book website.

To do many of the system administration tasks in this chapter, it is necessary to either have superuser or root user privileges on the system. That means you need to know the superuser password, which you (or a designated system administrator) can establish at installation of your Linux system.

Both the **su** and **sudo** commands are used to execute programs and other commands with root permissions. The root user has maximum permissions and can do anything to the system that a system administrator needs to. Normal users execute programs and commands with reduced permissions.

To execute something that requires maximum permissions, you'll have to first execute the **su** or **sudo** commands.

Using the **su** command makes you the superuser—or root user—when you execute it with no additional options. You are prompted to enter the root account's password. Also, the **su** command allows you to switch to any user account. If you execute the command **su**, you'll be prompted to enter the password, and the command shell current working directory will be your home directory.

Once you're done running commands in the root shell, you should type **exit** to leave the root shell and go back to limited-privileges mode.

The root shell is not the login shell.

In comparison, **sudo** runs a single command with root privileges. When you execute **sudo command**, the system prompts you for your current user account's password before running **command** as the root user. To use the **sudo** command, you must be part of the sudoers group. More information on the **sudo** command can be found in Section 17.9.2.1, and we give a much more extensive treatment of it in Chapter W26, Section W26.9.2.1 at the book website. We also provide an installation guide for Debian, Ubuntu, Linux Mint, and CentOS, that shows how to install and activate the **sudo** command on those systems, if it is not already installed and activated, in Appendix A, Section 17.9.2.1.

In our "learning-by-doing" approach, we have selected a common set of system administration tasks aimed specifically at an individual novice Linux user. The tasks we present fit our model of any general use case that novice would be working with. These tasks are as follows:

1. Do a fresh install (sometimes called a "bare-metal" install) of a 64-bit, X86 architecture version of Linux from DVD media, using a GUI installer onto a single hard disk desktop system, with a GUI desktop environment. Do a preliminary configuration of that system, using the default disk partitioning scheme, so that it is the only operating system on the hardware.

2. Illustrate how the system boots, starts up, and how to gracefully bring the system down, in the context of systemd.

3. Detail the basics of using systemd to manage system services.

4. Add additional users and groups to the system, and show how to design and maintain user accounts.

5. Adding hardware to the system, such as disk drives, using a recommended storage model.

6. Provide strategies, using the traditional and generic Linux commands, to backup and archive system files and user files.

7. Upgrade and maintain the operating system, and add/upgrade/remove user application package repository software, to both increase functionality by adding new software, and update existing software packages.

8. Monitor the performance of the system and tune it for optimal performance characteristics, given any use case.

9. Provide strategies for system security to harden the individual desktop computer.

10. Provide network connectivity strategies, both on a Local Area Network (LAN) and the Internet.

Besides the ten common tasks we have selected to present, there are numerous extensions and also many additional tasks that Linux system administration encompasses.

You must realize that we have listed all of the tasks earlier based upon the use case we target.

To extend some of the system administration topics covered in this book so that they are reflective of modern Linux systems, and to anticipate other use cases you might have, we provide some additional materials and references, as follows:

If you want to install Linux in a virtual environment so that it runs simultaneously with some other operating system on your computer hardware, we give specific instructions on how to do this in Chapter W23, titled "Virtualization Methodologies" at the book website. There we give the specifics of efficient ways of installing Linux as a "guest" operating system on other operating system hosts, using VirtualBox installed on the host. We also cover LXC/LXD container virtualization there.

In Section 17.8.1.2, we show the details of the use of Control Groups (cgroups) in systemd. cgroup techniques, executed on the command line, allow an administrator to more explicitly set and view the limits of the three major functions performed by the modern Linux kernel (or the Linux superkernel, systemd): concurrency, virtualization, and persistence.

At the book website, in Section W26.9.3, we fully illustrate the use of POSIX.1e and Network File System, version 4 (NFSv4) Access Control Lists (ACLs). We show the application of these facilities, which further extend the Discretionary Access Control (DAC) methods of the traditional Linux permissions model presented in Chapter 5. We apply ACLs to file system objects mounted as NFSv4 "shared" network drives. In addition, we provide an extended example of using ACLs on NFSv4 network drives, where the file system objects are Zettabyte File System (ZFS) formatted and controlled. This combination of NFSv4, ACLs, and ZFS provides the system administrator with a modern, feature-rich repertoire of techniques she can use in maintaining security and persistence of data on a Linux system.

We provide Chapter 18, which gives extensive details covering how a modern Linux system uses the Linux superkernel, systemd, to startup, and manage system services.

The additional materials and references we provide expedite all of the system administration techniques and methods shown in this chapter. They are found later in the sections of this chapter, in Chapter 18, in Appendix A, and in Chapters W26 and W27 at the book website.

17.1.1 System Administration the Easy Way

How can you do the majority of what is shown in this chapter in an easy way?

Use Webmin.

Webmin is a modern Linux Web browser-based, GUI system administration tool that allows you to do many of the system administration tasks shown in this chapter in a very intuitive, fast, and simple way. We strongly encourage you to first download and install Webmin on your Linux system, following the instructions we give in Section W26.1.2 of Appendix A or in Section W26.1.2 at the book website. Then, you can explore the Webmin facilities before you begin the rest of this chapter, and even go back to Webmin to find out how something which we show in the rest of this chapter can be done (or can't be done!) in Webmin.

17.2 Doing a Fresh Install from DVD ISO-Image Media and Preliminary System Configuration

This section assumes that you have already obtained the DVD ISO-image media for the Linux system you want to install and that your system is a 64-bit, X86 architecture computer that you are willing to use at least one entire hard disk to install the system on. If you have not obtained the ISO image, or do not have a 64-bit computer, you should follow the online documentation available at the websites listed in Section 17.2.2 to get the ISO image of the DVD media.

Since installation of a Linux system is highly variable between distributions and changes in significant ways as later upgrades and releases become available, we only provide a sample of Linux system installation. See Appendix A, Section 17.2, for specific installation instructions for Linux Mint, our default Linux distribution.

You can also use a release of the software system that is designed for a 32-bit machine, but be aware that some of the functionality we show in this chapter is limited to a 64-bit architecture machine.

"Obtaining the DVD ISO-image media" means downloading an ISO file for the release of the software you want to use, and then burning that ISO image to a DVD, or copying it to a USB thumb drive. We do *not* give instructions here for those processes.

At the time of writing of this book, the Linux systems we used to illustrate everything were at the following releases:

Debian 9.1, Ubuntu 16.04LTS, Linux Mint 18.2, CentOS 7.4

Be aware that some of the systems administration tasks illustrated for Linux are done in a different way in earlier (or later) releases of the software!

This section also assumes that you will *not* be *permanently* running the "live" version of the Linux system from persistent or nonpersistent media, like a USB thumb drive or CD/DVD. You are welcome to do that if you want to casually test drive Linux to get a feel for which of the "flavors" you like the most.

As stated in Section 17.1, if you want to install Linux in a virtual environment so that it runs simultaneously with some other operating system on your computer hardware, we give specific instructions on how to do this in Chapter W23, titled "Virtualization Methodologies" at the book website. There we give the specifics of installing Ubuntu Linux as a "guest" operating systems, using both LXC/LXD and VirtualBox.

17.2.1 Pre- and Postinstallation Considerations and Choices

All Linux systems have minimum hardware and driver requirements that can be determined before you begin an installation. These hardware requirements are listed in the documentation online for each of the systems.

Preinstallation Considerations

Once you have determined that your computer system hardware meets or exceeds those minimum requirements, the following listing of questions suggest very important considerations you can make over and above the default installation choices presented to you, before proceeding with the installation. Rather than viewing these suggestions as highly formal constraints, you should consider them as informally presented guides to how you can choose certain aspects of your Linux system installation, in the very subjective light of how you use your computer.

0. Given the way that you want to use your computer running Linux, does the hardware have an adequate amount of physical memory? This consideration is a performance concern. In any case, you should always first consider increasing the amount or size of physical memory

installed, if that is a viable option. As an almost necessary corequisite of considering memory size increases, the speed that the memory bus operates at is important as well.

What's the easiest way of finding out what the memory bus speed is? Refer to the motherboard information documents for your hardware. There are more complex, command line methods, but the most reliable and easiest way for beginners is to look at the documents.

For example, the memory modules that are actually installed may be rated to operate at a slower rate than what the bus can support, because at the time the system was built, that was all the builder could afford. So upgrading to faster memory modules, that take advantage of the higher rates of information transfer that the bus is capable of, leads to performance improvements. Other relevant, but more involved, memory considerations are the size of virtual memory, paging, and swap space.

1. Two of the most integral and important considerations are what persistent media data storage model are you going to use given your use case, and how many hard disks can or does your computer have physically connected to it?

With regard to the first consideration, are you going to need to attach virtual disk drives, via protocols such as NFSv4 or Internet Small Computer System Interface (iSCSI) (which we give basic details in Section 17.4)? Are you going to connect and work within the contexts of a Network-Attached Storage (NAS) or Storage Area Network (SAN) (which we also give basic details in Section 17.4)?

With regard to the second consideration, if you only have one hard disk drive, that will most likely be the bootable system disk and the user file system data disk. If it is possible to have two or more hard disk drives installed on the system, what strategies will you use to backup or "clone" the bootable system disk, and perhaps "mirror" all other user data on the multiple data disk hard drives?

Note: Our Data Storage Model Recommendation—If your hardware can support multiple internally mounted hard drives, or fast externally mounted hard drives, we recommend that you install the operating system on one internally mounted hard drive (preferably a solid-state drive [SSD]), and store all of the user data on another single disk or an array of hard disk drives. That way, if the operating system and its bootable hard disk drive become corrupted or unusable for some reason, your user data is on a separate hard disk or array of hard disks. This technique or storage model dovetails very well with the most practical methods of operating system upgrades. Using it, you can then simply replace the operating system hard disk and reinstall either the current version of the operating system or a newer version, without significantly impacting your data storage. It is then highly recommended and very easy to install the ZFS as a kernel-loadable module to construct the file system(s) on the data disks, and use the ZFS facilities that we show in Chapter W22, titled "ZFS Administration and Use" at the website. See Appendix A, Section W26.4.7 for information and installation instructions that are relevant for installation on Debian-family and CentOS systems. This will allow you to reattach the data hard drives to the new operating system and its hard disk drive. This is highly valuable not only for desktop computers but also server-class systems as well. The way that your data is deployed on your disks is a critical design consideration when you are building your system, and is highly dependent on the particular use case that is guiding it.

One caveat you must consider when using the earlier recommended storage model is some applications software, most prominently LXC/LXD containers, store required files in both the system area and in the data storage area. In Chapter W23, titled "Virtualization Methodologies" at the book website, we consider the case of LXC/LXD containers that are integrated with ZFS data storage. We supply some useful backup strategies for LXC/LXD containers that work with our recommended data storage mode.

Another caveat true at the time of writing of this book is that ZFS is not as easily installed on CentOS as it is on the three representative Debian-family systems we show.

2. Do you have a wireless connection to a local area network and the Internet? A wired connection through DHCP is automatically done in Linux during the default installs of all representative systems.

3. How many users are you going to initially establish at installation, and what are their user profiles going to be? For example, what users will have administrative privileges other than yourself, and what kind of security will each profile have? Integral with these questions are the considerations of the storage model you will establish after the initial installation of the operating system on a single hard disk, as defined in (1). For example, will all user's home directories be on a separate hard drive from the system disk? Also, what user groups are you going to establish at installation, and how are you going to manage groups in a postinstallation environment?

4. Who will be responsible for the ten items of system administration tasks listed earlier? For example, that will influence your disk management tasks concerning file systems for users and projects, according to the data storage model you employ.

5. What kind of software tools do you want to include for the kinds of tasks you and your user base will be doing? For example, during installation, you are able to add packages on top of the default package installation to help accomplish those tasks. What kind of software tools do you want to include for the kinds of tasks you and your user base will be doing? How are the user groups established in Item 3 going to have access privileges to this software? What are your policies with respect to group access to software tools, and how do you enforce this policy?

6. What kind of GUI windowing system do you want to install with the system? You have a choice at installation of Gnome, Cinnamon, Mate, and many others, and your previous experience and preferences with a particular style of desktop environment can be implemented at installation. Will you be doing a server install, based on your use case? A server install, and the management of a server system, involves another whole universe of considerations and design decisions.

Exercise 17.1

Make a detailed listing on paper of your answers to the earlier seven sets of questions before you begin to install your operating system. Then, read through the sections later referencing the procedures for actual installation, and for each of your answers, determine ahead of time how you will proceed through the procedures. This exercise is meant to serve as a "dry run" through any particular path you might take through the installation procedures.

Postinstallation Considerations

The highly practical things you do to your Linux system after it is installed successfully are *very* dependent upon your individual preinstallation considerations, and how you implemented those considerations in detail. As presented in the preinstallation considerations, you should view the suggested considerations here as informally presented guides to how you can choose certain postinstallation aspects of your Linux system.

Our minimal set of recommendations for postinstallation tasks is specific to our base Linux Mint system and its hardware configuration. They follow our preinstallation considerations.

0. Since we installed a desktop system that was going to heavily rely on a convenient and easy-to-use GUI environment, it was economically feasible to increase the amount of installed physical memory, to speed up this kind of usage. Additionally, this gave us increased speed of system operations that were compute-bound. See the output of the **inxi** command shown immediately following this recommendations listing, and In-Chapter Exercise 17.37, to gain more information and insight into how to view physical memory size on your Linux system.

   Your particular physical memory needs depend upon how you use your Linux system.

   The memory modules installed on our system were rated at the maximum speed allowed for the bus speed the system was capable of, as discovered in the motherboard documents.

   We went with the defaults affecting the size of virtual memory, paging, and swap space, although these may be important considerations for you.

1. Use the Software Manager to install absolutely essential applications that you need to use that are not already preinstalled by default (Firefox, LibreOffice). For us, that included, in priority order, opensshd, Filezilla, Gparted, VirtualBox, and Webmin. We also put desktop icons on our Cinnamon desktop for all of our important applications. We supply some of the installation instructions for this software in Appendix A.

2. Our highly recommended storage model dictates the use of ZFS, so we installed the zfsutils-linux package. The Linux Mint operating system was built on an SSD, and we added an additional spinning hard disk mirrored pair of vdevs for user data file storage immediately after installing zfsutils-linux.

3. Since printing documents was an essential operation that our system needed to perform, we attached and setup via Common UNIX Printing System (CUPS; as shown in detail on the website, in Chapter W26, Section W26.5, titled "CUPS Printing"), a laserprinter directed to the hardware of the computer with a USB connection.

4. We exposed our Linux Mint system on the Internet as a basic ssh server, so we immediately modified the sshd configuration file, and constructed ufw firewall rules, according to what we show in this chapter, to secure the system. In addition, we also installed the fail2ban package with its defaults to protect against brute-force attacks, mainly on port 22. If you are using a CentOS system, selinux is enabled and enforced by default, whereas it is not in the Debian family.

5. We minimally exposed our Linux Mint system as a Web server (using nginx and our custom programs as shown on the website), but went with the default security measures for those applications.

6. We needed only a sparse set of additional users and groups on the system, so we immediately set those up using the method shown in Section 17.3.4, and added to them as necessary. This method of adding new users and groups conforms to item 2. The primary advantage this confers is that users, and their data files, are on their own redundant pair of hard drives, discreet and maintained separately from the operating system installation. We feel that this model is useful for a single-user computer desktop system, a shared multiuser computer, or low-to-midlevel enterprise-level servers as well.

There are various Linux commands that survey system hardware, to provide you with a basis for doing performance tuning and postinstallation modification of the system. Some of these are lscpu, lshw, hwinfo, lspci, lspci, lsusb, inxi, lsblk, df, fdisk, mount, free, hdparm, and examining the relevant contents of the /proc directories. We detailed the use of some of these in previous chapters and illustrate the use of some of them in this chapter as well.

But, to get a very useful summary of the actual postinstallation configuration of their system, we have found that the Linux **inxi** command is the most useful and expedient way for beginners.

We show three uses of that command, on three different server-class machines: an HP ProLiant-MicroServer running Linux Mint, an IBM System x3650 -[7979E9U] running Debian 9, and a Dell PowerEdge T110 running CentOS 7.4.

Note that, on the Dell PowerEdge T110 running CentOS 7.4, we had to install **inxi** as superuser first, using the command **yum install inxi**.

We utilize various options that yield our specific required outputs for those machines, as follows:

```
$ sudo inxi -GSCMm -t c -P -x
[sudo] password for bob: qqqqq
System:    Host: bob-ProLiant-MicroServer Kernel: 4.4.0-89-generic x86_64 (64 bit
gcc: 5.4.0) Console: tty 0
           Distro: Linux Mint 18.2 Sonya
Machine:   System: HP product: ProLiant MicroServer serial: 5C7142P200
       Mobo: N/A model: N/A Bios: HP v: 041 date: 07/29/2011
CPU:       Dual core AMD Turion II Neo N40L (-MCP-) cache: 2048 KB
       flags: (lm nx sse sse2 sse3 sse4a svm) bmips: 5990
       clock speeds: max: 1500 MHz 1: 1000 MHz 2: 1300 MHz
```

```
Memory:      Array-1 capacity: 8 GB devices: 2 EC: Single-bit ECC
        Device-1: DIMM0 size: 4 GB speed: 1333 MHz type: Other part: N/A
        Device-2: DIMM1 size: 2 GB speed: 1333 MHz type: Other part: N/A
Graphics:  Card: Advanced Micro Devices [AMD/ATI] RS880M [Mobility Radeon HD
4225/4250] bus-ID: 01:05.0
        Display Server: X.org 1.18.4 drivers: ati,radeon (unloaded: fbdev,vesa)
        tty size: 157x44 Advanced Data: N/A for root out of X
Partition: ID-1: / size: 105G used: 34G (34%) fs: ext4 dev: /dev/sda1
        ID-2: swap-1 size: 6.31GB used: 0.00GB (0%) fs: swap dev: /dev/sda5
Processes: CPU: % used - Memory: MB / % used - Used/Total: 865.7/5841.4MB - top 5
            active
        1: cpu: 1.0% command: cinnamon pid: 2686 mem: 159.88MB (2.7%)
        2: cpu: 0.6% command: -bash pid: 17156 mem: 5.09MB (0.0%)
        3: cpu: 0.3% daemon: ~kworker/0:0~ pid: 16504 mem: 0.00MB (0.0%)
        4: cpu: 0.2% command: Xorg pid: 2175 mem: 53.16MB (0.9%)
        5: cpu: 0.1% command: sshd: pid: 17142 mem: 7.55MB (0.1%)
$

[root@debian]#  inxi -DRGSCMm -t c -P -x
System:     Host: debian9 Kernel: 4.9.0-3-amd64 x86_64 (64 bit gcc: 6.3.0)
        Desktop: Cinnamon 3.2.7 (Gtk 3.22.11-1) Distro: Debian GNU/Linux 9
        (stretch)
Machine:    Device: server System: IBM product: IBM System x3650 -[7979E9U]-
serial: 99BN632
        Mobo: IBM model: N/A BIOS: IBM v: -[GGE136AUS-1.09]- date: 02/07/2008
CPU(s):     2 Quad core Intel Xeon X5450s (-HT-MCP-SMP-) cache: 12288 KB
        flags: (lm nx sse sse2 sse3 sse4_1 ssse3 vmx) bmips: 47879
        clock speeds: max: 2988 MHz 1: 1992 MHz 2: 1992 MHz 3: 1992 MHz 4: 1992
        MHz
        5: 2988 MHz 6: 1992 MHz 7: 1992 MHz 8: 1992 MHz
Memory:     Used/Total: 962.1/24114.9MB
        Array-1 capacity: 63.5 GB devices: 12 EC: Multi-bit ECC
        Device-1: DIMM 1 size: 2 GB speed: 266 MHz type: DDR2 part: N/A
        Device-2: DIMM 4 size: 2 GB speed: 266 MHz type: DDR2 part: N/A
        Device-3: DIMM 2 size: 2 GB speed: 266 MHz type: DDR2 part: N/A
        Device-4: DIMM 5 size: 2 GB speed: 266 MHz type: DDR2 part: N/A
        Device-5: DIMM 3 size: 2 GB speed: 266 MHz type: DDR2 part: N/A
        Device-6: DIMM 6 size: 2 GB speed: 266 MHz type: DDR2 part: N/A
        Device-7: DIMM 7 size: 2 GB speed: 266 MHz type: DDR2 part: N/A
        Device-8: DIMM 10 size: 2 GB speed: 266 MHz type: DDR2 part: N/A
        Device-9: DIMM 8 size: 2 GB speed: 266 MHz type: DDR2 part: N/A
        Device-10: DIMM 11 size: 2 GB speed: 266 MHz type: DDR2 part: N/A
        Device-11: DIMM 9 size: 2 GB speed: 266 MHz type: DDR2 part: N/A
        Device-12: DIMM 12 size: 2 GB speed: 266 MHz type: DDR2 part: N/A
Graphics:  Card: Advanced Micro Devices [AMD/ATI] ES1000 bus-ID: 01:06.0
        Display Server: X.org 1.19.2 drivers: ati,vesa (unloaded:
        modesetting,fbdev,radeon)
        tty size: 92x24 Advanced Data: N/A for root
Drives:    HDD Total Size: 1068.1GB (1.1% used)
        ID-1: /dev/sda model: 2_1TB size: 1000.0GB temp: 68C
        ID-2: USB /dev/sdb model: KINGSTON_SV300S3 size: 60.0GB temp: 0C
        ID-3: USB /dev/sdc model: DT_101_G2 size: 8.0GB temp: 0C
Partition: ID-1: / size: 49G used: 4.9G (11%) fs: ext4 dev: /dev/sdb1
        ID-2: swap-1 size: 6.31GB used: 0.00GB (0%) fs: swap dev: /dev/sdb5
RAID:       No RAID data: /proc/mdstat missing-is md_mod kernel module loaded?
Processes: CPU: % used - Memory: MB / % used - Used/Total: 963.0/24114.9MB - top 5
active
        1: cpu: 14.3% command: cinnamon pid: 2199 mem: 355.28MB (1.4%)
        2: cpu: 8.8% command: Xorg pid: 2007 mem: 84.68MB (0.3%)
        3: cpu: 8.6% command: gnome-shell pid: 1742 mem: 151.65MB (0.6%)
        4: cpu: 0.3% daemon: ~kworker/6:2~ pid: 1224 mem: 0.00MB (0.0%)
        5: cpu: 0.2% command: nemo pid: 2232 mem: 51.88MB (0.2%)
[root@debian]#
```

```
[root@centos]# inxi -DGSCMm -t c -P -x
System:     Host: localhost.localdomain Kernel: 3.10.0-693.17.1.el7.x86_64 x86_64
       bits: 64 gcc: 4.8.5
       Console: tty 0 Distro: CentOS Linux release 7.4.1708 (Core)
Machine:    Device: server System: Dell product: PowerEdge T110 serial: 8GJC8P1
       Mobo: Dell model: 0V52N7 v: A01 serial: ..CN708210AN00ER.
       BIOS: Dell v: 1.5.2 date: 10/18/2010
CPU:        Dual core Intel Pentium G6950 (-MCP-) arch: Nehalem rev.5 cache: 3072
            KB
       flags: (lm nx sse sse2 sse3 ssse3 vmx) bmips: 11172
       clock speeds: max: 2793 MHz 1: 2793 MHz 2: 2793 MHz
Memory:     Used/Total: 298.2/3713.6MB
       Array-1 capacity: 16 GB devices: 4 EC: Multi-bit ECC
       Device-1: DIMM_A1 size: 1 GB speed: 1333 MHz type: DDR3 part:
       HMT112U7BFR8C-H9
       Device-2: DIMM_A2 size: 1 GB speed: 1333 MHz type: DDR3 part:
       HMT112U7TFR8C-H9
       Device-3: DIMM_A3 size: 1 GB speed: 1333 MHz type: DDR3 part:
       HMT112U7BFR8C-H9
       Device-4: DIMM_A4 size: 1 GB speed: 1333 MHz type: DDR3 part:
HMT112U7TFR8C-H9
Graphics:   Card: Matrox Systems MGA G200eW WPCM450 bus-ID: 05:03.0
       Display Server: N/A driver: mgag200
       tty size: 100x64 Advanced Data: N/A for root out of X
Drives:     HDD Total Size: 1060.2GB (0.6% used)
       ID-1: /dev/sda model: SPCC_Solid_State size: 60.0GB temp: 33C
       ID-2: /dev/sdb model: ST1000DM003 size: 1000.2GB temp: 20C
Partition: ID-1: / size: 35G used: 2.1G (6%) fs: xfs dev: /dev/dm-0
       ID-2: /boot size: 1014M used: 235M (24%) fs: xfs dev: /dev/sda1
       ID-3: /home size: 17G used: 60M (1%) fs: xfs dev: /dev/dm-2
       ID-4: swap-1 size: 4.16GB used: 0.00GB (0%) fs: swap dev: /dev/dm-1
Processes: CPU: % used - Memory: MB / % used - Used/Total: 298.0/3713.6MB - top 5
active
       1: cpu: 3.6% command: webmincron.pl pid: 12380 mem: 33.43MB (0.9%)
       2: cpu: 0.4% command: systemd pid: 1 mem: 6.66MB (0.1%)
       3: cpu: 0.2% command: python pid: 699 mem: 32.07MB (0.8%)
       4: cpu: 0.1% command: systemd-journald pid: 499 mem: 3.43MB (0.0%)
       5: cpu: 0.1% daemon: ~migration/1~ pid: 12 mem: 0.00MB (0.0%)
[root@centos]#
```

As you can see from the earlier output, **inxi** shows system hardware, central processing unit (CPU), drivers, Xorg, desktop for machines running a GUI desktop system, kernel information, disk information and statistics, in our cases the top five processes, Random Access Memory (RAM) usage, and other useful information about the three systems. Interestingly enough, the IBM System x3650 -[7979E9U] does *not* show any Redundant array of Inexpensive Disks (RAID) data available, although the two 1 TB drives are in a hardware RAID mirroring configuration. For more information about the **inxi** command, see the man page for **inxi** on your Linux system.

Exercise 17.2

Part a: Following up on your answer to In-Chapter Exercise 17.1, completely list and detail how you implemented your preinstallation considerations on a Linux system you have installed. Or, if you did not implement them because you didn't do an installation, how you *would* implement them on an actual hardware platform of your choice. Use our set of minimal recommendations for postinstallation tasks as a guide for how to fulfill the requirements of this exercise. In addition, you could consult with the person that actually installed your system on the hardware, if that person is available.

Part b: From looking at the output of the three **inxi** commands, which system has the slowest memory modules installed? Can you infer what the memory bus speed is from the speed of the CPU on that system? And what is Error Code Correction (ECC) memory? Why would these machines have that kind of memory? What is the file system type of the CentOS 7 computer? Given your Linux system and hardware, what options of the **inxi** command would be most useful and revealing for you to use?

17.2.2 Installation of Linux

Since we treat four representative Linux systems as models for various things we show throughout this chapter, the quickest way of installing any one of these systems is to follow the online instructions supplied with the software, for the version that you are installing. We supply installation instructions in Appendix A, Section 17.2, for Linux Mint, version 18.2. As stated in Section 17.1, we suggest limiting the installation to a single-disk system, where Linux is the only operating system on the disk. We also suggest using the 64-bit ISO-created DVD medium available at the following websites, for the particular version of Linux you want to install:

Debian: www.debian.org

Ubuntu: www.ubuntu.com

Linux Mint: www.linuxmint.com

CentOS: www.centos.org

The step-by-step installation process, guided by the instructions that are part of the documentation for the particular Linux system you want to install, and the choices you make during installation, have defaults for the simple installation we advise you to do. For example, for the four systems listed, we recommend that you use the default disk partition scheme when you are installing to a single hard drive system. Going with the defaults is particularly true if you do an installation of the GUI version of any of the earlier representative systems.

17.2.3 System Services Administration, Startup, and Shutdown Procedures

This section details the general procedures for starting up, managing the startup process, and gracefully shutting down Linux. It gives a brief overview of the steps that the system goes through in successfully starting up and shutting down. It then gives some further references to the basic but important systemd services administration utilities available to a system administrator. In those further references, we give examples of commands for manipulating and changing system services by enabling and running new services.

17.2.3.1 *The Boot and Startup Processes*

At its very simplest, a computer running a systemd-controlled Linux system goes from a powered-off state to full, normal operation in the sequence Booting, then Startup. This is shown in detail as

Booting

- The firmware on the main system board finds the bootable media.
- From the bootable media, the bootloader program loads the monolithic kernel from the boot media into RAM and starts it.

Startup

- The kernel uses the commands passed by the bootloader to find the main file system, at which point it can find and run systemd.
- systemd runs other programs to find the file systems and start network and all necessary local service processes.

The term "booting" is used here to mean bringing the operating system from a complete power-off condition to a point where systemd can take over. Then in "startup," systemd brings the computer to a steady-state fully normal operating condition.

Figure 17.1 illustrates the sequence of powering on, booting, startup, and shutdown on a typical modern computer running Linux. In Figure 17.1

- Power on means pushing the power on button on the hardware.
- Power-On Self-Test (POST) is hardware initialization done in conjunction with the bootloader programs.
- On our representative Linux systems, the Basic Input/Output System (BIOS) and the Unified Extensible Firmware Interface (UEFI) are the firmware systems that take over initialization of the booting sequence.
- The GRand Unified Bootloader, version 2 (GRUB2), is the bootloader program.
- The kernel is initiated, runs, and then executes /etc/systemd, which in turn brings the system to a normal operating state defined by the default Target. In a text-based server, this default Target is usually defined as multiuser.target. In a GUI environment, this default Target is usually defined as graphical.target.
- The shutdown procedure is illustrated in Section 17.2.3.2.

Further details of how a Linux system that is using systemd starts up are given in Section W26.2.3.1.1, titled "systemd Bootup," at the book website.

The BIOS booting component process, and its modern equivalent, the UEFI process, work according to specifications that control and define the firmware interface to the operating system. The operations of systemd when Linux starts up, and shuts down, are shown in detail in Chapter 18. As shown in Figure 17.1, several steps occur before systemd as a startup and initialization program takes over. POST, BIOS (or UEFI), GRUB2, and then the kernel, initialize hardware and prepare for the first process to be run from the root file system. After the root file system is found and mounted, systemd takes over and is

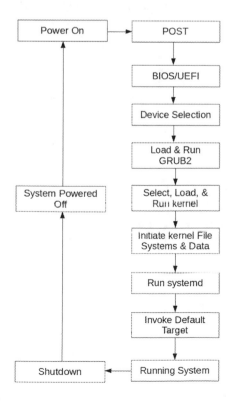

FIGURE 17.1 Booting, startup, and shutdown on a systemd Linux system.

then responsible for initializing all remaining hardware, mounting all necessary file systems, invoking the default systemd target, and managing the start of system services.

But then what does an ordinary user do when a normal boot and startup cannot be done at all, or normal shutdown cannot be achieved, for whatever reason?

For booting and startup, barring power or hardware-related issues exposed during POST, such as failed hard disk or memory, there are a couple of steps you can take to examine the possible problem. You can actually intercede at any of the steps illustrated in Figure 17.1.

You can attempt to boot the system into "rescue" mode, by trying to boot to the systemd emergency or rescue target states. This is shown in detail in Chapter W27, Sections 2.1 and 5.1.8.8, at the book website. This is similar to the legacy operation of entering "single-user" mode upon booting.

You can also use the bootloader program, GRUB2, to enter a "recovery" mode upon booting. On our representative Linux Mint system, when POST is completed successfully, and just before GRUB2 takes over the boot process, you can hold down the Shift key on the keyboard to enter a GRUB2 menu. On other Linux systems, that interrupt can be different.

Be aware that we do not give the details of GRUB2 configuration and modification here.

The advanced option of the GRUB2 menu includes choices for booting into alternative, or previous, "boot environments," and into recovery modes for those environments. These environments are usually named according to kernel version. So if you have upgraded the operating system kernel during the course of using your computer system, different environments with different kernels will be available to you. For an ordinary user, this choice for troubleshooting the system, by entering these system states during the boot stage, is a useful possibility that would help you recover data, logs, and other valuable information.

When a system hangs up during the shutdown procedure, a contemporary way of diagnosing the problem is to examine the systemd journal, with the **journalctl** command, as shown in Chapter W27, Sections 27.2.4 and 6.2. This way, if some process or program is preventing the system from reaching a powered-off condition, the journal logs will allow you to see what that process or program is. On shutdown, systemd attempts to stop all services, and unmount all file systems. The system is finally powered down. We give more details of how a Linux system using systemd shuts down in the next section.

Exercise 17.3

How can you boot into an alternative environment found on another boot medium on your computer system, and in which phase of the bootup process is this alternative method presented to you? Why would you want to do this? Make note of exactly how that is done on your system. Also, draw an arrow in Figure 17.1 that would specify what path would be followed if the system needs to be rebooted rather than powered off.

17.2.3.2 systemd and Traditional System Reboot or Shutdown

systemd is in charge of taking the system from the normal operating condition to a power-off state or to reboot the system. The relevant systemd **systemctl** commands that achieve shutting the system down or rebooting it, and briefly what they accomplish, are shown in Table 17.1. We describe these commands

TABLE 17.1

systemd Shutdown Commands

Command	Description
systemctl halt	Halts the system
systemctl poweroff	Powers off the system
systemctl reboot	Restarts the system
systemctl suspend	Suspends the system
systemctl hibernate	Hibernates the system
systemctl hybrid-sleep	Hibernates and suspends the system

and the systemd shutdown process further in Chapter 18, Section 2.3, and in Chapter W27, Section 5.1.10, at the book website. The systemd shutdown path through possible target states, from a normal operating condition to a powered-off state, is described in Figure 18.3.

Traditionally, the graceful shutdown or reboot procedures can be done graphically from the active window system or from the command line or console window, using the **halt** or **shutdown** commands and their options. The **shutdown** command has the advantage of allowing you to specify a time when shutdown processes are initiated.

Graceful shutdown procedures are generally done as follows:

1. Shutdown system and user processes
2. Flush system memory to disk
3. Unmount file systems
4. Power off

An example of using the **shutdown** command on Linux is as follows:

```
$ shutdown -h now
```

where **-h** means halt the processor and **now** is the time option and means immediately.

To get a more complete description of the Linux **shutdown** command, particularly the format of the time option, see the man page on your Linux system.

17.2.3.3 Preliminary Considerations When Managing System Services with systemd

The system service manager that is available in our base Linux systems is systemd. In fact, systemd is the system service manager for all distributions in the three major branches of Linux, and all important peripheral Linux distributions as well.

We detail Linux system service management with systemd completely in Chapter 18, and provide several examples there of how to effectively control operating system services in a modern Linux system.

We first pose a preliminary question about "system services" in Linux.

Question: What is a daemon?

Answer: Basically a daemon is a background, ongoing process that is not linked or controlled by a terminal. Particularly not connected in the usual way to standard output or stand error.

Question: Is a "service" a daemon?

Answer: Sometimes.

More accurately and generally, a service can be a process or collection of processes, the overall state of the system, or the state of a physical device, a virtual device, a dataset, etc. And as a provider of resources, or a collection of application capabilities, it can have more than one instance. For example, the many layered file systems present on the system, or the multiple means of remote login to the system.

17.2.3.4 Further References for System Service Management Using systemd

In this section, we provide further references for the use of systemd in Linux for service management, beyond what is presented in Chapter 18. Our primary reference is an example, detailed in Chapter W26 at the book website, of how to enable server side host services for a secure ftp daemon, known as vsftpd. That complete primary, basic example of using systemd to manage system services, is Example W26.1, titled "System Service Management Using systemd."

In addition, Chapter W23, titled "Virtualization Methodologies" at the book website, also shows examples of how you can implement the vsftpd service securely inside an LXC/LXD container, and within a VirtualBox Virtual Machine (VM).

As noted in the references, you should be cautious when executing these examples, particularly if you are not using one of the virtualization methodologies shown in Chapter W23 at the book website, and are on an insecure network connected to the Internet. You must have root privileges on the computer to execute the examples we show in the references as well.

Exercise 17.4
What is the danger of using ftp, telnet, rlogin, or rsh from a remote site on the Internet into your home computer?

17.3 User Administration

The two most important objectives of user administration are service and security. First, providing service to ensure that the user base has access to and can fully take advantage of the resources that a modern Linux system can provide. Second, securing the files and processes that the user base needs to utilize those resources. Later in this chapter, we go over some of the security methodologies that a system administrator can deploy to keep the system secure. In Chapter W23, titled "Virtualization Methodologies" at the book website, we also show security methods that utilize various virtualization technologies such as LXC/LXD containers and VirtualBox VM.

The traditional Linux technique for providing service and security to users is through access privileges, for an individual user, the group, or all others on the system, to specific objects on the system, such as files. We show how to set access privileges on files in Chapter 5, Sections 4 and 5. Designing and implementing user groups and access privileges for user groups are the most important parts of this technique.

Certainly, user account and group creation and configuration are the first steps in providing maximum service and security to the user base. That is true even if you are the only user of the system.

A very integral part of designing user accounts and groups is the detailed consideration given to the recommended user data storage model we show in Section 17.2.1, Item (1). This is done by securing user files on a second, possibly redundant, hard disk drive, so that they can be kept separate from the operating system boot disk itself. It can be practically achieved by properly creating user accounts so that they have their home directories on that second hard drive. As we show in Section 17.4, various ways of adding redundant persistent media very easily and effectively implement the model. In particular, Section 17.3.4 offers a general method of how user account and group creation and management can be done so that it conforms to the model.

We concentrate here on how to manage user accounts and groups on Linux with the adduser, addgroup, deluser, delgroup, newusers, and passwd commands. These activities are usually done a significant time after the initial installation of your Linux system, but can be done during installation and initialization of the system as well. Table 17.2 shows brief descriptions of these commands:

TABLE 17.2

Basic User Administration Commands

Command	Description
adduser	Add a new user account
deluser	Delete a user account
addgroup	Add a new group
delgroup	Delete a group
newusers	Update and create new users in batch mode
passwd	Manage user passwords
usermod	Modifies user accounts

The examples in this section illustrate

- simple cases of text-based user account creation and configuration,
- text-based group creation and configuration, and
- text-based deletion of a user account and a group.

Exercise 17.5

Make a table or chart of what users and groups need to be added to your system, and what their default account parameters and group memberships should be. What command can you use to identify all existing groups on the system?

17.3.1 Adding a User and Group in a Text-Based Interface on Linux

The commands adduser and addgroup, by default, add a "normal" (non-system level) user or group to the system using command line options and option arguments. They take default configuration information from /etc/adduser.conf. They are a more developed front-end for the older, low-level commands, such as useradd, groupadd, and usermod. adduser utilizes patented user ID (UID) and group ID (GID) values for creation of new accounts and groups, and by default creates a home directory from a skeletal configuration file, etc.

adduser itself is a script file that interfaces to the following commands and their functionality:

```
useradd, groupadd, passwd, gpasswd, usermod, chfn, chage, edquota.
```

An abbreviated listing of the syntax of the adduser and addgroup commands is as follows:

adduser, addgroup - add a user or group to the system

Syntax:
```
adduser  [options]  [--home  DIR]  [--shell  SHELL]  [--no-create-home]
[--uid ID] [--firstuid ID] [--lastuid ID] [--ingroup GROUP | --gid  ID]
[--disabled-password]       [--disabled-login]        [--gecos      GECOS]
[--add_extra_groups] [--encrypt-home] user

adduser --system [options] [--home DIR] [--shell  SHELL]  [--no-create-
home]  [--uid  ID]  [--group | --ingroup GROUP | --gid ID] [--disabled-
password] [--disabled-login] [--gecos GECOS] user

addgroup [options] [--gid ID] group

addgroup --system [options] [--gid ID] group

adduser [options] user group
```

Purpose: adduser and addgroup add users and groups to the system according to command line options and configuration information in /etc/adduser.conf. They are friendlier front ends to the low-level tools like useradd, groupadd, and usermod programs, by default choosing Debian policy conformant UID and GID values, creating a home directory with skeletal configuration, running a custom script, and other features. adduser and addgroup can be run in one of five modes

Output: New or modified user accounts or groups

Commonly used options/features:

[--quiet]

[--debug]

[--force-badname]

[--help|-h]

[--version]

[--conf FILE]

The basic, simple usage as superuser of the adduser command does the following five things:

```
$ sudo adduser username
```

1. Create the user named username.
2. Create the user's home directory (the default is /home/username), and copy the files from /etc/ skel into it.
3. Create a group with the same name as the user, and place the user in it.
4. Prompt for a password for the user (by default not shown on-screen), and its confirmation.
5. Prompt for personal information about the user, such as telephone, office contact, etc.

Given later are two simple examples of how to add a new user to Linux in a text-based, command line interface.

The first example illustrates how to interactively create a single user account from the command line as superuser. If you make a mistake in creating a user account, you can always remove the account immediately by using the **deluser** command, shown in Section 17.3.2.

Example 17.1 Linux adduser Command for a Single User Account

```
$ sudo adduser sarwar
[sudo] password for bob: qqq
Adding user `sarwar' ...
Adding new group `sarwar' (1002) ...
Adding new user `sarwar' (1002) with group `sarwar' ...
Creating home directory `/home/sarwar' ...
Copying files from `/etc/skel' ...
Enter new UNIX password: www
Retype new UNIX password: www
passwd: password updated successfully
Enter the new value, or press ENTER for the default
        Full Name []: ENTER
        Room Number []: ENTER
        Work Phone []: ENTER
        Home Phone []: ENTER
        Other []: ENTER
Is the information correct? [Y/n] Y
$
```

Exercise 17.6

Create a single new account on your Linux system, with the **adduser** command as shown earlier. Be sure to use the entries you made in your answer to In-Chapter Exercise 17.5 to override the defaults for user account configurations on your system.

The second example uses the command **newusers**, which allows "batch" mode creation of new users, in conjunction with a file that contains a listing of several user accounts that can be added all at one time.

A brief description of the newusers command is as follows:

newusers—update and create new users in batch mode

Syntax:

newusers [options] [file]

Purpose: The newusers command reads a file (or the standard input by default) and uses this information to update a set of existing users or to create new users. Each line is in the same format as the standard password file, as follows:

pw_name:pw_passwd:pw_uid:pw_gid:pw_gecos:pw_dir:pw_shell

---

where

pw_name is the name of the user you want to create,

pw_passwd is the password,

pw_uid is the UID,

pw_gid is the GID,

pw_gecos is an identifying comment,

pw_dir is the home directory path,

pw_shell is the new users default shell

Common Options:

-h, --help   Display help message and exit

-r, --system Create a system account

Example 17.2 Creating Several New Users in "Batch" Mode

1. Use your favorite text editor to create the following file, named "dev_grp.txt" in your home directory. Feel free to add more lines similar to the one shown to add more new users to the listing.

```
hassan:QQQ:2001:2001:CFO of Accounting:/home/hassan:/bin/bash
```

The seven fields of the earlier line, separated by colons (:), are the new user accounts name, password, UID, GID, GECOS commentary, default home directory, and default shell.

2. Use the following command to create the user(s) encoded in the file dev_grp.txt:

```
$ sudo newusers dev_grp.txt
[sudo] password for bob: qqq
$
```

3. Check the addition of the new user(s) using the following command:

```
$ more /etc/group
Output truncated…
hassan:x:2001:
$
```

Exercise 17.7

According to the table or chart of user accounts and their configuration requirements you did for In-Chapter Exercise 17.5 earlier, create a file of new user accounts. As shown in Example 17.2, for each user account you must have a single line in the file with seven colon-delimited fields that contain the configuration for each user. The table or chart design determines what is in the file.

17.3.2 Adding and Maintaining Groups in a Text-Based Interface

The easiest and most efficient way to create and manage user accounts and groups is by using the GUI-based Users and Groups Manager, accessed from the Cinnamon Desktop in Linux Mint via the menu choices Menu > System Settings > Users and Groups. But that method is somewhat limited in what it can accomplish. Therefore, it is recommended to use the addgroup and adduser command to manage groups.

Following is an example of how to add and manage groups.

Example 17.3 Adding and Managing Groups and Users from the Command Line

Objectives: To add a new group and manage those users that are in that group.

Prerequisites: Examples 17.1 and 17.2 having default groups available on your system.

Background: Text-based management of groups is accomplished with the addgroup command and its options.

Requirements: Do the following steps, in the order presented, to complete the requirements of this example

1. Create a group named "development," using the addgroup command as follows:

    ```
    $ sudo addgroup development
    [sudo] password for bob:
    Adding group `development' (GID 1003) ...
    Done.
    $
    ```

2. Add the user sarwar to the group named "development" using the adduser command as follows:

    ```
    $ sudo adduser sarwar development
    Adding user `sarwar' to group `development' ...
    Adding user sarwar to group development
    Done.
    $
    ```

3. List the groups on the system, and verify sarwar as a member of the group named "development":

    ```
    $ cd /etc
    /etc$ more group
    Output truncated...
    sarwar:x:1002:
    development:x:1003:sarwar
    /etc $
    ```

4. Remove the user sarwar from the group named "development":

    ```
    /etc $ sudo deluser sarwar development
    Removing user `sarwar' from group `development' ...
    Done.
    /etc $
    ```

5. Verify that sarwar has been removed from the group named "development":

    ```
    /etc $ more group
    Output truncated...
    sarwar:x:1002:
    development:x:1003:
    ```

Exercise 17.8

You have an air-gapped Linux computer in a room. You want only certain users in a defined group named "proj" to be able to access files in a directory named "proj1" in your account on the file system of that computer. Use the graphical User Manager to add those certain users, and yourself, to the group "proj." Create the "proj1" directory. Then, set the permission bits appropriately on your home directory, the "proj1" directory, and any files put in that directory, so that only those certain users can read, write, and execute the files in it. The room is open to the public, but to login to the computer, each individual user has to use her own password.

Exercise 17.9

There is another air-gapped computer in the room, older, but it still runs the latest Linux! Unfortunately, it does not have a graphics-capable monitor. It can only use a text, command line interface. Achieve the same results as Scenario 1, but by using only text-based commands. Name the group "project2" and the directory "proj2."

17.3.3 Deleting a User Account and Group from the Command Line

To effectively delete a user account or a group from the system using the command line, use the deluser or delgroup commands. A brief description of those commands is as follows:

deluser, delgroup - remove a user or group from the system

Syntax:
```
deluser [options][--force][--remove-home]  [--remove-all-files]
                  [--backup] [--backup-to DIR] user
deluser --group [options] group
delgroup [options] [--only-if-empty] group
deluser [options] user group
```
Purpose: deluser and delgroup remove users and groups from the system according to command line options and configuration information in /etc/deluser.conf and /etc/adduser.conf

Common Options:

[--conf FILE]   Use FILE instead of /etc/deluser.conf

--version   Display version and copyright information

Following is an example of deleting a user account named sarwar from the system, using the deluser command:

Example 17.4 User Account Deletion with deluser

Objectives: To delete a user account from the command line.

Prerequisites: Doing Examples 17.1–17.3, or having a user account on your system that you want to permanently delete.

Background: Similar to the adduser command, deluser deletes a user account from the system.

Requirements:

Do the following steps, in the order shown, to complete the requirements of this example:

1. To delete the user account sarwar we created in Example 17.1, type the following command:

```
$ sudo deluser sarwar
[sudo] password for bob: XXX
Removing user `sarwar' ...
Warning: group `sarwar' has no more members.
Done.
$
```

17.3.4 A Method of User and Group Creation Utilizing a Second Hard Disk

We provide a general, basic method here to allow you to create users and groups on your Linux system that conforms to a non-ZFS version of our recommended user data storage model from Section 17.2.1, Item (1).

The method comprises specific commands and techniques shown throughout this chapter. For further explanations of the commands listed here, consult the man pages on your system.

The method is as follows:

Step 1: A second persistent medium, such as an internally mounted SATA hard drive, an externally mounted USB-bus device, or some other medium connected that uses some other bus
architecture, must be added to your system. See the basic traditional approach of doing this in
Section 17.4.4.

In Section 17.4.4.1, Example 17.6, you create a new primary partition on that medium, and then
create an ext4 file system on that partition. You can also create "home" directories on that drive,
as implied in the conclusion of that example as well, that will become the home directories for
new users that you add to the system as needed.

Step 2: When you are using the adduser command as shown in Section 17.3.1 to add individual
new users, if you use the --home option, you can specify that the new user's home directory be the same directory you created in Step 1. To use a template for larger-scale account
creation, you can use the newuser command. You can also customize the default "skeleton"
file that is used by the adduser command so that it uniformly provisions newly created user
accounts, groups, etc., so that all new accounts have their home directories on the new hard
disk.

Step 3: If you manipulate the ownership permissions of the directories from Steps 1 and 2, the new
user will own and have access to the directories on the second hard disk. This is achieved by a
privileged user via the **chmod** and **chown** commands.

But you must also consider the impact of these permission changes on your system's security
model.

By tailoring home directory and file security so that it is in conformance with your security
model, via traditional file access permissions, or via ACLs as shown in Section W26.9.3 at the
book website, you can ensure that users and groups are limited and structured to conform to
both the security model and any performance use case(s) designed by you.

We pose a problem at the end of this chapter that asks you to deploy the earlier method
using your system's hardware capabilities, the details of your own security designs, and system-
specific parameters you must supply.

17.3.5 Basic Password Management

It is often necessary for the person responsible for user and group account management, and for obvious
security reasons, to make changes to passwords. This is accomplished most easily with the **passwd** command. We give a few basic examples of the use of this command. For a more complete description of the
passwd command, see the man page on your Linux system.

To reset the password for a user account, use the following command:

```
$ sudo passwd mansoor
[sudo] password for bob: QQQ
Enter new UNIX password: ZZZ
Retype new UNIX password: ZZZ
passwd: password updated successfully
$
```

To obtain a brief listing of the options available with the passwd command, use the following command:

```
$ sudo passwd -h mansoor
Usage: passwd [options] [LOGIN]

Options:
  -a, --all                 report password status on all accounts
  -d, --delete              delete the password for the named account
  -e, --expire              force expire the password for the named account
  -h, --help                display this help message and exit
```

```
-k, --keep-tokens              change password only if expired
-i, --inactive INACTIVE        set password inactive after expiration to INACTIVE
-l, --lock                     lock the password of the named account
-n, --mindays MIN_DAYS         set minimum number of days before password
                                              change to MIN_DAYS
-q, --quiet                    quiet mode
-r, --repository REPOSITORY    change password in REPOSITORY repository
-R, --root CHROOT_DIR          directory to chroot into
-S, --status                   report password status on the named account
-u, --unlock                   unlock the password of the named account
-w, --warndays WARN_DAYS       set expiration warning days to WARN_DAYS
-x, --maxdays MAX_DAYS         set maximum number of days before password
                                              change to MAX_DAYS
$
```

To see the status of a particular users' password, use the following command:

```
$ sudo passwd -S mansoor
mansoor P 11/15/2016 0 99999 7 -1
$
```

The output of the earlier command is described in the passwd man page as

The status information consists of seven fields. The first field is the user's login name. The second field indicates if the user account has a locked password (L), has no password (NP), or has a usable password (P). The third field gives the date of the last password change. The next four fields are the minimum age, maximum age, warning period, and inactivity period for the password. These ages are expressed in days.

17.3.6 Account Security

Once a user or group account has been added to the system, the next concern of a system administrator is how to maintain security, in terms of that user or group accounts' permissions to access and utilize system services and files in general. We show a method of doing this using ACLs in the Section W26.9.3, titled "Using Access Control Lists (ACLs) in Linux" at the book website. We discuss system security in general in Section 17.9 and follow that up with details in subsections.

As stated in Section 17.9.3, traditional Linux permissions, which define secure access to file objects like a regular file or a directory, are the permissions of read, write, and execute. Chapter 5, Section 5.4 has shown that form of file and directory permissions. Additionally, there are other advanced techniques for setting permissions, such as setting the setuid, setgid, and sticky bit. Beyond these, the ACL model gives users finer-grained control over file and directory object security. Every file object can be thought of as having associated with it an access ACL that controls the discretionary access to that object. For a directory, this ACL is referred to as a default ACL. Linux, by default, uses what is known as the POSIX.1e ACLs model.

17.4 File Systems, Connections to Persistent Media, and Adding Disks to Your System

Question: What is a computer file system?

Answer: A way of logically ordering data, so that if persistent, can be securely and easily located very quickly, and then accessed in a consistent way for use.

Providing data persistence, which is the third primary objective of the Linux operating system itself, in large part involves establishing and maintaining connections to persistent media, such as disk drives. These drives, and the file systems found on them, can be physically connected directly inside of the computer, such as with spinning hard disks, SSDs, or USB thumb drives. They can also be some form of remote virtual drive, such as a network-available drive, or remote volumes and complements of drives.

And in some cases, a file system may *not* make use of a persistent storage device or medium at all! At this higher level of abstraction, the file system can access, use, organize, and represent *any* form of data, whether it is persistent or volatile. We use the word volatile here to mean during the transient lifetime of some process or service. Of course, systemd and the Linux kernel control all processes and services running on the computer. *Pseudo* and special purpose file systems (sometimes called *synthetic file systems*), which can be thought of as virtual file systems, have this characteristic.

There are also established protocols for connecting to and accessing either physical, virtual, or pseudofile systems, in the locations where they may reside. Two examples of access protocols that establish and maintain the connections to virtual, network-available media and file systems are the NFSv4 and the iSCSI.

In this section, we first present an organizing scheme that you can use to think about types of drives and file systems. A file system may be nominally assigned according to what medium it exists on: a directly connected physical medium (such as SATA or Serial Attached Small Computer System Interface (SAS) hard disks), a virtual medium (such as NAS or SANs that use NFSv4 or iSCSI), or as a specialized pseudofile system that is not on a persistent medium at all (such as the cgroups or proc filesystems).

We then go on to give some particular examples of adding directly connected, physical disks to a Linux system.

Additionally, we suggest two approaches, when using the recommended storage model we give in Section 17.2.1, to adding persistent media to your Linux system. They are as follows:

1. Use the ZFS to integrate the medium as a virtual device into a new or existing pooled array of devices, where the pool can provide a variety of different levels of redundancy. These pools are known as "zpools." This is a simple method where, after placing a primary partition on the medium, it is automatically given a file system and mounted. This is shown in detail in Chapter W22, titled "ZFS Administration and Use" at the book website.

   At the time of the writing of this book, it was possible to install ZFS on the Debian-family of Linux (Debian, Ubuntu, and Linux Mint) and CentOS 7.4, in an easy and efficient manner.

   See Appendix A, Section 26.4.7, for detailed installation instructions for ZFS on Debian-family (Debian 9.1, Ubuntu 16.04, and Linux Mint 18.2) and CentOS 7.4.

2. A more traditional approach, where you first properly connect the device to the computer with cables, partition that medium, manually add a file system to it (typically the ext4 file system), and then finally create directories and files in the partition(s) on the media. The sections later detail this more traditional approach.

According to our organizing scheme, a file system can be separated into three hierarchically arranged layers that perform particular functions. These layers, arranged from farthest-to-nearest to the actual hardware of the drives in question, are as follows:

The Logical Layer: This layer is used for interaction with user application programs, and the processes they consist of, via Linux system calls. It provides the application programming interface (API) for file operations, for example system calls to OPEN, CLOSE, READ, etc., and connects with the layer below it for processing. The logical layer achieves efficient file access, logically organized directory operations, and provides user autonomy and security. In Figure 17.2, this layer is represented by the ISCSI and Process Inputs blocks at the top of the figure.

The Virtual Layer: This layer provides the interface mechanisms for maintaining multiple, simultaneously existing implementations of physical and virtual file systems on the same computer. For example, it makes possible mounting and transparently using NFSv4, btrfs, ext3, ext4, fat32, and ZFS at the same time on the same system, and operating with files from all of those implementations as if they were all of the same type.

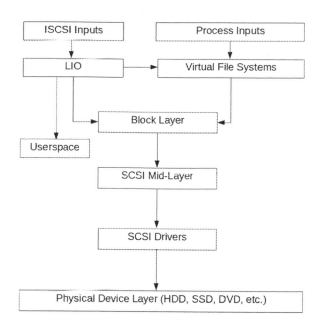

FIGURE 17.2	Linux file system storage layers.

The Linux IO (LIO) block shown in Figure 17.2 is known as an iSCSI "target," and represents virtual layer connectivity to persistent media from various network connections. These connections traditionally use high-speed Fibre Channel technology to create a Storage Area Network (SAN), but most importantly can also be achieved using Ethernet and TCP/IP.

A more detailed architectural scheme of classifying virtual layer file systems further separates that layer into block-based, network, stackable, pseudo, and special-purpose categories. In Figure 17.2, many of the different implementations of file systems, such as NFSv4, ext4, xfs, btrfs, initramfs, and the procfs, would be situated within the Virtual File Systems block shown in that diagram. Two very contemporary and extremely important virtual file systems are the cgroups file system (cgroupsfs) used by systemd, and the userland, block-based file system, ZFS.

The Physical Layer: This layer is concerned with the physical operation of the persistent storage device. It processes physical blocks being read or written. It handles buffering and memory management and is responsible for the physical placement of blocks in specific locations on the storage medium. The physical file system interacts with the device drivers or with the channels that physical devices communicate over. At a certain level of simplification, this layer can be schematically represented in Figure 17.2 by the integrated grouping of Block Layer, SCSI Layers, and the Physical Device Layer.

Furthermore, and as a integral part of the operation of the Linux kernel as it exists in a transient and volatile state as we have defined it, the pseudo and special-purpose file systems can be viewed as a series of conduits through which the entire system itself "flows." Thinking along these lines, when the kernel is in the CPU and attendant RAM, the kernel code itself is organized as a file system. The kernel (or systemd superkernel) can be viewed as a file system of volatile data structures that maintains the steady state of the hardware and software using these conduits exclusively; this achieves the overarching goals of virtualization, concurrency, user autonomy, security, and the necessary archival, long-term data persistence on the file systems that are established to do so.

Over the history of Linux, various file systems have been used to provide speed, efficiency, security, and utility to the ordinary user. The most contemporary and universal of these, across the three major branches of Linux, is the Linux Extended File System (ext). All of the representative flavors of Linux we show in this book use it as the default file system.

The 4th version of ext, ext4, is the current and most robust version so far. It has several features, such as large scalability, the ability to map to very large disk array sizes, and other very critical features, such as journaling. Following is a compact listing of some of the features of ext4:

It can support volumes with sizes up to 1 exabyte and files with sizes up to 16 terabytes.

It uses an "extents" mapping scheme, which replaces block mapping used by earlier versions of ext. An extent is a range of contiguous physical blocks.

It is backward-compatible with ext3 and ext2. Therefore, ext3 and ext2 can be mounted as ext4.

It delays block allocation until data is flushed to disk.

It has an unlimited number of subdirectories that can be created.

It has a multi block allocator that can make better choices about allocating files contiguously on disk.

It provides timestamps, measured in nanoseconds.

Most importantly, from the perspective of our recommended data storage model, adding a second hard drive, for example, allows you to keep the operating system and the user data on two different physical persistent media. That way, if the operating system fails and the system disk is corrupted and unusable, the data survives on the user data disk, and can very easily be recovered.

There are many traditional, legacy methods of achieving the objective of our data storage model, using facilities such as Linux disk and file maintenance commands, utilities such as mdadm (a software RAID manager), and Linux Volume Management (LVM). But from our perspective, the modern and contemporary way of implementing the recommended data storage model is using ZFS on redundant additional persistent media. With ZFS, you get bit-level data integrity, volume management, RAID capabilities at all levels, and a failure-proof backup strategy, all rolled into one utility.

There are some very important reasons for adding persistent media to your system, aside from conforming to our recommended storage model. Your hard disk may be running out of space, or beginning to show signs of failure. In a traditional legacy scheme, these situations might involve downtime of the system in order to correct the problem, whereas they can be handled by ZFS without bringing the system down.

Partitioning Schemes and Strategies: From all of your previous work in the Chapters 1–14, it should be evident to you that Linux organizes everything in files and uses a file system to organize those files. A disk partition can be most simply defined as a logical area of the disk that holds a file system.

Once you verify that your Linux system can recognize additional media, there are several reasons to adopt a particular partitioning scheme for a newly added disk drive. Creating multiple partitions on your hard drive avoids full disk problems by segregating directories into those partitions and gives the system administrator control over access to those directories and partitions. And you can maintain different file system structures simultaneously in different partitions. Even doing the initial installation of the system into a customized partitioning scheme allows you to install multiple operating systems on your computer.

When you are adding new disk drives, you have the option of using the traditional method of doing disk partitioning with command line utilities such as fdisk or gdisk, or the option of using a GUI-based method, with a utility such as Gparted. As we show in the later sections, you can implement our recommended data storage model with a traditional ext4 Linux file system scheme. We present an example detailing how to do this using the traditional ext4 scheme.

To help the administrator of a Linux system with the task of adding persistent media, the following sections will also address these general concerns:

- The availability of software device drivers for the new hardware to be added.
- How the hardware will be recognized, configured, and deployed on the system.
- Identification of possible paths to replacement and upgrading of existing hardware, within the context of using ZFS and our recommended data storage model.

As discussed in Item (1) of Section 17.2.1, our data storage model for desktop computers is capable of deploying two or more hard drives, and its implementation is made possible by what we show in the following subsections.

We show some further examples of additions of persistent media in Section W26.4.7 at the book website. Here, the primary example details ZFS-based additions done for SATA hard disks, whereas the secondary example details creating and managing RAID arrays in Linux using mdadm.

The additions in the traditional ext4-based example shown in Sections 17.4.3 and 17.4.4 will be done not only for USB thumb drives but can also be easily extended to other persistent media, such as internally or externally mounted SSD(s) or spinning hard disk(s).

SATA Hard Disks and USB Media: Generally, when you add a SATA hard disk some significant time after you have installed the operating system on the computer, you will want to partition it. You might even want to create a new partition table on it, create one or more partitions, and format those partitions using a standard Linux file system, such as ext4. We emphasize and encourage the use of the Gparted GUI-based application to do this.

When you add an external USB bus medium, such as a thumb drive or other form of persistent storage device, it is generally already formatted to the file system type known as FAT32 (in the case of most popular commercially available USB thumb drives) or to some other format depending on the media. Traditionally, you can then partition the disk using the fdisk command and add a file system to it with the mkfs command. We emphasize and encourage the use of fdisk, or its newer sister, gdisk, to do the partitioning.

Note that safe removal of USB media can be done manually in Linux. For example, using a desktop system GUI, right click on the USB thumb drive icon on the desktop, and make either the Eject or Safely Remove choices. Unmounting a USB thumb drive, or other USB-bus media, can also be done from the command line with the **umount** command.

When a USB thumb drive or other external medium is automatically mounted on Linux Mint, the path to it is **/media/your_home_dir/id**, where **your_home_dir** is the name of your home directory on the system, and **id** is the disk id number or identifier.

For example, when we added a 2-TB Seagate USB 3.0 SATA hard disk and create a single primary partition on it with Gparted, along with an ext4 file system automatically added to that partition at the same time, the newly mounted hard drive was accessed via the path /media/bob/Seagate_Backup_Drive_Plus.

17.4.1 Preliminary Considerations When Adding New Disk Drives

If you insert a USB thumb drive that you know is functioning properly into your computer, and it is *not* recognized, the chances are that your Linux system does not have a device driver available to enable communication between the computer and the thumb drive.

How do you know if a new disk drive is recognized, and most importantly, is usable on your system? There are at least three quick and easy ways to know whether the new disk drive is recognized.

1. If it is a USB thumb drive, and it is formatted to FAT32, Linux Mint will automount it, and an icon for it will open on the Cinnamon desktop (along with a file folder view of its contents).
2. In a terminal window, you can use the command **systemctl –f** and watch the screen display. It will show that a new device has been added, even though in the case of a USB thumb drive, it might be formatted to something other than FAT32!
3. Use the before-and-after technique shown in Section 17.4.2.

The same is true when you connect a SATA bus hard drive properly, but the probability of it not being recognized is much lower. The best and easiest thing to do in a case like this is to use another USB or SATA device. Linux has facilities to find and install device drivers on your system for a device, but this process is time consuming and may not be fruitful for the particular device in question. Also, it is possible to write a driver for your device, which is even more time consuming. The important thing here is not that the USB thumb drive is formatted to FAT32, but that a manufacturer has the device drivers available automatically when their device is inserted. This is not always true.

In many instances, it is important to know the physical device name, the instance name, and the logical device name of disk drives on your system, but practically speaking, for the Linux system administrator, easily finding out the logical device name of a disk drive is most important.

You may want to add a hard disk to your Linux system that has been used on another computer operating system previously. In that case, the primary and secondary examples we show can be deployed to repartition and prepare that hard disk for new use on your Linux system.

17.4.2 Five Quick and Easy Ways to Find Out the Logical Device Names of Disks

Before attaching a new disk drive to your Linux system, it is important to know how to determine, in a very quick and easy manner, what the currently installed logical device names of the disk drives actually attached and usable on your system are. What we mean by "attached and usable" is that the disk drive is properly connected and recognized by the system and has a device driver that the system can use to communicate with it.

Before and after: If you want to find out the logical device name of a new disk you want to add to the system, use one of the following methods to see what disks are on your system *before* you add the new disk, and then use the same method *after* the new disk has been added and note the difference. The different or new logical name that appears will be the logical device name of the new disk.

The five simple methods that follow show how to determine what disk drives are attached and usable on your system, and what are the logical device names of those and any others you might want to add to your system.

Method 1: Change your current working directory to /dev. Type ls. Hard drives, for example, show up in the ls listing as sda, sdb, etc. The full path to the first slice, or partition, on one of these disks is specified as /dev/sda1. A USB bus device, like a thumb drive, would show up in the ls listing as /dev/sdc, or whatever letter designation comes after the hard drives, and the full path to the first slice on it would be /dev/sdc1.

Method 2: Type **df -h** on the command line to find out the file system names and paths they are mounted at on your system. On our Linux Mint system, when we did this to see if a Lexar 60 Gb USB thumb drive was recently successfully attached to the system, this is the output

```
$ df -h
file system       Size     Used       Avail    Use%    Mounted on
udev              2.8G        0        2.8G      0%     /dev
tmpfs             584M      8.5M       576M      2%     /run
/dev/sda1         105G       90G       9.4G     91%     /
tmpfs             2.9G      1.5M       2.9G      1%     /dev/shm
tmpfs             5.0M        0        5.0M      0%     /run/lock
tmpfs             2.9G        0        2.9G      0%     /sys/fs/cgroup
cgmfs             100K        0        100K      0%     /run/cgmanager/fs
tmpfs             584M       80K       584M      1%     /run/user/1000
/dev/sdb1          60G      1.1G        59G      2%     /media/bob/Lexar
$
```

We address more details of the **df** command in Section 8.1.4.

Method 3: Very similar to using the **df** command, use the **lsblk -a** command. When we used this command and option on our Linux Mint system, after we had attached the Lexar 60 Gb USB thumb drive, we got the following output

```
$ lsblk -a
NAME       MAJ:MIN   RM   SIZE     RO   TYPE   MOUNTPOINT
loop1      7:1       0             0    loop
sdb        8:16      1    59.6G    0    disk
└─sdb1     8:17      1    59.6G    0    part   /media/bob/Lexar
loop6      7:6       0             0    loop
loop4      7:4       0             0    loop
```

```
sr0       11:0    1    1024M    0    rom
loop2     7:2     0             0    loop
loop0     7:0     0             0    loop
sda       8:0     0    111.8G   0    disk
├─sda2    8:2     0    1K       0    part
├─sda5    8:5     0    5.9G     0    part    [SWAP]
└─sda1    8:1     0    105.9G   0    part    /
loop7     7:7     0             0    loop
loop5     7:5     0             0    loop
loop3     7:3     0             0    loop
$
```

Method 4: First shown in Chapter 4, Section 4.7, use the **findmnt** command. When we used this on our Linux Mint system, we obtained the following output

```
$ findmnt
TARGET                              SOURCE        FSTYPE      OPTIONS
/                                   /dev/sda1     ext4        rw,relatime,errors=remount-ro...
├─/sys                              sysfs         sysfs       rw,nosuid,nodev,noexec,relatime...
│ ├─/sys/kernel/security            securityfs    security    rw,nosuid,nodev,noexec,relatime...
│ ├─/sys/fs/cgroup                  tmpfs         tmpfs       rw,mode=755
│ │ ├─/sys/fs/cgroup/systemd        cgroup        cgroup      rw,nosuid,nodev,noexec,relatime,...
│ │ ├─/sys/fs/cgroup/blkio          cgroup        cgroup      rw,nosuid,nodev,noexec,relatime,...
│ │ ├─/sys/fs/cgroup/freezer        cgroup        cgroup      rw,nosuid,nodev,noexec,relatime,...
│ │ ├─/sys/fs/cgroup/devices        cgroup        cgroup      rw,nosuid,nodev,noexec,relatime,...
│ │ ├─/sys/fs/cgroup/pids           cgroup        cgroup      rw,nosuid,nodev,noexec,relatime,...
│ │ ├─/sys/fs/cgroup/hugetlb        cgroup        cgroup      rw,nosuid,nodev,noexec,relatime,...
│ │ ├─/sys/fs/cgroup/memory         cgroup        cgroup      rw,nosuid,nodev,noexec,relatime,...
│ │ ├─/sys/fs/cgroup/net_cls,       cgroup        rw,nosuid,nodev,noexec,relatime,...
│ │ │ net_prio cgroup
│ │ ├─/sys/fs/cgroup/rdma           cgroup        cgroup      rw,nosuid,nodev,noexec,relatime,...
│ │ ├─/sys/fs/cgroup/               cgroup        cgroup      rw,nosuid,nodev,noexec,relatime,...
│ │ │ perf_event
│ │ ├─/sys/fs/cgroup/cpu,           cgroup        cgroup      rw,nosuid,nodev,noexec,relatime,...
│ │ │ cpuacct
│ │ └─/sys/fs/cgroup/cpuset         cgroup        cgroup      rw,nosuid,nodev,noexec,relatime,...
│ ├─/sys/fs/pstore                  pstore        pstore      rw,nosuid,nodev,noexec,relatime...
│ ├─/sys/fs/fuse/connections        fusectl       fusectl     rw,relatime
│ ├─/sys/kernel/config              configfs      configfs    rw,relatime
│ └─/sys/kernel/debug               debugfs       debugfs     rw,relatime
│   └─/sys/kernel/debug/            tracefs       tracefs     rw,relatime
│     tracing
├─/proc                             proc          proc        rw,nosuid,nodev,noexec,relatime...
│ └─/proc/sys/fs/binfmt_misc        systemd-1     autofs      rw,relatime,fd=24,pgrp=1,timeout...
│   └─/proc/sys/fs/binfmt_misc      binfmt_misc   binfmt_m    rw,relatime
├─/dev                              udev          devtmpfs    rw,nosuid,relatime,size=2926944k...
│ ├─/dev/pts                        devpts        devpts      rw,nosuid,noexec,relatime,gid=5,...
│ ├─/dev/shm                        tmpfs         tmpfs       rw,nosuid,nodev
│ ├─/dev/hugepages                  hugetlbfs     hugetlbf    rw,relatime,pagesize=2M...
│ └─/dev/mqueue                     mqueue        mqueue      rw,relatime
├─/run                              tmpfs         tmpfs       rw,nosuid,noexec,relatime,size=5...
│ ├─/run/lock                       tmpfs         tmpfs       rw,nosuid,nodev,noexec,relatime,...
│ ├─/run/cgmanager/fs               cgmfs         tmpfs       rw,relatime,size=100k,mode=755...
│ ├─/run/rpc_pipefs                 sunrpc        rpc_pipe    rw,relatime
│ └─/run/user/1000                  tmpfs         tmpfs       rw,nosuid,nodev,relatime,size=59...
│   └─/run/user/1000/gvfs           gvfsd-fuse    fuse.gvf    rw,nosuid,nodev,relatime,user_id...
└─/media/bob/Lexar                  /dev/sdb1     vfat        rw,nosuid,nodev,relatime,uid=100...
$
```

Notice that some of the file system types (FSTYPE) are shown as ext4, sysfs, tmpfs, cgroup, proc, vfat (for the a Lexar USB thumb drive mounted as /dev/sdb1 at /media/bob/Lexar).

Method 5: You can also very efficiently use the GUI-Based Gparted Partition Editor, as shown in Example 17.5. With Gparted, the installation of which is shown for all of our representative Linux systems in Appendix A, Section 17.4.3, you can easily find out the logical device names of disks on your system. In addition, with Gparted, you can use graphical editing methods to affect several important characteristics of the media, such as the format and partitioning of the drives.

Exercise 17.10

Insert a USB thumb drive into your computer and mount it if necessary. What command would you use to mount it? Use the **findmnt** command to find out its logical device name. What is the logical device name for this thumb drive? Along what path is it mounted on your Linux system? What are the uses and meanings of the other file system types shown, as output to the **findmnt** command? For example, are cgroup, proc,fuse.gvf, and tempfs logical, virtual, or physical file systems, and how exactly do these differ from ext4, or ZFS?

17.4.3 Adding a New Hard Disk to the System

The following example shows how to add a new hard disk to your system and partition it. We chose to use the GUI-based Gparted Partition editor, primarily because it is easy to use and can be quickly installed on Linux Mint.

We choose this partition editor rather than use the gdisk utility or the legacy fdisk command, because it has a GUI front end and is easy to use. This GUI technique not only lets you find out what the logical device names of disks are but also allows you to partition newly added disks and add a file system into them at the same time.

> **Example 17.5 Adding a New Disk and Using Gparted to "Slice" or Partition It**
>
> 1. Install the Gparted Partition Editor software if it is not already installed on your system. This is most efficiently done using a GUI-based software manager. We give instructions for the installation of Gparted from the command line with the Advanced Packaging Tool (APT) or Yellowdog Updater Modified (YUM) package management systems, in Appendix A, Section 17.4.3. Also, it would be efficient to place an icon for this software on your Linux desktop if you are using a GUI desktop management system.
> 2. Power down the system and properly connect the new disk to the system. Power the system backup.
> 3. Launch the Gparted Partition Editor, either graphically or from the command line.
> 4. The Gparted screen appears, as shown in Figure 17.3.
> 5. The current disks attached to the system appear in the menu bar at the upper right. Make note of all the complete paths to the current disks. For example, the boot or root disk might be designated as /dev/sda, as seen in Figure 17.3.
> 6. When the Gparted screen reappears, click the down-facing arrow shown in the menu bar in the upper right corner of the Gparted screen. Scroll in that bar until you reach the disk you just added to the system. If the disk drive you just added doesn't appear in the Gparted listing, you can't easily use that disk drive. If it does appear, continue to the next step. On our system, the new hard drive appeared as /dev/sdb.
> 7. Pick that new disk in the menu bar. It is shown in the main Gparted pane. Click on that disk in the main Gparted pane. You can now partition and format that new disk. In our example, it is shown as /dev/sdb, a new disk we inserted at Step 4. This is shown in Figure 17.4. Notice in that figure that the disk contains a single partition, /dev/sdb1, that has an ext4 file system on it, and no label.
> 8. From the pull-down menus at the top of the Gparted window, make the menu choice Partition>Delete, or click on the red X in the icon bar. This will delete the partition information on that disk. It is now a pending operation.
> 9. To execute the pending operation, make the pull-down menu choice Edit>Apply All Operations. In the warning window, click Apply. A window appears showing you the progress and hopefully successful application of the pending operation. Click close in that window when the operation is complete.
> 10. The new disk should now be unallocated. Click on its listing in the main Gparted pane. Make the pull-down menu choice Device>Create Partition Table. Change the Select new partition table type: to gpt. Click Apply in the warning window. Everything on that disk will be erased! When Gparted has created a new partition table, click on that disk again in the main Gparted pane.

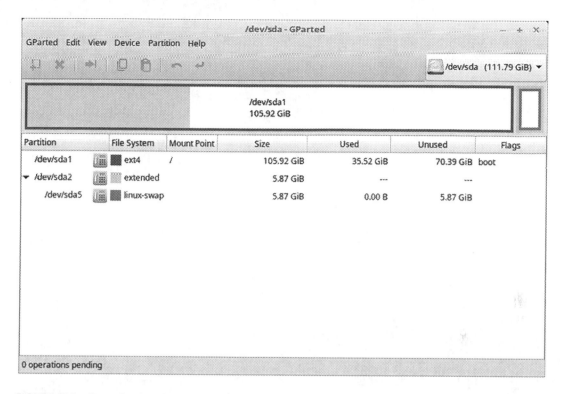

FIGURE 17.3 Gparted main window.

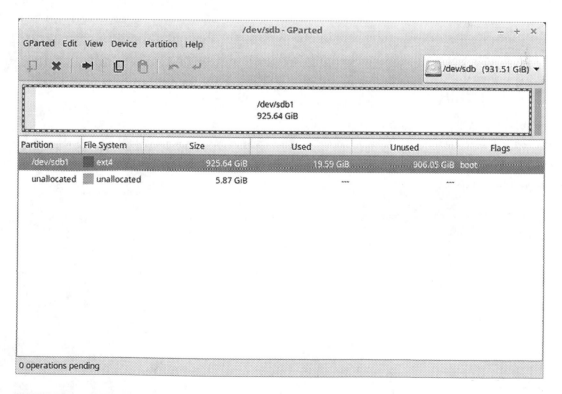

FIGURE 17.4 sdb Partition 1.

11. Make the pull-down menu choice Partition>New. The Create New Partition window appears on screen, as shown in Figure 17.5. The defaults for the new partition as seen in Figure 17.5 are to take the whole disk up with this partition, create it as a primary partition, and set the file system as ext4.

12. Add a label designation of your choice in the Label field. We have chosen to label this partition as zfs-prep, as seen in Figure 17.6, to prepare for placing a ZFS on it. Leave all of the other defaults in place. Click the Add button. Make the pull-down

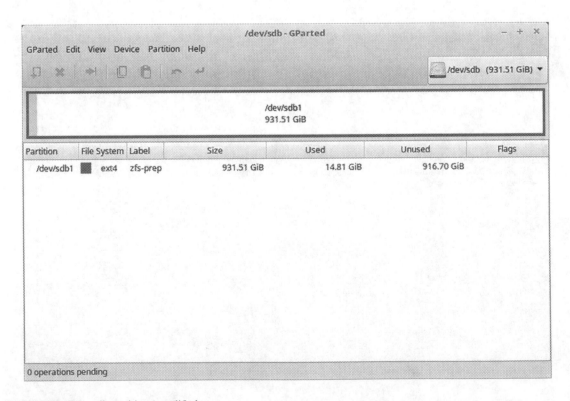

FIGURE 17.5   Create new partition window.

FIGURE 17.6   sdb partition 1 modified.

menu choice Edit>Apply All Operations. In the warning window, click Apply. Click the Close button when the Applying pending operations appears.

13. You now have a created a partition table on, partitioned and formatted a usable disk on our system, shown in Figure 17.6.
14. Quit Gparted by making the pull-down menu choice Gparted>Quit.
    Conclusion: We have added a new hard disk to our Linux system, and partitioned it with Gparted.

17.4.4 Adding Disks Using fdisk

A more traditional approach to adding a disk to the system a significant time after installation is covered in this section. We conform to our recommended storage model, as in the previous section where we implemented it for an additional hard disk added using ZFS. In this section, we use the fdisk command to partition additional disks. The example in this section uses an externally mounted USB thumb drive, which is more easily added, and is cheaper and much more available to the ordinary user. This example can be easily extended to internally mounted SATA bus drives or other bus architectures. In Chapter W22, titled "ZFS Administration and Use" at the book website, we give more examples of ZFS-based redundancy, to illustrate a similar level of this redundancy within ZFS.

17.4.4.1 Partitioning a Disk Using fdisk

fdisk is a command line, interactive, dialog-driven program used to create and manage partition tables on Linux. The following is a brief description of fdisk syntax and use:

fdisk - Create and manage partition tables

Syntax:
```
fdisk [options] device
              fdisk -l [device...]
```

Output: New or manipulated partitions on disk **device**

Common Interactive Commands:

N   Create a new partition. You enter a partition number, starting sector, and an ending sector. Both start and end sectors can be specified in absolute terms as sector numbers. Pressing the Enter key with no input specifies the default value, which is the start of the largest available block for the start sector and the end of the same block for the end sector

p   Display basic partition summary data. This includes partition numbers, starting and ending sector numbers, partition sizes, fdisk's partition types codes, and partition names

t   Change a single partition's type code. You enter the type code using a two-byte hexadecimal number

w   Write data. Use this command to save your changes

Example:
```
sudo fdisk /dev/sdb         As superuser, run fdisk on the disk /dev/sdb
```

Example 17.6 Placing a Single Partition on a USB Thumbdrive Using fdisk

Objectives: To practice placing a Linux partition, with an ext4 filesysem on that partition, on a USB thumb drive.

Prerequisites: Having a usable USB thumb drive for your system.

Requirements: Do the later steps in the sequence presented to complete the requirements of this example.

1. Having previously determined that the thumb drive is recognized on your system, and what its logical device name is (ours was /dev/sdb), insert it into a USB port. We used an 8-GB Kingston thumb drive; *very* reliable, and always automounts on Linux Mint and other NIX-like systems.

    On Linux Mint, the thumb drive was automounted. Use the following commands to unmount it:

    ```
    $ sudo umount /dev/sdb1
    ```

2. Use fdisk to partition the newly added USB thumb drive, making sure to first delete any partitions that are on it. For our Kingston USB thumb drive, there was only one default partition on it.

    ```
    $ sudo fdisk /dev/sdb
    Welcome to fdisk (util-linux 2.27.1).
    Changes will remain in memory only, until you decide to write
    them.
    Be careful before using the write command.
    Command (m for help):d
    d
    Selected partition 1
    Partition 1 has been deleted.
    Command (m for help):n
    Partition number (1-128, default 1): <Enter>
    First sector (34-15240542, default 2048): <Enter>
    Last sector, +sectors or +size{K,M,G,T,P} (2048-15240542,
    default 15240542): <Enter>

    Created a new partition 1 of type 'Linux file system' and of
    size 7.3 GiB.

    Command (m for help): p
    Disk /dev/sdb: 7.3 GiB, 7803174912 bytes, 15240576 sectors
    Units: sectors of 1 * 512 = 512 bytes
    Sector size (logical/physical): 512 bytes / 512 bytes
    I/O size (minimum/optimal): 512 bytes / 512 bytes
    Disklabel type: gpt
    Disk identifier: ABB88BEE-C60C-4B4F-B31B-12089C3DB19E
    Device      Start      End  Sectors  Size Type
    /dev/sdb1    2048 15240542 15238495  7.3G Linux file system

    Command (m for help): l
    1 EFI System               C12A7328-F81F-11D2-BA4B-00A0C93EC93B
    Output truncated...
    28 Linux RAID              A19D880F-05FC-4D3B-A006-743F0F84911E
    Output truncated...
    Command (m for help): t
    Selected partition 1
    Hex code (type L to list all codes): 83
    Changed type of partition 'Linux ' to 'Linux'.

    Command (m for help): p
    Disk /dev/sdb: 7.3 GiB, 7803174912 bytes, 15240576 sectors
    Units: sectors of 1 * 512 = 512 bytes
    Sector size (logical/physical): 512 bytes / 512 bytes
    I/O size (minimum/optimal): 512 bytes / 512 bytes
    Disklabel type: gpt
    Disk identifier: ABB88BEE-C60C-4B4F-B31B-12089C3DB19E
    ```

```
Device      Start      End  Sectors  Size Type
/dev/sdb1    2048 15240542 15238495  7.3G Linux

Command (m for help): w
The partition table has been altered.
Calling ioctl() to reread partition table.
Syncing disks.
$
```

3. In order for this partition to be usable, you must add a file system to it. We would like to add an ext4 file system to the partition. To do this, use the **mkfs** command as follows:

```
$ sudo mkfs -t ext4 /dev/sdb1
[sudo] password for bob:
mke2fs 1.42.13 (17-May-2015)
Creating file system with 1893320 4k blocks and 474208 inodes
file system UUID: b89135c2-c38e-45a3-85ff-375781777546
Superblock backups stored on blocks:
    32768, 98304, 163840, 229376, 294912, 819200, 884736, 1605632

Allocating group tables: done
Writing inode tables: done
Creating journal (32768 blocks): done
Writing superblocks and file system accounting information:
done

$
```

To use this partition for data, you can create directories and files in those directories.

Conclusion: Using the fdisk utility, and the **mkfs** command, we deleted any existing partitions on a USB thumb drive, and created a Linux partition on it with an ext4 file system.

Exercise 17.11

Do the steps of Example 17.6 to partition a USB thumb drive with a single primary partition, and then add an ext4 file system to it. Create directories on the thumb drive.

Exercise 17.12

Do the steps of Example 17.6 to partition an externally mounted hard drive in a USB enclosure, placing a single primary partition on it, and then add an ext4 file system. Create directories on the hard drive.

17.5 Configuring a Printer

We briefly discussed the Linux commands for printing files in Chapter 2, and in more detail in Chapter 6, Section 6.8. There are additional basic methods for adding and configuring a local printer, and printing documents on it, using the CUPS. We detail these methods extensively in Section W26.5 titled "What the Common Unix Printing System Accomplishes," at the book website. There, we also give some additional details of print commands such as lpadmin, lpc, lpinfo, lpmove, lpoptions, lpq, lp, lpr, lprm, and lpstat.

17.6 File System Backup Strategies and Techniques

The general necessity of backing up user and system files on a single-user Linux system, as a part of an ordinary user's routine operations, should be quite obvious. Also, for complex multiuser systems, it is a necessary procedure for anyone responsible for the administration of the system.

According to Linux system professionals, there is an easy-to-remember and important set of considerations you must make when backing up the system, as the system administrator. This set of considerations can be posed in simple question form as How, What, Why, When, Where, and Who? Some of the answers to these simple questions can be dovetailed together, and we provide a selected list of example answers as follows:

"How" means most importantly using what commands, utilities, applications, or combination of hardware and software to accomplish the backup and archive. These facilities are described in the later subsections. It also means incrementally, in a rolling fashion, or across the entire file system structure totally, using various strategies.

"What" means just some of the user files and user account files, all of them, only certain kinds of documents, the whole disk drive, multiple disk drives, system files, either all or a selected subset of the system files, etc.

"Why" means deciding on the relative importance of *"What"* you are backing up.

"When" means hourly, once a day, once a week, once a month, every time you save a particular file, and at what time exactly, like 3 A.M.

"Where" means on a local disk, to Dropbox/Google/Amazon cloud storage, to a USB thumb drive manually, to another computer or Network Attached Storage (NAS) on your home network automatically by cron, to another hard disk manually, totally, and incrementally, and to RAID of various levels or any variant and combination of the previous.

"Who" means by the only user on the system, by the initiator or executor, such as by cron automatically, by the designated system administrator, either manually or semiautomatically, by an automated process at Dropbox, Google, or Amazon.

17.6.1 A Strategic Synopsis and Overview of File Backup Facilities

There are several strategies that a single user, or system administrator, can use in confidently and efficiently backing up the file components of a Linux operating system. Table 17.3, and its included descriptions, give a basic overview of those strategies, and the facilities that implement them on a modern Linux system. In the sections that follow the table, we briefly give details of these facilities. The man pages on your system give more complete descriptions, along with command options, arguments, and option arguments for all of the Linux commands listed in the Table 17.3.

17.6.2 Linux Gnu tar

The Linux operating system has several utilities that allow you to archive your files and directories in a single file, and the **tar** command is the most popular, widely used, and traditional method that allows you to do this.

TABLE 17.3

Linux File Backup Facilities

Backup Facility	Description
tar	Command and options to pack a file or a directory hierarchy as an ordinary disk file for backup, archiving, or moving to another location or system. gtar is the Gnu version
cpio	Less popular than tar, but with much of the same functionality
rsync	A disk space-efficient command to copy files and directories
dd	A simple and abbreviated backup utility
zfs snapshot	Built-in commands and options in zfs that offer a variety of backup modes
Script files	Administrator or user-written shell scripts or other programming language backup systems, that can use all of the earlier commands in them
3rd party software	Many products, both local and online. Two examples that are most significant for ordinary use are Clonezilla and Filezilla

Contemporary tar on Linux is the Gnu version.

The tar (short for *tape archive*) utility was originally designed to save file systems on tape as a backup so that files could be recovered in the event of a system crash. It is primarily used now to pack a directory hierarchy as an ordinary disk file. That disk file can then be either saved for system backup purposes locally or remotely, or transmitted to someone via the Internet. It is also used commonly with a compression utility, such as gzip, via a command line option. Doing so saves disk space and transmission time. The saving in disk space results primarily from the fact that empty space within a cluster is not wasted. A brief description of the tar utility is as follows.

The Gnu version of tar has some important functional features and incorporates a more friendly syntax than a traditional UNIX tar. Therefore, for beginners, we only show the "long form" of the Gnu-style syntax because it is more intuitive and easy to understand. Once you get more familiar with tar, you may switch to the UNIX-style short form at your discretion.

System administrators normally use a cost-effective archival medium for archiving complete file system structures as backups, so that, when a system crashes for some reason, files can be recovered. Linux-based computer systems normally crash for reasons beyond the operating system's control, such as a disk head crash because of a power surge. Linux rarely causes a system to crash because it is a well-designed, coded, and tested operating system. In a typical installation, backup is done every day during off hours (late night or early morning) when the system is not normally in use.

The general syntax of the tar command, shown with Gnu-style syntax as opposed to traditional UNIX-style syntax, is as follows:

tar

Syntax:
`tar [operation mode] [operation mode options] [FILE…]`

Purpose: Archive (copy in a particular format) files to or, restore files from, an archival medium (which can be an ordinary file). Directories are by default archived and restored recursively

Output: Archived or restored files or directory structures

Main Operations and Operation Options in Gnu-Style Usage:

--append	Append files to the end of an archive
--concatenate	Append an archive to the end of another archive
--compare	Find differences between archive and file system
--create	Create a new tape and record archive files on it
--delete	Delete from the archive
--extract	Extract files from an archive
--help	Display a short option summary
--list	List the contents of an archive
--show-defaults	Show built-in defaults for options
--test-label	Test the archive volume label and then exit
--update	Append files that are newer than the versions in an archive
--usage	Display a list of available options
--version	Display program version and copyright information

Common Options:

preserve-permissions	Extract information about traditional file permissions
--acls	Enable POSIX.1e ACL support
--gzip	Filter the archive through gzip
--verbose	Verbosely list files processed
--file ARCHIVE.tar	Send archive to file named ARCHIVE.tar

Command Arguments:
     FILE...   Target, either an archive file, or file object to be archived

17.6.2.1 Archiving and Restoring Files Using tar

A normal Linux user can archive their own work if they want to. They would normally need to do this with file objects related to a project, so that they can transfer them to someone via e-mail, ssh, or via secondary storage media (USB thumb drive, DVD, or CD-ROM). Or perhaps they would want to retain their own file system objects as backups via the same methods.

The primary reason for making an archive is the convenience of dealing with (sending or receiving) a single file instead of a complete directory hierarchy. Without an archive, the sender might have to send several files and directories (a file structure) that the receiver would have to restore in their correct hierarchical structure. Without an archiving facility such as tar, depending on the size of the files and directory structure, the task of sending, receiving, and reconstructing the file structure can be time consuming.

In Chapter 7, Section 7.7, we discussed file compression by using the bzip2, gzip, gunzip, gzexe, xz, zcat, and zmore commands, and pointed out that compression saves disk space and transmission time. However, compressing small files normally does not result in much compression. Moreover, compressing files of one cluster in size (the minimum unit of disk storage; one or more sectors) or less does not help save disk space even if compression does result in smaller files, because the system ends up using one cluster to save the compressed file anyway. But if compression does result in a smaller file, you do save time in transmitting the compressed version. If the disk block size is 512 bytes and a cluster consists of more than two blocks, you can use the tar command to pack files together in one file, with a 512-byte tar header at the beginning of each file, as shown in Figure 17.7.

FIGURE 17.7   tar-packed file.

17.6.2.2 *Eight Easy* **tars**

In this section, we show eight of the simplest uses of the tar command. In subsequent sections, we expand upon these "easy tars" with more detailed example sessions run on our Linux system.

1. Creating an archive

    This is the simplest command to create an archive file from a directory.

    ```
    $ tar --create --verbose --file archive_name.tar directory_name
    ```

    where

    --create is the operation to create a new archive

    --verbose is the operation to list the files being processed

    --file is the operation that specifies that the file

    archive_name.tar is the name of the archive file and

    directory_name is the directory you want to archive

2. Creating an archive and preserving its ACL permissions

    A simple variation on creating an archive is using an option to preserve ACLs that the directory has. ACLs are detailed fully in Section W26.9.2 at the book website.

    ```
    $ tar --acls --create --verbose --file  archive_name.tar directory_name
    ```

    where

    --acls is the option to preserve the ACLs that the directory has

    --create is the operation to create a new archive

    --verbose is the operation to list the files being processed

    --file is the operation that specifies that the file archive_name.tar is the name of the archive
        file and

    directory_name is the directory you want to archive

3. Extracting an archive

    This is the simplest command to extract a directory from an archive file.

    ```
    $ tar --extract --verbose --file archive_name.tar
    ```

    where

    --extract is the operation to extract a directory_name from an archive file

    --verbose is the operation to list the files being processed

    --file is the operation that specifies that the file archive_name.tar is the name of the archive file

    In its simplest form, the directory which will be created from the archived file will have the same name as the archive file, minus the .tar extension.

4. Extracting an archive and restoring its ACL entries

    A simple variation on extracting an archive is using an option to restore the preserved ACLs that the archive file has (if any).

    ```
    $ tar --acls --extract --verbose --file  archive_name.tar
    ```

    where

    --acls is the option to restore the preserved the ACLs that the archived file has

    --extract is the operation to extract the archived file

    --verbose is the operation to list the files being processed

    --file is the operation that specifies that the file archive_name.tar is the name of the archive file

In this simplest form variation, the directory which will be created from the archived file will have the same name as the archive file, minus the .tar extension.

5. Listing the contents of an archive

This procedure allows you to view the contents of an archive file without extracting anything from it.

```
$ tar --list --verbose --file archive_name.tar
```

where

--list is the operation that lists the contents of the archive file

--verbose is the operation to list the files being processed

--file is the operation that specifies that the file archive_name.tar is the name of the archive file

6. Extracting a single file from an archive file

The following command extracts only a specific file from an archive file.

```
$ tar --extract --verbose --file archive_file.tar file_pathname
```

where

--extract is the operation to extract the single file

--verbose is the operation to list the files being processed

--file is the operation that specifies that the file archive_name.tar is the name of the archive file you are extracting from and

file_pathname is the complete pathname specification to the file you want to extract inside the archive

7. Extracting a single directory from an archive file

A variation on extracting just a single file from an archive, to extract a single directory, along with recursively extracting any of its subdirectories and files, is to specify the directory name that you want to extract as follows:

```
$ tar --extract --verbose --filef archive_file.tar directory_pathname
```

where

--extract is the operation to extract a single file

--verbose is the operation to list the files being processed

--file is the operation that specifies that the file archive_name.tar is the name of the archive file you are extracting from and

directory_pathname is the complete pathname specification to the directory you want to extract inside the archive

8. Adding a file or directory to an existing archive using --append option

To add files to an existing archive, do the following:

```
$ tar –append –verbose --file archive_name.tar file_or_directory_name
```

where

--append is the operation to add the single file or directory to the archive

--verbose is the operation to list the files or directories being processed

--file is the operation that specifies that the file archive_name.tar is the name of the archive file you are adding files or directories to and

file_or_directory_name is the complete pathname specification to the file or directory you want to add inside the archive

Exercise 17.13a and 17.13b

    A.   Referring to the abbreviated man page shown earlier, or to the man page for tar on your Linux system, give a command that would create a gzipped archive of a directory named ziptest.

    B.   Give a command that would extract the gzipped archive into another directory named ziptest2.

17.6.2.3 Creating an Archive with *tar*

As seen in the previous section, the most common use of the tar command is for archiving (also known as packing) a list of files and/or directories by using the --create or --append options. The --create option creates a new archive, whereas the --append option appends files at the end of the current archive.

In the example sessions presented in this and the following section, we use the directory structure shown in Figure 17.8. The following session shows that there are two directories under the **unix-book** directory called **current** and **final**. In addition, each of these directories contains six files (see Figure 17.8), displayed by the ls -l command.

```
$ cd unixbook
$ pwd
/users/sarwar/unixbook
$ ls -l
drwx------     2        sarwar 512      Jul   22      13:21     current
drwx------     2        sarwar 512      Jul   22      13:21     final
$ cd current
$ ls -l
-rw-------     1        sarwar 204288   Jul   19      13:06     ch07.doc
-rw-------     1        sarwar 87552    Jul   19      13:06     ch08.doc
-rw-------     1        sarwar 86016    Jul   19      13:06     ch09.doc
-rw-------     1        sarwar 121344   Jul   19      13:06     ch10.doc
-rw-------     1        sarwar 152576   Jul   19      13:06     ch11.doc
-rw-------     1        sarwar 347648   Jul   19      13:06     ch12.doc
$ cd ..
$ cd final
$ ls -l
-rw-------     1        sarwar 41984    Jul   19      13:06     ch1.doc
-rw-------     1        sarwar 54272    Jul   19      13:06     ch2.doc
```

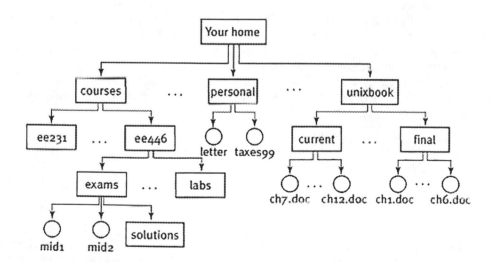

FIGURE 17.8   Directory structure.

```
-rw-------        1        sarwar  142848  Jul    19    13:06   ch3.doc
-rw-------        1        sarwar  86528   Jul    19    13:06   ch4.doc
-rw-------        1        sarwar  396288  Jul    19    13:06   ch5.doc
-rw-------        1        sarwar  334848  Jul    19    13:06   ch6.doc
$
```

If you want to create an archive of the unixbook directory on a disk file use the following command. Here we made **~/unixbook** our current directory and created a *tar archive* of this directory in a file called **unixbook.tar**. Note that **.tar** is not an extension required by the tar utility. We have used this extension because it allows us to identify tar archives by looking at the file name. With no such extension, we would have to use the file command to identify our tar archive files, as shown in the last command line in the session (in case you need reminding what the file command does).

```
$ cd ~/unixbook
$ tar --create --verbose --file unixbook.tar .
./final/ OK
./final/ch1.doc 41K
./final/ch2.doc 53K
./final/ch3.doc 140K
./final/ch4.doc 85K
./final/ch5.doc 387K
./final/ch6.doc 327K
./current/ OK
./current/ch07.doc 200K
./current/ch08.doc 86K
./current/ch09.doc 84K
./current/ch10.doc 119K
./current/ch11.doc 149K
./current/ch12.doc 340K

$ ls -l
drwx------        2        sarwar  512     Jul    22    13:21   current
drwx------        2        sarwar  512     Jul    22    13:21   final
-rw-------        1        sarwar  2064896 Jul    22    13:47   unixbook.tar
$ file unixbook.tar
unixbook.tar:              USTAR tar archive
$
```

You can also create the tar archive of the current directory by using the following command. The - argument informs tar that the archive is to be sent to standard output, which is redirected to the **unixbook.tar** file. As we discussed in Chapter 12, Section 3.4, the back quotes (grave accents) are used for command substitution—that is, to execute the find command and substitute its output for the command, including the back quotes. The output of the find . -print command (the names of all the files and directories in the current directory) are passed to the tar command as its parameters. These file and directory names are taken as the list of files to be archived by the tar command. Thus, the net effect of the command line is that a tar archive of the current directory is created in the archive file named **unixbook.tar**.

```
$ tar –create –verbose --file - `find . -print` > unixbook.tar
$
```

In the following in-chapter exercise, you will use the tar command with the c option to create an archive of a directory.

Exercise 17.14
Create a tar archive of a subdirectory hierarchy of your choice under your own home directory on your Linux system. What command line(s) did you use? What is the name of your archive file?

17.6.2.4 Restoring Archived Files with *tar*

You can restore (unpack) an archive by using the option --extract of the tar command. To restore the archive created in Section 17.6.2.3 and place it in a directory called **~/backups**, you can run the following command sequence. The cp command copies the archive file, assumed to be in your home directory, to the directory (**~/backups**) where the archived files are to be restored. The cd command is used to make the destination directory the current directory. Finally, the tar command is used to do the restoration. Notice that the destination directory is the current directory.

```
$ cp unixbook.tar ~/backups
$ cd ~/backups
$ tar --extract --verbose --file unixbook.tar
., 0 bytes, 0 tape blocks
./final, 0 bytes, 0 tape blocks
./final/ch1.doc, 41984 bytes, 82 tape blocks
./final/ch2.doc, 54272 bytes, 106 tape blocks
./final/ch3.doc, 142848 bytes, 279 tape blocks
./final/ch4.doc, 86528 bytes, 169 tape blocks
./final/ch5.doc, 396288 bytes, 774 tape blocks
./final/ch6.doc, 334848 bytes, 654 tape blocks
./current, 0 bytes, 0 tape blocks
./current/ch07.doc, 204288 bytes, 399 tape blocks
./current/ch08.doc, 87552 bytes, 171 tape blocks
./current/ch09.doc, 86016 bytes, 168 tape blocks
./current/ch10.doc, 121344 bytes, 237 tape blocks
./current/ch11.doc, 152576 bytes, 298 tape blocks x ./current/ch12.doc, 347648
bytes, 679 tape blocks
$ ls -l
drwx------      2      sarwar 512      Jul 22 13:21   current
drwx------      2      sarwar 512      Jul 22 13:21   final
-rw-------      1      sarwar 2064896 Jul 22 13:47   unixbook.tar
$
```

The **unixbook.tar** file remains intact after it has been unpacked. This result makes sense considering that the primary purpose of the tar archive is to backup files, and it should remain intact after restoration in case the system crashes after the file is restored but before it is archived again. After restoration of the **unixbook.tar** file, your directory structure looks like that shown in Figure 17.9.

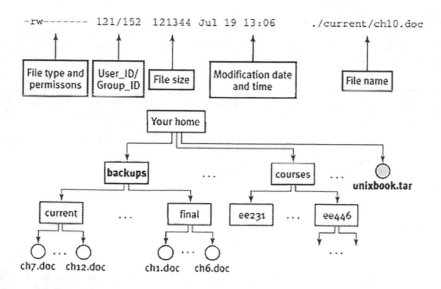

FIGURE 17.9 Restored archive file.

At times, you might need to restore a subset of files in a tar archive. System administrators often do this after a system crashes (usually caused by a disk head crash resulting from a power surge) and destroys some user files with it. In such cases, system administrators restore only those files from the backup archive that reside on the damaged portion of the disk. Selective restoration is possible with the --extract operation as long as the pathnames of the files to be restored are known. If you do not remember the pathnames of the files to be restored, you can use the --list operation to display the pathnames of files and directories on the archive file. The output of the tar command with the t option is in a format similar to the output of the ls -l command, as shown in the following session. As shown in the sample output, the first field specifies file type and access permissions, the second field specifies user ID/group ID of the owner of the file, the third field shows the file size in bytes, the next several fields show the time and date that the file was last modified, and the last field shows the pathname of the file stored in the archive.

If an archive contains a large number of files, you can pipe the output of the tar command with the --list operation added to the more command for page-by-page view. If you know the name of the file but not its pathname, you can pipe output of the tar command with the --list operation to the grep command. Files can also be restored, or their pathnames viewed, selectively. The following session illustrates these points.

```
$ tar --list --verbose --file unixbook.tar
drwx------     121/152 0       Jul  22   13:47   2015    ./
drwx------     121/152 0       Jul  22   13:21   2015    ./final/
-rw-------     121/152 41984   Jul  19   13:06   2015    ./final/ch1.doc
-rw-------     121/152 54272   Jul  19   13:06   2015    ./final/ch2.doc
-rw-------     121/152 142848  Jul  19   13:06   2015    ./final/ch3.doc
-rw-------     121/152 86528   Jul  19   13:06   2015    ./final/ch4.doc
-rw-------     121/152 396288  Jul  19   13:06   2015    ./final/ch5.doc
-rw-------     121/152 334848  Jul  19   13:06   2015    ./final/ch6.doc
drwx------     121/152 0       Jul  22   13:21   2015    ./current/
-rw-------     121/152 204288  Jul  19   13:06   2015    ./current/ch07.doc
-rw-------     121/152 87552   Jul  19   13:06   2015    ./current/ch08.doc
-rw-------     121/152 86016   Jul  19   13:06   2015    ./current/ch09.doc
-rw-------     121/152 121344  Jul  19   13:06   2015    ./current/ch10.doc
-rw-------     121/152 152576  Jul  19   13:06   2015    ./current/ch11.doc
-rw-------     121/152 347648  Jul  19   13:06   2015    ./current/ch12.doc

$ tar --list --verbose --file unixbook.tar | grep ch10.doc
-rw-------     121/152 121344 Jul 19 13:06   2015    ./current/ch10.doc
$
```

If you want to restore the file **ch10.doc** into the **~/unixbook/current** directory, you can use the following command sequence. Be sure that you give the pathname of the file to be restored, not just its name.

```
$ cd ~/unixbook
$ tar --extract --verbose --file ~/backups/unixbook.tar current/ch10.doc
$
```

The earlier output shows that the file **ch10.doc** has been restored in the **~/backups/current** directory. You can confirm this result by using the command

```
ls -l ~/backups/current
```

In the following in-chapter exercises, you will use the tar command with --list and --extract options to appreciate how attributes of the archived files can be viewed and how archived files can be restored.

Exercise 17.15

List the attributes of the files in the archive that you created in Exercise 17.14 and identify the sizes in bytes of all the files in it. What command did you use to do this?

Exercise 17.16

Copy the archive file that you created in Exercise 17.14 to a file called **mytar**. Extract it into a directory called **dir.backup** under your home directory. Show the commands you used for this task.

17.6.2.5 Copying Directory Hierarchies Locally and Remotely

You can use the tar command to copy one directory to another directory. You can also use the cp -r command to do so, but the disadvantage of using cp -r is that the traditional file access permissions and file modification times are not preserved. The access permissions of the copied files and directories are determined by the value of umask, and the modification time is set to current time. To make up for this shortcoming of cp -r, you can use the cp -p command to preserve file permissions.

With the tar command option --preserve-permissions, you can preserve the file objects traditional permissions when you extract the archive. Preservation of permissions when you extract an archive is not done by default, the permissions of extracted archive members is determined by the umask setting in the environment you are extracting to.

For example, in the following instructive command, tarring the contents of a directory, changing the current working directory, and untarring the original directory contents, while preserving the permissions of members, into this new current working directory is achieved.

```
$ mkdir pertest
$ cd pertest
~/pertest $ touch testfile2
~/pertest $ ls -la testfile2
-rw-r--r-- 1 bob bob 0 Oct 23 05:22 testfile2
~/pertest $ chmod u+x testfile2
~/pertest $ chmod g+w testfile2
~/pertest $ chmod o+wx testfile2
~/pertest $ ls -la testfile2
-rwxrw-rwx 1 bob bob 0 Oct 23 05:22 testfile2
~/pertest $ mkdir perback
~/pertest $ cd /home/bob/pertest; tar --create --verbose --file - .) | (cd /home/
         bob/perback; tar --extract --preserve-permissions --file -)
./
./testfile2
~/pertest $ cd perback
~/pertest/perback $ ls -la
total 8
drwxr-xr-x  2 bob bob 4096 Oct 23 06:03 .
drwxr-xr-x 45 bob bob 4096 Oct 23 05:27 ..
-rwxrw-rwx  1 bob bob    0 Oct 23 05:22 testfile2
~/pertest/perback $
```

The critical option is adding --preserve-permissions to the extract operation when the archive is extracted into /home/bob/perback.

Exercise 17.17

Do the earlier command line session on your system, but exclude the --preserve-permissions option in the extract operation. What are the permissions of the file testfile2 as it exists in the directory perback after execution of the tar commands?

The earlier command line session illustrates a very common use case for the tar command, that is

archive a local source directory,

change to a local (or remote) destination directory,

untar (unpack) the archived directory in this latter directory.

The entire task can be performed with one command by using command grouping and piping.

For example, in the following instructive command, tarring the contents of a directory, changing the current working directory, and untarring the original directory contents into this new current working directoy is achieved.

```
$ tar -create -verbose --file - . | (cd /home/bob/backup; tar -extract -verbose
  --file -)
```

where (in order from left to right):

> --create signals creation of an archive
>
> --verbose signals descriptive feedback on the process
>
> --file designates a file to be used as the archive
>
> - (the hyphen) designates the file as stdout. stdout will be piped to another program for further processing
>
> . represents the current working directory
>
> | designates the pipe or redirect of stdout to stdin of the program to the right of the |
>
> ( ) signals creation a new shell. Thus, the stdin of everything after the pipe is fed to a new shell.
>
> cd /home/bob/backup designates the destination of the backup
>
> ; signals a new command
>
> tar -extract -verbose --file designates verbose extraction into a file
>
> - (the hyphen) designates that stdin is used to create the archive file

In the following session, the **~/unixbook/examples** directory is copied to the **~/unixbook/examples. bak** directory. The tar command to the left of the pipe sends the archive to **stdout**, and the tar command to the right of the pipe unpacks the archive it receives at its **stdin**.

```
$ (cd ~/unixbook/examples; tar --create --verbose --file - .) |
          (cd ~/unixbook/examples.bak; tar --extract --verbose --file -)
 ./ OK
 ./Bshell/Domain, 14 bytes, 1 tape blocks a ./Bshell/IP, 18 bytes, 1 tape blocks
 ./Bshell/dns_demo1, 227 bytes, 1 tape blocks

Output truncated...

 ./Bshell/fs.csh, 1531 bytes, 3 tape blocks x ./Bshell/dir1, 0 bytes, 0 tape
blocks
 ./Bshell/copy, 2222 bytes, 5 tape blocks

$
```

The advantages of using the earlier multiple Linux command are that both cd and tar commands are available on all Linux (and even UNIX) systems.

Furthermore, the tar command can be used to copy directories from the local machine to a remote machine on a network.

To illustrate this, do the following on your Linux system. The assumption we make here is that you can rsh into another Linux or UNIX machine on your LAN and that you have an account on that machine. In our later sample session, that remote machine is 192.168.0.31- a UNIX TrueOS machine.

```
$ mkdir tartest
$ cd tartest
~/tartest $ touch testfile
~/tartest $ cd
$ (cd tartest; tar -create --file - .) | rsh 192.168.0.31 tar -extract --file -
The authenticity of host '192.168.0.31 (192.168.0.31)' can't be established.
```

```
ECDSA key fingerprint is SHA256:D8i5J7a2/dq3QAR846nGzmPaq5issWlyyBMPegpQMdM.
Are you sure you want to continue connecting (yes/no)? yes
Password for bob@trueos-6170: ZZZ
$
```

Exercise 17.18

What resulting files are found on both the computer you run the earlier command on, on 192.168.0.31, and in what directory on 192.168.0.31?

In the next instructive command line example, the **~/unixbook/examples** directory is copied to the **~/unixbook/examples.bak** directory on a remote machine, designated as trueos-6170 at IP address 192.168.0.8. The rsh command is used to execute the quoted command group on 192.168.0.8. The presumption in the following command is that you are opening a remote shell in an account on 192.168.0.8 that has the same name as your account on the client machine, and with login privileges on the remote machine 192.168.0.8. Also, that on 192.168.0.8, there already exists a directory named **unixbook/example.bak**. Because the tar command is not being run in verbose mode, the only feedback on the client machine will be the ssh authenticity prompt shown. You will have to supply your password, and then the command will complete silently, without any output on stdout of the client machine.

```
$ (cd ~/unixbook/examples; tar --create --file - .) | rsh 192.168.0.8
        "cd   ~/unixbook/example.bak; tar --extract --file -"
```

```
The authenticity of host '192.168.0.8(192.168.0.8)' can't be established.
ECDSA key fingerprint is SHA256:D8i5J7a2/dq3QAR846nGzmPaq5issWlyyBMPegpQMdM.
Are you sure you want to continue connecting (yes/no)? yes
Password for bob@trueos-6170: ZZZ
$
```

Exercise 17.19

What resulting files are found on both the computer you run the earlier command on, and on 192.168.0.8?

Exercise 17.20

What options would you have to add to the command to preserve the ACLs on the files being transmitted to 192.168.0.8. The assumption here is that, since the trueos-6170 system is a UNIX system that supports ACLs, nothing would have to be added to the tar command executed at that machine.

17.6.2.6 Extracting Software Distributions with *tar*

Companies often use the tar command to distribute their software because it results in a single file that the customer needs to copy, and it saves disk space compared with the unarchived directory hierarchies that might contain the software to be distributed. Also, most companies keep their distribution packs (in the tar format) on their Internet sites, where their customers can download them via the ftp command. Thus, the tar format also results in less "copying" time and reduced work on the part of the customer, who uses only one get (or mget) ftp subcommand, versus several sequences of commands if unarchived directory hierarchies have to be downloaded.

Because the sizes of software packages are large due to their graphical interfaces and multimedia formats, archives are typically compressed before they are put on secure ftp sites or websites. The users of the software need to uncompress and restore the archive file to expand the directory hierarchy of the package(s).

Now consider a file, **VirtualBox-4.3.16-95972-SunOS.tar.gz**. To restore this file after downloading it from the Internet, we can uncompress and untar it on our local Linux machine using one multiple Linux command, as follows:

```
$ gunzip -cd VirtualBox-4.3.16-95972-SunOS.tar.gz | tar --extract --verbose --file -
... [Output truncated]
$
```

If a software app is distributed on a secondary storage medium (USB thumb drive, DVD, or CD-ROM), you need to copy the appropriate files to the appropriate directory and repeat the earlier multiple command.

If you want to distribute some software app that you have written that is stored in a directory hierarchy, as shown earlier, you can create an archive of it and compress it at the same time using the tar command. This creates a tar archive in a compressed file that can be placed in an ftp repository, on a Web page, or sent as an e-mail attachment.

Archiving to a disk file is the most modern and cost-effective way to create software packages and apps for distribution and for creating backups.

The --gzip option can be used to generate the compressed version of the tar archive. The --gunzip option can be used to restore the compressed version of the tar archive. The use of these options eliminates the use of the gzip (or gunzip) utility to separately compress or uncompress an archive, and then the tar command; the two-step process can be performed by the tar command alone. In the following session, the first tar command generates a compressed tar archive of the current directory in the **~/unix-book/backups/ub.tar.gz** file, and the second tar command restores the compressed tar archive from the same file into the current directory.

```
$ tar --gzip --create --file ~/unixbook/backups/ub.tar.gz .
$ tar --gunzip --extract --verbose --file ~/unixbook/backups/ub.tar.gz
Output truncated ...
$
```

17.6.3 Other Linux Archiving and Backup Facilities

In addition to the traditional tar facility, there are several other facilities and methods a system administrator can use to archive and backup individual user accounts, files, file systems, and the entire system itself. As stated earlier, to get a more complete listing of the capabilities and options available for the command line facilities shown in this and all other sections, consult the man pages on your system for these commands. We briefly describe and give a simple example of some of the more modern and useful facilities and methods later. Also, in Chapter W22, titled "ZFS Management and Use" at the book website, we provide a methodology for doing archiving and backups in Linux using ZFS snapshots.

17.6.3.1 rsync

The rsync command is a space-efficient way to backup files and directories, particularly from one machine to another using **ssh** across a network. Its operation can also be automated via the use of systemd scheduling.

Its most important and defining feature is the data transfer "quick check" algorithm it uses, which reduces the amount of data transmitted. This is done by sending only the differences between the source files, and the same existing destination files (if any). If files need to be transferred, because there is a difference between source and destination, they are copied, or "synced," using this algorithm. The algorithm looks for files that have changed in size or modification time. Changes in the other possible preserved file object attributes (such as permissions or ACLs) are made on the destination file directly when the quick check indicates that the destination file needs to be updated in that way.

An abbreviated summary of the rsync man page is as follows:

rsync

> **Syntax**:
>   Local: rsync [OPTION...] SRC... [DEST]
> **Purpose**: To transfer files locally or remotely using an efficient and fast data transfer algorithm.
>   To copy locally, to/from (push/pull) another host over any remote shell, or to/from (push/pull) a remote **rsync** daemon
> **Remote**:
>   via remote shell:
>   Pull mode: rsync [OPTION...] [USER@]HOST:SRC... [DEST]

```
Push mode: rsync [OPTION...] SRC... [USER@]HOST:DEST
```

via rsync daemon:

```
Pull mode: rsync [OPTION...] [USER@]HOST::SRC... [DEST]
rsync [OPTION...] rsync://[USER@]HOST[:PORT]/SRC... [DEST]
Push mode: rsync [OPTION...] SRC... [USER@]HOST::DEST
rsync [OPTION...] SRC... rsync://[USER@]HOST[:PORT]/DEST
```

Output: Transferred files, either locally or remotely.
Common Options:

-a, --archive	archive mode
-A, --acls	preserve ACLs (implies -p)
--delete	delete extraneous files from destination dirs
-e, --rsh=COMMAND	specify the remote shell to use
--existing	skip creating new files on receiver
-h, --human-readable	output numbers in a human-readable format
-H, --hard-links	preserve hard links
-k, --copy-dirlinks	transform symlink to dir into referent dir
-K, --keep-dirlinks	treat symlinked dir on receiver as dir
-n, --dry-run	perform a trial run with no changes made
-p, --perms	preserve permissions
-q, --quiet	suppress nonerror messages
-r, --recursive	recurse into directories
--remove-source-files	sender removes synchronized files (nondir)
-v, --verbose	increase verbosity
-z, --compress	compress file data during the transfer

Command Arguments:

SRC	the source file or directory
DEST	the destination path

One of the basic assumptions we are making in the following set of examples is that the source is generally a file object that is changing over time, such as a directory where you are modifying files by adding or deleting them from it, or from its subdirectories. Also, the destination is a file object that is unchanging, or fixed over time, such as when you do a backup. This is usually the case when you are using an archiving utility, although as you have seen earlier with the tar command, the archive (or destination) can be added to or extracted from in various ways.

Generally, we do not cover more complex or advanced uses of rsync.

We have organized the rsync examples later based upon whether they apply where both source and destination are on a local machine or where either the source or destination are on a remote machine. The examples may also be divided into those that deal with files exclusively or those that deal with directories and files.

Local examples:

1. Using rsync to copy a single file named rsynctest from the source current working directory to the local destination directory /home/bob/USBint. The files are transferred in "archive" mode, done using the -a option. This ensures that symbolic links, devices, attributes, permissions, ownerships, etc. are preserved.

```
$ rsync -av rsynctest /home/bob/USBint
```

2. Copying a single local file in compressed form, with the operation feedback presented in human-readable form, from the source backup.tar (in the current working directory) to the destination directory /home/mansoor/backup1 which you have permissions on.

   ```
   $ rsync -zvh backup.tar /home/mansoor/backup1
   ```

3. Using rsync to copy an entire source directory contents from the directory named syncdir locally to a destination directory locally named /usr/home/bob/USBint. The slash at the end of the designated source means "copy the contents of this directory," not "copy the directory by name"

   ```
   $ rsync -av syncdir/ /usr/home/bob/USBint
   ```

4. Copying a single source directory /home/bob/backups in compressed form, with the operation feedback presented in human-readable form, to the destination directory /home/mansoor/back-ups which you have permissions on

   ```
   $ rsync -avzh /home/bob/backups /home/mansoor/backups/
   ```

5. Copying a single source directory /home/bob/backups, and its files and subdirectory structure, in compressed form, to the destination directory /home/mansoor/backups which you have per-missions on

   ```
   $ rsync -zvr /home/bob/backups /home/mansoor/backups
   ```

6. Doing a dry-run, where you remove the source files and compress the transmitted file object

   ```
   $ rsync --dry-run --remove-source-files -zvh backup.tar /tmp/backups/backup.
     tar
   ```

Remote examples:

7. Copying a single directory, named Music, from the local machine to a remote machine at 192.168.0.8 into mansoor's home

   ```
   $ rsync -avz Music mansoor@192.168.0.8:/home/mansoor/
   ```

8. Copying an entire local directory, named syncdir2, using ssh, to a remote machine at 192.168.0.7 and into the OS X-style directory /Users/b/Linux3e

   ```
   $ rsync -av -e ssh syncdir2 bob@192.168.0.7:/Users/b/Linux3e
   ```

9. Copying a remote directory to a local directory

   ```
   $ rsync -avzh root@192.168.0.100:/home/bob/backups /home/mansoor/temp
   ```

10. Copying just the contents of the remote directory /home/mansoor/Linux2e
    updating (sometimes called "syncing") with the --existing option, only the existing files at the destination. If the source has new files, which are not on the destination, these new files are not put on the destination.

    ```
    $ rsync -avz --existing mansoor@192.168.1.2:/home/mansoor/Linux2e/ .
    ```

11. Copying a single local file, named backup.tar, using ssh, to a remote machine at 192.168.0.8

    ```
    $ rsync -avzhe ssh backup.tar bob@192.168.0.8:/home/bob/backups/
    ```

12. Copying a single remote file to the current working directory using ssh

    ```
    $ rsync -avzhe ssh bob@192.168.0.100:/usr/home/bob/systemctl.log .
    ```

13. Copying only the local files in /home/bob/files to a remote machine with an IP address of 192.168.0.8. With the --delete option present, extraneous files on the destination that no longer exist in the source directory are deleted. The -v option ensures that what is being transferred is output. rsync using ssh on an ephemeral port 22

    ```
    $ rsync -av --delete -e 'ssh -p 32000' /home/bob/files/ bob@192.168.0.8:/home/
    bob/files/
    ```

Exercise 17.21

What is the basic assumption in all of the examples earlier where you copy to a remote host?

Exercise 17.22

Do all of the earlier 13 examples locally on your system, and remotely on your LAN, substitute file names, directories, account names, and IP addresses as necessary.

17.6.3.2 Script Files for Backup and Recovery

User-written script files can be deployed by the system administrator to quickly and efficiently do backup and archiving. Whether they are coded in Python (or other scripting language) or in a shell programming language that embeds any of the earlier command line facilities, they can be used to facilitate and automate the backup, recovery, and archiving of files, directories, or file systems. Additionally, there are several online sources for backup and archiving script files available, and we will not justify the need for planning, coding, or maintaining such script files versus using an online, ready-made program.

Following is an example of a Bash shell script file using a legacy application of **tar**.

Example 17.7 Simple Bash Shell Example for Automating tar Backups

Objective: The following Bash shell script will backup a directory, /home/bob/bashtest, to another directory previously created named simple_backup in a compressed format.

Procedures: Do the following steps, in the order presented, to complete the requirements of this example.

1. Create a subdirectory of your home directory named bashtest.
2. Create another subdirectory of your home directory named simple_backup.
3. Use an editor of your choice to create the following file and name that file backup.bash.

```
#!/bin/bash
# To backup additional directories, put more pathnames in the
following command,
# separated by spaces, such as /home/bob /home/sarwar etc.
backup_source=/home/bob/bashtest
backup_destination=/home/bob/simple_backup
filename=back1.tgz
echo "Backing up your bashtest directory"
tar --create --verbose --gzip --file $backup_destination/$filename
$backup_source
echo "Backup Complete"
```

4. Execute the Bash script with the command ./backup.bash

Exercise 17.23

Give the commands necessary to automate the earlier script file using cron, to run the backup. sh script at 3:00 A.M. every week. Feel free to modify time, day, and source and destination directories and files, so that you can use the script file to backup information important to you on your Linux system.

17.6.3.3 Software for Backup and Recovery: The "Zillas" and Ghost

There are many commercial software and hardware packages that can be deployed to backup and archive your system. One very easy-to-use commercial product that allows you to clone entire hard disks in a "broadcast" fashion over a network, from a server to one or more machines, is Norton Ghost. But the two most readily available, useful, and free software facilities that can do a variety of file system backup and recovery, disk recovery, and disk cloning operations are FileZilla and Clonezilla.

17.6.3.3.1 FileZilla

FileZilla is nominally a graphics-based ftp client and server program that can use ssh as the tunnel or conduit between systems. It has a number of useful functions and menu choices that allow the system administrator to successfully, confidentially, and efficiently backup and restore single files or directories, via a network globally. It is most useful for backing up and restoring single-user files and directories.

It is not a replacement or substitute for the command line facilities shown in the preceding sections. Figure 17.10 illustrates the screen display and menus available in the "client" version of FileZilla.

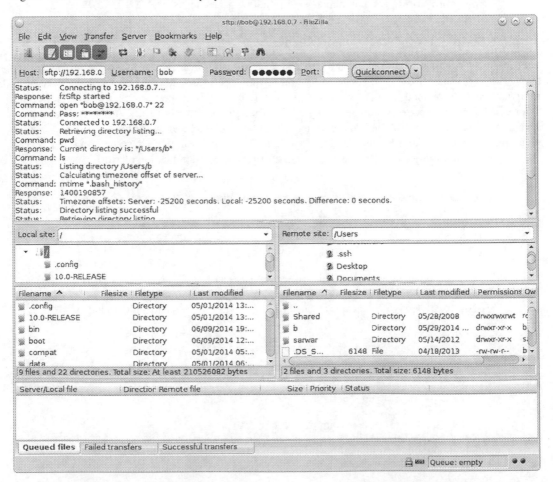

FIGURE 17.10   FileZilla main window.

Both client and server, in our case a local machine running Linux Mint, and a remote Linux machine, must have ssh communications protocol enabled between them. You can have login access to an account on the remote server or you can anonymously login as well if that is enabled.

After launching FileZilla on the client, to login to a remote host server, you need to supply the IP address of the server, the login name and password, and the port number (22 for ssh). Once you have successfully logged in, the local machine's directory and file structure is shown on the left side of the figure. The remote machine's directory structure is shown on the right side of the figure. To transfer files or directories between machines, you simply drag and drop between the appropriate panes on the left or right. If you are overwriting previously transferred files or directories, the FileZilla default is to give you the chance to overwrite or rename the files being transferred.

There are a number of other menu choices at the top of the FileZilla screen that allow you to affect preferences, set bookmarks, etc. For example, via the menu choice Manage Bookmarks and the Site Manager, you can automatically make multiple local directories and remote directories available for ssh transfer as soon as you log in to the remote server sites.

17.6.3.3.2 Clonezilla

Clonezilla is the most useful and readily available tool for the ordinary single computer user to do partition and whole-disk cloning. The "live" version of it can be deployed to do nonbroadcast cloning of entire disks, including the boot sector.

This technique allows an ordinary user or a system administrator to take fast snapshots of entire disks, thus greatly enhancing and facilitating any backup strategy employed.

As an example, on a large single disk, Clonezilla is capable of quickly duplicating the entire disk so that the clone can be used in exactly the same way as the original disk. So, instead of using any of the other backup schemes shown earlier, at any single instant in time, you can create an exact copy of any of the disks on your system. Of course, you would have to weigh the merits of this versus any of the other backup strategies.

In Linux, cloning the entire system disk, including the boot partition, is possible with Clonezilla.

This would allow you to have a bootable duplicate of your system disk, possibly with all of your user data files on it as well. In CloneZilla, the source disk can be an internal hard drive, and the target clone can be an externally mounted hard disk in a USB-connected enclosure. This is a very useful procedure if you are running Linux on a laptop computer that is only capable of having one internal hard drive.

17.6.3.4 Using git and GitHub to Backup File Systems

The git command and GitHub are useful, efficient, and powerful means you can use to create a distributed version-control system for your software projects. They can also be used as a cloud-based storage and archiving facility, particularly if you follow the storage model we recommend in this chapter. That is, if your Linux operating system is installed on an SSD, and all of your user data files are stored on another persistent medium, such as additional SSDs or hard disks.

Of course, using a private GitHub repository for backup and archiving is not free. In addition, there are limits to the amount of data that you can store on GitHub. Therefore, this method of backup and archiving works best for an ordinary user in a public repository at GitHub, with small user data storage requirements. Also, if you use another data storage model that we recommend, the mechanics of doing GitHub archiving and backup would be dependent on that data storage model.

17.7 Software Updates and Operating System Upgrades (On Debian-Family and CentOS)

Question: Why would the two most significant, postinstallation tasks that an ordinary user on a single-user system needs to perform be as follows?

   a. The addition, removal, or update of software applications as packages on the system.

   b. The upgrade of the operating system itself to a newer, currently stable release?

Answer:

   a. The standard installation of the operating system does *not* include software that the user
       needs, given the use cases she puts the system to. Changing conditions in the use cases
       require removal of software packages as well. Finally, as newer and improved versions
       of installed packages become available, there must be a quick and easy way to update the
       packages to incorporate those improvements.

   b. Purely from a performance and security viewpoint, upgrades of the system happen very
       regularly, and even an ordinary user should take advantage of these improvements via
       upgrades.

Question: What is the difference between "updating" and "upgrading" a Linux system?

Answer: Generally in the literature, the two terms are often confused for one another. But from
   our perspective in this section (and throughout this chapter and book), we define the two terms
   as follows:

Updating is the process of adding or replacing application software packages. For example, using the
package management system to add the Gnu C compiler to the system or bringing the nginx Web server
application from an earlier-installed version to the latest stable release of it.

Upgrading is the process of replacing the entire operating system with a newer, stable release. The
upgrade can be to a newer minor release or to a new major release. For example, upgrading CentOS 7.3
to CentOS 7.4 is considered a major release upgrade or upgrading Linux Mint 17.3 to Linux Mint 18 is
also considered a major release upgrade.

In this section, we show the basics of how these two fundamental tasks can be done on our representative Linux systems: Debian, Ubuntu, Linux Mint, and CentOS.

For software application addition, removal, or updates, we limit the discussion to covering the
Advanced Package Tool (APT) for the Debian/Ubuntu/Linux Mint family, and the Yellowdog Updater,
Modified(YUM) for CentOS. Both of these package mangers can be used very effectively in a text-
based, Command User Interface (CUI) environment. That is how we approach them in this book. That
does not preclude the use of GUI-based front ends to package managers, such as Synaptic.

For upgrading the operating system itself, we show you a basic set of procedures in Section 17.7.4 that
suits the variable conditions of upgrading an operating system.

Furthermore, adding the Gnu C Compiler (gcc) and the requisite attendant libraries for it is our primary example of adding a software application package on our representative systems. We show this in
detail in Appendix A, Sections 17.7.2.2 and 17.7.2.3.

We illustrate gcc installation in Appendix A because it is a critical procedure to execute before you can
do any of the work in Chapters 15 and 16 on system programming.

If you have done your own installation of the operating system using the techniques we show in Section
17.2.2, it is only installed by default on one of the four representative systems, with the necessary libraries.

If you have not done the installation of the operating system on your computer system yourself, be
aware that the installation of gcc may have already been done for you by the administrator of your system!

Exercise 17.24

   In the light of what we have said about updating and upgrading your Linux system, list applications that you want to install, given the use case(s) you undertake on your system. Then, as you
   go through the rest of this section, use the appropriate package manager to install these.
   The best way for you to get more basic help on the package management commands we show
   in this section is to read the man pages on your system.

17.7.1 Preliminary Storage Model Suggestions

We previously provided an ordinary user with a suggested storage model, where the operating system
is on a single disk drive (which we call the "system disk"), and the user data is on one or more other
ZFS vdevs. One of the most useful and practical advantages this model gives an ordinary user is the

ability to add or update software applications and upgrade the operating system itself in a reliable and time-efficient way. The software applications and operating system, since they exist on the system disk, can be incrementally updated or upgraded independently of the user data. So for example, if you want to go from an earlier major release of the operating system to the latest release, you can do a clean, "bare metal" install of the new release onto the system disk and then simply reattach the user data to it. We have found that this storage model and this method of update/upgrade are safest and most reliable, even for an ordinary user, i.e. it doesn't break the system.

If you choose to keep the older major release of the operating system and just want to incrementally update software and upgrade the operating system, the sections later will give you a basic overview of how to do that.

There is a wide selection of software programs available via the use of the facilities we show here. These programs are found in software archives known as "repositories." The repositories contain collections of "packages," and each package contains bundled-together software and dependencies that are all automatically installed together. When downloaded and installed on your system by package management facilities such as apt or yum (the two package management systems we illustrate), that software becomes available to a user without the need to do anything else.

For software that provides a system service, such as sshd, or a Web server like nginx, the package manager not only downloads and installs all necessary components and their dependencies but also uses systemd facilities to start the service and make sure it automatically starts at every subsequent system boot.

Of all the chapters and sections in this book, this section is probably the most sensitive to, and dependent upon, explicit instructions for the particular version and release of the operating system you are working with.

Exercise 17.25

Determine exactly how your Linux system deploys a storage model. For example, what are the paths to a user's home directory? How is the file system structure segregated on the disk drives? How is this different from the storage model we suggest? How can you migrate the storage model that now exists on your Linux system to conform to our suggested model?

17.7.2 Using APT

For beginners, a CUI approach to package management at first might not seem to be the quickest and easiest way to accomplish one of the most significant postinstallation tasks. GUI approaches have some tremendous advantages, not only in terms of saving time interacting with the Linux system itself. But we feel the explicit control you get by using APT, YUM, on the command line outweigh the initial advantages of a GUI. Of all of the tools available for APT, apt-get is the most useful for a beginner.

17.7.2.1 Using apt-get

The APT is used for all aspects of managing packages. The most used (and useful) of its package management utilities is the CUI apt-get tool. Following is an abbreviated man page description of apt-get:

apt-get: APT package handling utility—CUI version

```
Syntax:
   apt-get [-asqdyfmubV] [-o=config_string] [-c=config_file]
                [-t=target_release] [-a=architecture] {update | upgrade |
                dselect-upgrade | dist-upgrade |
                install pkg [{=pkg_version_number | /target_release}]... |
                remove pkg... | purge pkg... |
                source pkg [{=pkg_version_number | /target_release}]... |
                build-dep pkg [{=pkg_version_number | /target_release}]... |
                download pkg [{=pkg_version_number | /target_release}]... |
                check | clean | autoclean | autoremove | {-v | --version} |
                {-h | --help}}
```

> **Purpose**: apt-get is the command line tool for handling packages, and may be considered the user's "back-end" to other tools using the APT library. Several "front-end" interfaces exist, such as Aptitude and Synaptic
>
> **Output**: Added, removed, or updated software packages on the system
>
> **Common Options and Commands**:
>
> *Note*: Unless the -h, or --help option is given, one of the commands later must be present. Options:
>
> -d, --download-only
>
> > Download only; package files are only retrieved and not unpacked or installed. Configuration Item: APT::Get::Download-Only
>
> --only-upgrade
>
> > Do not install new packages; when used in conjunction with install, only upgrade will install upgrades for already installed packages and ignore requests to install new packages
>
> -q, --quiet
>
> > Quiet; produces output suitable for logging, omitting progress indicators
>
> -y, --yes, --assume-yes
>
> > Automatic yes to prompts; assume "yes" as answer to all prompts and run noninteractively
>
> Commands:
>
> install
>
> > install is followed by one or more packages desired for installation or upgrading. The /etc/apt/sources.list file is used to locate the desired packages
>
> Remove
>
> > remove is identical to install, except that packages are removed instead of installed. Removing a package leaves its configuration files on the system
>
> Update
>
> > update is used to resynchronize the package index files from their sources. The indexes of available packages are fetched from the location(s) specified in /etc/apt/sources.list
>
> Upgrade
>
> > upgrade is used to install the newest versions of all packages currently installed on the system from the sources enumerated in /etc/apt/sources.list

We cover the following functions in APT for our representative Debian-family Linux systems (at the time this book was written, those are Debian 9.1, Ubuntu 16.04, and Linux Mint 18.2):

Installing, reinstalling, removing, and searching for packages.

Upgrading and updating packages.

In the Operations later, when we refer to package_name, this represents any specific named package.

Operation 1. To use the apt-cache package query function to list available packages on your system, use the following command:

```
[root@Linux_Mint] # apt-cache pkgnames
```

Although this listing can be quite long, it is useful to find out if a package has been installed on your system. For example, in the gcc example we show, before you try to install gcc, it is a good idea to see if it has already been installed!

Operation 2. To install a package, use the following command:

```
[root@Linux_Mint] # apt-get install package_name
```

Operation 3. To remove a package, or the package and its configuration files, use the following commands:

```
[root@Linux_Mint] # apt-get remove package_name
[root@Linux_Mint] # apt-get purge package_name
```

Operation 4. To update the package index files from the local sources.list directories (the metadata for packages) or upgrade all Debian packages to their latest versions, use the following commands:

```
[root@Linux_Mint] # apt-get update
[root@Linux_Mint] # apt-get upgrade
```

Operation 5. To individually update an installed package, by removing it and then reinstalling it, use the following command:

```
[root@Linux_Mint] # apt-get  --reinstall install package_name
```

Operation 6. To show metadata associated with an installed package on the system, use the following command:

```
[root@Linux_Mint] # apt-cache show package_name
```

For example, on our Debian-family Linux Mint system, the output of this command shows the Gnu C Compiler (gcc) metadata information

```
[root@Linux_Mint] # apt-cache show gcc
```

Package: gcc

Priority: optional

Section: devel

Installed-Size: 44

Maintainer: Ubuntu Developers <ubuntu-devel-discuss@lists.ubuntu.com>

Original-Maintainer: Debian GCC Maintainers <debian-gcc@lists.debian.org>

Architecture: amd64

Source: gcc-defaults (1.150ubuntu1)

Version: 4:5.3.1-1ubuntu1

Provides: c-compiler

Depends: cpp (>= 4:5.3.1-1ubuntu1), gcc-5 (>= 5.3.1-3~)

Recommends: libc6-dev | libc-dev

Suggests: gcc-multilib, make, manpages-dev, autoconf, automake, libtool, flex, bison, gdb, gcc-doc

Output truncated...

Again, see Sections 17.7.2.2 and 17.7.2.3 in Appendix A for Debian-family and CentOS installation instructions for gcc.

Exercise 17.26

First, install gcc on your Linux system, as detailed in this section, and following the instructions in Appendix A. Then, if they haven't already been installed by your system administrator, use APT on your Linux system to install the following application packages: openssh-server, nginx, vsftpd. Finally, test each of these applications. Do these applications restart after a system reboot? How can you know that?

Exercise 17.27

Use APT on your Linux system to install application packages that fit your particular use case(s). Go back to your answer to In-Chapter Exercise 24 for a preliminary listing of these.

17.7.3 YUM

YUM is a text command-based package-management facility for Red Hat-family Linux system that are using the RedHat Package Manager (RPM). Though YUM has a text-based command-based interface, there are several other tools available that do provide GUIs to YUM.

YUM allows automatic updates, and package and dependency management, on RPM-based Linux distributions. Like APT, YUM works with highly customizable software repositories that you can establish as defaults, which can be accessed locally, or over a network connection.

YUM can perform operations, dependent upon your configuration of repositories, such as

Installing packages

Deleting packages

Updating existing installed packages

Listing available packages

Listing installed packages

Besides the distributions that use YUM directly, openSUSE Linux (a flavor that is representative of the third major branch of Linux, Slackware) has added support for YUM repositories in YaST (the package manager used in openSUSE), and the Open Build Service repositories use the YUM XML repository for the formatting of package metadata.

17.7.3.1 YUM Package Installation Procedures

Preliminary Operation minus 1

Before we could use YUM on our representative Linux CentOS 7.4 system, we had to make sure that the network connection through the available Network Interface Card (NIC) on the computer was turned "On." This was accomplished via the Gnome Desktop menu choices, in descending order, Applications>System Tools>Settings>Network. After we made the slider-switch choice to turn the NIC "On," we verified that the network was reachable (and our NIC was automatically assigned an IP address by the DHCP server on our network) by using the `ip addr show` command in a terminal window. There was no need to modify the firewalld settings in CentOS 7.4, since the firewall (although active) was not blocking Internet http traffic by default.

Preliminary Operation 0.

Another preliminary operation we did on our CentOS 7.4 system was to enable the Extra Packages for Enterprise Linux repository, so we could use YUM to install a few specific packages that we wanted to from this community-based "extras" repository. The commands to accomplish this are as follows:

```
[root@CentOS_7 ~]# systemctl stop packagekit
[root@CentOS_7 ~]# systemctl disable packagekit
[root@CentOS_7 ~]# yum install epel-release
```

The first two commands circumvented a bug with a Python program holding the lock on the YUM program and may not be necessary on later releases of CentOS beyond version 7.4. The first systemctl command stopped the offending service, and the second systemctl command disabled it for subsequent reboots of the system. See more on the systemd systemctl command in Chapter 18 on systemd.

In the following ten operations, package_name refers to a particular named package, such as nginx. Also, all commands are executed as the root user.

Operation 1. To start with, check all of the enabled YUM repositories on your system, using the following command:

```
[root@CentOS_7 ~]# yum repolist
```

Operation 2. To list all enabled and disabled YUM repositories on your system, use the following command:

```
[root@CentOS_7 ~]# yum repolist all
```

If you have done Preliminary Operation 0, epel (and its metalink) should be in that listing!

Operation 3. To list all available packages in the YUM database, use the following command:

```
[root@CentOS_7 ~]# yum list | less
```

Operation 4. To list all installed packages on your system, use the following command:

```
[root@CentOS_7 ~]# yum list installed | less
```

Operation 5. To search for a particular package amongst those available at your default repositories, use the following command:

```
[root@CentOS_7 ~]# yum search package_name
```

Operation 6. To install a package that you know the exact name of, just run the following command. As noted, YUM will automatically find and install all required dependencies for that package!

```
[root@CentOS_7 ~]# yum install package_name
```

Operation 7. If you want YUM to do the install without confirmation, use the following command:

```
[root@CentOS_7 ~]# yum -y install package_name
```

Operation 8. To remove a package that you know the exact name of, use the following command:

```
[root@CentOS_7 ~]# yum remove package_name
```

or without confirmation:

```
[root@CentOS_7 ~]# yum -y remove package_name
```

Operation 9. To update a package you already have on your system that you know the name of, use the following command:

```
[root@CentOS_7 ~]# yum update package_name
```

Operation 10. To check that your system is up to date, particularly with respect to security packages, and then actually update the packages, use the following commands:

```
[root@CentOS_7 ~]# yum check-update
[root@CentOS_7 ~]# yum update
```

Again, see Appendix A, Section 17.7.3.2, for details of the installation of gcc using YUM on CentOS.

Exercise 17.28

First, install gcc on your Linux system, as detailed in this section. Then, if they haven't already been installed by your system administrator, use YUM on your Linux system to install the following application packages: openssh-server, nginx, vsftpd. Finally, test each of these applications. Do these applications restart after a system reboot? How can you know that?

Exercise 17.29

Use YUM on your Linux system to install application packages that fit your particular use case(s). Go back to your answer to In-Chapter Exercise 24 for a preliminary listing of these.

17.7.4 Upgrading the Operating System

Similar to our recommendations for the exact, specific details of initially installing our four representative Linux systems, the minor or major release upgrades of these systems is dependent on the specific system you are dealing with, and the versions of that system you are trying to upgrade between.

Although the Debian family (Debian, Ubuntu, Linux Mint) has a very similar upgrade procedure, via APT, for its members, and CentOS has a similar upgrade procedure within the members of the RedHat family, there is enough variability that presenting the exact, specific details of upgrading these systems is best left to consulting relevant online sources at the time you want to do the upgrades.

When we wrote this book, the upgrade procedure for the four representative systems could be summarized as follows:

For the Debian family, the following APT commands worked for minor release upgrades:

```
[root@Debian] # apt update
[root@Debian] # apt upgrade
[root@Debian] # apt dist-upgrade
```

The major upgrade procedure is best done by consulting the online documentation at the time you want to do the upgrade.

For CentOS, the following YUM commands worked for minor and major release upgrades (remember to check that your NIC was turned on as shown in Operation minus one in Section 17.7.3.1!):

To check the release your system is currently at

```
[root@CentOS] # cat /etc/release-redhat
[root@CentOS] # yum check-update
[root@CentOS] # yum update
```

To recheck the release your system is now at

```
[root@CentOS] # cat /etc/release-redhat
```

Exercise 17.30

Create a document that details exactly what steps you must take to upgrade your Linux system, serving for both minor and major releases of the software. The scheduling of your procedural steps should reflect, for example, a weekly or monthly basis for minor release changes, or at some unspecified interval for major release changes.

Then, implement the procedures that your document specifies.

17.8 System and Software Performance Monitoring and Adjustment

The most important considerations the Linux system manager (and for the ordinary, single user on their own computer) has to make when dealing with system performance revolves around CPU process management, memory management, disk usage/management, and network performance. Table 17.4 lists the controlling facilities and functions Linux provides for system tuning and performance monitoring, most notably with the kernel-loadable Z File System component added, if that is the case.

TABLE 17.4

Performance Tuning Functions

System Component	Control Facility
CPU	Nice numbers
	Process priorities
	Cgroup management
	Batch queues
	Scheduler parameters
Memory	Process resource limits
	Cgroup management
	Memory management parameters
	Paging space
ZFS vdevs, zpools, and I/O (if installed)	ZFS pool and file system organization
	ZFS deduplication, efficiency, optimization
	I/O parameters
	Zvol creation and administration
Network I/O	Network memory buffers
	Network-related parameters
	Network infrastructure

The commands that implement and affect some of the functions from Table 17.4 are as follows. We also include commands that allow you to monitor these functions as well. We cover some of these commands in the following sections.

```
df, du, free, journalctl, nice, pgrep, ps, renice, systemctl, systemd-cgls,
systemd-run, top
```

17.8.1 Application-Level Process/Thread Resource Management

The basic reason for monitoring system resources, such as process CPU usage and system memory, in terms of how the system is performing given its designed use case(s), is to allow you to observe, assess, and then gain control, at the application level, of how programs are effectively using major components of system resources.

For example, if a particular standard application (such as a Web server which you have installed from an online repository) or even a user-written program's processes are consuming too much of CPU or memory resources, according to criteria you determine given the use case(s) your system is designed for, you can scale back that task's operation to better serve your user base. On the other hand, if those same processes are being underserved or starved for resources, for whatever reason, you can scale up that task's operation as well. There may be some situations where underservice may be caused by hardware malfunctions, or illegal intrusion, but we do not cover those situations in this section.

The key is to observe and assess the usage of system resources in a contemporary way on a modern Linux system and then take some actions based on those observations and assessments. Those actions are of course guided by how you have strategically designed and implemented your administrative plan, given the use case(s) of your system.

Another way of assessing process activity, for example, in terms of scheduling processes and process groups through the CPU, is to examine the proc file system. We show the basics of proc in Chapter 10, Section 8.

17.8.1.1 Traditional Process Control

Traditionally, the most important, complete, and readily available display of system process activity is given by the ps and top (or htop) commands, which allow you to monitor running processes. For a more complete description of the ps and top commands, with several examples, see Chapter 10, Sections 5.1–5.3. See Section W26.8 at the book website, for a more detailed description, with command line examples, of traditional process control.

17.8.1.2 systemd Cgroups, Affecting the Limits of CPU Scheduling

With the adoption of systemd into the Linux kernel in all major branches of Linux, including all of our representative systems, several new and many updated features have been introduced. These features completely change the startup phase of booting the system, for example. But most importantly, they radically modify how both system and applications services (such as sshd), and even common userland processes, are managed. One of these radically modified and updated features, known as Control Groups (cgroups), is completely reorganized with respect to service and process/thread management. Processes and threads, as well as system services, are grouped together in a unified manner. This moves system resource management, such as what percentage of CPU time a process, thread, or service is allocated, from the process level to the application level. This is achieved by mapping cgroup structure to the systemd "unit tree." We sketch systemd cgroups impact on the performance scheduling of processes and threads through the CPU. In the next section, we focus on how cgroups impact memory management to affect system performance.

In Chapter 18, we detail all aspects of systemd. We give a more complete treatment of cgroups in Chapter W27, Section 6.1 at the book website. We have also given an exposition of Linux CPU scheduling component of the kernel, and its algorithm, in Chapter 10, Sections 10.2 and 10.5.1.

cgroups is a mechanism for organizing services and processes/threads in a hierarchic tree, or "unit tree," of numbered groups to facilitate resource management. The structure of this tree is graphically illustrated using the systemd-cgls command. We show how the systemd-cgls command can be used to view and assess the cgroups strucure. Then, control is exercised on these numbered groups to affect resources like CPU useage, memory management, network bandwidth, and other system service-related activities. This control can be done on a transient, or temporary, basis using the systemd-run command. Or it can be done on a persistent or permanent basis by making critical changes to the systemd units file that all permanent, start-at-boot-time services have. We organize our presentation in this section around these two bases.

Before the introduction of cgroups and systemd in the Linux kernel, all processes received similar amounts of system resources that the administrator could modify to enhance performance (per the use case(s) the system operated under by design) with process *nice* values. This is shown, for example, in Chapter 10. The drawback of this legacy approach was that, for an application, such as a Web server or database management program, which may have a large number of subprocesses or threads running at the same time, there was an inadequate mechanism available to give priority to these groups of tasks. Another task or set of tasks that was more use case, mission-critical, but had a much smaller number of subtasks, had to share resources equally with the Web server or database management program.

systemd cgroups have a tremendous advantage over the legacy approach. For example, using systemd cgroups, all the processes started by the nginx Web server will have their own cgroup numbers. This allows you to control nginx at the application level, and all of its "worker children," with the systemd systemctl command. We completely detail the systemctl command in Chapter 18, Section 5.

We give a more in-depth explanation of the hierarchic cgoups unit tree in Chapter W27, Section 6.1 at the book website. In that section, we also show the utility of the systemd-cgls command, and how it depicts in a easy-to-read, graphical tree format all of the cgroups and their hierarchic structuring on your Linux system.

The key takeaway here, and something that is often misunderstood, is that if you run some task, and nothing else of higher "priority" (in terms of cgroup-levied resource controls) is running at the same time, that task gets 100% of the resources (or nearly so, since the system needs to have a small chunk of resources to keep itself running!).

17.8.1.2.1 Creating a Transient or Persistent Control Group

As we detail in Chapter W27, Section 6.1 at the book website, the place where cgroup resources are managed is the systemd "unit" file. Placing this management resource facility at the level of unit files allows their operations to be managed with systemd's command line utilities. Depending on the kind of unit file, your resource management settings can be *transient* or *persistent*.

To create a **transient cgroup** for a service via its unit file, start the service with the systemd-run command. You can then modify resources on an application per the design of your system use case(s). It is worth noting here that system programming applications can create transient cgroups by using cgroup API calls to **systemd**. A transient unit is removed from the system as soon as the service is stopped using the systemctl command, or the system is rebooted (since there is no permanent unit file created for the service using systemd-run.)

To create a **persistent cgroup** for a service service, create/edit its unit configuration file. The configuration can be preserved after system reboot, so it can be used to manage services that are started automatically.

We present Examples 17.8 and 17.9, with relevant background information, on these two methods of service creation.

Creating a Transient Cgroup with systemd-run *and Changing its CPUShare*

To create a transient cgroup and run a command in that cgroup, the systemd-run command is used. The general form of the systemd-run command, with some specific options, is as follows:

```
systemd-run --unit=name --slice=slice_name --property prop_name= val command
```

where

name is your name for the unit. If --unit is not specified, a nondescriptive unit name will be chosen automatically. It is highly recommended to always include this option and choose some descriptive name you can easily recognize.

slice_name is one of the standard, existing slices on your system you want the unit to operate under. You can list slices on your system using the systemctl -t slice command. Or you can even create a new slice by designating a new, unique name. By default, services and scopes are created in the **system. slice**.

prop_name is an allowed property you want to initially set for the unit, and Val is the value you want to assign to that property. In this section, we use a separate systemctl command to do this to change a unit CPUShares. But it can also be done using this option of the systemd-run command.

command is the Linux command, and its valid options, option arguments, and command arguments that will execute in the service unit when it is started. The command should always be at the end of the systemd-run command to avoid confusing systemd-run options with the command you want to run in the unit!

See the systemd-run man page on your system for more information on options and their arguments.

To set a property, such as the % share of the CPU that a service gets, you use the systemctl command. The general syntax for setting a property of a service with systemctl is

```
systemctl set-property service_name property=property_value
```

where

service name is the name of the service you want to modify the property of,
property is the valid name of a property, and
property_value is an allowable value of the given property.
See the man page for systemctl on your Linux system for more information.

Example 17.8 Starting a New Service with systemd-run and Allocating CPU Resources

Objectives. The following example allows you to start three transient services in the user. slice, and allocate a larger percentage of the CPU to one of them. Therefore, one of them will get more time in the CPU. The program that will be running in each of the services will be the top command, with the -b batch option.

Prerequisites: That you have superuser privilege on the system.

Requirements: Do the following steps, in the order presented later.

1. Execute the following systemd-run command, with the arguments shown:

```
$ systemd-run --unit=test  --slice=user.slice top -b
==== AUTHENTICATING FOR org.freedesktop.systemd1.manage-units
===
Authentication is required to manage system services or other
units.
Authenticating as: bob,,, (bob)
Password: zzzzz
==== AUTHENTICATION COMPLETE ===
Running as unit test.service.
$
```

2. Check the status of the service.

```
$ systemctl status test.service
• test.service - /usr/bin/top -b
  Loaded: loaded
Transient: yes
  Drop-In: /run/systemd/system/test.service.d
        └─50-CPUShares.conf, 50-Description.conf, 50-ExecStart.
conf, 50-Slice
   Active: active (running) since Mon 2017-08-14 15:18:23 PDT;
      5min ago
Main PID: 18522 (top)
   Tasks: 1 (limit: 512)
     CPU: 924ms
   CGroup: /user.slice/test.service
        └─18522 /usr/bin/top -b

Aug 14 15:23:31 bob-PowerEdge-T110 top[18522]:    752 root
0 -20        0
Aug 14 15:23:31 bob-PowerEdge-T110 top[18522]:    753 root
0 -20        0
Aug 14 15:23:31 bob-PowerEdge-T110 top[18522]:    754 root
0 -20        0
Output truncated...
$
```

3. Run two more similar transient units to provide a comparison test for the relative
   setting of CPUShares.

```
$ systemd-run --unit=test2  --slice=user.slice top -b
Output truncated…
$ systemd-run --unit=test 3 --slice=user.slice top -b
Output truncated…
$
```

4. Use the systemctl command, and the set-property subcommand, to give the user.
   slice and test.service a higher CPU share. Then, reload the systemd daemon.

```
$ systemctl set-property user.slice CPUShares=3000
Output truncated...
$ systemctl set-property test.service CPUShares=3000
Output truncated…
$ systemctl daemon-reload
Output-truncated…
$
```

Exercise 17.31
Use the top command on your system to verify that test.service is actually receiving
more of a percentage of the CPU time than both services test2, test3, and other

user.slice services. How much more % time on every cycle through the top command output?

Exercise 17.32

If the default CPU allocation number for a process is 1024, what do numbers like 3000 (shown in the earlier systemctl set-property command) or 100, mean, in terms of relative percentage utilization of the CPU? Correlate your answer to this last question to the output of the top command you used in Exercise 17.31. Why did we set the CPUShares of user.slice to 3000 also?

5. Finally, to stop the three transient services started earlier, use the following commands:

```
$ sudo systemctl stop test.service
[sudo] password for bob: zzzzz
$ sudo systemctl stop test2.service
$ sudo systemctl stop test3.service
$
```

Conclusion: We used systemd-run to execute a set of three transient services, then gave one of those services a higher CPU priority. We had to use the systemctl daemon-reload command to basically reload the systemd manager configuration, reload all service unit files and other units, and redo the systemd dependency tree. This forces the CPUShares changes we made to take affect.

Persistent Cgroups and Setting Their Resources

Example 17.9 Creating Persistent Cgroups and Modifying Their CPUShare and CPUQuota

Objectives: To create persistent, or permanent, cgroup, by constructing a service unit file for it in the /etc/systemd/system directory. This is the standard location for user-installed and user-defined services in systemd, as shown in Chapter 18. The basic objectives of the following example will be to create a simple, user-defined service, and balance its CPU usage with the same systemctl set-property command we have shown for transient services in Example 17.8.

Prerequisites: Completion of Example 17.8, and having superuser privilege on the system.

Requirements: Do the following steps, in the order presented later.

1. Use your favorite text editor as the root, or superuser, to create and save the following text file, named test4.service, exactly as shown, in /etc/systemd/system:
[Unit]
Description=A test service that uses the sha256sum command
After=remote-fs.target nss-lookup.target
[Service]
ExecStart=/usr/bin/sha256sum /dev/zero
ExecStop=/bin/kill -WINCH ${MAINPID}
[Install]
WantedBy=multiuser.target
The structure and components of the earlier file are detailed completely in Chapter 18.

2. Now we can start this service, and make it persistent, using the systemctl command, and its unit subcommand. We can also use the ps and systemctl command, and its unit subcommand status, to monitor valuable information about test4.service

```
$ sudo systemctl start test4.service
[sudo] password for bob: zzzzz
$ sudo systemctl enable test4.service
Created symlink from /etc/systemd/system/multi-user.target.
wants….
Output truncated...
```

```
$ sudo systemctl status test4.service
[sudo] password for bob: zzzzz
• test4.service – A test service that uses the sha256sum command
  Loaded: loaded (/etc/systemd/system/test4.service; enabled;
    vendor preset: enabled)
  Drop-In: /etc/systemd/system/test4.service.d
        └─50-CPUShares.conf
  Active: active (running) since Mon 2017-08-14 19:30:49 PDT;
    5min ago
 Main PID: 2089 (sha256sum)
     CPU: 5min 12.188s
  CGroup: /system.slice/test4.service
       └─2089 /usr/bin/sha256sum /dev/zero
Aug 14 19:30:49 bob-Inspiron-6000 systemd[1]: Started test4
Output truncated…
$
$ ps -p 2089 -o pid,comm,cputime,%cpu
  PID COMMAND              TIME %CPU
2089 sha256sum            00:02:10 99.6
$
```

The valuable information we can see from the systemctl status and ps commands is that this service is consuming 99.6% of the CPU resource on our system. Also, it's pid is 2089, it is running the command sha256sum, and it is in the system.slice cgroup "block."

3. There are two approaches we can take to "throttle" back the amount of CPU resource test4.service consumes. First, we can decrease the CPUShare for test4.service, to "throttle" it back to a more acceptable value. This change will be persistent across system reboots. We can set the value of CPUShares for test4.service to 100 (the value we will set in the command later), which will be effective upon all subsequent restarts of that service. Second, we can set a maximum quota on CPU consumption for test4.service, by changing its CPUQuota property to some lower percentage value.

   To make a change in CPUShares or CPUQuota for test4.service that is transient or temporary (nonpersistent across restarts or reboots), add the --runtime option to the following systemctl set-property commands.

   To throttle back test4.service using CPUShares, use the following command

   ```
   $ sudo systemctl set-property test4.service CPUShares=100
   ```

   To throttle back test4.service using CPUQuota, use the following command

   ```
   $ sudo systemctl set-property test4.service CPUQuota=20%
   ```

   To make these changes effective, use the following commands:

   ```
   $ sudo systemctl daemon-reload
   $ sudo systemctl restart test4.service
   ```

4. Now let's see what the sha256sum process has as a CPUShares value. On our Linux Mint system, the following command shows this:

   ```
   $ cat /sys/fs/cgroup/cpu/system.slice/test4.service/cpu.shares
   100
   $
   ```

Exercise 17.33

Use the top command to verify the CPU % usage of test4.service after just limiting its CPUShare to 100. What is the CPUShare on your system and why? Is this percentage in agreement with the ps command percentage shown in the earlier session? What does the top command reveal about %CPU usage for test4.service after you have throttled it back with the CPUQuota property value of 20% that we show. Do both methods yield similar results? Why or why not?

Exercise 17.34

Create and start another system.slice service with the systemd-run command, named test5.service, that executes the md5sum command. Start it and make it persistent. Then, use the top command to verify the % CPU of both test4.service and test5.service. What are their relative % CPU usages on your system and why?

Exercise 17.35

How do you stop test4.service and test5.service from running? Look ahead to Chapter 18 to learn how to completely delete them from your system.

5. To cleanup your system to stop and remove the test4.service, use the following commands:

```
$ sudo systemctl stop test4.service
$ sudo rm -r /etc/systemd/system/test4*
$
```

Conclusion: We have successfully created a simple, yet persistent, user-defined service and balanced its CPU usage with the systemctl command.

17.8.1.3 Managing Memory

systemd allows you to easily control and set limits on a unit's memory usage. This can be critical in cases where there is a minimum amount of installed memory on the computer hardware. To see what percentage of memory a unit is consuming, use the top command, and examine the column output for the Memory Usage Variable (%MEM). Given the use case(s) your system has been designed for, if an application or service is consuming too much memory, there are two ways you can limit its memory usage.

The first way is to use the systemctl set-properties command, with the property MemoryLimit assigned a value, for a particular service. The following command sets MemoryLimit for the unit test4.service that we created in Section 17.8.1.2.1 to 500 MB:

```
$ sudo systemctl set-property test4.service MemoryLimit=500M
```

The limit specifies how much process and kernel memory can be used by tasks in this unit. The limit is a memory size in bytes. If the numerical value is followed by K, M, G, or T, the specified memory size is parsed as Kilobytes, Megabytes, Gigabytes, or Terabytes (with the base 1024), respectively. The special value "infinity" means that no memory limit is applied.

The earlier systemctl set-properties command will create a MemoryLimit.conf file as /etc/systemd/system/test4.service.d/MemoryLimit.conf, which can be edited in the second way later.

The second way is to put the following directive in the [Service] section of the unit's configuration file:

```
MemoryLimit=value
```

Replace value with a limit on maximum memory usage of the processes executed in the cgroup. Use suffixes K, M, G, or T to identify Kilobyte, Megabyte, Gigabyte, or Terabyte as the unit of measurement. Also, the MemoryAccounting parameter has to be enabled for the unit.

The MemoryLimit parameter controls the memory.limit_in_bytes control group parameter.

To assign a 500 MB memory limit to the test4 service, modify the MemoryLimit setting in the /etc/systemd/system/test4.service.d/MemoryLimit.conf file to include the following:

```
[Service]
MemoryLimit=500M
```

To apply the changes in either of the two ways, reload systemd's configuration and restart the service.

```
$ sudo systemctl daemon-reload
$ sudo systemctl restart test4.service
```

Exercise 17.36
Set the MemoryLimit value for selected services on your Linux system, either higher or lower, depending upon specific use-case requirements.

Exercise 17.37
Use the Linux **free** command on your system, and its options, to gauge the distribution of system memory. Do the quantities you get agree with the amount of installed system memory you have?
From the output the **free** commands, what can you infer about the installed memory quantity on your system, in GB? See the man page for the **free** command on your system to get more information about its usage and options.

17.8.1.4 Assessment of System Disk Usage

Along with CPU usage and memory management, the ability to observe, assess, and modify the resources of persistent storage media, particularly hard disk drives, is critical to a system administrators task. Even for an ordinary user, disk space is a precious commodity, regardless of the availability and lower cost of terabyte-capacity hard drives. Consider a computer system that can only physically support one hard drive internally on the fastest bus architecture it has or a system where an ordinary user is storing large files, such as video or other types of media file.

Traditionally, the Linux **du** command allowed you to get a text-based summary of disk usage for files and directories in a variety of formats. An administrator could then take action to trim hard disk usage according to the use case of the system. For more information on the du command, see the man page for du on your Linux system.

The Linux **df** command (df is an abbreviation for disk free) allows you to display the amount of available disk space for file systems on which the invoking user has read access. df is typically invoked using the -h option, which gives a human-readable output format. For example, on our Linux Mint system, the following command gives information about disk and file system layout:

```
$ df -h
file system   Size    Used    Avail.   Use%    Mounted on
udev          3.8G    0       3.8G     0%.     /dev
tmpfs         767M    9.4M    757M     2%      /run
/dev/sda2     51G     17G     32G      34%     /
tmpfs         3.8G    1.1M    3.8G     1%      /dev/shm
tmpfs         5.0M    4.0K    5.0M     1%      /run/lock
tmpfs         3.8G    0       3.8G     0%      /sys/fs/cgroup
/dev/sda1     511M    3.4M    508M     1%      /boot/efi
cgmfs         100K    0       100K     0%      /run/cgmanager/fs
tmpfs         767M    56K     766M     1%      /run/user/1000
/dev/sdb1     7.3G    119M    7.1G     2%      /media/bob/07D5-128C
$
```

More contemporary graphical methods can be used to view and assess disk usage on our four representative Linux systems. For example, on Debian-family Linux systems such as our Linux Mint system, you can use the gnome disk usage analyzer, named "Baobob." The same graphical tool is available by default on CentOS. It displays either a bar or sector chart representation of relative disk usage, for any attached persistent medium. This tool is very useful when you want to get a picture of where, and what, your disks are being used for. Figure 17.11 shows a typical disk capacity display of the root directory on a Linux Mint system. Similar tools are found on all of our representative systems.

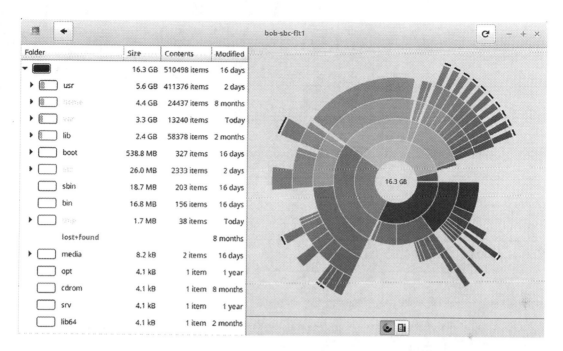

FIGURE 17.11   Disk usage analyzer display of the root directory.

Another useful summary tool, similar to Gparted, is the gnome-disk-utility, launched from the Preferences Menu in Linux Mint.

17.8.2 Network Configuration with the `ip` Command

The most important and useful command for the system administrator when doing a postinstall network configuration is the `ip` command. The following sections illustrate the basic usage of this command, and give a use-case example that illustrates how to assign more than one IP address to a NIC. A NIC is what the system hardware uses to interface with a network.

With the new toolkit, it is as easy as with the old to add new ip addresses:

```
ip addr add 192.168.1.1/24 dev eth0
```

17.8.2.1 Basic `ip` Command Syntax, Options, and Operations

An abbreviated version of the man page for the `ip` command is as follows:

`ip`

```
Syntax: ip [ Options] Object { Command | help }
        ip [ -force ] -batch filename
     where

     Object   { link | addr | addrlabel | route | rule | neigh | ntable | tunnel
                 | tuntap | mad-
          dress | mroute | mrule | monitor | xfrm | netns | l2tp | tcp_metrics }
     Options: { -V[ersion] | -h[uman-readable] | -s[tatistics] | -r[esolve] |
                 -f[amily] { inet |
```

```
        inet6 | ipx | dnet | link } | -o[neline] | -n[etns] name | -a[ll] | -c[olor]
    }
```
Command:

Specifies the action to perform on the object. The set of possible actions depends on the object type

For most objects, the commands possible are **add**, **delete,** and **show**. Some objects do not allow all of these operations or have some additional commands. The help command is available for all objects. It prints out a list of available commands and argument syntax conventions. See the syntax and output of the first example later

If no command is given, usually the **show** command is the default, if that option can be applied to the objects specified

Purpose: Show or manipulate routing, devices, policy routing, and tunnels

Output: Output or modified routing, devices, policy routing, or tunnels

Commonly used options/features:

    addr   IP address

To view an abbreviated syntax display of help on the ip route command, type the following:

```
$ ip addr help
Usage: ip address {add|change|replace} IFADDR dev IFNAME [ LIFETIME ]
                                       [ CONFFLAG-LIST ]
       ip address del IFADDR dev IFNAME [mngtmpaddr]
       ip address {show|save|flush} [ dev IFNAME ] [ scope SCOPE-ID ]
                    [ to PREFIX ] [ FLAG-LIST ] [ label LABEL ] [up]
       ip address {showdump|restore}
IFADDR := PREFIX | ADDR peer PREFIX
       [ broadcast ADDR ] [ anycast ADDR ]
       [ label IFNAME ] [ scope SCOPE-ID ]
Output truncated...
$
```

From the earlier listing, you can see the Commands available to work on the Object route, i.e. add, change, replace, del(ete), show, save, flush, and showdump.

To see the status of the specific NIC named enp2s0, type the following:

```
$ ip addr show enp2s0
2: enp2s0: <BROADCAST,MULTICAST,UP,LOWER_UP> mtu 1500 qdisc mq state UP group
default qlen 1000
    link/ether e4:11:5b:12:c2:77 brd ff:ff:ff:ff:ff:ff
    inet 192.168.0.6/24 brd 192.168.0.255 scope global enp2s0
       valid_lft forever preferred_lft forever
    inet6 fe80::e611:5bff:fe12:c277/64 scope link
       valid_lft forever preferred_lft forever
```

To see the status of all the network interfaces attached or defined on the system, type the following:

```
$ ip addr show
1: lo: <LOOPBACK,UP,LOWER_UP> mtu 65536 qdisc noqueue state UNKNOWN group default
qlen 1
    link/loopback 00:00:00:00:00:00 brd 00:00:00:00:00:00
    inet 127.0.0.1/8 scope host lo
       valid_lft forever preferred_lft forever
    inet6 ::1/128 scope host
       valid_lft forever preferred_lft forever
```

```
2: enp2s0: <BROADCAST,MULTICAST,UP,LOWER_UP> mtu 1500 qdisc mq state UP group
default qlen 1000
    link/ether e4:11:5b:12:c2:77 brd ff:ff:ff:ff:ff:ff
    inet 192.168.0.6/24 brd 192.168.0.255 scope global enp2s0
       valid_lft forever preferred_lft forever
    inet6 fe80::e611:5bff:fe12:c277/64 scope link
       valid_lft forever preferred_lft forever
```

To display link characteristics of enp2s0, type the following:

```
$ ip link show dev enp2s0
2: enp2s0: <BROADCAST,MULTICAST,UP,LOWER_UP> mtu 1500 qdisc mq state UP mode
DEFAULT group default qlen 1000
    link/ether e4:11:5b:12:c2:77 brd ff:ff:ff:ff:ff:ff
$
```

17.8.2.2 Use Case Example: Assigning Several IP Addresses to a NIC

This section and its example show how to assign several IP addresses to one NIC, without using the deprecated ifconfig command or its obsolete "alias" notation.

Question: Why would you want to do this?

Answer: You want to have your favorite Web server, nginx, listen on several IP addresses of your LAN, but you only have one NIC attached to the system. You proceed with the following commands, and then modify the nginx configuration to do so. We do not show the nginx configuration changes here.

Example 17.10 Assigning Several IP Addresses to a NIC

Objectives: To use the Linux ip command to assign several IP addresses to your system's NIC.

Prerequisites: All of Chapter 11 on Networking and Internetworking, and relevant sections of Chapter W20 on Interprocess Communication at the book website. Having super-user privileges on your system.

Requirements: Do the following steps, in the order presented.

1. If you need an additional IP address temporarily, you can add it to any network interface by using the following general format command:

```
$ sudo ip address add <ip-address>/<netmask> dev <interface>
An actual example would be as follows:
$ sudo ip address add 192.168.0.100/24 dev enp2s0
[sudo] password for bob: QQQ
$
```

The earlier command would add an additional IP address to the network interface, using a 24 bit netmask, to whatever current IP addresses are already assigned to the nic enp2s0.

You can check the result with the following command:

```
$ ip address show enp2s0
2: enp2s0: <BROADCAST,MULTICAST,UP,LOWER_UP> mtu 1500 qdisc mq state
UP group default qlen 1000
    link/ether e4:11:5b:12:c2:77 brd ff:ff:ff:ff:ff:ff
    inet 192.168.0.6/24 brd 192.168.0.255 scope global enp2s0
          valid_lft forever preferred_lft forever
    inet 192.168.0.100/24 scope global secondary enp2s0
          valid_lft forever preferred_lft forever
    inet6 fe80::e611:5bff:fe12:c277/64 scope link
          valid_lft forever preferred_lft forever
```

You can delete this address again using the following command:

```
$ sudo ip address del 192.168.0.100/24 dev enp2s0
```

These changes are lost when you reboot your machine.

2. To make the additional IP address on the nic permanent, you can edit the file /etc/network/interfaces, and add the earlier IP address to the nic, while retaining the defaults found in that file. This is what the interfaces file would look like with the earlier assignment added:

```
# interfaces(5) file used by ifup(8) and ifdown(8)
auto lo
iface lo inet loopback
# The primary network interface, the default
auto enp2s0
iface enp2s0 inet dhcp
# The address added to the default
iface enp2s0 inet static
     address 192.168.0.100/24
```

The earlier file shows you can retain the default settings for network interfaces, and add additional ones.

To return to the default network interface settings for the nic enp2s0, simply edit the /etc/network/interfaces file and remove the add IP address lines.

3. This step brings up a very interesting point. We wish to have the new settings take effect, without rebooting the machine. In the earlier two steps, we were in reality accessing the system via ssh. So, to accomplish having the new settings take effect, it was necessary for us to execute the following multiple command. This is because, for us, the first part of it, sudo ifdown enp2s0, would drop our ssh session connection. As a multiple command (two commands on the same command line), the ssh-session is maintained, and we are not logged off the system. And in addition, the settings take effect without rebooting the system.

To have these settings take effect without a reboot, we use the commands ifdown/ifup in the following multiple command:

```
$ sudo ifdown enp2s0 && sudo ifup enp2s0
Killed old client process
Internet Systems Consortium DHCP Client 4.3.3
Copyright 2004-2015 Internet Systems Consortium.
All rights reserved.
For info, please visit https://www.isc.org/software/dhcp/
Listening on LPF/enp2s0/e4:11:5b:12:c2:77
Sending on   LPF/enp2s0/e4:11:5b:12:c2:77
Sending on   Socket/fallback
DHCPRELEASE on enp2s0 to 192.168.0.1 port 67 (xid=0x7dc0aeea)
RTNETLINK answers: No such process
RTNETLINK answers: Cannot assign requested address
Internet Systems Consortium DHCP Client 4.3.3
Copyright 2004-2015 Internet Systems Consortium.
All rights reserved.
For info, please visit https://www.isc.org/software/dhcp/
Listening on LPF/enp2s0/e4:11:5b:12:c2:77
Sending on   LPF/enp2s0/e4:11:5b:12:c2:77
Sending on   Socket/fallback
DHCPDISCOVER on enp2s0 to 255.255.255.255 port 67 interval 3
(xid=0xc6039266)
DHCPDISCOVER on enp2s0 to 255.255.255.255 port 67 interval 8
(xid=0xc6039266)
DHCPREQUEST of 192.168.0.6 on enp2s0 to 255.255.255.255 port 67
(xid=0x669203c6)
```

```
DHCPOFFER of 192.168.0.6 from 192.168.0.1
DHCPACK of 192.168.0.6 from 192.168.0.1
bound to 192.168.0.6 -- renewal in 288071 seconds.
$
```

Exercise 17.38

How would you delete the assignment of the address 192.168.0.100 to enp2s0 during this session?

Exercise 17.39

How would you ensure that the deletion persisted between system boots?

Conclusion: We assigned several IP addresses on LAN to one NIC on our Linux system.

17.9 System Security

There are two meanings of the word "security" when referring to computers. We will consider only one of them.

The first meaning denotes reliability. In other words, the files and data on the disks can be relied upon to be persistent over time. That is the objective of not only having persistent media, but also of backing up the operating system itself, and the user files and data. For example, if the operating system crashes and cannot be resurrected, the user files and data are archived, secure, and can be recovered in total.

The second meaning denotes free from malicious intrusion by agents (such as robotic cracker programs) or objects (such as the processes that are the basis of program execution) that do not have the authority to access a specific part of the system or any of its components, files, or data. The reasons for implementing system security in this second sense, both for an ordinary, single-user system and in a multiuser environment, should be obvious. Especially in the light of contemporary privacy issues, the widespread hacking and system penetration forces that are prevalent on the Internet currently.

We deal here primarily with the second meaning.

Fundamentally according to this second meaning, there are at least four places where you can situate system security measures as we show them in this section. They are as follows:

1. At the Process Level—The process model of the Linux system makes this site most important. This model mandates security implemented on processes, via authentication and credentialing techniques. Processes that are the basis for program execution in Linux and that bridge kernel and user spaces. Examples of these measures are as follows:
   - Setting traditional UNIX permission bits, the special permission bits SUID, SGID, and the sticky bit detailed in Chapter 5. The setting of permission bits, as detailed in Chapter 5, implies that the processes that shell commands create are the real objects that the permission bits that files possess are aimed at, or applied to. This is true of everything that creates processes in Linux, which means user-written programs and all other applications programs and system programs as well.
   - Deployment of ACL's and extended POSIX1.e ACL's, as detailed in this chapter.
   - Use of Linux-specific "capabilities" applied to processes, as detailed in this chapter.
   - Namespace isolation of processes, using the Linux kernel namespace API, and the six major implementations of it. We give a brief description of this in Chapter W26, Section 9.9 at the book website.

2. At the Physical Level—When someone sits directly in front of the computer and tries to log in and use it. One of the ways that can be accomplished is through the use of password protection into user accounts. On a public Linux computer system, this is the most valuable way to maintain general access security and user/account security. There are also many security techniques in this place that limit physical access to the hardware of the system, such as locking the door

to the room you keep the machine in or having protocols in place for limiting physical access to the machine.

3. At the Network Level—By placing safeguards on the computer's network connection. This is accomplished using various forms of not only monitoring with systemd journalctl, and using the many forms of intrusion detection, but also with intrusion prevention systems (IPSs) we illustrate later.

4. At the Persistent Media Level—On the persistent media attached, either physically or virtually, to the computer itself. These techniques are applied to the system disk and other attached media such as user data disks or mounted Network File Systems (NFS). Examples of this are the use of virtualized "sandboxed" machines on traditional volumes, or on ZFS volumes, our recommended user data storage model from Section 17.2.1, and the various domains of disk data encryption.

Furthermore, the earlier sites use the following specific techniques. We will describe some of the important ones in the following sections. A further elaboration of some of these techniques is given in Chapter W26, Sections 9.2 and 9.3, titled "Access Control Credentials: Discretionary (DAC), Mandatory (MAC), and Role-Based (RBAC)" and "Using Access Control Lists (ACLs) in Linux" at the book website.

- Password-based authentication
- Access control, either DAC, MAC, or RBAC
- The sudo command
- Setting POSIX1.e ACLs and extended ACLs for files and directories
- Intrusion Detection and Intrusion Prevention
- Security Software
- ufw System Firewall
- Process Credentialing
- User namespace process isolation
- Whole disk, partition, directory, or file-level data encryption

17.9.1 Password-Based Authentication

The first line of defense in system security, and the technique employed almost universally across many types of computer system, is password-based authentication. The Linux system compares a user-entered password at login for a user ID, compares the password to a previously established and stored one held in a password file for that user, and based on the comparison authenticates the user. The ID not only determines whether the user can gain access to the system itself but also determines what privileges the user has. For example, superuser privilege. Also, the ID is used in DAC, as shown in Chapter 9, Section 2. The password file and a hash/salt scheme using a SHA256 hashing algorithm for encrypting the password work together to authenticate a user ID. The password file on the system, in /etc/passwd, holds user information and works in conjunction with the /etc/shadow file to authenticate a user ID.

Exercise 17.40

Examine the contents of the /etc/passwd file on your Linux system. With reference to your username entry in that file, how many colon-delimited fields are there, what are the meanings of those colon-delimited fields?

17.9.2 Access Control Credentials: DAC, MAC, and RBAC

The nomenclature we use in this section is important if you want to understand the different types of access control. When we talk about access control via security checks, here are the important terms

Objects: A fundamental component of an executing Linux kernel are the programs and processes it maintains. Objects are the entities that are targeted, or worked on, by the processes of a program. For example, processes themselves can be objects or the processes that are generated by executing instances of a program.

Files/inodes are another form of object, particularly the executable form of file objects, and the data structure(s) holding their information. This should not be confused with file system objects, which we have referred to as either an ordinary file or a directory.

Object Ownership: Indicates the owning user and group.

Object Context: Security checks done when objects are acted on.

Subjects: An object that is acted upon by another object. Processes are active subjects, such as those processes that are created by an exec() or fork() system call from some originating process.

Subject Context: Security checks done when an active subject performs its operations.

Action: What a subject does to an object. This includes reading, writing, creating, and deleting files; forking or signaling.

Permissions: Security checks when a subject acts upon an object. Taking the subject context, the object context, and the action, and searching one or more sets of permissions to see whether the subject is granted or denied permission to act in the desired manner on the object, given those contexts. In simple terms, match subject and object permissions, and let the subject act or not on the object.

There are three basic "classes" of permissions are

1. DAC:

    Sometimes the object will include sets of rules as part of its description. This is an "Access Control List" or "ACL." A Linux file may supply more than one ACL. A traditional Linux file, for example, includes a permissions mask that is an abbreviated ACL with three fixed classes of subject ("user," "group," and "other"), each of which may be granted certain privileges ("read," "write," and "execute"—whatever those map to for the object in question). Linux file permissions do not allow the arbitrary specification of subjects, however, and so are of limited use.

    A Linux file might also support a POSIX.1e ACL, or in the case of ZFS, an NFSv4 ACL. This is a list of rules that grants various permissions to arbitrary subjects.

2. Mandatory access control (MAC):

    The system as a whole may have one or more sets of permissions that get applied to all subjects and objects, regardless of their source. Security Extended Linux is an example of this.

3. Role-Based Access Control (RBAC):

    Rather than use the user ID to determine what access rights users and groups have on the system, the RBAC model grants access based on the role or roles that a user assumes. The classic RBAC example is the use of the **su** or **sudo** commands to grant an unprivileged user root privileges. Another example can be found in ZFS, when you execute the **zfs** command, and your action is checked to see that the subject issuing the command has the role privilege, *even if the user is root*.

    These classes of access control policies determine what action is allowed on what object, under what circumstances (DAC, MAC, or RBAC) and by what subject.

    A permission in the traditional UNIX/Linux sense, for example, is read, write, or execute privilege. A subject, for example, can be thought of as an executing process. Most importantly, an object is a Linux process (detailed in Chapter 10), since everything done on files and the data in them, is done through active processes on the system.

Exercise 17.41
    What major apparatus controls the execution of processes in Linux?

Exercise 17.42
> Give other examples of DAC, MAC, and RBAC.
> On the command line, an ordinary user or the system administrator is able to implement resource use restrictions and privileges by controlling process credential assignments, exercised on subject executable image files, via the **chmod** command. On Linux, an ordinary unprivileged user can be given the required privileged role with the **su** or **sudo** command. Then, as root, she can issue a privileged **chmod**, **chown**, and **chgrp** to grant or modify file and directory access permissions, and use the DAC, MAC, or RBAC methods.

Types of Credentials

We are concerned with the three basic types of credentials that the Linux kernel supports. These are as follows, with major references in the text shown:

Traditional UNIX Credentials—See Chapter 5 for a more complete exposition of these.

1. Real UID
2. Real GID

UID and GID are assigned to most Linux objects. These in large part define the object context of that object, with processes included in this assignment.

3. Effective (EUID), Saved (SUID), and File System (FSID) UID
4. Effective (EGID), Saved (SGID), and (FSGID) GID
5. Supplementary groups

The additional credentials used by processes are EUID/EGID/GROUPS, and are used as the subject context, and real UID/GID will be used as the object context.

Access Control Lists

ACLs provide the ordinary, unprivileged user with the ability to set finer access controls on directories and files than the traditional Linux permissions, whether they are used on ext4 or ZFS file systems. Two different basic types of ACL apply to files and directories. An ACL that defines the current access permissions of files and directories is called an *access* ACL. An ACL, which can only logically be set on a directory, and that defines the permissions that a directory object inherits from its parent directory at the time of its creation, is called a *default* ACL. Additional basic types of ACL are *minimal* and *extended* ACLs. ACL permissions that can be equivalent to the traditional file mode permissions are called *minimal ACLs*. Minimal ACLs have three entries, which can be the same as the traditional file permissions. ACLs with more than three entries are called extended ACLs. Extended ACLs also contain a mask entry and may contain any number of named user and named group entries.

See Chapter W26, Section 9.3 and its subsections, at the book website, for more information about POSIX.1e and NFSV4 ACL's.

Capabilities—See the man page for capabilities on your system

1. Set of permitted capabilities
2. Set of inheritable capabilities
3. Set of effective capabilities
4. Capability bounding set

These are most pertinent when they apply to processes, which are the active elements in system operation. They are privileged permissions exercised in a "finer-grained" context. Finer-grained is used here to mean a more specific, targeted privilege. These are applied to a process or processes that ordinarily, via the traditional model, could only be granted a blanket, all-or-nothing scheme of privileges. Putting a user in the sudoers file is an example of this traditional model's application.

Capabilities are controlled by changes in the traditional Linux permissions but can also be set more finely and viewed directly by the capset and getcap system calls.

See Chapter W26, Section 9.8.3, at the book website, for a further exposition of Linux process capabilities.

Exercise 17.43

How (and why) do you apply Linux capabilities on the command line? Provide an example of this that you have implemented and tested on your Linux system.

17.9.2.1 sudo

On Linux, programs, commands, and files are traditionally accessed through user and group permissions. Each user has a unique identifier, given either as a username or UID. Users belong to unique groups, given either as a group name or a GID. Specific users and groups have permission to access available programs, commands, and files.

The **sudo** program, or command, allows a single command to be run as root, or even as some other user. Only the system administrator, or root user, can utilize a policy listing file (named sudoers) that contains commands that each user can execute. So the administrator controls what users have what privileges on the system. When any user needs to run a command that requires root permissions, that user types **sudo command** in a console terminal, allowing them to run command with root privilege. Then, **sudo** consults its permissions list in the policy listing file. If the user has permission to run that command, it runs the command. If the user does not have permission to run the command, **sudo** denies execution. Running **sudo** does not require knowing root's password, but by default requires the user's own password to execute successfully.

For a more detailed description of the **sudo** command (with numerous examples of enabling and using it on our representative Linux systems), see Chapter W26, Sections W26.9.2.1.1–W26.9.2.1.3 at the book website. We also give installation instructions on how to install the sudo command, for Debian-family and CentOS systems, in Section 17.9.2.1 of Appendix A.

The **su** command (an abbreviation for "switch user") allows a user to switch roles and become the superuser on the system without logging off from their own account. You must know the password of the root account if you want to assume the role of root. On our representative Linux systems, this command is restricted, and you are encouraged to include users that need to have higher levels of privilege on the system in the sudoers file. That way they can execute privileged operations on a per-command basis.

Exercise 17.44

Are you included in the sudoers group and file on your Linux system? How did you find out if you are?

Exercise 17.45

Enter some unprivileged user on your Linux system into the sudoers file. Then have them test the **sudo** command and its operation as we have shown it in this chapter and Chapter 18.

Exercise 17.46

Given any of the representative systems we illustrate in this book (or one you are using), is the **su** command available for your use on your Linux system? Why or why not?

17.9.3 IDS and IPS

A conceptual layout of how malicious activity from outside of the operating system interfaces through the components of a Linux system in given in Figure 17.12. It is important to realize that Figure 17.12 does not specify the multitude of types of attack that can intrude upon your system from a LAN or the Internet, but does show the arrangement of system components that these attacks can target. It also does not show where in the software, kernel, or hardware any defensive or preventative mechanisms are placed.

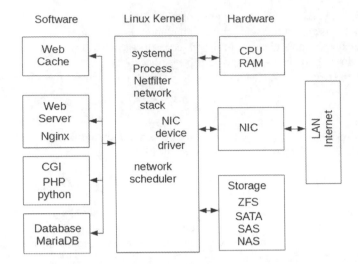

FIGURE 17.12 Routes of attack and Linux system components.

Exercise 17.47

Where would a Russian bot be situated in Figure 17.12?

Exercise 17.48

What is the Red Team Field Manual? What is the Blue Team Field Manual? What is Pen testing? Why would you need to either or both of these manuals in the context of Pen testing?

An Intrusion Detection System (IDS) is usually a software application that monitors a network or systems for unauthorized activity or user protocol violations. It is applied via a software-specific interface, and most importantly, through log file monitoring.

A wide variety of free or commercial IDSs, from antivirus software to hierarchical systems that monitor the traffic of a backbone network, are available. They can be classified as Network IDSs (NIDS) or as Host IDSs (HIDS). An IDS that monitors operating system files is an example of HIDS. An IDS that analyzes incoming network traffic is an example of NIDS. Furthermore, IDS can be classified by detection strategy. The most well-known types of detection strategies are signature-based detection (recognizing bad patterns, such as malware) and anomaly-based detection (detecting deviations from a predefined model of "good" traffic, which often uses a form of Artificial Intelligence. IDSs that have some predetermined and structured response protocol are referred to as IPS.

Typical network IDSs try to detect malicious activity, such as brute-force attacks, denial-of-service attacks, port scans, or attempts to crack into computers by monitoring network traffic.

Following is a partial listing of some free IDSs available for installation, but not installed by default, on our Linux Mint system:

- The Advanced Intrusion Detection Environment (AIDE) is a Linux Mint HIDS that can monitor and analyse the internals of the operating system.
- Snort is a Linux Mint NIDS that performs packet logging and real-time traffic analysis on IP networks.
- fail2ban is an example of a Linux Mint IPS. One of the many configurable actions it takes based upon malicious activity coming into your system from a LAN or Internet is to lock out IP addresses that attempt to login via ssh more than five times with the wrong password. This prevents brute-force attacks through port 22.

These three IDSs are available through the Linux Mint Software Manager. The best way to find out more about the particulars of installing and configuring them is by reading their online documentation and tutorials.

Exercise 17.49

Install one or all of the earlier IDS packages on your Linux system, using the package management system available. How do you deploy fail2ban, and how do you control its environment? For example, how do you "unblock" an IP address using fail2ban? Test this on your Linux system.

Exercise 17.50

What are the capabilities and uses of Wireshark? Give a description of it in terms of the IDSs we described earlier, and then install Wireshark and use it on your Linux system.

17.9.4 Linux Security Software

All larger Linux computer server systems, as well as individual desktop systems used by an ordinary user, can have several types of network-based and host-based security software available to detect malicious activity, protect systems and data, and support intrusion detection and the appropriate responses to them. They can be organized into the following categories:

- IDS and IPS
- Remote Access Software
- Web Proxies
- Vulnerability Management Software
- Authentication Servers
- Routers
- Firewalls
- Network Quarantine Servers

We covered some of the details of Intrusion Detection and also give a more complete description of Firewalls in the next section.

17.9.4.1 System Firewall

A firewall is a facility that prevents unauthorized access to or from a private network or a computer. Firewalls can be implemented in either hardware, software, or a combination of both. Firewalls are primarily used to prevent unauthorized access to a private network or intranet, from the Internet. All traffic entering or leaving a single computer or intranet passes through the firewall, which examines each message and blocks those that do not meet the specified security criteria or *firewall rules*.

In addition to limiting access to your computer and network, a firewall is also useful for allowing remote access to a private network through secure authentication certificates and logins. A common practice is to let a stand-alone computer serve as a single hardware firewall to a private network.

Hardware firewalls can be purchased as a stand-alone product. They can also be found integral to broadband routers, such as the Actiontec PK5000. Most hardware firewalls will have a minimum of four network ports to connect other computers, but for larger networks, business networking firewall solutions are available.

In Linux, if the firewall is active (enabled, or running), all incoming traffic is usually *blocked* until you specify a TCP or UDP port that traffic is allowed to come in on. Also, usually all outgoing traffic on all ports is allowed. In the following sections, we describe the Uncomplicated FireWall (ufw) for the Debian-family of Linux (including Debian, Ubuntu, and Linux Mint), and also firewalld for CentOS.

To determine whether the firewall is active, use the following systemd command on any of our representative Linux systems:

```
$ systemctl status firewall_name
```

where firewall_name is ufw on the Debian-family and firewalld on CentOS.

When we executed this command on our Debian-family Linux Mint system, we got the following output:

```
$ systemctl status ufw
● ufw.service - Uncomplicated firewall
   Loaded: loaded (/lib/systemd/system/ufw.service; enabled; vendor preset:
enabled)
   Active: active (exited) since Fri 2017-10-13 11:32:43 PDT; 9h ago
  Process: 317 ExecStart=/lib/ufw/ufw-init start quiet (code=exited, status=0/
SUCCESS)
 Main PID: 317 (code=exited, status=0/SUCCESS)
   CGroup: /system.slice/ufw.service

Warning: Journal has been rotated since unit was started. Log output is incomplete
or unavailable.
$
```

Be very careful when adding custom rules or modifying the firewall, it may endanger your system's security!

Exercise 17.51

What firewall is available on your Linux system, and is it active? Are there any firewall rules in effect by default, and what are they? How did you find this out?

17.9.4.2 *ufw and Netfilter Interface in Debian Linux Family*

The Debian family of Linux uses ufw to protect the system. When we installed our systems according to the recommendations in Section 17.2, by default ufw was *not* installed on Debian, but was installed and active on Ubuntu and Linux Mint. Incoming traffic was blocked on Ubuntu, but not on Linux Mint. Outgoing traffic on both of the systems was installed and was *not* blocked. The default configuration file for its rules is located in /etc/ufw/ufw.conf. The easiest way for you to find out more about text-based modifications to firewall rules is to see the **ufw** man page on your Linux system. We provide an extensive section giving the basic details of using the command line method of modifying firewall rules in Chapter W26, Section 9.6.2, titled "Linux Uncomplicated Firewall (ufw)" at the book website.

It is not absolutely necessary to change the ufw firewall rules, unless your security model, and its impact on your particular use case, warrants firewall rule customization. We do provide a general overview of ufw in the next section.

17.9.4.2.1 *Linux ufw*

The default firewall configuration utility for Linux Mint is the ufw. Unlike other firewall rule-based systems, such as the iptables firewall configuration utility, ufw is a much easier way to create an IPv4 or IPv6 host-based firewall. A "rules-based system" means that you create syntactically correct rules to control network connection access to your system. ufw's most basic application is to allow or deny access on ports or deny them from specific IP addresses that you know are problematic security risks.

ufw is based upon the Netfilter interface to the Linux kernel, and particularly, the filter table operations and protocols found in that interface. The rule format is also similar to the Packet Filter syntax in OpenBSD UNIX.

17.9.4.3 *firewalld in CentOS*

firewalld is the default firewall management tool for our CentOS system. Similar to ufw, it provides extensive security features by acting as a front-end for the iptables packet filtering system provided by the Linux kernel. The code for firewalld is written in Python.

firewalld supports IPv4 and IPv6 networks, and most notably is capable of managing separate firewall "zones." Each zone can have different configurations of rules. You can configure the Network Manager on CentOS to automatically switch zone configurations.

Very importantly, in the light of systemd cgroups control, system services and applications can use the D-Bus interface to interact with firewalld. The D-Bus interface, or Desktop Bus, is a software-based bus architecture used for Interprocess Communication and Remote Procedure Calls. It allows communication between processes running on the same system.

firewalld supports timed rules, meaning the number of connections (or "hits") to a Web server, for example, can be limited. firewalld can't limit incoming hits from known security risk IPs or URLs, similar to applications such as fail2ban. Fail2ban uses hit counting and rejection-of-service on a per source IP address to prevent brute-force and distributed denial-of-service attacks. That is a valid design constraint when you are formulating your postinstallation software requirements, as we did in Section 17.2.1.

For more in-depth information on firewalld, see Chapter W26, Section 9.7 at the book website.

The following Linux command shows that firewalld is active (by default) on our CentOS system:

```
[root@CentOS ~]# systemctl status firewalld
• firewalld.service - firewalld - dynamic firewall daemon
   Loaded: loaded (/usr/lib/systemd/system/firewalld.service; enabled; vendor
preset: enabled)
   Active: active (running) since Fri 2017-10-13 16:40:39 EDT; 1min 25s ago
     Docs: man:firewalld(1)
 Main PID: 685 (firewalld)
   CGroup: /system.slice/firewalld.service
         └─685 /usr/bin/python -Es /usr/sbin/firewalld --nofork --nopid
[root@CentOS ~]#
```

Exercises 17.52

What do you think is the biggest advantage of having a record of, or "logging", ufw or firewalld events?

17.9.5 Persistent Media Security

There are two basic strategies a system administrator can take to secure the persistent media, such as spinning and SSD hard disks, including USB-mounted media, and also to harden the file system structure on those media. The actual techniques of these two strategies overlap considerably with the other system administration tasks we show in the other sections of this chapter.

An example of an overlapping technique is found in the addition of hard disks to the system, and perhaps using ZFS to provide redundancy on these disks per the recommended storage model that we have shown in Section 17.4. This allows a system administrator to segregate the user files and other components of the system. That segregation provides a way of isolating those resources, and therefore allows them to be more securely accessed. Adding additional disks addresses the administrator's responsibility to backup important data, most critically user data files.

The first strategy involves providing additional physical media, or additional partitions on existing media, to accommodate user data files, or other components of the Linux file system.

The second strategy involves designing and implementing file and directory access permissions on those additional partitions or media, so that process authentication through the various forms of process credentialing.

17.9.5.1 Persistent Media Allocations for/home

According to the recommended user data storage model we provided in Section 17.2.1, you should put the /home subdirectory, where all of your user directories are located, on its own physical medium. That is a traditional system administration technique of designing your file system structure. When you add mount options to a single formatted partition on this medium, you can do the following:

- Set the nosuid option to prevent SUID and SGID permission-enabled executable programs running from there. Programs that need SUID and SGID permissions should not be stored in /home, as a preventative measure.

- Set the nodev option so no device file located there will be recognized. Device files should be stored in /dev, not in /home.
- Optionally set the noexec option, so no executable programs which are stored in /home can be run.

Exercise 17.53

You are tasked by your boss to migrate all of the /home directories of all users on your system off their default location on a single-disk system (where the system is also installed on that disk), to a multidisk installation on the same computer system. Sketch a detailed plan of exactly how you would do this, using the migration of a single user as an example. Show all commands you would use to achieve the migration, and successfully delete the old /home directory for that users account.

17.9.5.2 Securing the File System

Another important part of securing your Linux system is setting proper file system security.

Setting permissions on files and directories was covered in Chapter 5. POSIX1e ACLs are covered in detail in Section W26.9.3.1, titled "Linux POSIX.1e ACL Model Details", found at the book website.

17.9.6 Process Credentials

The multiprogramming model that we detail in Chapter 10 requires that processes, whether they are generated by shell built-in or external commands, by user-written programs, or by any system programs that use the fork, exec, execve system calls, have their credentials authenticated before they can make use of objects such as system resources, other processes, sockets, or files/inodes.

Linux systems assign credentials to processes, which associate the process with a specific user and a specific group. These credentials are essential in a multiuser and multiprogramming system because they determine what each process can or cannot do in user space and in kernel space, maintaining the autonomy and the security of each user's personal data and the stability of the system.

The use of credentials is applied in the process data structure and in the shared resources the processes are trying to access, similar to a key and a set of tumblers in a lock. Files are the critical resource on the system. Thus, in the default Linux ext4 file system, each file is owned by a specific user and belongs to a group of users. The owner of a file determines what kind of operation is allowed on that file, distinguishing among herself, the file's user group, and all other users on the system. When a process tries to make use of a file, the Linux Virtual File System always validates whether the access is allowed or not, according to the permissions established by the file owner and the process credentials. Clearly, process credentials and file access permissions are integrated and inseparable.

This authentication coupling is based upon the forms of file permissions we detail in Chapter 5, and we have shown that the **chmod** command is the instrument for granting these permissions to files and directories.

17.9.6.1 File Permission-Based Mechanisms

In the file-based permission mechanisms shown in Chapter 5, those forms involve managing the basic permission bits on files and directories, and the SetUID (SUID), the SetGID (SGID), and sticky bit permissions. For example, setting permissions using DAC on the /usr/bin/passwd and /usr//bin/sudo commands. Each of these files should maintain their SUID permissions, as shown in the following output:

```
$ ls -la /usr/bin/passwd
-rwsr-xr-x 1 root root 54256 Mar 29  2016 /usr/bin/passwd
$ ls -la /usr/bin/sudo
-rwsr-xr-x 1 root root 136808 Aug 17 06:20 /usr/bin/sudo
$
```

The commands passwd and sudo are SUID-capable programs. Even though these commands are ostensibly run with root user privelege, a nonprivileged user can only change their own

password with the passwd command, and can only assume root privileges with sudo. That is, if they are given privelege to do so in the /etc/sudoers file.

The find command allows you to search your system to see if there are any hidden or otherwise inappropriate SUID and SGID commands on your system. Here is an example:

```
$ sudo find / -perm /6000 -ls
  2104368    4 drwxrwsr-x    2 root    mail    4096 Jun 28  2016 /var/mail
  2104383    4 drwxr-sr-x   30 man     root    4096 Dec 25 07:41 /var/cache/man
  2235989    4 drwxr-sr-x    ? man     root    4096 Dec 25 07:41 /var/cache/man/tr
  2104517    4 drwxr-sr-x    2 man     root    4096 Dec 25 07:41 /var/cache/man/ja
  2104518    4 drwxr-sr-x    2 man     root    4096 Dec 25 07:41 /var/cache/man/ko
  3300204    4 drwxr-sr-x    2 man     root    4096 Dec 25 07:41 /var/cache/man/ro
  2104520    4 drwxr-sr-x    2 man     root    4096 Dec 25 07:41 /var/cache/man/pl
  2104519    4 drwxr-sr-x    2 man     root    4096 Dec 25 07:41 /var/cache/man/nl
Output truncated...
```

The find command reveals SetUID and SetGID commands that unprivileged users can run to assume the privileged role.

Exercise 17.54

Use the find command to list any hidden or inappropriate SUID and SGID commands on your Linux system.

17.9.7 Disk Encryption

When the ordinary user is considering disk encryption for security, the critical determination to make is the scope of encryption she wants to implement. For example, she can encrypt the entire system disk she is installing Linux on at the time she installs the system. Or she can do a postinstallation encryption of an entire disk, selected partitions, selected directories, or individual files. What governs the scope of encryption is up to the user, and depends a great deal on the particular use case. It is critical to know why you want to do encryption when considering all of the possible strategies of encrypting your persistent media.

In Chapter W26, Sections 9.7.2, titled "Whole Disk Encryption at Installation," through 9.7.5, titled "Encrypting a USB Thumb Drive," at the book website, we detail a large-scope encryption strategy. In those sections, we show how to encrypt the entire system disk at installation of Linux Mint and also an entire single partition on a USB thumb drive. We also cover a smaller-scope encryption strategy of encrypting a directory and a particular important file in your home directory in those sections.

17.9.7.1 The Meaning of Encryption

Encryption uses a very developed science, known as *cryptography*, to implement data hiding.

Cryptography is used on a Linux system to encode data to hide it from unprivileged users, and then decrypt, or decode the data for privileged users. On a Linux system, the following are some of the objects that can be possible targets for encryption:

Individual files
Volumes or individual partitions
Web page connections
Network connections
Backup file objects on additional hard disks or other media
Compressed directories or files

Encryption/decryption basically uses a variety of mathematical algorithms to treat the earlier targets. These algorithms are called *cryptographic ciphers*. The important terms to know when dealing with cryptography are *plain text* and *cypher text*. Plain text is the unencrypted or decrypted format, and cypher text is the encrypted format.

The details of encryption algorithms, their mathematics and complexity, are not as useful to the ordinary user as knowing where and when to apply them. That is, at the system, whole disk, partition, directory, or individual file levels. And also knowing why you want to use them, and carefully reasoning and designing your strategies for implementing these algorithms with the common command line tools, as shown in Chapter W26, Sections 9.7.2–9.7.5 at the book website.

Exercise 17.55

Which of the encryption strategies, entire disk, selected partitions, selected directories, or individual files, would you deploy on your Linux system, and for what specific reasons?

Exercise 17.56

After deciding upon one, or even multiple encryption strategies on your Linux system that you would want to use, look at Chapter W26, Sections 9.7.2, titled "Whole Disk Encryption at Installation," through 9.7.5, titled "Encrypting a USB Thumb Drive," at the book website, and implement those strategies.

17.10 Virtualization Methodologies

A virtual environment for a computer program, and for an operating system, can be defined as a shell within which the program functions autonomously. And as we have shown earlier, one of the parts of looking at an operating system at a certain level of abstraction is the ability of the operating system to present the illusion of virtual environments. A multitasking, multiprogramming operating system, where all users are presented with this illusion that each individual user is working autonomously on her own discreet computer (when in fact many people are also working on the same hardware platform), is a basic underlying aspect of operating system virtualization. In many respects, and at some other level of abstraction, systemd can be thought of as a virtualizing program. It virtualizes the Linux kernel and gives you the illusion that you are working directly with it, when in fact you are working in the systemd environment. That way of looking at the systemd "superkernel," and the Linux kernel itself, most importantly assumes that the function of these two programs is to help maintain the steady state of the hardware and software as the power is turned on, the system is running, or it becomes necessary to have the power turned off. In Chapter 18, we give a complete and detailed description of systemd and its operation.

In Chapter W23, titled "Virtualization Methodologies," at the book website, we show two popular and important facilities for creating a virtual operating system environment within a Linux host environment: LXC/LXD containers and VirtualBox VMs. These facilities provide extensions of some of the topics we covered in this chapter and also extend many of the topics covered in all of the previous chapters in this book.

What differentiates these two facilities is that, for the first one, all virtual environments are running under the same kernel on one host machine. In the second one, any number of different kernels can be running simultaneously on one host machine. That means that, LXC/LXD containers, you can be running many different distributions of Linux at the same time on one machine. With VirtualBox, you can be running Ubuntu, Linux Mint, or CentOS, or any number of other supported operating systems at the same time on one host machine.

There are other similar virtualization facilities available for Linux. Docker is another container application suite that allows you to "spin up" container instances that are lightweight, in terms of the disk space size they occupy. A typical example of one of these lightweight Docker container instances is the Web server software nginx. In terms of disk space occupied in general, the lightest weight facility is a Docker container, the middleweight is an LXC/LXD container, and the heavyweight is a VirtualBox guest.

Both LXC/LXD and Docker rely upon the namespaces system programming API, and the clone(2) system call. We briefly mention Linux namespaces and their uses in Chapter W27, Section 6.1.3, at the book website.

17.10.1 Virtualization Applications

The important application of LXC/LXD containers and VirtualBox, in the context of system administration, is to provide a measure of system security. For example, it is possible with both facilities to isolate a system service or application program in a guest operating environment, completely autonomous from the host operating system. A service like a Secure Shell (SSH) server can be run inside of an LXC/LXD container instance, or a VirtualBox guest, and anything that intrudes upon that server and its system space does not intrude upon the host operating system space. If LAN or Internet traffic to and from the server is compromised in any way, the server can be stopped, restarted, or even deleted, without affecting the host operating system. Another example would be if a faulty, bug-ridden application program were run in an LXC/LXD container instance, or VirtualBox guest environment, it could bring the guest operating system kernel to a halt without affecting in any way the host operating system. Those two example cases are probably the most useful aspects of maintaining a virtual environment, but there are others.

In addition, another application of these methodologies is to allow you to use and experiment with different operating systems on a single piece of hardware. You can play with these other operating systems in a "sandbox" that is isolated from your host system and can access that host system *simultaneously* with the virtualized environments.

We emphasize the word simultaneously, because it is possible to create different boot environments on one computer and boot into each of these environments sequentially, but not simultaneously.

We encourage you to go through all of the examples shown in Chapter W23, titled "Virtualization Methodologies" at the book website, to gain a better understanding of virtualization methodologies as they currently exist in Linux.

Exercise 17.57

What is kernel-based VM (KVM) and how does it compare to the virtualization applications we describe in this section? Which cloud-based system deploys KVM? Answer this question in terms of type of virtualization, size of a VM instances, ease of use, and speed of execution of VM instances. Compare these KVM criteria to LXC/LXD containers, VirtualBox VM's, and Google Cloud and Amazon EC2 instances.

Summary

In this chapter, we use a "learning-by-doing" approach to accomplish the following common system administration tasks

1. Do a fresh install of a 64-bit, X86 architecture version from DVD media using a GUI installer onto a single hard disk desktop system, with a Cinnamon GUI desktop. Do a preliminary configuration of that system.
2. Illustrate booting strategies and how to gracefully bring the system down.
3. Detail the basics of using systemd to manage system services.
4. Add additional users and groups to the system and show how to design and maintain user accounts.
5. Adding persistent media to the system in particular disk drives. We also established a framework for connecting and maintaining the file system on that media, which classified the file system as either existing on a physical medium (hard disks physically connected to the computer), a virtual medium (NFSv4, iSCSI), or as a specialized pseudofile system that was not on a medium at all (cgroups, proc).

6. Provide strategies using the traditional and generic Linux commands to backup and archive the system files and user files.

7. Update and maintain the operating system and add/upgrade/remove user application package repository software to both to increase functionality and upgrade existing packages.

8. Monitor the performance of the system and tune it for optimal performance characteristics.

9. Provide strategies for system security to harden the individual desktop computer.

10. Provide network connectivity strategies, both on a LAN and the Internet.

Questions and Problems

1. Write a brief outline of how you installed your version of the Linux operating system on your computer and detail exactly what special procedures you used for your particular installation, and how they may have differed from the ones presented in the current installation procedures for your version of the software. If you didn't do the installation, find out from the system administrator how the installation was done, and why it was done in that way. If you did a server install, explain how and why that was done, given the particular use case that the system had to conform to.

2. Do the following steps to complete the requirements for this problem:

   a. If you have not already done so, download, install, and test the vsftpd service on your system, as shown in Appendix A, 17.2.3.4 vsftpd Installation, Example 17.1.

   b. Use the **adduser** command, as shown in Example 17.1, to create a new user account on your Linux system with the following configuration

```
$ sudo adduser ftp2
Adding user 'ftp2' ...
Adding new group 'ftp2' (1004) ...
Adding new user 'ftp2' (1003) with group 'ftp2' ...
Creating home directory '/home/ftp2' ...
Copying files from '/etc/skel' ...
Enter new UNIX password: YYY
Retype new UNIX password: YYY
passwd: password updated successfully
Changing the user information for ftp2
Enter the new value, or press ENTER for the default
        Full Name []:
        Room Number []:
        Work Phone []:
        Home Phone []:
        Other []:
Is the information correct? [Y/n] Y
$
```

   c. Test your new user account locally on your LAN by using the command **ftp 0** with username **ftp2**. Test it from the Internet. Put files in the users account and retrieve files from that account locally from another account and from the Internet.

3. Given the steps needed to accomplish user account management shown earlier in Section 17.3, make a table or chart of what users and groups need to be added to your system, and what their default account parameters and group memberships should be. Then, use the methods shown in Sections 17.3 to accomplish user account creation, modification, and deletion from the command line. What command can you use to identify all existing groups on the system? Use the batch mode account creation technique shown in Section 17.3 to implement the users and groups from the table or chart you created.

4. Using the facilities available both in systemd and CUPS, add a printer to your Linux system with a direct USB connection. In detail, list all the steps necessary to get the printer to actually work given your installation type and your specific Linux system.

5. What printer commands did you use to test the addition of the printer you added to your system in Problem 17.4? In other words, what commands did you use to actually print documents on the printer you added?

6. What is the meaning of the term archive?

7. What is the tar command used for? Write a short, explicit, and articulate summary report that gives all its uses.

8. You want to create a tar archive of a project that contains several directories, subdirectories, and files, and save the archive on a USB thumb drive mounted on your system so that you can distribute the archive to your friends. (a) What is the pathname to a USB thumb drive mounted on your system? (b) How would you designate a USB thumb drive as the destination from where the tar archive would be created, as an argument to the tar command?

9. Give a command line example of creating a tar archive of your current working directory.

10. Give the explicit commands for compressing and keeping a tar archive of a "backups" directory in your home directory.

11. Give commands for restoring the backup file in Problem 17.8 in a directory named ~/backups.

12. Give a command line example of copying your home directory to another directory called home.back, so that access privileges and file modify time are preserved.

13. Why is the tar command preferred over the **cp -r** command for creating backup copies of directory hierarchies?

14. Suppose that you download a file, Linuxbook.tar.Z, from an ftp site. Give the sequence of commands for restoring this archive and installing it in a directory named ~/Linuxbook.

15. Use the **tar** command to create a compressed archive of a directory of your choice in a new directory you create named mybackups under your home directory. Name the compressed archive **something.tar.gz** where something is the name of the directory you chose to backup. Show the command lines that you used to perform these tasks.

16. Use the **tar** command to restore the compressed tar archive ~/mybackups/something.tar.gz you produced in Problem 17.15, into a new directory named mirrors under your home directory. Show the command lines that you used to perform these tasks.

17. Use the latest version of CloneZilla Live to make a bootable clone of your Linux system disk. The source and target disks for cloning can be either both internally mounted, or in the case of a laptop computer, internally mounted for the source, and externally mounted in a USB or E-SATA enclosure for the target. The instructions for using CloneZilla to do this procedure are found online at the CloneZilla website. Make sure the target disk has large enough capacity to achieve the cloning!

    To test the clone, gracefully shut down your system and remove the original source system disk. Then, replace it with the cloned target and restart the system.

18. This problem is specifically targeted at Ubuntu and Linux Mint systems. Successful completion of it will allow you to add the Universe and Multiverse Repositories in those systems.

    Add one or all of these additional software repositories to either your /etc/apt/sources.list file or to a separate .list file in the /etc/apt/sources.list.d/ directory

    deb http://us.archive.ubuntu.com/ubuntu/ xenial universe
    deb-src http://us.archive.ubuntu.com/ubuntu/ xenial universe
    deb http://us.archive.ubuntu.com/ubuntu/ xenial-updates universe
    deb-src http://us.archive.ubuntu.com/ubuntu/ xenial-updates universe
    deb http://us.archive.ubuntu.com/ubuntu/ xenial multiverse
    deb-src http://us.archive.ubuntu.com/ubuntu/ xenial multiverse

deb http://us.archive.ubuntu.com/ubuntu/ xenial-updates multiverse

deb-src http://us.archive.ubuntu.com/ubuntu/ xenial-updates multiverse

You should replace "us." by another country code, referring to a mirror server in your region. Don't forget to retrieve the updated package lists with the command sudo apt-get update.

19. Execute the steps of the user and group creation method shown in Section 17.3.4 in detail for your Linux system, according to the following constraints:

   a. Your system capabilities to accommodate additional persistent media,

   b. whether or not you want to use a ZFS-based or traditional partitioning/file system creation/ file system mounting-based approach,

   c. your security model, in terms of how it isolates users and groups from one another,

   d. your system performance model as it affects users and groups,

   e. how many users and groups you plan to accommodate.

   For example, you can add one or more additional hard drives to your system in a ZFS-based individual or multiple virtual pools, limit the privileges you want ordinary users to have, allow them to run their own executable programs from their home directories, and have a small-scale set of user accounts and groups.

Advanced Problems and Projects

20. Using the systemd cgroups methods of Sections 17.8.1.2 and 17.8.1.3, and the constraints imposed by your particular use case, first examine and assess the most critical and important applications default use of the system resources of CPU and memory on your Linux system. Then, give the applications that you believe warrant it greater "weight" in accessing the CPU and memory than what they have by default. For example, if you are running a server that uses nginx to serve Web pages, how can you give the nginx application and its worker processes more weight in accessing the CPU and system memory? How does the number of CPUs and CPU cores that your computer has affect your redistribution of CPU usage, and how can you modify that redistribution to take advantage of those CPUs and cores? That might include taking advantage of CPU affinity, if the application you want to give more weight to is a system program written and developed by you.

21. After downloading and installing Webmin, as shown in Appendix A, W26.1.2, test a connection to Webmin, both from the local machine, and over an intranet or the Internet from another computer system. What URL did you use to gain access to Webmin on the local machine's Web browser? What URL did you use to gain access to Webmin from another machine on an intranet? What URL did you use to gain access to Webmin from another machine on the Internet? List several precautionary steps you could take to ensure that no one else on the Internet can log into your system's Webmin interface?

22. After reading through Chapter W26, Section 4.8, titled "Creating and Managing RAID Arrays in Linux," found at the book website, complete all of the steps shown in Example W26.8b "Adding a RAID1 Mirror Using mdadm" on your Linux system. Use the appropriate package management commands at Step 1 to install mdadm on your Linux system, as shown in Appendix A, W26.4.8.

23.

   a. Examine the cups.service systemd service unit file in /lib/systemd/system and write a short paragraph-long description, in your own words, on why CUPS is inactive when no powered-on printers are attached to the system. This question assumes that CUPS is installed on your system, according to the requirements shown in Chapter W26, Section 5, at the book website, and that it is enabled at system boot.

b. How would you make CUPS available even though no printer is plugged in or attached?

24. After reading through and executing the commands shown in Chapter W26, Section 8, titled "Traditional Process Control and Monitoring," at the book website, answer the following questions:

a. Use the top command on your system to list the top processes running. What are they?

b. $ top -d 10

```
top - 06:00:36 up 1 day, 15:40,  1 user,  load average: 0.12, 0.20, 0.19
Tasks: 258 total,   1 running, 257 sleeping,  0 stopped,  0 zombie
%Cpu(s):  0.5 us,  0.4 sy,  0.0 ni, 99.1 id,  0.1 wa,  0.0 hi,  0.0 si,  0.0 st
KiB Mem :  5981784 total,  3483416 free,  749348 used,  1749020 buff/cache
KiB Swap:  6157308 total,  6157308 free,       0 used.  4876824 avail Mem

PID   USER   PR  NI   VIRT     RES     SHR    S  %CPU  %MEM  TIME+    COMMAND
15159 root   20   0   0        0       0      S  0.5   0.0   0:14.54  kworker/0:0
2228  root   20   0   399344   92040   55144  S  0.4   1.5   25:22.85 Xorg
2570  bob    20   0   1572696  193004  69664  S  0.3   3.2   31:44.13 cinnamon
7     root   20   0   0        0       0      S  0.2   0.0   3:35.24  rcu_sched
15079 bob    20   0   485236   36100   27676  S  0.2   0.6   0:03.94  gnome-term+
1     root   20   0   185604   6160    3896   S  0.0   0.1   0:07.46  systemd
Output truncated...
```

In the earlier top display, what are the top processes running, and why?

25. After reading through and executing the commands shown in Chapter W26, Sections 9.2.1–9.2.1.2 at the book website, put a particular, single unprivileged user on your Linux system into the sudoers file. Give that user access according to your particular system security model.

26. After reading through and executing the commands shown in Chapter W26, Sections 9.2.1–9.2.1.2 at the book website, answer the following question:

For User Specifications 1–8 in the sample sudoers file found in Section 9.2.1.2, describe briefly what privileges are given by each entry.

27. Execute all use cases shown in Chapter W26, Section 9.6.2.1, on the book website, as they apply on your Linux system, and note the results.

28. Is it possible to directly run a non-Linux operating system inside of an LXC/LXD container? Give a couple of example container instances, or templates, that are available for LXC/LXD.

29. Is it possible to run more than one non-Linux guest machine in VirtualBox?

30. Define *synthetic file system* in your own words and as verbosely as possible.

Then answer the following questions:

Do the contents of the pseudo and special-purpose file systems, in the virtual layer we have partitioned the Linux storage scheme into in Section 17.4, exist across boots, or are the data structures that define them, and their specific content, created anew each time the system boots and is restarted? Are their data structures and content fixed immutably at the time the system is built, or does their content vary across time, and how?

In your answers to the earlier questions, list several examples of pseudo and special-purpose file systems, what function(s) they perform, how you can discover which parts of them are either volatile, persistent, or both to whatever degree, and how you can know which ones are fixed, and how, when the system is built, if that's the case.

Projects

Project 1

Do a "bare metal" install of your Linux system, so that the entire system disk is encrypted at installation. For example, we show this procedure in Chapter W26, Section 9.7.2, at the book website, for Linux Mint.

What advantage does this confer in terms of system security, especially in terms of combating a specific form of security breach?

Project 2

On your Linux system, mirror an internally mounted SATA hard disk that is *not* the system disk onto a USB thumb drive. The SATA hard disk can be used as the first vdev you initially create a zpool on, and the USB thumb drive can be the new vdev you attach to the zpool as a mirror. Following are some advisories about how to complete this problem successfully:

a. Make sure you know the logical device name of your system disk, such as /dev/sda.
   Don't do this problem by mistake using that drive or you will render your system disk unusable!

b. Make sure you have installed ZFS on your Linux system, as shown in Appendix A, W26.4.7!

c. You have to unmount the SATA hard drive and the thumb drive before you can use the zpool command on them.

d. First create a zpool on the SATA hard drive first and name that pool sata_usb_test.

e. You should make sure that the thumb drive has the same capacity as the SATA hard drive.

f. Create zfs datasets on the mirrored pair, but be sure to delete the datasets and destroy the zpool before removing the thumb drive from the machine.

Project 3

On your Linux system, mirror two USB thumb drives. The first USB thumb drive can be used as the first vdev you initially create a zpool on, and the second USB thumb drive can be the new vdev you attach (with the zpool attach command) to the zpool as a mirror. Following are some advisories about how to complete this problem successfully:

a. Make sure you have installed ZFS on your Linux system, as shown in Appendix A, W26.4.7!

b. You have to unmount the thumb drives before you can use the zpool command on them.

c. You should make sure that each thumb drive has the same capacity as the other thumb drive.

d. Initially, create a zpool on the first thumb drive.

e. Create zfs datasets on the mirrored pair, but be sure to delete the datasets and destroy the zpool before removing the thumb drives from the machine.

Project 4

The assumption in this project is that the source directories and files that you want to back up are created and maintained on a second persistent media device, such as another internal SATA or external USB hard disk, or USB thumb drives. *Not on the system disk*. This is congruent with the recommended storage model in this chapter! Therefore, you do not need to use any script file or the Linux file copying commands to achieve archiving of selected directories or files.

Use the RAID capabilities provided by mdadm to accomplish "automatic" backup of the source disks and partitions on them. See Chapter W26, Section 4.8 at the book website, for an introduction to mdadm.

In particular, use the method of RAID 1, mirroring, presented for mdadm in Chapter W26, Example W26.8b at the book website, to achieve the automatic backup of directories and files which you want backed up. Present the exact and specific commands that you used to provide the tactics you employ using mdadm, in a verbose, explicit, and articulate report format.

Project 5

What is Linux LVM, and why would you want to deploy it on your Linux system? Is it possible to install the system itself with LVM as the default? Is LVM already installed on your Linux system?

After answering these questions, add a hard disk to your system, and manage it with LVM as a volume in conjunction with the original storage complement on your system. What advantages does this confer, given your particular use case(s), and how does this method of volume management compare with ZFS?

Project 6

Do all of the examples and in-chapter exercises shown in Sections W26.5–W26.5.5 at the book website on CUPS printing.

Project 7

In preparation for this project, read through what is presented in Section W26.9.9 at the book website. Then execute all of the steps shown in Example W26.28 on your Linux system.

Project 8

Install KVM on your Linux system, and use it to create and launch a VM image of an operating system of your choice.

Project 9

Besides the ten common system administration tasks we have selected to present in this chapter, there are numerous extensions, and also many additional tasks that Linux system administration encompass. A large variety of these tasks are further detailed at the book website.

From the following list of additional tasks, choose one or several of the items shown, and either fully implement the details of that item, or items, on your Linux system. The item(s) you select might be very relevant to the particular way you use your Linux system. If you choose *not* to implement any of the items, do a personalized research report that would allow you to achieve implementation of particular items of interest to you at some future date, given how your use case might change beyond what it currently is.

1. Creating, configuring, and running a "live" version of your Linux operating system from a persistent USB thumb drive, or installing and running over a LAN or from a server.
2. Doing an advanced, text-based installation of a server with a discretionary higher level of complex configuration than what we have illustrated in this chapter.
3. Maintaining a large user base across multiple machines and networks, possibly using ACLs.
4. Configuring customized partitions onto which the operating system is installed, or using higher levels of RAIDZ, LVM, or multipartitioned, multiboot environment disks with multiple operating systems on them. Possibly using Zvols as a means of partitioning your storage media.
5. Using commercial backup software.
6. Hot-swapping SAS hard drives and software.
7. Hand-building the operating system or applications software systems from GitHub source, or using advanced package repository resources and configurations.
8. Using the Linux system exclusively to stream media via NAS to multiple displays.
9. Doing the common system administration tasks we illustrated in this chapter on OSX, iOS, UNIX, or Android operating systems.
10. Write and administer, or install, malware-fighting programs.

11. Implementing a secure full LAMP stack, in preparation for Internet facing your system(s).

12. Create a hubless SAN using iSCSI over Ethernet.

Project 10

Following up on your answers to In-Chapter Exercises 17.1 and 17.2, completely list and detail what the preinstallation considerations for three distinct and personally-chosen, use cases of a Linux hardware platform would be. For example, a public-facing Web server that uses Docker containers running nginx. Then, implement one of those use cases on actual hardware. Use our set of minimal recommendations for postinstallation tasks as a guide to begin to fulfill the requirements of this project.

Looking for more? Visit our sites for additional readings, recommended resources, and exercises.

CRC Press e-Resource: https://www.crcpress.com/9781138710085

Authors' GitHub: https://github.com/bobk48/linuxthetextbook

18

---

systemd

OBJECTIVES

- To describe the uses of systemd for an ordinary user
- To give an overview of systemd components and commands
- To describe, in basic terms, how systemd starts up Linux, starts system services, and shuts down
- To define and explain systemd units and unit files, with an emphasis on service unit files
- To define and explain systemd target unit files, and their purposes and uses
- To show examples of systemd service management and the systemctl command
- To show how to harness user-written programs to systemd service management
- To detail some uses of the systemd journal for system logging
- To supply numerous systemd command examples
- To create and add a new systemd-controlled service from scratch
- To cover the following commands and primitives:

    `systemctl, journalctl, systemd unit files, systemd targets`

18.1 Introduction: Why You Should Know about and Use systemd

For an ordinary user of Linux on a desktop computer, since systemd works completely automatically without that user doing anything, why know about and use it?

We are not talking about why a system administrator, a higher level IT specialist, a developer, or a systems programmer would want or know or use systemd.

Three preliminary examples answer the earlier question. They are typical scenarios a beginner, or even a more experienced, Linux user might experience when first using systemd.

Primary Example 1: You want to ssh into a newly installed Linux system from another machine on your LAN or intranet. When you attempt that, you get some obscure error message that denies you access via ssh to the new machine. Out of the multitude of debugging scenarios you could go through to find out why you are getting the error, you could use one single systemd command while sitting at the new machine to find out whether the sshd daemon is running or even installed. Then, proceed to solve your problem quickly and easily from there.

Primary Example 2: A new app you have installed refuses to run on your Linux system. Out of the multitude of debugging scenarios you could go through to find out why that app is misbehaving, you could use a single systemd command to check the most recent record of system activity and then solve the problem quickly and easily from there.

Primary Example 3: Your Linux system is *very* slow to respond to even simple commands in a terminal, and you suspect a piece of hardware is failing. You can use systemd to analyze the overall operations of the system quickly and easily by switching to an emergency or rescue system state using the "targets" available to enter a debugging mode of system operation.

We show the procedures for doing these Primary Examples in the sections of this chapter that follow.

Incidentally, another reason why an ordinary user needs to know about systemd is because it has replaced the Berkeley Software Distribution (BSD)-style and System V-style init and service management mechanisms in the kernel of the three major branches of Linux: Debian (of which Ubuntu and Linux Mint are derivatives), Slackware (openSuse), and Redhat (RHEL, CentOS).

As the second component of the system administration division of this book, this chapter is a combined short, basic reference and tutorial on systemd for the ordinary user. There are at least four ways to approach the material presented here, and, as with everything else in this book, your personal approach depends on what and how much you want to learn. They are as follows:

1. Read and use the reference sections provided at the beginning of the chapter.
2. Do all of the examples in the tutorial sections, answer the in-chapter exercise questions, and then selectively answer, solve, or execute the Questions/Problems/Projects at the end of the chapter.
3. Complete 1 and 2, and then go on to the advanced material provided in Chapter W27 at the book website. We provide many more topics, exercises, examples, questions/problems/projects, which extend what is shown in this chapter and in Chapter W27 at the book website.
4. Do some combination of all the three (1–3) ways .

From that personal perspective, we cumulatively show the following:

- How systemd is structured, from the top down, i.e., from the user to the kernel.
- Introduce (using many command line examples) its most important commands.
- Give basic and extended, typical use cases for applying those important commands.

systemd is, on the surface, a combined Linux initialization program and service manager, that also includes many additional important system state control programs and utilities.

But in reality, systemd is a "superkernel," which uses the Linux kernel itself as a subsidiary program to maintain the steady state of the hardware and software components of a Linux computer. systemd then supervises and manages everything that the kernel does to optimize operation in that steady-state condition, whether it be start-up, process management and control, Interprocess Communication (IPC), hardware services, network operations, shutdown, and a multitude of other operations.

And if you have gone through Chapter 17 and Chapter W23 at the book website, you must realize that the virtualization technologies we cover there are "superkernels" as well, forming layers on top of the Linux kernel. Those virtualization technologies in turn manage system start-up, disks, file systems, and entire virtualized machines. At a certain level of abstraction, operating system level and full virtualization are just instances of a superkernel.

We have suggested that the most important features of the Linux kernel, and of an operating system in general, are to provide virtualization, concurrency, and persistence. systemd provides and efficiently manages all of those features. And, in the context of those features, systemd permeates and threads through every one of them.

Exercise 18.1

Describe in your own words why you would want to, or need to, use systemd on your Linux system.

Question: What is service management?

Answer: System service management basically means controlling which service daemons run, how they run and consume system resources such as central processing unit (CPU), and when they run.

systemd runs and controls all of the Linux systems we illustrate this book with Debian, Ubuntu, Linux Mint, and CentOS. In this chapter, we cover systemd in much more depth and extend what we presented in Chapter 17, Section 2.3.4.

18.1.1 Some Prerequisites

Although much of the material in this chapter can be done without having done anything else in this book, there are some things which we assume that you have done in preparation for your work here.

The following is a list of those prerequisites, from specific to general knowledge of material covered:

1. Knowing the structure of a Linux command and having basic mastery of file maintenance commands.
2. Being familiar with the materials presented in the core chapters of this book, Chapters 3–11.
3. Basic and advanced Bash shell programming as shown in Chapters 12 and 13.
4. Having read through and referring to Chapter 17 on system administration.

An overview of the top-down architecture of the components of systemd is given in Figure 18.1.

You should first understand and follow these instructive points we make about Figure 18.1

0. Things (from the point of view of the user typing Utility Commands available at the top level of Figure 18.1) work in a descending order, from highest to lowest layers.
1. Starting at the top layer, the Utility Commands are what you can type at the Linux command line. These commands implement the user-side, front end to systemd. The most prominent and important one, the systemctl command, is detailed in Section 18.3.
2. The next layer contains Daemons and Targets. systemd Daemons are what control all other Linux daemons, and services, including systemd itself. Targets are basically operating states you can put Linux in.

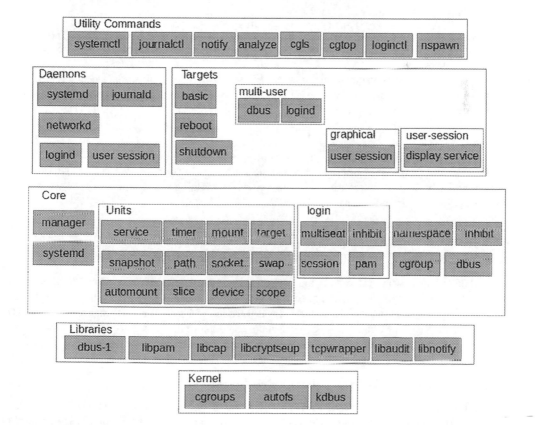

FIGURE 18.1 systemd architecture.

3. The next layer contains the Core features, most prominently Units. Units are what describe and define services, affecting everything in Linux service management. Units affect everything in Linux.

4. The next layer, Libraries, contains what the upper layers call upon to get their work done.

5. The lowest layer, the Kernel, is where cgroups and other kernel modules do all of the essential control. That is why cgroups, and their application to systemd, are one of the most important features of systemd for the system programmer. We cover cgroups from the user's perspective in Chapter 17, Section 8.1.2.

For example, a user executes a correctly formulated systemd command on the command line. This command is passed through the Daemons and Targets layer and affects appropriate modules in that layer. These modules in turn affect Core layer modules, such as invoking or executing a Target Unit. This invocation or execution uses the library modules in the lowest layer to change the state of the Kernel at that lowest layer.

Units, Targets, and the systemctl command are the most important items in the context of the diagram, particularly for a novice user trying to understand systemd.

18.1.2 Where to Get Further Help and Documentation

Documentation for systemd exists right on your Linux system!

We encourage you to read through and generally refer to the following man pages on your system (or on the Internet). Starting from the top, and arranged in order of detail, you can use these man pages to gain further and more complete information about the specific topics that we cover in this chapter:

```
systemd.unit, systemd.target, systemctl,
systemd.special, systemd.directives, systemd.service, journalctl
systemd.socket, systemd.device, systemd.mount, systemd.automount, systemd.swap,
systemd.path, systemd.timer, systemd.scope, systemd.slice, systemd.snapshot
```

These man pages can serve as an electronic reference library, directly on your Linux system, for systemd. For example, if you don't remember the exact meanings of, and how to apply, the arguments to the systemctl command, you can refer to its man page on your system to refresh your memory!

Exercise 18.2

Carefully read through the man pages on your system for systemd.unit, systemd.target, and systemctl, and then write a brief, paragraph-long summary, in your own words, that will assist you with each topic. Then refer to these summaries when doing the remainder of the work we show in this chapter and in Chapter W27 at the book website.

18.2 System Start-Up, Initialization, and Shutdown Using systemd

To begin to understand what systemd is, and part of what it achieves as a "superkernel," we examine the first and last major tasks that systemd accomplishes for the Linux operating system—start-up and shutdown. The first of these tasks is what systemd was introduced for: to more effectively and efficiently replace the older UNIX and Linux init systems. But, as you explore more of this chapter, you will come to understand that systemd is much more than this.

18.2.1 An Overview of the Start-Up and Shutdown Processes

The basic details of how a Linux system that is using systemd boots and starts up are given in Chapter 17, Section 2.5. In this section, we further expand upon the particulars of the start-up phase, with regard to the systemd-specific steps involved.

As seen in Chapter 17, several processes come into play even before systemd as an initialization program takes over during system start-up. Figure 17.1 illustrates booting with POST, BIOS or UEFI, GRUB, and then start-up, where the kernel initializes hardware and prepares for the first process to be run from the root file system. After the root file system is found and mounted, systemd takes over and is then responsible for initializing all remaining hardware, mounting all necessary file systems, managing the start of system services, and allowing fine-grained control over all processes.

On shutdown, systemd stops all services and unmounts all file systems (detaching the storage programs handling them). The system is then either rebooted or finally powered down.

Exercise 18.3

What process is Process Identification number 1 (PID 1) on your system, and how did you find this out?

Exercise 18.4

Describe, in your own words, the differences between the boot process and the start-up process in Linux. Refer specifically to the steps the system goes through to achieve the states that it is in during those general processes.

18.2.2 systemd Start-Up: Targets, Target States, and Target Units

After the boot phase is over, systemd is the controlling component of the start-up phase, and is responsible for the completion of start-up. It does this by initializing the required file systems, services, and drivers that are necessary for the normal operating condition of the system. With systemd, this process is split up into "run-time" steps (operations that happen as the system is actually executing in real time), whose objectives are to reach required, or requested, "target states," formerly known as "run levels" in UNIX and Linux jargon. The entire process is done as much as possible in parallel. This parallelism achieves a significantly faster start-up time, a hallmark of systemd-controlled Linux. The start-up procedure also has no fixed path through it, so that the order in which "target units" are reached is determined when the system is actually started up. We give a complete definition, and provide examples, of targets, target states, and target units, in the sections that follow.

"Dependencies" are sequentially ordered relations between target units. When systemd takes over the start-up phase, it first activates target units that are upstream dependencies of a target state known as the default.target, and then proceeds to activate all other downstream dependencies. default.target is an alias of graphical.target or multi-user.target, depending on whether the system by default has been configured for a GUI (on a desktop system, for example) or only for a text console, CUI (on a server system, for example). Generally, desktop users prefer the GUI environment because of its graphical interactive operations and window system. Servers generally have a CUI because of the low memory and resource overhead involved in operating at that target state. It is important to realize that both use cases have very specific performance requirements, and therefore different target unit activation requirements as well.

Figure 18.2 shows the ordering of "milestone" target unit descriptions reached according to whether a GUI graphical.target or CUI multi-user.target state is ultimately reached. The arrows describe which units are called in what order, from top to bottom.

In terms of target unit dependencies, targets higher up in the diagram are the dependencies of targets lower down in the diagram.

Some representative milestone target units, in the order that they are called, are as follows: local_fs_pre.target, sysint.target, basic.target, and graphical.target. Other target units, such as timers.target, are called by basic.target, and can affect system state when their "dependencies," i.e., the resources and other target states that depend on them, require it.

Exercise 18.5

Why do you think that multi-user.target is a necessary prerequisite of achieving the graphical. target state of the system?

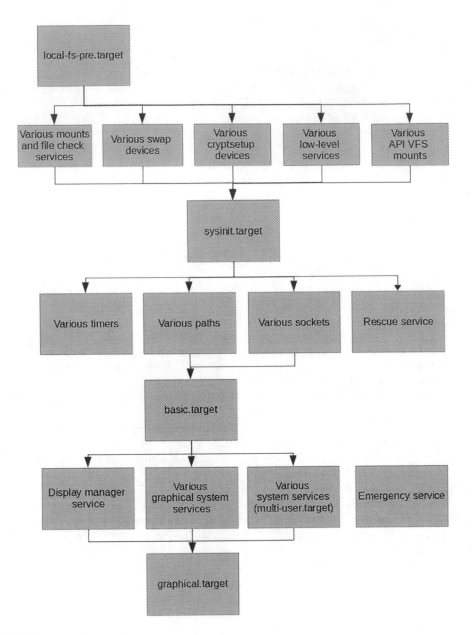

FIGURE 18.2   systemd boot targets.

18.2.3 systemd-Controlled Shutdown

As introduced in Chapter 17, Section 2.3.2, systemd-controlled shutdown can follow several paths, depending upon which target units are "pulled in" to achieve the desired form of powering off or rebooting of the system. Referring to Figure 18.3, these paths flow from shutdown.target and umount.target to the poweroff.target. The specific systemctl commands that initiate procedures along these these paths were shown in Table 17.1.

Figure 18.3 shows both the common targets and services involved in a shutdown, reboot, and power off. In the case of kexec, which enables you to load and boot into another kernel from the currently running kernel, the reboot is done very quickly without waiting for the whole BIOS boot process to finish.

We give additional information on shutdown targets in Chapter W27, Sections 4 and 5, at the book website.

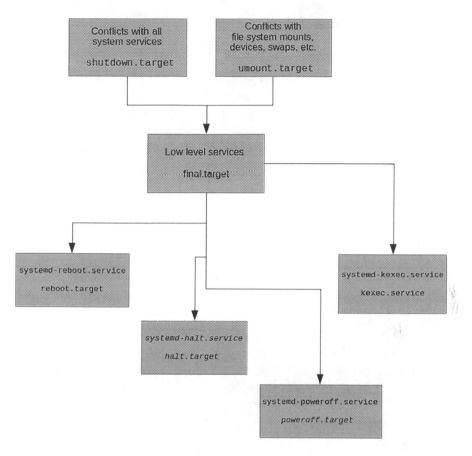

FIGURE 18.3   systemd shutdown path.

18.3 systemd System Service Management Basics Using the systemctl Command

After the boot and start-up processes are over and the system is in a steady state (its normal operating condition), the main purpose of systemd is to maintain that steady state by efficiently and expediently managing system services and processes. This section details that operation, and gives several use-case examples of the most important utility command for system service management: systemctl.

The general form of the systemctl command is as follows:

Syntax:
```
systemctl [OPTIONS...] COMMAND [Name...]
```
Purpose: systemctl is the controlling command for the state of the "systemd" system and service manager

Output: Modified system state control

Commonly used options/features:

-t, --type=argument	where argument should be a comma-separated list of unit types, such as service and socket
-a, --all	used with the list-units command, to show inactive units and units which are following other units
--after	used with list-dependencies, to show the units that are ordered before the specified unit

Common commands and usage:

Systemctl	List all the service unit files and other units
systemctl start	Start a service (not persistent)
systemctl stop Name	Stop a service (not persistent)
systemctl restart Name	Stop and then restart a service
systemctl reload Name	Reloads the config file without interrupting pending operations
systemctl restart Name	Restarts if the service is already running
systemctl status Name	Tells whether a service is currently running
systemctl enable Name	Turn the service on, for start at next boot, or other trigger
systemctl disable Name	Turn the service off for the next reboot or any other trigger
systemctl is-enabled Name	Check whether a service is configured to start
systemctl list-unit-files	Lists unit files
systemctl daemon-reload	Restart the systemd management program itself

Exercise 18.6

Use the systemctl command on your Linux system to determine how many units are loaded. How many of those units are inactive? What command did you use to view the names of all loaded but inactive units?

In many illustrations in the later sections, and in Chapter W27 at the book website, we show the general or generic format of the systemctl command, with its possible options and subcommands. You can identify these as generically formatted by noting that the command line prompt is absent, and the command argument is a generic name.service. For example:

```
sudo systemctl start name.service
```

We also show many systemctl commands, prefaced with the Bash $ command line prompt, that shows specific options and arguments as they are present on our representative Linux systems. For example:

```
$ sudo systemctl set-default multi-user.target
```

Also notice that we execute all systemctl commands as an ordinary user using sudo! That's because many of the systemctl commands we illustrate later change system state, requiring you to have sudo privilege on your Linux system. So to avoid having to differentiate commands that do or do not require that privilege, we use the **sudo** command in a blanket fashion.

18.3.1 How to Use systemctl to Manage systemd Services and Units

As a command directly available to the user (i.e., at the top of Figure 18.1), the following sections illustrate some of the important uses to which the systemctl command can be put. These sections will further expand your understanding of how extensive, and pervasive, systemd is in terms of maintaining the steady state of the kernel, and of the entire software/hardware ensemble that is a Linux computer.

Table 18.1 contains a summary of the most-often used systemctl subcommands and a brief description of each.

18.3.1.1 Service Management

One of the fundamental purposes of an initialization, and start-up system such as systemd, is to initialize the components that must be started after the Linux kernel is booted. These components are traditionally known as "userland" components. The initialization system is also used to manage services and daemons

TABLE 18.1

systemctl Commands

Command and Argument	Description
list-units [Pattern]	List units [according to Pattern]. If more than one Pattern is specified, only units matching one of them is shown
list-unit-files [Pattern]	List installed unit files [according to Pattern] and their state
list-timers [Pattern]	List timer units showing the time they elapse next
list-dependencies [Name]	Recursively list units required and wanted by unit [Name]
set-default Name	Set the default target the system boots into
start Pattern	Activate pattern units
stop Pattern	Deactivate pattern units
daemon-reload	Reload the systemd manager configuration
restart Pattern	Restart units specified in Pattern. If those units are not yet running, they will be started
isolate Name	Start the unit Name, and its dependencies, and stop all other units
get-default	Lists the default target the system boots into. This lists the target unit name that default.target is symlinked to
is-active Pattern	Check whether any of the units specified by Pattern are active (running)
status [Pattern]	Show run-time status information about one or more units, and most recent log data from the journal
enable Name	Units will automatically start upon next reboot
reenable Name	Resets the symlinks to allow you to disable and then enable a unit
disable Name	Disables Name units. Removes all symlinks to the Name unit files from the unit configuration directory. Units will *not* automatically start on the next reboot.
is-enabled Name	Gives the enabled status of Name
emergency	Enter emergency target state
rescue	Enter rescue target state
poweroff	Shut down and power off the computer
reboot	Shut down and reboot the computer
halt	Shut down and halt the system

Note: Anything enclosed in [ ] is optional.

at any point while the system is running in its normal operating condition. In addition, this service management function is extended to encompass graceful and orderly reboot or shutdown of the system.

In systemd, the objects to be managed are "units," which are resources that systemd is most naturally designed to manage. Units are categorized by the type of resource they represent, and they are defined with files, known as *unit files*. The type of each unit can be inferred from the suffix on the end of the file name for that unit.

For service management tasks, the unit will be a service unit, which has unit files with a suffix of .service. However, for most service management commands you execute, you can actually leave off the .service suffix, because systemd by default operates on a service when using service management commands.

18.3.1.2 *Starting and Stopping Services*

To start a systemd service, by executing instructions in the service's unit file, use the start command, as follows. If you are running as a nonroot user, you will have to use sudo, because as we mentioned earlier, starting a service will change the state of the operating system:

```
sudo systemctl start name.service
```

To stop a currently running service, you can use the stop command, as follows:

```
sudo systemctl stop name.service
```

18.3.1.3 Restarting and Reloading

To restart a running service, you can use the restart command, as follows:

```
sudo systemctl restart name.service
```

If the application is able to reload its configuration files (without restarting), you can use the reload command to initiate that process, as follows:

```
sudo systemctl reload name.service
```

If you are unsure whether the service can reload its configuration file, you can use the reload-or-restart command. This will reload the configuration in-place, if available. Otherwise, it will restart the service, so the new configuration is invoked, as follows:

```
sudo systemctl reload-or-restart name.service
```

18.3.1.4 Enabling and Disabling Services

The earlier commands are useful for starting or stopping services during the current session. To tell systemd to start services automatically at boot and start-up, you must enable them.

To start a service at boot and start-up, use the enable command, as follows:

```
sudo systemctl enable name.service
```

This will create a symbolic link from the system's copy of the service file (universally across all of our representative Linux systems in /lib/systemd/system or /etc/systemd/system) into the location on disk where systemd looks for autostart files (traditionally /etc/systemd/system/some_target.target.wants.) To change your mind and disable the service from starting automatically, use the following command:

```
sudo systemctl disable name.service
```

This will remove the symbolic link that indicated that the service should be started automatically.

Keep in mind that enabling a service does not start it in the current session.

If you wish to start the service and enable it at boot and start-up, you will have to issue both the start and enable commands.

> **Exercise 18.7**
> To practice the earlier systemctl commands, install a program using the methods of Chapter 17 which you find useful. Then
> a. Check if it has actually been installed as a systemd-controlled service.
> b. If it has been installed as a service, start, restart, stop, and check the status of the service.
> c. Enable the service to start at system boot.

18.3.1.5 Checking the Status of Services

To check the status of a service on your system, you can use the systemctl status command. Following we present the command and its output, executed on our Linux system, for the installed nginx Web server service:

```
$ sudo systemctl status nginx.service
[sudo] password for bob: zzzzz
```

• nginx.service - A high performance web server and a reverse proxy server
   Loaded: loaded (/lib/systemd/system/nginx.service; enabled; vendor preset:
enabled)
   Active: active (running) since Thu 2017-08-10 13:49:52 PDT; 1 weeks 5 days ago
  Main PID: 1826 (nginx)
    CGroup: /system.slice/nginx.service
            ├─1826 nginx: master process /usr/sbin/nginx -g daemon on; master_process on
            ├─1827 nginx: worker process
            └─1828 nginx: worker process

Aug 10 13:49:51 bob-ProLiant-MicroServer systemd[1]: Starting A high performance
web server and a reverse proxy server
Aug 10 13:49:52 bob-ProLiant-MicroServer systemd[1]: Started A high performance
web server and a reverse proxy server
lines 1-11/11 (END)

The earlier command shows you the service state (loaded and active), the cgroup hierarchy, and the first few journald log entry lines for the nginx Web server service.

It gives you an overview of the current status of the application, notifying you of any problems and any actions that may be required.

There are also methods for checking for specific states. For instance, to check to see if a unit is currently active (or running), you can use the is-active command, as follows:

```
sudo systemctl is-active name.service
```

This will return the current unit state, which can be active or inactive. Also, if the unit state is "enabled," this means that it starts up at system boot.

This brings up a very practical possibility: how can you quickly know whether a service has actually been installed on your system? That is, either by default, by you, using the package management system commands, or built by hand, and then installed? You can use the systemctl status command, and if the output yields something like the following (for the Webmin service in this case), you know it has *not* been installed, or possibly has been incorrectly installed!

```
$ sudo systemctl status webmin
• webmin.service
   Loaded: not-found (Reason: No such file or directory)
   Active: inactive (dead)
$
```

To see if the unit is enabled, you can use the systemctl is-enabled command, as follows:

```
sudo systemctl is-enabled name.service
```

This will show whether the service is enabled or disabled.

A third status report you can get is to see if the unit is in a failed state. This indicates that there was a problem starting the unit:

```
sudo systemctl is-failed name.service
```

This will return active if it is running properly or failed if an error occurred. If the unit was intentionally stopped, it may return unknown or inactive. An exit status of "0" indicates that a failure occurred and an exit status of "1" indicates any other status.

Exercise 18.8

Use the systemctl options shown in this section to see if the sshd service is installed, started, or active on your Linux system.

18.3.1.6 How to Check System State

The earlier commands are used for effectively managing single services, but they are not very helpful for exploring the current state of the entire system and its services. There are a number of systemctl subcommands and their options that give you this kind of information.

Listing Current Units

To see a list of all of the active units that systemd knows about, we can use the list-units command:

```
$ sudo systemctl list-units
```

This will show you a list of all of the units that systemd currently has active on the system.
  The output has the following columns:

1. UNIT: The systemd unit name.
2. LOAD: Whether the unit's configuration has been read properly by systemd. The configuration of loaded units is kept in memory.
3. ACTIVE: A summary statement about whether the unit is active. This is a basic way to tell if the unit has started successfully or not.
4. SUB: This is a lower-level state that indicates more detailed information about the unit. This often varies by unit type, state, and the actual method in which the unit runs.
5. DESCRIPTION: A short textual description of what the unit is/does.

Since the list-units command shows only active units by default, all of the earlier entries will show "loaded" in the LOAD column and "active" in the ACTIVE column. This display is the default behavior of systemctl, when it is invoked without additional subcommands. It is the same as using systemctl with no arguments.
  systemd has a feature that tracks each service, whether it started up successfully at system start-up, whether it exited with a nonzero exit code, whether it timed out, or whether it terminated abnormally (with a segmentation fault), both during start-up and at run time. Typing systemctl at the shell prompt allows you to see the state of all services, for example:

```
$ sudo systemctl
```

The ACTIVE column, which shows you the high-level state of a service, whether it is active (i.e., running), inactive (i.e., not running), or in any other state, is critical. An item in the list that is marked maintenance and highlighted in red tells you about a service that failed to run or perhaps encountered a problem.
  In the following case, for the zfs-import-cache.service we have on our Linux system, we can find out what specifically happened to zfs-import-cache, with the systemctl status command:

```
$ sudo systemctl status zfs-import-cache.service
• zfs-import-cache.service - Import ZFS pools by cache file
   Loaded: loaded (/lib/systemd/system/zfs-import-cache.service; static; vendor
preset: enabled)
   Active: failed (Result: exit-code) since Thu 2016-07-14 18:42:38 PDT; 13h ago
  Process: 611 ExecStart=/sbin/zpool import -c /etc/zfs/zpool.cache -aN
(code=exited, status=1/FAILURE)
  Process: 512 ExecStartPre=/sbin/modprobe zfs (code=exited, status=0/SUCCESS)
 Main PID: 611 (code=exited, status=1/FAILURE)

Jul 14 18:42:36 bob-ProLiant-MicroServer systemd[1]: Starting Import ZFS pools by
cache file...
Jul 14 18:42:38 bob-ProLiant-MicroServer zpool[611]: cannot import 'test1': no
such pool or dataset
Jul 14 18:42:38 bob-ProLiant-MicroServer zpool[611]:          Destroy and re-create
the pool from
```

```
Jul 14 18:42:38 bob-ProLiant-MicroServer zpool[611]:          a backup source.
Jul 14 18:42:38 bob-ProLiant-MicroServer systemd[1]: zfs-import-cache.service:
Main process exited, code=exited, status=1/FAILURE
Jul 14 18:42:38 bob-ProLiant-MicroServer systemd[1]: Failed to start Import ZFS
pools by cache file.
Jul 14 18:42:38 bob-ProLiant-MicroServer systemd[1]: zfs-import-cache.service:
Unit entered failed state.
Jul 14 18:42:38 bob-ProLiant-MicroServer systemd[1]: zfs-import-cache.service:
Failed with result 'exit-code'.
```

This shows us that zfs-import-cache.service terminated during run-time and tells us the error condition: the process exited with an exit status of 611.

To use systemctl to output many other forms of information, you add additional options to the command. To see all of the units that systemd has loaded (or attempted to load), regardless of whether they are currently active, you can use the --all flag, like this:

```
$ sudo systemctl list-units --all
```

This will show any unit that systemd loaded or attempted to load, regardless of its current state on the system. Some units become inactive after running, and some units that systemd attempted to load may have not been found on disk.

You can use other options to filter these results. For example, we can use the --state= option and arguments to indicate the LOAD, ACTIVE, or SUB states that we wish to see. You will have to keep the --all flag so that systemctl allows nonactive units to be displayed, as follows:

```
$ sudo systemctl list-units --all --state=inactive
```

Another common filter is the --type= option. We can tell systemctl to only display units of the type we are interested in. For example, to see only active service units, we can use the following:

```
systemctl list-units --type=service
```

Listing All Unit Files

The list-units command only displays units that systemd has attempted to read and load into memory. Since systemd will only read units that it thinks it needs, this will not necessarily include all of the available units on the system. To see every available unit file within the systemd paths, including those that systemd has not attempted to load, you can use the list-unit-files command instead:

```
$ sudo systemctl list-unit-files
```

Units are representations of resources that systemd knows about. Since systemd has not necessarily read all of the unit definitions according to this view, it only presents information about the files themselves. The output has two columns: the unit file and the state.

The state will be "enabled," "disabled," "static," or "masked." Static means that the unit file does not contain an "install" section, which is used to enable a unit. These units cannot be enabled. This means that the unit performs a one-time-only action, or is used only as a dependency of another unit and should not be run by itself.

Exercise 18.9

To get practice with the service management commands presented in the all of the sub-sections of Section 18.3.1 execute as many of them on your system as you can. Instead of using the arguments to the commands shown, use units and services of your own choice.

Exercise 18.10

What systemctl command allows you to list the service unit files on your system?

18.4 Using systemctl to View the System State

As we have previously shown, systemd units, in particular service units, that can be manipulated in logical groups are known as target units or simply targets. This section shows how systemctl can be used to view the system state, so that subsequently you can adjust various aspects of service units or other target units.

18.4.1 systemd Service Viewing

systemd allows you to analyze the system startup process. You can list all services and their status. In terms of efficiency, systemd also allows you to scan the startup procedure to find out how much time each service took during start-up.

To review the complete list of active services that have been started since the system last stated, enter the following command:

```
$ sudo systemctl
```

To get information on a specific service, use the following command:

```
systemctl status name.service
```

To show what services failed to start, add the --failed option to the systemctl command:

```
$ sudo systemctl --failed
```

To debug system startup times, you can use the systemd-analyze command. It shows the total startup time, and depending on the options you use, a list of services ordered by startup time. We illustrate the output of this command on our Linux Mint system, as follows:

```
$ sudo systemd-analyze critical-chain
The time after the unit is active or started is printed after the "@" character.
The time the unit takes to start is printed after the "+" character.

graphical.target @12.759s
└─multi-user.target @12.759s
  └─virtualbox.service @12.562s +197ms
    └─network-online.target @12.554s
      └─NetworkManager-wait-online.service @4.369s +8.184s
        └─NetworkManager.service @3.991s +377ms
          └─dbus.service @3.912s
            └─basic.target @3.899s
              └─paths.target @3.899s
                └─acpid.path @3.899s
                  └─sysinit.target @3.894s
                    └─apparmor.service @3.668s +224ms
                      └─local-fs.target @3.666s
                        └─run-cgmanager-fs.mount @4.159s
                          └─local-fs-pre.target @619ms
                            └─keyboard-setup.service @262ms +356ms
                              └─system.slice @251ms
                                └─-.slice @220ms
```

The command can also generate a Scalable Vector Graphics (SVG) file, showing the time services took to start in relation to other services. We illustrate this case on a Linux Mint system as follows:

```
$ sudo systemd-analyze plot > /home/bob/mint18-startup.svg
```

The preceding command lets you review the services that started and the time it took to start them in a graphical plot file named mint18startup.svg, saved in /home/bob.

Exercise 18.11

Use the systemctl status command on your Linux system to find out the status of a particular service of interest to you.

Exercise 18.12

Use the systemd-analyze command to produce an .svg plot of the startup of your system.

18.5 Creating and Adding a New systemd-Controlled Service

In this section, we begin by showing some simple examples of how to create a "new-style" systemd daemon. New style, as we further describe later, means the daemon used the functionality of systemd, as opposed to the "old-style" daemon illustrated in Chapter W21 at the book website.

We then provide a complete and practical extended example of using systemd: applying some of the basic systemd facilities shown in the previous sections to a simple, static-page Web server daemon program in C. This program is designed so that it will log messages to the systemd journal and can be controlled as a systemd service.

18.5.1 New-Style Daemons and Services

To begin this section, it is important to present a few easy-to-understand examples of new-style systemd daemons.

Modern daemons, written specifically to be implemented as systemd services exclusively, should follow a simple scheme, which we call "new-style" daemons. We leave it to you to contrast and compare them to the "old-style" Linux daemons that run under legacy BSD-init and System V-init. We show introductory, simple new-style daemons, and a more involved one, and make some observations about them to give you a feeling for how systemd can simplify, as well as add more functionality, to your interactions with systemd, and its service management in Linux.

A new-style daemon in systemd can be as simple as a Bash shell script or a short Python script. It can also be as complex as a complete, multimodule, C program, with thousands of lines of code in it.

To control it with systemd, you can simply use the systemd-run command, or you can install it in a more complex way by adding it as a service, with attendant timer, socket, or other unit files. These procedures are explained in the sections that follow.

18.5.1.1 A systemd-run Command Example

In this example, we use the systemd-run command to execute a nonprivileged user Bash shell script and illustrate that this daemon can be controlled and monitored as a systemd service. Do the following steps in the order presented to gain an appreciation of these points:

1. Create the following executable Bash shell script, named test-daemon.bash, in your home directory, using nano, or your favorite text editor:

```
#!/bin/bash
i=0
while sleep 10
do
        echo "<$i>This message is coming from a new-style daemon"
        i=$(((i+1)%8))
done
```

2. Execute the following command, some of the details of which we gave in Chapter 17, Section 8.1.2.1, on creating transient cgroups:

```
$ systemd-run --user --unit=test-daemon ./test-daemon.sh
Running as unit test-daemon.service.
```

3. Check the status of the daemon with the following command:

```
$ systemctl --user status test-daemon
• test-daemon.service - /home/bob/./test-daemon.sh
     Loaded: loaded
  Transient: yes
    Drop-In: /run/user/1000/systemd/user/test-daemon.service.d
             └─50-Description.conf, 50-ExecStart.conf
     Active: active (running) since Sat 2016-08-06 09:04:09 PDT; 20s ago
   Main PID: 2252 (test-daemon.sh)
     CGroup: /user.slice/user-1000.slice/user@1000.service/test-daemon.service
             ├─2252 /bin/sh /home/bob/./test-daemon.sh
             └─2255 sleep 10

Aug 06 09:04:09 bob systemd[1152]: Started /home/bob/./test-daemon.sh.
Aug 06 09:04:19 bob test-daemon.sh[2252]: This message is coming from a new-
style daemon
Aug 06 09:04:29 bob test-daemon.sh[2252]: This message is coming from a new-
style daemon
Output truncated…
```

4. Finally, to stop the daemon from running, use the following command:

```
$ systemctl --user stop test-daemon
```

Exercise 18.13

Use the four steps in Section 18.5.1.1 to convert another personally meaningful and useful Bash script of your choosing into a systemd-controlled service. Then, use the service, either temporarily or permanently, as your use case dictates.

18.5.1.2 Bourne-Again Daemon

Here is the same Bash shell script that we use a more "permanent" and privileged user technique to make into a service. Do the following steps in the order shown to get an appreciation of these points:

1. Create the following Bash shell script in /usr/local/bin. Name it new-style.bash, using your favorite text editor. Be sure to have root privileges when you do this, so you can save and execute this file using the sudo command in Step 2:

```
#!/bin/bash
i=0
while sleep 10
do
        echo "<$i>This message is coming from a new-style daemon" > /dev/pts/0 &
        i=$(((i+1)%8))
done
```

2. While the present working directory is /usr/local/bin, execute the script as root with the command

```
$ sudo ./new-style.bash
```

Note its output on stdout and then terminate the execution of the script file with Ctrl-C.

Exercise 18.14

You can use the journalctl –r command to see what the latest entries to the systemd journal are. Is any of the output of this command showing output from new-style.bash? Why or why not?

3. Edit the file new-style.bash and delete the **> /dev/pts/0 &** from the line that begins with echo. Resave it.

4. Create the following service file, named new-style.service, in /etc/systemd/system, using your favorite text editor
   ```
   [Service]
   ExecStart=/usr/local/bin/new-style.bash
   ```
5. Start the service with the following command
   ```
   $ sudo systemctl start new-style
   ```

Let the service run for a minute or two.

6. Show the status of the running daemon with the following command:
   ```
   $ sudo systemctl status new-style
   ```
7. Give the following command to stop the service. You may have to open another terminal window to do this
   ```
   $ sudo systemctl stop new-style
   ```

Exercise 18.15

Use the journalctl –r command to see what the latest entries to the systemd journal are. Is any of the output of this command showing output from new-style.bash?

Exercise 18.16

Use the seven steps in Section 18.5.1.1 to create another Bash script of your choice into a systemd-controlled service. Then, use the service, either temporarily or permanently, as your use case dictates.

18.5.2 An Old-Style Simple Daemon Web Server Example

We will take a daemon Web server program, named webserver2, and put it under the control of systemd as a service. The C code for this program may be found GitHub site for this book. We leave it to you to examine, and most importantly, analyze the logic and details of that code, to gain an understanding of what it does, and how it accomplishes what it does. The way the program is made into a systemd-controlled service is found in Section 18.5.3.

The program has the distinguishing characteristic of logging messages to the systemd journal. These messages will contain both information about the incoming requests to the server and the outgoing server content, as well as errors generated by the requests. We illustrate many of the uses of the systemd journal in Chapter W27, Section 6.2, at the book website.

Note that to compile C programs on your Linux system, you may have to install the Gnu compiler (gcc), depending upon which Linux system you have installed. We give detailed instructions for installation of gcc with APT (using the apt-get command) and YUM (using the yum command) in Appendix A, Sections 17.7.2.2 and 17.2.3.2. That installation procedure is dependent upon the details of using the package management system on your Linux system. On the representative Linux systems we use in this book (Debian, Ubuntu, Linux Mint, and CentOS), we use the installation of gcc as an illustration of how to use the package management system on those systems.

For example, in Debian-family Linux Mint, to use the gcc compiler and its systemd library modules, you first must obtain and install the build-essential and libsystemd-dev packages. Alternatively, in a Linux Mint Cinnamon GUI environment, you can use the Software Manager to install both gcc and build-essential on your system. An alternative command line way (shown in Appendix A) to obtain and install build-essential is by using the following command:

```
$ sudo apt-get install build-essential
```

To compile the example weberver2 source code on our Linux Mint system, we used the following command and compilation flags:

```
$ gcc webserver2.c -o webserver2 `pkg-config --cflags --libs libsystemd`
```

18.5.2.1 How to Run *webserver2* before You Make It a Service

You must have a valid index.html file in the directory you are running webserver2 from. We do not provide you with or illustrate the creation on an index.html file, but leave that up to you.

Once you have compiled webserver2 correctly, you can run it by typing the following command:

```
$ ./webserver2 port_number /pathname_to_web_directory
```

where

port_number is the port you want it to listen on, such as 8083,

/pathname_to_web_directory is the path where the executable binary image, and an index.html file, is located, such as /home/bob

The command to run it on our Linux Mint system is

```
$ ./webserver2 8083 /home/bob/index.html
```

Please realize that webserver2 is a static page server—it will send back to the requesting agent any page ending with .htm, .html, .jpg, .jpeg, .gif, .png, .ico, and .zip. It does not support running Common Gateway Interface (CGI) scripts, Perl, or Hypertext Preprocessor (PHP) extensions.

To test webserver2, set your Web browser to the IP address of your machine (or localhost 127.0.0.1) and the port_number you chose. For example, to view our index.html file, we set our URL as follows:

```
192.168.0.6:8083
```

Whatever content is defined in your index.html will now appear in your Web browser.

When you are done serving index.html from webserver2, you can stop it by using the kill command, for example:

```
$ sudo kill -9 13325
```

where

13325 is the PID of the webserver2 process, as shown by the **ps -C webserver2** command.

Before going on to the next section, you should kill the webserver2 daemon.

18.5.2.2 *Webserver* Example Program Logic Model

Following the basic design and implementation of the various daemon programs illustrated in Chapter W21, Sections 3–13, found at the book website, our example webserver2 daemon contains the system calls and program structure that is very much like the classic Linux daemon model. Refer to Figure W21.3 for a basic description of the "old-style" Linux daemon model.

Exercise 18.17

What are the differences, at the gross level of program structure, between the code of the webserver2.c program and what is shown in Figure W21.3?

One of the major differences between our webserver2 code and the Figure W21.3 model is the maintenance of program robustness. That means error checking and handling, and system logging, at many of the 11 steps shown in that figure.

As a particular difference, the umask is not set to 0 in webserver2. Why?

As stated earlier, one notable exception to the model in our example webserver2 daemon is that there is a separate module that does all of the message logging, including errors, to the systemd journal. So for example, whenever an exception is generated by incoming requests to the webserver, those errors are logged in the systemd journal using the sd_print_journal system call. Also, Web page requests and sends are logged to the systemd journal, allowing a system administrator to monitor the operation of webserver2 as a service, and also to examine possible intrusion alerts via Hypertext Transfer Protocol gets and posts, and the error messages printed to the journal.

We formulate a problem at the end of this chapter that has you compare the system calls made in Chapter W21 at the book website, Sections 3–13, and summarized in Figure W21.3, to the example webserver2 system calls.

18.5.2.3 *Using the Journal Application Programming Interface (API) in webserver2*

In this subsection, we show what native API system calls the systemd journal provides for logging. We also illustrate the use of these native API system calls in the webserver2 example program.

The following example code fragment shows the use of using the journal's logging API call sd_journal_print:

```
#include <systemd/sd-journal.h>

int main(int argc, char *argv[]) {
        sd_journal_print(LOG_NOTICE, "systemd rocks!");
        return 0;
}
```

As noted earlier, this program must be compiled on a Linux Mint system with the following compiler directives. Also note that the package libsystemd-dev must first be installed (most expediently with the Software Manager) on a Linux Mint system as well.

`pkg-config --cflags --libs libsystemd`

The journal's native logging APIs provide specifically structured log messages to be sent from a program of interest, perhaps which you have written and wish to monitor for those specific structures, to the journal. This specifically structured log data is searchable with the journalctl command and is available to other ancillary programs. A system administrator would use the systemd journal logs, the ancillary programs, and other intrusion detection techniques to maintain the security of the system. The journal logs with specifically structured data attached would allow the system administrator to track down issues beyond what may be found in the human-readable log message text. Here's a basic example of how to do that with the system call sd_journal_send():

```
#include <systemd/sd-journal.h>
#include <unistd.h>
#include <stdlib.h>

int main(int argc, char *argv[]) {
        sd_journal_send("MESSAGE=Hello World!",
                        "MESSAGE_ID=52fb62f99e2c49d89cfbf9d6de5e3555",
                        "PRIORITY=5",
                        "HOME=%s", getenv("HOME"),
                        "TERM=%s", getenv("TERM"),
                        "PAGE_SIZE=%li", sysconf(_SC_PAGESIZE),
                        "N_CPUS=%li", sysconf(_SC_NPROCESSORS_ONLN),
                        NULL);
        return 0;
}
```

The preceding program will write a log message to the journal with six additional metadata-structured fields attached. We used sd_journal_send() to accomplish this, and it is similar to sd_journal_print(), but takes a NULL terminated list of format strings each followed by its arguments. The format strings must include the field name and a "=" before the values.

The structured message included seven fields. The first three are well-known fields:

MESSAGE= is the actual human readable message part of the structured message.

PRIORITY= is the numeric message priority value formatted as an integer string.

MESSAGE_ID= is a 128-bit ID that identifies our specific message call, formatted as hexadecimal string.

See the man page for systemd.journal-fields for a complete list of the currently well-known fields.

In the webserver2 example, we use the following code structure within the logging module to print logging information to the journal. Various error conditions, and most importantly, requests/sends to the daemon (via the case condition LOG), are logged to the journal in this way.

```
switch (type) {
      case ERROR: sd_journal_print(LOG_NOTICE, "ERROR: %s:%s Errno=%d exiting pid=%d",
                s1, s2, errno,getpid());
            break;
      case FORBIDDEN:
            (void)write(socket_fd, "HTTP/1.1 403 Forbidden\nContent-Length:
            185\nConnection: close\nContent-Type: text/html\n\n<html><head>\n<title>403
            Forbidden</title>\n</head><body>\n<h1>Forbidden</h1>\nThe requested URL,
            file type or operation is not allowed on this simple static file webserver.\
            n</body></html>\n",271);
            sd_journal_print(LOG_NOTICE, "FORBIDDEN: %s:%s",s1, s2);
            break;
      case NOTFOUND:
            (void)write(socket_fd, "HTTP/1.1 404 Not Found\nContent-Length:
            136\nConnection: close\nContent-Type: text/html\n\n<html><head>\n<title>404
            Not Found</title>\n</head><body>\n<h1>Not Found</h1>\n
            The requested URL was not found on this server.\n</body></html>\n",224);
            /*(void)sprintf(logbuffer,"NOT FOUND: %s:%s",s1, s2); */
            sd_journal_print(LOG_NOTICE, "NOT FOUND: %s:%s",s1, s2);
            break;
      case LOG: sd_journal_print(LOG_NOTICE, "INFO: %s:%s:%d",s1, s2,socket_fd);
            break;
      }
      if(type == ERROR || type == NOTFOUND || type == FORBIDDEN) exit(3);
}
```

18.5.3 Making a Daemon a Service: A Simple Three-Step Example

Following is an example of how a system program in C, "webserver2," that is shown in Section 18.5.2, can be made into a systemd-controlled service. The simplest way to do this is by creating a unit file for it, starting the service, and then, if desired, enabling the service to start upon all subsequent reboots.

The steps necessary to do this for a generic program, and particularly for our webserver2 example, are given as follows:

Step 1: Create the service unit file, named **your_name.service**, with your favorite text editor. Save the service unit file in /etc/systemd/system.
The following are the contents of that service unit file:
[Unit]
Description=A simple description
After=network.target

```
[Service]
ExecStart=your_path your_port your_directory
Type=forking

[Install]
WantedBy=multi-user.target
```

A line-by-line explanation of the components of this simple service unit file is as follows:

The [Unit] directive Description= documents what this service does. Additionally, the documentation string you put here appears in the systemctl status report of this service.

The [Unit] After= directive instructs systemd to start this unit when the network.target state needs it.

The [Service] directive ExecStart= supplies your_path, which is the pathname to the location of the executable program on your system, including the actual name of the program; in terms of the particulars of this argument for the webserver2 example, your_port is the port number you wish the program to listen on, and your_directory is the directory on your system where the program has its Web pages (such as index.html), and any other attendant files.

The [Service] Type= directive specified as forked means systemd will wait until your program forks. For example, the Web server daemon creates its control socket after starting, but before forking. So, without the forking directive, systemd may start something that tries to connect to the socket before the Web server daemon creates it, thus yielding a dead service.

The [Install] directive WantedBy= specifies multi-user.target as the state that is reached before this daemon service is started.

Step 2: Start the service with the command `systemctl start your_name.service`.

You can check the status of the service with the following command:

`$ systemctl status your_name.service`

To stop the service at any time, give the command

`$ systemctl stop your_name.service`

Step 3: To ensure that the service starts every time the system boots, type the command:

`$ systemctl enable your_name.service`

After you do this, a symbolic link /etc/systemd/system/multiuser.target.wants/your_name.service, linking to the actual unit file in /etc/systemd/system, will be created. It tells systemd to "pull in" the unit when starting the multi-user.target.

If you want to discontinue having this service start at boot, you can give the command

`$ systemctl disable your_name.service`

This will remove the symlink.

18.5.3.1 Making webserver2 a systemd Service

The particulars of doing the three steps in Section 18.5.3 to make our webserver2 example a service, and most importantly, to use the service, are as follows:

Step 1: Make sure webserver2 is not running as instructed in Section 18.5.2.1.

Step 2: The name of the executable program is webserver2 and the directory containing it and attendant code (such as an index.html file, etc.) is /home/bob/webserver2.

Step 3: The unit file you create with your text editor should be named webserver2.service and be saved in /etc/systemd/system. The [Service] ExecStart= directive should be

```
ExecStart=/home/bob/webserver2 8081 /home/bob/webserver2
```

Step 4: Start the service with the command:

```
$ sudo systemctl start webserver2.service
```
To see the status of the service with the command:

```
$ sudo systemctl status webserver2.service
```
To stop the service with the command:

```
$ sudo systemctl stop webserver2.service
```

Step 5: To make sure that the service starts automatically on every subsequent reboot of the system, use the command:

```
$ sudo systemctl enable webserver2.service
```
To stop the service from automatically starting on every subsequent reboot of the system, use the command:

```
$ sudo systemctl disable webserver2.service
```

Step 6: To connect to the server and view the Web pages you put in /home/bob/webserver2, point a Web browser to the IP address, port, and file location of the machine your started service is on, such as

```
192.168.0.6:8081/home/bob/webserver2/index.html
```

Exercise 18.18

Is it possible to run the nonservice activated webserver2 that you compiled and executed in Section 18.5.2.1 multiple times, so that each instance of it listens on a different port on your Linux system? Is it possible to serve different Web pages from each of these multiple instances of webserver2? Is it possible to run multiple webserver2-based systemd services on your Linux system, with each service listening at a different port, and serving different Web pages? Write a description of how each of these scenarios would work. Compare this to file locking on daemons as shown in Chapter W21 at the book website.

Summary

This chapter was a combined basic reference and tutorial on systemd for the ordinary user. From the user perspective, we illustrated how systemd is structured from the top down. We defined the fundamental terms unit and target. We then introduced, through many examples, its most important command, systemctl. Finally, we gave extended and typical use cases for applying those commands. We used a basic Web server program as an example of a systemd-managed service. We covered the following systemd topics, using many text-based commands:

System start-up, initialization, and shutdown

Units and unit files, with an emphasis on service unit files

Target unit files, and their purposes and uses

System service management basics and the systemctl command

Other important user commands, for example, the journalctl command

Creating, adding, and controlling a new systemd service from scratch

Furthermore, we defined systemd as a combined Linux initialization program and service manager, which also contains and implements many additional important system state control programs and utilities.

We have suggested that systemd is not only an initialization program and service manager but is also a "superkernel," which uses the Linux kernel itself as a subsidiary program. The Linux kernel, at a certain level of abstraction, maintains the steady state of the hardware and software components of the computer for systemd. Extending that abstraction, systemd supervises and manages everything that the kernel does to optimize operation in that steady-state condition, whether it be start-up, process control and IPC, hardware services, network operations, shutdown, and many other facilities.

We concluded the chapter by showing, with some extended examples, how the ordinary user can add and control a new systemd service.

Questions and Problems

1. Answer the following as completely as you can

   a. Is a systemd unit a target, or can it be a target?

   b. Is a systemd target a unit, or can it be a unit?

   This question is aimed at getting you to think about the conceptual organization of systemd as shown in Figure 18.1, as well as the details of what units and targets are.

2. How can you easily verify that the milestone boot targets shown in Figure 18.2 are reached in the order local-fs-pre.target, sysinit.target, basic.target, multi-user.target, and graphical.target shown in that figure, when your system boots? The basic assumption here is that you are booting into a Linux desktop environment, but you can also describe verification on the basis of using a CUI text-only set of systemd commands.

3. What systemctl command can be used to change your Linux system from the final target state graphical.target to multi-user.target, and then change the final target state back to graphical. target. Test these operations on your Linux system.

4. Take a Bash script file that is useful to you, and turn it into a systemd-controlled service which executes at an appropriately timed interval. Place the script file in your home directory on the system and set permissions on it properly. Start, stop, restart, and edit the operation of the service with the appropriate systemctl commands.

5. After examining the code and logic of execution of the daemon examples presented in Section 18.5, compare the daemon-creating system calls in those, to those presented in the C code from Chapter W21, Section 13, at the book website, named "test_server.c" (summarized in Figure W21.3). What differences do you notice? What system calls from the Chapter W21 test_server are *not* present in the example programs or vice versa? Why?

6. Modify the daemon webserver2 example presented in Section 18.5 so that it writes other fields of interest that might be useful for you, to the journal using the sd_journal_send systemd function.

7. Compare the information and logs found in /var/log to the contents of the journal, excluding the directory /var/log/journal if that exists. Prepare a report of the significant similarities and differences that you notice, particularly with respect to searching, rotating, and retention of the logs.

8. Download, compile, and execute the webserver2 program code found at the GitHub site for this book on your Linux system. Then, make modifications of your own choosing to the example webserver2 program, by using the editing techniques found in Section 18.3. Be sure to restart the webserver2 service after your changes are made. Then, reverse the changes to the webserver2 service by using the techniques shown in Section 18.3.

9. Download, compile, and execute the webserver2 program code found at the GitHub site for this book on your Linux system. Then, after carefully examining the error-handling capabilities of the program code in detail, design a test suite that generates error conditions from the command line or from your Web browser that would test all of the error-handling conditions in the program code as given to you. Make a complete and detailed list of the systemd journal entries and any other feedback that the webserver2 program provides via its error-handling capabilities, as output of your testing results.

10. Part 1.

Write a simple Bash script file that can repeatedly back up a single source directory under your home directory, to a target directory that is either on a USB-mounted thumbdrive on the system, or to a remote computer on your LAN. Use the tar command and appropriate options, to compress the target-archived directory, or use the scp or rsync commands to create an uncompressed backup on the target. Be sure to name the backup on the target with some meaningful label, to differentiate sequential backups that you make, and make sure you test this script file to see that it works correctly!

In order to help you complete this problem and to refresh your memory about how to do copying operations with the scp, ssh, and rsync commands presented in previous chapters, we repeat some examples of those copying operations here.

Following is a condensed collection of examples that use scp, ssh/dd, and rsync to copy files and directories locally and remotely. It is understood in these examples that you have permission access to the local files and directories and that you have permission access to the remote locations as well.

1. An example of using secure copy (scp) to copy all the files in the directory webserver2 to a remote directory of the same name.

   $ scp webserver2/*.* bob@192.168.0.25:/home/bob/webserver2/

2. An example of extracting the remote file backup.tar file at /home/bob in 512-byte blocks and streaming it through dd to the current working directory on the system you typed this command.

   $ ssh bob@192.168.0.13:/home/bob "dd if=backup.tar ibs=512" | tar xvBf –

3. An example of using rsync to copy an entire directory named syncdir in the current working directory locally to a mounted USB thumbdrive named USBint

   $ rsync -av syncdir /media/bob/USBint')

4. An example of using rsync to copy an entire directory named syncdir2 in the current working directory in push mode, remotely to an OS X machine

   $ rsync -av -e ssh syncdir2 bob@192.168.0.7:/Users/b/unix3e

5. An example of using rsync to copy a file named rsynctest from the current working directory to the local destination on a thumbdrive named USBint is

   $ rsync -av rsynctest /home/bob/USBint

Part 2.

To make the earlier script file practical and useful for you, make it a systemd service that can be run on some regular basis (to be determined by you), using the examples from Sections 18.5. So, for example, you could time the service to run daily at 9:00 P.M. Make sure to enable this service so it is persistent across boots!

11. Examine the man page for the systemctl command on your system, and write a short report that describes its syntax, purpose, options and option arguments, and subcommands that pertain to units and unit files.

12. It would be helpful if you could "rewind" the system into a previously saved boot environment. How would that be possible, using what you know of systemd and the Linux facilities we have shown in all previous chapters? Sketch a way of doing that.

13. Write a brief description of why you would change the system state into the following targets: poweroff.target, rescue.target, multi-user.target, graphical.target, reboot. target, emergency.target.

14. Use the systemd-run command to run the program webserver2, and all of the script files shown in this chapter, one at a time, as transient scope units.

15. Determine which systemd commands must be run as root, using the sudo command, and which do not require root privilege. Produce a table of those commands. Explain exactly what your strategy was in finding out the information that allowed you to answer this question.

Advanced Questions and Problems

16. Install systemadm on your computer, and using the techniques of Chapter W27, Section W27.8, at the book website, start, stop, and get the status of the webserver2 service that we worked with in the sections of this chapter. Then determine which method (either command line or GUI) you prefer to manage services with.

17. What is the snapshot feature of systemd, and how and why do you use it? What other Linux facilities, programs, and systems have similar capabilities, and how does their implementations compare in detail to systemd snapshots?

18. In terms of starting up your Linux system, what does systemd do when the target units in Figure 18.2 *cannot* be reached, for whatever reason? What strategies can be deployed, and what specific systemd tactics can be used, to start up the system when these targets cannot be reached? In other words, what path of execution can be followed, and why? Does the system immediately enter into rescue or emergency mode? What are systemd's error-handling capabilities, as far as system start-up go? How did you find this out? And most importantly, how can you use your answers to these questions as a troubleshooting guide on your Linux system, when the requisite target states cannot be reached?

Projects

Project 1

To practice with changing the system state to rescue and emergency modes, use the commands in Chapter W27, Sections W27.5.1.8.8 and W27.5.1.8.9, at the book website, to achieve that on your system. And when your system goes into those states, carefully explore and document the capabilities of rescue- and emergency-mode operations. For example, what file system commands and facilities do you have available to you in those modes? Can you mount external media while the system is in those modes, so that you can save and backup important user data files? It would be very useful to become familiar with these two system state targets (by experimenting with bringing your Linux system into these states!), so that if an emergency happens, you can recover from it.

Project 2

Follow the instructions in Chapter W27, Section W27.7.1.3, at the website, "A Python Webserver as a New-Style Daemon," and create, start, and test the service named simple5.service that those instructions specify. It would be useful to look through Chapter W19 at the book website on Python before doing this project, but it is not mandatory to do so.

Then do the following:

Modify the Python script simple5.py, so that the Web server is exposed on port 8091. Also, modify the contents of the built-in html code in that script file so that it is customized to your liking. Add things like more text, images, links to other pages, etc. Name the Python script file simple6.py, and complete the six steps shown in Section W27.7.1.3 to make it into a systemd-controlled service. Then run the services simple5 and simple6 simultaneously. Test both services by browsing with your favorite Web browser to the ports 8090 and 8091 on the system you installed these two services on.

Project 3

Take the following Python rolling-backup program script, and use the methods of Chapter W27, Sections W27.6.3.1 and W27.6.3.2, at the book website, to run it on your Linux system. As shown in those sections, implement the Python script as a clock-time-based or calendar-based systemd service. Modify the Python code so that the service runs at time intervals that are useful for you.

Also, make sure you change the source and target directories in the Python code so that they suit your specific needs, and work with the file system structure of your Linux system. Refer to Chapter W19 at the book website on Python for the pertinent details of how to execute/implement this Python script. For those familiar with Python programming, use any enhancements to it that you feel would make the script file more "Pythonic."

```python
#!/usr/local/bin/python
import os
import shutil
target = "/home/bob/USBint/" #Target directory, or where you are backing up to
i = 1
while i <= 5:
    temp_path = target + str(i) + "/"
    if not os.path.exists( temp_path ):
        try:
            os.makedirs( temp_path )
            print "Created   " + temp_path
        except:
            print " Could not create   " + temp_path
    i = i + 1
print "Deleting the oldest archive"
shutil.rmtree( target + "5" )
print "Recycle the backups"
os.rename( target + "4", target + "5" )
os.rename( target + "3", target + "4" )
os.rename( target + "2", target + "3" )
# Do the backups
os.system('cp -a ' + target + "1" + " " + target + "2")
os.system('rsync -av /home/bob/python/' + " " + target + "1") #Source directory
```

Project 4

Convert the Python script file shown in Project 3 to a Bash script file that does exactly the same thing. Then, as specified in Project 3, use systemd to enable the Bash script to run at a time interval that is useful for you. Also, change the source and target directories to suit your specific use-case needs, and Linux system configuration as well.

Looking for more? Visit our sites for additional readings, recommended resources, and exercises.

CRC Press e-Resource: https://www.crcpress.com/9781138710085.

Authors' GitHub: https://github.com/bobk48/linuxthetextbook.

Appendix A: Installation Instructions

Who this is for: We provide the following installation instructions for critical software on our representative Linux systems to the readers who do *not* have that software already installed.

Specifically, those readers would not only be the system administrator on the computer that is running Linux (therefore, they would have Root privileges!) but also be the person that installed the operating system itself on the hardware. Our principal target is a user that wants to install Linux, and this critical software, on her own PC.

As specified in the Preface, to make it easy for you to differentiate which system we are providing instructions for "whenever we illustrate a Linux command, by default we preface the command with only the Bash shell prompt "$," and we use the output that a Debian-family distribution would give. That output is generally uniform across Debian, Ubuntu, and Linux Mint, and also CentOS.

Whenever the syntax, execution, and output of a command significantly differs from the default on the Debian-family distributions, particularly as seen in Appendix A on Debian and CentOS, we preface the command with a shell prompt as follows:

```
[root@debian]#      for Debian
[root@centos]#      for CentOS"
```

Disclaimer: At the time of writing of this book, all of these installation instructions were tested on Debian 9.1, Ubuntu 16.04, Linux Mint 18.2, and CentOS 7.4. Over time, some of the procedures and steps we show may change slightly to accommodate bug fixes or other upgrades in any of these representative systems. But in general, they will remain the same for major new releases of the operating system.

Index of References in the Printed Book Text and at the Book Website

(Book website references have the letter "W" as a prefix, as indicated in Preface)

Installation Instructions

W25.1.2 emacs Installation

On our four representative Linux systems, emacs was *not* preinstalled.

Debian-family Installation of emacs
Type the following command to install emacs:

```
$ sudo apt-get install emacs24
```

CentOS Installation of emacs
As with all of the following installations on CentOS, there were some preliminary steps that needed to be taken before we could use Yellow Dog Updater, Modified (YUM). On our CentOS 7.4 system, we had to first stop and disable the packagekit app that is holding the lock on YUM. Then we could install the epel "extras" package, in which emacs is found. Use the following commands to do this, and then to install emacs with YUM:

```
[root@centos]# systemctl stop packagekit
[root@centos]# systemctl disable packagekit
[root@centos]# yum install epel-release
Output truncated...
[root@centos]# yum install emacs
Output truncated...
[root@centos]#
```

W25.5.7 Version Control with git
Installing git on Debian-family-

```
$ sudo apt install git
[sudo] password for bob: zzz
Reading package lists... Done
Building dependency tree
Reading state information... Done
The following additional packages will be installed:
  git-man liberror-perl
Suggested packages:
  git-daemon-run | git-daemon-sysvinit git-doc git-el git-email git-gui gitk
  gitweb git-arch git-cvs git-mediawiki git-svn
The following NEW packages will be installed:
  git git-man liberror-perl
0 upgraded, 3 newly installed, 0 to remove and 111 not upgraded.
Need to get 3,760 kB of archives.
After this operation, 25.6 MB of additional disk space will be used.
Do you want to continue? [Y/n] Y
Output Truncated...
Setting up git (1:2.7.4-0ubuntu1) ...
$
```

On CentOS (Note that all of the following commands are run as root!)-

1. You have to stop and disable the packagekit app so that it releases its lock on YUM. Do this with the following commands:
   ```
   [root@centos] # systemctl stop packagekit
   [root@centos] # systemctl disable packagekit
   ```
2. Now you can install git with the following command:
   ```
   [root@centos] # yum install git -y
   Loaded plugins: fastestmirror, langpacks
   Loading mirror speeds from cached hostfile
    * base: mirror.web-ster.com
    * epel: mirrors.develooper.com
    * extras: mirror.cs.pitt.edu
    * updates: centos.mirror.constant.com
   Resolving Dependencies
   --> Running transaction check
   ```

```
---> Package git.x86_64 0:1.8.3.1-12.el7_4 will be installed
Output truncated...
```

3. Test to see that the version shown in Step 2 was loaded and installed with the following command:
   [root@centos] # **git --version**

W26.1.2 Webmin Download and Installation

For Webmin Installation on Debian-family-

Example W26.1 Webmin Installation

Step 1. Add the official repository. You can do this by appending these two lines to your /etc/apt/sources.list file, using the following commands:

$ **sudo vi /etc/apt/sources.list**

Then, add the two lines shown here:

deb http://download.webmin.com/download/repository sarge contrib

deb http://webmin.mirror.somersettechsolutions.co.uk/repository sarge contrib

Save and exit the file.

Step 2. Fetch and install the Gnu Privacy Guard (GPG) key for Webmin, using the following commands:

$ **sudo wget http://www.webmin.com/jcameron-key.asc**
$ **sudo apt-key add jcameron-key.asc**

Step 3. Install Webmin with the following commands:

```
$ sudo apt-get update
$ sudo apt-get install webmin -y
Reading package lists... Done
Building dependency tree
Reading state information... Done
The following additional packages will be installed:
  apt-show-versions libauthen-pam-perl
The following NEW packages will be installed:
  apt-show-versions libauthen-pam-perl webmin
0 upgraded, 3 newly installed, 0 to remove and 45 not upgraded.
Need to get 15.3 MB of archives.
Output truncated...
```

For Webmin Installation on CentOS 7 (Note that all the following commands are executed as root!)-

1. Update all the currently installed software to the latest version available using the command:
   [root@centos] # **yum -y update**
2. Use a Web browser, go to the Webmin download page, and check for the Webmin RedHat Package Manager (RPM) package. The RPM package works for any RedHat, Fedora, or CentOS system. To download the package, use wget on the command line. For our CentOS system, we used the following command:
   [root@centos] # **wget http://prdownloads.sourceforge.net/webadmin/webmin-1.831-1.noarch.rpm**

To proceed with the installation, you need to make sure that all dependencies are installed on your CentOS Virtual Private Server (VPS). If they are not installed, you can install them using the following command:

[root@centos] # **yum -y install perl perl-Net-SSLeay openssl perl-IO-Tty**

Once the dependencies are installed, you can install Webmin using the following command:

```
[root@centos] # rpm -U webmin-1.831-1.noarch.rpm
```

3. To start Webmin, use the following command:

```
[root@centos] # service webmin start
```

To enable Webmin on system boot, use the following command:

```
[root@centos] # chkconfig webmin on
```

4. Before you can login to Webmin via a Web browser, you need to first adjust the public zone of firewalld with the following commands:

```
[root@centos] # firewall-cmd -permanent -zone=public -add-port=10000/tcp
[root@centos]  #firewall-cmd -reload
```

5. To access Webmin, open your favorite Web browser, enter HTTPS as protocol, enter your server IP address and use 10000 as a port number.
   https://YOUR-IP-ADDRESS:10000

You can accept the exception to the SSL certification. Then login to Webmin with the username set to root and root's password.

17.2 Linux System Installation

Linux Mint 18 Cinnamon Installation from a DVD-created ISO File in Ten Easy Steps

We are assuming that you want to install a 64-bit version of Linux Mint onto computer hardware that has an Internet connection, where it will be the only operating system on the hard drive, and you want to dedicate the entire hard drive to Linux Mint, so the hardware boots into Linux Mint.

1. Download the ISO image from www.linuxmint.com/download.php
   Choose the edition you want, in our case 64-bit Cinnamon Desktop version.
2. "Burn" the ISO image to a DVD or to a USB thumbdrive. We do not give instructions here for that operation.
3. Boot your computer to the DVD or USB thumbdrive, however, that is achieved with your hardware. Generally, that is done by pressing a function key on the keyboard, and selecting the boot device order.
4. At this point, Linux Mint runs from the DVD or USB thumbdrive, and you can test drive Linux Mint. Double click on the icon on the Linux Mint Cinnamon Desktop labeled "Install Linux Mint."
5. Make your language choice and choose Install Third Party Software if you desire. Make sure your computer hardware has enough disk drive space available and is connected to the Internet, as asked on the next installation screen.
6. **IMPORTANT** We assume you will be using and dedicating the entire hard drive of your computer to Linux Mint installation, so choose "Erase entire disk and install Linux Mint" on the next installation screen. If you want to do something else, such as dual boot the hardware into multiple operating systems, we do *not* give instructions for that here.
7. Click on the "Install Now" button, and verify, by clicking "Continue," that you want to write the designated installation to the disks and partitions presented in the next installation screen.
8. Choose your location in the world on the map presented.
9. Select the keyboard layout you desire. In general, you will not have to touch the keyboard mapping.

10. On the "Who are you?" screen, type in your name and a username, and most importantly, choose a password you can remember! This will give you root privileges on the newly installed Linux Mint system and put you automatically into the sudoers group. Click on the "Continue" button to install Linux Mint.

17.2.3.4 vsftpd Installation, From Example 17.1

The package management system command (either APT, YUM, or one used on your Linux system) used to install the vsftpd package is dependent upon what Linux system you are using. See Section 17.7 for details of package management.

For Debian, Ubuntu, and Linux Mint-

1. Download and install the vsftpd server with the following command:

```
$ sudo apt-get install vsftpd
[sudo] password for bob: QQQ
Reading package lists... Done
Building dependency tree
Reading state information... Done
The following packages were automatically installed and are no longer required:
  liblxc1 lxc-common lxcfs uidmap
Use 'sudo apt autoremove' to remove them.
The following NEW packages will be installed:
  vsftpd
0 upgraded, 1 newly installed, 0 to remove and 321 not upgraded.
Output truncated…
Processing triggers for ureadahead (0.100.0-19) ...
Processing triggers for systemd (229-4ubuntu4) ...
$
```

2. Use the systemctl command to check the status of the vsftpd service, using the following command:

```
$ systemctl status vsftpd.service
● vsftpd.service - vsftpd FTP server
   Loaded: loaded (/lib/systemd/system/vsftpd.service; enabled; vendor preset:
enabled)
   Active: active (running) since Sat 2016-07-02 12:10:11 PDT; 2min 15s ago
 Main PID: 8780 (vsftpd)
   CGroup: /system.slice/vsftpd.service
           └─8780 /usr/sbin/vsftpd /etc/vsftpd.conf

Jul 02 12:10:10 bob-VirtualBox systemd[1]: Starting vsftpd FTP server...
Jul 02 12:10:11 bob-VirtualBox systemd[1]: Started vsftpd FTP server.
$
```

Notice from the preceding output that the command from Step 1 is not only downloaded but also installed and started, the vsftpd service.

3. To check to see that vsftpd is actually accessible on your system from another host, from another host machine on your network, use the ftp command to connect to the vsftpd server on your machine. Substitute the IP address of the machine you want to ftp into for the IP 192.168.0.30 shown in this command:

```
# ftp 192.168.0.30
Connected to 192.168.0.30.
220 (vsFTPd 3.0.3)
Name (192.168.0.30:bob): bob
331 Please specify the password.
```

```
Password: QQQ
230 Login successful.
Remote system type is UNIX.
Using binary mode to transfer files.
...
ftp>
```

4. Terminate the connection with the following ftp command:

```
ftp> exit
221 Goodbye.
#
```

For CentOS 7.5 (All commands are run as root!)-

On our CentOS 7.5 system, we had to disable and stop the packagekit app to free the lock on YUM. We did this with the following commands:

```
[root@centos]# systemctl stop packagekit
[root@centos]# systemctl disable packagekit
[root@centos] #
```

We also used the **ip** command to determine the IP address of the machine on our local intranet that we were installing vsftpd on. In Step 5, this IP address was entered as 192.168.0.43.

1. Execute the following command to install the vsftpd package with YUM:

```
[root@centos] # yum install vsftpd ftp -y
Loaded plugins: fastestmirror, langpacks
Loading mirror speeds from cached hostfile
 * base: repo1.ash.innoscale.net
 * epel: mirror.ancl.hawaii.edu
 * extras: mirror.cs.pitt.edu
 * updates: mirror.cloud-bricks.net
Resolving Dependencies
--> Running transaction check
---> Package ftp.x86_64 0:0.17-67.el7 will be installed
---> Package vsftpd.x86_64 0:3.0.2-22.el7 will be installed
Output Truncated...
[root@centos] #
```

2. To configure vsftpd, you have to edit vsftpd configuration file /etc/vsftpd/vsftpd.conf as follows. If you want to use nano to do this instead of vi, you are welcome to do so.
   ```
   [root@centos] # vi /etc/vsftpd/vsftpd.conf
   ```
   At the following lines in that file, make the changes as shown.

   ```
   ...
   ## Disable anonymous login ##
   anonymous_enable=NO

   ## Uncomment ascii upload and download directive##
   ascii_upload_enable=YES
   ascii_download_enable=YES
   ```

   Save the file and exit the editor.

3. To enable the service at all subsequent boots and to start the vsftpd service, use the following commands:

   ```
   [root@centos] # systemctl enable vsftpd
   [root@centos] # systemctl start vsftpd
   ```

4. To allow the ftp service in through the firewall on port 21, use the following commands:

```
[root@centos] # firewall-cmd --permanent –add-port=21/tcp
success
[root@centos] # firewall-cmd --permanent –add-service=ftp
success
[root@centos] # firewall-cmd –reload
success
```

5. To connect to FTP server using an already-created user named mansoor, use the following commands, not run as root:

```
[root@centos] # exit

$ ftp 192.168.0.43
Connected to 192.168.0.43 (192.168.0.43).
220 (vsFTPd 3.0.2)
Name (192.168.0.43:bob): mansoor
331 Please specify the password.
Password: QQQ
230 Login successful.
Remote system type is UNIX.
Using binary mode to transfer files.
ftp> exit
221 Goodbye.
$
```

17.4.3 Installing Gparted from the Command Line
   For Debian-family (Ubuntu shown)-

```
$ sudo apt-get install gparted
[sudo] password for bob:
Reading package lists... Done
Building dependency tree
Reading state information... Done
The following packages were automatically installed and are no longer required:
  libllvm3.8 libmircommon5 libpango1.0-0 libpangox-1.0-0 libsnapd-glib1
snapd-login-service
  ubuntu-core-launcher
Use 'sudo apt autoremove' to remove them.
The following additional packages will be installed:
  libgtkmm-2.4-1v5 libparted-fs-resize0
Suggested packages:
  xfsprogs reiserfsprogs reiser4progs jfsutils kpartx dmraid dmsetup gpart
libparted-dev
The following NEW packages will be installed:
  gparted libgtkmm-2.4-1v5 libparted-fs-resize0
0 upgraded, 3 newly installed, 0 to remove and 78 not upgraded.
Need to get 1,129 kB of archives.
After this operation, 6,973 kB of additional disk space will be used.
Do you want to continue? [Y/n] y
Output truncated...
Setting up gparted (0.25.0-1) ...
Processing triggers for libc-bin (2.23-0ubuntu9) ...
$
```

For CentOS 7

   At the time of writing of this book, on our CentOS 7.4 system, we had to first stop and disable the packagekit app that is holding the lock on YUM. Then, we could install the epel "extras" package in which gparted is found. Use the following commands to do this, and then install gparted:

```
[root@centos]# systemctl stop packagekit
[root@centos]# systemctl disable packagekit
[root@centos]# yum install epel-release
Output truncated...
[root@centos]#
[root@centos] # yum install gparted
Loaded plugins: fastestmirror, langpacks
epel/x86_64/metalink
| 12 kB   00:00:00      epel
| 4.3 kB  00:00:00      (1/3): epel/x86_64/group_gz
| 261 kB  00:00:10      (2/3): epel/x86_64/updateinfo
| 839 kB  00:00:18      (3/3): epel/x86_64/primary_db
| 6.1 MB  00:00:19      Loading mirror speeds from cached hostfile
 * base: mirror.cogentco.com
 * epel: mirror.sfo12.us.leaseweb.net
 * extras: centos.sonn.com
 * updates: mirrors.seas.harvard.edu
Resolving Dependencies
--> Running transaction check
---> Package gparted.x86_64 0:0.19.1-6.el7 will be installed
Output truncated...
Installed:
  gparted.x86_64 0:0.19.1-6.el7
Dependency Installed:
  polkit-gnome.x86_64 0:0.106-0.1.20170423gita0763a2.el7
Complete!
[root@centos]#
```

W26.4.7 Installing ZFS

For Ubuntu 16.04 and Linux Mint 18.2-

From Example W26.8a Adding a New Hard Drive Using ZFS

Update your package manager, and download and install the zfsutils-linux on your system, using the following commands:

```
$ sudo apt update
Output truncated...
$ sudo apt upgrade
Output truncated...
$ sudo apt install zfsutils-linux
Output truncated...
```

1. Once the zfsutils-linux is installed, list the zpools on your system with the following command, to test that the installation has been successful:

```
$ sudo zpool list
[sudo] password for bob: zzz
no pools available
$
```

For Debian 9.1 and later

Because of our installation type, we needed to keep the Debian 9.1 installation DVD mounted on our computer to execute the following steps. Of course, some of these steps may not be necessary if there are bug fixes in later versions of Debian!

Step 1. You must enable the contrib, nonfree repositories available, by using the sed editor to find and replace the word "main" in the file /etc/apt/sources.list:

```
[root@debian]# sed -i 's/main/main contrib non-free/g' /etc/apt/sources.list
```

Then you can update your Debian systems' packages.

```
[root@debian]# apt-get update
```

Step 2. ZFS Installation

Install Linux headers. We install ZFS using dkms for our installation. The symbolic link is used here because of a bug inside the zfs distribution in Debian that doesn't look for rm binary correctly.

```
[root@debian]# apt -y install linux-headers-$(uname -r)
[root@debian]# ln -s /bin/rm /usr/bin/rm
[root@debian]# apt-get -y install zfs-dkms
```

Step 3. ZFS Start and Restart

At this point, zfs-dkms is installed. To start ZFS, use the following commands:

```
[root@debian]# /sbin/modprobe zfs
[root@debian]# systemctl restart zfs-import-cache
[root@debian]# systemctl restart zfs-import-scan
[root@debian]# systemctl restart zfs-mount
[root@debian]# systemctl restart zfs-share
```

Step 4. Create a zpool before reboot. Another caveat, otherwise zfs will not load the proper modules if there are no zpools! Reboot after these commands, and you'll have working ZFS.

```
[root@debian]# truncate -s 100M /test1
[root@debian]# zpool create mypool /test1
[root@debian]# zpool scrub mypool
[root@debian]# zpool status
pool: mypool
 state: ONLINE
  scan: scrub repaired 0 in 0h0m with 0 errors on Sun Dec 24 08:11:27 2017
config:

        NAME          STATE     READ WRITE CKSUM
        mypool        ONLINE       0   0   0
          /test1   ONLINE    0   0   0

errors: No known data errors
[root@debian]#
```

For CentOS 7.4-

Step 0. Stop and disable packagekit, so that YUM is unlocked, with the following commands:

```
[root@centos] #  systemctl stop packagekit
[root@centos] #  systemctl disable packagekit
```

Step 1. Verify that you have the 7.4 release with the following command:

```
[root@centos] # cat /etc/redhat-release
```

CentOS Linux release 7.4.1708 (Core)

If you have an earlier release of CentOS, such as Release 7.3, just substitute 7.3 for 7.4 in the following command. We have not tested this install on Release 6 or lower!

Step 2. Install the ZFS repository package for CentOS 7.4:

```
[root@centos] # yum -y install http://download.zfsonlinux.org/epel/zfs-release.
                el7_4.noarch.rpm
Output truncated...
```

Step 3. Modify your /etc/yum.repos.d/zfs.repo with a text editor of your choice.

To install the kABI-recommended packages so ZFS won't have to be rebuilt every time the kernel is updated, do this:

Edit the /etc/yum.repos.d/zfs.repo as root (so your changes will be written and saved!) and disable the [zfs] repository and enable the [zfs-kmod]. The first two sections of your file should look like this (changes shown in bold):

```
[zfs]
name=ZFS on Linux for EL7 - dkms
baseurl=http://download.zfsonlinux.org/epel/7.3/$basearch/
enabled=0
metadata_expire=7d
gpgcheck=1
gpgkey=file:///etc/pki/rpm-gpg/RPM-GPG-KEY-zfsonlinux
[zfs-kmod]
name=ZFS on Linux for EL7 - kmod
baseurl=http://download.zfsonlinux.org/epel/7.3/kmod/$basearch/
enabled=1
metadata_expire=7d
gpgcheck=1
gpgkey=file:///etc/pki/rpm-gpg/RPM-GPG-KEY-zfsonlinux
```

Step 4. Install ZFS with the following command:

```
[root@centos] # yum -y install zfs
Output truncated...
```

Step 5. Reboot your system. Then, to start ZFS, use the following command:

```
[root@centos] # modprobe zfs
[root@centos] #
```

Step 6. According to the methodology shown in the section entitled "ZFS Administration and Use" at the GitHub site, create a psuedovirtual device (vdev) in /root as a file named zfs01.img, with the following commands:

You may wish at this time to use an actual physical device for a vdev, such as a hard disk or Solid State Drive (SSD) on the Serial ATA (SATA) bus, or USB thumbdrive(s), as shown in the methodologies illustrated in the section titled "ZFS Administration and Use" at the GitHub site.

```
[root@centos] # cd /root
[root@centos] # truncate -s 2G zfs01.img
```

Step 7. Create a ZFS pool using our psuedo file-as-a-vdev:

```
[root@centos] # zpool create first-zfs-pool /root/zfs01.img
[root@centos] #
```

The file was used a virtual device, a zpool was built on it, and it was mounted as /first-zfs-pool.

Step 8. Check that the pool was created with the following df command:

```
[root@centos] # df -hT
Filesystem                      Type       Size   Used   Avail   Use%   Mounted on
/dev/mapper/centos-root         xfs         18G   4.3G     13G    25%   /
devtmpfs                        devtmpfs   905M      0    905M     0%   /dev
tmpfs                           tmpfs      920M      0    920M     0%   /dev/shm
tmpfs                           tmpfs      920M   8.8M    911M     1%   /run
tmpfs                           tmpfs      920M      0    920M     0%   /sys/fs/cgroup
/dev/sda1                       xfs       1014M   289M    726M    29%   /boot
tmpfs                           tmpfs      184M   8.0K    184M     1%   /run/user/42
```

```
tmpfs                          tmpfs    184M   0    184M   0%   /run/user/1000
first-zfs-pool                 zfs      1.9G   0    1.9G   0%   /first-zfs-pool
```

Step 9. Enable ZFS startup after reboots using the following command:

```
[root@centos] # systemctl preset zfs-import-cache zfs-import-scan zfs-mount
               zfs-share zfs-zed zfs.target
```

Step 10. Reboot the system to verify that ZFS starts up and you still have the zpool you created in Step 6.

Step 11. After reboot, check that ZFS and your psuedo-vdev filesystem is running and available, with the following command:

```
[root@centos] # df -hT | egrep 'zfs|Filesystem'
Filesystem          Type    Size  Used Avail Use% Mounted on
first-zfs-pool      zfs     1.9G   0   1.9G  0%   /first-zfs-pool
[root@centos] #
```

W26.4.8 mdadm Installation, From Example W26.8b

For Debian-family

1. Install mdadm to your system if it is not there already:

```
$ sudo apt-get install mdadm
Reading package lists... Done
Building dependency tree
Reading state information... Done
Suggested packages:
  default-mta | mail-transport-agent
The following NEW packages will be installed:
  mdadm
0 upgraded, 1 newly installed, 0 to remove and 54 not upgraded.
Need to get 394 kB of archives.
After this operation, 1,201 kB of additional disk space will be used.
...
W: mdadm: /etc/mdadm/mdadm.conf defines no arrays.
$
```

For CentOS 7-

On out CentOS 7.4 system, mdadm was already installed by default. So we just started the mdadm monitor and enable it when the system reboots using the following commands:

```
[root@centos] # systemctl start mdmonitor
[root@centos] # systemctl enable mdmonitor
```

17.7.2.2 gcc Installation with apt-get

For Debian-family gcc Installation-

As shown in Chapters 15 and 16, the Gnu C Compiler (gcc) is a critical development tool, both for developing system programs, and also generally for writing other application C program code. As we show in this example, it can be installed on our representative Linux systems.

To use apt to install the Gnu C Compiler, and other attendant libraries, on the Debian family of Linux, do the following steps according to the system you want to use gcc on:

On Debian, starting with version 9.1

Step 1. The simplest way to determine whether gcc has already been installed is to see if the build-essential package has been installed already on your system. To do this, you can try to compile a C program (developed and tested on another system) that you know can compile and run, and if you get an error that the command gcc does not exist on your Debian system, that usually means gcc is not installed.

You can also use the apt-cache pkgnames command shown previously to determine whether the build-essential package has already been installed by a system administrator! If it hasn't, proceed to step 2.

Step 2. Use the following command to install the build-essential package:

```
[root@debian] # apt-get update && apt-get install build-essential
```

In the middle of installing build-essential, you may have to insert the Debian DVD installation media for the installation to continue. That of course depends on the exact details of how you have set up repositories on your system at the time you, or a system administrator, installed the system! We show repository management details for Debian, Ubuntu, Linux Mint, and CentOS at the book website, in Chapter W26, Section 7.3.

Step 3. You can now compile C programs with the standard libraries.

On Ubuntu starting with version 16.04

Step 0. As of version 16.04, the gcc and build-essential packages are already installed, if you have done a default installation of the system. You can now compile C programs with standard libraries.

On Linux Mint starting with version 18

Step 1. The simplest way to determine whether gcc has already been installed is to see whether the build-essential package has been installed already on your system. To do this, you can try to compile a C program (developed and tested on another system) that you know can compile and run, and if you get an error that the command gcc does not exist on your Linux Mint system, that usually means gcc is not installed.

You can also use the apt-cache pkgnames command shown earlier to determine whether the build-essential package has already been installed by a system administrator! If it hasn't, proceed to Step 2.

Step 2. Use the following command to install the build-essential package:

```
[root@Linux_Mint] # apt-get update && apt-get install build-essential
```

Step 3. You can now compile C programs with standard libraries.

19.7.3.2 A gcc Installation Example with YUM

The following example details the installation of the Gnu C Compiler (gcc), and its attendant libraries, on our representative Linux system, CentOS 7.4.

Step 0. First, we had to stop the packagekit app, and disable it so that the YUM lock was released. Then we had to install the epel "extras" package. The following commands achieve these two objectives:

```
[root@centos]# systemctl stop packagekit
[root@centos]# systemctl disable packagekit
[root@centos]# yum install epel-release
Output truncated...
```

Step 1. Use YUM to list the groups installed on your system.

```
[root@centos]# yum group list
```

Step 2. If you don't see the Development Group listed onscreen, install it with the following command:

```
[root@centos]# yum group install "Development Tools"
```

Step 3. After successfully doing Step 2, check the version number of gcc with the following command:

```
[root@centos]# gcc –version
```

On our CentOS 7.4 system, the version number was 4.8.5

You are now ready to compile and test the C programs in Chapters 15 and 16, and the example C programs in Chapters W20 and W21.

17.9.2.1 sudo Installation

For Debian-family Installation (also works for Ubuntu and Linux Mint *if* sudo is not installed)-

Debian—As root, use the following command:

`[root@debian] # `**`apt-get install sudo`**

Then give yourself, or a designated user, sudo privilege, using the following command:

`[root@debian] # `**`visudo`**

Your default text editor will then open (nano on our Debian 9.1 system), and you will be able to enter the following line into the section entitled "# User privilege specification" of the sudoers file:

```
username            All-(ALL:ALL)           ALL
```

where username is the username you want to give blanket sudo privilege to.

Save the file and exit the editor.

Ubuntu—As of the time of the writing of this book, sudo was already installed, and the default user was in the sudoers group. This can be checked with the command **id -Gn**

Linux Mint—As of the time of the writing of this book, sudo was already installed, and the default user was in the sudoers group. This can be checked with the command **id -Gn**

For CentOS Installation

As of the time of the writing of this book, sudo was already installed, but as root, you must give yourself (or a designated user) sudo privilege. To give yourself, or a designated user, sudo privilege, use the following command:

`[root@centos] # `**`visudo`**

Your default text editor will then open, and you will be able to enter the following line into the section titled "# User privilege specification" of the sudoers file:

```
username            ALL=(ALL:ALL)           ALL
```

where username is the username you want to give blanket sudo privilege to.

Save the file, and exit the editor.

Appendix B: Books for Further Reference

To successfully use all of the information we present in this book, we encourage the reader to refer to the following printed book sources to extend and amplify the material in the topics we have chosen. This can be done generally from the primary/secondary sources we list and selectively from the additional topic areas.

Each entry is formatted in the following way

Author(s), **Title, (edition)**, Date of Publication, Publisher

A brief description and pertinent subject references in the cited work.

Applicable to Chapters/Sections: Chapters and sections you can apply the information in the cited work to in *Linux: The Textbook*, 2nd edition.

Be aware that later editions of the cited works may be available in the future, but the later editions will be enhancements to the editions we list here.

General References, and System, Kernel, or Network Programming

Primary Sources

Arpaci-Dusseau, **Operating Systems: Three Easy Pieces**, 2014, Arpaci-Dusseau Books
An excellent synoptic overview of operating systems, cast in terms of virtualization, concurrency, and persistence.
Applicable to Chapters/Sections: All

Kerrisk, **The Linux Programming Interface**, 2010, No Starch Press
The Bible of Linux system programming. Specifically, Chapters 2–9 offer background material. Also see Chapter 12 on System and Process Information, and any other chapter you need to extend your knowledge of system programming in Linux with. No systemd, but look for that in the next edition!
Applicable to Chapters/Sections: 15, 16, W20, W21

Negus, **Linux Bible, 9th edition**, 2015, Wiley
The best general reference to everything in Linux.
Applicable to Chapters/Sections: All

Stevens, Rago, **Advanced Programming in the UNIX Environment, 3rd edition**, 2013, Addison-Wesley
The Old Testament of NIX-like system programming. Very instructive to compare the system calls and chapters in this book to the Kerrisk book on Linux.
Applicable to Chapters/Sections: 15, 16, W20, W21

Das, **Your UNIX/Linux: The Ultimate Guide, 3rd edition**, 2012, McGraw-Hill
A great general reference for all things Linux, including chapters on system programming.
Applicable to Chapters/Sections: All except 18

Sobell, Helmke, **A Practical Guide to Linux Commands, Editors, & Shell Programming, 4th edition**, 2017, Addison-Wesley
A very comprehensive and thorough guide to core Linux commands and shell programming.
Applicable to Chapters/Sections: 1–14, W25

Barrett, **Linux Pocket Guide, 3rd edition**, 2016, O'Reilly Media
A very handy command dictionary and guide to scripting in Bash.
Applicable to Chapters/Sections: 1–14

Tanenbaum, Bos, **Modern Operating Systems, 4th edition**, 2014, Pearson Education
A good overview of operating system principles.
Applicable to Chapters/Sections: 1, 2, 15, 16, W20, W21

Kurose, Ross, **Computer Networking A Top-Down Approach, 7th edition**, 2016, Pearson

Secondary Sources

Stevens, **UNIX Network Programming Interprocess Communication, 2nd edition**, 1999, Addison-Wesley
A great general reference on UNIX IPC system programming.
Applicable to Chapters/Sections: 15, 16, W20, W21

Stevens, Fenner, Rudoff, **UNIX Network Programming Sockets Networking API, 3rd edition**, 2004, Addison-Wesley
Extensive coverage of the sockets API.
Applicable to Chapters/Sections: 15, 16, W20, W21

Glass, Ables, **Linux for Programmers and Users, 3rd edition**, 2003, Pearson Education
Applicable to Chapters/Sections: 1–14

Silberschatz, Galvin, Gagne, **Operating System Concepts, 9th edition**, 2012, Wiley
Comparison of the design of Linux with other OS's.

Silberschatz, Galvin, Gagne, **Operating System Concepts Essentials, 2nd Edition**, 2013, Wiley
Presents a subset of chapters in **Operating System Concepts, 9th edition**.

Boyce, **Linux Networking Cookbook**, 2016, Packt Publishing

Cobbaut, **Mastering Linux- Fundamentals**, 2016, Samurai Media Ltd.

Duntemann, **Assembly Language Step by Step Programming with Linux, 3rd edition**, 2009, Wiley

Shotts, **The Linux Command Line: A Complete Introduction, 1st edition**, 2012, No Starch Press
Applicable to Chapters/Sections: 1–14

Bovet, Cesati, **Understanding the Linux Kernel, 3rd edition**, 2006, O'Reilly Media
Applicable to Chapters/Sections: 15, 16, W20, W21

Sheer, **Linux Rute User's Tutorial and Exposition (RUTE)**, 2001, Prentice Hall

Love, **Linux Kernel Development, 3rd edition**, 2010, Addison-Wesley Professional

Matthew, Stones, **Beginning Linux Programming, 4th edition**, 2007, Wrox
Applicable to Chapters/Sections: 15, 16, W20, W21

Mauerer, **Professional Linux Kernel Architecture**, 2008, Wrox

Fox, **Linux with Operating System Concepts**, 2014, Taylor & Francis/CRC Press

Petersen, **Linux: The Complete Reference, 6th edition**, 2007, McGraw-Hill Education

Certification and Competency Exams

Brunson, **Walberg, CompTIA Linux+/LPIC-1 Certification Guide, 1st edition**, 2015, Pearson IT Certification
   Applicable to Chapters/Sections: All

Tracy, **LPI Linux Essentials Certification All-in-One Exam Guide, 2nd edition**, 2015, McGraw-Hill
   Applicable to Chapters/Sections: All

Jang, Orsaria, **RHCSA/RHCE Red Hat Linux Certification Study Guide, 7th edition** (Exams EX200 & EX300), 2016, McGraw-Hill
   Certification preparation for administrators of Red Hat and CentOS systems.
   Applicable to Chapters/Sections: All

C Language Programming

Kernighan, Ritchie, **C Programming Langue, 2nd edition**, 1988, Prentice Hall

Prata, **C Primer Plus, 6th edition**, 2013, Addison-Wesley Professional

Prinz, Crawford, **C in a Nutshell**, 2006, O'Reilly Media

Press, Flannery, Teukolsky, Vetterling, **Numerical Recipes in C**, 1988, Cambridge University Press

git

Loeliger, McCullough, **Version Control with Git, 2nd edition**, 2012, O'Reilly Media
   Refer to Chapters 1–7 for basic information, and Chapter 12 on remote repositories.
   Applicable to Chapters/Sections: W24, Section 5.7

Chacon, Straub, **Pro Git, 2nd edition**, 2014, Apress
   Refer to Chapters 1–3 for basic information.
   Applicable to Chapters/Sections: W24, Section 5.7

POSIX, POSIX Threads

Lewine, **POSIX Programmers Guide**, 1991, O'Reilly & Associates

Butenhof, **Programming with POSIX Threads**, 1997, Addison-Wesley Professional

Nichols, Buttlar, Farrell, **Pthreads Programming**, 1998, O'Reilly Media

Python

Lutz, **Programming Python, 4th edition**, 2011, O'Reilly Media
  Refer to Chapter 2, pages 75–102 on the sys and os modules. Refer to Part III., Chapters 8 and 9 on tkinter and GUI programming.
  Applicable to Chapters/Sections: W19, W26, W27

Lutz, **Learning Python, 5th edition**, 2013, O'Reilly Media
  The best overview and tutorial on Python. Parts I, II, and III Cover Getting Started, Types and Operations, and Statements Syntax thoroughly.
  Applicable to Chapters/Sections: W19, W26, W27

Lutz, **Python Pocket Reference, 5th edition**, 2014, O'Reilly Media
  Refer to Pages 67–104 for statement syntax, pages 173–231 for module descriptions.
  Applicable to Chapters/Sections: W19, W26, W27

Sweigart, **Automate the Boring Stuff with Python Practical Programming for Total Beginners, 1st edition**, 2015, No Starch Press
  Applicable to Chapters/Sections: W19, W26, W27

Guttag, **Introduction to Computation and Programming Using Python**, 2013, MIT Press
  An entire course run on programming and Python.
  Applicable to Chapters/Sections: W19, W26, W27

Beazley, **Python Essential Reference, 4th edition**, 2009, Pearson Education
  Refer to Part I, Chapters 1–11 for a complete specification of the syntax of the language.
  Applicable to Chapters/Sections: W19, W26, W27

Shell Scripts and Programming

Kochan, Wood, **Shell Programming in Unix, Linux and OS X: The Fourth Edition of Unix Shell Programming, 4th edition**, 2016, Addison-Wesley
  Applicable to Chapters/Sections: 12, 13

Newham, Rosenblatt, **Learning the bash Shell: Unix Shell Programming, 3rd edition**, 2016, O'Reilly
  Applicable to Chapters/Sections: 12, 13

Blum, Bresnahan, **Linux Command Line and Shell Scripting Bible, 3rd edition**, 2015, Wiley
  An excellent source for Linux shells and the art of shell script programming. Numerous shell scripting examples.
  Applicable to Chapters/Sections: 1–14

Quigley, **UNIX Shells by Example, 4th edition**, 2005, Pearson Education/Prentice Hall
  Covers all Linux shells and extensive programming in each.
  Applicable to Chapters/Sections: All, particularly 12, 13

ssh

Stahnke, **Pro OpenSSH**, 2006, Apress
  Refer to Parts I and II for introduction and command line options/setup.
  Applicable to Chapters/Sections: Sections 11.8.5–11.8.8, 17, W26

Barrett, Silverman, Byrnes, **SSH The Secure Shell: The Definitive Guide, 2nd edition**, 2005, O'Reilly Media
    Refer to Chapters 1–3 for Introduction, Basic Client Use, and Inside SSH.
    Applicable to Chapters/Sections: Sections 11.8.5–11.8.8, 17, W26

System Administration

Soyinka, **Linux Administration: A Beginner's Guide, 7th edition**, 2015, McGraw-Hill
    The best general reference on Linux system administration. Refer to Chapter 10, Virtual Filesystems. Covers systemd.
    Applicable to Chapters/Sections: 17, 18, W26, W27

Nemeth, Snyder, Hein, Whaley, Mackin, **UNIX and Linux System Administration Handbook, 5th edition**, 2017, Addison-Wesley Professional
    Very comprehensive, covers FreeBSD 11.0, Debian, Ubuntu, RHEL/CentOS. systemd permeates everything.
    Applicable to Chapters/Sections: 17, 18, W26, W27

Ubuntu

Helmke, **Ubuntu Unleashed 2017 Edition, 12th edition**, 2016, Sams
    Applicable to Chapters/Sections: All

vi and emacs

Robbins, Hannah, Lamb, **Learning the vi and Vim Editors, 7th edition**, 2008, O'Reilly Media
    Applicable to Chapters/Sections: W25

Cameron, Elliott, Loy, Raymond, Rosenblatt, **Learning GNU Emacs, 3rd edition**, 2005, O'Reilly Media
    Refer to Chapters 1–4 for basics, Chapters 6 and 10 for customization.
    Applicable to Chapters/Sections: W25

ZFS

Lucas, Jude, **FreeBSD Mastery ZFS**, 2015, Tilted Windmill Press
    A guide to practical applications of basic ZFS.
    Applicable to Chapters/Sections: W22

Ancillary Books

Hennessy, Patterson, **Computer Architecture: A Quantitative Approach, 5th edition**, 2014, Morgan Kaufmann
    Hardware-oriented, descriptions of instruction set architectures for multiprocessor and parallel computers.

Null, Lobur, **The Essentials of Computer Organization and Architecture, 4th edition**, 2014, Jones & Bartlett Learning

Pierce, **Types and Programming Languages**, 2002, MIT Press
Describes OOP, functional programming.

Abelson, Sussman, **Structure and Interpretation of Computer Programs, 2nd edition**, 1996, MIT Press
   How to think about writing computer programs. Can be used very successfully as a pseudocoding language for Bash scripts, C, or Python programs. Requires a scheme interpreter on your computer or tablet.

McKusick, Neville-Neil, Watson, **The Design and Implementation of the FreeBSD Operating System, 2nd edition**, 2015, Addison-Wesley
   Useful to compare FreeBSD and system-level theory and its Linux equivalent. Very instructive to compare Figure 7.1 Kernel I/O, page 315, with the Linux Storage Stack Diagram, for kernel V4.10,
   www.thomas-krenn.com/en/wiki/Linux_Storage_Stack_Diagram

Bryant, O'Hallaron, **Computer Systems: A Programmer's Perspective, 3rd edition**, 2016, Pearson

Sarwar, Koretsky, **UNIX: The Textbook, 3rd edition**, 2016, Taylor & Francis/CRC Press
   Everything in Linux: The Textbook, 2nd edition, minus systemd, but plus ZFS as the root file system!
   Applicable to Chapters/Sections: All

Index